Graduate Texts in Mathematics 113

SPRINGER STUDY EDITION

Ioannis Karatzas   Steven E. Shreve

Brownian Motion and Stochastic Calculus

Second Edition

With 10 Illustrations

Springer

Ioannis Karatzas
Departments of Mathematics
and Statistics
Columbia University
New York, NY 10027
USA
ik@math.columbia.edu

Steven E. Shreve
Department of Mathematical Sciences
Carnegie Mellon University
Pittsburgh, PA 15213
USA
shreve@cmu.edu

Mathematics Subject Classification (2000): 60G07, 60H05

Library of Congress Cataloging-in-Publication Data
Karatzas, Ioannis.
Brownian motion and stochastic calculus / Ioannis Karatzas, Steven E. Shreve.—2nd ed.
p. cm. — (Graduate texts in mathematics; 113)
Includes bibliographical references and index.
ISBN 978-0-387-97655-6 ISBN 978-1-4612-0949-2 (eBook)
DOI 10.1007/978-1-4612-0949-2
1. Brownian motion processes. 2. Stochastic analysis.
I. Shreve, Steven E. II. Title. III. Series.
QH274.75.K37 1991
530.4′75—dc20 91-22775

The present volume is the corrected softcover second edition of the previously published hardcover first edition (**ISBN 978-0-387-97655-6**).

ISBN 978-0-387-97655-6 Printed on acid-free paper.

9

springeronline.com

To Eleni and Dot

Preface

Two of the most fundamental concepts in the theory of stochastic processes are the *Markov property* and the *martingale property*.* This book is written for readers who are acquainted with both of these ideas in the discrete-time setting, and who now wish to explore stochastic processes in their continuous-time context. It has been our goal to write a systematic and thorough exposition of this subject, leading in many instances to the frontiers of knowledge. At the same time, we have endeavored to keep the mathematical prerequisites as low as possible, namely, knowledge of measure-theoretic probability and some familiarity with discrete-time processes. The vehicle we have chosen for this task is *Brownian motion*, which we present as the canonical example of both a Markov process and a martingale. We support this point of view by showing how, by means of stochastic integration and random time change, all continuous-path martingales and a multitude of continuous-path Markov processes can be represented in terms of Brownian motion. This approach forces us to leave aside those processes which do not have continuous paths. Thus, the Poisson process is not a primary object of study, although it is developed in Chapter 1 to be used as a tool when we later study passage times and local time of Brownian motion.

The text is organized as follows: Chapter 1 presents the basic properties of martingales, as they are used throughout the book. In particular, we generalize from the discrete to the continuous-time context the martingale convergence theorem, the optional sampling theorem, and the Doob–Meyer decomposition. The latter gives conditions under which a submartingale can be written

* According to M. Loève, "martingales, Markov dependence and stationarity are the only three dependence concepts so far isolated which are sufficiently general and sufficiently amenable to investigation, yet with a great number of deep properties" (*Ann. Probab.* **1** (1973), p. 6).

as the sum of a martingale and an increasing process, and associates to every martingale with continuous paths a "quadratic variation process." This process is instrumental in the construction of stochastic integrals with respect to continuous martingales.

Chapter 2 contains three different constructions of Brownian motion, as well as discussions of the Markov and strong Markov properties for continuous-time processes. These properties are motivated by d-dimensional Brownian motion, but are developed in complete generality. This chapter also contains a careful discussion of the various filtrations commonly associated with Brownian motion. In Section 2.8 the strong Markov property is applied to a study of one-dimensional Brownian motion on a half-line, and on a bounded interval with absorption and reflection at the endpoints. Many densities involving first passage times, last exit times, absorbed Brownian motion, and reflected Brownian motion are explicitly computed. Section 2.9 is devoted to a study of sample path properties of Brownian motion. Results found in most texts on this subject are included, and in addition to these, a complete proof of the Lévy modulus of continuity is provided.

The theory of stochastic integration with respect to continuous martingales is developed in Chapter 3. We follow a middle path between the original constructions of stochastic integrals with respect to Brownian motion and the more recent theory of stochastic integration with respect to right-continuous martingales. By avoiding discontinuous martingales, we obviate the need to introduce the concept of predictability and the associated, highly technical, measure-theoretic machinery. On the other hand, it requires little extra effort to consider integrals with respect to continuous martingales rather than merely Brownian motion. The remainder of Chapter 3 is a testimony to the power of this more general approach; in particular, it leads to strong theorems concerning representations of continuous martingales in terms of Brownian motion (Section 3.4). In Section 3.3 we develop the chain rule for stochastic calculus, commonly known as Itô's formula. The Girsanov Theorem of Section 3.5 provides a method of changing probability measures so as to alter the drift of a stochastic process. It has become an indispensable method for constructing solutions of stochastic differential equations (Section 5.3) and is also very important in stochastic control (e.g., Section 5.8) and filtering. Local time is introduced in Sections 3.6 and 3.7, and it is shown how this concept leads to a generalization of the Itô formula to convex but not necessarily differentiable functions.

Chapter 4 is a digression on the connections between Brownian motion, Laplace's equation, and the heat equation. Sharp existence and uniqueness theorems for both these equations are provided by probabilistic methods; applications to the computation of boundary crossing probabilities are discussed, and the formulas of Feynman and Kac are established.

Chapter 5 returns to our main theme of stochastic integration and differential equations. In this chapter, stochastic differential equations are driven by Brownian motion and the notions of *strong* and *weak* solutions are presented. The basic Itô theory for strong solutions and some of its ramifications, including comparison and approximation results, are offered in Section 5.2, whereas Section 5.3 studies weak solutions in the spirit of Yamada & Watanabe. Essentially equivalent to the search for a weak solution is the search for a solution to the "Martingale Problem" of Stroock & Varadhan. In the context of this martingale problem, a full discussion of existence, uniqueness, and the strong Markov property for solutions of stochastic differential equations is given in Section 5.4. For one-dimensional equations it is possible to provide a complete characterization of solutions which exist only up to an "explosion time," and this is set forth in Section 5.5. This section also presents the recent and quite striking results of Engelbert & Schmidt concerning existence and uniqueness of solutions to one-dimensional equations. This theory makes substantial use of the local time material of Sections 3.6, 3.7 and the martingale representation results of Subsections 3.4.A,B. By analogy with Chapter 4, we discuss in Section 5.7 the connections between solutions to stochastic differential equations and elliptic and parabolic partial differential equations. Applications of many of the ideas in Chapters 3 and 5 are contained in Section 5.8, where we discuss questions of option pricing and optimal portfolio/consumption management. In particular, the Girsanov theorem is used to remove the difference between average rates of return of different stocks, a martingale representation result provides the optimal portfolio process, and stochastic representations of solutions to partial differential equations allow us to recast the optimal portfolio and consumption management problem in terms of two linear parabolic partial differential equations, for which explicit solutions are provided.

Chapter 6 is for the most part derived from Paul Lévy's profound study of Brownian excursions. Lévy's intuitive work has now been formalized by such notions as filtrations, stopping times, and Poisson random measures, but the remarkable fact remains that he was able, 40 years ago and working without these tools, to penetrate into the *fine structure of the Brownian path* and to inspire all the subsequent research on these matters until today. In the spirit of Lévy's work, we show in Section 6.2 that when one travels along the Brownian path with a clock run by the local time, the number of excursions away from the origin that one encounters, whose duration exceeds a specified number, has a Poisson distribution. Lévy's heuristic construction of Brownian motion from its excursions has been made rigorous by other authors. We do not attempt such a construction here, nor do we give a complete specification of the distribution of Brownian excursions; in the interest of intelligibility, we content ourselves with the specification of the distribution for the durations of the excursions. Sections 6.3 and 6.4 derive distributions for functionals of Brownian motion involving its local time; we present, in particular, a Feynman–Kac result for the so-called "elastic" Brownian motion, the formulas of D. Williams and H. Taylor, and the Ray–Knight description of

Brownian local time. An application of this theory is given in Section 6.5, where a one-dimensional stochastic control problem of the "bang-bang" type is solved.

The writing of this book has become for us a monumental undertaking involving several people, whose assistance we gratefully acknowledge. Foremost among these are the members of our families, Eleni, Dot, Andrea, and Matthew, whose support, encouragement, and patience made the whole endeavor possible. Parts of the book grew out of notes on lectures given at Columbia University over several years, and we owe much to the audiences in those courses. The inclusion of several exercises, the approaches taken to a number of theorems, and several citations of relevant literature resulted from discussions and correspondence with F. Baldursson, A. Dvoretzky, W. Fleming, O. Kallenberg, T. Kurtz, S. Lalley, J. Lehoczky, D. Stroock, and M. Yor. We have also taken exercises from Mandl, Lánská & Vrkoč (1978), and Ethier & Kurtz (1986). As the project proceeded, G.-L. Xu, Z.-L. Ying, and Th. Zariphopoulou read large portions of the manuscript and suggested numerous corrections and improvements. Careful reading by Daniel Ocone and Manfred Schäl revealed minor errors in the first printing, and these have been corrected. Others, including F. Åkesson, S. Dayanik, B. Doytchinov, H.J. Engelbert, R. Höhnle, C. Hou, A. Karolik, W. Nichols, L. Nielsen, D. Ocone, N. Vaillant and H. Wang found errors and/or contributed ideas, which have resulted in improvements in subsequent printings. However, our greatest single debt of gratitude goes to Marc Yor, who read much of the near-final draft and offered substantial mathematical and editorial comments on it. The typing was done tirelessly, cheerfully, and efficiently by Stella DeVito and Doodmatie Kalicharan; they have our most sincere appreciation.

We are grateful to Sanjoy Mitter and Dimitri Bertsekas for extending to us the invitation to spend the critical initial year of this project at the Massachusetts Institute of Technology. During that time the first four chapters were essentially completed, and we were partially supported by the Army Research Office under grant DAAG-299-84-K-0005. Additional financial support was provided by the National Science Foundation under grants DMS-84-16736 and DMS-84-03166 and by the Air Force Office of Scientific Research under grants AFOSR 82-0259, AFOSR 85-0360, and AFOSR 86-0203.

Ioannis Karatzas
Steven E. Shreve

Contents

Suggestions for the Reader

We use a hierarchical numbering system for equations and statements. The k-th equation in Section j of Chapter i is labeled $(j.k)$ at the place where it occurs and is cited as $(j.k)$ within Chapter i, but as $(i.j.k)$ outside Chapter i. A definition, theorem, lemma, corollary, remark, problem, exercise, or solution is a "statement," and the k-th statement in Section j of Chapter i is labeled $j.k$ *Statement* at the place where it occurs, and is cited as *Statement* $j.k$ within Chapter i but as *Statement* $i.j.k$ outside Chapter i.

This book is intended as a text and can be used in either a one-semester or a two-semester course, or as a text for a special topic seminar. The accompanying figure shows dependences among sections, and in some cases among subsections. In a one-semester course, we recommend inclusion of Chapter 1 and Sections 2.1, 2.2, 2.4, 2.5, 2.6, 2.7, §2.9.A, B, E, Sections 3.2, 3.3, 5.1, 5.2, and §5.6.A, C. This material provides the basic theory of stochastic integration, including the Itô calculus and the basic existence and uniqueness results for strong solutions of stochastic differential equations. It also contains matters of interest in engineering applications, namely, Fisk–Stratonovich integrals and approximation of stochastic differential equations in §3.3.A and 5.2.D, and Gauss–Markov processes in §5.6.A. Progress through this material can be accelerated by omitting the proof of the Doob–Meyer Decomposition Theorem 1.4.10 and the proofs in §2.4.D. The statements of Theorem 1.4.10, Theorem 2.4.20, Definition 2.4.21, and Remark 2.4.22 should, however, be retained. If possible in a one-semester course, and certainly in a two-semester course, one should include the topic of weak solutions of stochastic differential equations. This is accomplished by covering §3.4.A, B, and Sections 3.5, 5.3, and 5.4. Section 5.8 serves as an introduction to *stochastic control*, and so we recommend adding §3.4.C, D, E, and Sections 5.7, and 5.8 if time permits. In either a one- or two-semester course, Section 2.8 and part or all of Chapter 4 may be included according to time and interest. The material on *local time* and its applications in Sections 3.6, 3.7, 5.5, and in Chapter 6 would normally be the subject of a special topic course with advanced students.

The text contains about 175 "problems" and over 100 "exercises." The former are assignments to the reader to fill in details or generalize a result, and these are often quoted later in the text. We judge approximately two-thirds of these problems to be nontrivial or of fundamental importance, and solutions for such problems are provided at the end of each chapter. The exercises are also often significant extensions of results developed in the text, but these will not be needed later, except perhaps in the solution of other exercises. Solutions for the exercises are not provided. There are some exercises for which the solution we know violates the dependencies among sections shown in the figure, but such violations are pointed out in the offending exercises, usually in the form of a hint citing an earlier result.

Interdependence of the Chapters

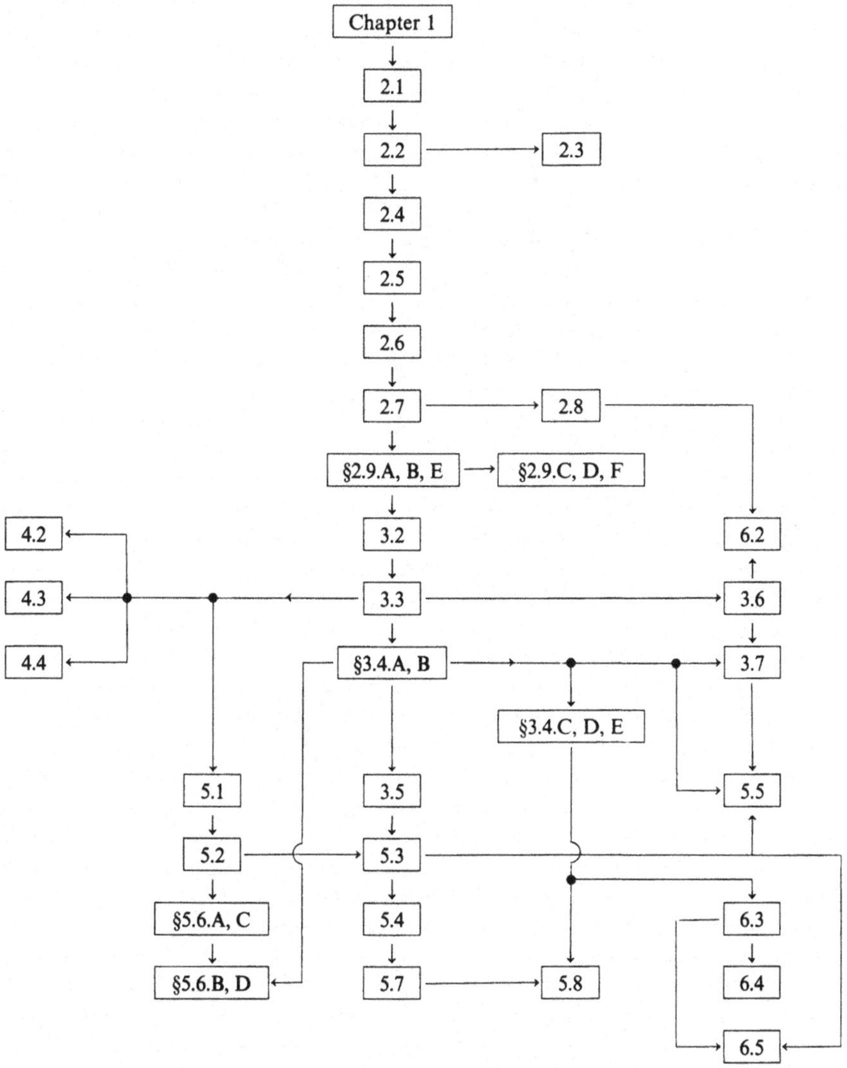

Frequently Used Notation

I. General Notation

Let a and b be real numbers.
(1) $\triangleq$ means "is defined to be."
(2) $a \wedge b \triangleq \min\{a, b\}$.
(3) $a \vee b \triangleq \max\{a, b\}$.
(4) $a^+ \triangleq \max\{a, 0\}$.
(5) $a^- \triangleq \max\{-a, 0\}$.

II. Sets and Spaces

(1) $\mathbb{N}_0 \triangleq \{0, 1, 2, \ldots\}$.
(2) Q is the set of rational numbers.
(3) Q^+ is the set of nonnegative rational numbers.
(4) $\mathbb{R}^d$ is the d-dimensional Euclidean space; $\mathbb{R}^1 = \mathbb{R}$.
(5) $B_r \triangleq \{x \in \mathbb{R}^d;\ \|x\| < r\}$ (p. 240).
(6) $(\mathbb{R}^d)^{[0,\infty)}$ is the set of functions from $[0, \infty)$ to $\mathbb{R}^d$ (pp. 49, 76).
(7) $C[0, \infty)^d$ is the subspace of $(\mathbb{R}^d)^{[0,\infty)}$ consisting of continuous functions; $C[0, \infty)^1 = C[0, \infty)$ (pp. 60, 64).
(8) $D[0, \infty)$ is the subspace of $\mathbb{R}^{[0,\infty)}$ consisting of functions which are right continuous and have left-limits (p. 409).
(9) $C^k(E)$, $C_b^k(E)$, $C_0^k(E)$: See Remark 4.1, p. 312.
(10) $C^{1,2}([0, T) \times E)$, $C^{1,2}((0, T) \times E)$: See Remark 4.1, p. 312.
(11) $\mathscr{L}$, $\mathscr{L}(M)$, $\mathscr{L}^*$, $\mathscr{L}^*(M)$: See pp. 130–131.
(12) $\mathscr{P}$, $\mathscr{P}(M)$, $\mathscr{P}^*$, $\mathscr{P}^*(M)$: See pp. 146–147.
(13) $\mathscr{M}_2(\mathscr{M}_2^c)$: The space of (continuous) square-integrable martingales (p. 30).
(14) $\mathscr{M}^{loc}(\mathscr{M}^{c,loc})$: The space of (continuous) local martingales (p. 36).

III. Functions

(1) $\operatorname{sgn}(x) = \begin{cases} 1; & x > 0, \\ -1; & x \le 0. \end{cases}$

(2) $1_A(x) \triangleq \begin{cases} 1; & x \in A, \\ 0; & x \notin A. \end{cases}$

(3) $p(t; x, y) \triangleq \dfrac{1}{\sqrt{2\pi t}} e^{-(x-y)^2/2t}$; $t > 0$, $x, y \in \mathbb{R}$ (p. 52).

(4) $p_{\pm}(t; x, y) \triangleq p(t; x, y) \pm p(t; x, -y)$; $t > 0$, $x, y \in \mathbb{R}$ (p. 97).

(5) $[\![t]\!]$ is the largest integer less than or equal to the real number t.

IV. σ-Fields

(1) $\mathscr{B}(U)$: The smallest σ-field containing all open sets of the topological space U (p. 1).

(2) $\mathscr{B}_t(C[0, \infty))$, $\mathscr{B}_t(C[0, \infty)^d)$. See pp. 60, 307.

(3) $\sigma(\mathscr{G})$: The smallest σ-field containing the collection of sets $\mathscr{G}$.

(4) $\sigma(X_s)$: The smallest σ-field with respect to which the random variable X_s is measurable.

(5) $\sigma(X_s; 0 \le s \le t)$: The smallest σ-field with respect to which the random variable X_s is measurable, $\forall\, s \in [0, t]$.

(6) $\mathscr{F}_t^X \triangleq \sigma(X_s; 0 \le s \le t)$, $\mathscr{F}_\infty \triangleq \sigma(\bigcup_{t \ge 0} \mathscr{F}_t)$: See p. 3.

(7) $\mathscr{F}_{t+} \triangleq \bigcap_{\varepsilon > 0} \mathscr{F}_{t+\varepsilon}$, $\mathscr{F}_{t-} \triangleq \sigma(\bigcup_{s<t} \mathscr{F}_s)$: See p. 4.

(8) $\mathscr{F}_T$: The σ-field of events determined prior to the stopping time T; see p. 8.

(9) $\mathscr{F}_{T+}$: The σ-field of events determined immediately after the optional time T; see p. 10.

(10) $\mathscr{F} \otimes \mathscr{G} \triangleq \sigma(A \times B; A \in \mathscr{F}, B \in \mathscr{G}\}$: The product σ-field formed from the σ-fields $\mathscr{F}$ and $\mathscr{G}$.

V. Operations on Functions

(1) $\Delta \triangleq \sum_{i=1}^{d} \dfrac{\partial^2}{\partial x_i^2}$: The Laplacian (p. 240).

(2) $\mathscr{A}$, $\mathscr{A}_t$: Second order differential operators; see pp. 281, 311.

VI. Operations on Processes

(1) θ_s, θ_S: Shift operator at the deterministic time s and the random time S; see pp. 77, 83.

(2) $I_t^M(X) \triangleq \int_0^t X_s\,dM_s$: The stochastic integral of X with respect to M. See p. 141 for $M \in \mathcal{M}_2^c$, $X \in \mathcal{L}^*(M)$; see p. 147 for $M \in \mathcal{M}^{c,loc}$, $X \in \mathcal{P}^*(M)$.
(3) $M_t^* \triangleq \max_{0 \le s \le t} |M_s|$: See p. 163 for $M \in \mathcal{M}^{c,loc}$.
(4) $\langle X \rangle$: The quadratic variation process of $X \in \mathcal{M}_2$ (p. 31) or $X \in \mathcal{M}^{c,loc}$ (p. 36).
(5) $\langle X, Y \rangle$: The cross-variation process of X, Y in $\mathcal{M}_2$ (p. 31) or in $\mathcal{M}^{c,loc}$ (p. 36).
(6) $\|X\|_t$, $\|X\|$: See p. 37 for $X \in \mathcal{M}_2$.

VII. Miscellaneous

(1) $m_T(X, \delta) \triangleq \sup\{|X_s - X_t|; 0 \le s < t \le T, t - s \le \delta\}$; See p. 33.
(2) $m^T(\omega, \delta) \triangleq \max\{|\omega(s) - \omega(t)|; 0 \le s < t \le T, t - s \le \delta\}$: See p. 62.
(3) $\bar{D}$: The closure of the set $D \subset \mathbb{R}^d$.
(4) D^c: The complement of the set D.
(5) ∂D: The boundary of the set $D \subset \mathbb{R}^d$.
(6) $\tau_D \triangleq \inf\{t \ge 0; W_t \in D^c\}$: The first time the Brownian motion W exits from the set $D \subset \mathbb{R}^d$ (p. 240).
(7) $T_b \triangleq \inf\{t \ge 0; W_t = b\}$: The first time the one-dimensional Brownian motion W reaches the level $b \in \mathbb{R}$ (p. 79).
(8) $\Gamma_+(t) \triangleq \int_0^t 1_{(0,\infty)}(W_s)\,ds$: The occupation time by Brownian motion of the positive half-line (p. 273).
(9) $P_n \xrightarrow{w} P$: Weak convergence of the sequence of probability measures $\{P_n\}_{n=1}^\infty$ to the probability measure P (p. 60).
(10) $X_n \xrightarrow{\mathcal{D}} X$: Convergence in distribution of the sequence of random variables $\{X_n\}_{n=1}^\infty$ to the random variable X (p. 61).
(11) P^x: Probability measure corresponding to Brownian motion (p. 72) or a Markov process (p. 74) with initial position $x \in \mathbb{R}^d$.
(12) P^μ: Probability measure corresponding to Brownian motion (p. 72) or a Markov process (p. 74) with initial distribution μ.
(13) $\mathcal{N}_t^\mu$, $\mathcal{N}^\mu$: Collections of P^μ-negligible sets (p. 89).
(14) $I(\sigma)$, $Z(\sigma)$: See pp. 331, 332.
(15) I_d: The $(d \times d)$ identity matrix.
(16) meas: Lebesgue measure on the real line (p. 105).

CHAPTER 1

Martingales, Stopping Times, and Filtrations

1.1. Stochastic Processes and σ-Fields

A *stochastic process* is a mathematical model for the occurrence, at each moment after the initial time, of a random phenomenon. The randomness is captured by the introduction of a measurable space $(\Omega, \mathscr{F})$, called the *sample space*, on which probability measures can be placed. Thus, a stochastic process is a collection of random variables $X = \{X_t; 0 \le t < \infty\}$ on $(\Omega, \mathscr{F})$, which take values in a second measurable space $(S, \mathscr{S})$, called the *state space*. For our purposes, the state space $(S, \mathscr{S})$ will be the d-dimensional Euclidean space equipped with the σ-field of Borel sets, i.e., $S = \mathbb{R}^d$, $\mathscr{S} = \mathscr{B}(\mathbb{R}^d)$, where $\mathscr{B}(U)$ will always be used to denote the smallest σ-field containing all open sets of a topological space U. The index $t \in [0, \infty)$ of the random variables X_t admits a convenient interpretation as *time*.

For a fixed sample point $\omega \in \Omega$, the function $t \mapsto X_t(\omega)$; $t \ge 0$ is the *sample path* (realization, trajectory) of the process X associated with ω. It provides the mathematical model for a random experiment whose outcome can be observed continuously in time (e.g., the number of customers in a queue observed and recorded over a period of time, the trajectory of a molecule subjected to the random disturbances of its neighbors, the output of a communications channel operating in noise).

Let us consider two stochastic processes X and Y defined on the same probability space $(\Omega, \mathscr{F}, P)$. When they are regarded as functions of t and ω, we would say X and Y were the same if and only if $X_t(\omega) = Y_t(\omega)$ for all $t \ge 0$ and all $\omega \in \Omega$. However, in the presence of the probability measure P, we could weaken this requirement in at least three different ways to obtain three related concepts of "sameness" between two processes. We list them here.

1.1 Definition. Y is a *modification* of X if, for every $t \geq 0$, we have $P[X_t = Y_t] = 1$.

1.2 Definition. X and Y have *the same finite-dimensional distributions* if, for any integer $n \geq 1$, real numbers $0 \leq t_1 < t_2 < \cdots < t_n < \infty$, and $A \in \mathscr{B}(\mathbb{R}^{nd})$, we have:

$$P[(X_{t_1}, \ldots, X_{t_n}) \in A] = P[(Y_{t_1}, \ldots, Y_{t_n}) \in A].$$

1.3 Definition. X and Y are called *indistinguishable* if almost all their sample paths agree:

$$P[X_t = Y_t; \forall\, 0 \leq t < \infty] = 1.$$

The third property is the strongest; it implies trivially the first one, which in turn yields the second. On the other hand, two processes can be modifications of one another and yet have completely different sample paths. Here is a standard example:

1.4 Example. Consider a positive random variable T with a continuous distribution, put $X_t \equiv 0$, and let $Y_t = \begin{Bmatrix} 0; & t \neq T \\ 1; & t = T \end{Bmatrix}$. Y is a modification of X, since for every $t \geq 0$ we have $P[Y_t = X_t] = P[T \neq t] = 1$, but on the other hand: $P[Y_t = X_t; \forall\, t \geq 0] = 0$.

A positive result in this direction is the following.

1.5 Problem. Let Y be a modification of X, and suppose that both processes have a.s. right-continuous sample paths. Then X and Y are indistinguishable.

It does not make sense to ask whether Y is a modification of X, or whether Y and X are indistinguishable, unless X and Y are defined on the same probability space and have the same state space. However, if X and Y have the same state space but are defined on different probability spaces, we can ask whether they have the same finite-dimensional distributions.

1.2′ Definition. Let X and Y be stochastic processes defined on probability spaces $(\Omega, \mathscr{F}, P)$ and $(\tilde{\Omega}, \tilde{\mathscr{F}}, \tilde{P})$, respectively, and having the same state space $(\mathbb{R}^d, \mathscr{B}(\mathbb{R}^d))$. X and Y have *the same finite-dimensional distributions* if, for any integer $n \geq 1$, real numbers $0 \leq t_1 < t_2 < \cdots < t_n < \infty$, and $A \in \mathscr{B}(\mathbb{R}^{nd})$, we have

$$P[(X_{t_1}, \ldots, X_{t_n}) \in A] = \tilde{P}[(Y_{t_1}, \ldots, Y_{t_n}) \in A].$$

Many processes, including d-dimensional Brownian motion, are defined in terms of their finite-dimensional distributions irrespective of their probability space. Indeed, in Chapter 2 we will construct a standard d-dimensional Brownian motion B on a *canonical* probability space and then state that any process, on any probability space, which has state space $(\mathbb{R}^d, \mathscr{B}(\mathbb{R}^d))$ and the same finite-dimensional distributions as B, is a standard d-dimensional Brownian motion.

For technical reasons in the theory of Lebesgue integration, probability measures are defined on σ-fields and random variables are assumed to be measurable with respect to these σ-fields. Thus, implicit in the statement that a random process $X = \{X_t; 0 \le t < \infty\}$ is a collection of $(\mathbb{R}^d, \mathscr{B}(\mathbb{R}^d))$-valued random variables on $(\Omega, \mathscr{F})$, is the assumption that each X_t is $\mathscr{F}/\mathscr{B}(\mathbb{R}^d)$-measurable. However, X is really a function of the pair of variables (t, ω), and so, for technical reasons, it is often convenient to have some joint measurability properties.

1.6 Definition. The stochastic process X is called *measurable* if, for every $A \in \mathscr{B}(\mathbb{R}^d)$, the set $\{(t, \omega); X_t(\omega) \in A\}$ belongs to the product σ-field $\mathscr{B}([0, \infty)) \otimes \mathscr{F}$; in other words, if the mapping

$$(t, \omega) \mapsto X_t(\omega): ([0, \infty) \times \Omega, \mathscr{B}([0, \infty)) \otimes \mathscr{F}) \to (\mathbb{R}^d, \mathscr{B}(\mathbb{R}^d))$$

is measurable.

It is an immediate consequence of Fubini's theorem that the trajectories of such a process are Borel-measurable functions of $t \in [0, \infty)$, and provided that the components of X have defined expectations, then the same is true for the function $m(t) = EX_t$; here, E denotes expectation with respect to a probability measure P on $(\Omega, \mathscr{F})$. Moreover, if X takes values in $\mathbb{R}$ and I is a subinterval of $[0, \infty)$ such that $\int_I E|X_t|\,dt < \infty$, then

$$\int_I |X_t|\,dt < \infty \text{ a.s. } P, \quad \text{and} \quad \int_I EX_t\,dt = E\int_I X_t\,dt.$$

There is a very important, nontechnical reason to include σ-fields in the study of stochastic processes, and that is to keep track of information. The temporal feature of a stochastic process suggests a flow of time, in which, at every moment $t \ge 0$, we can talk about a *past*, *present*, and *future* and can ask how much an observer of the process knows about it at present, as compared to how much he knew at some point in the past or will know at some point in the future. We equip our sample space $(\Omega, \mathscr{F})$ with a *filtration*, i.e., a nondecreasing family $\{\mathscr{F}_t; t \ge 0\}$ of sub-σ-fields of $\mathscr{F}$: $\mathscr{F}_s \subseteq \mathscr{F}_t \subseteq \mathscr{F}$ for $0 \le s < t < \infty$. We set $\mathscr{F}_\infty = \sigma(\bigcup_{t \ge 0} \mathscr{F}_t)$.

Given a stochastic process, the simplest choice of a filtration is that generated by the process itself, i.e.,

$$\mathscr{F}_t^X \triangleq \sigma(X_s; 0 \le s \le t),$$

the smallest σ-field with respect to which X_s is measurable for every $s \in [0, t]$.

We interpret $A \in \mathscr{F}_t^X$ to mean that by time t, an observer of X knows whether or not A has occurred. The next two exercises illustrate this point.

1.7 Exercise. Let X be a process, *every* sample path of which is RCLL (i.e., *r*ight-*c*ontinuous on $[0, \infty)$ with finite *l*eft-hand *l*imits on $(0, \infty)$). Let A be the event that X is continuous on $[0, t_0)$. Show that $A \in \mathscr{F}_{t_0}^X$.

1.8 Exercise. Let X be a process whose sample paths are RCLL *almost surely*, and let A be the event that X is continuous on $[0, t_0)$. Show that A can fail to be in $\mathscr{F}_{t_0}^X$, but if $\{\mathscr{F}_t; t \geq 0\}$ is a filtration satisfying $\mathscr{F}_t^X \subseteq \mathscr{F}_t$, $t \geq 0$, and $\mathscr{F}_{t_0}$ contains all P-null sets of $\mathscr{F}$, then $A \in \mathscr{F}_{t_0}$.

Let $\{\mathscr{F}_t; t \geq 0\}$ be a filtration. We define $\mathscr{F}_{t-} \triangleq \sigma(\bigcup_{s<t} \mathscr{F}_s)$ to be the *σ-field of events strictly prior to $t > 0$* and $\mathscr{F}_{t+} \triangleq \bigcap_{\varepsilon>0} \mathscr{F}_{t+\varepsilon}$ to be the *σ-field of events immediately after $t \geq 0$*. We decree $\mathscr{F}_{0-} \triangleq \mathscr{F}_0$ and say that the filtration $\{\mathscr{F}_t\}$ is *right-* (*left-*)*continuous* if $\mathscr{F}_t = \mathscr{F}_{t+}$ (resp., $\mathscr{F}_t = \mathscr{F}_{t-}$) holds for every $t \geq 0$.

The concept of measurability for a stochastic process, introduced in Definition 1.6, is a rather weak one. The introduction of a filtration $\{\mathscr{F}_t\}$ opens up the possibility of more interesting and useful concepts.

1.9 Definition. The stochastic process X is *adapted* to the filtration $\{\mathscr{F}_t\}$ if, for each $t \geq 0$, X_t is an $\mathscr{F}_t$-measurable random variable.

Obviously, every process X is adapted to $\{\mathscr{F}_t^X\}$. Moreover, if X is adapted to $\{\mathscr{F}_t\}$ and Y is a modification of X, then Y is also adapted to $\{\mathscr{F}_t\}$ provided that $\mathscr{F}_0$ contains all the P-negligible sets in $\mathscr{F}$. Note that this requirement is not the same as saying that $\mathscr{F}_0$ is complete, since some of the P-negligible sets in $\mathscr{F}$ may not be in the completion of $\mathscr{F}_0$.

1.10 Exercise. Let X be a process with every sample path LCRL (i.e., *l*eft-*c*ontinuous on $(0, \infty)$ with finite *r*ight-hand *l*imits on $[0, \infty)$), and let A be the event that X is continuous on $[0, t_0]$. Let X be adapted to a right-continuous filtration $\{\mathscr{F}_t\}$. Show that $A \in \mathscr{F}_{t_0}$.

1.11 Definition. The stochastic process X is called *progressively measurable* with respect to the filtration $\{\mathscr{F}_t\}$ if, for each $t \geq 0$ and $A \in \mathscr{B}(\mathbb{R}^d)$, the set $\{(s, \omega); 0 \leq s \leq t, \omega \in \Omega, X_s(\omega) \in A\}$ belongs to the product σ-field $\mathscr{B}([0, t]) \otimes \mathscr{F}_t$; in other words, if the mapping $(s, \omega) \mapsto X_s(\omega)$: $([0, t] \times \Omega, \mathscr{B}([0, t]) \otimes \mathscr{F}_t) \to (\mathbb{R}^d, \mathscr{B}(\mathbb{R}^d))$ is measurable, for each $t \geq 0$.

The terminology here comes from Chung & Doob (1965), which is a basic reference for this section and the next. Evidently, any progressively measurable process is measurable and adapted; the following theorem of Chung & Doob (1965) provides the extent to which the converse is true.

1.12 Proposition. *If the stochastic process X is measurable and adapted to the filtration $\{\mathscr{F}_t\}$, then it has a progressively measurable modification.*

The reader is referred to the book of Meyer (1966), p. 68, for the (lengthy, and rather demanding) proof of this result. It will be used in this text only in a tangential fashion. Nearly all processes of interest are either right- or left-continuous, and for them the proof of a stronger result is easier and will now be given.

1.13 Proposition. *If the stochastic process X is adapted to the filtration $\{\mathscr{F}_t\}$ and every sample path is right-continuous or else every sample path is left-continuous, then X is also progressively measurable with respect to $\{\mathscr{F}_t\}$.*

PROOF. We treat the case of right-continuity. With $t > 0$, $n \geq 1$, $k = 0, 1, \ldots, 2^n - 1$, and $0 \leq s \leq t$, we define:

$$X_s^{(n)}(\omega) = X_{(k+1)t/2^n}(\omega) \quad \text{for} \quad \frac{kt}{2^n} < s \leq \frac{k+1}{2^n}t,$$

as well as $X_0^{(n)}(\omega) = X_0(\omega)$. The so-constructed map $(s, \omega) \mapsto X_s^{(n)}(\omega)$ from $[0, t] \times \Omega$ into $\mathbb{R}^d$ is demonstrably $\mathscr{B}([0, t]) \otimes \mathscr{F}_t$-measurable. Besides, by right-continuity we have: $\lim_{n\to\infty} X_s^{(n)}(\omega) = X_s(\omega)$, $\forall (s, \omega) \in [0, t] \times \Omega$. Therefore, the (limit) map $(s, \omega) \mapsto X_s(\omega)$ is also $\mathscr{B}([0, t]) \otimes \mathscr{F}_t$-measurable. □

1.14 Remark. If the stochastic process X is right- or left-continuous, but not necessarily adapted to $\{\mathscr{F}_t\}$, then the same argument shows that X is measurable.

A *random time* T is an $\mathscr{F}$-measurable random variable, with values in $[0, \infty]$.

1.15 Definition. If X is a stochastic process and T is a random time, we define the function X_T on the event $\{T < \infty\}$ by

$$X_T(\omega) \triangleq X_{T(\omega)}(\omega).$$

If $X_\infty(\omega)$ is defined for all $\omega \in \Omega$, then X_T can also be defined on Ω, by setting $X_T(\omega) \triangleq X_\infty(\omega)$ on $\{T = \infty\}$.

1.16 Problem. If the process X is measurable and the random time T is finite, then the function X_T is a random variable.

1.17 Problem. Let X be a measurable process and T a random time. Show that the collection of all sets of the form $\{X_T \in A\}$ and $\{X_T \in A\} \cup \{T = \infty\}$; $A \in \mathscr{B}(\mathbb{R})$, forms a sub-$\sigma$-field of $\mathscr{F}$. We call this the *σ-field generated by X_T*.

We shall devote our next section to a very special and extremely useful class of random times, called *stopping times*. These are of fundamental importance in the study of stochastic processes, since they constitute our most effective tool in the effort to "tame the continuum of time," as Chung (1982) puts it.

1.2. Stopping Times

Let us keep in mind the interpretation of the parameter t as time, and of the σ-field $\mathscr{F}_t$ as the accumulated information up to t. Let us also imagine that we are interested in the occurrence of a certain phenomenon: an earthquake with intensity above a certain level, a number of customers exceeding the safety requirements of our facility, and so on. We are thus forced to pay particular attention to the instant $T(\omega)$ at which the phenomenon manifests itself *for the first time*. It is quite intuitive then that the event $\{\omega; T(\omega) \le t\}$, which occurs if and only if the phenomenon has appeared prior to (or at) time t, should be part of the information accumulated by that time.

We can now formulate these heuristic considerations as follows:

2.1 Definition. Let us consider a measurable space $(\Omega, \mathscr{F})$ equipped with a filtration $\{\mathscr{F}_t\}$. A random time T is a *stopping time* of the filtration, if the event $\{T \le t\}$ belongs to the σ-field $\mathscr{F}_t$, for every $t \ge 0$. A random time T is an *optional time* of the filtration, if $\{T < t\} \in \mathscr{F}_t$, for every $t \ge 0$.

2.2 Problem. Let X be a stochastic process and T a stopping time of $\{\mathscr{F}_t^X\}$. Suppose that for some pair $\omega, \omega' \in \Omega$, we have $X_t(\omega) = X_t(\omega')$ for all $t \in [0, T(\omega)] \cap [0, \infty)$. Show that $T(\omega) = T(\omega')$.

2.3 Proposition. *Every random time equal to a nonnegative constant is a stopping time. Every stopping time is optional, and the two concepts coincide if the filtration is right-continuous.*

Proof. The first statement is trivial; the second is based on the observation $\{T < t\} = \bigcup_{n=1}^{\infty} \{T \le t - (1/n)\} \in \mathscr{F}_t$, because if T is a stopping time, then $\{T \le t - (1/n)\} \in \mathscr{F}_{t-(1/n)} \subseteq \mathscr{F}_t$ for $n \ge 1$. For the third claim, suppose that T is an optional time of the right-continuous filtration $\{\mathscr{F}_t\}$. Since for every positive integer m, we have $\{T \le t\} = \bigcap_{n=m}^{\infty} \{T < t + (1/n)\}$, we deduce that $\{T \le t\} \in \mathscr{F}_{t+(1/m)}$; whence $\{T \le t\} \in \mathscr{F}_{t+} = \mathscr{F}_t$. □

2.4 Corollary. *T is an optional time of the filtration $\{\mathscr{F}_t\}$ if and only if it is a stopping time of the* (*right-continuous!*) *filtration $\{\mathscr{F}_{t+}\}$.*

2.5 Example. Consider a stochastic process X with right-continuous paths, which is adapted to a filtration $\{\mathscr{F}_t\}$. Consider a subset $\Gamma \in \mathscr{B}(\mathbb{R}^d)$ of the state space of the process, and define the *hitting time*

$$H_\Gamma(\omega) = \inf\{t \geq 0; X_t(\omega) \in \Gamma\}.$$

We employ the standard convention that the infimum of the empty set is infinity.

2.6 Problem. If the set Γ in Example 2.5 is open, show that H_Γ is an optional time.

2.7 Problem. If the set Γ in Example 2.5 is closed and the sample paths of the process X are continuous, then H_Γ is a stopping time.

Let us establish some simple properties of stopping times.

2.8 Lemma. *If T is optional and θ is a positive constant, then $T + \theta$ is a stopping time.*

PROOF. If $0 \leq t < \theta$, then $\{T + \theta \leq t\} = \varnothing \in \mathscr{F}_t$. If $t \geq \theta$, then

$$\{T + \theta \leq t\} = \{T \leq t - \theta\} \in \mathscr{F}_{(t-\theta)+} \subseteq \mathscr{F}_t. \qquad \square$$

2.9 Lemma. *If T, S are stopping times, then so are $T \wedge S$, $T \vee S$, $T + S$.*

PROOF. The first two assertions are trivial. For the third, start with the decomposition, valid for $t > 0$:

$$\{T + S > t\} = \{T = 0, S > t\} \cup \{0 < T < t, T + S > t\}$$
$$\cup \{T > t, S = 0\} \cup \{T \geq t, S > 0\}.$$

The first, third, and fourth events in this decomposition are in $\mathscr{F}_t$, either trivially or by virtue of Proposition 2.3. As for the second event, we rewrite it as:

$$\bigcup_{\substack{r \in Q^+ \\ 0<r<t}} \{t > T > r, S > t - r\},$$

where Q^+ is the set of rational numbers in $[0, \infty)$. Membership in $\mathscr{F}_t$ is now obvious. $\square$

2.10 Problem. Let T, S be optional times; then $T + S$ is optional. It is a stopping time, if one of the following conditions holds:

(i) $T > 0, S > 0$;
(ii) $T > 0$, T is a stopping time.

2.11 Lemma. *Let $\{T_n\}_{n=1}^\infty$ be a sequence of optional times; then the random times*

$$\sup_{n\geq 1} T_n, \quad \inf_{n\geq 1} T_n, \quad \overline{\lim_{n\to\infty}}\, T_n, \quad \underline{\lim}_{n\to\infty}\, T_n$$

are all optional. Furthermore, if the T_n's are stopping times, then so is $\sup_{n\geq 1} T_n$.

PROOF. Obvious, from Corollary 2.4 and from the identities

$$\left\{\sup_{n\geq 1} T_n \leq t\right\} = \bigcap_{n=1}^{\infty} \{T_n \leq t\} \quad \text{and} \quad \left\{\inf_{n\geq 1} T_n < t\right\} = \bigcup_{n=1}^{\infty} \{T_n < t\}. \qquad \square$$

How can we measure the information accumulated up to a stopping time T? In order to broach this question, let us suppose that an event A is part of this information, i.e., that the occurrence or nonoccurrence of A has been decided by time T. Now if by time t one observes the value of T, which can happen only if $T \leq t$, then one must also be able to tell whether A has occurred. In other words, $A \cap \{T \leq t\}$ and $A^c \cap \{T \leq t\}$ must both be $\mathscr{F}_t$-measurable, and this must be the case for any $t \geq 0$. Since

$$A^c \cap \{T \leq t\} = \{T \leq t\} \cap (A \cap \{T \leq t\})^c,$$

it is enough to check only that $A \cap \{T \leq t\} \in \mathscr{F}_t$, $t \geq 0$.

2.12 Definition. Let T be a stopping time of the filtration $\{\mathscr{F}_t\}$. The *σ-field $\mathscr{F}_T$ of events determined prior to the stopping time* T consists of those events $A \in \mathscr{F}$ for which $A \cap \{T \leq t\} \in \mathscr{F}_t$ for every $t \geq 0$.

2.13 Problem. Verify that $\mathscr{F}_T$ is actually a σ-field and T is $\mathscr{F}_T$-measurable. Show that if $T(\omega) = t$ for some constant $t \geq 0$ and every $\omega \in \Omega$, then $\mathscr{F}_T = \mathscr{F}_t$.

2.14 Exercise. Let T be a stopping time and S a random time such that $S \geq T$ on Ω. If S is $\mathscr{F}_T$-measurable, then it is also a stopping time.

2.15 Lemma. *For any two stopping times T and S, and for any $A \in \mathscr{F}_S$, we have $A \cap \{S \leq T\} \in \mathscr{F}_T$. In particular, if $S \leq T$ on Ω, we have $\mathscr{F}_S \subseteq \mathscr{F}_T$.*

PROOF. It is not hard to verify that, for every stopping time T and positive constant t, $T \wedge t$ is an $\mathscr{F}_t$-measurable random variable. With this in mind, the claim follows from the decomposition:

$$A \cap \{S \leq T\} \cap \{T \leq t\} = [A \cap \{S \leq t\}] \cap \{T \leq t\} \cap \{S \wedge t \leq T \wedge t\},$$

which shows readily that the left-hand side is an event in $\mathscr{F}_t$. $\square$

2.16 Lemma. *Let T and S be stopping times. Then $\mathscr{F}_{T\wedge S} = \mathscr{F}_T \cap \mathscr{F}_S$, and each of the events*

$$\{T < S\}, \{S < T\}, \{T \leq S\}, \{S \leq T\}, \{T = S\}$$

belongs to $\mathscr{F}_T \cap \mathscr{F}_S$.

PROOF. For the first claim we notice from Lemma 2.15 that $\mathscr{F}_{T\wedge S} \subseteq \mathscr{F}_T \cap \mathscr{F}_S$. In order to establish the opposite inclusion, let us take $A \in \mathscr{F}_S \cap \mathscr{F}_T$ and observe that

$$A \cap \{S \wedge T \le t\} = A \cap [\{S \le t\} \cup \{T \le t\}]$$
$$= [A \cap \{S \le t\}] \cup [A \cap \{T \le t\}] \in \mathscr{F}_t,$$

and therefore $A \in \mathscr{F}_{S \wedge T}$.

From Lemma 2.15 we have $\{S \le T\} \in \mathscr{F}_T$, and thus $\{S > T\} \in \mathscr{F}_T$. On the other hand, consider the stopping time $R = S \wedge T$, which, again by virtue of Lemma 2.15, is measurable with respect to $\mathscr{F}_T$. Therefore, $\{S < T\} = \{R < T\} \in \mathscr{F}_T$. Interchanging the roles of S, T we see that $\{T > S\}$, $\{T < S\}$ belong to $\mathscr{F}_S$, and thus we have shown that both these events belong to $\mathscr{F}_T \cap \mathscr{F}_S$. But then the same is true for their complements, and consequently also for $\{S = T\}$. □

2.17 Problem. Let T, S be stopping times and Z an integrable random variable. We have

(i) $E[Z|\mathscr{F}_T] = E[Z|\mathscr{F}_{S \wedge T}]$, P-a.s. on $\{T \le S\}$
(ii) $E[E(Z|\mathscr{F}_T)|\mathscr{F}_S] = E[Z|\mathscr{F}_{S \wedge T}]$, P-a.s.

Now we can start to appreciate the usefulness of the concept of stopping time in the study of stochastic processes.

2.18 Proposition. *Let $X = \{X_t, \mathscr{F}_t; 0 \le t < \infty\}$ be a progressively measurable process, and let T be a stopping time of the filtration $\{\mathscr{F}_t\}$. Then the random variable X_T of Definition 1.15, defined on the set $\{T < \infty\} \in \mathscr{F}_T$, is $\mathscr{F}_T$-measurable, and the "stopped process" $\{X_{T \wedge t}, \mathscr{F}_t; 0 \le t < \infty\}$ is progressively measurable.*

Proof. For the first claim, one has to show that for any $B \in \mathscr{B}(\mathbb{R}^d)$ and any $t \ge 0$, the event $\{X_T \in B\} \cap \{T \le t\}$ is in $\mathscr{F}_t$; but this event can also be written in the form $\{X_{T \wedge t} \in B\} \cap \{T \le t\}$, and so it is sufficient to prove the progressive measurability of the stopped process.

To this end, one observes that the mapping $(s, \omega) \mapsto (T(\omega) \wedge s, \omega)$ of $[0, t] \times \Omega$ into itself is $\mathscr{B}([0, t]) \otimes \mathscr{F}_t$-measurable. Besides, by the assumption of progressive measurability, the mapping

$$(s, \omega) \mapsto X_s(\omega): ([0, t] \times \Omega, \mathscr{B}([0, t]) \otimes \mathscr{F}_t) \to (\mathbb{R}^d, \mathscr{B}(\mathbb{R}^d))$$

is measurable, and therefore the same is true for the composite mapping

$$(s, \omega) \mapsto X_{T(\omega) \wedge s}(\omega): ([0, t] \times \Omega, \mathscr{B}([0, t]) \otimes \mathscr{F}_t) \to (\mathbb{R}^d, \mathscr{B}(\mathbb{R}^d)). \qquad \square$$

2.19 Problem. Under the same assumptions as in Proposition 2.18, and with $f(t, x): [0, \infty) \times \mathbb{R}^d \to \mathbb{R}$ a bounded, $\mathscr{B}([0, \infty)) \otimes \mathscr{B}(\mathbb{R}^d)$-measurable function, show that the process $Y_t = \int_0^t f(s, X_s)\, ds; t \ge 0$ is progressively measurable with respect to $\{\mathscr{F}_t\}$, and Y_T is an $\mathscr{F}_T$-measurable random variable.

2.20 Definition. Let T be an optional time of the filtration $\{\mathscr{F}_t\}$. The *σ-field $\mathscr{F}_{T+}$ of events determined immediately after the optional time T* consists of those events $A \in \mathscr{F}$ for which $A \cap \{T \le t\} \in \mathscr{F}_{t+}$ for every $t \ge 0$.

2.21 Problem. Verify that the class $\mathscr{F}_{T+}$ is indeed a σ-field with respect to which T is measurable, that it coincides with $\{A \in \mathscr{F}; A \cap \{T < t\} \in \mathscr{F}_t, \forall t \ge 0\}$, and that if T is a stopping time (so that both $\mathscr{F}_T, \mathscr{F}_{T+}$ are defined), then $\mathscr{F}_T \subseteq \mathscr{F}_{T+}$.

2.22 Problem. Verify that analogues of Lemmas 2.15 and 2.16 hold if T and S are assumed to be optional and $\mathscr{F}_T$, $\mathscr{F}_S$ and $\mathscr{F}_{T \wedge S}$ are replaced by $\mathscr{F}_{T+}$, $\mathscr{F}_{S+}$ and $\mathscr{F}_{(T \wedge S)+}$, respectively. Prove that if S is an optional time and T is a *positive* stopping time with $S \le T$, and $S < T$ on $\{S < \infty\}$, then $\mathscr{F}_{S+} \subseteq \mathscr{F}_T$.

2.23 Problem. Show that if $\{T_n\}_{n=1}^\infty$ is a sequence of optional times and $T = \inf_{n \ge 1} T_n$, then $\mathscr{F}_{T+} = \bigcap_{n=1}^\infty \mathscr{F}_{T_n+}$. Besides, if each T_n is a positive stopping time and $T < T_n$ on $\{T < \infty\}$, then we have $\mathscr{F}_{T+} = \bigcap_{n=1}^\infty \mathscr{F}_{T_n}$.

2.24 Problem. Given an optional time T of the filtration $\{\mathscr{F}_t\}$, consider the sequence $\{T_n\}_{n=1}^\infty$ of random times given by

$$T_n(\omega) = \begin{cases} T(\omega); & \text{on } \{\omega; T(\omega) = +\infty\} \\ \dfrac{k}{2^n}; & \text{on } \left\{\omega; \dfrac{k-1}{2^n} \le T(\omega) < \dfrac{k}{2^n}\right\} \end{cases}$$

for $n \ge 1, k \ge 1$. Obviously $T_n \ge T_{n+1} \ge T$, for every $n \ge 1$. Show that each T_n is a stopping time, that $\lim_{n\to\infty} T_n = T$, and that for every $A \in \mathscr{F}_{T+}$ we have $A \cap \{T_n = (k/2^n)\} \in \mathscr{F}_{k/2^n}$; $n, k \ge 1$.

We close this section with a statement about the set of jumps for a stochastic process whose sample paths do not admit discontinuities of the second kind.

2.25 Definition. A filtration $\{\mathscr{F}_t\}$ is said to satisfy the *usual conditions* if it is right-continuous and $\mathscr{F}_0$ contains all the P-negligible events in $\mathscr{F}$.

2.26 Proposition. *If the process X has RCLL paths and is adapted to the filtration $\{\mathscr{F}_t\}$ which satisfies the usual conditions, then there exists a sequence $\{T_n\}_{n=1}^\infty$ of stopping times of $\{\mathscr{F}_t\}$ which exhausts the jumps of X, i.e.,*

(2.1)

$$\{(t,\omega) \in (0,\infty) \times \Omega; X_t(\omega) \ne X_{t-}(\omega)\} \subseteq \bigcup_{n=1}^\infty \{(t,\omega) \in [0,\infty) \times \Omega; T_n(\omega) = t\}.$$

The proof of this result is based on the powerful "section theorems" of the general theory of processes. It can be found in Dellacherie (1972), p. 84, or Elliott (1982), p. 61. Note that our definition of the terminology "$\{T_n\}_{n=1}^\infty$ exhausts the jumps of X" as set forth in (2.1) is a bit different from that found on p. 60 of Elliott (1982). However, the proofs in the cited references justify our version of Proposition 2.26.

1.3. Continuous-Time Martingales

We assume in this section that the reader is familiar with the concept and basic properties of martingales in discrete time. An excellent presentation of this material can be found in Chung (1974, §§9.3 and 9.4, pp. 319–341) and we shall cite from this source frequently. Alternative references are Ash (1972) and Billingsley (1979). The purpose of this section is to extend the discrete-time results to continuous-time martingales.

The standard example of a continuous-time martingale is one-dimensional Brownian motion. This process can be regarded as the continuous-time version of the one-dimensional symmetric random walk, as we shall see in Chapter 2. Since we have not yet introduced Brownian motion, we shall take instead the compensated Poisson process as a continuing example developed in the problems throughout this section. The compensated Poisson process is a martingale which will serve us later in the construction of Poisson random measures, a tool necessary for the treatment of passage and local times of Brownian motion.

In this section we shall consider exclusively real-valued processes $X = \{X_t; 0 \leq t < \infty\}$ on a probability space $(\Omega, \mathscr{F}, P)$, adapted to a given filtration $\{\mathscr{F}_t\}$ and such that $E|X_t| < \infty$ holds for every $t \geq 0$.

3.1 Definition. The process $\{X_t, \mathscr{F}_t; 0 \leq t < \infty\}$ is said to be a *submartingale* (respectively, a *supermartingale*) if, for every $0 \leq s < t < \infty$, we have, a.s. P: $E(X_t|\mathscr{F}_s) \geq X_s$ (respectively, $E(X_t|\mathscr{F}_s) \leq X_s$).

We shall say that $\{X_t, \mathscr{F}_t; 0 \leq t < \infty\}$ is a *martingale* if it is both a submartingale and a supermartingale.

3.2 Problem. Let $T_1, T_2, \ldots$ be a sequence of independent, exponentially distributed random variables with parameter $\lambda > 0$:

$$P[T_i \in dt] = \lambda e^{-\lambda t}\, dt, \quad t \geq 0.$$

Let $S_0 = 0$ and $S_n = \sum_{i=1}^n T_i$; $n \geq 1$. (We may think of S_n as the time at which the n-th customer arrives in a queue, and of the random variables T_i, $i = 1, 2, \ldots$ as the interarrival times.) Define a continuous-time, integer-valued RCLL process

$$N_t = \max\{n \geq 0;\ S_n \leq t\}; \quad 0 \leq t < \infty. \tag{3.1}$$

(We may regard N_t as the number of customers who arrive up to time t.)

(i) Show that for $0 \leq s < t$ we have

$$P[S_{N_s+1} > t | \mathscr{F}_s^N] = e^{-\lambda(t-s)}, \quad \text{a.s. } P.$$

(*Hint*: Choose $\tilde{A} \in \mathscr{F}_s^N$ and a nonnegative integer n. Show that there exists an event $A \in \sigma(T_1, \ldots, T_n)$ such that $A \cap \{N_s = n\} = \tilde{A} \cap \{N_s = n\}$, and use the independence between T_{n+1} and the pair $(S_n, 1_A)$ to establish

$$\int_{\tilde{A} \cap \{N_s = n\}} P[S_{n+1} > t | \mathscr{F}_s^N] \, dP = e^{-\lambda(t-s)} P[\tilde{A} \cap \{N_s = n\}].)$$

(ii) Show that for $0 \leq s < t$, $N_t - N_s$ is a Poisson random variable with parameter $\lambda(t-s)$, independent of $\mathscr{F}_s^N$. (*Hint*: With $\tilde{A} \in \mathscr{F}_s^N$ and $n \geq 0$ as before, use the result in (i) to establish

$$\int_{\tilde{A} \cap \{N_s = n\}} P[N_t - N_s \leq k | \mathscr{F}_s^N] \, dP$$

$$= P[\tilde{A} \cap \{N_s = n\}] \cdot \sum_{j=0}^{k} e^{-\lambda(t-s)} \frac{(\lambda(t-s))^j}{j!}$$

for every integer $k \geq 0$.)

3.3 Definition A *Poisson process with intensity* $\lambda > 0$ is an adapted, integer-valued RCLL process $N = \{N_t, \mathscr{F}_t; 0 \leq t < \infty\}$ such that $N_0 = 0$ a.s., and for $0 \leq s < t$, $N_t - N_s$ is independent of $\mathscr{F}_s$ and is Poisson distributed with mean $\lambda(t-s)$.

We have demonstrated in Problem 3.2 that the process $N = \{N_t, \mathscr{F}_t^N; 0 \leq t < \infty\}$ of (3.1) is Poisson. Given a Poisson process N with intensity λ, we define the *compensated Poisson process*

$$M_t \triangleq N_t - \lambda t, \mathscr{F}_t; \quad 0 \leq t < \infty.$$

Note that the filtrations $\{\mathscr{F}_t^M\}$ and $\{\mathscr{F}_t^N\}$ agree.

3.4 Problem. Prove that a compensated Poisson process $\{M_t, \mathscr{F}_t; t \geq 0\}$ is a martingale.

3.5 Remark. The reader should notice the decomposition $N_t = M_t + A_t$ of the (submartingale) Poisson process as the sum of the martingale M and the increasing function $A_t = \lambda t$, $t \geq 0$. A general result along these lines, due to P. A. Meyer, will be the object of the next section (Theorem 4.10).

A. Fundamental Inequalities

Consider a submartingale $\{X_t; 0 \leq t < \infty\}$, and an integrable, $\mathscr{F}_\infty$-measurable random variable X_∞; we recall here that $\mathscr{F}_\infty = \sigma(\bigcup_{t \geq 0} \mathscr{F}_t)$. If we also have, for every $0 \leq t < \infty$,

$$E(X_\infty | \mathscr{F}_t) \geq X_t \quad \text{a.s. } P,$$

then we say that "$\{X_t, \mathscr{F}_t; 0 \le t \le \infty\}$ is a submartingale with last element X_∞". We have a similar convention in the (super)martingale case.

A straightforward application of the conditional Jensen inequality (Chung (1974), Thm. 9.1.4) yields the following result.

3.6 Proposition. *Let $\{X_t, \mathscr{F}_t; 0 \le t < \infty\}$ be a martingale (respectively, submartingale), and $\varphi: \mathbb{R} \to \mathbb{R}$ a convex (respectively, convex nondecreasing) function, such that $E|\varphi(X_t)| < \infty$ holds for every $t \ge 0$. Then $\{\varphi(X_t), \mathscr{F}_t; 0 \le t < \infty\}$ is a submartingale.*

The method used to prove Jensen's inequality and Proposition 3.6 extends to the vector situation of the next problem.

3.7 Problem. Let $\{X_t = (X_t^{(1)}, \ldots, X_t^{(d)}), \mathscr{F}_t; 0 \le t < \infty\}$ be a vector of martingales, and $\varphi: \mathbb{R}^d \to \mathbb{R}$ a convex function with $E|\varphi(X_t)| < \infty$ valid for every $t \ge 0$. Then $\{\varphi(X_t), \mathscr{F}_t; 0 \le t < \infty\}$ is a submartingale; in particular $\{\|X_t\|, \mathscr{F}_t; 0 \le t < \infty\}$ is a submartingale.

Let $X = \{X_t; 0 \le t < \infty\}$ be a real-valued stochastic process. Consider two numbers $\alpha < \beta$ and a finite subset F of $[0, \infty)$. We define the *number of upcrossings* $U_F(\alpha, \beta; X(\omega))$ of the interval $[\alpha, \beta]$ by the restricted sample path $\{X_t; t \in F\}$ as follows. Set

$$\tau_1(\omega) = \min\{t \in F; X_t(\omega) \le \alpha\},$$

and define recursively for $j = 1, 2, \ldots$

$$\sigma_j(\omega) = \min\{t \in F; t \ge \tau_j(\omega), X_t(\omega) > \beta\},$$

$$\tau_{j+1}(\omega) = \min\{t \in F; t \ge \sigma_j(\omega), X_t(\omega) < \alpha\}.$$

The convention here is that the minimum of empty set is $+\infty$, and we denote by $U_F(\alpha, \beta; X(\omega))$ the largest integer j for which $\sigma_j(\omega) < \infty$. If $I \subset [0, \infty)$ is not necessarily finite, we define

$$U_I(\alpha, \beta; X(\omega)) = \sup\{U_F(\alpha, \beta; X(\omega)); F \subseteq I, F \text{ is finite}\}.$$

The *number of downcrossings* $D_I(\alpha, \beta; X(\omega))$ is defined similarly.

The following theorem extends to the continuous-time case certain well-known results of discrete martingales.

3.8 Theorem. *Let $\{X_t, \mathscr{F}_t; 0 \le t < \infty\}$ be a submartingale whose every path is right-continuous, let $[\sigma, \tau]$ be a subinterval of $[0, \infty)$, and let $\alpha < \beta$, $\lambda > 0$ be real numbers. We have the following results:*

(i) First submartingale inequality:

$$\lambda \cdot P\left[\sup_{\sigma \le t \le \tau} X_t \ge \lambda\right] \le E(X_\tau^+).$$

(ii) Second submartingale inequality:

$$\lambda \cdot P\left[\inf_{\sigma \le t \le \tau} X_t \le -\lambda\right] \le E(X_\tau^+) - E(X_\sigma).$$

(iii) Upcrossings and downcrossings inequalities:

$$EU_{[\sigma,\tau]}(\alpha, \beta; X(\omega)) \le \frac{E(X_\tau^+) + |\alpha|}{\beta - \alpha}, \quad ED_{[\sigma,\tau]}(\alpha, \beta; X(\omega)) \le \frac{E(X_\tau - \alpha)^+}{\beta - \alpha}.$$

(iv) Doob's maximal inequality:

$$E\left(\sup_{\sigma \le t \le \tau} X_t\right)^p \le \left(\frac{p}{p-1}\right)^p E(X_\tau^p), \quad p > 1,$$

provided $X_t \ge 0$ a.s. P *for every* $t \ge 0$, *and* $E(X_\tau^p) < \infty$.

(v) Regularity of the paths: *Almost every sample path* $\{X_t(\omega); 0 \le t < \infty\}$ *is bounded on compact intervals; is free of discontinuities of the second kind, i.e., admits left-hand limits everywhere on* $(0, \infty)$; *and if the filtration* $\{\mathscr{F}_t\}$ *satisfies the usual conditions, then the jumps are exhausted by a sequence of stopping times* (*Proposition* 2.26).

PROOF. Let the finite set F consist of σ, τ, and a finite subset of $[\sigma, \tau] \cap Q$. We obtain from Theorem 9.4.1 of Chung (1974): $\mu P[\max_{t \in F} X_t > \mu] \le E(X_\tau^+)$ as well as: $\mu P[\min_{t \in F} X_t < -\mu] \le E(X_\tau^+) - E(X_\sigma)$. By considering an increasing sequence $\{F_n\}_{n=1}^\infty$ of finite sets whose union is the whole of $([\sigma, \tau] \cap Q) \cup \{\sigma, \tau\}$, we may replace F by this union in the preceding inequalities. The right-continuity of sample paths implies then $\mu P[\sup_{\sigma \le t \le \tau} X_t > \mu] \le E(X_\tau^+)$ and $\mu P[\inf_{\sigma \le t \le \tau} X_t < -\mu] \le E(X_\tau^+) - E(X_\sigma)$. Finally, we let $\mu \uparrow \lambda$ to obtain (i) and (ii).

Being the limit of random variables of the form $U_F(\alpha, \beta; X(\omega))$ with finite F, $U_{[\sigma,\tau]}(\alpha, \beta; X(\omega))$ is measurable. We obtain (iii), (iv) from Theorems 9.4.2, 9.5.4 in Chung (1974) (see also Meyer (1966), pp. 93–94). For (v), we note first that the boundedness of (almost all) sample paths on the compact interval $[0, n]$, $n \ge 1$, follows directly from (i), (ii); second, we consider the events

$$A_{\alpha,\beta}^{(n)} \triangleq \{\omega \in \Omega; U_{[0,n]}(\alpha, \beta; X(\omega)) = \infty\}, \quad n \ge 1, \alpha < \beta.$$

By virtue of (iii), these have zero probability, and the same is true for the union

$$A^{(n)} = \bigcup_{\substack{\alpha < \beta \\ \alpha, \beta \in Q}} A_{\alpha,\beta}^{(n)},$$

which includes the set

$$\left\{\omega \in \Omega; \varliminf_{s \uparrow t} X_s(\omega) < \varlimsup_{s \uparrow t} X_s(\omega), \text{ for some } t \in [0, n]\right\}.$$

Consequently, for every $\omega \in \Omega \backslash A^{(n)}$, the left limit $X_{t-}(\omega) = \lim_{s \uparrow t} X_s(\omega)$ exists for all $0 < t \le n$. This is true for every $n \ge 1$, so the preceding left limit exists for every $0 < t < \infty$, $\omega \in (\bigcup_{n=1}^\infty A^{(n)})^c$. □

3.9 Problem. Let N be a Poisson process with intensity λ.

(a) For any $c > 0$,

$$\overline{\lim_{t\to\infty}} P\left[\sup_{0\le s\le t} (N_s - \lambda s) \ge c\sqrt{\lambda t}\right] \le \frac{1}{c\sqrt{2\pi}}$$

(b) For any $c > 0$,

$$\overline{\lim_{t\to\infty}} P\left[\inf_{0\le s\le t} (N_s - \lambda s) \le -c\sqrt{\lambda t}\right] \le \frac{1}{c\sqrt{2\pi}}.$$

(c) For $0 < \sigma < \tau$, we have

$$E\left[\sup_{\sigma\le t\le\tau} \left(\frac{N_t}{t} - \lambda\right)^2\right] \le \frac{4\tau\lambda}{\sigma^2}.$$

(*Hint*: Use Stirling's approximation to show that $\lim_{t\to\infty}(1/\sqrt{\lambda t})E(N_t - \lambda t)^+ = 1/\sqrt{2\pi}$.)

3.10 Remark. From Problem 3.9 (a) and (b), we see that for each $c > 0$, there exists $T_c > 0$ such that

$$P\left[\left|\frac{N_t}{t} - \lambda\right| \ge c\sqrt{\frac{\lambda}{t}}\right] \le \frac{3}{c\sqrt{2\pi}}, \quad \forall t \ge T_c.$$

From this we can conclude the *weak law of large numbers* for Poisson processes: $(N_t/t) \to \lambda$, in probability as $t \to \infty$. In fact, by choosing $\sigma = 2^n$ and $\tau = 2^{n+1}$ in Problem 3.9 (c) and using Čebyšev's inequality, one can show

$$P\left[\sup_{2^n\le t\le 2^{n+1}} \left|\frac{N_t}{t} - \lambda\right| \ge \varepsilon\right] \le \frac{8\lambda}{\varepsilon^2 2^n}$$

for every $n \ge 1$, $\varepsilon > 0$. Then by a Borel-Cantelli argument (see Chung (1974), Theorems 4.2.1, 4.2.2), we obtain the *strong law of large numbers* for Poisson processes: $\lim_{t\to\infty}(N_t/t) = \lambda$, a.s. P.

The following result from the discrete-parameter theory will be used repeatedly in the sequel; it is contained in the proof of Theorem 9.4.7 in Chung (1974), but it deserves to be singled out and reviewed.

3.11 Problem. Let $\{\mathscr{F}_n\}_{n=1}^\infty$ be a decreasing sequence of sub-σ-fields of $\mathscr{F}$ (i.e., $\mathscr{F}_{n+1} \subseteq \mathscr{F}_n \subseteq \mathscr{F}$, $\forall n \ge 1$), and let $\{X_n, \mathscr{F}_n; n \ge 1\}$ be a *backward submartingale*; i.e., $E|X_n| < \infty$, X_n is $\mathscr{F}_n$-measurable, and $E(X_n|\mathscr{F}_{n+1}) \ge X_{n+1}$ a.s. P, for every $n \ge 1$. Then $l \triangleq \lim_{n\to\infty} E(X_n) > -\infty$ implies that the sequence $\{X_n\}_{n=1}^\infty$ is uniformly integrable.

3.12 Remark. If $\{X_t, \mathscr{F}_t; 0 \le t < \infty\}$ is a submartingale and $\{t_n\}_{n=1}^\infty$ is a nonincreasing sequence of nonnegative numbers, then $\{X_{t_n}, \mathscr{F}_{t_n}; n \ge 1\}$ is a backward submartingale.

It was supposed in Theorem 3.8 that the submartingale X has right-continuous sample paths. It is of interest to investigate conditions under which we may assume this to be the case.

3.13 Theorem. *Let $X = \{X_t, \mathscr{F}_t; 0 \le t < \infty\}$ be a submartingale, and assume the filtration $\{\mathscr{F}_t\}$ satisfies the usual conditions. Then the process X has a right-continuous modification if and only if the function $t \mapsto EX_t$ from $[0, \infty)$ to $\mathbb{R}$ is right-continuous. If this right-continuous modification exists, it can be chosen so as to be RCLL and adapted to $\{\mathscr{F}_t\}$, hence a submartingale with respect to $\{\mathscr{F}_t\}$.*

The proof of Theorem 3.13 requires the following proposition, which we establish first.

3.14 Proposition. *Let $X = \{X_t, \mathscr{F}_t; 0 \le t < \infty\}$ be a submartingale. We have the following:*

(i) *There is an event $\Omega^* \in \mathscr{F}$ with $P(\Omega^*) = 1$, such that for every $\omega \in \Omega^*$:*

$$\text{the limits } X_{t+}(\omega) \triangleq \lim_{\substack{s \downarrow t \\ s \in Q}} X_s(\omega), \quad X_{t-} \triangleq \lim_{\substack{s \uparrow t \\ s \in Q}} X_s(\omega)$$

exist for all $t \ge 0$ (respectively, $t > 0$).

(ii) *The limits in (i) satisfy*

$$E(X_{t+} | \mathscr{F}_t) \ge X_t \quad \text{a.s. } P, \forall\, t \ge 0.$$

$$E(X_t | \mathscr{F}_{t-}) \ge X_{t-} \quad \text{a.s. } P, \forall\, t > 0.$$

(iii) *$\{X_{t+}, \mathscr{F}_{t+}; 0 \le t < \infty\}$ is a submartingale with P-almost every path RCLL.*

PROOF.

(i) We wish to imitate the proof of (v), Theorem 3.8, but because we have not assumed right-continuity of sample paths, we may not use (iii) of Theorem 3.8 to argue that the events $A^{(n)}_{\alpha,\beta}$ appearing in that proof have probability zero. Thus, we alter the definition slightly by considering the submartingale X evaluated only at rational times, and setting

$$A^{(n)}_{\alpha,\beta} = \{\omega \in \Omega; U_{[0,n] \cap Q}(\alpha, \beta; X(\omega)) = \infty\}, \quad n \ge 1, \alpha < \beta,$$

$$A^{(n)} = \bigcup_{\substack{\alpha < \beta \\ \alpha, \beta \in Q}} A^{(n)}_{\alpha,\beta}.$$

Then each $A^{(n)}_{\alpha,\beta}$ has probability zero, as does each $A^{(n)}$. The conclusions follow readily.

(ii) Let $\{t_n\}_{n=1}^{\infty}$ be a sequence of rational numbers in (t, ∞), monotonically decreasing to $t \ge 0$ as $n \to \infty$. Then $\{X_{t_n}, \mathscr{F}_{t_n}; n \ge 1\}$ is a backward submartingale, and the sequence $\{E(X_{t_n})\}_{n=1}^{\infty}$ is decreasing and bounded below by $E(X_t)$. Problem 3.11 tells us that $\{X_{t_n}\}_{n=1}^{\infty}$ is a uniformly integrable sequence. From the submartingale property we have $\int_A X_t \, dP \le \int_A X_{t_n} \, dP$, for every $n \ge 1$ and $A \in \mathscr{F}_t$; uniform integrability renders almost sure into L^1-convergence (Chung (1974), Theorem 4.5.4), and by letting $n \to \infty$ we obtain $\int_A X_t \, dP \le \int_A X_{t+} \, dP$, for every $A \in \mathscr{F}_t$. The first inequality in (ii) follows.

Now take a sequence $\{t_n\}_{n=1}^\infty$ in $(0, t) \cap Q$, monotonically increasing to $t > 0$. According to the submartingale property $E[X_t | \mathscr{F}_{t_n}] \ge X_{t_n}$ a.s. We may let $n \to \infty$ and use Lévy's theorem (Chung (1974), Theorem 9.4.8) to obtain the second inequality in (ii).

(iii) Take a monotone decreasing sequence $\{s_n\}_{n=1}^\infty$ of rational numbers, with $0 \le s < s_n < t$ holding for every $n \ge 1$, and $\lim_{n\to\infty} s_n = s$. According to the first part of (ii), $E(X_{t+} | \mathscr{F}_{s_n}) \ge X_{s_n}$ a.s. Letting $n \to \infty$ and using Lévy's theorem again, we obtain the submartingale property $E(X_{t+} | \mathscr{F}_{s+}) \ge X_{s+}$ a.s. It is not difficult to show, using (i) and Theorem 3.8(v), that P-almost every path $t \mapsto X_{t+}$ is RCLL. □

Proof of Theorem 3.13. Assume that the function $t \mapsto EX_t$ is right-continuous; we show that $\{X_{t+}, \mathscr{F}_t; 0 \le t < \infty\}$ as defined in Proposition 3.14 is a modification of X. The former process is adapted because of the right-continuity of $\{\mathscr{F}_t\}$. Given $t \ge 0$, let $\{q_n\}_{n=1}^\infty$ be a sequence of rational numbers with $q_n \downarrow t$. Then $\lim_{n\to\infty} X_{q_n} = X_{t+}$, a.s., and uniform integrability implies that $EX_{t+} = \lim_{n\to\infty} EX_{q_n}$. By assumption, $\lim_{n\to\infty} EX_{q_n} = EX_t$, and Proposition 3.14 (ii) gives $X_{t+} \ge X_t$, a.s. It follows that $X_{t+} = X_t$, a.s.

Conversely, suppose that $\{\tilde{X}_t; 0 \le t < \infty\}$ is a right-continuous modification of X. Fix $t \ge 0$ and let $\{t_n\}_{n=1}^\infty$ be a sequence of numbers with $t_n \downarrow t$. We have $P[X_t = \tilde{X}_t, X_{t_n} = \tilde{X}_{t_n}; n \ge 1] = 1$ and $\lim_{n\to\infty} \tilde{X}_{t_n} = \tilde{X}_t$, a.s. Therefore, $\lim_{n\to\infty} X_{t_n} = X_t$ a.s., and the uniform integrability of $\{X_{t_n}\}_{n=1}^\infty$ implies that $EX_t = \lim_{n\to\infty} EX_{t_n}$. The right-continuity of the function $t \mapsto EX_t$ follows. □

B. Convergence Results

For the remainder of this section, we deal only with right-continuous processes, usually imposing no condition on the filtrations $\{\mathscr{F}_t\}$. Thus, the description *right-continuous* in phrases such as "right-continuous martingale" refers to the sample paths and not the filtration. It will be obvious that the assumption of right-continuity can be replaced in these results by the assumption of right-continuity for P-almost every sample path.

3.15 Theorem (Submartingale Convergence). *Let $\{X_t, \mathscr{F}_t; 0 \le t < \infty\}$ be a right-continuous submartingale and assume $C \triangleq \sup_{t\ge 0} E(X_t^+) < \infty$. Then $X_\infty(\omega) \triangleq \lim_{t\to\infty} X_t(\omega)$ exists for a.e. $\omega \in \Omega$, and $E|X_\infty| < \infty$.*

Proof. From Theorem 3.8 (iii) we have for any $n \ge 1$ and real numbers $\alpha < \beta$: $EU_{[0,n]}(\alpha, \beta; X(\omega)) \le (E(X_n^+) + |\alpha|)/(\beta - \alpha)$, and by letting $n \to \infty$ we obtain, thanks to the monotone convergence theorem:

$$EU_{[0,\infty)}(\alpha, \beta; X(\omega)) \le \frac{C + |\alpha|}{\beta - \alpha}.$$

The events $A_{\alpha,\beta} \triangleq \{\omega; U_{[0,\infty)}(\alpha, \beta; X(\omega)) = \infty\}$, $-\infty < \alpha < \beta < \infty$, are thus P-negligible, and the same is true for the event $A = \bigcup_{\substack{\alpha<\beta\\ \alpha,\beta \in Q}} A_{\alpha,\beta}$, which contains the set $\{\omega; \overline{\lim}_{t\to\infty} X_t(\omega) > \underline{\lim}_{t\to\infty} X_t(\omega)\}$.

Therefore, for every $\omega \in \Omega \backslash A$, $X_\infty(\omega) = \lim_{t\to\infty} X_t(\omega)$ exists. Moreover,

$$E|X_t| = 2E(X_t^+) - E(X_t) \le 2C - EX_0$$

shows that the assumption $\sup_{t\ge0} E(X_t^+) < \infty$ is equivalent to the apparently stronger one $\sup_{t\ge0} E|X_t| < \infty$, which in turn forces the integrability of X_∞, by Fatou's lemma. □

3.16 Problem. Let $\{X_t, \mathscr{F}_t; 0 \le t < \infty\}$ be a right-continuous, nonnegative supermartingale; then $X_\infty(\omega) = \lim_{t\to\infty} X_t(\omega)$ exists for P-a.e. $\omega \in \Omega$, and $\{X_t, \mathscr{F}_t; 0 \le t \le \infty\}$ is a supermartingale.

3.17 Definition. A right-continuous, nonnegative supermartingale $\{Z_t, \mathscr{F}_t; 0 \le t < \infty\}$ with $\lim_{t\to\infty} E(Z_t) = 0$ is called a *potential*.

Problem 3.16 guarantees that a potential $\{Z_t, \mathscr{F}_t; 0 \le t < \infty\}$ has a last element Z_∞, and $Z_\infty = 0$ a.s. P.

3.18 Exercise. Suppose that the filtration $\{\mathscr{F}_t\}$ satisfies the usual conditions. Then every right-continuous, uniformly integrable supermartingale $\{X_t, \mathscr{F}_t; 0 \le t < \infty\}$ admits the *Riesz decomposition* $X_t = M_t + Z_t$, a.s. P, as the sum of a right-continuous, uniformly integrable martingale $\{M_t, \mathscr{F}_t; 0 \le t < \infty\}$ and a potential $\{Z_t, \mathscr{F}_t; 0 \le t < \infty\}$.

3.19 Problem. The following three conditions are equivalent for a nonnegative, right-continuous submartingale $\{X_t, \mathscr{F}_t; 0 \le t < \infty\}$:

(a) it is a uniformly integrable family of random variables;
(b) it converges in L^1, as $t \to \infty$;
(c) it converges P a.s. (as $t \to \infty$) to an integrable random variable X_∞, such that $\{X_t, \mathscr{F}_t; 0 \le t \le \infty\}$ is a submartingale.

Observe that the implications (a) ⇒ (b) ⇒ (c) hold without the assumption of nonnegativity.

3.20 Problem. The following four conditions are equivalent for a right-continuous martingale $\{X_t, \mathscr{F}_t; 0 \le t < \infty\}$:

(a), (b) as in Problem 3.19;
(c) it converges P a.s. (as $t \to \infty$) to an integrable random variable X_∞, such that $\{X_t, \mathscr{F}_t; 0 \le t \le \infty\}$ is a martingale;

(d) there exists an integrable random variable Y, such that $X_t = E(Y|\mathscr{F}_t)$ a.s. P, for every $t \geq 0$.

Besides, if (d) holds and X_∞ is the random variable in (c), then

$$E(Y|\mathscr{F}_\infty) = X_\infty \quad \text{a.s. } P.$$

3.21 Problem. Let $\{N_t, \mathscr{F}_t; 0 \leq t < \infty\}$ be a Poisson process with parameter $\lambda > 0$. For $u \in \mathbb{C}$ and $i = \sqrt{-1}$, define the process

$$X_t = \exp[iuN_t - \lambda t(e^{iu} - 1)]; \quad 0 \leq t < \infty.$$

(i) Show that $\{\text{Re}(X_t), \mathscr{F}_t; 0 \leq t < \infty\}$, $\{\text{Im}(X_t), \mathscr{F}_t; 0 \leq t < \infty\}$ are martingales.
(ii) Consider X with $u = -i$. Does this martingale satisfy the equivalent conditions of Problem 3.20?

C. The Optional Sampling Theorem

What can happen if one samples a martingale at random, instead of fixed, times? For instance, if X_t represents the fortune, at time t, of an indefatigable gambler (who plays continuously!) engaged in a "fair" game, can he hope to improve his expected fortune by judicious choice of the time to quit? If no clairvoyance into the future is allowed (in other words, if our gambler is restricted to quit at stopping times), and if there is any justice in the world, the answer should be "no." Doob's *optional sampling theorem* tells us under what conditions we can expect this to be true.

3.22 Theorem (Optional Sampling). *Let $\{X_t, \mathscr{F}_t; 0 \leq t \leq \infty\}$ be a right-continuous submartingale with a last element X_∞, and let $S \leq T$ be two optional times of the filtration $\{\mathscr{F}_t\}$. We have*

$$E(X_T|\mathscr{F}_{S+}) \geq X_S \quad \text{a.s. } P.$$

If S is a stopping time, then $\mathscr{F}_S$ can replace $\mathscr{F}_{S+}$ above. In particular, $EX_T \geq EX_0$, and for a martingale with a last element we have $EX_T = EX_0$.

Proof. Consider the sequence of random times

$$S_n(\omega) = \begin{cases} S(\omega) & \text{if} \quad S(\omega) = +\infty \\ \dfrac{k}{2^n} & \text{if} \quad \dfrac{k-1}{2^n} \leq S(\omega) < \dfrac{k}{2^n}, \end{cases}$$

and the similarly defined sequences $\{T_n\}$. These were shown in Problem 2.24 to be stopping times. For every fixed integer $n \geq 1$, both S_n and T_n take on a countable number of values and we also have $S_n \leq T_n$. Therefore, by the "discrete" optional sampling Theorem 9.3.5 in Chung (1974) we have

$\int_A X_{S_n}\,dP \le \int_A X_{T_n}\,dP$ for every $A \in \mathscr{F}_{S_n}$, and a fortiori for every $A \in \mathscr{F}_{S+} = \bigcap_{n=1}^{\infty} \mathscr{F}_{S_n}$, by virtue of Problem 2.23. If S is a stopping time, then $S \le S_n$ implies $\mathscr{F}_S \subseteq \mathscr{F}_{S_n}$ as in Lemma 2.15, and the preceding inequality also holds for every $A \in \mathscr{F}_S$.

It is checked similarly that $\{X_{S_n}, \mathscr{F}_{S_n}; n \ge 1\}$ is a backward submartingale, with $\{E(X_{S_n})\}_{n=1}^{\infty}$ decreasing and bounded below by $E(X_0)$. Therefore, the sequence of random variables $\{X_{S_n}\}_{n=1}^{\infty}$ is uniformly integrable (Problem 3.11), and the same is of course true for $\{X_{T_n}\}_{n=1}^{\infty}$. The process is right-continuous, so $X_T(\omega) = \lim_{n\to\infty} X_{T_n}(\omega)$ and $X_S(\omega) = \lim_{n\to\infty} X_{S_n}(\omega)$ hold for a.e. $\omega \in \Omega$. It follows from uniform integrability that X_T, X_S are integrable, and that $\int_A X_S\,dP \le \int_A X_T\,dP$ holds for every $A \in \mathscr{F}_{S+}$. □

3.23 Problem. Establish the optional sampling theorem for a right-continuous submartingale $\{X_t, \mathscr{F}_t; 0 \le t < \infty\}$ and optional times $S \le T$ under either of the following two conditions:

(i) T is a *bounded* optional time (there exists a number $a > 0$, such that $T \le a$);
(ii) there exists an integrable random variable Y, such that $X_t \le E(Y|\mathscr{F}_t)$ a.s. P, for every $t \ge 0$.

3.24 Problem. Suppose that $\{X_t, \mathscr{F}_t; 0 \le t < \infty\}$ is a right-continuous submartingale and $S \le T$ are stopping times of $\{\mathscr{F}_t\}$. Then

(i) $\{X_{T\wedge t}, \mathscr{F}_t; 0 \le t < \infty\}$ is a submartingale;
(ii) $E[X_{T\wedge t}|\mathscr{F}_S] \ge X_{S\wedge t}$ a.s. P, for every $t \ge 0$.

3.25 Problem. A submartingale of constant expectation, i.e., with $E(X_t) = E(X_0)$ for every $t \ge 0$, is a martingale.

3.26 Problem. A right-continuous process $X = \{X_t, \mathscr{F}_t; 0 \le t < \infty\}$ with $E|X_t| < \infty$; $0 \le t < \infty$ is a submartingale if and only if for every pair $S \le T$ of bounded stopping times of the filtration $\{\mathscr{F}_t\}$ we have

$$E(X_T) \ge E(X_S). \tag{3.2}$$

3.27 Problem. Let T be a bounded stopping time of the filtration $\{\mathscr{F}_t\}$, which satisfies the usual conditions, and define $\tilde{\mathscr{F}}_t = \mathscr{F}_{T+t}$; $t \ge 0$. Then $\{\tilde{\mathscr{F}}_t\}$ also satisfies the usual conditions.

(i) If $X = \{X_t, \mathscr{F}_t; 0 \le t < \infty\}$ is a right-continuous submartingale, then so is $\tilde{X} = \{\tilde{X}_t \triangleq X_{T+t} - X_T, \tilde{\mathscr{F}}_t; 0 \le t < \infty\}$.
(ii) If $\tilde{X} = \{\tilde{X}_t, \tilde{\mathscr{F}}_t; 0 \le t < \infty\}$ is a right-continuous submartingale with $\tilde{X}_0 = 0$, a.s. P, then $X = \{X_t \triangleq \tilde{X}_{(t-T)\vee 0}, \mathscr{F}_t; 0 \le t < \infty\}$ is also a submartingale.

3.28 Problem. Let $Z = \{Z_t, \mathscr{F}_t; 0 \le t < \infty\}$ be a continuous, nonnegative martingale with $Z_\infty \triangleq \lim_{t\to\infty} Z_t = 0$, a.s. P. Then for every $s \ge 0$, $b > 0$:

(i) $P\left[\sup_{t>s} Z_t \geq b | \mathscr{F}_s\right] = \frac{1}{b} Z_s$, a.s. on $\{Z_s < b\}$.

(ii) $P\left[\sup_{t\geq s} Z_t \geq b\right] = P[Z_s \geq b] + \frac{1}{b} E[Z_s 1_{\{Z_s<b\}}]$.

3.29 Problem. Let $\{X_t, \mathscr{F}_t; 0 \leq t < \infty\}$ be a continuous, nonnegative supermartingale and $T = \inf\{t \geq 0; X_t = 0\}$. Show that

$$X_{T+t} = 0; \quad 0 \leq t < \infty \quad \text{holds a.s. on } \{T < \infty\}.$$

3.30 Exercise. Suppose that the filtration $\{\mathscr{F}_t\}$ satisfies the usual conditions and let $X^{(n)} = \{X_t^{(n)}, \mathscr{F}_t; 0 \leq t < \infty\}, n \geq 1$ be an increasing sequence of right-continuous supermartingales, such that the random variable $\xi_t \triangleq \lim_{n\to\infty} X_t^{(n)}$ is nonnegative and integrable for every $0 \leq t < \infty$. Then there exists an RCLL supermartingale $X = \{X_t, \mathscr{F}_t; 0 \leq t < \infty\}$ which is a modification of the process $\xi = \{\xi_t, \mathscr{F}_t; 0 \leq t < \infty\}$.

1.4. The Doob-Meyer Decomposition

This section is devoted to the decomposition of certain submartingales as the summation of a martingale and an increasing process (Theorem 4.10, already presaged by Remark 3.5). We develop first the necessary discrete-time results.

4.1 Definition. Consider a probability space $(\Omega, \mathscr{F}, P)$ and a random sequence $\{A_n\}_{n=0}^{\infty}$ adapted to the discrete filtration $\{\mathscr{F}_n\}_{n=0}^{\infty}$. The sequence is called *increasing*, if for P-a.e. $\omega \in \Omega$ we have $0 = A_0(\omega) \leq A_1(\omega) \leq \cdots$, and $E(A_n) < \infty$ holds for every $n \geq 1$.

An increasing sequence is called *integrable* if $E(A_\infty) < \infty$, where $A_\infty = \lim_{n\to\infty} A_n$. An arbitrary random sequence $\{\xi_n\}_{n=0}^{\infty}$ is called *predictable* for the filtration $\{\mathscr{F}_n\}_{n=0}^{\infty}$, if for every $n \geq 1$ the random variable ξ_n is $\mathscr{F}_{n-1}$-measurable. Note that if $A = \{A_n, \mathscr{F}_n; n = 0, 1, \ldots\}$ is predictable with $E|A_n| < \infty$ for every n, and if $\{M_n, \mathscr{F}_n; n = 0, 1, \ldots\}$ is a bounded martingale, then the *martingale transform* of A by M defined by

$$Y_0 = 0 \quad \text{and} \quad Y_n = \sum_{k=1}^{n} A_k(M_k - M_{k-1}); \quad n \geq 1, \tag{4.1}$$

is itself a martingale. This martingale transform is the discrete-time version of the stochastic integral with respect to a martingale, defined in Chapter 3. A fundamental property of such integrals is that they are martingales when parametrized by their upper limit of integration.

Let us recall from Chung (1974), Theorem 9.3.2 and Exercise 9.3.9, that any submartingale $\{X_n, \mathscr{F}_n; n = 0, 1, \ldots\}$ admits the *Doob decomposition* $X_n = M_n + A_n$ as the summation of a martingale $\{M_n, \mathscr{F}_n\}$ and an increasing sequence $\{A_n, \mathscr{F}_n\}$. It suffices for this to take $A_0 = 0$ and $A_{n+1} = A_n - X_n + E(X_{n+1}|\mathscr{F}_n) = \sum_{k=0}^{n} [E(X_{k+1}|\mathscr{F}_k) - X_k]$, for $n \geq 0$. This increasing sequence is actually predictable, and with this proviso the Doob decomposition of a submartingale is unique.

We shall try in this section to extend the Doob decomposition to suitable continuous-time submartingales. In order to motivate the developments, let us discuss the concept of predictability for stochastic sequences in some further detail.

4.2 Definition. An increasing sequence $\{A_n, \mathscr{F}_n; n = 0, 1, \ldots\}$ is called *natural* if for every bounded martingale $\{M_n, \mathscr{F}_n; n = 0, 1, \ldots\}$ we have

$$E(M_n A_n) = E \sum_{k=1}^{n} M_{k-1}(A_k - A_{k-1}), \quad \forall n \geq 1. \tag{4.2}$$

A simple rewriting of (4.1) shows that an increasing sequence A is natural if and only if the martingale transform $Y = \{Y_n\}_{n=0}^{\infty}$ of A by every bounded martingale M satisfies $EY_n = 0$, $n \geq 0$. It is clear then from our discussion of martingale transforms that every predictable increasing sequence is natural. We now prove the equivalence of these two concepts.

4.3 Proposition. *An increasing random sequence A is predictable if and only if it is natural.*

PROOF. Suppose that A is natural and M is a bounded martingale. With $\{Y_n\}_{n=0}^{\infty}$ defined by (4.1), we have

$$E[A_n(M_n - M_{n-1})] = EY_n - EY_{n-1} = 0, \quad n \geq 1.$$

It follows that

$$\begin{aligned} E[M_n\{A_n - E(A_n|\mathscr{F}_{n-1})\}] &= E[(M_n - M_{n-1})A_n] \\ &\quad + E[M_{n-1}\{A_n - E(A_n|\mathscr{F}_{n-1})\}] \\ &\quad - E[(M_n - M_{n-1})E(A_n|\mathscr{F}_{n-1})] = 0 \end{aligned} \tag{4.3}$$

for every $n \geq 1$. Let us take an arbitrary but fixed integer $n \geq 1$, and show that the random variable A_n is $\mathscr{F}_{n-1}$-measurable. Consider (4.3) for this fixed integer, with the martingale M given by

$$M_k = \begin{cases} \operatorname{sgn}[A_n - E(A_n|\mathscr{F}_{n-1})], & k = n, \\ M_n, & k > n, \\ E(M_n|\mathscr{F}_k), & k = 0, 1, \ldots, n. \end{cases}$$

We obtain $E|A_n - E(A_n|\mathscr{F}_{n-1})| = 0$, whence the desired conclusion. □

From now on we shall revert to our filtration $\{\mathscr{F}_t\}$ parametrized by $t \in [0, \infty)$ on the probability space $(\Omega, \mathscr{F}, P)$. Let us consider a process $A = \{A_t; 0 \le t < \infty\}$ adapted to $\{\mathscr{F}_t\}$. By analogy with Definitions 4.1 and 4.2, we have the following:

4.4 Definition. An adapted process A is called *increasing* if for P-a.e. $\omega \in \Omega$ we have

(a) $A_0(\omega) = 0$
(b) $t \mapsto A_t(\omega)$ is a nondecreasing, right-continuous function,

and $E(A_t) < \infty$ holds for every $t \in [0, \infty)$. An increasing process is called *integrable* if $E(A_\infty) < \infty$, where $A_\infty = \lim_{t\to\infty} A_t$.

4.5 Definition. An increasing process A is called *natural* if for every bounded, right-continuous martingale $\{M_t, \mathscr{F}_t; 0 \le t < \infty\}$ we have

$$E\int_{(0,t]} M_s\, dA_s = E\int_{(0,t]} M_{s-}\, dA_s, \quad \text{for every } 0 < t < \infty. \tag{4.4}$$

4.6 Remarks.

(i) If A is an increasing and X a measurable process, then with $\omega \in \Omega$ fixed, the sample path $\{X_t(\omega); 0 \le t < \infty\}$ is a measurable function from $[0, \infty)$ into $\mathbb{R}$. It follows that the Lebesgue-Stieltjes integrals

$$I_t^{\pm}(\omega) \triangleq \int_{(0,t]} X_s^{\pm}(\omega)\, dA_s(\omega)$$

are well defined. If X is progressively measurable (e.g., right-continuous and adapted), and if $I_t = I_t^+ - I_t^-$ is well defined and finite for all $t \ge 0$, then I is right-continuous and progressively measurable.

(ii) Every continuous, increasing process is natural. Indeed then, for P-a.e. $\omega \in \Omega$ we have

$$\int_{(0,t]} (M_s(\omega) - M_{s-}(\omega))\, dA_s(\omega) = 0 \quad \text{for every } 0 < t < \infty,$$

because every path $\{M_s(\omega); 0 \le s < \infty\}$ has only countably many discontinuities (Theorem 3.8(v)).

(iii) It can be shown that every natural increasing process is adapted to the filtration $\{\mathscr{F}_{t-}\}$ (see Liptser & Shiryaev (1977), Theorem 3.10), provided that $\{\mathscr{F}_t\}$ satisfies the usual conditions.

4.7 Lemma. *If A is an increasing process and $\{M_t, \mathscr{F}_t; 0 \le t < \infty\}$ is a bounded, right-continuous martingale, then*

$$E(M_t A_t) = E\int_{(0,t]} M_s\, dA_s. \tag{4.5}$$

In particular, condition (4.4) *in Definition* 4.5 *is equivalent to*

$$E(M_t A_t) = E\int_{(0,t]} M_{s-}\,dA_s. \tag{4.4$'$}$$

PROOF. Consider a partition $\Pi = \{t_0, t_1, \ldots, t_n\}$ of $[0,t]$, with $0 = t_0 \le t_1 \le \cdots \le t_n = t$, and define

$$M_s^\Pi = \sum_{k=1}^{n} M_{t_k} 1_{(t_{k-1}, t_k]}(s).$$

The martingale property of M yields

$$E\int_{(0,t]} M_s^\Pi\,dA_s = E\sum_{k=1}^{n} M_{t_k}(A_{t_k} - A_{t_{k-1}}) = E\left[\sum_{k=1}^{n} M_{t_k}A_{t_k} - \sum_{k=1}^{n-1} M_{t_{k+1}}A_{t_k}\right]$$

$$= E(M_t A_t) - E\sum_{k=1}^{n-1} A_{t_k}(M_{t_{k+1}} - M_{t_k}) = E(M_t A_t).$$

Now let $\|\Pi\| \triangleq \max_{1\le k\le n}(t_k - t_{k-1}) \to 0$, so $M_s^\Pi \to M_s$, and use the bounded convergence theorem for Lebesgue-Stieltjes integration to obtain (4.5). □

The following concept is a strengthening of the notion of uniform integrability for submartingales.

4.8 Definition. Let us consider the class $\mathscr{S}(\mathscr{S}_a)$ of all stopping times T of the filtration $\{\mathscr{F}_t\}$ which satisfy $P(T < \infty) = 1$ (respectively, $P(T \le a) = 1$ for a given finite number $a > 0$). The right-continuous process $\{X_t, \mathscr{F}_t; 0 \le t < \infty\}$ is said to be *of class D*, if the family $\{X_T\}_{T\in\mathscr{S}}$ is uniformly integrable; *of class DL*, if the family $\{X_T\}_{T\in\mathscr{S}_a}$ is uniformly integrable, for every $0 < a < \infty$.

4.9 Problem. Suppose $X = \{X_t, \mathscr{F}_t; 0 \le t < \infty\}$ is a right-continuous submartingale. Show that under any one of the following conditions, X is of class DL.

(a) $X_t \ge 0$ a.s. for every $t \ge 0$.
(b) X has the special form

$$X_t = M_t + A_t, \quad 0 \le t < \infty \tag{4.6}$$

suggested by the Doob decomposition, where $\{M_t, \mathscr{F}_t; 0 \le t < \infty\}$ is a martingale and $\{A_t, \mathscr{F}_t; 0 \le t < \infty\}$ is an increasing process.

Show also that if X is a uniformly integrable martingale, then it is of class D.

The celebrated theorem which follows asserts that membership in DL is also a sufficient condition for the decomposition of the semimartingale X in the form (4.6).

4.10 Theorem (Doob-Meyer Decomposition). *Let $\{\mathscr{F}_t\}$ satisfy the usual conditions (Definition 2.25). If the right-continuous submartingale $X = \{X_t, \mathscr{F}_t; 0 \le t < \infty\}$ is of class DL, then it admits the decomposition* (4.6) *as the summa-*

tion of a right-continuous martingale $M = \{M_t, \mathscr{F}_t; 0 \le t < \infty\}$ *and an increasing process* $A = \{A_t, \mathscr{F}_t; 0 \le t < \infty\}$. *The latter can be taken to be natural; under this additional condition, the decomposition* (4.6) *is unique (up to indistinguishability). Further, if X is of class D, then M is a uniformly integrable martingale and A is integrable.*

PROOF. For *uniqueness*, let us assume that X admits both decompositions $X_t = M'_t + A'_t = M''_t + A''_t$, where M' and M'' are martingales and A', A'' are natural increasing processes. Then $\{B_t \triangleq A'_t - A''_t = M''_t - M'_t, \mathscr{F}_t; 0 \le t < \infty\}$ is a martingale (of bounded variation), and for every bounded and right-continuous martingale $\{\xi_t, \mathscr{F}_t\}$ we have

$$E[\xi_t(A'_t - A''_t)] = E\int_{(0,t]} \xi_{s-}\, dB_s = \lim_{n\to\infty} E\sum_{j=1}^{m_n} \xi_{t_{j-1}^{(n)}}[B_{t_j^{(n)}} - B_{t_{j-1}^{(n)}}],$$

where $\Pi_n = \{t_0^{(n)}, \ldots, t_{m_n}^{(n)}\}$, $n \ge 1$ is a sequence of partitions of $[0, t]$ with $\|\Pi_n\| = \max_{1 \le j \le m_n}(t_j^{(n)} - t_{j-1}^{(n)})$ converging to zero as $n \to \infty$. But now

$$E[\xi_{t_{j-1}^{(n)}}(B_{t_j^{(n)}} - B_{t_{j-1}^{(n)}})] = 0, \quad \text{and thus} \quad E[\xi_t(A'_t - A''_t)] = 0.$$

For an arbitrary bounded random variable ξ, we can select $\{\xi_t, \mathscr{F}_t\}$ to be a right-continuous modification of $\{E[\xi|\mathscr{F}_t], \mathscr{F}_t\}$ (Theorem 3.13); we obtain $E[\xi(A'_t - A''_t)] = 0$ and therefore $P(A'_t = A''_t) = 1$, for every $t \ge 0$. The right-continuity of A' and A'' now gives us their indistinguishability.

For the *existence* of the decomposition (4.6) on $[0, \infty)$, with X of class DL, it suffices to establish it on every finite interval $[0, a]$; by uniqueness, we can then extend the construction to the entire of $[0, \infty)$. Thus, for fixed $0 < a < \infty$, let us select a right-continuous modification of the nonpositive submartingale

$$Y_t \triangleq X_t - E[X_a|\mathscr{F}_t], \quad 0 \le t \le a.$$

Consider the partitions $\Pi_n = \{t_0^{(n)}, t_1^{(n)}, \ldots, t_{2^n}^{(n)}\}$ of the interval $[0, a]$ of the form $t_j^{(n)} = (j/2^n)a$, $j = 0, 1, \ldots, 2^n$. For every $n \ge 1$, we have the Doob decomposition

$$Y_{t_j^{(n)}} = M_{t_j^{(n)}}^{(n)} + A_{t_j^{(n)}}^{(n)}, \quad j = 0, 1, \ldots, 2^n$$

where the predictable increasing sequence $A^{(n)}$ is given by

$$\begin{aligned} A_{t_j^{(n)}}^{(n)} &= A_{t_{j-1}^{(n)}}^{(n)} + E[Y_{t_j^{(n)}} - Y_{t_{j-1}^{(n)}}|\mathscr{F}_{t_{j-1}^{(n)}}] \\ &= \sum_{k=0}^{j-1} E[Y_{t_{k+1}^{(n)}} - Y_{t_k^{(n)}}|\mathscr{F}_{t_k^{(n)}}], \quad j = 1, \ldots, 2^n. \end{aligned}$$

Notice also that because $M_a^{(n)} = -A_a^{(n)}$, we have

$$Y_{t_j^{(n)}} = A_{t_j^{(n)}}^{(n)} - E[A_a^{(n)}|\mathscr{F}_{t_j^{(n)}}], \quad j = 0, 1, \ldots, 2^n. \tag{4.7}$$

We now show that *the sequence* $\{A_a^{(n)}\}_{n=1}^{\infty}$ *is uniformly integrable*. With $\lambda > 0$, we define the random times

$$T_\lambda^{(n)} = a \wedge \min\{t_{j-1}^{(n)};\, A_{t_j^{(n)}}^{(n)} > \lambda \text{ for some } j,\ 1 \le j \le 2^n\}.$$

We have $\{T_\lambda^{(n)} \le t_{j-1}^{(n)}\} = \{A_{t_j^{(n)}}^{(n)} > \lambda\} \in \mathscr{F}_{t_{j-1}^{(n)}}$ for $j = 1, \ldots, 2^n$, and $\{T_\lambda^{(n)} < a\} = \{A_a^{(n)} > \lambda\}$. Therefore, $T_\lambda^{(n)} \in \mathscr{S}_a$. On each set $\{T_\lambda^{(n)} = t_j^{(n)}\}$, we have $E[A_a^{(n)} | \mathscr{F}_{t_j^{(n)}}] = E[A_a^{(n)} | \mathscr{F}_{T_\lambda^{(n)}}]$, so (4.7) implies

$$Y_{T_\lambda^{(n)}} = A_{T_\lambda^{(n)}}^{(n)} - E[A_a^{(n)} | \mathscr{F}_{T_\lambda^{(n)}}] \le \lambda - E[A_a^{(n)} | \mathscr{F}_{T_\lambda^{(n)}}] \tag{4.8}$$

on $\{T_\lambda^{(n)} < a\}$. Thus

$$\int_{\{A_a^{(n)} > \lambda\}} A_a^{(n)}\, dP \le \lambda P[T_\lambda^{(n)} < a] - \int_{\{T_\lambda^{(n)} < a\}} Y_{T_\lambda^{(n)}}\, dP. \tag{4.9}$$

Replacing λ by $\lambda/2$ in (4.8) and integrating the equality over the $\mathscr{F}_{T_{\lambda/2}^{(n)}}$-measurable set $\{T_{\lambda/2}^{(n)} < a\}$, we obtain

$$\begin{aligned} -\int_{\{T_{\lambda/2}^{(n)} < a\}} Y_{T_{\lambda/2}^{(n)}}\, dP &= \int_{\{T_{\lambda/2}^{(n)} < a\}} (A_a^{(n)} - A_{T_{\lambda/2}^{(n)}}^{(n)})\, dP \\ &\ge \int_{\{T_\lambda^{(n)} < a\}} (A_a^{(n)} - A_{T_{\lambda/2}^{(n)}}^{(n)})\, dP \ge \frac{\lambda}{2} P[T_\lambda^{(n)} < a], \end{aligned}$$

and thus (4.9) leads to

$$\int_{\{A_a^{(n)} > \lambda\}} A_a^{(n)}\, dP \le -2 \int_{\{T_{\lambda/2}^{(n)} < a\}} Y_{T_{\lambda/2}^{(n)}}\, dP - \int_{\{T_\lambda^{(n)} < a\}} Y_{T_\lambda^{(n)}}\, dP. \tag{4.10}$$

The family $\{X_T\}_{T \in \mathscr{S}_a}$ is uniformly integrable by assumption, and thus so is $\{Y_T\}_{T \in \mathscr{S}_a}$. But

$$P[T_\lambda^{(n)} < a] = P[A_a^{(n)} > \lambda] \le \frac{E(A_a^{(n)})}{\lambda} = -\frac{E(Y_0)}{\lambda},$$

so

$$\sup_{n \ge 1} P[T_\lambda^{(n)} < a] \to 0 \quad \text{as } \lambda \to \infty.$$

Since the sequence $\{Y_{T_c^{(n)}}\}_{n=1}^\infty$ is uniformly integrable for every $c > 0$, it follows from (4.10) that the sequence $\{A_a^{(n)}\}_{n=1}^\infty$ is also uniformly integrable.

By the Dunford-Pettis compactness criterion (Meyer (1966), p. 20, or Dunford & Schwartz (1963), p. 294), uniform integrability of the sequence $\{A_a^{(n)}\}_{n=1}^\infty$ guarantees the existence of an integrable random variable A_a, as well as of a subsequence $\{A_a^{(n_k)}\}_{k=1}^\infty$ which converges to A_a weakly in L^1:

$$\lim_{k \to \infty} E(\xi A_a^{(n_k)}) = E(\xi A_a)$$

for every bounded random variable ξ. To simplify typography we shall assume henceforth that the preceding subsequence has been relabeled and shall denote it simply as $\{A_a^{(n)}\}_{n=1}^\infty$. By analogy with (4.7), we define the process $\{A_t, \mathscr{F}_t\}$ as a right-continuous modification of

$$A_t = Y_t + E(A_a | \mathscr{F}_t); \quad 0 \le t \le a. \tag{4.11}$$

4.11 Problem. Show that if $\{A^{(n)}\}_{n=1}^{\infty}$ is a sequence of integrable random variables on a probability space $(\Omega, \mathscr{F}, P)$ which converges weakly in L^1 to an integrable random variable A, then for each σ-field $\mathscr{G} \subset \mathscr{F}$, the sequence $E[A^{(n)}|\mathscr{G}]$ converges to $E[A|\mathscr{G}]$ weakly in L^1.

Let $\Pi = \bigcup_{n=1}^{\infty} \Pi_n$. For $t \in \Pi$, we have from Problem 4.11 and a comparison of (4.7) and (4.11) that $\lim_{n\to\infty} E(\xi A_t^{(n)}) = E(\xi A_t)$ for every bounded random variable ξ. For $s, t \in \Pi$ with $0 \le s < t \le a$, and any bounded and nonnegative random variable ξ, we obtain $E[\xi(A_t - A_s)] = \lim_{n\to\infty} E[\xi(A_t^{(n)} - A_s^{(n)})] \geqq 0$, and by selecting $\xi = 1_{\{A_s > A_t\}}$ we get $A_s \le A_t$, a.s. P. Because Π is countable, for a.e. $\omega \in \Omega$ the function $t \mapsto A_t(\omega)$ is nondecreasing on Π, and right-continuity shows that it is nondecreasing on $[0, a]$ as well. It is trivially seen that $A_0 = 0$, a.s. P. Further, for any bounded and right-continuous martingale $\{\xi_t, \mathscr{F}_t\}$, we have from (4.2) and Proposition 4.3:

$$\begin{aligned} E(\xi_a A_a^{(n)}) &= E \sum_{j=1}^{2^n} \xi_{t_{j-1}^{(n)}} [A_{t_j^{(n)}}^{(n)} - A_{t_{j-1}^{(n)}}^{(n)}] \\ &= E \sum_{j=1}^{2^n} \xi_{t_{j-1}^{(n)}} [Y_{t_j^{(n)}} - Y_{t_{j-1}^{(n)}}] \\ &= E \sum_{j=1}^{2^n} \xi_{t_{j-1}^{(n)}} [A_{t_j^{(n)}} - A_{t_{j-1}^{(n)}}], \end{aligned}$$

where we are making use of the fact that both sequences $\{A_t - Y_t, \mathscr{F}_t\}$ and $\{A_t^{(n)} - Y_t, \mathscr{F}_t\}$, for $t \in \Pi_n$, are martingales. Letting $n \to \infty$ one obtains by virtue of (4.5):

$$E \int_{(0,a]} \xi_s \, dA_s = E \int_{(0,a]} \xi_{s-} \, dA_s,$$

as well as $E \int_{(0,t]} \xi_s \, dA_s = E \int_{(0,t]} \xi_{s-} \, dA_s$, $\forall t \in [0, a]$, if one remembers that $\{\xi_{s \wedge t}, \mathscr{F}_s; 0 \le s \le a\}$ is also a (bounded) martingale (cf. Problem 3.24). Therefore, the process A defined in (4.11) is natural increasing, and (4.6) follows with $M_t = E[X_a - A_a | \mathscr{F}_t]$, $0 \le t \le a$.

Finally, if the submartingale X is of class D it is uniformly integrable, hence it possesses a last element X_∞ to which it converges both in L^1 and almost surely as $t \to \infty$ (Problem 3.19). The reader will have no difficulty repeating the preceding argument, with $a = \infty$, and observing that $E(A_\infty) < \infty$. □

Much of this book is devoted to the presentation of Brownian motion as the typical continuous martingale. To develop this theme, we must specialize the Doob-Meyer result just proved to *continuous submartingales*, where we discover that continuity and a bit more implies that both processes in the decomposition also turn out to be continuous. This fact will allow us to conclude that the quadratic variation process for a continuous martingale (Section 5) is itself continuous.

4.12 Definition. A submartingale $\{X_t, \mathscr{F}_t; 0 \le t < \infty\}$ is called *regular* if for every $a > 0$ and every nondecreasing sequence of stopping times $\{T_n\}_{n=1}^\infty \subseteq \mathscr{S}_a$ with $T = \lim_{n\to\infty} T_n$, we have $\lim_{n\to\infty} E(X_{T_n}) = E(X_T)$.

4.13 Problem. Verify that a continuous, nonnegative submartingale is regular.

4.14 Theorem. *Suppose that $X = \{X_t; 0 \le t < \infty\}$ is a right-continuous submartingale of class DL with respect to the filtration $\{\mathscr{F}_t\}$, which satisfies the usual conditions, and let $A = \{A_t; 0 \le t < \infty\}$ be the natural increasing process in the Doob-Meyer decomposition* (4.6). *The process A is continuous if and only if X is regular.*

PROOF. Continuity of A yields the regularity of X quite easily, by appealing to the optional sampling theorem for bounded stopping times (Problem 3.23(i)).

Conversely, let us suppose that X is regular; then for any sequence $\{T_n\}_{n=1}^\infty$ as in Definition 4.12, we have by optional sampling: $\lim_{n\to\infty} E(A_{T_n}) = \lim_{n\to\infty} E(X_{T_n}) - E(M_T) = E(A_T)$, and therefore $A_{T_n(\omega)}(\omega) \uparrow A_{T(\omega)}(\omega)$ except for ω in a P-null set which may depend on T.

To remove this dependence on T, let us consider the same sequence $\{\Pi_n\}_{n=1}^\infty$ of partitions of the interval $[0, a]$ as in the proof of Theorem 4.10, and select a number $\lambda > 0$. For each interval $(t_j^{(n)}, t_{j+1}^{(n)}), j = 0, 1, \ldots, 2^n - 1$ we consider a right-continuous modification of the martingale

$$\xi_t^{(n)} = E[\lambda \wedge A_{t_{j+1}^{(n)}} | \mathscr{F}_t], \quad t_j^{(n)} < t \le t_{j+1}^{(n)}.$$

This is possible by virtue of Theorem 3.13. The resulting process $\{\xi_t^{(n)}; 0 \le t \le a\}$ is right-continuous on $(0, a)$ except possibly at the points of the partition, and dominates the increasing process $\{\lambda \wedge A_t; 0 \le t \le a\}$; in particular, the two processes agree a.s. at the points $t_1^{(n)}, \ldots, t_{2^n}^{(n)}$. Because A is a natural increasing process, we have from (4.4)

$$E\int_{(t_j^{(n)}, t_{j+1}^{(n)}]} \xi_s^{(n)}\, dA_s = E\int_{(t_j^{(n)}, t_{j+1}^{(n)}]} \xi_{s-}^{(n)}\, dA_s; \quad j = 0, 1, \ldots, 2^n - 1,$$

and by summing over j, we obtain

$$E\int_{(0,t]} \xi_s^{(n)}\, dA_s = E\int_{(0,t]} \xi_{s-}^{(n)}\, dA_s, \tag{4.12}$$

for any $0 \le t \le a$. Now the process

$$\eta_t^{(n)} = \begin{cases} \xi_{t+}^{(n)} - (\lambda \wedge A_t), & 0 \le t < a, \\ 0, & t = a, \end{cases}$$

is right-continuous and adapted to $\{\mathscr{F}_t\}$; therefore, for any $\varepsilon > 0$ the random time

$$T_n(\varepsilon) = a \wedge \inf\{0 \le t \le a; \eta_t^{(n)} > \varepsilon\} = a \wedge \inf\{0 \le t \le a; \xi_t^{(n)} - (\lambda \wedge A_t) > \varepsilon\}$$

is an optional time of the right-continuous filtration $\{\mathscr{F}_t\}$, hence a stopping time in $\mathscr{S}_a$ (cf. Problem 2.6 and Proposition 2.3). Further, defining for each $n \geq 1$ the function $\varphi_n : [0, a] \to \Pi_n$ by $\varphi_n(t) = t_{j+1}^n$ for $t_j^{(n)} \leq t < t_{j+1}^{(n)}$ and $\varphi_n(a) = a$, we have $\varphi_n(T_n(\varepsilon)) \in \mathscr{S}_a$. Because $\xi^{(n)}$ is decreasing in n, the increasing limit $T_\varepsilon = \lim_{n\to\infty} T_n(\varepsilon)$ exists a.s., is a stopping time in $\mathscr{S}_a$, and we also have

$$T_\varepsilon = \lim_{n\to\infty} \varphi_n(T_n(\varepsilon)) \quad \text{a.s. } P.$$

By optional sampling we obtain now

$$\begin{aligned} E[\xi_{T_n(\varepsilon)+}^{(n)}] &= \sum_{j=0}^{2^n-2} E[E(\lambda \wedge A_{t_{j-1}}^{(n)} | \mathscr{F}_{T_n(\varepsilon)}) 1_{\{t_j^{(n)} \leq T_n(\varepsilon) < t_{j-1}^{(n)}\}}] \\ &\quad + E[E(\lambda \wedge A_a | \mathscr{F}_{T_n(\varepsilon)}) 1_{\{t_{2^n-1}^{(n)} \leq T_n(\varepsilon) \leq a\}}] \\ &= E[\lambda \wedge A_{\varphi_n(T_n(\varepsilon))}], \end{aligned}$$

where we set $\xi_{a+}^{(n)} = \xi_a^{(n)}$. Therefore

$$\begin{aligned} E[(\lambda \wedge A_{\varphi_n(T_n(\varepsilon))}) - (\lambda \wedge A_{T_n(\varepsilon)})] &= E[\xi_{T_n(\varepsilon)+}^{(n)} - (\lambda \wedge A_{T_n(\varepsilon)})] \\ &= E[1_{\{T_n(\varepsilon) < a\}}(\xi_{T_n(\varepsilon)^{(n)}+} - (\lambda \wedge A_{T_n(\varepsilon)}))] \\ &\geq \varepsilon P[T_n(\varepsilon) < a]. \end{aligned}$$

We employ now the regularity of X to conclude that for every $\varepsilon > 0$,

$$P[Q_n > \varepsilon] = P[T_n(\varepsilon) < a] \leq \frac{1}{\varepsilon} E[(\lambda \wedge A_{\varphi_n(T_n(\varepsilon))}) - (\lambda \wedge A_{T_n(\varepsilon)})] \to 0$$

as $n \to \infty$, where $Q_n \triangleq \sup_{0 \leq t \leq a} |\xi_t^{(n)} - (\lambda \wedge A_t)|$. Therefore, this last sequence of random variables converges to zero in probability, and hence also almost surely along a (relabeled) subsequence. We apply this observation to (4.12), along with the monotone convergence theorem for Lebesgue-Stieltjes integration, to obtain

$$E \int_{(0,t]} (\lambda \wedge A_s)\, dA_s = E \int_{(0,t]} (\lambda \wedge A_{s-})\, dA_s, \quad 0 \leq t < \infty,$$

which yields the continuity of the path $t \mapsto \lambda \wedge A_t(\omega)$ for every $\lambda > 0$, and hence the continuity of $t \mapsto A_t(\omega)$ for P-a.e. $\omega \in \Omega$. □

4.15 Problem. Let $X = \{X_t, \mathscr{F}_t; 0 \leq t < \infty\}$ be a continuous, nonnegative process with $X_0 = 0$ a.s., and $A = \{A_t, \mathscr{F}_t; 0 \leq t < \infty\}$ any continuous, increasing process for which

$$E(X_T) \leq E(A_T) \tag{4.13}$$

holds for every bounded stopping time T of $\{\mathscr{F}_t\}$. Introduce the process $V_t \triangleq \max_{0 \leq s \leq t} X_s$, consider a continuous, increasing function F on $[0, \infty)$ with $F(0) = 0$, and define $G(x) \triangleq 2F(x) + x\int_x^\infty u^{-1}\, dF(u)$; $0 < x < \infty$.

Establish the inequalities

$$P[V_T \geq \varepsilon] \leq \frac{E(A_T)}{\varepsilon}; \quad \forall \varepsilon > 0 \tag{4.14}$$

$$P[V_T \geq \varepsilon, A_T < \delta] \leq \frac{E(\delta \wedge A_T)}{\varepsilon}; \quad \forall \varepsilon > 0, \delta > 0 \tag{4.15}$$

$$EF(V_T) \leq EG(A_T) \tag{4.16}$$

for *any* stopping time T of $\{\mathscr{F}_t\}$.

4.16 Remark. If the process X of Problem 4.15 is a submartingale, then A can be taken as the continuous, increasing process in the Doob-Meyer decomposition (4.6) of Theorems 4.10 and 4.14.

4.17 Remark. The corollary

$$P[V_T \geq \varepsilon] \leq \frac{E(\delta \wedge A_T)}{\varepsilon} + P[A_T \geq \delta] \tag{4.15$'$}$$

of (4.15) is very useful in the limit theory of continuous-time martingales; it is known as the *Lenglart inequality*. We shall use it to establish convergence results for martingales (Problem 5.25) and stochastic integrals (Proposition 3.2.26). On the other hand, it follows easily from (4.16) that

$$E(V_T^p) \leq \frac{2-p}{1-p} E(A_T^p); \quad 0 < p < 1 \tag{4.17}$$

holds for any stopping time T of $\{\mathscr{F}_t\}$.

1.5. Continuous, Square-Integrable Martingales

In order to appreciate Brownian motion properly, one must understand the role it plays as the canonical example of various classes of processes. One such class is that of continuous, square-integrable martingales. Throughout this section, we have a fixed probability space $(\Omega, \mathscr{F}, P)$ and a filtration $\{\mathscr{F}_t\}$ which satisfies the usual conditions (Definition 2.25).

5.1 Definition. Let $X = \{X_t, \mathscr{F}_t; 0 \leq t < \infty\}$ be a right-continuous martingale. We say that X is *square-integrable* if $EX_t^2 < \infty$ for every $t \geq 0$. If, in addition, $X_0 = 0$ a.s., we write $X \in \mathscr{M}_2$ (or $X \in \mathscr{M}_2^c$, if X is also continuous).

5.2 Remark. Although we have defined $\mathscr{M}_2^c$ so that its members have every sample path continuous, the results which follow are also true if we assume only that P-almost every sample path is continuous.

For any $X \in \mathscr{M}_2$, we observe that $X^2 = \{X_t^2, \mathscr{F}_t; 0 \leq t < \infty\}$ is a nonnegative submartingale (Proposition 3.6), hence of class DL, and so X^2 has a unique Doob-Meyer decomposition (Theorem 4.10, Problem 4.9):

$$X_t^2 = M_t + A_t; \quad 0 \le t < \infty$$

where $M = \{M_t, \mathscr{F}_t; 0 \le t < \infty\}$ is a right-continuous martingale and $A = \{A_t, \mathscr{F}_t; 0 \le t < \infty\}$ is a natural increasing process. We normalize these processes so that $M_0 = A_0 = 0$, a.s. P. If $X \in \mathscr{M}_2^c$, then A and M are *continuous* (Theorem 4.14 and Problem 4.13); recall Definitions 4.4 and 4.5 for the terms *increasing* and *natural.*

5.3 Definition. For $X \in \mathscr{M}_2$, we define the *quadratic variation* of X to be the process $\langle X \rangle_t \triangleq A_t$, where A is the natural increasing process in the Doob-Meyer decomposition of X^2. In other words, $\langle X \rangle$ is that unique (up to indistinguishability) adapted, natural increasing process, for which $\langle X \rangle_0 = 0$ a.s. and $X^2 - \langle X \rangle$ is a martingale.

5.4 Example. Consider a Poisson process $\{N_t, \mathscr{F}_t; 0 \le t < \infty\}$ as in Definition 3.3 and assume that the filtration $\{\mathscr{F}_t\}$ satisfies the usual conditions (this can be accomplished, for instance, by "augmentation" of $\{\mathscr{F}_t^N\}$; cf. Remark 2.7.10). It is easy to verify that the martingale $M_t = N_t - \lambda t$, $\mathscr{F}_t$ of Problem 3.4 is in $\mathscr{M}_2$, and $\langle M \rangle_t = \lambda t$.

If we take two elements X, Y of $\mathscr{M}_2$, then both processes $(X + Y)^2 - \langle X + Y \rangle$ and $(X - Y)^2 - \langle X - Y \rangle$ are martingales, and therefore so is their difference $4XY - [\langle X + Y \rangle - \langle X - Y \rangle]$.

5.5 Definition. For any two martingales X, Y in $\mathscr{M}_2$, we define their *cross-variation process* $\langle X, Y \rangle$ by

$$\langle X, Y \rangle_t \triangleq \tfrac{1}{4}[\langle X + Y \rangle_t - \langle X - Y \rangle_t]; \quad 0 \le t < \infty,$$

and observe that $XY - \langle X, Y \rangle$ is a martingale. Two elements X, Y of $\mathscr{M}_2$ are called *orthogonal* if $\langle X, Y \rangle_t = 0$, a.s. P, holds for every $0 \le t < \infty$.

The uniqueness argument in Theorem 4.10 also shows that $\langle X, Y \rangle$ is, up to indistinguishability, the only process of the form $A = A^{(1)} - A^{(2)}$ with $A^{(j)}$ adapted and natural increasing ($j = 1, 2$), such that $XY - A$ is a martingale. In particular, $\langle X, X \rangle = \langle X \rangle$. For continuous X and Y, we give a different uniqueness argument in Theorem 5.13.

5.6 Remark. In view of the identities

$$\begin{aligned} E[(X_t - X_s)(Y_t - Y_s)|\mathscr{F}_s] &= E[X_t Y_t - X_s Y_s|\mathscr{F}_s] \\ &= E[\langle X, Y \rangle_t - \langle X, Y \rangle_s|\mathscr{F}_s], \end{aligned}$$

valid P a.s. for every $0 \le s < t < \infty$, the orthogonality of X, Y in $\mathscr{M}_2$ is equivalent to the statements "XY is a martingale" or "the increments of X and Y over $[s, t]$ are conditionally uncorrelated, given $\mathscr{F}_s$".

5.7 Problem. Show that $\langle \cdot, \cdot \rangle$ is a bilinear form on $\mathscr{M}_2$, i.e., for any members X, Y, Z of $\mathscr{M}_2$ and real numbers α, β, we have

(i) $\langle \alpha X + \beta Y, Z\rangle = \alpha\langle X, Z\rangle + \beta\langle Y, Z\rangle$,
(ii) $\langle X, Y\rangle = \langle Y, X\rangle$,
(iii) $|\langle X, Y\rangle|^2 \leqq \langle X\rangle\langle Y\rangle$.
(iv) For P-a.e. $\omega \in \Omega$,

$$\check{\xi}_t(\omega) - \check{\xi}_s(\omega) \leqq \tfrac{1}{2}[\langle X\rangle_t(\omega) - \langle X\rangle_s(\omega) + \langle Y\rangle_t(\omega) - \langle Y\rangle_s(\omega)];$$
$$0 \leq s < t < \infty,$$

where $\check{\xi}_t$ denotes the total variation of $\xi \triangleq \langle X, Y\rangle$ on $[0, t]$.

The use of the term *quadratic variation* in Definition 5.3 may appear to be unfounded. Indeed, a more conventional use of this term is the following. Let $X = \{X_t; 0 \leq t < \infty\}$ be a process, fix $t > 0$, and let $\Pi = \{t_0, t_1, \ldots, t_m\}$, with $0 = t_0 \leq t_1 \leq t_2 \leq \cdots \leq t_m = t$, be a partition of $[0, t]$. Define the *p-th variation* $(p > 0)$ *of X over the partition* Π to be

$$V_t^{(p)}(\Pi) = \sum_{k=1}^{m} |X_{t_k} - X_{t_{k-1}}|^p.$$

Now define the mesh of the partition Π as $\|\Pi\| = \max_{1 \leq k \leq m}|t_k - t_{k-1}|$. If $V_t^{(2)}(\Pi)$ converges in some sense as $\|\Pi\| \to 0$, the limit is entitled to be called the quadratic variation of X on $[0, t]$. Our justification of Definition 5.3 for *continuous* martingales (on which we shall concentrate from now on) is the following result:

5.8 Theorem. *Let X be in $\mathscr{M}_2^c$. For partitions Π of $[0, t]$, we have $\lim_{\|\Pi\| \to 0} V_t^{(2)}(\Pi) = \langle X\rangle_t$ (in probability); i.e., for every $\varepsilon > 0, \eta > 0$ there exists $\delta > 0$ such that $\|\Pi\| < \delta$ implies*

$$P[|V_t^{(2)}(\Pi) - \langle X\rangle_t| > \varepsilon] < \eta.$$

The proof of Theorem 5.8 proceeds through two lemmas. The key fact employed here is that, *when squaring sums of martingale increments and taking the expectation, one can neglect the cross-product terms.* More precisely, if $X \in \mathscr{M}_2$ and $0 \leq s < t \leq u < v$, then

$$E[(X_v - X_u)(X_t - X_s)] = E\{E[X_v - X_u|\mathscr{F}_u](X_t - X_s)\} = 0.$$

We shall apply this fact to both martingales $X \in \mathscr{M}_2$ and $X^2 - \langle X\rangle$. In the latter case, we note that because

$$E[(X_v - X_u)^2|\mathscr{F}_t] = E[X_v^2 - 2X_u E[X_v|\mathscr{F}_u] + X_u^2|\mathscr{F}_t]$$
$$= E[X_v^2 - X_u^2|\mathscr{F}_t] = E[\langle X\rangle_v - \langle X\rangle_u|\mathscr{F}_t],$$

the terms $X_v^2 - \langle X\rangle_v - (X_u^2 - \langle X\rangle_u)$ and $(X_v - X_u)^2 - (\langle X\rangle_v - \langle X\rangle_u)$ have the same conditional expectation given $\mathscr{F}_t$, namely zero, and thus the expectation of products of such terms over nonoverlapping intervals is still zero.

5.9 Lemma. *Let $X \in \mathcal{M}_2$ satisfy $|X_s| \le K < \infty$ for all $s \in [0,t]$, a.s. P. Let $\Pi = \{t_0, t_1, \ldots, t_m\}$, with $0 = t_0 \le t_1 \le \cdots \le t_m = t$, be a partition of $[0,t]$. Then $E[V_t^{(2)}(\Pi)]^2 \le 6K^4$.*

PROOF. Using the martingale property, we have for $0 \le k \le m-1$,

$$E\left[\sum_{j=k+1}^{m} (X_{t_j} - X_{t_{j-1}})^2 \,\middle|\, \mathscr{F}_{t_k}\right] = E\left[\sum_{j=k+1}^{m} (X_{t_j}^2 - 2X_{t_{j-1}} E(X_{t_j}|\mathscr{F}_{t_{j-1}}) + X_{t_{j-1}}^2) | \mathscr{F}_{t_k}\right]$$

$$= E\left[\sum_{j=k+1}^{m} (X_{t_j}^2 - X_{t_{j-1}}^2) \,\middle|\, \mathscr{F}_{t_k}\right]$$

$$\le E[X_{t_m}^2 | \mathscr{F}_{t_k}] \le K^2$$

so

$$E\left[\sum_{k=1}^{m-1} \sum_{j=k+1}^{m} (X_{t_j} - X_{t_{j-1}})^2 (X_{t_k} - X_{t_{k-1}})^2\right]$$

$$= E\left[\sum_{k=1}^{m-1} (X_{t_k} - X_{t_{k-1}})^2 \sum_{j=k+1}^{m} E[(X_{t_j} - X_{t_{j-1}})^2 | \mathscr{F}_{t_k}]\right] \tag{5.1}$$

$$\le K^2 E\left[\sum_{k=1}^{m-1} (X_{t_k} - X_{t_{k-1}})^2\right] \le K^4.$$

We also have

$$E\left[\sum_{k=1}^{m} (X_{t_k} - X_{t_{k-1}})^4\right] \le 4K^2 E\left[\sum_{k=1}^{m} (X_{t_k} - X_{t_{k-1}})^2\right] \le 4K^4. \tag{5.2}$$

Inequalities (5.1) and (5.2) imply

$$E[V_t^{(2)}(\Pi)]^2 = E\left[\sum_{k=1}^{m} (X_{t_k} - X_{t_{k-1}})^4\right]$$

$$+ 2E\left[\sum_{k=1}^{m-1} \sum_{j=k+1}^{m} (X_{t_j} - X_{t_{j-1}})^2 (X_{t_k} - X_{t_{k-1}})^2\right]$$

$$\le 6K^4. \qquad \square$$

5.10 Lemma. *Let $X \in \mathcal{M}_2^c$ satisfy $|X_s| \le K < \infty$ for all $s \in [0,t]$, a.s. P. For partitions Π of $[0,t]$, we have*

$$\lim_{\|\Pi\| \to 0} EV_t^{(4)}(\Pi) = 0.$$

PROOF. For any partition Π, we may write

$$V_t^{(4)}(\Pi) \le V_t^{(2)}(\Pi) \cdot m_t^2(X; \|\Pi\|),$$

where

$$m_t(X;\delta) \triangleq \sup\{|(X_r - X_s)|; 0 \le r < s \le t, s - r \le \delta\} \tag{5.3}$$

is measurable because the supremum can be restricted to rational s and r. The

Hölder inequality implies

$$EV_t^{(4)}(\Pi) \le (E[V_t^{(2)}(\Pi)]^2)^{1/2}(Em_t^4(X; \|\Pi\|))^{1/2}.$$

As $\|\Pi\|$ approaches zero, the first factor on the right-hand side remains bounded and the second term tends to zero, by the uniform continuity on $[0, t]$ of the sample paths of X and by the bounded convergence theorem.

□

PROOF OF THEOREM 5.8. We consider first the bounded case: $|X_s| \le K < \infty$ and $\langle X_s \rangle \le K$ hold for all $s \in [0, t]$, a.s. P. For any partition $\Pi = \{t_0, t_1, \ldots, t_m\}$ as earlier we may write (see the discussion preceding Lemma 5.9 and relation (5.3)):

$$\begin{aligned}
E(V_t^{(2)}(\Pi) - \langle X \rangle_t)^2 &= E\left[\sum_{k=1}^{m} \{(X_{t_k} - X_{t_{k-1}})^2 - (\langle X \rangle_{t_k} - \langle X \rangle_{t_{k-1}})\}\right]^2 \\
&= \sum_{k=1}^{m} E[(X_{t_k} - X_{t_{k-1}})^2 - (\langle X \rangle_{t_k} - \langle X \rangle_{t_{k-1}})]^2 \\
&\le 2 \sum_{k=1}^{m} E[(X_{t_k} - X_{t_{k-1}})^4 + (\langle X \rangle_{t_k} - \langle X \rangle_{t_{k-1}})^2] \\
&\le 2EV_t^{(4)}(\Pi) + 2E[\langle X \rangle_t \cdot m_t(\langle X \rangle; \|\Pi\|)].
\end{aligned}$$

As the mesh of Π approaches zero, the first term on the right-hand side of this inequality converges to zero because of Lemma 5.10; the second term does as well, by the bounded convergence theorem and the sample path uniform continuity of $\langle X \rangle$. Convergence in L^2 implies convergence in probability, so this proves the theorem for martingales which are uniformly bounded.

Now suppose $X \in \mathcal{M}_2^c$ is not necessarily bounded. We use the technique of *localization* to reduce this case to the one already studied. Let us define a sequence of stopping times (Problem 2.7) for $n = 1, 2, \ldots$ by

$$T_n = \inf\{t \ge 0; |X_t| \ge n \text{ or } \langle X \rangle_t \ge n\}.$$

Now $X_t^{(n)} \triangleq X_{t \wedge T_n}$ is a bounded martingale relative to the filtration $\{\mathscr{F}_t\}$ (Problem 3.24), and likewise $\{X_{t \wedge T_n}^2 - \langle X \rangle_{t \wedge T_n}, \mathscr{F}_t; 0 \le t < \infty\}$ is a bounded martingale. From the uniqueness of the Doob-Meyer decomposition, we see that

$$\langle X^{(n)} \rangle_t = \langle X \rangle_{t \wedge T_n}. \tag{5.4}$$

Therefore, for partitions Π of $[0, t]$, we have

$$\lim_{\|\Pi\| \to 0} E\left[\sum_{k=1}^{m} (X_{t_k \wedge T_n} - X_{t_{k-1} \wedge T_n})^2 - \langle X \rangle_{t \wedge T_n}\right]^2 = 0.$$

Since $T_n \uparrow \infty$ a.s., we have for any fixed t that $\lim_{n\to\infty} P[T_n < t] = 0$. These facts can be used to prove the desired convergence of $V_t^{(2)}(\Pi)$ to $\langle X \rangle_t$ in probability. □

5.11 Problem. Let $\{X_t, \mathscr{F}_t; 0 \le t < \infty\}$ be a continuous process with the property that for each fixed $t > 0$ and for some $p > 0$,

$$\lim_{\|\Pi\|\to 0} V_t^{(p)}(\Pi) = L_t \quad \text{(in probability)},$$

where L_t is a random variable taking values in $[0, \infty)$ a.s. Show that for $q > p$, $\lim_{\|\Pi\|\to 0} V_t^{(q)}(\Pi) = 0$ (in probability), and for $0 < q < p$, $\lim_{\|\Pi\|\to 0} V_t^{(q)}(\Pi) = \infty$ (in probability) on the event $\{L_t > 0\}$.

5.12 Problem. Let X be in $\mathscr{M}_2^c$, and T be a stopping time of $\{\mathscr{F}_t\}$. If $\langle X \rangle_T = 0$, a.s. P, then we have $P[X_{T\wedge t} = 0;\ \forall 0 \le t < \infty] = 1$.

The conclusion to be drawn from Theorem 5.8 and Problems 5.11 and 5.12 is that for continuous, square-integrable martingales, quadratic variation is the "right" variation to study. All variations of higher order are zero, and, except in trivial cases where the martingale is a.s. constant on an initial interval, all variations of lower order are infinite with positive probability. Thus, the sample paths of continuous, square-integrable martingales are quite different from "ordinary" continuous functions. Being of unbounded first variation, they cannot be differentiable, nor is it possible to define integrals of the form $\int_0^t Y_s(\omega)\, dX_s(\omega)$ with respect to $X \in \mathscr{M}_2^c$ in a pathwise (i.e., for every or P-almost every $\omega \in \Omega$), Lebesgue-Stieltjes sense. We shall return to this problem of the definition of stochastic integrals in Chapter 3, where we shall give Itô's construction and change-of-variable formula; the latter is the counterpart of the chain rule from classical calculus, adapted to account for the unbounded first, but bounded second, variation of such processes.

It is also worth noting that for $X \in \mathscr{M}_2^c$, the process $\langle X \rangle$, being monotone, is its own first variation process and has quadratic variation zero. Thus, an integral of the form $\int Y_t d\langle X \rangle_t$ is defined in a pathwise, Lebesgue-Stieltjes sense (Remark 4.6 (i)).

We discuss now the cross-variation between two *continuous*, square-integrable martingales.

5.13 Theorem. *Let $X = \{X_t, \mathscr{F}_t; 0 \le t < \infty\}$ and $Y = \{Y_t, \mathscr{F}_t; 0 \le t < \infty\}$ be members of $\mathscr{M}_2^c$. There is a unique (up to indistinguishability) $\{\mathscr{F}_t\}$-adapted, continuous process of bounded variation $\{A_t, \mathscr{F}_t; 0 \le t < \infty\}$ satisfying $A_0 = 0$ a.s. P, such that $\{X_t Y_t - A_t, \mathscr{F}_t; 0 \le t < \infty\}$ is a martingale. This process is given by the cross-variation $\langle X, Y \rangle$ of Definition 5.5.*

PROOF. Clearly, $A = \langle X, Y \rangle$ enjoys the stated properties (continuity is a consequence of Theorem 4.14 and Problem 4.13). This shows existence of A. To prove uniqueness, suppose there exists another process B satisfying the conditions on A. Then

$$M \triangleq (XY - A) - (XY - B) = B - A$$

is a continuous martingale with finite first variation. If we define

$$T_n = \inf\{t \geq 0: |M_t| = n\},$$

then $\{M_t^{(n)} \triangleq M_{t \wedge T_n}, \mathscr{F}_t; 0 \leq t < \infty\}$ is a continuous, bounded (hence square-integrable) martingale, with finite first variation on every interval $[0, t]$. It follows from (5.4) and Problem 5.11 that

$$\langle M \rangle_{t \wedge T_n} = \langle M^{(n)} \rangle_t = 0 \quad \text{a.s.,} \quad t \geq 0.$$

Problem 5.12 shows that $M^{(n)} \equiv 0$ a.s.; and since $T_n \uparrow \infty$ as $n \to \infty$, we conclude that $M \equiv 0$, a.s. P. □

5.14 Problem. Show that for $X, Y \in \mathscr{M}_2^c$ and $\Pi = \{t_0, t_1, \ldots, t_m\}$ a partition of $[0, t]$,

$$\lim_{\|\Pi\| \to 0} \sum_{k=1}^{m} (X_{t_k} - X_{t_{k-1}})(Y_{t_k} - Y_{t_{k-1}}) = \langle X, Y \rangle_t \quad \text{(in probability).}$$

Twice in this section we have used the technique of localization, once in the proof of Theorem 5.8 to extend a result about bounded martingales to square-integrable ones, and again in the proof of Theorem 5.13 to apply a result about square-integrable martingales to a continuous martingale which was not necessarily square-integrable. The next definitions and problems develop this idea formally.

5.15 Definition. Let $X = \{X_t, \mathscr{F}_t; 0 \leq t < \infty\}$ be a (continuous) process. If there exists a nondecreasing sequence $\{T_n\}_{n=1}^{\infty}$ of stopping times of $\{\mathscr{F}_t\}$, such that $\{X_t^{(n)} \triangleq X_{t \wedge T_n}, \mathscr{F}_t; 0 \leq t < \infty\}$ is a martingale for each $n \geq 1$ and $P[\lim_{n \to \infty} T_n = \infty] = 1$, then we say that X is a (continuous) *local martingale*; if, in addition, $X_0 = 0$ a.s., we write $X \in \mathscr{M}^{\text{loc}}$ (respectively, $X \in \mathscr{M}^{c,\text{loc}}$ if X is continuous).

5.16 Remark. Every martingale is a local martingale (cf. Problem 3.24(i)), but the converse is not true. We shall encounter continuous, local martingales which are integrable, or even *uniformly integrable*, but fail to be martingales (cf. Exercises 3.3.36, 3.3.37, 3.5.18 (iii)).

5.17 Problem. Let X, Y be in $\mathscr{M}^{c,\text{loc}}$. Then there is a unique (up to indistinguishability) adapted, continuous process of bounded variation $\langle X, Y \rangle$ satisfying $\langle X, Y \rangle_0 = 0$ a.s. P, such that $XY - \langle X, Y \rangle \in \mathscr{M}^{c,\text{loc}}$. If $X = Y$, we write $\langle X \rangle = \langle X, X \rangle$, and this process is nondecreasing.

5.18 Definition. We call the process $\langle X, Y \rangle$ of Problem 5.17 the *cross-variation* of X and Y, in accordance with Definition 5.5. We call $\langle X \rangle$ the *quadratic variation* of X.

5.19 Problem.

(i) A local martingale of class DL is a martingale.
(ii) A nonnegative local martingale is a supermartingale.
(iii) If $M \in \mathscr{M}^{c,\text{loc}}$ and S is a stopping time of $\{\mathscr{F}_t\}$, then $E(M_S^2) \leq E\langle M \rangle_S$, where $M_\infty^2 \triangleq \underline{\lim}_{t \to \infty} M_t^2$.

We shall show in Theorem 3.3.16 that one-dimensional Brownian motion $\{B_t, \mathscr{F}_t; 0 \le t < \infty\}$ is the unique member of $\mathscr{M}^{c,\text{loc}}$ whose quadratic variation at time t is t; i.e., $B_t^2 - t$ is a martingale. We shall also show that d-dimensional Brownian motion $\{(B_t^{(1)}, \ldots, B_t^{(d)}), \mathscr{F}_t; 0 \le t < \infty\}$ is characterized by the condition

$$\langle B^{(i)}, B^{(j)} \rangle_t = \delta_{ij} t, \quad t \ge 0,$$

where δ_{ij} is the Kronecker delta.

5.20 Exercise. Suppose $X \in \mathscr{M}_2$ has stationary, independent increments, and $\{\mathscr{F}_t\}$ is the filtration generated by X. Then $\langle X \rangle_t = t(EX_1^2)$, $t \ge 0$.

5.21 Exercise. Employ the localization technique used in the solution of Problem 5.17 to show that the conclusions of Theorem 5.8 and of Problem 5.12 hold for every $X \in \mathscr{M}^{c,\text{loc}}$. In particular, every $X \in \mathscr{M}^{c,\text{loc}}$ of bounded first variation is identically equal to zero.

We close this section by imposing a metric structure on $\mathscr{M}_2$ and discussing the nature of both $\mathscr{M}_2$ and its subspace $\mathscr{M}_2^c$ under this metric.

5.22 Definition. For any $X \in \mathscr{M}_2$ and $0 \le t < \infty$, we define

$$\|X\|_t \triangleq \sqrt{E(X_t^2)}.$$

We also set

$$\|X\| \triangleq \sum_{n=1}^{\infty} \frac{\|X\|_n \wedge 1}{2^n}.$$

Let us observe that the function $t \mapsto \|X\|_t$ on $[0, \infty)$ is nondecreasing, because X^2 is a submartingale. Further, $\|X - Y\|$ is a pseudo-metric on $\mathscr{M}_2$, which becomes a metric if we identify indistinguishable processes. Indeed, suppose that for $X, Y \in \mathscr{M}_2$ we have $\|X - Y\| = 0$; this implies $X_n = Y_n$ a.s. P, for every $n \ge 1$, and thus $X_t = E(X_n | \mathscr{F}_t) = E(Y_n | \mathscr{F}_t) = Y_t$ a.s. P, for every $0 \le t \le n$. Since X and Y are right-continuous, they are indistinguishable (Problem 1.5).

5.23 Proposition. *Under the preceding metric, $\mathscr{M}_2$ is a complete metric space, and $\mathscr{M}_2^c$ a closed subspace of $\mathscr{M}_2$.*

Proof. Let us consider a Cauchy sequence $\{X^{(n)}\}_{n=1}^{\infty} \subseteq \mathscr{M}_2$: $\lim_{n,m\to\infty} \|X^{(n)} - X^{(m)}\| = 0$. For each fixed t, $\{X_t^{(n)}\}_{n=1}^{\infty}$ is Cauchy in $L^2(\Omega, \mathscr{F}_t, P)$, and so has an L^2-limit X_t. For $0 \le s < t < \infty$ and $A \in \mathscr{F}_s$, we have from L^2-convergence and the Cauchy-Schwarz inequality that $\lim_{n\to\infty} E[1_A(X_s^{(n)} - X_s)] = 0$, $\lim_{n\to\infty} E[1_A(X_t^{(n)} - X_t)] = 0$. Therefore, $E[1_A X_t^{(n)}] = E[1_A X_s^{(n)}]$ implies $E[1_A X_t] = E[1_A X_s]$, and X is seen to be a martingale; we can choose a right-continuous modification so that $X \in \mathscr{M}_2$. We have $\lim_{n\to\infty} \|X^{(n)} - X\| = 0$.

To show that $\mathcal{M}_2^c$ is closed, let $\{X^{(n)}\}_{n=1}^\infty$ be a sequence in $\mathcal{M}_2^c$ with limit X in $\mathcal{M}_2$. We have by the first submartingale inequality of Theorem 3.8:

$$P\left[\sup_{0\le t\le T}|X_t^{(n)}-X_t|\ge\varepsilon\right]\le\frac{1}{\varepsilon^2}E|X_T^{(n)}-X_T|^2=\frac{1}{\varepsilon^2}\|X^{(n)}-X\|_T^2\to 0$$

as $n\to\infty$. Along an appropriate subsequence $\{n_k\}_{k=1}^\infty$ we must have

$$P\left[\sup_{0\le t\le T}|X_t^{(n_k)}-X_t|\ge\frac{1}{k}\right]\le\frac{1}{2^k};\quad k\ge 1$$

and the Borel-Cantelli lemma implies that $X_t^{(n_k)}$ converges to X_t, uniformly on $[0,T]$, almost surely. The continuity of X follows. □

5.24 Problem. Let $M=\{M_t,\mathscr{F}_t;0\le t<\infty\}$ be a process in $\mathcal{M}_2\cup\mathcal{M}^{c,\text{loc}}$ and assume that its quadratic variation process $\langle M\rangle$ is integrable: $E\langle M\rangle_\infty<\infty$. Then:

(i) M is a martingale, and M and the submartingale M^2 are both uniformly integrable; in particular, $M_\infty=\lim_{t\to\infty}M_t$ exists a.s. P, and $EM_\infty^2=E\langle M\rangle_\infty$;
(ii) we may take a right-continuous modification of $Z_t=E(M_\infty^2|\mathscr{F}_t)-M_t^2$; $t\ge 0$, which is a potential.

5.25 Problem. Let $M\in\mathcal{M}^{c,\text{loc}}$ and show that for any stopping time T of $\{\mathscr{F}_t\}$,

$$P\left[\max_{0\le t\le T}|M_t|\ge\varepsilon\right]\le\frac{E(\delta\wedge\langle M\rangle_T)}{\varepsilon^2}+P[\langle M\rangle_T\ge\delta], \tag{5.5}$$

$\forall\,\varepsilon>0,\delta>0$. In particular, for a sequence $\{M^{(n)}\}_{n=1}^\infty\subseteq\mathcal{M}^{c,\text{loc}}$ we have

$$\langle M^{(n)}\rangle_T\xrightarrow[n\to\infty]{P}0\quad\Rightarrow\quad\max_{0\le t\le T}|M_t^{(n)}|\xrightarrow[n\to\infty]{P}0. \tag{5.6}$$

5.26 Problem. Let $\{M_t,\mathscr{F}_t;0\le t<\infty\}$ and $\{N_t,\mathscr{G}_t;0\le t<\infty\}$ on $(\Omega,\mathscr{F},P)$ be continuous local martingales relative to their respective filtrations, and assume that $\mathscr{F}_\infty$ and $\mathscr{G}_\infty$ are independent. With $\mathscr{H}_t\triangleq\sigma(\mathscr{F}_t\cup\mathscr{G}_t)$, show that $\{M_t,\mathscr{H}_t;0\le t<\infty\}$, $\{N_t,\mathscr{H}_t;0\le t<\infty\}$ and $\{M_tN_t,\mathscr{H}_t;0\le t<\infty\}$ are local martingales. If we define $\tilde{\mathscr{H}}_t=\bigcap_{s>t}\sigma(\mathscr{H}_s\cup\mathscr{N})$, where $\mathscr{N}$ is the collection of P-negligible events in $\mathscr{F}$, then the filtration $\{\tilde{\mathscr{H}}_t\}$ satisfies the usual conditions, and relative to it the processes M, N and MN are still local martingales. In particular, $\langle M,N\rangle\equiv 0$.

1.6. Solutions to Selected Problems

1.8. We first construct an example with $A\notin\mathscr{F}_{t_0}^X$. The collection of sets of the form $\{(X_{t_1},X_{t_2},\ldots)\in B\}$, where $B\in\mathscr{B}(\mathbb{R}^d)\otimes\mathscr{B}(\mathbb{R}^d)\otimes\cdots$ and $0\le t_1<t_2<\cdots\le t_0$, forms a σ-field, and each such set is in $\mathscr{F}_{t_0}^X$. Indeed, every set in $\mathscr{F}_{t_0}^X$ has such a representation; cf. Doob (1953), p. 604. Choose $\Omega = [0, 2)$, $\mathscr{F} = \mathscr{B}([0, 2))$, and $P(F) = \text{meas}(F \cap [0, 1])$; $F \in \mathscr{F}$, where *meas* stands for "Lebesgue measure." For $\omega \in [0, 1]$, define $X_t(\omega) = 0$, $\forall t \geq 0$; for $\omega \in (1, 2)$, define $X_t(\omega) = 0$ if $t \neq \omega$, $X_\omega(\omega) = 1$. Choose $t_0 = 2$. If $A \in \mathscr{F}_{t_0}^X$, then for some $B \in \mathscr{B}(\mathbb{R}) \otimes \mathscr{B}(\mathbb{R}) \otimes \cdots$ and some sequence $\{t_k\}_{k=1}^\infty \subseteq [0, 2]$, we have $A = \{(X_{t_1}, X_{t_2}, \ldots) \in B\}$. Choose $\bar{t} \in (1, 2)$, $\bar{t} \notin \{t_1, t_2, \ldots\}$. Since $\omega = \bar{t}$ is not in A and $X_{t_k}(\bar{t}) = 0$, $k = 1, 2, \ldots$, we see that $(0, 0, \ldots) \notin B$. But $X_{t_k}(\omega) = 0$, $k = 1, 2, \ldots$, for all $\omega \in [0, 1]$; we conclude that $[0, 1] \cap A = \varnothing$, which contradicts the definition of A and the construction of X.

We next show that if $\mathscr{F}_{t_0}^X \subseteq \mathscr{F}_{t_0}$ and $\mathscr{F}_{t_0}$ contains all P-null sets of $\mathscr{F}$, then $A \in \mathscr{F}_{t_0}$. Let $N \subset \Omega$ be the set on which X is not RCLL. Then

$$A = \left(\bigcup_{n=1}^\infty A_n\right)^c \cap N^c,$$

where

$$A_n = \bigcap_{m=1}^\infty \bigcup_{\substack{q_1, q_2 \in \mathbb{Q} \cap [0, t_0) \\ |q_1 - q_2| < (1/m)}} \left\{|X_{q_1} - X_{q_2}| > \frac{1}{n}\right\}.$$

2.6. Try to argue the validity of the identity $\{H_\Gamma < t\} = \bigcup_{\substack{s \in \mathbb{Q} \\ 0 \leq s < t}} \{X_s \in \Gamma\}$, for any $t > 0$. The inclusion $\supseteq$ is obvious, even for sets which are not open. Use right-continuity, and the fact that Γ is open, to go the other way.

2.7. (Wentzell (1981)): For $x \in \mathbb{R}^d$, let $\rho(x, \Gamma) = \inf\{\|x - y\|; y \in \Gamma\}$, and consider the nested sequence of open neighborhoods of Γ given by $\Gamma_n = \{x \in \mathbb{R}^d; \rho(x, \Gamma) < (1/n)\}$. By virtue of Problem 2.6, the times $T_n \triangleq H_{\Gamma_n}$; $n \geq 1$, are optional. They form a nondecreasing sequence, dominated by $H = H_\Gamma$, with limit $T \triangleq \lim_{n\to\infty} T_n \leq H$, and we have the following dichotomy:

On $\{H = 0\}$: $T_n = 0, \forall n \geq 1$.
On $\{H > 0\}$: there exists an integer $k = k(\omega) \geq 1$ such that
$$T_n = 0; \quad \forall 1 \leq n < k, \quad \text{and} \quad 0 < T_n < T_{n+1} < H; \quad \forall n \geq k.$$

We shall show that $T = H$, and for this it suffices to establish $T \geq H$ on $\{H > 0, T < \infty\}$.

On the indicated event we have, by continuity of the sample paths of X: $X_T = \lim_{n\to\infty} X_{T_n}$ and $X_{T_m} \in \partial\Gamma_m \subseteq \Gamma_n$; $\forall m > n \geq k$. Now we can let $m \to \infty$, to obtain $X_T \in \Gamma_n$; $\forall n \geq k$, and thus $X_T \in \bigcap_{n=1}^\infty \Gamma_n = \Gamma$. We conclude with the desired result $H \leq T$. The conclusion follows now from $\{H \leq t\} = \bigcap_{n=1}^\infty \{T_n < t\}$, valid for $t > 0$, and $\{H = 0\} = \{X_0 \in \Gamma\}$.

2.17. For every $A \in \mathscr{F}_T$ we know that $A \cap \{T \leq S\}$ belongs to both $\mathscr{F}_T$ (Lemma 2.16) and $\mathscr{F}_S$ (Lemma 2.15), and therefore also to $\mathscr{F}_{T \wedge S} = \mathscr{F}_T \cap \mathscr{F}_S$. Consequently, $\int_A 1_{\{T \leq S\}} E(Z|\mathscr{F}_{T\wedge S})\,dP = \int_{A \cap \{T \leq S\}} Z\,dP = \int_{A \cap \{T \leq S\}} E(Z|\mathscr{F}_T)\,dP = \int_A 1_{\{T \leq S\}} E(Z|\mathscr{F}_T)\,dP$, so (i) follows.

For claim (ii) we conclude from (i) that

$$\begin{aligned} 1_{\{T \leq S\}} E[E(Z|\mathscr{F}_T)|\mathscr{F}_S] &= E[1_{\{T \leq S\}} E(Z|\mathscr{F}_T)|\mathscr{F}_S] = E[1_{\{T \leq S\}} E(Z|\mathscr{F}_{S\wedge T})|\mathscr{F}_S] \\ &= 1_{\{T \leq S\}} E[Z|\mathscr{F}_{S\wedge T}], \end{aligned}$$

which proves the desired result on the set $\{T \leq S\}$. Interchanging the roles of S and T and replacing Z by $E(Z|\mathscr{F}_T)$, we can also conclude from (i) that

$$1_{\{S<T\}}E[E(Z|\mathscr{F}_T)|\mathscr{F}_S] = 1_{\{S<T\}}E[E(Z|\mathscr{F}_T)|\mathscr{F}_{S\wedge T}]$$
$$= 1_{\{S<T\}}E[Z|\mathscr{F}_{S\wedge T}].$$

2.22. We discuss only the second claim, following Chung (1982). For any $A \in \mathscr{F}_{S+}$, we have

$$A = \left(\bigcup_{r\in Q}[A\cap\{S<r<T\}\right)\cup[A\cap\{S=\infty\}].$$

Now $A\cap\{S<r<T\} = A\cap\{S<r\}\cap\{T>r\}$ is an event in $\mathscr{F}_T$, as is easily verified, because $A\cap\{S<r\}\in\mathscr{F}_r$. On the other hand, $A\cap\{S=\infty\} = [A\cap\{S=\infty\}]\cap\{T=\infty\}$ is easily seen to be in $\mathscr{F}_T$. It follows that $A\in\mathscr{F}_T$.

2.23. T is an optional time, by Lemma 2.11, and so $\mathscr{F}_{T+}$ is defined and contained in $\mathscr{F}_{T_n+}$ for every $n\geq 1$. Therefore, $\mathscr{F}_{T+}\subseteq\bigcap_{n=1}^{\infty}\mathscr{F}_{T_n+}$. To go the other way, consider an event A such that $A\cap\{T_n<t\}\in\mathscr{F}_t$, for every $n\geq 1$ and $t\geq 0$. Obviously then, $A\cap\{T<t\} = A\cap(\bigcup_{n=1}^{\infty}\{T_n<t\}) = \bigcup_{n=1}^{\infty}(A\cap\{T_n<t\})\in\mathscr{F}_t$, and thus $A\in\mathscr{F}_{T+}$. The second claim is justified similarly, using Problem 2.22.

3.2. (i) Fix $s\geq 0$ and a nonnegative integer n. Consider the "trace" σ-field $\mathscr{G}$ of all sets obtained by intersecting the members of $\mathscr{F}_s^N$ with the set $\{N_s=n\}$. Consider also the similar trace σ-field $\mathscr{H}$ of $\sigma(T_1,\ldots,T_n)$ on $\{N_s=n\}$. A generating family for $\mathscr{G}$ is the collection of sets of the form $\{N_{t_1}\leq n_1,\ldots, N_{t_k}\leq n_k, N_s=n\}$, where $0\leq t_1\leq\cdots\leq t_k\leq s$, and each such set is a member of $\mathscr{H}$. A generating family for $\mathscr{H}$ is the collection of all sets of the form $\{S_1\leq t_1,\ldots,S_n\leq t_n, N_s=n\}$, where $0\leq t_1\leq\cdots\leq t_{n-1}\leq s$, and each such set is a member of $\mathscr{G}$. It follows that $\mathscr{G}=\mathscr{H}$.

Therefore, for every $\tilde{A}\in\mathscr{F}_s^N$ there exists $A\in\sigma(T_1,\ldots,T_n)$ such that $\tilde{A}\cap\{N_s=n\} = A\cap\{N_s=n\}$. Using the independence of T_{n+1} and $(S_n, 1_A)$ we obtain

$$\begin{aligned}
&\int_{\tilde{A}\cap\{N_s=n\}} P[S_{n+1}>t|\mathscr{F}_s^N]\,dP\\
&= P[\{S_{n+1}>t\}\cap A\cap\{S_n\leq s<S_{n+1}\}]\\
&= P[\{S_n+T_{n+1}>t\}\cap A\cap\{S_n\leq s\}]\\
&= \int_{t-s}^{\infty} P[\{S_n>t-u\}\cap A\cap\{S_n\leq s\}]\lambda e^{-\lambda u}\,du\\
&= e^{-\lambda(t-s)}\int_0^{\infty} P[\{S_n+u>s\}\cap A\cap\{S_n\leq s\}]\lambda e^{-\lambda u}\,du\\
&= e^{-\lambda(t-s)}P[\{S_n+T_{n+1}>s\}\cap A\cap\{S_n\leq s\}]\\
&= e^{-\lambda(t-s)}P[\tilde{A}\cap\{N_s=n\}].
\end{aligned}$$

Summation over $n\geq 0$ yields $\int_{\tilde{A}} P[S_{N_s+1}>t|\mathscr{F}_s^N]\,dP = e^{-\lambda(t-s)}P(\tilde{A})$ for every $\tilde{A}\in\mathscr{F}_s^N$.

(ii) For an arbitrary but fixed $k\geq 1$, the random variable $Y_k \triangleq S_{n+k+1}-S_{n+1} = \sum_{j=n+2}^{n+k+1}T_j$ is independent of $\sigma(T_1,\ldots,T_{n+1})$; it has the gamma density $P[Y_k\in du] = [(\lambda u)^{k-1}/(k-1)!]\lambda e^{-\lambda u}\,du;\ u>0$, for which one checks easily the identity:

$$P[Y_k > \theta] = \sum_{j=0}^{k-1} \frac{(\lambda\theta)^j}{j!} e^{-\lambda\theta}; \quad k \geq 1, \theta > 0.$$

We have, as in (i),

$$\begin{aligned}
&\int_{\tilde{A}\cap\{N_s=n\}} P[N_t - N_s \leq k | \mathscr{F}_s^N]\, dP \\
&= P[\{N_t \leq n + k\} \cap \tilde{A} \cap \{N_s = n\}] \\
&= P[\{S_{n+k+1} > t\} \cap A \cap \{N_s = n\}] \\
&= P[\{S_{n+1} + Y_k > t\} \cap A \cap \{N_s = n\}] \\
&= \int_0^\infty P[\{S_{n+1} + u > t\} \cap A \cap \{N_s = n\}] \cdot P(Y_k \in du) \\
&= P[\tilde{A} \cap \{N_s = n\}] \cdot P(Y_k > t - s) \\
&\quad + \int_0^{t-s} P[\{S_{n+1} > t - u\} \cap \tilde{A} \cap \{N_s = n\}] P(Y_k \in du) \\
&= P[\tilde{A} \cap \{N_s = n\}] \left(\sum_{j=0}^{k-1} e^{-\lambda(t-s)} \frac{(\lambda(t-s))^j}{j!} \right. \\
&\quad \left. + \int_0^{t-s} e^{-\lambda(t-s-u)} P(Y_k \in du) \right) \\
&= P[\tilde{A} \cap \{N_s = n\}] \cdot \sum_{j=0}^{k} e^{-\lambda(t-s)} \frac{(\lambda(t-s))^j}{j!}.
\end{aligned}$$

Adding up over $n \geq 0$ we obtain

$$\int_{\tilde{A}} P[N_t - N_s \leq k | \mathscr{F}_s^N]\, dP = P(\tilde{A}) \sum_{j=0}^{k} e^{-\lambda(t-s)} \frac{(\lambda(t-s))^j}{j!}$$

for every $\tilde{A} \in \mathscr{F}_s^N$ and $k \geq 1$, and both assertions follow.

3.7. Let $\{h_\alpha; \alpha \in A\}$ be a collection of linear functions from $\mathbb{R}^d \to \mathbb{R}$ for which $\varphi = \sup_{\alpha \in A} h_\alpha$. Then for $0 \leq s < t$ we have

$$E[\varphi(X_t)|\mathscr{F}_s] \geq E[h_\alpha(X_t)|\mathscr{F}_s] = h_\alpha(X_s), \quad \forall \alpha \in A.$$

Taking the supremum over α, we obtain the submartingale inequality for $\varphi(X)$. Now $\|\cdot\|$ is convex and $E\|X_t\| \leq E(|X_t^{(1)}| + \cdots + |X_t^{(d)}|) < \infty$, so $\|X\|$ is a submartingale.

3.11. Thanks to the Jensen inequality (as in Proposition 3.6) we have that $\{X_n^+, \mathscr{F}_n; n \geq 1\}$ is also a backward submartingale, and so with $\lambda > 0$: $\lambda \cdot P[|X_n| > \lambda] \leq E|X_n| = -E(X_n) + 2E(X_n^+) \leq -l + 2E(X_1^+) < \infty$. It follows that $\sup_{n\geq 1} P[|X_n| > \lambda]$ converges to zero as $\lambda \to \infty$, and by the submartingale property:

$$\int_{\{X_n^+ > \lambda\}} X_n^+\, dP \leq \int_{\{X_n^+ > \lambda\}} X_1^+\, dP \leq \int_{\{|X_n| > \lambda\}} X_1^+\, dP.$$

Therefore, $\{X_n^+\}_{n=1}^\infty$ is a uniformly integrable sequence. On the other hand,

$$0 \ge \int_{\{X_n < -\lambda\}} X_n\,dP = E(X_n) - \int_{\{X_n \ge -\lambda\}} X_n\,dP \ge E(X_n) - \int_{\{X_n \ge -\lambda\}} X_m\,dP$$

$$= E(X_n) - E(X_m) + \int_{\{X_n < -\lambda\}} X_m\,dP, \quad \text{for } n > m.$$

Given $\varepsilon > 0$, we can certainly choose m so large that $0 \le E(X_m) - E(X_n) \le \varepsilon/2$ holds for every $n > m$, and for that m we select $\lambda > 0$ in such a way that

$$\sup_{n>m} \int_{\{X_n < -\lambda\}} |X_m|\,dP < \frac{\varepsilon}{2}.$$

Consequently, for these choices of m and λ we have:

$$\sup_{n>m} \int_{\{X_n^- > \lambda\}} X_n^-\,dP < \varepsilon, \quad \text{and thus } \{X_n^-\}_{n=1}^\infty \text{ is also uniformly integrable.}$$

3.19. (a) ⇒ (b): Uniform integrability allows us to invoke the submartingale convergence Theorem 3.15, to establish the existence of an almost sure limit X_∞ for $\{X_t; 0 \le t < \infty\}$ as $t \to \infty$, and to convert almost sure convergence into L^1-convergence.

(b) ⇒ (c): Let X_∞ be the L^1-limit of $\{X_t; 0 \le t < \infty\}$. For $0 \le s < t$ and $A \in \mathscr{F}_s$ we have $\int_A X_s\,dP \le \int_A X_t\,dP$, and letting $t \to \infty$ we obtain the submartingale property $\int_A X_s\,dP \le \int_A X_\infty\,dP$; $0 \le s < \infty$, $A \in \mathscr{F}_s$.

(c) ⇒ (a): For $0 \le t < \infty$ and $\lambda > 0$, we have $\int_{\{X_t > \lambda\}} X_t\,dP \le \int_{\{X_t > \lambda\}} X_\infty\,dP$, which converges to zero, uniformly in t, as $\lambda \uparrow \infty$, because $P[X_t > \lambda] \le (1/\lambda)EX_t \le (1/\lambda)EX_\infty$.

3.20. Apply Problem 3.19 to the submartingales $\{X_t^\pm, \mathscr{F}_t; 0 \le t < \infty\}$ to obtain the equivalence of (a), (b), and (c). The latter obviously implies (d), which in turn gives (a). If (d) holds, then $\int_A Y\,dP = \int_A X_t\,dP$, $\forall A \in \mathscr{F}_t$. Letting $t \to \infty$, we obtain $\int_A Y\,dP = \int_A X_\infty\,dP$. The collection of sets $A \in \mathscr{F}_\infty$ for which this equality holds is a monotone class containing the field $\bigcup_{t \ge 0} \mathscr{F}_t$. Consequently, the equality holds for every $A \in \mathscr{F}_\infty$, which gives $E(Y|\mathscr{F}_\infty) = X_\infty$, a.s.

3.26. The necessity of (3.2) follows from the version of the optional sampling theorem for bounded stopping times (Problem 3.23 (i)). For sufficiency, consider $0 \le s < t < \infty$, $A \in \mathscr{F}_s$ and define the stopping time $S(\omega) \triangleq s1_A(\omega) + t1_{A^c}(\omega)$. The condition $E(X_t) \ge E(X_S)$ is tantamount to the submartingale property $E[X_t 1_A] \ge E[X_s 1_A]$.

3.27. By assumption, each $\tilde{\mathscr{F}}_t$ contains the P-negligible events of $\mathscr{F}$. For the right-continuity of $\{\tilde{\mathscr{F}}_t\}$, select a sequence $\{t_n\}_{n=1}^\infty$ strictly decreasing to t; according to Problem 2.23,

$$\tilde{\mathscr{F}}_{t+} \triangleq \bigcap_{n=1}^{\infty} \tilde{\mathscr{F}}_{t_n} = \bigcap_{n=1}^{\infty} \mathscr{F}_{T+t_n} = \mathscr{F}_{(T+t)+},$$

and the latter agrees with $\mathscr{F}_{T+t} = \tilde{\mathscr{F}}_t$ under the assumption of right-continuity of $\{\mathscr{F}_t\}$ (Definition 2.20).

(i) With $0 \le s < t < \infty$, Problem 3.23 implies

$$E[\tilde{X}_t|\tilde{\mathscr{F}}_s] = E[X_{T+t} - X_T|\mathscr{F}_{T+s}] \ge X_{T+s} - X_T = \tilde{X}_s, \quad \text{a.s. } P.$$

(ii) Let $S_1 \le S_2$ be bounded stopping times of $\{\mathscr{F}_t\}$, and set $\tau_j = (S_j - T) \vee 0$. By Lemma 2.16, $\{\tau_j \le t\} = \{S_j \le T + t\} \in \tilde{\mathscr{F}}_t$ for all $t \ge 0$, so $\tau_1 \le \tau_2$ are bounded stopping times of $\{\tilde{\mathscr{F}}_t\}$. Furthermore, $0 \le E\tilde{X}_{\tau_j} = E\tilde{X}^+_{\tau_j} - E\tilde{X}^-_{\tau_j}$ and $E\tilde{X}^+_{\tau_j} \le E\tilde{X}^+_{S_j} < \infty$, so $E|\tilde{X}^+_{\tau_j}| < \infty$. According to Problem 3.26,

$$EX_{S_2} = E\tilde{X}_{\tau_2} \ge E\tilde{X}_{\tau_1} = EX_{S_1}.$$

Another application of Problem 3.26 shows that X is a submartingale.

3.28. (Robbins and Siegmund (1970)): With the stopping time

$$T = \inf\{t \in [s, \infty); Z_t = b\},$$

the process $\{Z_{T\wedge t}, \mathscr{F}_t; 0 \le t < \infty\}$ is a martingale (Problem 3.24 (i)). It follows that for every $A \in \mathscr{F}_s$, $t \ge s$:

$$\begin{aligned}\int_{A\cap\{Z_s<b\}} Z_s\,dP &= \int_{A\cap\{Z_s<b\}} Z_{T\wedge t}\,dP \\ &= b\cdot P[A \cap \{Z_s < b, T \le t\}] + \int_{A\cap\{Z_s<b\}} Z_t 1_{\{T>t\}}\,dP.\end{aligned}$$

The integrand $Z_t 1_{\{T>t\}}$ is dominated by b, and converges to zero as $t \to \infty$ by assumption; it develops then from the dominated convergence theorem that

$$\begin{aligned}\int_{A\cap\{Z_s<b\}} Z_s\,dP &= b\cdot P[A \cap \{Z_s < b, T < \infty\}] \\ &= b\int_{A\cap\{Z_s<b\}} P[T<\infty|\mathscr{F}_s]\,dP,\end{aligned}$$

establishing the first conclusion. The second follows readily.

4.9. (a) According to Problem 3.23 (i) we have

$$\int_{\{X_T>\lambda\}} X_T\,dP \le \int_{\{X_T>\lambda\}} X_a\,dP \quad \text{and} \quad P[X_T > \lambda] \le \frac{E(X_T)}{\lambda} \le \frac{E(X_a)}{\lambda}$$

for every $a > 0$, $\lambda > 0$, $T \in \mathscr{S}_a$, and therefore

$$\lim_{\lambda\to\infty} \sup_{T\in\mathscr{S}_a} \int_{\{X_T>\lambda\}} X_T\,dP = 0.$$

(b) It suffices to show the uniform integrability of $\{M_T\}_{T\in\mathscr{S}_a}$. Once again, Problem 3.23 (i) yields $M_T = E(M_a|\mathscr{F}_T)$ a.s. P, for every $T \in \mathscr{S}_a$, and the claim then follows easily, just as in the implication (d) $\Rightarrow$ (a) of Problem 3.20.

This latter problem, coupled with Theorem 3.22, yields the representation $X_T = E(X_\infty|\mathscr{F}_T)$ a.s. P, $\forall\, T \in \mathscr{S}$ for every uniformly integrable martingale X, which is thus shown to be of class D.

4.11. For an arbitrary bounded random variable ξ on $(\Omega, \mathscr{F}, P)$,

$$E[\xi E(A^{(n)}|\mathscr{G})] = E[E(\xi|\mathscr{G})\cdot E(A^{(n)}|\mathscr{G})] = E[A^{(n)}E(\xi|\mathscr{G})],$$

which converges to $E[A\cdot E(\xi|\mathscr{G})] = E[\xi E(A|\mathscr{G})]$.

4.15. Define the stopping times $H_\varepsilon = \inf\{t \geq 0; X_t \geq \varepsilon\}$, $S = \inf\{t \geq 0; A_t \geq \delta\}$ (Problem 2.7) and $T_n = T \wedge n \wedge H_\varepsilon$. We have

$$\varepsilon P[V_{T_n} \geq \varepsilon] \leq E[X_{T_n} 1_{\{V_{T_n} \geq \varepsilon\}}] \leq E(X_{T_n}) \leq E(A_{T_n}) \leq E(A_T)$$

and (4.14) follows because $T_n \uparrow T \wedge H_\varepsilon$, a.s. as $n \to \infty$. On the other hand, we have

$$P[V_T \geq \varepsilon, A_T < \delta] \leq P[V_{T \wedge S} \geq \varepsilon] \leq \frac{E(A_{S \wedge T})}{\varepsilon} = \frac{E(\delta \wedge A_T)}{\varepsilon}$$

thanks to (4.14), and (4.15) follows (adapted from Lenglart (1977)). From the identity $F(x) = \int_0^\infty 1_{\{x \geq u\}} dF(u)$, the Fubini theorem, and (4.15)′ we have

$$\begin{aligned} EF(V_T) &= \int_0^\infty P(V_T \geq u)\, dF(u) \leq \int_0^\infty \left\{ \frac{E(u \wedge A_T)}{u} + P(A_T \geq u) \right\} dF(u) \\ &= \int_0^\infty \left[2P(A_T \geq u) + \frac{1}{u} E(A_T 1_{\{A_T < u\}}) \right] dF(u) \\ &= E\left[2F(A_T) + A_T \int_{A_T}^\infty \frac{1}{u}\, dF(u) \right] = EG(A_T) \end{aligned}$$

(taken from Revuz & Yor (1987); see also Burkholder (1973), p. 26).

5.11. Let $\Pi = \{t_0, \ldots, t_m\}$, with $0 = t_0 \leq t_1 \leq \cdots \leq t_m = t$, be a partition of $[0, t]$. For $q > p$, we have

$$V_t^{(q)}(\Pi) \leq V_t^{(p)}(\Pi) \cdot \max_{1 \leq k \leq m} |X_{t_k} - X_{t_{k-1}}|^{q-p}.$$

As $\|\Pi\| \to 0$, the first term on the right-hand side has a finite limit in probability, and the second term converges to zero in probability. Therefore, the product converges to zero in probability. For the second assertion, suppose that $0 < q < p$, $P(L_t > 0) > 0$ and assume that $V_t^{(q)}(\Pi)$ does not tend to ∞ (in probability) as $\|\Pi\| \to 0$. Then we can find $\delta > 0$, $0 < K < \infty$, and a sequence of partitions $\{\Pi_n\}_{n=1}^\infty$ such that $P(A_n) \geq \delta P(L_t > 0)$, where

$$A_n = \{L_t > 0, V_t^{(q)}(\Pi_n) \leq K\}; \quad n \geq 1.$$

Consequently, with $\Pi_n = \{t_0^{(n)}, \ldots, t_{m_n}^{(n)}\}$, we have

$$V_t^{(p)}(\Pi_n) \leq K m_t^{p-q}(X; \|\Pi_n\|) \quad \text{on } A_n; \quad n \geq 1.$$

This contradicts the fact that $V_t^{(p)}(\Pi_n)$ converges (in probability) to the positive random variable L_t on $\{L_t > 0\}$.

5.12. Because $\langle X \rangle$ is continuous and nondecreasing, we have $P[\langle X \rangle_{T \wedge t} = 0;\ 0 \leq t < \infty] = 1$. An application of the optional sampling theorem to the continuous martingale $M \triangleq X^2 - \langle X \rangle$ yields (Problem 3.23 (i)): $0 = EM_{T \wedge t} = E[X_{T \wedge t}^2 - \langle X \rangle_{T \wedge t}] = EX_{T \wedge t}^2$, which implies $P[X_{T \wedge t} = 0] = 1$, for every $0 \leq t < \infty$. The conclusion follows now by continuity.

5.17. There are sequences $\{S_n\}$, $\{T_n\}$ of stopping times such that $S_n \uparrow \infty$, $T_n \uparrow \infty$, and $X_t^{(n)} \triangleq X_{t \wedge S_n}$, $Y_t^{(n)} \triangleq Y_{t \wedge T_n}$ are $\{\mathscr{F}_t\}$-martingales. Define

$$R_n \triangleq S_n \wedge T_n \wedge \inf\{t \geq 0: |X_t| = n \text{ or } |Y_t| = n\},$$

and set $\tilde{X}_t^{(n)} = X_{t \wedge R_n}$, $\tilde{Y}_t^{(n)} = Y_{t \wedge R_n}$. Note that $R_n \uparrow \infty$ a.s. Since $\tilde{X}_t^{(n)} = X_{t \wedge R_n}^{(n)}$, and likewise for $\tilde{Y}^{(n)}$, these processes are also $\{\mathscr{F}_t\}$-martingales (Problem 3.24), and are in $\mathscr{M}_2^c$ because they are bounded. For $m > n$, $\tilde{X}_t^{(n)} = \tilde{X}_{t \wedge R_n}^{(m)}$ and so

$$(\tilde{X}_t^{(n)})^2 - \langle \tilde{X}^{(m)} \rangle_{t \wedge R_n} = (\tilde{X}_{t \wedge R_n}^{(m)})^2 - \langle \tilde{X}^{(m)} \rangle_{t \wedge R_n}$$

is a martingale. This implies $\langle \tilde{X}^{(n)} \rangle_t = \langle \tilde{X}^{(m)} \rangle_{t \wedge R_n}$. We can thus decree $\langle X \rangle_t \triangleq \langle \tilde{X}^{(n)} \rangle_t$ whenever $t \le R_n$ and be assured that $\langle X \rangle$ is well defined. The process $\langle X \rangle$ is adapted, continuous, and nondecreasing and satisfies $\langle X \rangle_0 = 0$ a.s. Furthermore,

$$X_{t \wedge R_n}^2 - \langle X \rangle_{t \wedge R_n} = (\tilde{X}_t^{(n)})^2 - \langle \tilde{X}^{(n)} \rangle_t$$

is a martingale for each n, so $X^2 - \langle X \rangle \in \mathscr{M}^{c,\text{loc}}$. As in Theorem 5.13, we may now take $\langle X, Y \rangle = \frac{1}{4}[\langle X + Y \rangle - \langle X - Y \rangle]$.

As for the question of uniqueness, suppose both A and B satisfy the conditions required of $\langle X, Y \rangle$. Then $M \triangleq XY - A$ and $N \triangleq XY - B$ are in $\mathscr{M}^{c,\text{loc}}$, so just as before we can construct a sequence $\{R_n\}$ of stopping times with $R_n \uparrow \infty$ such that $M_t^{(n)} \triangleq M_{t \wedge R_n}$ and $N_t^{(n)} \triangleq N_{t \wedge R_n}$ are in $\mathscr{M}_2^c$. Consequently $M_t^{(n)} - N_t^{(n)} = B_{t \wedge R_n} - A_{t \wedge R_n} \in \mathscr{M}_2^c$, and being of bounded variation this process must be identically zero (see the proof of Theorem 5.13). It follows that $A = B$.

5.24. If $M \in \mathscr{M}_2$, then $E(M_t^2) = E\langle M \rangle_t \le E\langle M \rangle_\infty$; $\forall 0 \le t < \infty$. If $M \in \mathscr{M}^{c,\text{loc}}$, Problem 5.19 (iii) gives $E(M_S^2) \le E\langle M \rangle_S \le E\langle M \rangle_\infty < \infty$ for every stopping time S; it follows that the family $\{M_S\}_{S \in \mathscr{S}}$ is uniformly integrable, i.e., that M is of class D and therefore a martingale (Problem 5.19 (i)).

In either case, therefore, M is a uniformly integrable martingale; Problem 3.20 now shows that $M_\infty = \lim_{t \to \infty} M_t$ exists, and that $E(M_\infty | \mathscr{F}_t) = M_t$ holds a.s. P, for every $t \ge 0$. Fatou's lemma now yields

$$E(M_\infty^2) = E\left(\lim_{t \to \infty} M_t^2\right) \le \lim_{t \to \infty} E(M_t^2) = \lim_{t \to \infty} E\langle M \rangle_t = E\langle M \rangle_\infty, \tag{6.1}$$

and Jensen's inequality: $M_t^2 \le E(M_\infty^2 | \mathscr{F}_t)$, a.s. P, for every $t \ge 0$. It follows that the nonnegative submartingale M^2 has a last element, i.e., that $\{M_t^2, \mathscr{F}_t; 0 \le t \le \infty\}$ is a submartingale. Problem 3.19 shows that this submartingale is uniformly integrable, and (6.1) holds with equality. Finally, $Z_t = E(M_\infty^2 | \mathscr{F}_t) - M_t^2$ is now seen to be a (right-continuous, by appropriate choice of modification) nonnegative supermartingale, with $E(Z_t) = E(M_\infty^2) - E(M_t^2)$ converging to zero as $t \to \infty$.

5.25. Problem 5.19 (iii) allows us to apply Remarks 4.16, 4.17 with $X = M^2$, $A = \langle M \rangle$.

1.7. Notes

Sections 1.1, 1.2: These two sections could have been lumped together under the rubric "Fields, Optionality, and Measurability" after the manner of Chung & Doob (1965). Although slightly dated, this article still makes excellent reading. Good accounts of this material in book form have been written by Meyer [(1966); Chapter IV], Dellacherie [(1972); Chapter III and to a lesser extent Chapter IV], Dellacherie & Meyer [(1975/1980); Chapter IV], and Chung [(1982); Chapter 1]. These sources provide material on the classification of stopping times as "predictable," "accessible," and "totally inaccessible," as well as corresponding notions of measurability for stochastic processes, which we need not broach here.

A new notion of "sameness" between two stochastic processes, called *synonimity* has been introduced by Aldous. It was expounded by Hoover (1984) and was found to be useful in the study of martingales.

A deep result of Dellacherie [(1972), p. 51] is the following: for every progressively measurable process X and $\Gamma \in \mathscr{B}(\mathbb{R})$, the hitting time H_Γ of Example 2.5 is a stopping time of $\{\mathscr{F}_t\}$, provided that this filtration is right-continuous and that each σ-field $\mathscr{F}_t$ is complete.

Section 1.3: The term *martingale* was introduced in probability theory by J. Ville (1939). The concept had been created by P. Lévy back in 1934, in an attempt to extend the Kolmogorov inequality and the law of large numbers beyond the case of independence. Lévy's zero-one law (Theorem 9.4.8 and Corollary in Chung (1974)) is the first martingale convergence theorem. The classic text, Doob (1953), introduced, for the first time, an impressively complete theory of the subject as we know it today. For the foundations of the discrete-parameter case there is perhaps no better source than the relevant sections in Chapter 9 of Chung (1974) that we have already mentioned; fuller accounts are Neveu (1975), Chow & Teicher (1978), Chapter 11, and Hall & Heyde (1980). Other books which contain material on the continuous-parameter case include Meyer [(1966); Chapters V, VI], Dellacherie & Meyer [(1975/1980); Chapters V–VIII], Liptser & Shiryaev [(1977), Chapters 2, 3] and Elliott [(1982), Chapters 3, 4].

Section 1.4: Theorem 4.10 is due to P. A. Meyer (1962, 1963); its proof was later simplified by K. M. Rao (1969). Our account of this theorem, as well as that of Theorem 4.14, follows closely Ikeda & Watanabe (1981).

For any *nonnegative* submartingale X satisfying the conditions of Theorem 3.13, Krylov (1990) shows the existence of an *increasing* process D such that

$$X_t = X_0 + E(D_t | \mathscr{F}_t), \qquad \text{a.s.},$$

holds for every fixed $t \in [0, \infty)$, and uses this representation to obtain a simple derivation of the Doob-Meyer decomposition (4.6) for such X.

The Doob-Meyer decomposition $X = M + A$ of Theorem 4.10 remains valid for a general right-continuous submartingale X (not necessarily of class DL), but now with M a *local* martingale; see, for example, Protter (1990), Theorem 7, p. 94.

Section 1.5: The study of square-integrable martingales began with Fisk (1966) and continued with the seminal article by Kunita & Watanabe (1967). Theorem 5.8 is due to Fisk (1966). In (5.6), the opposite implication is also true; see Lemma A.1 in Pitman & Yor (1986).

CHAPTER 2

Brownian Motion

2.1. Introduction

Brownian movement is the name given to the irregular movement of pollen, suspended in water, observed by the botanist Robert Brown in 1828. This random movement, now attributed to the buffeting of the pollen by water molecules, results in a dispersal or *diffusion* of the pollen in the water. The range of application of Brownian motion as defined here goes far beyond a study of microscopic particles in suspension and includes modeling of stock prices, of thermal noise in electrical circuits, of certain limiting behavior in queueing and inventory systems, and of random perturbations in a variety of other physical, biological, economic, and management systems. Furthermore, integration with respect to Brownian motion, developed in Chapter 3, gives us a unifying representation for a large class of martingales and diffusion processes. Diffusion processes represented this way exhibit a rich connection with the theory of partial differential equations (Chapter 4 and Section 5.7). In particular, to each such process there corresponds a second-order parabolic equation which governs the transition probabilities of the process.

The history of Brownian motion is discussed more extensively in Section 11; see also Chapters 2–4 in Nelson (1967).

1.1 Definition. A (*standard, one-dimensional*) *Brownian motion* is a continuous, adapted process $B = \{B_t, \mathscr{F}_t; 0 \le t < \infty\}$, defined on some probability space $(\Omega, \mathscr{F}, P)$, with the properties that $B_0 = 0$ a.s. and for $0 \le s < t$, the increment $B_t - B_s$ is independent of $\mathscr{F}_s$ and is normally distributed with mean zero and variance $t - s$. We shall speak sometimes of a Brownian motion $B = \{B_t, \mathscr{F}_t; 0 \le t \le T\}$ on $[0, T]$, for some $T > 0$, and the meaning of this terminology is apparent.

If B is a Brownian motion and $0 = t_0 < t_1 < \cdots < t_n < \infty$, then the increments $\{B_{t_j} - B_{t_{j-1}}\}_{j=1}^n$ are independent and the distribution of $B_{t_j} - B_{t_{j-1}}$ depends on t_j and t_{j-1} only through the difference $t_j - t_{j-1}$; to wit, it is normal with mean zero and variance $t_j - t_{j-1}$. We say that the process B has *stationary, independent increments.* It is easily verified that B is a square-integrable martingale and $\langle B \rangle_t = t$, $t \geq 0$.

The *filtration* $\{\mathscr{F}_t\}$ is a part of the definition of Brownian motion. However, if we are given $\{B_t; 0 \leq t < \infty\}$ but no filtration, and if we know that B has stationary, independent increments and that $B_t = B_t - B_0$ is normal with mean zero and variance t, then $\{B_t, \mathscr{F}_t^B; 0 \leq t < \infty\}$ is a Brownian motion (Problem 1.4). Moreover, if $\{\mathscr{F}_t\}$ is a "larger" filtration in the sense that $\mathscr{F}_t^B \subseteq \mathscr{F}_t$ for $t \geq 0$, and if $B_t - B_s$ is independent of $\mathscr{F}_s$ whenever $0 \leq s < t$, then $\{B_t, \mathscr{F}_t; 0 \leq t < \infty\}$ is also a Brownian motion.

It is often interesting, and sometimes necessary, to work with a filtration $\{\mathscr{F}_t\}$ which is larger than $\{\mathscr{F}_t^B\}$. For instance, we shall see in Example 5.3.5 that the stochastic differential equation (5.3.1) does not have a solution, unless we take the driving process W to be a Brownian motion with respect to a filtration which is *strictly larger* than $\{\mathscr{F}_t^W\}$. The desire to have existence of solutions to stochastic differential equations is a major motivation for allowing $\{\mathscr{F}_t\}$ in Definition 1.1 to be strictly larger than $\{\mathscr{F}_t^B\}$.

The first problem one encounters with Brownian motion is its *existence.* One approach to this question is to write down what the finite-dimensional distributions of this process (based on the stationarity, independence, and normality of its increments) must be, and then construct a probability measure and a process on an appropriate measurable space in such a way that we obtain the prescribed finite-dimensional distributions. This direct approach is the one most often used to construct a Markov process, but is rather lengthy and technical; we spell it out in Section 2. A more elegant approach for Brownian motion, which exploits the *Gaussian* property of this process, is based on Hilbert space theory and appears in Section 3; it is close in spirit to Wiener's (1923) original construction, which was modified by Lévy (1948) and later further simplified by Ciesielski (1961). Nothing in the remainder of the book depends on Section 3; however, Theorems 2.2 and 2.8 as well as Problem 2.9 will be useful in later developments.

Section 4 provides yet another proof for the existence of Brownian motion, this time based on the idea of the weak limit of a sequence of random walks. The properties of the space $C[0, \infty)$ developed in this section will be used extensively throughout the book.

Section 5 defines the *Markov property,* which is enjoyed by Brownian motion. Section 6 presents the *strong Markov property,* and, using a proof based on the optional sampling theorem for martingales, shows that Brownian motion is a strong Markov process. In Section 7 we discuss various choices of the filtration for Brownian motion. The central idea here is augmentation of the filtration generated by the process, in order to obtain a right-continuous filtration. Developing this material in the context of strong Markov processes requires no additional effort, and we adopt this level of generality.

Sections 8 and 9 are devoted to properties of Brownian motion. In Section 8 we compute distributions of a number of elementary Brownian functionals; among these are first passage times, last exit times, and time and level of the maximum over a fixed time-interval. Section 9 deals with almost sure properties of the Brownian sample path. Here we discuss its growth as $t \to \infty$, its oscillations near $t = 0$ (law of the iterated logarithm), its nowhere differentiability and nowhere monotonicity, and the topological perfectness of the set of times when the sample path is at the origin.

We conclude this introductory section with the Dynkin system theorem (Ash (1972), p. 169). This result will be used frequently in the sequel whenever we need to establish that a certain property, known to hold for a collection of sets closed under intersection, also holds for the σ-field generated by this collection. Our first application of this result occurs in Problem 1.4.

1.2 Definition. A collection $\mathscr{D}$ of subsets of a set Ω is called a *Dynkin system* if

(i) $\Omega \in \mathscr{D}$,
(ii) $A, B \in \mathscr{D}$ and $B \subseteq A$ imply $A \backslash B \in \mathscr{D}$,
(iii) $\{A_n\}_{n=1}^{\infty} \subseteq \mathscr{D}$ and $A_1 \subseteq A_2 \subseteq \cdots$ imply $\bigcup_{n=1}^{\infty} A_n \in \mathscr{D}$.

1.3 Dynkin System Theorem. *Let $\mathscr{C}$ be a collection of subsets of Ω which is closed under pairwise intersection. If $\mathscr{D}$ is a Dynkin system containing $\mathscr{C}$, then $\mathscr{D}$ also contains the σ-field $\sigma(\mathscr{C})$ generated by $\mathscr{C}$.*

1.4 Problem. Let $X = \{X_t; 0 \leq t < \infty\}$ be a stochastic process for which $X_0, X_{t_1} - X_{t_0}, \ldots, X_{t_n} - X_{t_{n-1}}$ are independent random variables, for every integer $n \geq 1$ and indices $0 = t_0 < t_1 < \cdots < t_n < \infty$. Then for any fixed $0 \leq s < t < \infty$, the increment $X_t - X_s$ is independent of $\mathscr{F}_s^X$.

2.2. First Construction of Brownian Motion

A. The Consistency Theorem

Let $\mathbb{R}^{[0,\infty)}$ denote the set of all real-valued functions on $[0, \infty)$. An *n-dimensional cylinder set* in $\mathbb{R}^{[0,\infty)}$ is a set of the form

$$C \triangleq \{\omega \in \mathbb{R}^{[0,\infty)}; (\omega(t_1), \ldots, \omega(t_n)) \in A\}, \tag{2.1}$$

where $t_i \in [0, \infty)$, $i = 1, \ldots, n$, and $A \in \mathscr{B}(\mathbb{R}^n)$. Let $\mathscr{C}$ denote the field of all cylinder sets (of all finite dimensions) in $\mathbb{R}^{[0,\infty)}$, and let $\mathscr{B}(\mathbb{R}^{[0,\infty)})$ denote the smallest σ-field containing $\mathscr{C}$.

2.1 Definition. Let T be the set of finite sequences $\underset{\sim}{t} = (t_1, \ldots, t_n)$ of distinct, nonnegative numbers, where the length n of these sequences ranges over the set of positive integers. Suppose that for each $\underset{\sim}{t}$ of length n, we have a probability measure $Q_{\underset{\sim}{t}}$ on $(\mathbb{R}^n, \mathscr{B}(\mathbb{R}^n))$. Then the collection $\{Q_{\underset{\sim}{t}}\}_{\underset{\sim}{t} \in T}$ is called a *family of finite-dimensional distributions*. This family is said to be *consistent* provided that the following two conditions are satisfied:

(a) if $\underset{\sim}{s} = (t_{i_1}, t_{i_2}, \ldots, t_{i_n})$ is a permutation of $\underset{\sim}{t} = (t_1, t_2, \ldots, t_n)$, then for any $A_i \in \mathscr{B}(\mathbb{R})$, $i = 1, \ldots, n$, we have

$$Q_{\underset{\sim}{t}}(A_1 \times A_2 \times \cdots \times A_n) = Q_{\underset{\sim}{s}}(A_{i_1} \times A_{i_2} \times \cdots \times A_{i_n});$$

(b) if $\underset{\sim}{t} = (t_1, t_2, \ldots, t_n)$ with $n \geq 1$, $\underset{\sim}{s} = (t_1, t_2, \ldots, t_{n-1})$, and $A \in \mathscr{B}(\mathbb{R}^{n-1})$, then

$$Q_{\underset{\sim}{t}}(A \times \mathbb{R}) = Q_{\underset{\sim}{s}}(A).$$

If we have a probability measure P on $(\mathbb{R}^{[0,\infty)}, \mathscr{B}(\mathbb{R}^{[0,\infty)}))$, then we can define a family of finite-dimensional distributions by

$$(2.2) \qquad Q_{\underset{\sim}{t}}(A) = P[\omega \in \mathbb{R}^{[0,\infty)}; (\omega(t_1), \ldots, \omega(t_n)) \in A],$$

where $A \in \mathscr{B}(\mathbb{R}^n)$ and $\underset{\sim}{t} = (t_1, \ldots, t_n) \in T$. This family is easily seen to be consistent. We are interested in the converse of this fact, because it will enable us to construct a probability measure P from the finite-dimensional distributions of Brownian motion.

2.2 Theorem (Daniell (1918), Kolmogorov (1933)). *Let $\{Q_{\underset{\sim}{t}}\}$ be a consistent family of finite-dimensional distributions. Then there is a probability measure P on $(\mathbb{R}^{[0,\infty)}, \mathscr{B}(\mathbb{R}^{[0,\infty)}))$, such that* (2.2) *holds for every $\underset{\sim}{t} \in T$.*

PROOF. We begin by defining a set function Q on the field of cylinders $\mathscr{C}$. If C is given by (2.1) and $\underset{\sim}{t} = (t_1, t_2, \ldots, t_n) \in T$, we set

$$(2.3) \qquad Q(C) = Q_{\underset{\sim}{t}}(A), \quad C \in \mathscr{C}.$$

2.3 Problem. The set function Q is well defined and finitely additive on $\mathscr{C}$, with $Q(\mathbb{R}^{[0,\infty)}) = 1$.

We now prove the countable additivity of Q on $\mathscr{C}$, and we can then draw on the Carathéodory extension theorem to assert the existence of the desired extension P of Q to $\mathscr{B}(\mathbb{R}^{[0,\infty)})$. Thus, suppose $\{B_k\}_{k=1}^{\infty}$ is a sequence of disjoint sets in $\mathscr{C}$ with $B \triangleq \bigcup_{k=1}^{\infty} B_k$ also in $\mathscr{C}$. Let $C_m = B \setminus \bigcup_{k=1}^{m} B_k$, so

$$Q(B) = Q(C_m) + \sum_{k=1}^{m} Q(B_k).$$

Countable additivity will follow from

$$(2.4) \qquad \lim_{m \to \infty} Q(C_m) = 0.$$

Now $Q(C_m) = Q(C_{m+1}) + Q(B_{m+1}) \geq Q(C_{m+1})$, so the limit in (2.4) exists. Assume that this limit is equal to $\varepsilon > 0$, and note that $\bigcap_{m=1}^{\infty} C_m = \emptyset$.

From $\{C_m\}_{m=1}^\infty$ we may construct another sequence $\{D_m\}_{m=1}^\infty$ which has the properties $D_1 \supseteq D_2 \supseteq \cdots$, $\bigcap_{m=1}^\infty D_m = \varnothing$, and $\lim_{m\to\infty} Q(D_m) = \varepsilon > 0$. Furthermore, each D_m has the form

$$D_m = \{\omega \in \mathbb{R}^{[0,\infty)}; (\omega(t_1), \ldots, \omega(t_m)) \in A_m\}$$

for some $A_m \in \mathscr{B}(\mathbb{R}^m)$, and the finite sequence $\underset{\sim}{t}_m \triangleq (t_1, \ldots, t_m) \in T$ is an extension of the finite sequence $\underset{\sim}{t}_{m-1} \triangleq (t_1, \ldots, t_{m-1}) \in T$, $m \geq 2$. This may be accomplished as follows. Each C_k has a form

$$C_k = \{\omega \in \mathbb{R}^{[0,\infty)}; (\omega(t_1), \ldots, \omega(t_{m_k})) \in A_{m_k}\}; \quad A_{m_k} \in \mathscr{B}(\mathbb{R}^{m_k}),$$

where $\underset{\sim}{t}_{m_k} = (t_1, \ldots, t_{m_k}) \in T$. Since $C_{k+1} \subseteq C_k$, we can choose these representations so that $\underset{\sim}{t}_{m_{k+1}}$ is an extension of $\underset{\sim}{t}_{m_k}$, and $A_{m_{k+1}} \subseteq A_{m_k} \times \mathbb{R}^{m_{k+1}-m_k}$. Define

$$D_1 = \{\omega; \omega(t_1) \in \mathbb{R}\}, \ldots, D_{m_1-1} = \{\omega; (\omega(t_1), \ldots, \omega(t_{m_1-1})) \in \mathbb{R}^{m_1-1}\}$$

and $D_{m_1} = C_1$, as well as

$$D_{m_1+1} = \{\omega; (\omega(t_1), \ldots, \omega(t_{m_1}), \omega(t_{m_1+1})) \in A_{m_1} \times \mathbb{R}\}, \ldots,$$

$$D_{m_2-1} = \{\omega; (\omega(t_1), \ldots, \omega(t_{m_1}), \omega(t_{m_1+1}), \ldots, \omega(t_{m_2-1})) \in A_{m_1} \times \mathbb{R}^{m_2-m_1-1}\}$$

and $D_{m_2} = C_2$. Continue this process, and note that by construction $\bigcap_{m=1}^\infty D_m = \bigcap_{m=1}^\infty C_m = \varnothing$.

2.4 Problem. We say that $A \in \mathscr{B}(\mathbb{R}^n)$ is *regular* if for every probability measure Q on $(\mathbb{R}^n, \mathscr{B}(\mathbb{R}^n))$ and for every $\varepsilon > 0$, there is a closed set F and an open set G such that $F \subseteq A \subseteq G$ and $Q(G \setminus F) < \varepsilon$. Show that every set in $\mathscr{B}(\mathbb{R}^n)$ is regular. (*Hint*: Show that the collection of regular sets is a σ-field containing all closed sets.)

According to Problem 2.4, there exists for each m a closed set $F_m \subseteq A_m$ such that $Q_{\underset{\sim}{t}_m}(A_m \setminus F_m) < \varepsilon/2^m$. By intersecting F_m with a sufficiently large closed sphere centered at the origin, we obtain a compact set K_m such that, with

$$E_m \triangleq \{\omega \in \mathbb{R}^{[0,\infty)}; (\omega(t_1), \ldots, \omega(t_m)) \in K_m\},$$

we have $E_m \subseteq D_m$ and

$$Q(D_m \setminus E_m) = Q_{\underset{\sim}{t}_m}(A_m \setminus K_m) < \frac{\varepsilon}{2^m}.$$

The sequence $\{E_m\}$ may fail to be nonincreasing, so we define

$$\tilde{E}_m = \bigcap_{k=1}^m E_k,$$

and we have

$$\tilde{E}_m = \{\omega \in \mathbb{R}^{[0,\infty)}; (\omega(t_1), \ldots, \omega(t_m)) \in \tilde{K}_m\},$$

where

$$\tilde{K}_m = (K_1 \times \mathbb{R}^{m-1}) \cap (K_2 \times \mathbb{R}^{m-2}) \cap \cdots \cap (K_{m-1} \times \mathbb{R}) \cap K_m,$$

which is compact. We can bound $Q_{\underset{\sim}{t}_m}(\tilde{K}_m)$ away from zero, since

$$\begin{aligned} Q_{\underset{\sim}{t}_m}(\tilde{K}_m) &= Q(\tilde{E}_m) = Q(D_m) - Q(D_m \backslash \tilde{E}_m) \\ &= Q(D_m) - Q\left(\bigcup_{k=1}^{m} (D_m \backslash E_k)\right) \\ &\geq Q(D_m) - Q\left(\bigcup_{k=1}^{m} (D_k \backslash E_k)\right) \\ &\geq \varepsilon - \sum_{k=1}^{m} \frac{\varepsilon}{2^k} > 0. \end{aligned}$$

Therefore, $\tilde{K}_m$ is nonempty for each m, and we can choose $(x_1^{(m)}, \ldots, x_m^{(m)}) \in \tilde{K}_m$. Being contained in the compact set $\tilde{K}_1$, the sequence $\{x_1^{(m)}\}_{m=1}^{\infty}$ must have a convergent subsequence $\{x_1^{(m_k)}\}_{k=1}^{\infty}$ with limit x_1. But $\{(x_1^{(m_k)}, x_2^{(m_k)})\}_{k=2}^{\infty}$ is contained in $\tilde{K}_2$, so it has a convergent subsequence with limit (x_1, x_2). Continuing this process, we can construct $(x_1, x_2, \ldots) \in \mathbb{R} \times \mathbb{R} \times \cdots$, such that $(x_1, \ldots, x_m) \in \tilde{K}_m$ for each m. Consequently, the set

$$S = \{\omega \in \mathbb{R}^{[0,\infty)} : \omega(t_i) = x_i, i = 1, 2, \ldots\}$$

is contained in each $\tilde{E}_m$, and hence in each D_m. This contradicts the fact that $\bigcap_{m=1}^{\infty} D_m = \varnothing$. We conclude that (2.4) holds. □

Our aim is to construct a probability measure P on $(\Omega, \mathscr{F}) \triangleq (\mathbb{R}^{[0,\infty)}, \mathscr{B}(\mathbb{R}^{[0,\infty)}))$ so that the process $B = \{B_t, \mathscr{F}_t^B; 0 \leq t < \infty\}$ defined by $B_t(\omega) \triangleq \omega(t)$, the *coordinate mapping process*, is almost a standard, one-dimensional Brownian motion under P. We say "almost" because we leave aside the requirement of sample path continuity for the moment and concentrate on the finite-dimensional distributions. Recalling the discussion following Definition 1.1, we see that whenever $0 = s_0 < s_1 < s_2 < \cdots < s_n$, the cumulative distribution function for $(B_{s_1}, \ldots, B_{s_n})$ must be

$$\begin{aligned} (2.5) \qquad & F_{(s_1,\ldots,s_n)}(x_1, \ldots, x_n) \\ &= \int_{-\infty}^{x_1} \int_{-\infty}^{x_2} \cdots \int_{-\infty}^{x_n} p(s_1; 0, y_1) p(s_2 - s_1; y_1, y_2) \ldots \\ &\quad \ldots p(s_n - s_{n-1}; y_{n-1}, y_n)\, dy_n \ldots dy_2\, dy_1 \end{aligned}$$

for $(x_1, \ldots, x_n) \in \mathbb{R}^n$, where p is the Gaussian kernel

$$(2.6) \qquad p(t; x, y) \triangleq \frac{1}{\sqrt{2\pi t}} e^{-(x-y)^2/2t}, \quad t > 0,\ x, y \in \mathbb{R}.$$

The reader can verify (and should, if he has never done so!) that (2.5) is equivalent to the statement that the increments $\{B_{s_j} - B_{s_{j-1}}\}_{j=1}^{n}$ are independent, and $B_{s_j} - B_{s_{j-1}}$ is normally distributed with mean zero and variance $s_j - s_{j-1}$.

Now let $\underset{\sim}{t} = (t_1, t_2, \ldots, t_n)$, where the t_j are not necessarily ordered but are distinct. Let the random vector $(B_{t_1}, B_{t_2}, \ldots, B_{t_n})$ have the distribution determined by (2.5) (where the t_j must be ordered from smallest to largest to obtain $(s_1, \ldots, s_n)$ appearing in (2.5)). For $A \in \mathscr{B}(\mathbb{R}^n)$, let $Q_{\underline{t}}(A)$ be the probability under this distribution that $(B_{t_1}, B_{t_2}, \ldots, B_{t_n})$ is in A. This defines a family of finite-dimensional distributions $\{Q_{\underline{t}}\}_{\underline{t} \in T}$.

2.5 Problem. Show that the just defined family $\{Q_{\underline{t}}\}_{\underline{t} \in T}$ is consistent.

2.6 Corollary to Theorem 2.2. *There is a probability measure P on $(\mathbb{R}^{[0,\infty)}, \mathscr{B}(\mathbb{R}^{[0,\infty)}))$, under which the coordinate mapping process*

$$B_t(\omega) = \omega(t); \quad \omega \in \mathbb{R}^{[0,\infty)}, \; t \geq 0,$$

has stationary, independent increments. An increment $B_t - B_s$, where $0 \leq s < t$, is normally distributed with mean zero and variance $t - s$.

B. The Kolmogorov-Čentsov Theorem

Our construction of Brownian motion would now be complete, were it not for the fact that we have built the process on the sample space $\mathbb{R}^{[0,\infty)}$ of all real-valued functions on $[0, \infty)$ rather than on the space $C[0, \infty)$ of continuous functions on this half-line. One might hope to overcome this difficulty by showing that the probability measure P in Corollary 2.6 assigns measure one to $C[0, \infty)$. However, as the next problem shows, $C[0, \infty)$ is not in the σ-field $\mathscr{B}(\mathbb{R}^{[0,\infty)})$, so $P(C[0, \infty))$ is not defined. This failure is a manifestation of the fact that the σ-field $\mathscr{B}(\mathbb{R}^{[0,\infty)})$ is, quite uncomfortably, "too small" for a space as big as $\mathbb{R}^{[0,\infty)}$; no set in $\mathscr{B}(\mathbb{R}^{[0,\infty)})$ can have restrictions on uncountably many coordinates. In contrast to the space $C[0, \infty)$, it is not possible to determine a function in $\mathbb{R}^{[0,\infty)}$ by specifying its values at only countably many coordinates. Consequently, the next theorem takes a different approach, which is to construct a continuous modification of the coordinate mapping process in Corollary 2.6.

2.7 Exercise. Show that the only $\mathscr{B}(\mathbb{R}^{[0,\infty)})$-measurable set contained in $C[0, \infty)$ is the empty set. (*Hint*: A typical set in $\mathscr{B}(\mathbb{R}^{[0,\infty)})$ has the form

$$E = \{\omega \in \mathbb{R}^{[0,\infty)}; (\omega(t_1), \omega(t_2), \ldots) \in A\},$$

where $A \in \mathscr{B}(\mathbb{R} \times \mathbb{R} \times \cdots)$).

2.8 Theorem (Kolmogorov, Čentsov (1956a)). *Suppose that a process $X = \{X_t; 0 \leq t \leq T\}$ on a probability space $(\Omega, \mathscr{F}, P)$ satisfies the condition*

$$E|X_t - X_s|^\alpha \leq C|t - s|^{1+\beta}, \quad 0 \leq s, t \leq T, \tag{2.7}$$

for some positive constants α, β, and C. Then there exists a continuous modification $\tilde{X} = \{\tilde{X}_t; 0 \leq t \leq T\}$ of X, which is locally Hölder-continuous with exponent γ

for every $\gamma \in (0, \beta/\alpha)$, *i.e.*,

$$P\left[\omega; \sup_{\substack{0<t-s<h(\omega) \\ s,t\in[0,T]}} \frac{|\tilde{X}_t(\omega) - \tilde{X}_s(\omega)|}{|t-s|^\gamma} \le \delta\right] = 1, \tag{2.8}$$

where $h(\omega)$ *is an a.s. positive random variable and* $\delta > 0$ *is an appropriate constant.*

PROOF. For notational simplicity, we take $T = 1$. Much of what follows is a consequence of the Čebyšev inequality. First, for any $\varepsilon > 0$, we have

$$P[|X_t - X_s| \ge \varepsilon] \le \frac{E|X_t - X_s|^\alpha}{\varepsilon^\alpha} \le C\varepsilon^{-\alpha}|t-s|^{1+\beta},$$

and so $X_s \to X_t$ in probability as $s \to t$. Second, setting $t = k/2^n$, $s = (k-1)/2^n$, and $\varepsilon = 2^{-\gamma n}$ (where $0 < \gamma < \beta/\alpha$) in the preceding inequality, we obtain

$$P[|X_{k/2^n} - X_{(k-1)/2^n}| \ge 2^{-\gamma n}] \le C2^{-n(1+\beta-\alpha\gamma)},$$

and consequently,

$$\begin{aligned} P\left[\max_{1\le k\le 2^n} |X_{k/2^n} - X_{(k-1)/2^n}| \ge 2^{-\gamma n}\right] &= P\left[\bigcup_{k=1}^{2^n} |X_{k/2^n} - X_{(k-1)/2^n}| \ge 2^{-\gamma n}\right] \\ &\le C2^{-n(\beta-\alpha\gamma)}. \end{aligned} \tag{2.9}$$

The last expression is the general term of a convergent series; by the Borel-Cantelli lemma, there is a set $\Omega^* \in \mathscr{F}$ with $P(\Omega^*) = 1$ such that for each $\omega \in \Omega^*$,

$$\max_{1\le k\le 2^n} |X_{k/2^n}(\omega) - X_{(k-1)/2^n}(\omega)| < 2^{-\gamma n}, \quad \forall\, n \ge n^*(\omega), \tag{2.10}$$

where $n^*(\omega)$ is a positive, integer-valued random variable.

For each integer $n \ge 1$, let us consider the partition $D_n = \{(k/2^n); k = 0, 1, \ldots, 2^n\}$ of $[0, 1]$, and let $D = \bigcup_{n=1}^\infty D_n$ be the set of dyadic rationals in $[0, 1]$. We shall fix $\omega \in \Omega^*$, $n \ge n^*(\omega)$, and show that for every $m > n$, we have

$$|X_t(\omega) - X_s(\omega)| \le 2 \sum_{j=n+1}^{m} 2^{-\gamma j}; \quad \forall\, t, s \in D_m, 0 < t - s < 2^{-n}. \tag{2.11}$$

For $m = n + 1$, we can only have $t = (k/2^m)$, $s = ((k-1)/2^m)$, and (2.11) follows from (2.10). Suppose (2.11) is valid for $m = n+1, \ldots, M-1$. Take $s < t$, $s, t \in D_M$, consider the numbers $t^1 = \max\{u \in D_{M-1}; u \le t\}$ and $s^1 = \min\{u \in D_{M-1}; u \ge s\}$, and notice the relationships $s \le s^1 \le t^1 \le t$, $s^1 - s \le 2^{-M}$, $t - t^1 \le 2^{-M}$. From (2.10) we have $|X_{s^1}(\omega) - X_s(\omega)| \le 2^{-\gamma M}$, $|X_t(\omega) - X_{t^1}(\omega)| \le 2^{-\gamma M}$, and from (2.11) with $m = M - 1$,

$$|X_{t^1}(\omega) - X_{s^1}(\omega)| \le 2 \sum_{j=n+1}^{M-1} 2^{-\gamma j}.$$

We obtain (2.11) for $m = M$.

We can show now that $\{X_t(\omega); t \in D\}$ is uniformly continuous in t for every $\omega \in \Omega^*$. For any numbers s, $t \in D$ with $0 < t - s < h(\omega) \triangleq 2^{-n^*(\omega)}$, we select $n \geq n^*(\omega)$ such that $2^{-(n+1)} \leq t - s < 2^{-n}$. We have from (2.11)

(2.12) $$|X_t(\omega) - X_s(\omega)| \leq 2 \sum_{j=n+1}^{\infty} 2^{-\gamma j} \leq \delta |t - s|^\gamma, \quad 0 < t - s < h(\omega),$$

where $\delta = 2/(1 - 2^{-\gamma})$. This proves the desired uniform continuity.

We define $\tilde{X}$ as follows. For $\omega \notin \Omega^*$, set $\tilde{X}_t(\omega) = 0$, $0 \leq t \leq 1$. For $\omega \in \Omega^*$ and $t \in D$, set $\tilde{X}_t(\omega) = X_t(\omega)$. For $\omega \in \Omega^*$ and $t \in [0, 1] \cap D^c$, choose a sequence $\{s_n\}_{n=1}^{\infty} \subseteq D$ with $s_n \to t$; uniform continuity and the Cauchy criterion imply that $\{X_{s_n}(\omega)\}_{n=1}^{\infty}$ has a limit which depends on t but not on the particular sequence $\{s_n\}_{n=1}^{\infty} \subseteq D$ chosen to converge to t, and we set $\tilde{X}_t(\omega) = \lim_{s_n \to t} X_{s_n}(\omega)$. The resulting process $\tilde{X}$ is thereby continuous; indeed, $\tilde{X}$ satisfies (2.12), so (2.8) is established.

To see that $\tilde{X}$ is a modification of X, observe that $\tilde{X}_t = X_t$ a.s. for $t \in D$; for $t \in [0, 1] \cap D^c$ and $\{s_n\}_{n=1}^{\infty} \subseteq D$ with $s_n \to t$, we have $X_{s_n} \to X_t$ in probability and $X_{s_n} \to \tilde{X}_t$ a.s., so $\tilde{X}_t = X_t$ a.s. □

2.9 Problem. A *random field* is a collection of random variables $\{X_t; t \in \mathscr{A}\}$, where $\mathscr{A}$ is a partially ordered set. Suppose $\{X_t; t \in [0, T]^d\}$, $d \geq 2$, is a random field satisfying

(2.13) $$E|X_t - X_s|^\alpha \leq C \|t - s\|^{d+\beta}$$

for some positive constants α, β, and C. Show that the conclusion of Theorem 2.8 holds, with (2.8) replaced by

(2.14) $$P\left[\omega; \sup_{\substack{0 < \|t-s\| < h(\omega) \\ s,t \in [0,T]^d}} \frac{|\tilde{X}_t(\omega) - \tilde{X}_s(\omega)|}{\|t - s\|^\gamma} \leq \delta\right] = 1.$$

2.10 Problem. Show that if $B_t - B_s$, $0 \leq s < t$, is normally distributed with mean zero and variance $t - s$, then for each positive integer n, there is a positive constant C_n for which

$$E|B_t - B_s|^{2n} = C_n |t - s|^n.$$

2.11 Corollary to Theorem 2.8. *There is a probability measure P on $(\mathbb{R}^{[0,\infty)}, \mathscr{B}(\mathbb{R}^{[0,\infty)}))$, and a stochastic process $W = \{W_t, \mathscr{F}_t^W; t \geq 0\}$ on the same space, such that under P, W is a Brownian motion.*

Proof. According to Theorem 2.8 and Problem 2.10, there is for each $T > 0$ a modification W^T of the process B in Corollary 2.6 such that W^T is continuous on $[0, T]$. Let

$$\Omega_T = \{\omega; W_t^T(\omega) = B_t(\omega) \text{ for every rational } t \in [0, T]\},$$

so $P(\Omega_T) = 1$. On $\tilde{\Omega} \triangleq \bigcap_{T=1}^{\infty} \Omega_T$, we have for positive integers T_1 and T_2,

$$W_t^{T_1}(\omega) = W_t^{T_2}(\omega), \text{ for every rational } t \in [0, T_1 \wedge T_2].$$

Since both processes are continuous on $[0, T_1 \wedge T_2]$, we must have $W_t^{T_1}(\omega) = W_t^{T_2}(\omega)$ for every $t \in [0, T_1 \wedge T_2]$, $\omega \in \tilde{\Omega}$. Define $W_t(\omega)$ to be this common value. For $\omega \notin \tilde{\Omega}$, set $W_t(\omega) = 0$ for all $t \geq 0$. □

2.12 Remark. Actually, for P-a.e. $\omega \in \mathbb{R}^{[0,\infty)}$, the Brownian sample path $\{W_t(\omega); 0 \leq t < \infty\}$ is locally Hölder-continuous with exponent γ, for every $\gamma \in (0, 1/2)$. This is a consequence of Theorem 2.8 and Problem 2.10.

2.3. Second Construction of Brownian Motion

This section provides a succinct, self-contained construction of Brownian motion. It may be omitted without loss of continuity.

Let us suppose that $\{B_t, \mathscr{F}_t; t \geq 0\}$ is a Brownian motion, fix $0 \leq s < t < \infty$, and set $\theta \triangleq (t + s)/2$; then, conditioned on $B_s = x$ and $B_t = z$, the random variable B_θ is normal with mean $\mu \triangleq (x + z)/2$ and variance $\sigma^2 \triangleq (t - s)/4$. To verify this, observe that the known distribution and independence of the increments B_s, $B_\theta - B_s$, and $B_t - B_\theta$ lead to the joint density

$$P[B_s \in dx, B_\theta \in dy, B_t \in dz] = p(s; 0, x) p\left(\frac{t-s}{2}; x, y\right) p\left(\frac{t-s}{2}; y, z\right) dx\, dy\, dz$$

$$= p(s; 0, x) p(t - s; x, z) \cdot \frac{1}{\sigma\sqrt{2\pi}} \exp\left\{-\frac{(y - \mu)^2}{2\sigma^2}\right\} dx\, dy\, dz$$

in the notation of (2.6), after a bit of algebra. Dividing by

$$P[B_s \in dx, B_t \in dz] = p(s; 0, x) p(t - s; x, z)\, dx\, dz,$$

we obtain

$$(3.1) \qquad P[B_{(t+s)/2} \in dy \mid B_s = x, B_t = z] = \frac{1}{\sigma\sqrt{2\pi}} e^{-(y-\mu)^2/2\sigma^2}\, dy.$$

The simple form of this conditional distribution for $B_{(t+s)/2}$ suggests that we can construct Brownian motion on some finite time-interval, say $[0, 1]$, by interpolation. Once we have completed the construction on $[0, 1]$, a simple "patching together" of a sequence of such Brownian motions will result in a Brownian motion defined for all $t \geq 0$.

To carry out this program, we begin with a countable collection $\{\xi_k^{(n)}; k \in I(n), n = 0, 1, \ldots\}$ of independent, standard (zero mean and unit variance) normal random variables on a probability space $(\Omega, \mathscr{F}, P)$. Here $I(n)$ is the set of odd integers between 0 and 2^n; i.e., $I(0) = \{1\}$, $I(1) = \{1\}$, $I(2) = \{1, 3\}$, etc. For each $n \geq 0$, we define a process $B^{(n)} = \{B_t^{(n)}; 0 \leq t \leq 1\}$ by recursion and linear interpolation, as follows. For $n \geq 1$, $B_{k/2^{n-1}}^{(n)}$ will agree with $B_{k/2^{n-1}}^{(n-1)}$, $k = 0, 1, \ldots, 2^{n-1}$. Thus, for each value of n, we need only specify $B_{k/2^n}^{(n)}$ for $k \in I(n)$. We set

$$B_0^{(0)} = 0, \quad B_1^{(0)} = \xi_1^{(0)}.$$

If the values of $B_{k/2^{n-1}}^{(n-1)}$, $k = 0, 1, \ldots, 2^{n-1}$ have been specified (so $B_t^{(n-1)}$ is defined for $0 \le t \le 1$ by piecewise-linear interpolation) and $k \in I(n)$, we denote $s = (k-1)/2^n$, $t = (k+1)/2^n$, $\mu = \frac{1}{2}(B_s^{(n-1)} + B_t^{(n-1)})$, and $\sigma^2 = (t-s)/4 = 1/2^{n+1}$ and set, in accordance with (3.1),

$$B_{k/2^n}^{(n)} \equiv B_{(t+s)/2}^{(n)} \triangleq \mu + \sigma \xi_k^{(n)}.$$

We shall show that, almost surely, $B_t^{(n)}$ converges uniformly in t to a continuous function B_t, and $\{B_t, \mathscr{F}_t^B; 0 \le t \le 1\}$ is a Brownian motion.

Our first step is to give a more convenient representation for the processes $B^{(n)}$, $n = 0, 1, \ldots$. We define the *Haar functions* by $H_1^{(0)}(t) = 1$, $0 \le t \le 1$, and for $n \ge 1$, $k \in I(n)$,

$$H_k^{(n)}(t) = \begin{cases} 2^{(n-1)/2}, & \dfrac{k-1}{2^n} \le t < \dfrac{k}{2^n}, \\ -2^{(n-1)/2}, & \dfrac{k}{2^n} \le t < \dfrac{k+1}{2^n}, \\ 0, & \text{otherwise.} \end{cases}$$

We define the *Schauder functions* by

$$S_k^{(n)}(t) = \int_0^t H_k^{(n)}(u)\,du, \quad 0 \le t \le 1, n \ge 0, k \in I(n).$$

Note that $S_1^{(0)}(t) = t$, and for $n \ge 1$ the graphs of $S_k^{(n)}$ are little tents of height $2^{-(n+1)/2}$ centered at $k/2^n$ and nonoverlapping for different values of $k \in I(n)$. It is clear that $B_t^{(0)} = \xi_1^{(0)} S_1^{(0)}(t)$, and by induction on n, it is easily verified that

$$B_t^{(n)}(\omega) = \sum_{m=0}^{n} \sum_{k \in I(m)} \xi_k^{(m)}(\omega) S_k^{(m)}(t), \quad 0 \le t \le 1, n \ge 0. \tag{3.2}$$

3.1 Lemma. *As $n \to \infty$, the sequence of functions $\{B_t^{(n)}(\omega); 0 \le t \le 1\}$, $n \ge 0$, given by* (3.2) *converges uniformly in t to a continuous function $\{B_t(\omega); 0 \le t \le 1\}$, for a.e. $\omega \in \Omega$.*

PROOF. Define $b_n = \max_{k \in I(n)} |\xi_k^{(n)}|$. For $x > 0$

$$P[|\xi_k^{(n)}| > x] = \sqrt{\frac{2}{\pi}} \int_x^\infty e^{-u^2/2}\,du \tag{3.3}$$

$$\le \sqrt{\frac{2}{\pi}} \int_x^\infty \frac{u}{x} e^{-u^2/2}\,du = \sqrt{\frac{2}{\pi}} \frac{e^{-x^2/2}}{x},$$

which gives

$$P[b_n > n] = P\left[\bigcup_{k \in I_n} \{|\xi_k^{(n)}| > n\}\right] \le 2^n P[|\xi_1^{(n)}| > n] \le \sqrt{\frac{2}{\pi}} \frac{2^n e^{-n^2/2}}{n}, \quad n \ge 1.$$

Now $\sum_{n=1}^{\infty} 2^n e^{-n^2/2}/n < \infty$, so the Borel-Cantelli lemma implies that there is a set $\tilde{\Omega}$ with $P(\tilde{\Omega}) = 1$ such that for each $\omega \in \tilde{\Omega}$ there is an integer $n(\omega)$ satisfying $b_n(\omega) \le n$ for all $n \ge n(\omega)$. But then

$$\sum_{n=n(\omega)}^{\infty} \sum_{k \in I(n)} |\xi_k^{(n)} S_k^{(n)}(t)| \le \sum_{n=n(\omega)}^{\infty} n 2^{-(n+1)/2} < \infty;$$

so for $\omega \in \tilde{\Omega}$, $B_t^{(n)}(\omega)$ converges uniformly in t to a limit $B_t(\omega)$. Continuity of $\{B_t(\omega); 0 \le t \le 1\}$ follows from the uniformity of the convergence. □

Under the inner product $\langle f, g \rangle = \int_0^1 f(t)g(t)\,dt$, $L^2[0,1]$ is a Hilbert space, and the Haar functions $\{H_k^{(n)}; k \in I(n), n \ge 0\}$ form a complete, orthonormal system (see, e.g., Kaczmarz & Steinhaus (1951), but also Exercise 3.3 later). The Parseval equality

$$\langle f, g \rangle = \sum_{n=0}^{\infty} \sum_{k \in I(n)} \langle f, H_k^{(n)} \rangle \langle g, H_k^{(n)} \rangle,$$

applied to $f = 1_{[0,t]}$ and $g = 1_{[0,s]}$ yields

$$\sum_{n=0}^{\infty} \sum_{k \in I(n)} S_k^{(n)}(t) S_k^{(n)}(s) = s \wedge t; \quad 0 \le s, t \le 1. \tag{3.4}$$

3.2 Theorem. *With $\{B_t^{(n)}\}_{n=1}^{\infty}$ defined by (3.2) and $B_t = \lim_{n\to\infty} B_t^{(n)}$, the process $\{B_t, \mathscr{F}_t^B; 0 \le t \le 1\}$ is a Brownian motion on $[0,1]$.*

PROOF. It suffices to prove that, for $0 = t_0 < t_1 < \cdots < t_n \le 1$, the increments $\{B_{t_j} - B_{t_{j-1}}\}_{j=1}^n$ are independent, normally distributed, with mean zero and variance $t_j - t_{j-1}$. For this, we show that for $\lambda_j \in \mathbb{R}$, $j = 1, \ldots, n$ and $i = \sqrt{-1}$,

$$E\left[\exp\left\{i \sum_{j=1}^n \lambda_j (B_{t_j} - B_{t_{j-1}})\right\}\right] = \prod_{j=1}^n \exp\left\{-\frac{1}{2}\lambda_j^2 (t_j - t_{j-1})\right\}. \tag{3.5}$$

Set $\lambda_{n+1} = 0$. Using the independence and standard normality of the random variables $\{\xi_k^{(n)}\}$, we have from (3.2)

$$\begin{aligned}
&E\left[\exp\left\{-i \sum_{j=1}^n (\lambda_{j+1} - \lambda_j) B_{t_j}^{(M)}\right\}\right] \\
&= E\left[\exp\left\{-i \sum_{m=0}^{M} \sum_{k \in I(m)} \xi_k^{(m)} \sum_{j=1}^n (\lambda_{j+1} - \lambda_j) S_k^{(m)}(t_j)\right\}\right] \\
&= \prod_{m=0}^{M} \prod_{k \in I(m)} E\left[\exp\left\{-i\xi_k^{(m)} \sum_{j=1}^n (\lambda_{j+1} - \lambda_j) S_k^{(m)}(t_j)\right\}\right] \\
&= \prod_{m=0}^{M} \prod_{k \in I(m)} \exp\left[-\frac{1}{2}\left\{\sum_{j=1}^n (\lambda_{j+1} - \lambda_j) S_k^{(m)}(t_j)\right\}^2\right] \\
&= \exp\left[-\frac{1}{2} \sum_{j=1}^n \sum_{i=1}^n (\lambda_{j+1} - \lambda_j)(\lambda_{i+1} - \lambda_i) \sum_{m=0}^{M} \sum_{k \in I(m)} S_k^{(m)}(t_i) S_k^{(m)}(t_j)\right].
\end{aligned}$$

Letting $M \to \infty$ and using (3.4), we obtain

$$
\begin{aligned}
E\left[\exp\left\{i \sum_{j=1}^{n} \lambda_j (B_{t_j} - B_{t_{j-1}})\right\}\right] &= E\left[\exp\left\{-i \sum_{j=1}^{n} (\lambda_{j+1} - \lambda_j) B_{t_j}\right\}\right] \\
&= \exp\left\{-\sum_{j=1}^{n-1} \sum_{i=j+1}^{n} (\lambda_{j+1} - \lambda_j)(\lambda_{i+1} - \lambda_i) t_j - \frac{1}{2} \sum_{j=1}^{n} (\lambda_{j+1} - \lambda_j)^2 t_j\right\} \\
&= \exp\left\{-\sum_{j=1}^{n-1} (\lambda_{j+1} - \lambda_j)(-\lambda_{j+1}) t_j - \frac{1}{2} \sum_{j=1}^{n} (\lambda_{j+1} - \lambda_j)^2 t_j\right\} \\
&= \exp\left\{\frac{1}{2} \sum_{j=1}^{n-1} (\lambda_{j+1}^2 - \lambda_j^2) t_j - \frac{1}{2} \lambda_n^2 t_n\right\} \\
&= \prod_{j=1}^{n} \exp\left\{-\frac{1}{2} \lambda_j^2 (t_j - t_{j-1})\right\}. \qquad \square
\end{aligned}
$$

3.3 Exercise. Prove Theorem 3.2 without resort to the Parseval identity (3.4), by completing the following steps.

(a) The increments $\{B_{k/2^n}^{(n)} - B_{(k-1)/2^n}^{(n)}\}_{k=1}^{2^n}$ are independent, normal random variables with mean zero and variance $1/2^n$.

(b) If $0 = t_0 < t_1 < \cdots < t_n \le 1$ and each t_j is a dyadic rational, then the increments $\{B_{t_j} - B_{t_{j-1}}\}_{j=1}^{n}$ are independent, normal random variables with mean zero and variance $(t_j - t_{j-1})$.

(c) The assertion in (b) holds even if $\{t_j\}_{j=1}^{n}$ is not contained in the set of dyadic rationals.

3.4 Corollary. *There is a probability space $(\Omega, \mathscr{F}, P)$ and a stochastic process $B = \{B_t, \mathscr{F}_t^B; 0 \le t < \infty\}$ on it, such that B is a standard, one-dimensional Brownian motion.*

PROOF. According to Theorem 3.2, there is a sequence $(\Omega_n, \mathscr{F}_n, P_n)$, $n = 1, 2, \ldots$ of probability spaces together with a Brownian motion $\{X_t^{(n)}; 0 \le t \le 1\}$ on each space. Let $\Omega = \Omega_1 \times \Omega_2 \times \cdots$, $\mathscr{F} = \mathscr{F}_1 \otimes \mathscr{F}_2 \otimes \cdots$, and $P = P_1 \times P_2 \times \cdots$. Define B on Ω recursively by

$$
\begin{aligned}
B_t &= X_t^{(1)}, \quad 0 \le t \le 1, \\
B_t &= B_n + X_{t-n}^{(n+1)}, \quad n \le t \le n+1.
\end{aligned}
$$

This process is clearly continuous, and the increments are easily seen to be independent and normal with zero mean and the proper variances. $\square$

2.4. The Space $C[0, \infty)$, Weak Convergence, and the Wiener Measure

The sample spaces for the Brownian motions we built in Sections 2 and 3 were, respectively, the space $\mathbb{R}^{[0,\infty)}$ of all real-valued functions on $[0, \infty)$ and a space Ω rich enough to carry a countable collection of independent, standard normal random variables. The "canonical" space for Brownian motion, the one most convenient for many future developments, is $C[0,\infty)$, the space of all continuous, real-valued functions on $[0,\infty)$ with metric

$$\rho(\omega_1,\omega_2) \triangleq \sum_{n=1}^{\infty} \frac{1}{2^n} \max_{0 \le t \le n} (|\omega_1(t) - \omega_2(t)| \wedge 1). \tag{4.1}$$

In this section, we show how to construct a measure, called *Wiener measure*, on this space so that the coordinate mapping process is Brownian motion. This construction is given as the proof of Theorem 4.20 (Donsker's invariance principle) and involves the notion of weak convergence of random walks to Brownian motion.

4.1 Problem. Show that ρ defined by (4.1) is a metric on $C[0,\infty)$ and, under ρ, $C[0,\infty)$ is a complete, separable metric space.

4.2 Problem. Let $\mathscr{C}(\mathscr{C}_t)$ be the collection of finite-dimensional cylinder sets of the form (2.1); i.e.,

$$C = \{\omega \in C[0,\infty); (\omega(t_1),\ldots,\omega(t_n)) \in A\}; \quad n \ge 1,\ A \in \mathscr{B}(\mathbb{R}^n), \tag{2.1$'$}$$

where, for all $i = 1, \ldots, n$, $t_i \in [0,\infty)$ (respectively, $t_i \in [0,t]$). Denote by $\mathscr{G}(\mathscr{G}_t)$ the smallest σ-field containing $\mathscr{C}(\mathscr{C}_t)$.

Show that $\mathscr{G} = \mathscr{B}(C[0,\infty))$, the Borel σ-field generated by the open sets in $C[0,\infty)$, and that $\mathscr{G}_t = \varphi_t^{-1}(\mathscr{B}(C[0,\infty))) \triangleq \mathscr{B}_t(C[0,\infty))$, where φ_t: $C[0,\infty) \to C[0,\infty)$ is the mapping $(\varphi_t\omega)(s) = \omega(t \wedge s)$; $0 \le s < \infty$.

Whenever X is a random variable on a probability space $(\Omega, \mathscr{F}, P)$ with values in a measurable space $(S, \mathscr{B}(S))$, i.e., the function $X: \Omega \to S$ is $\mathscr{F}/\mathscr{B}(S)$-measurable, then X induces a probability measure PX^{-1} on $(S, \mathscr{B}(S))$ by

$$PX^{-1}(B) = P\{\omega \in \Omega; X(\omega) \in B\}, \quad B \in \mathscr{B}(S). \tag{4.2}$$

An important special case of (4.2) occurs when $X = \{X_t; 0 \le t < \infty\}$ is a continuous stochastic process on $(\Omega, \mathscr{F}, P)$. Such an X can be regarded as a random variable on $(\Omega, \mathscr{F}, P)$ with values in $(C[0,\infty), \mathscr{B}(C[0,\infty)))$, and PX^{-1} is called the *law of* X. The reader should verify that the law of a continuous process is determined by its finite-dimensional distributions.

A. Weak Convergence

The following concept is of fundamental importance in probability theory.

4.3 Definition. Let (S, ρ) be a metric space with Borel σ-field $\mathscr{B}(S)$. Let $\{P_n\}_{n=1}^{\infty}$ be a sequence of probability measures on $(S, \mathscr{B}(S))$, and let P be another measure on this space. We say that $\{P_n\}_{n=1}^{\infty}$ *converges weakly to* P and write $P_n \xrightarrow{w} P$, if and only if

$$\lim_{n\to\infty} \int_S f(s)\,dP_n(s) = \int_S f(s)\,dP(s)$$

for every bounded, continuous real-valued function f on S.

It follows, in particular, that the *weak limit* P is a probability measure, and that it is unique.

4.4 Definition. Let $\{(\Omega_n, \mathscr{F}_n, P_n)\}_{n=1}^\infty$ be a sequence of probability spaces, and on each of them consider a random variable X_n with values in the metric space (S, ρ). Let $(\Omega, \mathscr{F}, P)$ be another probability space, on which a random variable X with values in (S, ρ) is given. We say that $\{X_n\}_{n=1}^\infty$ *converges to X in distribution*, and write $X_n \xrightarrow{\mathscr{D}} X$, if the sequence of measures $\{P_n X_n^{-1}\}_{n=1}^\infty$ converges weakly to the measure PX^{-1}.

Equivalently, $X_n \xrightarrow{\mathscr{D}} X$ if and only if

$$\lim_{n\to\infty} E_n f(X_n) = Ef(X)$$

for every bounded, continuous real-valued function f on S, where E_n and E denote expectations with respect to P_n and P, respectively.

Recall that if S in Definition 4.4 is $\mathbb{R}^d$, then $X_n \xrightarrow{\mathscr{D}} X$ if and only if the sequence of characteristic functions $\varphi_n(u) \triangleq E_n \exp\{i(u, X_n)\}$ converges to $\varphi(u) \triangleq E\exp\{i(u, X)\}$, for every $u \in \mathbb{R}^d$. This is the so-called Cramér-Wold device (Theorem 7.7 in Billingsley (1968)).

The most important example of convergence in distribution is that provided by the central limit theorem. In the Lindeberg-Lévy form used here, the theorem asserts that if $\{\xi_n\}_{n=1}^\infty$ is a sequence of independent, identically distributed random variables with mean zero and variance σ^2, then $\{S_n\}$ defined by

$$S_n = \frac{1}{\sigma\sqrt{n}} \sum_{k=1}^n \xi_k$$

converges in distribution to a standard normal random variable. It is this fact which dictates that *a properly normalized sequence of random walks will converge in distribution to Brownian motion* (the invariance principle of Subsection D).

4.5 Problem. Suppose $\{X_n\}_{n=1}^\infty$ is a sequence of random variables taking values in a metric space (S_1, ρ_1) and converging in distribution to X. Suppose (S_2, ρ_2) is another metric space, and $\varphi: S_1 \to S_2$ is continuous. Show that $Y_n \triangleq \varphi(X_n)$ converges in distribution to $Y \triangleq \varphi(X)$.

B. Tightness

The following theorem is stated without proof; its special case $S = \mathbb{R}$ is used to prove the central limit theorem. In the form provided here, a proof can be found in several sources, for instance Billingsley (1968), pp. 35–40, or Parthasarathy (1967), pp. 47–49.

4.6 Definition. Let (S, ρ) be a metric space and let Π be a family of probability measures on $(S, \mathscr{B}(S))$. We say that Π is *relatively compact* if every sequence of elements of Π contains a weakly convergent subsequence. We say that Π is *tight* if for every $\varepsilon > 0$, there exists a compact set $K \subseteq S$ such that $P(K) \geq 1 - \varepsilon$, for every $P \in \Pi$.

If $\{X_\alpha\}_{\alpha \in A}$ is a family of random variables, each one defined on a probability space $(\Omega_\alpha, \mathscr{F}_\alpha, P_\alpha)$ and taking values in S, we say that this family is *relatively compact* or *tight* if the family of induced measures $\{P_\alpha X_\alpha^{-1}\}_{\alpha \in A}$ has the appropriate property.

4.7 Theorem (Prohorov (1956)). *Let Π be a family of probability measures on a complete, separable metric space S. This family is relatively compact if and only if it is tight.*

We are interested in the case $S = C[0, \infty)$. For this case, we shall provide a characterization of tightness (Theorem 4.10). To do so, we define for each $\omega \in C[0, \infty)$, $T > 0$, and $\delta > 0$ the *modulus of continuity* on $[0, T]$:

$$m^T(\omega, \delta) \triangleq \max_{\substack{|s-t| \leq \delta \\ 0 \leq s, t \leq T}} |\omega(s) - \omega(t)|. \tag{4.3}$$

4.8 Problem. Show that $m^T(\omega, \delta)$ is continuous in $\omega \in C[0, \infty)$ under the metric ρ of (4.1), is nondecreasing in δ, and $\lim_{\delta \downarrow 0} m^T(\omega, \delta) = 0$ for each $\omega \in C[0, \infty)$.

We shall need the following version of the Arzelà-Ascoli theorem.

4.9 Theorem. *A set $A \subseteq C[0, \infty)$ has compact closure if and only if the following two conditions hold:*

$$\sup_{\omega \in A} |\omega(0)| < \infty, \tag{4.4}$$

$$\lim_{\delta \downarrow 0} \sup_{\omega \in A} m^T(\omega, \delta) = 0 \qquad \text{for every } T > 0. \tag{4.5}$$

PROOF. Assume that the closure of A, denoted by $\bar{A}$, is compact. Since $\bar{A}$ is contained in the union of the open sets

$$G_n = \{\omega \in C[0, \infty); |\omega(0)| < n\}, \qquad n = 1, 2, \ldots$$

it must be contained in some particular G_n, and (4.4) follows. For $\varepsilon > 0$, let $K_\delta = \{\omega \in \bar{A}; m^T(\omega, \delta) \geq \varepsilon\}$. Each K_δ is closed (Problem 4.8) and is contained in $\bar{A}$, so each K_δ is compact. Problem 4.8 implies $\bigcap_{\delta > 0} K_\delta = \varnothing$, so for some $\delta(\varepsilon) > 0$, we must have $K_{\delta(\varepsilon)} = \varnothing$. This proves (4.5).

We now assume (4.4), (4.5) and prove the compactness of $\bar{A}$. Since $C[0, \infty)$ is a metric space, it suffices to prove that every sequence $\{\omega_n\}_{n=1}^{\infty} \subseteq A$ has a convergent subsequence. We fix $T > 0$ and note that for some $\delta_1 > 0$, we have $m^T(\omega, \delta_1) \leq 1$ for each $\omega \in A$; so for fixed integer $m \geq 1$ and $t \in (0, T]$ with $(m-1)\delta_1 < t \leq m\delta_1 \wedge T$, we have from (4.5):

$$|\omega(t)| \leq |\omega(0)| + \sum_{k=1}^{m-1} |\omega(k\delta_1) - \omega((k-1)\delta_1)| + |\omega(t) - \omega((m-1)\delta_1)|$$

$$\leq |\omega(0)| + m.$$

It follows that for each $r \in Q^+$, the set of nonnegative rationals, $\{\omega_n(r)\}_{n=1}^{\infty}$ is bounded. Let $\{r_0, r_1, r_2, \ldots\}$ be an enumeration of Q^+. Then choose $\{\omega_n^{(0)}\}_{n=1}^{\infty}$, a subsequence of $\{\omega_n\}_{n=1}^{\infty}$ with $\omega_n^{(0)}(r_0)$ converging to a limit denoted $\omega(r_0)$. From $\{\omega_n^{(0)}\}_{n=1}^{\infty}$, choose a further subsequence $\{\omega_n^{(1)}\}_{n=1}^{\infty}$ such that $\omega_n^{(1)}(r_1)$ converges to a limit $\omega(r_1)$. Continue this process, and then let $\{\tilde{\omega}_n\}_{n=1}^{\infty} = \{\omega_n^{(n)}\}_{n=1}^{\infty}$ be the "diagonal sequence." We have $\tilde{\omega}_n(r) \to \omega(r)$ for each $r \in Q^+$.

Let us note from (4.5) that for each $\varepsilon > 0$, there exists $\delta(\varepsilon) > 0$ such that $|\tilde{\omega}_n(s) - \tilde{\omega}_n(t)| \leq \varepsilon$ whenever $0 \leq s, t \leq T$ and $|s - t| \leq \delta(\varepsilon)$. The same inequality, therefore, holds for ω when we impose the additional condition $s, t \in Q^+$. It follows that ω is uniformly continuous on $[0, T] \cap Q^+$ and so has an extension to a continuous function, also called ω, on $[0, T]$; furthermore, $|\omega(s) - \omega(t)| \leq \varepsilon$ whenever $0 \leq s, t \leq T$ and $|s - t| \leq \delta(\varepsilon)$. For n sufficiently large, we have that whenever $t \in [0, T]$, there is some $r_k \in Q^+$ with $k \leq n$ and $|t - r_k| \leq \delta(\varepsilon)$. For sufficiently large $M \geq n$, we have $|\tilde{\omega}_m(r_j) - \omega(r_j)| \leq \varepsilon$ for all $j = 0, 1, \ldots, n$ and $m \geq M$. Consequently,

$$|\tilde{\omega}_m(t) - \omega(t)| \leq |\tilde{\omega}_m(t) - \tilde{\omega}_m(r_k)| + |\tilde{\omega}_m(r_k) - \omega(r_k)| + |\omega(r_k) - \omega(t)|$$

$$\leq 3\varepsilon, \quad \forall m \geq M, 0 \leq t \leq T.$$

We can make this argument for any $T > 0$, so $\{\tilde{\omega}_n\}_{n=1}^{\infty}$ converges uniformly on bounded intervals to the function $\omega \in C[0, \infty)$. □

4.10 Theorem. *A sequence $\{P_n\}_{n=1}^{\infty}$ of probability measures on $(C[0, \infty), \mathscr{B}(C[0, \infty)))$ is tight if and only if*

(4.6) $$\lim_{\lambda \uparrow \infty} \sup_{n \geq 1} P_n[\omega; |\omega(0)| > \lambda] = 0,$$

(4.7) $$\lim_{\delta \downarrow 0} \sup_{n \geq 1} P_n[\omega; m^T(\omega, \delta) > \varepsilon] = 0; \quad \forall T > 0, \varepsilon > 0.$$

PROOF. Suppose first that $\{P_n\}_{n=1}^{\infty}$ is tight. Given $\eta > 0$, there is a compact set K with $P_n(K) \geq 1 - \eta$, for every $n \geq 1$. According to Theorem 4.9, for sufficiently large $\lambda > 0$, we have $|\omega(0)| \leq \lambda$ for all $\omega \in K$; this proves (4.6). According to the same theorem, if T and ε are also given, then there exists δ_0 such that $m^T(\omega, \delta) \leq \varepsilon$ for $0 < \delta < \delta_0$ and $\omega \in K$. This gives us (4.7).

Let us now assume (4.6) and (4.7). Given a positive integer T and $\eta > 0$, we choose $\lambda > 0$ so that

$$\sup_{n\geq 1} P_n[\omega; |\omega(0)| > \lambda] \leq \eta/2^{T+1}.$$

We choose $\delta_k > 0$, $k = 1, 2, \ldots$ such that

$$\sup_{n\geq 1} P_n\left[\omega; m^T(\omega, \delta_k) > \frac{1}{k}\right] \leq \eta/2^{T+k+1}.$$

Define the closed sets

$$A_T = \left\{\omega; |\omega(0)| \leq \lambda, m^T(\omega, \delta_k) \leq \frac{1}{k}, k = 1, 2, \ldots\right\}, \quad A = \bigcap_{T=1}^{\infty} A_T,$$

so $P_n(A_T) \geq 1 - \sum_{k=0}^{\infty} \eta/2^{T+k+1} = 1 - \eta/2^T$ and $P_n(A) \geq 1 - \eta$, for every $n \geq 1$. By Theorem 4.9, A is compact, so $\{P_n\}_{n=1}^{\infty}$ is tight. □

4.11 Problem. Let $\{X^{(m)}\}_{m=1}^{\infty}$ be a sequence of continuous stochastic processes $X^{(m)} = \{X_t^{(m)}; 0 \leq t < \infty\}$ on $(\Omega, \mathscr{F}, P)$, satisfying the following conditions:

(i) $\sup_{m\geq 1} E|X_0^{(m)}|^{\nu} \triangleq M < \infty$,
(ii) $\sup_{m\geq 1} E|X_t^{(m)} - X_s^{(m)}|^{\alpha} \leq C_T|t - s|^{1+\beta}; \quad \forall T > 0$ and $0 \leq s, t \leq T$

for some positive constants α, β, ν (universal) and C_T (depending on $T > 0$).

Show that the probability measures $P_m \triangleq P(X^{(m)})^{-1}; m \geq 1$ induced by these processes on $(C[0, \infty), \mathscr{B}(C[0, \infty)))$ form a tight sequence. (*Hint*: Follow the technique of proof in the Kolmogorov-Čentsov Theorem 2.8, to verify the conditions (4.6), (4.7) of Theorem 4.10).

4.12 Problem. Suppose $\{P_n\}_{n=1}^{\infty}$ is a sequence of probability measures on $(C[0, \infty), \mathscr{B}(C[0, \infty)))$ which converges weakly to a probability measure P. Suppose, in addition, that $\{f_n\}_{n=1}^{\infty}$ is a uniformly bounded sequence of real-valued, continuous functions on $C[0, \infty)$ converging to a continuous function f, the convergence being uniform on compact subsets of $C[0, \infty)$. Then

$$\lim_{n\to\infty} \int_{C[0,\infty)} f_n(\omega)\, dP_n(\omega) = \int_{C[0,\infty)} f(\omega)\, dP(\omega). \tag{4.8}$$

4.13 Remark. Theorems 4.9, 4.10 and Problems 4.11, 4.12 have natural extensions to $C[0, \infty)^d$, the space of continuous, $\mathbb{R}^d$-valued functions on $[0, \infty)$. The proofs of these extensions are the same as for the one-dimensional case.

C. Convergence of Finite-Dimensional Distributions

Suppose that X is a continuous process on some $(\Omega, \mathscr{F}, P)$. For each ω, the function $t \mapsto X_t(\omega)$ is a member of $C[0, \infty)$, which we denote by $X(\omega)$. Since $\mathscr{B}(C[0, \infty))$ is generated by the one-dimensional cylinder sets and $X_t(\cdot)$ is $\mathscr{F}$-measurable for each fixed t, the random function $X: \Omega \to C[0, \infty)$ is $\mathscr{F}/\mathscr{B}(C[0, \infty))$-measurable. Thus, if $\{X^{(n)}\}_{n=1}^{\infty}$ is a sequence of continuous processes (with each $X^{(n)}$ defined on a perhaps distinct probability space

$(\Omega_n, \mathscr{F}_n, P_n)$), we can ask whether $X^{(n)} \xrightarrow{\mathscr{D}} X$ in the sense of Definition 4.4. We can also ask whether the finite-dimensional distributions of $\{X^{(n)}\}_{n=1}^{\infty}$ converge to those of X, i.e., whether

$$(X_{t_1}^{(n)}, X_{t_2}^{(n)}, \ldots, X_{t_d}^{(n)}) \xrightarrow{\mathscr{D}} (X_{t_1}, X_{t_2}, \ldots, X_{t_d}).$$

The latter question is considerably easier to answer than the former, since the convergence in distribution of finite-dimensional random vectors can be resolved by studying characteristic functions.

For any finite subset $\{t_1, \ldots, t_d\}$ of $[0, \infty)$, let us define the *projection mapping* $\pi_{t_1,\ldots,t_d}: C[0, \infty) \to \mathbb{R}^d$ as

$$\pi_{t_1,\ldots,t_d}(\omega) = (\omega(t_1), \ldots, \omega(t_d)).$$

If the function $f: \mathbb{R}^d \to \mathbb{R}$ is bounded and continuous, then the composite mapping $f \circ \pi_{t_1,\ldots,t_d}: C[0, \infty) \to \mathbb{R}$ enjoys the same properties; thus, $X^{(n)} \xrightarrow[n\to\infty]{\mathscr{D}} X$ implies

$$\begin{aligned}\lim_{n\to\infty} E_n f(X_{t_1}^{(n)}, \ldots, X_{t_d}^{(n)}) &= \lim_{n\to\infty} E_n(f \circ \pi_{t_1,\ldots,t_d})(X^{(n)}) \\ &= E(f \circ \pi_{t_1,\ldots,t_d})(X) = Ef(X_{t_1}, \ldots, X_{t_d}).\end{aligned}$$

In other words, if the sequence of processes $\{X^{(n)}\}_{n=1}^{\infty}$ converges in distribution to the process X, then all finite-dimensional distributions converge as well. The converse holds in the presence of tightness (Theorem 4.15), but not in general; this failure is illustrated by the following exercise.

4.14 Exercise. Consider the sequence of (nonrandom) processes

$$X_t^{(n)} = nt \cdot 1_{[0,1/2n]}(t) + (1 - nt) \cdot 1_{(1/2n, 1/n]}(t);$$

$0 \le t < \infty, n \ge 1$ and let $X_t = 0, t \ge 0$. Show that all finite-dimensional distributions of $X^{(n)}$ converge weakly to the corresponding finite-dimensional distributions of X, but the sequence of processes $\{X^{(n)}\}_{n=1}^{\infty}$ does *not* converge in distribution to the process X.

4.15 Theorem. *Let $\{X^{(n)}\}_{n=1}^{\infty}$ be a tight sequence of continuous processes with the property that, whenever $0 \le t_1 < \cdots < t_d < \infty$, then the sequence of random vectors $\{(X_{t_1}^{(n)}, \ldots, X_{t_d}^{(n)})\}_{n=1}^{\infty}$ converges in distribution. Let P_n be the measure induced on $(C[0, \infty), \mathscr{B}(C[0, \infty)))$ by $X^{(n)}$. Then $\{P_n\}_{n=1}^{\infty}$ converges weakly to a measure P, under which the coordinate mapping process $W_t(\omega) \triangleq \omega(t)$ on $C[0, \infty)$ satisfies*

$$(X_{t_1}^{(n)}, \ldots, X_{t_d}^{(n)}) \xrightarrow{\mathscr{D}} (W_{t_1}, \ldots, W_{t_d}), \quad 0 \le t_1 < \cdots < t_d < \infty, d \ge 1.$$

PROOF. Every subsequence $\{\tilde{X}^{(n)}\}$ of $\{X^{(n)}\}$ is tight, and so has a further subsequence $\{\hat{X}^{(n)}\}$ such that the measures induced on $C[0, \infty)$ by $\{\hat{X}^{(n)}\}$ converge weakly to a probability measure P, by the Prohorov theorem 4.7. If a different subsequence $\{\check{X}^{(n)}\}$ induces measures on $C[0, \infty)$ converging to a probability measure Q, then P and Q must have the same finite-dimensional distributions, i.e.,

$$P[\omega \in C[0,\infty); (\omega(t_1),\ldots,\omega(t_d)) \in A] = Q[\omega \in C[0,\infty); (\omega(t_1),\ldots,\omega(t_d)) \in A],$$
$$0 \le t_1 < t_2 < \cdots < t_d < \infty, \quad A \in \mathscr{B}(\mathbb{R}^d), \quad d \ge 1.$$

This means $P = Q$.

Suppose the sequence of measures $\{P_n\}_{n=1}^\infty$ induced by $\{X^{(n)}\}_{n=1}^\infty$ did not converge weakly to P. Then there must be a bounded, continuous function $f: C[0,\infty) \to \mathbb{R}$ such that $\lim_{n\to\infty} \int f(\omega) P_n(d\omega)$ does not exist, or else this limit exists but is different from $\int f(\omega) P(d\omega)$. In either case, we can choose a subsequence $\{\tilde{P}_n\}_{n=1}^\infty$ for which $\lim_{n\to\infty} \int f(\omega)\tilde{P}_n(d\omega)$ exists but is different from $\int f(\omega) P(d\omega)$. This subsequence can have no further subsequence $\{\hat{P}_n\}_{n=1}^\infty$ with $\hat{P}_n \xrightarrow{w} P$, and this violates the conclusion of the previous paragraph. □

We shall need the following result.

4.16 Problem. Let $\{X^{(n)}\}_{n=1}^\infty$, $\{Y^{(n)}\}_{n=1}^\infty$, and X be random variables with values in a separable metric space (S, ρ); we assume that for each $n \ge 1$, $X^{(n)}$ and $Y^{(n)}$ are defined on the same probability space. If $X^{(n)} \xrightarrow{\mathscr{D}} X$ and $\rho(X^{(n)}, Y^{(n)}) \to 0$ in probability, as $n \to \infty$, then $Y^{(n)} \xrightarrow{\mathscr{D}} X$ as $n \to \infty$.

D. The Invariance Principle and the Wiener Measure

Let us consider now a sequence $\{\xi_j\}_{j=1}^\infty$ of independent, identically distributed random variables with mean zero and variance σ^2, $0 < \sigma^2 < \infty$, as well as the sequence of partial sums $S_0 = 0$, $S_k = \sum_{j=1}^k \xi_j$, $k \ge 1$. A continuous-time process $Y = \{Y_t; t \ge 0\}$ can be obtained from the sequence $\{S_k\}_{k=0}^\infty$ by linear interpolation; i.e.,

(4.9) $$Y_t = S_{[\![t]\!]} + (t - [\![t]\!])\xi_{[\![t]\!]+1}, \quad t \ge 0,$$

where $[\![t]\!]$ denotes the greatest integer less than or equal to t. Scaling appropriately both time and space, we obtain from Y a sequence of processes $\{X^{(n)}\}$:

(4.10) $$X_t^{(n)} = \frac{1}{\sigma\sqrt{n}} Y_{nt}, \quad t \ge 0.$$

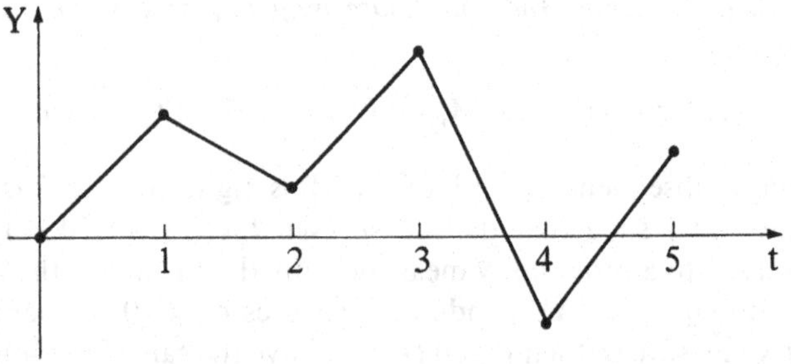

Note that with $s = k/n$ and $t = (k+1)/n$, the increment $X_t^{(n)} - X_s^{(n)} = (1/\sigma\sqrt{n})\xi_{k+1}$ is independent of $\mathscr{F}_s^{X^{(n)}} = \sigma(\xi_1, \dots, \xi_k)$. Furthermore, $X_t^{(n)} - X_s^{(n)}$ has zero mean and variance $t - s$. This suggests that $\{X_t^{(n)}; t \geq 0\}$ is approximately a Brownian motion. We now show that, even though the random variables ξ_j are not necessarily normal, the central limit theorem dictates that the limiting distributions of the increments of $X^{(n)}$ are normal.

4.17 Theorem. *With $\{X^{(n)}\}$ defined by (4.10) and $0 \leq t_1 < \cdots < t_d < \infty$, we have*

$$(X_{t_1}^{(n)}, \dots, X_{t_d}^{(n)}) \xrightarrow{\mathscr{D}} (B_{t_1}, \dots, B_{t_d}) \quad \text{as } n \to \infty,$$

where $\{B_t, \mathscr{F}_t^B; t \geq 0\}$ is a standard, one-dimensional Brownian motion.

PROOF. We take the case $d = 2$; the other cases differ from this one only by being notationally more cumbersome. Set $s = t_1$, $t = t_2$. We wish to show

$$(X_s^{(n)}, X_t^{(n)}) \xrightarrow{\mathscr{D}} (B_s, B_t).$$

Since

$$\left| X_t^{(n)} - \frac{1}{\sigma\sqrt{n}} S_{[\![tn]\!]} \right| \leq \frac{1}{\sigma\sqrt{n}} |\xi_{[\![tn]\!]+1}|,$$

we have by the Čebyšev inequality,

$$P\left[\left| X_t^{(n)} - \frac{1}{\sigma\sqrt{n}} S_{[\![tn]\!]} \right| > \varepsilon \right] \leq \frac{1}{\varepsilon^2 n} \to 0$$

as $n \to \infty$. It is clear then that

$$\left\| (X_s^{(n)}, X_t^{(n)}) - \frac{1}{\sigma\sqrt{n}} (S_{[\![sn]\!]}, S_{[\![tn]\!]}) \right\| \to 0 \quad \text{in probability,}$$

so, by Problem 4.16, it suffices to show

$$\frac{1}{\sigma\sqrt{n}} (S_{[\![sn]\!]}, S_{[\![tn]\!]}) \xrightarrow{\mathscr{D}} (B_s, B_t).$$

From Problem 4.5 we see that this is equivalent to proving

$$\frac{1}{\sigma\sqrt{n}} \left(\sum_{j=1}^{[\![sn]\!]} \xi_j, \sum_{j=[\![sn]\!]+1}^{[\![tn]\!]} \xi_j \right) \xrightarrow{\mathscr{D}} (B_s, B_t - B_s).$$

The independence of the random variables $\{\xi_j\}_{j=1}^{\infty}$ implies

$$\lim_{n\to\infty} E\left[\exp\left\{ \frac{iu}{\sigma\sqrt{n}} \sum_{j=1}^{[\![sn]\!]} \xi_j + \frac{iv}{\sigma\sqrt{n}} \sum_{j=[\![sn]\!]+1}^{[\![tn]\!]} \xi_j \right\} \right]$$

$$(4.11) \qquad = \lim_{n\to\infty} E\left[\exp\left\{ \frac{iu}{\sigma\sqrt{n}} \sum_{j=1}^{[\![sn]\!]} \xi_j \right\} \right] \cdot \lim_{n\to\infty} E\left[\exp\left\{ \frac{iv}{\sigma\sqrt{n}} \sum_{j=[\![sn]\!]+1}^{[\![tn]\!]} \xi_j \right\} \right],$$

provided both limits on the right-hand side exist. We deal with $\lim_{n\to\infty}\cdot E[\exp\{(iu/\sigma\sqrt{n})\sum_{j=1}^{[\![sn]\!]}\xi_j\}]$; the other limit can be treated similarly. Since

$$\left|\frac{1}{\sigma\sqrt{n}}\sum_{j=1}^{[\![sn]\!]}\xi_j - \frac{\sqrt{s}}{\sigma\sqrt{[\![sn]\!]}}\sum_{j=1}^{[\![sn]\!]}\xi_j\right| \to 0 \quad \text{in probability,}$$

and, by the central limit theorem, $(\sqrt{s}/\sigma\sqrt{[\![sn]\!]})\sum_{j=1}^{[\![sn]\!]}\xi_j$ converges in distribution to a normal random variable with mean zero and variance s, we have

$$\lim_{n\to\infty} E\left[\exp\left\{\frac{iu}{\sigma\sqrt{n}}\sum_{j=1}^{[\![sn]\!]}\xi_j\right\}\right] = e^{-u^2 s/2}.$$

Similarly,

$$\lim_{n\to\infty} E\left[\exp\left\{\frac{iv}{\sigma\sqrt{n}}\sum_{j=[\![sn]\!]+1}^{[\![tn]\!]}\xi_j\right\}\right] = e^{-v^2(t-s)/2}.$$

Substitution of these last two equations into (4.11) completes the proof. □

Actually, the sequence $\{X^{(n)}\}$ of linearly interpolated and normalized random walks in (4.10) converges to Brownian motion *in distribution*. For the tightness required to carry out such an extension (recall Theorem 4.15), we shall need two auxiliary results.

4.18 Lemma. *Set $S_k = \sum_{j=1}^{k}\xi_j$, where $\{\xi_j\}_{j=1}^{\infty}$ is a sequence of independent, identically distributed random variables, with mean zero and finite variance $\sigma^2 > 0$. Then, for any $\varepsilon > 0$,*

$$\lim_{\delta\downarrow 0}\overline{\lim_{n\to\infty}}\frac{1}{\delta}P\left[\max_{1\le j\le [\![n\delta]\!]+1}|S_j| > \varepsilon\sigma\sqrt{n}\right] = 0.$$

Proof. By the central limit theorem, we have for each $\delta > 0$ that $(1/\sigma\sqrt{[\![n\delta]\!]+1})S_{[\![n\delta]\!]+1}$ converges in distribution to a standard normal random variable Z, whence $(1/\sigma\sqrt{n\delta})S_{[\![n\delta]\!]+1} \xrightarrow{\mathscr{D}} Z$. Fix $\lambda > 0$ and let $\{\varphi_k\}_{k=1}^{\infty}$ be a sequence of bounded, continuous functions on $\mathbb{R}$ with $\varphi_k \downarrow 1_{(-\infty,-\lambda]\cup[\lambda,\infty)}$. We have for each i,

$$\overline{\lim_{n\to\infty}} P[|S_{[\![n\delta]\!]+1}| \ge \lambda\sigma\sqrt{n\delta}] \le \lim_{n\to\infty} E\varphi_k\left(\frac{1}{\sigma\sqrt{n\delta}}S_{[\![n\delta]\!]+1}\right) = E\varphi_k(Z).$$

Let $k\to\infty$ to conclude

$$\text{(4.12)}\quad \overline{\lim_{n\to\infty}} P[|S_{[\![n\delta]\!]+1}| \ge \lambda\sigma\sqrt{n\delta}] \le P[|Z|\ge\lambda] \le \frac{1}{\lambda^3}E|Z|^3, \quad \lambda > 0.$$

We now define $\tau = \min\{j\ge 1; |S_j| > \varepsilon\sigma\sqrt{n}\}$. With $0 < \delta < \varepsilon^2/2$, we have (imitating the proof of the Kolmogorov inequality; e.g., Chung (1974), p. 116):

$$(4.13)\qquad P\left[\max_{0 \le j \le [\![n\delta]\!]+1} |S_j| > \varepsilon\sigma\sqrt{n}\right] \le P[|S_{[\![n\delta]\!]+1}| \ge \sigma\sqrt{n}(\varepsilon - \sqrt{2\delta})] + \sum_{j=1}^{[\![n\delta]\!]} P[|S_{[\![n\delta]\!]+1}| < \sigma\sqrt{n}(\varepsilon - \sqrt{2\delta}) | \tau = j] P[\tau = j].$$

But if $\tau = j$, then $|S_{[\![n\delta]\!]+1}| < \sigma\sqrt{n}(\varepsilon - \sqrt{2\delta})$ implies $|S_j - S_{[\![n\delta]\!]+1}| > \sigma\sqrt{2n\delta}$. By the Čebyšev inequality, the probability of this event is bounded above by

$$\frac{1}{2n\delta\sigma^2} E[(S_j - S_{[\![n\delta]\!]+1})^2 | \tau = j] = \frac{1}{2n\delta\sigma^2} E\left(\sum_{i=j+1}^{[\![n\delta]\!]+1} \xi_i^2\right) \le \frac{1}{2}, \quad 1 \le j \le [\![n\delta]\!].$$

Returning to (4.13), we may now write

$$P\left[\max_{0 \le j \le [\![n\delta]\!]+1} |S_j| > \varepsilon\sigma\sqrt{n}\right] \le P[|S_{[\![n\delta]\!]+1}| \ge \sigma\sqrt{n}(\varepsilon - \sqrt{2\delta})] + \frac{1}{2} P[\tau \le [\![n\delta]\!]]$$
$$\le P[|S_{[\![n\delta]\!]+1}| \ge \sigma\sqrt{n}(\varepsilon - \sqrt{2\delta})] + \frac{1}{2} P\left[\max_{0 \le j \le [\![n\delta]\!]+1} |S_j| > \varepsilon\sigma\sqrt{n}\right],$$

from which follows

$$P\left[\max_{0 \le j \le [\![n\delta]\!]+1} |S_j| > \varepsilon\sigma\sqrt{n}\right] \le 2P[|S_{[\![n\delta]\!]+1}| \ge \sigma\sqrt{n}(\varepsilon - \sqrt{2\delta})].$$

Setting $\lambda = (\varepsilon - \sqrt{2\delta})/\sqrt{\delta}$ in (4.12), we see that

$$\overline{\lim_{n\to\infty}} \frac{1}{\delta} P\left[\max_{0 \le j \le [\![n\delta]\!]+1} |S_j| > \varepsilon\sigma\sqrt{n}\right] \le \frac{2\sqrt{\delta}}{(\varepsilon - \sqrt{2\delta})^3} E|Z|^3,$$

and letting $\delta \downarrow 0$ we obtain the desired result. □

4.19 Lemma. *Under the assumptions of Lemma* 4.18, *we have for any* $T > 0$,

$$\lim_{\delta \downarrow 0} \overline{\lim_{n\to\infty}} P\left[\max_{\substack{1 \le j \le [\![n\delta]\!]+1 \\ 0 \le k \le [\![nT]\!]+1}} |S_{j+k} - S_k| > \varepsilon\sigma\sqrt{n}\right] = 0.$$

Proof. For $0 < \delta \le T$, let $m = m(\delta) \ge 2$ be the unique integer satisfying $T/m < \delta \le T/(m-1)$. Since

$$\lim_{n\to\infty} \frac{[\![nT]\!] + 1}{[\![n\delta]\!] + 1} = \frac{T}{\delta} < m,$$

we have $[\![nT]\!] + 1 < ([\![n\delta]\!] + 1)m$ for sufficiently large n. For such a large

n, suppose $|S_{j+k} - S_k| > \varepsilon\sigma\sqrt{n}$ for some k, $0 \le k \le [\![nT]\!] + 1$, and some j, $1 \le j \le [\![n\delta]\!] + 1$. There exists then a unique integer p, $0 \le p \le m - 1$, such that

$$([\![n\delta]\!] + 1)p \le k < ([\![n\delta]\!] + 1)(p + 1).$$

There are two possibilities for $k + j$. One possibility is that

$$([\![n\delta]\!] + 1)p \le k + j \le ([\![n\delta]\!] + 1)(p + 1),$$

in which case either $|S_k - S_{([\![n\delta]\!]+1)p}| > \frac{1}{3}\varepsilon\sigma\sqrt{n}$, or else $|S_{k+j} - S_{([\![n\delta]\!]+1)p}| > \frac{1}{3}\varepsilon\sigma\sqrt{n}$. The second possibility is that

$$([\![n\delta]\!] + 1)(p + 1) < k + j < ([\![n\delta]\!] + 1)(p + 2),$$

in which case either $|S_k - S_{([\![n\delta]\!]+1)p}| > \frac{1}{3}\varepsilon\sigma\sqrt{n}$, $|S_{([\![n\delta]\!]+1)p} - S_{([\![n\delta]\!]+1)(p+1)}| > \frac{1}{3}\varepsilon\sigma\sqrt{n}$, or else $|S_{([\![n\delta]\!]+1)(p+1)} - S_{k+j}| > \frac{1}{3}\varepsilon\sigma\sqrt{n}$. In conclusion, we see that

$$\left\{\max_{\substack{1 \le j \le [\![n\delta]\!]+1 \\ 0 \le k \le [\![nT]\!]+1}} |S_{j+k} - S_k| > \varepsilon\sigma\sqrt{n}\right\}$$
$$\subseteq \bigcup_{p=0}^{m} \left\{\max_{1 \le j \le [\![n\delta]\!]+1} |S_{j+p([\![n\delta]\!]+1)} - S_{p([\![n\delta]\!]+1)}| > \frac{1}{3}\varepsilon\sigma\sqrt{n}\right\}.$$

But

$$P\left[\max_{1 \le j \le [\![n\delta]\!]+1} |S_{j+p([\![n\delta]\!]+1)} - S_{p([\![n\delta]\!]+1)}| > \frac{1}{3}\varepsilon\sigma\sqrt{n}\right]$$
$$= P\left[\max_{1 \le j \le [\![n\delta]\!]+1} |S_j| > \frac{1}{3}\varepsilon\sigma\sqrt{n}\right],$$

and thus:

$$P\left[\max_{\substack{1 \le j \le [\![n\delta]\!]+1 \\ 0 \le k \le [\![nT]\!]+1}} |S_{j+k} - S_k| > \varepsilon\sigma\sqrt{n}\right] \le (m + 1)P\left[\max_{1 \le j \le [\![n\delta]\!]+1} |S_j| > \frac{1}{3}\varepsilon\sigma\sqrt{n}\right].$$

Since $m \le (T/\delta) + 1$, we obtain the desired conclusion from Lemma 4.18. □

We are now in a position to establish the main result of this section, namely the convergence in distribution of the sequence of normalized random walks in (4.10) to Brownian motion. This result is also known as the *invariance principle*.

4.20 Theorem (The Invariance Principle of Donsker (1951)). *Let* $(\Omega, \mathscr{F}, P)$ *be a probability space on which is given a sequence* $\{\xi_j\}_{j=1}^{\infty}$ *of independent, identically distributed random variables with mean zero and finite variance* $\sigma^2 > 0$. *Define* $X^{(n)} = \{X_t^{(n)}; t \ge 0\}$ *by (4.10), and let* P_n *be the measure induced by* $X^{(n)}$ *on* $(C[0, \infty), \mathscr{B}(C[0, \infty)))$. *Then* $\{P_n\}_{n=1}^{\infty}$ *converges weakly to a measure* P_*, *under which the coordinate mapping process $W_t(\omega) \triangleq \omega(t)$ on $C[0, \infty)$ is a standard, one-dimensional Brownian motion.*

PROOF. In light of Theorems 4.15 and 4.17, it remains to show that $\{X^{(n)}\}_{n=1}^{\infty}$ is tight. For this we use Theorem 4.10, and since $X_0^{(n)} = 0$ a.s. for every n, we need only establish, for arbitrary $\varepsilon > 0$ and $T > 0$, the convergence

$$\lim_{\delta \downarrow 0} \sup_{n \geq 1} P\left[\max_{\substack{|s-t| \leq \delta \\ 0 \leq s,t \leq T}} |X_s^{(n)} - X_t^{(n)}| > \varepsilon\right] = 0. \tag{4.14}$$

We may replace $\sup_{n \geq 1}$ in this expression by $\overline{\lim}_{n\to\infty}$, since for a finite number of integers n we can make the probability appearing in (4.14) as small as we choose, by reducing δ. But

$$P\left[\max_{\substack{|s-t| \leq \delta \\ 0 \leq s,t \leq T}} |X_s^{(n)} - X_t^{(n)}| > \varepsilon\right] = P\left[\max_{\substack{|s-t| \leq n\delta \\ 0 \leq s,t \leq nT}} |Y_s - Y_t| > \varepsilon\sigma\sqrt{n}\right],$$

and

$$\max_{\substack{|s-t| \leq n\delta \\ 0 \leq s,t \leq nT}} |Y_s - Y_t| \leq \max_{\substack{|s-t| \leq [\![n\delta]\!]+1 \\ 0 \leq s,t \leq [\![nT]\!]+1}} |Y_s - Y_t| \leq \max_{\substack{1 \leq j \leq [\![n\delta]\!]+1 \\ 0 \leq k \leq [\![nT]\!]+1}} |S_{j+k} - S_k|,$$

where the last inequality follows from the fact that Y is piecewise linear and changes slope only at integer values of t. Now (4.14) follows from Lemma 4.19. □

4.21 Definition. The probability measure P_* on $(C[0, \infty), \mathscr{B}(C[0, \infty)))$, under which the coordinate mapping process $W_t(\omega) \triangleq \omega(t)$, $0 \leq t < \infty$, is a standard, one-dimensional Brownian motion, is called *Wiener measure.*

4.22 Remark. A standard, one-dimensional Brownian motion defined on any probability space can be thought of as a random variable with values in $C[0, \infty)$; regarded this way, Brownian motion induces the Wiener measure on $(C[0, \infty), \mathscr{B}(C[0, \infty)))$. For this reason, we call $(C[0, \infty), \mathscr{B}(C[0, \infty)), P_*)$, where P_* is Wiener measure, the *canonical probability space* for Brownian motion.

2.5. The Markov Property

In this section we define the notion of a d-dimensional Markov process and cite d-dimensional Brownian motion as an example. There are several equivalent statements of the Markov property, and we spend some time developing them.

A. Brownian Motion in Several Dimensions

5.1 Definition. Let d be a positive integer and μ a probability measure on $(\mathbb{R}^d, \mathscr{B}(\mathbb{R}^d))$. Let $B = \{B_t, \mathscr{F}_t; t \geq 0\}$ be a continuous, adapted process with values in $\mathbb{R}^d$, defined on some probability space $(\Omega, \mathscr{F}, P)$. This process is called a *d-dimensional Brownian motion with initial distribution μ*, if

(i) $P[B_0 \in \Gamma] = \mu(\Gamma), \quad \forall \Gamma \in \mathscr{B}(\mathbb{R}^d)$;
(ii) for $0 \leq s < t$, the increment $B_t - B_s$ is independent of $\mathscr{F}_s$ and is normally distributed with mean zero and covariance matrix equal to $(t-s)I_d$, where I_d is the $(d \times d)$ identity matrix.

If μ assigns measure one to some singleton $\{x\}$, we say that B is a *d-dimensional Brownian motion starting at x*.

Here is one way to construct a d-dimensional Brownian motion with initial distribution μ. Let $X(\omega_0) = \omega_0$ be the identity random variable on $(\mathbb{R}^d, \mathscr{B}(\mathbb{R}^d), \mu)$, and for each $i = 1, \ldots, d$, let $\tilde{B}^{(i)} = \{\tilde{B}_t^{(i)}, \mathscr{F}_t^{\tilde{B}(i)}; t \geq 0\}$ be a standard, one-dimensional Brownian motion on some $(\Omega^{(i)}, \mathscr{F}^{(i)}, P^{(i)})$. On the product space

$$(\mathbb{R}^d \times \Omega^{(1)} \times \cdots \times \Omega^{(d)}, \mathscr{B}(\mathbb{R}^d) \otimes \mathscr{F}^{(1)} \otimes \cdots \otimes \mathscr{F}^{(d)}, \mu \times P^{(1)} \times \cdots \times P^{(d)}),$$

define

$$B_t(\omega) \triangleq X(\omega_0) + (\tilde{B}_t^{(1)}(\omega_1), \ldots, \tilde{B}_t^{(d)}(\omega_d)),$$

and set $\mathscr{F}_t = \mathscr{F}_t^B$. Then $B = \{B_t, \mathscr{F}_t; t \geq 0\}$ is the desired object.

There is a second construction of d-dimensional Brownian motion with initial distribution μ, a construction which motivates the concept of *Markov family*, to be introduced in this section. Let $P^{(i)}$, $i = 1, \ldots, d$ be d copies of Wiener measure on $(C[0,\infty), \mathscr{B}(C[0,\infty)))$. Then $P^0 \triangleq P^{(1)} \times \cdots \times P^{(d)}$ is a measure, called *d-dimensional Wiener measure*, on $(C[0,\infty)^d, \mathscr{B}(C[0,\infty)^d))$. Under P^0, the coordinate mapping process $B_t(\omega) \triangleq \omega(t)$ together with the filtration $\{\mathscr{F}_t^B\}$ is a d-dimensional Brownian motion starting at the origin. For $x \in \mathbb{R}^d$, we define the probability measure P^x on $(C[0,\infty)^d, \mathscr{B}(C[0,\infty)^d))$ by

$$P^x(F) = P^0(F - x), \quad F \in \mathscr{B}(C[0,\infty)^d), \tag{5.1}$$

where $F - x = \{\omega \in C[0,\infty)^d; \omega(\cdot) + x \in F\}$. Under P^x, $B \triangleq \{B_t, \mathscr{F}_t^B; t \geq 0\}$ is a d-dimensional Brownian motion starting at x. Finally, for a probability measure μ on $(\mathbb{R}^d, \mathscr{B}(\mathbb{R}^d))$, we define P^μ on $\mathscr{B}(C[0,\infty)^d)$ by

$$P^\mu(F) = \int_{\mathbb{R}^d} P^x(F)\mu(dx). \tag{5.2}$$

Problem 5.2 shows that such a definition is possible.

5.2 Problem. Show that for each $F \in \mathscr{B}(C[0,\infty)^d)$, the mapping $x \mapsto P^x(F)$ is $\mathscr{B}(\mathbb{R}^d)/\mathscr{B}([0,1])$-measurable. (*Hint*: Use the Dynkin System Theorem 1.3.)

5.3 Proposition. The coordinate mapping process $B = \{B_t, \mathscr{F}_t^B; t \geq 0\}$ on $(C[0,\infty)^d, \mathscr{B}(C[0,\infty)^d), P^\mu)$ is a d-dimensional Brownian motion with initial distribution μ.

5.4 Problem. Give a careful proof of Proposition 5.3 and the assertions preceding Problem 5.2.

5.5 Problem. Let $\{B_t = (B_t^{(1)}, \ldots, B_t^{(d)}), \mathscr{F}_t; 0 \leq t < \infty\}$ be a d-dimensional Brownian motion. Show that the processes

$$M_t^{(i)} \triangleq B_t^{(i)} - B_0^{(i)}, \mathscr{F}_t; \quad 0 \leq t < \infty, 1 \leq i \leq d$$

are continuous, square-integrable martingales, with $\langle M^{(i)}, M^{(j)} \rangle_t = t\delta_{ij}$; $1 \leq i, j \leq d$. Furthermore, the vector of martingales $M = (M^{(1)}, \ldots, M^{(d)})$ is independent of $\mathscr{F}_0$.

5.6 Definition. Given a metric space (S, ρ), we denote by $\overline{\mathscr{B}(S)}^\mu$ the completion of the Borel σ-field $\mathscr{B}(S)$ (generated by the open sets) with respect to the finite measure μ on $(S, \mathscr{B}(S))$. The *universal σ-field* is $\mathscr{U}(S) \triangleq \bigcap_\mu \overline{\mathscr{B}(S)}^\mu$, where the intersection is over all finite measures (or, equivalently, all probability measures) μ. A $\mathscr{U}(S)/\mathscr{B}(\mathbb{R})$-measurable, real-valued function is said to be *universally measurable*.

5.7 Problem. Let (S, ρ) be a metric space and let f be a real-valued function defined on S. Show that f is universally measurable if and only if for every finite measure μ on $(S, \mathscr{B}(S))$, there is a Borel-measurable function $g_\mu: S \to \mathbb{R}$ such that $\mu\{x \in S; f(x) \neq g_\mu(x)\} = 0$.

5.8 Definition. A *d-dimensional Brownian family* is an adapted, d-dimensional process $B = \{B_t, \mathscr{F}_t; t \geq 0\}$ on a measurable space $(\Omega, \mathscr{F})$, and a family of probability measures $\{P^x\}_{x \in \mathbb{R}^d}$, such that

(i) for each $F \in \mathscr{F}$, the mapping $x \mapsto P^x(F)$ is universally measurable;
(ii) for each $x \in \mathbb{R}^d$, $P^x[B_0 = x] = 1$;
(iii) under each P^x, the process B is a d-dimensional Brownian motion starting at x.

We have already seen how to construct a family of probability measures $\{P^x\}$ on the canonical space $(C[0,\infty)^d, \mathscr{B}(C[0,\infty)^d))$ so that the coordinate mapping process, relative to the filtration it generates, is a Brownian motion starting at x under any P^x. With $\mathscr{F} = \mathscr{B}(C[0,\infty)^d)$, Problem 5.2 shows that the universal measurability requirement (i) of Definition 5.8 is satisfied. Indeed, for this canonical example of a d-dimensional Brownian family, the mapping $x \mapsto P^x(F)$ is actually Borel-measurable for each $F \in \mathscr{F}$. The reason we formulate Definition 5.8 with the weaker measurability condition is to allow expansion of $\mathscr{F}$ to a larger σ-field (see Remark 7.16).

B. Markov Processes and Markov Families

Let us suppose now that we observe a Brownian motion with initial distribution μ up to time s, $0 \le s < t$. In particular, we see the value of B_s, which we call y. Conditioned on these observations, what is the probability that B_t is in some set $\Gamma \in \mathscr{B}(\mathbb{R}^d)$? Now $B_t = (B_t - B_s) + B_s$, and the increment $B_t - B_s$ is independent of the observations up to time s and is distributed just as B_{t-s} is under P^0. On the other hand, B_s does depend on the observations; indeed, we are conditioning on $B_s = y$. It follows that the sum $(B_t - B_s) + B_s$ is distributed as B_{t-s} is under P^y. Two points then become clear. First, knowledge of the whole past up to time s provides no more useful information about B_t than knowing the value of B_s; in other words,

$$P^\mu[B_t \in \Gamma | \mathscr{F}_s] = P^\mu[B_t \in \Gamma | B_s], \quad 0 \le s < t, \Gamma \in \mathscr{B}(\mathbb{R}^d). \tag{5.3}$$

Second, conditioned on $B_s = y$, B_t is distributed as B_{t-s} is under P^y; i.e.,

$$P^\mu[B_t \in \Gamma | B_s = y] = P^y[B_{t-s} \in \Gamma], \quad 0 \le s < t, \Gamma \in \mathscr{B}(\mathbb{R}^d). \tag{5.4}$$

5.9 Problem. Make the preceding discussion rigorous by proving the following. If X and Y are d-dimensional random vectors on $(\Omega, \mathscr{F}, P)$, $\mathscr{G}$ is a sub-σ-field of $\mathscr{F}$, X is independent of $\mathscr{G}$ and Y is $\mathscr{G}$-measurable, then for every $\Gamma \in \mathscr{B}(\mathbb{R}^d)$:

$$P[X + Y \in \Gamma | \mathscr{G}] = P[X + Y \in \Gamma | Y], \quad \text{a.s. } P; \tag{5.5}$$

$$P[X + Y \in \Gamma | Y = y] = P[X + y \in \Gamma], \quad \text{for } PY^{-1}\text{-a.e. } y \in \mathbb{R}^d \tag{5.6}$$

in the notation of (4.2).

5.10 Definition. Let d be a positive integer and μ a probability measure on $(\mathbb{R}^d, \mathscr{B}(\mathbb{R}^d))$. An adapted, d-dimensional process $X = \{X_t, \mathscr{F}_t; t \ge 0\}$ on some probability space $(\Omega, \mathscr{F}, P^\mu)$ is said to be a *Markov process with initial distribution* μ if

(i) $P^\mu[X_0 \in \Gamma] = \mu(\Gamma)$, $\forall\, \Gamma \in \mathscr{B}(\mathbb{R}^d)$;
(ii) for $s, t \ge 0$ and $\Gamma \in \mathscr{B}(\mathbb{R}^d)$,

$$P^\mu[X_{t+s} \in \Gamma | \mathscr{F}_s] = P^\mu[X_{t+s} \in \Gamma | X_s], \quad P^\mu\text{-a.s.}$$

Our experience with Brownian motion indicates that it is notationally and conceptually helpful to have a whole family of probability measures, rather than just one. Toward this end, we define the concept of a Markov family.

5.11 Definition. Let d be a positive integer. A d-dimensional *Markov family* is an adapted process $X = \{X_t, \mathscr{F}_t; t \ge 0\}$ on some $(\Omega, \mathscr{F})$, together with a family of probability measures $\{P^x\}_{x \in \mathbb{R}^d}$ on $(\Omega, \mathscr{F})$, such that

(a) for each $F \in \mathscr{F}$, the mapping $x \mapsto P^x(F)$ is universally measurable;
(b) $P^x[X_0 = x] = 1$, $\forall\, x \in \mathbb{R}^d$;

(c) for $x \in \mathbb{R}^d$, $s, t \geq 0$ and $\Gamma \in \mathscr{B}(\mathbb{R}^d)$,

$$P^x[X_{t+s} \in \Gamma | \mathscr{F}_s] = P^x[X_{t+s} \in \Gamma | X_s], \quad P^x\text{-a.s;}$$

(d) for $x \in \mathbb{R}^d$, $s, t \geq 0$ and $\Gamma \in \mathscr{B}(\mathbb{R}^d)$,

$$P^x[X_{t+s} \in \Gamma | X_s = y] = P^y[X_t \in \Gamma], \quad P^x X_s^{-1}\text{-a.e. } y$$

in the notation of (4.2).

The following statement is a consequence of Problem 5.9 and the discussion preceding it.

5.12 Theorem. *A d-dimensional Brownian motion is a Markov process. A d-dimensional Brownian family is a Markov family.*

C. Equivalent Formulations of the Markov Property

The Markov property, encapsulated by conditions (c) and (d) of Definition 5.11, can be reformulated in several equivalent ways. Some of these formulations amount to incorporating (c) and (d) into a single condition; others replace the evaluation of X at the single time $s + t$ by its evaluation at multiple times after s. The bulk of this subsection presents those formulations of the Markov property which we shall find most convenient in the sequel.

Given an adapted process $X = \{X_t, \mathscr{F}_t; t \geq 0\}$ and a family of probability measures $\{P^x\}_{x \in \mathbb{R}^d}$ on $(\Omega, \mathscr{F})$, such that condition (a) of Definition 5.11 is satisfied, we can define a collection of operators $\{U_t\}_{t \geq 0}$ which map bounded, Borel-measurable, real-valued functions on $\mathbb{R}^d$ into bounded, universally measurable, real-valued functions on the same space. These are given by

$$(U_t f)(x) \triangleq E^x f(X_t). \tag{5.7}$$

In the case where f is the indicator of $\Gamma \in \mathscr{B}(\mathbb{R}^d)$, we have $E^x f(X_t) = P^x[X_t \in \Gamma]$, and the universal measurability of $U_t f$ follows directly from Definition 5.11 (a); for an arbitrary, Borel-measurable function f, the universal measurability of $U_t f$ is then a consequence of the bounded convergence theorem.

5.13 Proposition. *Conditions* (c) *and* (d) *of Definition* 5.11 *can be replaced by*:

(e) *For* $x \in \mathbb{R}^d$, $s, t \geq 0$ *and* $\Gamma \in \mathscr{B}(\mathbb{R}^d)$,

$$P^x[X_{s+t} \in \Gamma | \mathscr{F}_s] = (U_t 1_\Gamma)(X_s), \quad P^x\text{-a.s.}$$

PROOF. First, let us assume that (c), (d) hold. We have from the latter:

$$P^x[X_{t+s} \in \Gamma | X_s = y] = (U_t 1_\Gamma)(y) \quad \text{for } P^x X_s^{-1}\text{-a.e. } y \in \mathbb{R}^d.$$

If the function $\alpha(y) \triangleq (U_t 1_\Gamma)(y)$: $\mathbb{R}^d \to [0, 1]$ were $\mathscr{B}(\mathbb{R}^d)$-measurable, as is the case for Brownian motion, we would then be able to conclude that, for all

$x \in \mathbb{R}^d$, $s \geq 0$: $P^x[X_{t+s} \in \Gamma | X_s] = \alpha(X_s)$, a.s. P^x, and from condition (c): $P^x[X_{t+s} \in \Gamma | \mathscr{F}_s] = \alpha(X_s)$, a.s. P^x, which would then establish (e).

However, we only know that $U_t 1_\Gamma(\cdot)$ is universally measurable. This means (from Problem 5.7) that, for given $s, t \geq 0, x \in \mathbb{R}^d$, there exists a Borel-measurable function $g: \mathbb{R}^d \to [0, 1]$ such that

$$(5.8) \qquad (U_t 1_\Gamma)(y) = g(y), \quad \text{for } P^x X_s^{-1}\text{-a.e. } y \in \mathbb{R}^d,$$

whence

$$(5.9) \qquad (U_t 1_\Gamma)(X_s) = g(X_s), \quad \text{a.s. } P^x.$$

One can then repeat the preceding argument with g replacing the function α.

Second, let us assume that (e) holds; then for any given $s, t \geq 0$ and $x \in \mathbb{R}^d$, (5.9) gives

$$(5.10) \qquad P^x[X_{t+s} \in \Gamma | \mathscr{F}_s] = g(X_s), \quad \text{a.s. } P^x.$$

It follows that $P^x[X_{t+s} \in \Gamma | \mathscr{F}_s]$ has a $\sigma(X_s)$-measurable version, and this establishes (c). From the latter and (5.10) we conclude

$$P^x[X_{t+s} \in \Gamma | X_s = y] = g(y) \quad \text{for } P^x X_s^{-1}\text{-a.e. } y \in \mathbb{R}^d,$$

and this in turn yields (d), thanks to (5.8). □

5.14 Remark on Notation. For given $\omega \in \Omega$, we denote by $X_{s+\cdot}(\omega)$ the function $t \mapsto X_{s+t}(\omega)$. Thus, $X_{s+\cdot}$ is a measurable mapping from $(\Omega, \mathscr{F})$ into $((\mathbb{R}^d)^{[0,\infty)}, \mathscr{B}((\mathbb{R}^d)^{[0,\infty)}))$, the space of all $\mathbb{R}^d$-valued functions on $[0, \infty)$ equipped with the smallest σ-field containing all finite-dimensional cylinder sets.

5.15 Proposition. *For a Markov family* $X, (\Omega, \mathscr{F}), \{P^x\}_{x \in \mathbb{R}^d}$, *we have*:

(c′) *For* $x \in \mathbb{R}^d$, $s \geq 0$ *and* $F \in \mathscr{B}((\mathbb{R}^d)^{[0,\infty)})$,

$$P^x[X_{s+\cdot} \in F | \mathscr{F}_s] = P^x[X_{s+\cdot} \in F | X_s], \quad P^x\text{-a.s;}$$

(d′) *For* $x \in \mathbb{R}^d$, $s \geq 0$ *and* $F \in \mathscr{B}((\mathbb{R}^d)^{[0,\infty)})$,

$$P^x[X_{s+\cdot} \in F | X_s = y] = P^y[X_\cdot \in F], \quad P^x X_s^{-1}\text{-a.e. } y.$$

(*Note*: If $\Gamma \in \mathscr{B}(\mathbb{R}^d)$ and $F = \{\omega \in (\mathbb{R}^d)^{[0,\infty)}; \omega(t) \in \Gamma\}$, for fixed $t \geqq 0$, then (c′) and (d′) reduce to (c) and (d), respectively, of Definition 5.11.)

PROOF. The collection of all sets $F \in \mathscr{B}((\mathbb{R}^d)^{[0,\infty)})$ for which (c′) and (d′) hold forms a Dynkin system; so by Theorem 1.3, it suffices to prove (c′) and (d′) for finite-dimensional cylinder sets of the form

$$F = \{\omega \in (\mathbb{R}^d)^{[0,\infty)}; \omega(t_0) \in \Gamma_0, \ldots, \omega(t_{n-1}) \in \Gamma_{n-1}, \omega(t_n) \in \Gamma_n\},$$

where $0 = t_0 < t_1 < \cdots < t_n$, $\Gamma_i \in \mathscr{B}(\mathbb{R}^d)$, $i = 0, 1, \ldots, n$, and $n \geq 0$. For such an F, condition (c′) becomes

$$(5.11)\qquad P^x[X_s\in\Gamma_0,\ldots,X_{s+t_{n-1}}\in\Gamma_{n-1},X_{s+t_n}\in\Gamma_n|\mathscr{F}_s]$$
$$= P^x[X_s\in\Gamma_0,\ldots,X_{s+t_{n-1}}\in\Gamma_{n-1},X_{s+t_n}\in\Gamma_n|X_s],\quad P^x\text{-a.s.}$$

We prove this statement by induction on n. For $n = 0$, it is obvious. Assume it true for $n-1$. A consequence of this assumption is that for any bounded, Borel-measurable $\varphi: \mathbb{R}^{dn}\to\mathbb{R}$,

$$(5.12)\quad E^x[\varphi(X_s,\ldots,X_{s+t_{n-1}})|\mathscr{F}_s] = E^x[\varphi(X_s,\ldots,X_{s+t_{n-1}})|X_s],\quad P^x\text{-a.s.}$$

Now (c) implies that

$$(5.13)\qquad P^x[X_s\in\Gamma_0,\ldots,X_{s+t_{n-1}}\in\Gamma_{n-1},X_{s+t_n}\in\Gamma_n|\mathscr{F}_s]$$
$$= E^x[1_{\{X_s\in\Gamma_0,\ldots,X_{s+t_{n-1}}\in\Gamma_{n-1}\}}P^x[X_{s+t_n}\in\Gamma_n|\mathscr{F}_{s+t_{n-1}}]|\mathscr{F}_s]$$
$$= E^x[1_{\{X_s\in\Gamma_0,\ldots,X_{s+t_{n-1}}\in\Gamma_{n-1}\}}P^x[X_{s+t_n}\in\Gamma_n|X_{s+t_{n-1}}]|\mathscr{F}_s].$$

Any $\sigma(X_{s+t_{n-1}})$-measurable random variable can be written as a Borel-measurable function of $X_{s+t_{n-1}}$ (Chung (1974), p. 299), and so there exists a Borel-measurable function $g:\mathbb{R}^d\to[0,1]$, such that $P^x[X_{s+t_n}\in\Gamma_n|X_{s+t_{n-1}}] = g(X_{s+t_{n-1}})$, a.s. P^x. Setting $\varphi(x_0,\ldots,x_{n-1})\triangleq 1_{\Gamma_0}(x_0)\ldots 1_{\Gamma_{n-1}}(x_{n-1})g(x_{n-1})$, we can use (5.12) to replace $\mathscr{F}_s$ by $\sigma(X_s)$ in (5.13) and then, reversing the previous steps, to obtain (5.11). The proof of (d′) is similar, although notationally more complex. □

It happens sometimes, for a given process $X = \{X_t, \mathscr{F}_t; t\ge 0\}$ on a measurable space $(\Omega,\mathscr{F})$, that one can construct a family of so-called *shift operators* $\theta_s:\Omega\to\Omega$, $s\ge 0$, such that each θ_s is $\mathscr{F}/\mathscr{F}$-measurable and

$$(5.14)\qquad X_{s+t}(\omega) = X_t(\theta_s\omega);\quad \forall\,\omega\in\Omega,\quad s,t\ge 0.$$

The most obvious examples occur when Ω is either $(\mathbb{R}^d)^{[0,\infty)}$ of Remark 5.14 or $C[0,\infty)^d$ of Remark 4.13, $\mathscr{F}$ is the smallest σ-field containing all finite-dimensional cylinder sets, and X is the coordinate mapping process $X_t(\omega) = \omega(t)$. We can then define $\theta_s\omega = \omega(s+\cdot)$, i.e.,

$$(5.15)\qquad (\theta_s\omega)(t) = \omega(s+t),\quad t\ge 0.$$

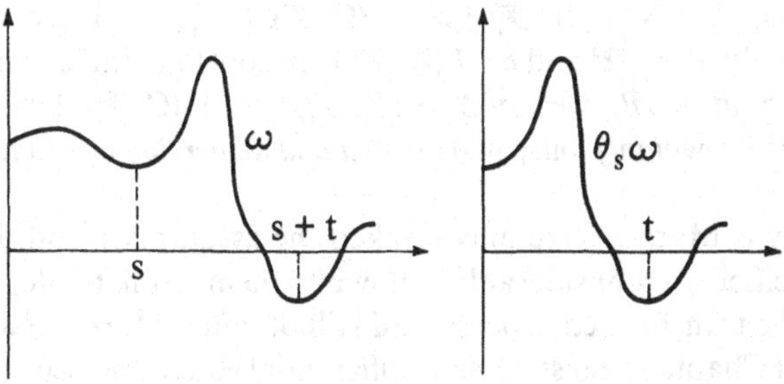

When the shift operators exist, then the function $X_{s+\cdot}(\omega)$ of Remark 5.14 is none other than $X_\cdot(\theta_s\omega)$, so $\{X_{s+\cdot}\in F\} = \theta_s^{-1}\{X_\cdot \in F\}$. As F ranges over $\mathscr{B}((\mathbb{R}^d)^{[0,\infty)})$, $\{X_\cdot \in F\}$ ranges over $\mathscr{F}_\infty^X$. Thus, (c′) and (d′) can be reformulated as follows: for every $F \in \mathscr{F}_\infty^X$ and $s \ge 0$,

$$\text{(c}''\text{)} \qquad P^x[\theta_s^{-1}F \mid \mathscr{F}_s] = P^x[\theta_s^{-1}F \mid X_s], \quad P^x\text{-a.s.}$$

$$\text{(d}''\text{)} \qquad P^x[\theta_s^{-1}F \mid X_s = y] = P^y[F], \quad P^xX_s^{-1}\text{-a.e. } y.$$

In a manner analogous to what was achieved in Proposition 5.13, we can capture both (c″) and (d″) in the requirement that for every $F \in \mathscr{F}_\infty^X$ and $s \ge 0$,

$$\text{(e}''\text{)} \qquad P^x[\theta_s^{-1}F \mid \mathscr{F}_s] = P^{X_s}(F), \quad P^x\text{-a.s.}$$

Since (e″) is often given as the primary defining property for a Markov family, we state a result about its equivalence to our definition.

5.16 Theorem. *Let $X = \{X_t, \mathscr{F}_t; t \ge 0\}$ be an adapted process on a measurable space $(\Omega, \mathscr{F})$, let $\{P^x\}_{x\in\mathbb{R}^d}$ be a family of probability measures on $(\Omega, \mathscr{F})$, and let $\{\theta_s\}_{s\ge 0}$ be a family of $\mathscr{F}/\mathscr{F}$-measurable shift-operators satisfying* (5.14). *Then X, $(\Omega, \mathscr{F})$, $\{P^x\}_{x\in\mathbb{R}^d}$ is a Markov family if and only if* (a), (b), *and* (e″) *hold.*

5.17 Exercise. Suppose that X, $(\Omega, \mathscr{F})$, $\{P^x\}_{x\in\mathbb{R}^d}$ is a Markov family with shift-operators $\{\theta_s\}_{s\ge 0}$. Use (c″) to show that for every $x \in \mathbb{R}^d$, $s \ge 0$, $G \in \mathscr{F}_s$ and $F \in \mathscr{F}_\infty^X$,

$$\text{(c}'''\text{)} \qquad P^x[G \cap \theta_s^{-1}F \mid X_s] = P^x[G \mid X_s]P^x[\theta_s^{-1}F \mid X_s], \quad P^x\text{-a.s.}$$

We may interpret this equation as saying the "past" G and the "future" $\theta_s^{-1}F$ are conditionally independent, given the "present" X_s. Conversely, show that (c‴) implies (c″).

We close this section with additional examples of Markov families.

5.18 Problem. Suppose $X = \{X_t, \mathscr{F}_t; t \ge 0\}$ is a Markov process on $(\Omega, \mathscr{F}, P)$ and $\varphi: [0,\infty) \to \mathbb{R}^d$ and $\Psi: [0,\infty) \to L(\mathbb{R}^d, \mathbb{R}^d)$, the space of linear transformations from $\mathbb{R}^d$ to $\mathbb{R}^d$, are given (nonrandom) functions with $\varphi(0) = 0$ and $\Psi(t)$ nonsingular for every $t \ge 0$. Set $Y_t = \varphi(t) + \Psi(t)X_t$. Then $Y = \{Y_t, \mathscr{F}_t; t \ge 0\}$ is also a Markov process.

5.19 Definition. Let $B = \{B_t, \mathscr{F}_t; t \ge 0\}$, $(\Omega, \mathscr{F})$, $\{P^x\}_{x\in\mathbb{R}^d}$ be a d-dimensional Brownian family. If $\mu \in \mathbb{R}^d$ and $\sigma \in L(\mathbb{R}^d, \mathbb{R}^d)$ are constant and σ is nonsingular, then with $Y_t \triangleq \mu t + \sigma B_t$, we say $Y = \{Y_t, \mathscr{F}_t; t \ge 0\}$, $(\Omega, \mathscr{F})$, $\{P^{\sigma^{-1}x}\}_{x\in\mathbb{R}^d}$ is a *d-dimensional Brownian family with drift μ and dispersion coefficient σ.*

This family is Markov. We may weaken the assumptions on the drift and diffusion coefficients considerably, allowing them both to depend on the location of the transformed process, and still obtain a Markov family. This is the subject of Chapter 5 on stochastic differential equations; see, in particular, Theorem 5.4.20 and Remark 5.4.21.

5.20 Definition. A *Poisson family with intensity* $\lambda > 0$ is a process $N = \{N_t, \mathscr{F}_t; t \geq 0\}$ on a measurable space $(\Omega, \mathscr{F})$ and a family of probability measures $\{P^x\}_{x \in \mathbb{R}}$, such that

(i) for each $E \in \mathscr{F}$, the mapping $x \mapsto P^x(E)$ is universally measurable;
(ii) for each $x \in \mathbb{R}$, $P^x[N_0 = x] = 1$;
(iii) under each P^x, the process $\{\tilde{N}_t = N_t - N_0, \mathscr{F}_t: t \geq 0\}$ is a Poisson process with intensity λ.

5.21 Exercise. Show that a Poisson family with intensity $\lambda > 0$ is a Markov family. Show furthermore that, in the notation of Definition 5.20 and under any P^x, the σ-fields $\mathscr{F}_\infty^{\tilde{N}}$ and $\mathscr{F}_0$ are independent.

Standard, one-dimensional Brownian motion is both a martingale and a Markov process. There are many Markov processes, such as Brownian motion with nonzero drift and the Poisson process, which are not martingales. There are also martingales which do not enjoy the Markov property.

5.22 Exercise. Construct a martingale which is not a Markov process.

2.6. The Strong Markov Property and the Reflection Principle

Part of the appeal of Brownian motion lies in the fact that the distribution of certain of its functionals can be obtained in closed form. Perhaps the most fundamental of these functionals is the *passage time* T_b to a level $b \in \mathbb{R}$, defined by

$$T_b(\omega) = \inf\{t \geq 0; B_t(\omega) = b\}. \tag{6.1}$$

We recall that a passage time for a continuous process is a stopping time (Problem 1.2.7).

We shall first obtain the probability density function of T_b by a heuristic argument, based on the so-called *reflection principle* of Désiré André (Lévy (1948), p. 293). A rigorous presentation of this argument requires use of the *strong Markov property* for Brownian motion. Accordingly, after some motivational discussion, we define the concept of a *strong Markov family* and prove that any Brownian family is strongly Markovian. This will allow us to place the heuristic argument on firm mathematical ground.

A. The Reflection Principle

Here is the argument of Désiré André. Let $\{B_t, \mathscr{F}_t; 0 \leq t < \infty\}$ be a standard, one-dimensional Brownian motion on $(\Omega, \mathscr{F}, P^0)$. For $b > 0$, we have

$$P^0[T_b < t] = P^0[T_b < t, B_t > b] + P^0[T_b < t, B_t < b].$$

Now $P^0[T_b < t, B_t > b] = P^0[B_t > b]$. On the other hand, if $T_b < t$ and $B_t < b$, then sometime before time t the Brownian path reached level b, and then in the remaining time it traveled from b to a point c less than b. Because of the symmetry with respect to b of a Brownian motion starting at b, the "probability" of doing this is the same as the "probability" of traveling from b to the point $2b - c$. The heuristic rationale here is that, for every path which crosses level b and is found at time t at a point below b, there is a "shadow path" (see figure) obtained from reflection about the level b which exceeds this level at time t, and these two paths have the same "probability." Of course, the actual probability for the occurrence of any particular path is zero, so this argument is only heuristic; even if the probability in question were positive, it would not be entirely obvious how to derive the type of "symmetry" claimed here from the definition of Brownian motion. Nevertheless, this argument leads us to the correct equation

$$P^0[T_b < t, B_t < b] = P^0[T_b < t, B_t > b] = P^0[B_t > b],$$

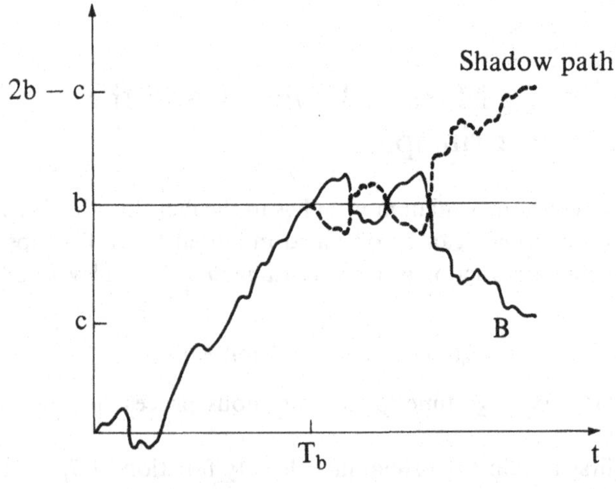

which then yields

$$P^0[T_b < t] = 2P^0[B_t > b] = \sqrt{\frac{2}{\pi}} \int_{bt^{-1/2}}^{\infty} e^{-x^2/2}\,dx. \tag{6.2}$$

Differentiating with respect to t, we obtain the density of the passage time

$$P^0[T_b \in dt] = \frac{|b|}{\sqrt{2\pi t^3}} e^{-b^2/2t}\,dt; \quad t > 0. \tag{6.3}$$

The preceding reasoning is based on the assumption that Brownian motion "starts afresh" (in the terminology of Itô & McKean (1974)) at the stopping time T_b, i.e., that the process $\{B_{t+T_b} - B_{T_b}; 0 \le t < \infty\}$ is Brownian motion, independent of the σ-field $\mathscr{F}_{T_b}$. If T_b were replaced by a nonnegative constant, it would not be hard to show this; if T_b were replaced by an arbitrary random time, the statement would be false (cf. Exercise 6.1). The fact that this "starting afresh" actually takes place at stopping times such as T_b is a consequence of the *strong Markov property* for Brownian motion.

6.1 Exercise. Let $\{B_t, \mathscr{F}_t; t \ge 0\}$ be a standard, one-dimensional Brownian motion. Give an example of a random time S with $P[0 \le S < \infty] = 1$, such that with $W_t \triangleq B_{S+t} - B_S$, the process $W = \{W_t, \mathscr{F}_t^W; t \ge 0\}$ is *not* a Brownian motion.

B. Strong Markov Processes and Families

6.2 Definition. Let d be a positive integer and μ a probability measure on $(\mathbb{R}^d, \mathscr{B}(\mathbb{R}^d))$. A progressively measurable, d-dimensional process $X = \{X_t, \mathscr{F}_t; t \ge 0\}$ on some $(\Omega, \mathscr{F}, P^\mu)$ is said to be a *strong Markov process with initial distribution* μ if

(i) $P^\mu[X_0 \in \Gamma] = \mu(\Gamma)$, $\forall\, \Gamma \in \mathscr{B}(\mathbb{R}^d)$;
(ii) for any optional time S of $\{\mathscr{F}_t\}$, $t \ge 0$ and $\Gamma \in \mathscr{B}(\mathbb{R}^d)$,

$$P^\mu[X_{S+t} \in \Gamma \mid \mathscr{F}_{S+}] = P^\mu[X_{S+t} \in \Gamma \mid X_S], \quad P^\mu\text{-a.s. on } \{S < \infty\}.$$

6.3 Definition. Let d be a positive integer. A d-dimensional *strong Markov family* is a progressively measurable process $X = \{X_t, \mathscr{F}_t; t \ge 0\}$ on some $(\Omega, \mathscr{F})$, together with a family of probability measures $\{P^x\}_{x \in \mathbb{R}^d}$ on $(\Omega, \mathscr{F})$, such that:

(a) for each $F \in \mathscr{F}$, the mapping $x \mapsto P^x(F)$ is universally measurable;
(b) $P^x[X_0 = x] = 1$, $\forall\, x \in \mathbb{R}^d$;
(c) for $x \in \mathbb{R}^d$, $t \ge 0$, $\Gamma \in \mathscr{B}(\mathbb{R}^d)$, and any optional time S of $\{\mathscr{F}_t\}$,

$$P^x[X_{S+t} \in \Gamma \mid \mathscr{F}_{S+}] = P^x[X_{S+t} \in \Gamma \mid X_S], \quad P^x\text{-a.s. on } \{S < \infty\};$$

(d) for $x \in \mathbb{R}^d$, $t \ge 0$, $\Gamma \in \mathscr{B}(\mathbb{R}^d)$, and any optional time S of $\{\mathscr{F}_t\}$,

$$P^x[X_{S+t} \in \Gamma \mid X_S = y] = P^y[X_t \in \Gamma], \quad P^x X_S^{-1}\text{-a.e. } y.$$

6.4 Remark. In Definitions 6.2, 6.3, $\{X_{S+t} \in \Gamma\} \triangleq \{S < \infty, X_{S+t} \in \Gamma\}$ and $P^x X_S^{-1}(\mathbb{R}^d) = P^x(S < \infty)$. The probability appearing on the right-hand side of Definition 6.2(ii) and Definition 6.3(c) is conditioned on the σ-field generated by X_S as defined in Problem 1.1.17. The reader may wish to verify in this connection that for any progressively measurable process X,

$$P^\mu[X_{S+t} \in \Gamma \mid \mathscr{F}_{S+}] = P^\mu[X_{S+t} \in \Gamma \mid X_S] = 0, \quad P^\mu\text{-a.s. on } \{S = \infty\},$$

and so the restriction $S < \infty$ in these conditions is unnecessary.

6.5 Remark. An optional time of $\{\mathscr{F}_t\}$ is a stopping time of $\{\mathscr{F}_{t+}\}$ (Corollary 1.2.4). Because of the assumption of progressive measurability, the random variable X_S appearing in Definitions 6.2 and 6.3 is $\mathscr{F}_{S+}$-measurable (Proposition 1.2.18). Moreover, if S is a stopping time of $\{\mathscr{F}_t\}$, then X_S is $\mathscr{F}_S$-measurable. In this case, we can take conditional expectations with respect to $\mathscr{F}_S$ on both sides of (c) in Definition 6.3, to obtain

$$P^x[X_{S+t}\in\Gamma|\mathscr{F}_S] = P^x[X_{S+t}\in\Gamma|X_S], \quad P^x\text{-a.s. on } \{S<\infty\}.$$

Setting S equal to a constant $s \geq 0$, we obtain condition (c) of Definition 5.11. Thus, every strong Markov family is a Markov family. Likewise, every strong Markov process is a Markov process. However, not every Markov family enjoys the strong Markov property; a counterexample to this effect, involving a progressively measurable process X, appears in Wentzell (1981), p. 161.

Whenever S is an optional time of $\{\mathscr{F}_t\}$ and $u > 0$, then $S + u$ is a stopping time of $\{\mathscr{F}_t\}$ (Problem 1.2.10). This fact can be used to replace the constant s in the proof of Proposition 5.15 by the optional time S, thereby obtaining the following result.

6.6 Proposition. *For a strong Markov family* $X = \{X_t, \mathscr{F}_t; t \geq 0\}$, $(\Omega, \mathscr{F})$, $\{P^x\}_{x\in\mathbb{R}^d}$, *we have*

(c′) *for* $x\in\mathbb{R}^d$, $F\in\mathscr{B}((\mathbb{R}^d)^{[0,\infty)})$, *and any optional time* S *of* $\{\mathscr{F}_t\}$,

$$P^x[X_{S+\cdot}\in F|\mathscr{F}_{S+}] = P^x[X_{S+\cdot}\in F|X_S], \quad P^x\text{-a.s. on } \{S<\infty\};$$

(d′) *for* $x\in\mathbb{R}^d$, $F\in\mathscr{B}((\mathbb{R}^d)^{[0,\infty)})$, *and any optional time* S *of* $\{\mathscr{F}_t\}$,

$$P^x[X_{S+\cdot}\in F|X_S = y] = P^y[X_\cdot\in F], \quad P^xX_S^{-1}\text{-a.e. } y.$$

Using the operators $\{U_t\}_{t\geq 0}$ in (5.7), conditions (c) and (d) of Definition 6.3 can be combined.

6.7 Proposition. *Let* $X = \{X_t, \mathscr{F}_t; t \geq 0\}$ *be a progressively measurable process on* $(\Omega, \mathscr{F})$, *and let* $\{P^x\}_{x\in\mathbb{R}^d}$ *be a family of probability measures satisfying* (a) *and* (b) *of Definition 6.3. Then* X, $(\Omega, \mathscr{F})$, $\{P^x\}_{x\in\mathbb{R}^d}$ *is strong Markov if and only if for any* $\{\mathscr{F}_t\}$*-optional time* S, $t \geq 0$, *and* $x\in\mathbb{R}^d$, *one of the following equivalent conditions holds*:

(e) *for any* $\Gamma\in\mathscr{B}(\mathbb{R}^d)$,

$$P^x[X_{S+t}\in\Gamma|\mathscr{F}_{S+}] = (U_t 1_\Gamma)(X_S), \quad P^x\text{-a.s. on } \{S<\infty\};$$

(e′) *for any bounded, continuous* $f\colon \mathbb{R}^d \to \mathbb{R}$,

$$E^x[f(X_{S+t})|\mathscr{F}_{S+}] = (U_t f)(X_S), \quad P^x\text{-a.s. on } \{S<\infty\}.$$

PROOF. The proof that (e) is equivalent to (c) and (d) is the same as the proof of the analogous equivalence for Markov families given in Proposition 5.13. Since any bounded, continuous real-valued function on $\mathbb{R}^d$ is the pointwise limit of a bounded sequence of linear combinations of indicators of Borel sets, (e′) follows from (e) and the bounded convergence theorem. On the other hand, if (e′) holds and $\Gamma \subseteq \mathbb{R}^d$ is closed, then 1_Γ is the pointwise limit of $\{f_n\}_{n=1}^\infty$, where $f_n(x) = [1 - n\rho(x,\Gamma)] \vee 0$ and $\rho(x,\Gamma) = \inf\{\|x-y\|; y \in \Gamma\}$. Each f_n is bounded and continuous, so (e) holds for closed sets Γ. The collection of sets $\Gamma \in \mathscr{B}(\mathbb{R}^d)$ for which (e) holds forms a Dynkin system, so, by Theorem 1.3, (e) holds for all $\Gamma \in \mathscr{B}(\mathbb{R}^d)$. □

6.8 Remark. If $X = \{X_t, \mathscr{F}_t; t \geq 0\}$, $(\Omega, \mathscr{F})$, $\{P^x\}_{x \in \mathbb{R}^d}$ is a strong Markov family and μ is a probability measure on $(\mathbb{R}^d, \mathscr{B}(\mathbb{R}^d))$, we can define a probability measure P^μ by (5.2) for every $F \in \mathscr{F}$, and then X on $(\Omega, \mathscr{F}, P^\mu)$ is a strong Markov process with initial distribution μ. Condition (ii) of Definition 6.2 can be verified upon writing condition (e) in integrated form:

$$\int_F (U_t 1_\Gamma)(X_S)\,dP^x = P^x[X_{S+t} \in \Gamma, F]; \quad F \in \mathscr{F}_{S+},$$

and then integrating both sides with respect to μ. Similarly, if X, $(\Omega, \mathscr{F})$, $\{P^x\}_{x \in \mathbb{R}^d}$ is a Markov family, then X on $(\Omega, \mathscr{F}, P^\mu)$ is a Markov process with initial distribution μ.

It is often convenient to work with *bounded* optional times only. The following problem shows that stating the strong Markov property in terms of such optional times entails no loss of generality. We shall use this fact in our proof that Brownian families are strongly Markovian.

6.9 Problem. Let S be an optional time of the filtration $\{\mathscr{F}_t\}$ on some $(\Omega, \mathscr{F}, P)$.

(i) Show that if Z_1 and Z_2 are integrable random variables and $Z_1 = Z_2$ on some $\mathscr{F}_{S+}$-measurable set A, then

$$E[Z_1 | \mathscr{F}_{S+}] = E[Z_2 | \mathscr{F}_{S+}], \quad \text{a.s. on } A.$$

(ii) Show under the conditions of (i) that if s is a positive constant, then

$$E[Z_1 | \mathscr{F}_{S+}] = E[Z_2 | \mathscr{F}_{(S \wedge s)+}], \quad \text{a.s. on } \{S \leq s\} \cap A.$$

(*Hint*: Use Problem 1.2.17(i)).

(iii) Show that if (e) (or (e′)) in Proposition 6.7 holds for every bounded optional time S of $\{\mathscr{F}_t\}$, then it holds for every optional time.

Conditions (e) and (e′) are statements about the conditional distribution of X at a single time $S + t$ after the optional time S. If there are shift operators $\{\theta_s\}_{s \geq 0}$ satisfying (5.14), then for any random time S we can define the *random shift* $\theta_S: \{S < \infty\} \to \Omega$ by

$$\theta_S = \theta_s \quad \text{on } \{S = s\}.$$

In other words, θ_S is defined so that whenever $S(\omega) < \infty$, then

$$X_{S(\omega)+t}(\omega) = X_t(\theta_S(\omega)).$$

In particular, we have $\{X_{S+\cdot} \in E\} = \theta_S^{-1}\{X_\cdot \in E\}$, and (c′) and (d′) are, respectively, equivalent to the statements: for every $x \in \mathbb{R}^d$, $F \in \mathscr{F}_\infty^X$, and any optional time S of $\{\mathscr{F}_t\}$,

(c″) $$P^x[\theta_S^{-1}F | \mathscr{F}_{S+}] = P^x[\theta_S^{-1}F | X_S], \quad P^x\text{-a.s. on } \{S < \infty\};$$

(d″) $$P^x[\theta_S^{-1}F | X_S = y] = P^y(F), \quad P^x X_S^{-1}\text{-a.e. } y.$$

Both (c″) and (d″) can be captured by the single condition:

(e″) for $x \in \mathbb{R}^d$, $F \in \mathscr{F}_\infty^X$, and any optional time S of $\{\mathscr{F}_t\}$,

$$P^x[\theta_S^{-1}F | \mathscr{F}_{S+}] = P^{X_S}(F), \quad P^x\text{-a.s. on } \{S < \infty\}.$$

Since (e″) is often given as the primary defining property for a strong Markov family, we summarize this discussion with a theorem.

6.10 Theorem. *Let $X = \{X_t, \mathscr{F}_t; t \geq 0\}$ be a progressively measurable process on $(\Omega, \mathscr{F})$, let $\{P^x\}_{x \in \mathbb{R}^d}$ be a family of probability measures on $(\Omega, \mathscr{F})$, and let $\{\theta_s\}_{s \geq 0}$ be a family of $\mathscr{F}/\mathscr{F}$-measurable shift operators satisfying* (5.14). *Then $X, (\Omega, \mathscr{F}), \{P^x\}_{x \in \mathbb{R}^d}$ is a strong Markov family if and only if* (a), (b), *and* (e″) *hold.*

6.11 Problem. Show that (e″) is equivalent to the following condition:

(e‴) For all $x \in \mathbb{R}^d$, any bounded, $\mathscr{F}_\infty^X$-measurable random variable Y, and any optional time S of $\{\mathscr{F}_t\}$, we have

$$E^x[Y \circ \theta_S | \mathscr{F}_{S+}] = E^{X_S}(Y), \quad P^x\text{-a.s. on } \{S < \infty\}.$$

(*Note*: If we write this equation with the arguments filled in, it becomes

$$E^x[Y \circ \theta_S | \mathscr{F}_{S+}](\omega) = \int_\Omega Y(\omega') P^{X_{S(\omega)}(\omega)}(d\omega'), \quad P^x\text{-a.e. } \omega \in \{S < \infty\},$$

where $(Y \circ \theta_S)(\omega'') \triangleq Y(\theta_{S(\omega'')}(\omega''))$.)

C. The Strong Markov Property for Brownian Motion

The discussion on the strong Markov property for Brownian motion will require some background material on regular conditional probabilities.

6.12 Definition. Let X be a random variable on a probability space $(\Omega, \mathscr{F}, P)$ taking values in a complete, separable metric space $(S, \mathscr{B}(S))$. Let $\mathscr{G}$ be a sub-σ-field of $\mathscr{F}$. A *regular conditional probability of X given $\mathscr{G}$* is a function $Q: \Omega \times \mathscr{B}(S) \to [0, 1]$ such that

(i) for each $\omega \in \Omega$, $Q(\omega; \cdot)$ is a probability measure on $(S, \mathscr{B}(S))$,
(ii) for each $E \in \mathscr{B}(S)$, the mapping $\omega \mapsto Q(\omega; E)$ is $\mathscr{G}$-measurable, and
(iii) for each $E \in \mathscr{B}(S)$, $P[X \in E | \mathscr{G}](\omega) = Q(\omega; E)$, P-a.e. ω.

Under the conditions of Definition 6.12 on X, $(\Omega, \mathscr{F}, P)$, $(S, \mathscr{B}(S))$, and $\mathscr{G}$, *a regular conditional probability for X given $\mathscr{G}$ exists* (Ash (1972), pp. 264–265, or Parthasarathy (1967), pp. 146–150). One consequence of this fact is that the conditional characteristic function of a random vector can be used to determine its conditional distribution, in the manner outlined by the next lemma.

6.13 Lemma. *Let X be a d-dimensional random vector on $(\Omega, \mathscr{F}, P)$. Suppose $\mathscr{G}$ is a sub-σ-field of $\mathscr{F}$ and suppose that for each $\omega \in \Omega$, there is a function $\varphi(\omega; \cdot): \mathbb{R}^d \to \mathbb{C}$ such that for each $u \in \mathbb{R}^d$,*

$$\varphi(\omega; u) = E[e^{i(u,X)} | \mathscr{G}](\omega), \quad P\text{-a.e. } \omega.$$

If, for each ω, $\varphi(\omega; \cdot)$ is the characteristic function of some probability measure P^ω on $(\mathbb{R}^d, \mathscr{B}(\mathbb{R}^d))$, i.e.,

$$\varphi(\omega; u) = \int_{\mathbb{R}^d} e^{i(u,x)} P^\omega(dx),$$

where $i = \sqrt{-1}$, then for each $\Gamma \in \mathscr{B}(\mathbb{R}^d)$, we have

$$P[X \in \Gamma | \mathscr{G}](\omega) = P^\omega(\Gamma), \quad P\text{-a.e. } \omega.$$

PROOF. Let Q be a regular conditional probability for X given $\mathscr{G}$, so for each fixed $u \in \mathbb{R}^d$ we can build up from indicators to show that

$$(6.4) \qquad \varphi(\omega; u) = E[e^{i(u,X)} | \mathscr{G}](\omega) = \int_{\mathbb{R}^d} e^{i(u,x)} Q(\omega; dx), \quad P\text{-a.e. } \omega.$$

The set of ω for which (6.4) fails may depend on u, but we can choose a countable, dense subset D of $\mathbb{R}^d$ and an event $\tilde{\Omega} \in \mathscr{F}$ with $P(\tilde{\Omega}) = 1$, so that (6.4) holds for every $\omega \in \tilde{\Omega}$ and $u \in D$. Continuity in u of both sides of (6.4) allows us to conclude its validity for every $\omega \in \tilde{\Omega}$ and $u \in \mathbb{R}^d$. Since a measure is uniquely determined by its characteristic function, we must have $P^\omega = Q(\omega; \cdot)$ for P-a.e. ω, and the result follows. □

Recall that a d-dimensional random vector N has a *d-variate normal distribution* with mean $\mu \in \mathbb{R}^d$ and $(d \times d)$ covariance matrix Σ if and only if it has characteristic function

$$(6.5) \qquad Ee^{i(u,N)} = e^{i(u,\mu) - (u, \Sigma u)/2}; \quad u \in \mathbb{R}^d.$$

Suppose $B = \{B_t, \mathscr{F}_t; t \geq 0\}$, $(\Omega, \mathscr{F})$, $\{P^x\}_{x \in \mathbb{R}^d}$ is a d-dimensional Brownian family. Choose $u \in \mathbb{R}^d$ and define the complex-valued process

$$M_t \triangleq \exp\left[i(u, B_t) + \frac{t}{2}\|u\|^2\right], \quad t \geq 0.$$

We denote the real and imaginary parts of this process by R_t and I_t, respectively.

6.14 Lemma. *For each $x \in \mathbb{R}^d$, the processes $\{R_t, \mathscr{F}_t; t \geq 0\}$ and $\{I_t, \mathscr{F}_t; t \geq 0\}$ are martingales on $(\Omega, \mathscr{F}, P^x)$.*

PROOF. For $0 \leq s < t$, we have

$$E^x[M_t|\mathscr{F}_s] = E^x\left[M_s \exp\left(i(u, B_t - B_s) + \frac{t-s}{2}\|u\|^2\right)\middle|\mathscr{F}_s\right]$$
$$= M_s E^x\left[\exp\left(i(u, B_t - B_s) + \frac{t-s}{2}\|u\|^2\right)\right] = M_s,$$

where we have used the independence of $B_t - B_s$ and $\mathscr{F}_s$, as well as (6.5). Taking real and imaginary parts, we obtain the martingale property for $\{R_t, \mathscr{F}_t; t \geq 0\}$ and $\{I_t, \mathscr{F}_t; t \geq 0\}$. □

6.15 Theorem. *A d-dimensional Brownian family is a strong Markov family. A d-dimensional Brownian motion is a strong Markov process.*

PROOF. We verify that a Brownian family $B = \{B_t, \mathscr{F}_t; t \geq 0\}, (\Omega, \mathscr{F}), \{P^x\}_{x \in \mathbb{R}^d}$ satisfies condition (e) of Proposition 6.7. Thus, let S be an optional time of $\{\mathscr{F}_t\}$. In light of Problem 6.9, we may assume that S is bounded. Fix $x \in \mathbb{R}^d$. The optional sampling theorem (Theorem 1.3.22 and Problem 1.3.23 (i)) applied to the martingales of Lemma 6.14 yields, for P^x-a.e. $\omega \in \Omega$:

$$E^x[\exp(i(u, B_{S+t}))|\mathscr{F}_{S+}](\omega) = \exp\left[i(u, B_{S(\omega)}(\omega)) - \frac{t}{2}\|u\|^2\right].$$

Comparing this to (6.5), we see that the conditional distribution of B_{S+t}, given $\mathscr{F}_{S+}$, is normal with mean $B_{S(\omega)}(\omega)$ and covariance matrix tI_d. This proves (e). □

We can carry this line of argument a bit further to obtain a related result.

6.16 Theorem. *Let S be an a.s. finite optional time of the filtration $\{\mathscr{F}_t\}$ for the d-dimensional Brownian motion $B = \{B_t, \mathscr{F}_t; t \geq 0\}$. Then with $W_t \triangleq B_{S+t} - B_S$, the process $W = \{W_t, \mathscr{F}_t^W; t \geq 0\}$ is a d-dimensional Brownian motion, independent of $\mathscr{F}_{S+}$.*

PROOF. We show that for every $n \geqq 1$, $0 \leq t_0 \leq \cdots \leq t_n < \infty$, and $u_1, \ldots, u_n \in \mathbb{R}^d$, we have a.s. P:

$$(6.6) \quad E\left[\exp\left(i\sum_{k=1}^{n}(u_k, W_{t_k} - W_{t_{k-1}})\right)\middle|\mathscr{F}_{S+}\right] = \prod_{k=1}^{n}\exp\left[-\frac{1}{2}(t_k - t_{k-1})\|u_k\|^2\right];$$

thus, according to Lemma 6.13 and (6.5), not only are the increments $\{W_{t_k} - W_{t_{k-1}}\}_{k=1}^n$ independent normal random vectors with mean zero and covariance matrices $(t_k - t_{k-1})I_d$, but they are also independent of the σ-field $\mathscr{F}_{S+}$. This substantiates the claim of the theorem.

We prove (6.6) for bounded, optional times S of $\{\mathscr{F}_t\}$; the argument given in Solution 6.9 can be used to extend this result to a.s. finite S. Assume (6.6) holds for some n, and choose $0 \le t_0 \le \cdots \le t_n \le t_{n+1}$. Applying the equality in the proof of Theorem 6.15 to the optional time $S + t_n$, we have

$$(6.7)\quad \begin{aligned} &E\{\exp[i(u_{n+1}, W_{t_{n+1}} - W_{t_n})] | \mathscr{F}_{(S+t_n)+}\} \\ &= E\{\exp[i(u_{n+1}, B_{S+t_{n+1}})] | \mathscr{F}_{(S+t_n)+}\} \cdot \exp[-i(u_{n+1}, B_{S+t_n})] \\ &= \exp[-\tfrac{1}{2}(t_{n+1} - t_n)\|u_{n+1}\|^2], \quad P\text{-a.s.} \end{aligned}$$

Therefore,

$$\begin{aligned} &E\left[\exp\left(i\sum_{k=1}^{n+1}(u_k, W_{t_k} - W_{t_{k-1}})\right)\bigg|\mathscr{F}_{S+}\right] \\ &= E\Bigg[\exp\left(i\sum_{k=1}^{n}(u_k, W_{t_k} - W_{t_{k-1}})\right) \\ &\qquad \cdot E\{\exp(i(u_{n+1}, W_{t_{n+1}} - W_{t_n})) | \mathscr{F}_{(S+t_n)+}\}\bigg|\mathscr{F}_{S+}\Bigg] \\ &= \exp\left[-\frac{1}{2}(t_{n+1} - t_n)\|u_{n+1}\|^2\right] E\left[\exp\left(i\sum_{k=1}^{n}(u_k, W_{t_k} - W_{t_{k-1}})\right)\bigg|\mathscr{F}_{S+}\right] \\ &= \prod_{k=1}^{n+1}\exp\left[-\frac{1}{2}(t_k - t_{k-1})\|u_k\|^2\right], \quad P\text{-a.s.,} \end{aligned}$$

which completes the induction step. The proof that (6.6) holds for $n = 1$ is obtained by setting $t_n = 0$ in (6.7). □

In order to present a rigorous derivation of the density (6.3) for the passage time T_b in (6.1), a slight extension of the strong Markov property for right-continuous processes will be needed.

6.17 Proposition. *Let $X = \{X_t, \mathscr{F}_t; t \ge 0\}$, $(\Omega, \mathscr{F})$, $\{P^x\}_{x \in \mathbb{R}^d}$ be a strong Markov family, and the process X be right-continuous. Let S be an optional time of $\{\mathscr{F}_t\}$ and T an $\mathscr{F}_{S+}$-measurable random time satisfying $T(\omega) \ge S(\omega)$ for all $\omega \in \Omega$. Then, for any $x \in \mathbb{R}^d$ and any bounded, continuous $f: \mathbb{R}^d \to \mathbb{R}$,*

$$(6.8)\quad E^x[f(X_T)|\mathscr{F}_{S+}](\omega) = (U_{T(\omega)-S(\omega)}f)(X_{S(\omega)}(\omega)), \quad P^x\text{-a.e. } \omega \in \{T < \infty\}.$$

PROOF. For $n \ge 1$, let

$$T_n = \begin{cases} S + \dfrac{1}{2^n}([\![2^n(T - S)]\!] + 1), & \text{if } T < \infty, \\ \infty, & \text{if } T = \infty, \end{cases}$$

so that $T_n = S + k2^{-n}$ when $(k-1)2^{-n} \le T - S < k2^{-n}$. We have $T_n \downarrow T$ on $\{T < \infty\}$. From (e′) we have for $k \ge 0$,

$$E^x[f(X_{S+k2^{-n}})|\mathscr{F}_{S+}] = (U_{k/2^n}f)(X_S), \quad P^x\text{-a.s. on } \{S < \infty\},$$

and Problem 6.9 (i) then implies

$$E^x[f(X_{T_n})|\mathscr{F}_{S+}](\omega) = (U_{T_n(\omega)-S(\omega)}f)(X_{S(\omega)}(\omega)), \quad P^x\text{-a.e. } \omega \in \{T < \infty\}.$$

The bounded convergence theorem for conditional expectations and the right-continuity of X imply that the left-hand side converges to $E^x[f(X_T)|\mathscr{F}_{S+}](\omega)$ as $n \to \infty$. Since $(U_t f)(y) = E^y f(X_t)$ is right-continuous in t for every $y \in \mathbb{R}^d$, the right-hand side converges to $(U_{T(\omega)-S(\omega)}f)(X_{S(\omega)}(\omega))$. □

6.18 Corollary. *Under the conditions of Proposition* 6.17, (6.8) *holds for every bounded, $\mathscr{B}(\mathbb{R}^d)/\mathscr{B}(\mathbb{R})$-measurable function f. In particular, for any $\Gamma \in \mathscr{B}(\mathbb{R}^d)$ we have for P^x-a.e. $\omega \in \{T < \infty\}$:*

$$P^x[X_T \in \Gamma|\mathscr{F}_{S+}](\omega) = (U_{T(\omega)-S(\omega)}1_\Gamma)(X_{S(\omega)}(\omega)).$$

PROOF. Approximate the indicator of a closed set Γ by bounded, continuous functions as in the proof of Proposition 6.7. Then prove the result for any $\Gamma \in \mathscr{B}(\mathbb{R}^d)$, and extend to bounded, Borel-measurable functions. □

6.19 Proposition. *Let $\{B_t, \mathscr{F}_t; t \ge 0\}$ be a standard, one-dimensional Brownian motion, and for $b \ne 0$, let T_b be the first passage time to b as in* (6.1). *Then T_b has the density given by* (6.3).

PROOF. Because $\{-B_t, \mathscr{F}_t; t \ge 0\}$ is also a standard, one-dimensional Brownian motion, it suffices to consider the case $b > 0$. In Corollary 6.18 set $S = T_b$,

$$T = \begin{cases} t & \text{if } S < t, \\ \infty & \text{if } S \ge t, \end{cases}$$

and $\Gamma = (-\infty, b)$. On the set $\{T < \infty\} = \{S < t\}$, we have $B_{S(\omega)}(\omega) = b$ and $(U_{T(\omega)-S(\omega)}1_\Gamma)(B_{S(\omega)}(\omega)) = \frac{1}{2}$. Therefore,

$$P^0[T_b < t, B_t < b] = \int_{\{T_b < t\}} P^0[B_T \in \Gamma|\mathscr{F}_{S+}]\,dP^0 = \frac{1}{2}P^0[T_b < t].$$

It follows that

$$\begin{aligned} P^0[T_b < t] &= P^0[T_b < t, B_t > b] + P^0[T_b < t, B_t < b] \\ &= P^0[B_t > b] + \tfrac{1}{2}P^0[T_b < t], \end{aligned}$$

and (6.2) is proved. □

6.20 Remark. It follows from (6.2), by letting $t \to \infty$, that the passage times are almost surely finite: $P^0[T_b < \infty] = 1$.

6.21 Problem. Recall the notions of Poisson family and Poisson process from Definitions 5.20 and 1.3.3, respectively. Show that the former is a strong Markov family, and the latter is a strong Markov process.

2.7. Brownian Filtrations

In Section 1 we made a point of defining Brownian motion $B = \{B_t, \mathscr{F}_t; t \geq 0\}$ with a filtration $\{\mathscr{F}_t\}$ which is possibly larger than $\{\mathscr{F}_t^B\}$ and anticipated some of the reasons that mandate this generality. One reason is related to the fact that, although the filtration $\{\mathscr{F}_t^B\}$ is left-continuous, it fails to be right-continuous (Problem 7.1). Some of the developments in later chapters require either right or two-sided continuity of the filtration $\{\mathscr{F}_t\}$, and so in this section we construct filtrations with these properties.

Let us recall the basic definitions from Section 1.1. For a filtration $\{\mathscr{F}_t; t \geq 0\}$ on the measurable space $(\Omega, \mathscr{F})$, we set $\mathscr{F}_{t+} = \bigcap_{\varepsilon>0} \mathscr{F}_{t+\varepsilon}$ for $t \geq 0$, $\mathscr{F}_{t-} = \sigma(\bigcup_{s<t} \mathscr{F}_s)$ for $t > 0$, $\mathscr{F}_{0-} = \mathscr{F}_0$, and $\mathscr{F}_\infty = \sigma(\bigcup_{t\geq 0} \mathscr{F}_t)$. We say that $\{\mathscr{F}_t\}$ is *right-* (respectively, *left-*) *continuous* if $\mathscr{F}_{t+} = \mathscr{F}_t$ (respectively, $\mathscr{F}_{t-} = \mathscr{F}_t$) holds for every $0 \leq t < \infty$. When $X = \{X_t, \mathscr{F}_t^X; t \geq 0\}$ is a process on $(\Omega, \mathscr{F})$, then left-continuity of $\{\mathscr{F}_t^X\}$ at some fixed $t > 0$ can be interpreted to mean that X_t can be discovered by observing X_s, $0 \leq s < t$. Right-continuity means intuitively that if X_s has been observed for $0 \leq s \leq t$, then nothing more can be learned by peeking infinitesimally far into the future. We recall here that $\mathscr{F}_t^X = \sigma(X_s; 0 \leq s \leq t)$.

7.1 Problem. Let $\{X_t, \mathscr{F}_t^X; 0 \leq t < \infty\}$ be a d-dimensional process.

(i) Show that the filtration $\{\mathscr{F}_{t+}^X\}$ is right-continuous.
(ii) Show that if X is left-continuous, then the filtration $\{\mathscr{F}_t^X\}$ is left-continuous.
(iii) Show by example that, even if X is continuous, $\{\mathscr{F}_t^X\}$ can fail to be right-continuous and $\{\mathscr{F}_{t+}^X\}$ can fail to be left-continuous.

We shall need to develop the important notions of *completion* and *augmentation* of σ-fields, in the context of a process $X = \{X_t, \mathscr{F}_t^X; 0 \leq t < \infty\}$ with initial distribution μ on the space $(\Omega, \mathscr{F}_\infty^X, P^\mu)$, where $P^\mu[X_0 \in \Gamma] = \mu(\Gamma)$; $\Gamma \in \mathscr{B}(\mathbb{R}^d)$. We start by setting, for $0 \leq t \leq \infty$,

$$\mathscr{N}_t^\mu \triangleq \{F \subseteq \Omega; \exists\, G \in \mathscr{F}_t^X \text{ with } F \subseteq G,\ P^\mu(G) = 0\}.$$

$\mathscr{N}_\infty^\mu$ will be called "the collection of *P^μ-null sets*" and denoted simply by $\mathscr{N}^\mu$.

7.2 Definition. For any $0 \leq t < \infty$, we define

(i) the *completion*: $\mathscr{F}_t^\mu \triangleq \sigma(\mathscr{F}_t^X \cup \mathscr{N}_t^\mu)$, and
(ii) the *augmentation*: $\mathscr{F}_t^\mu \triangleq \sigma(\mathscr{F}_t^X \cup \mathscr{N}^\mu)$

of the σ-field $\mathscr{F}_t^X$ under P^μ. For $t = \infty$ the two concepts agree, and we set simply $\mathscr{F}^\mu \triangleq \sigma(\mathscr{F}_\infty^X \cup \mathscr{N}^\mu)$.

The *augmented filtration* $\{\mathscr{F}_t^\mu\}$ possesses certain desirable properties, which will be used frequently in the sequel and are developed in the ensuing problems and propositions.

7.3 Problem. For any sub-σ-field $\mathscr{G}$ of $\mathscr{F}_\infty^X$, define $\mathscr{G}^\mu = \sigma(\mathscr{G} \cup \mathscr{N}^\mu)$ and

$$\mathscr{H} = \{F \subseteq \Omega; \exists\, G \in \mathscr{G} \text{ such that } F \triangle G \in \mathscr{N}^\mu\}.$$

Show that $\mathscr{G}^\mu = \mathscr{H}$. We now extend P^μ by defining $P^\mu(F) \triangleq P^\mu(G)$ whenever $F \in \mathscr{G}^\mu$, and $G \in \mathscr{G}$ is chosen to satisfy $F \triangle G \in \mathscr{N}^\mu$. Show that the probability space $(\Omega, \mathscr{G}^\mu, P^\mu)$ is *complete*:

$$F \in \mathscr{G}^\mu,\ P^\mu(F) = 0,\ D \subseteq F \quad \Rightarrow \quad D \in \mathscr{G}^\mu.$$

7.4 Problem. From Definition 7.2 we have $\bar{\mathscr{F}}_t^\mu \subseteq \mathscr{F}_t^\mu$, for every $0 \le t < \infty$. Show by example that the inclusion can be strict.

7.5 Problem. Show that the σ-field $\mathscr{F}^\mu$ of Definition 7.2 agrees with

$$\mathscr{F}_\infty^\mu \triangleq \sigma\left(\bigcup_{t \ge 0} \mathscr{F}_t^\mu\right).$$

7.6 Problem. If the process X has left-continuous paths, then the filtration $\{\mathscr{F}_t^\mu\}$ is left-continuous.

A. Right-Continuity of the Augmented Filtration for a Strong Markov Process

We are ready now for the key result of this section.

7.7 Proposition. *For a d-dimensional strong Markov process $X = \{X_t, \mathscr{F}_t^X; t \ge 0\}$ with initial distribution μ, the augmented filtration $\{\mathscr{F}_t^\mu\}$ is right-continuous.*

Proof. Let $(\Omega, \mathscr{F}_\infty^X, P^\mu)$ be the probability space on which X is defined. Fix $s \ge 0$ and consider the degenerate, $\{\mathscr{F}_t^X\}$-optional time $S = s$. With $0 \le t_0 < t_1 < \cdots < t_n \le s < t_{n+1} < \cdots < t_m$ and $\Gamma_0, \ldots, \Gamma_m$ in $\mathscr{B}(\mathbb{R}^d)$, the strong Markov property gives

$$\begin{aligned} &P^\mu[X_{t_0} \in \Gamma_0, \ldots, X_{t_m} \in \Gamma_m | \mathscr{F}_{s+}^X] \\ &\quad = 1_{\{X_{t_0} \in \Gamma_0, \ldots, X_{t_n} \in \Gamma_n\}} P^\mu[X_{t_{n+1}} \in \Gamma_{n+1}, \ldots, X_{t_m} \in \Gamma_m | X_s], \end{aligned}$$

P^μ-a.s. It is now evident that $P^\mu[X_{t_0} \in \Gamma_0, \ldots, X_{t_m} \in \Gamma_m | \mathscr{F}_{s+}^X]$ has an $\mathscr{F}_s^X$-

measurable version. The collection of all sets $F \in \mathscr{F}_\infty^X$ for which $P^\mu[F|\mathscr{F}_{s+}^X]$ has an $\mathscr{F}_s^X$-measurable version is a Dynkin system. We conclude from the Dynkin System Theorem 1.3 that, for every $F \in \mathscr{F}_\infty^X$, the conditional probability $P^\mu[F|\mathscr{F}_{s+}^X]$ has an $\mathscr{F}_s^X$-measurable version.

Let us take now $F \in \mathscr{F}_{s+}^X \subseteq \mathscr{F}_\infty^X$; we have $P^\mu[F|\mathscr{F}_{s+}^X] = 1_F$, a.s. P^μ, so 1_F has an $\mathscr{F}_s^X$-measurable version which we denote by Y. Because $G \triangleq \{Y = 1\} \in \mathscr{F}_s^X$ and $F \triangle G \subseteq \{1_F \neq Y\} \in \mathscr{N}^\mu$, we have $F \in \mathscr{F}_s^\mu$ and consequently $\mathscr{F}_{s+}^X \subseteq \mathscr{F}_s^\mu$; $s \geq 0$.

Now let us suppose that $F \in \mathscr{F}_{s+}^\mu$; for every integer $n \geq 1$ we have $F \in \mathscr{F}_{s+(1/n)}^\mu$, as well as a set $G_n \in \mathscr{F}_{s+(1/n)}^X$ such that $F \triangle G_n \in \mathscr{N}^\mu$. We define $G \triangleq \bigcap_{m=1}^\infty \bigcup_{n=m}^\infty G_n$, and since $G = \bigcap_{m=M}^\infty \bigcup_{n=m}^\infty G_n$ for any positive integer M, we have $G \in \mathscr{F}_{s+}^X \subseteq \mathscr{F}_s^\mu$. To prove that $F \in \mathscr{F}_s^\mu$, it suffices to show $F \triangle G \in \mathscr{N}^\mu$. Now

$$G \backslash F \subseteq \left(\bigcup_{n=1}^\infty G_n\right) \backslash F = \bigcup_{n=1}^\infty (G_n \backslash F) \in \mathscr{N}^\mu.$$

On the other hand,

$$F \backslash G = F \cap \left(\bigcap_{m=1}^\infty \bigcup_{n=m}^\infty G_n\right)^c = F \cap \left(\bigcup_{m=1}^\infty \bigcap_{n=m}^\infty G_n^c\right)$$

$$= \bigcup_{m=1}^\infty \left[F \cap \left(\bigcap_{n=m}^\infty G_n^c\right)\right] \subseteq \bigcup_{m=1}^\infty (F \cap G_m^c) = \bigcup_{m=1}^\infty (F \backslash G_m) \in \mathscr{N}^\mu.$$

It follows that $F \in \mathscr{F}_s^\mu$, so $\mathscr{F}_{s+}^\mu \subseteq \mathscr{F}_s^\mu$ and right-continuity is proved. □

7.8 Corollary. *For a d-dimensional, left-continuous strong Markov process $X = \{X_t, \mathscr{F}_t^X; t \geq 0\}$ with initial distribution μ, the augmented filtration $\{\mathscr{F}_t^\mu\}$ is continuous.*

7.9 Theorem. *Let $B = \{B_t, \mathscr{F}_t^B; t \geq 0\}$ be a d-dimensional Brownian motion with initial distribution μ on $(\Omega, \mathscr{F}_\infty^B, P^\mu)$. Relative to the filtration $\{\mathscr{F}_t^\mu\}$, $\{B_t, t \geq 0\}$ is still a d-dimensional Brownian motion.*

PROOF. Augmentation of σ-fields does not disturb the assumptions of Definition 5.1.

7.10 Remark. Consider a Poisson process $\{N_t, \mathscr{F}_t^N; 0 \leq t < \infty\}$ as in Definition 1.3.3 and denote by $\{\mathscr{F}_t\}$ the augmentation of $\{\mathscr{F}_t^N\}$. In conjunction with Problems 6.21 and 7.3, Proposition 7.7 shows that $\{\mathscr{F}_t\}$ satisfies the usual conditions; furthermore, $\{N_t, \mathscr{F}_t; 0 \leq t < \infty\}$ is a Poisson process.

Since *any* d-dimensional Brownian motion is strongly Markov (Theorem 6.15), the augmentation of the filtration in Theorem 7.9 does not affect the strong Markov property. This raises the following general question. Suppose $\{X_t, \mathscr{F}_t^X; t \geq 0\}$ is a d-dimensional, strong Markov process with initial distribution μ on $(\Omega, \mathscr{F}_\infty^X, P^\mu)$. Is the process $\{X_t, \mathscr{F}_t^\mu; t \geq 0\}$ also strongly Markov? In other words, is it true, for every optional time S of $\{\mathscr{F}_t^\mu\}$, $t \geq 0$ and $\Gamma \in \mathscr{B}(\mathbb{R}^d)$, that

$$(7.1) \qquad P^\mu[X_{S+t} \in \Gamma | \mathscr{F}_{S+}^\mu] = P^\mu[X_{S+t} \in \Gamma | X_S], \quad P^\mu\text{-a.s. on } \{S < \infty\}?$$

Although the answer to this question is affirmative, phrased in this generality the question is not as important as it might appear. In each particular case, some technique must be used to prove that $\{X_t, \mathscr{F}_t^X; t \geq 0\}$ is strongly Markov in the first place, and this technique can usually be employed to establish the strong Markov property for $\{X_t, \mathscr{F}_t^\mu; t \geq 0\}$ as well. Theorems 7.9 and 6.15 exemplify this kind of argument for d-dimensional Brownian motion. Nonetheless, the interested reader can work through the following series of exercises to verify that (7.1) is valid in the generality claimed.

In Exercises 7.11–7.13, $X = \{X_t, \mathscr{F}_t^X; 0 \leq t < \infty\}$ is a strong Markov process with initial distribution μ on $(\Omega, \mathscr{F}_\infty^X, P^\mu)$.

7.11 Exercise. Show that any optional time S of $\{\mathscr{F}_t^\mu\}$ is also a stopping time of this filtration, and for each such S there exists an optional time T of $\{\mathscr{F}_t^X\}$ with $\{S \neq T\} \in \mathscr{N}^\mu$. Conclude that $\mathscr{F}_{S+}^\mu = \mathscr{F}_S^\mu = \mathscr{F}_T^\mu$, where $\mathscr{F}_T^\mu$ is defined to be the collection of sets $A \in \mathscr{F}^\mu$ satisfying $A \cap \{T \leq t\} \in \mathscr{F}_t^\mu$, $\forall 0 \leq t < \infty$.

7.12 Exercise. Suppose that T is an optional time of $\{\mathscr{F}_t^X\}$. For fixed positive integer n, define

$$T_n = \begin{cases} T, & \text{on } \{T = \infty\} \\ \dfrac{k}{2^n}, & \text{on } \left\{\dfrac{k-1}{2^n} \leq T < \dfrac{k}{2^n}\right\}. \end{cases}$$

Show that T_n is a stopping time of $\{\mathscr{F}_t^X\}$, and $\mathscr{F}_T^\mu \subseteq \sigma(\mathscr{F}_{T_n}^X \cup \mathscr{N}^\mu)$. Conclude that $\mathscr{F}_T^\mu \subseteq \sigma(\mathscr{F}_{T+}^X \cup \mathscr{N}^\mu)$. (*Hint*: Use Problems 1.2.23 and 1.2.24.)

7.13 Exercise. Establish the following proposition: if for each $t \geq 0$, $\Gamma \in \mathscr{B}(\mathbb{R}^d)$, and optional time T of $\{\mathscr{F}_t^X\}$, we have the strong Markov property

$$(7.2) \qquad P^\mu[X_{T+t} \in \Gamma | \mathscr{F}_{T+}^X] = P^\mu[X_{T+t} \in \Gamma | X_T], \quad P^\mu\text{-a.s. on } \{T < \infty\},$$

then (7.1) holds for every optional time S of $\{\mathscr{F}_t^\mu\}$.

This completes our discussion of the augmentation of the filtration generated by a strong Markov process. At first glance, augmentation appears to be a rather artificial device, but in retrospect it can be seen to be more useful and natural than merely completing each σ-field $\mathscr{F}_t^X$ with respect to P^μ. It is more natural because it involves only one collection of P^μ-null sets, the collection we called $\mathscr{N}^\mu$, rather than a separate collection for each $t \geq 0$. It is more useful because completing each σ-field $\mathscr{F}_t^X$ need not result in a right-continuous filtration, as the next problem demonstrates.

7.14 Problem. Let $\{B_t; t \geq 0\}$ be the coordinate mapping process on $(C[0,\infty), \mathscr{B}(C[0,\infty)))$, P^0 be Wiener measure, and $\bar{\mathscr{F}}_t$ denote the completion of $\mathscr{F}_t^B$ under P^0. Consider the set

$$F = \{\omega \in C[0,\infty); \omega \text{ is constant on } [0,\varepsilon] \text{ for some } \varepsilon > 0\}.$$

Show that: (i) $P^0(F) = 0$, (ii) $F \in \mathscr{F}_{0+}^B$, and (iii) $F \notin \bar{\mathscr{F}}_0$.

B. A "Universal" Filtration

The difficulty with the filtration $\{\mathscr{F}_t^\mu\}$, obtained for a strong Markov process with initial distribution μ, is its dependence on μ. In particular, such a filtration is inappropriate for a strong Markov family, where there is a continuum of initial conditions. We now construct a filtration which is well suited for this case.

Let $\{X_t, \mathscr{F}_t^X; t \geq 0\}$, $(\Omega, \mathscr{F}_\infty^X)$, $\{P^x\}_{x \in \mathbb{R}^d}$ be a d-dimensional, strong Markov family. For each probability measure μ on $(\mathbb{R}^d, \mathscr{B}(\mathbb{R}^d))$, we define P^μ as in (5.2):

$$P^\mu(F) = \int_{\mathbb{R}^d} P^x(F)\mu(dx), \quad \forall F \in \mathscr{F}_\infty^X,$$

and we construct the augmented filtration $\{\mathscr{F}_t^\mu\}$ as before. We define

$$\tilde{\mathscr{F}}_t \triangleq \bigcap_\mu \mathscr{F}_t^\mu, \quad 0 \leq t \leq \infty, \tag{7.3}$$

where the intersection is over all probability measures μ on $(\mathbb{R}^d, \mathscr{B}(\mathbb{R}^d))$. Note that $\mathscr{F}_t^X \subseteq \tilde{\mathscr{F}}_t \subseteq \mathscr{F}_t^\mu, 0 \leq t \leq \infty$ for any probability measure μ on $(\mathbb{R}^d, \mathscr{B}(\mathbb{R}^d))$; therefore, if $\{X_t, \mathscr{F}_t^X; t \geq 0\}$ and $\{X_t, \mathscr{F}_t^\mu; t \geq 0\}$ are both strongly Markovian under P^μ, then so is $\{X_t, \tilde{\mathscr{F}}_t; t \geq 0\}$. Because the order of intersection is interchangeable and $\{\mathscr{F}_t^\mu\}$ is right-continuous, we have

$$\tilde{\mathscr{F}}_{t+} = \bigcap_{s>t} \bigcap_\mu \mathscr{F}_s^\mu = \bigcap_\mu \bigcap_{s>t} \mathscr{F}_s^\mu = \bigcap_\mu \mathscr{F}_t^\mu = \tilde{\mathscr{F}}_t.$$

Thus $\{\tilde{\mathscr{F}}_t\}$ is also right-continuous.

7.15 Theorem. *Let $B = \{B_t, \mathscr{F}_t^B; t \geq 0\}$ $(\Omega, \mathscr{F}_\infty^B)$, $\{P^x\}_{x \in \mathbb{R}^d}$ be a d-dimensional Brownian family. Then $\{B_t, \tilde{\mathscr{F}}_t; t \geq 0\}$, $(\Omega, \tilde{\mathscr{F}}_\infty)$, $\{P^x\}_{x \in \mathbb{R}^d}$ is also a Brownian family.*

PROOF. It is easily verifed that, under each P^x, $\{B_t, \tilde{\mathscr{F}}_t; t \geq 0\}$ is a d-dimensional Brownian motion starting at x. It remains only to establish the universal measurability condition (i) of Definition 5.8. Fix $F \in \tilde{\mathscr{F}}_\infty$. For each probability measure μ on $(\mathbb{R}^d, \mathscr{B}(\mathbb{R}^d))$, we have $F \in \mathscr{F}^\mu$, so there is some $G \in \mathscr{F}_\infty^B$ with $F \triangle G \in \mathscr{N}^\mu$. Let $N \in \mathscr{F}_\infty^B$ satisfy $F \triangle G \subseteq N$ and $P^\mu(N) = 0$. The functions $g(x) \triangleq P^x(G)$ and $n(x) \triangleq P^x(N)$ are universally measurable by assumption. Furthermore,

$$\int_{\mathbb{R}^d} n(x)\mu(dx) = P^\mu(N) = 0,$$

so $n = 0$, μ-a.e. The nonnegative functions $h_1(x) \triangleq P^x(F\backslash G)$ and $h_2(x) \triangleq P^x(G\backslash F)$ are dominated by n, so h_1 and h_2 are zero μ-a.e., and hence h_1 and h_2 are measurable with respect to $\mathscr{B}(\mathbb{R}^d)^\mu$, the completion of $\mathscr{B}(\mathbb{R}^d)$ under μ. Set $f(x) \triangleq P^x(F)$. We have $f(x) = g(x) + h_1(x) - h_2(x)$, so f is also $\mathscr{B}(\mathbb{R}^d)^\mu$-measurable. This is true for every μ; thus, f is universally measurable. □

7.16 Remark. In Theorem 7.15, even if the mapping $x \mapsto P^x(F)$ is Borel-measurable for each $F \in \mathscr{F}_\infty^B$ (c.f. Problem 5.2), we can conclude only its universal measurability for each $F \in \tilde{\mathscr{F}}_\infty$. This explains why Definition 5.8 was designed with a condition of universal rather than Borel-measurability.

C. The Blumenthal Zero-One Law

We close this section with a useful consequence of the results concerning augmentation.

7.17 Theorem (Blumenthal (1957) Zero-One Law). *Let $\{B_t, \tilde{\mathscr{F}}_t; t \geq 0\}$, $(\Omega, \mathscr{F})$, $\{P^x\}_{x \in \mathbb{R}^d}$ be a d-dimensional Brownian family, where $\tilde{\mathscr{F}}_t$ is given by (7.3). If $F \in \tilde{\mathscr{F}}_0$, then for each $x \in \mathbb{R}^d$ we have either $P^x(F) = 0$ or $P^x(F) = 1$.*

PROOF. For $F \in \tilde{\mathscr{F}}_0$ and each $x \in \mathbb{R}^d$, there exists $G \in \mathscr{F}_0^B$ such that $P^x(F \,\Delta\, G) = 0$. But G must have the form $G = \{B_0 \in \Gamma\}$ for some $\Gamma \in \mathscr{B}(\mathbb{R}^d)$, so

$$P^x(F) = P^x(G) = P^x\{B_0 \in \Gamma\} = 1_\Gamma(x). \qquad \square$$

7.18 Problem. Show that, with probability one, a standard, one-dimensional Brownian motion changes sign infinitely many times in any time-interval $[0, \varepsilon]$, $\varepsilon > 0$.

7.19 Problem. Let $\{W_t, \mathscr{F}_t; 0 \leq t < \infty\}$ be a standard, one-dimensional Brownian motion on $(\Omega, \mathscr{F}, P)$, and define

$$S_b = \inf\{t \geq 0; W_t > b\}; \quad b \geq 0.$$

(i) Show that for each $b \geq 0$, $P[T_b \neq S_b] = 0$.
(ii) Show that if L is a finite, nonnegative random variable on $(\Omega, \mathscr{F}, P)$ which is independent of $\mathscr{F}_\infty^W$, then $\{T_L \neq S_L\} \in \mathscr{F}$ and $P[T_L \neq S_L] = 0$.

2.8. Computations Based on Passage Times

In order to motivate the strong Markov property in Section 2.6, we derived the density for the first passage time of a one-dimensional Brownian motion from the origin to $b \neq 0$. In this section we obtain a number of distributions related to this one, including the distribution of *reflected Brownian motion, Brownian motion on* $[0, a]$ *absorbed at the endpoints*, the *time and value of the maximum of Brownian motion* on a fixed time interval, and the *time of the last exit of Brownian motion* from the origin before a fixed time. Although derivations of all of these distributions can be based on the strong Markov property and the reflection principle, we shall occasionally provide arguments based on the optional sampling theorem for martingales. The former method yields densities, whereas the latter yields Laplace transforms of densities (moment generating functions). The reader should be acquainted with both methods.

A. Brownian Motion and Its Running Maximum

Throughout this section, $\{W_t, \mathscr{F}_t; 0 \le t < \infty\}$, $(\Omega, \mathscr{F})$, $\{P^x\}_{x \in \mathbb{R}}$ will be a one-dimensional Brownian family. We recall from (6.1) the *passage times*

$$T_b = \inf\{t \ge 0; W_t = b\}; \quad b \in \mathbb{R},$$

and define the *running maximum* (or maximum-to-date)

$$M_t = \max_{0 \le s \le t} W_s. \tag{8.1}$$

8.1 Proposition. *We have for* $t > 0$ *and* $a \le b$, $b \ge 0$:

$$P^0[W_t \in da, M_t \in db] = \frac{2(2b - a)}{\sqrt{2\pi t^3}} \exp\left\{-\frac{(2b - a)^2}{2t}\right\} da\, db. \tag{8.2}$$

PROOF. For $a \le b$, $b \ge 0$, the symmetry of Brownian motion implies that

$$\begin{aligned}(U_{t-s} 1_{(-\infty, a]})(b) &\triangleq P^b[W_{t-s} \le a] = P^b[W_{t-s} \ge 2b - a] \\ &\triangleq (U_{t-s} 1_{[2b-a, \infty)})(b); \quad 0 \le s \le t.\end{aligned}$$

Corollary 6.18 then yields

$$\begin{aligned}P^0[W_t \le a | \mathscr{F}_{T_b+}] &= (U_{t-T_b} 1_{(-\infty, a]})(b) = (U_{t-T_b} 1_{[2b-a, \infty)})(b) \\ &= P^0[W_t \ge 2b - a | \mathscr{F}_{T_b+}], \quad \text{a.s. } P^0 \text{ on } \{T_b \le t\}.\end{aligned}$$

Integrating both sides of this equation over $\{T_b \le t\}$ and noting that $\{T_b \le t\} = \{M_t \ge b\}$, we obtain

$$\begin{aligned}P^0[W_t \le a, M_t \ge b] &= P^0[W_t \ge 2b - a, M_t \ge b] \\ &= P^0[W_t \ge 2b - a] = \frac{1}{\sqrt{2\pi t}} \int_{2b-a}^{\infty} e^{-x^2/2t}\, dx.\end{aligned}$$

Differentiation leads to (8.2). □

8.2 Problem. Show that for $t > 0$, $b > 0$,

$$(8.3)\quad P^0[M_t \in db] = P^0[|W_t| \in db] = P^0[M_t - W_t \in db] = \sqrt{\frac{2}{\pi t}} e^{-b^2/2t} db,$$

$$(8.3)'\quad P^0\left[\max_{0 \le u \le t} |W_u| \ge b\right] \le 4P^0[W_t \ge b] \le \sqrt{\frac{t}{2\pi}} \frac{4}{b} e^{-b^2/2t}.$$

8.3 Remark. From (8.3) we see that

$$(8.4)\quad P^0[T_b \le t] = P^0[M_t \ge b] = \frac{2}{\sqrt{2\pi}} \int_{b/\sqrt{t}}^{\infty} e^{-x^2/2} dx; \quad b > 0.$$

By differentiation, we recover the passage time density (6.3):

$$(8.5)\quad P^0[T_b \in dt] = \frac{b}{\sqrt{2\pi t^3}} e^{-b^2/2t} dt; \quad b > 0, t > 0.$$

For future reference, we note that this density has Laplace transform

$$(8.6)\quad E^0 e^{-\alpha T_b} = e^{-b\sqrt{2\alpha}}; \quad b > 0, \alpha > 0.$$

By letting $t \uparrow \infty$ in (8.4) or $\alpha \downarrow 0$ in (8.6), we see that $P^0[T_b < \infty] = 1$. It is clear from (8.5), however, that $E^0 T_b = \infty$.

8.4 Exercise. Derive (8.6) (and consequently (8.5)) by applying the optional sampling theorem to the $\{\mathscr{F}_t\}$-martingale

$$(8.7)\quad X_t = \exp\{\lambda W_t - \tfrac{1}{2}\lambda^2 t\}; \quad 0 \le t < \infty,$$

with $\lambda = \sqrt{2\alpha} > 0$.

The following simple proposition will be extremely helpful in our study of local time in Section 6.2.

8.5 Proposition. *The process of passage times $T = \{T_a, \mathscr{F}_{T_a+}; 0 \le a < \infty\}$ has the property that, under P^0 and for $0 \le a < b$, the increment $T_b - T_a$ is independent of $\mathscr{F}_{T_a+}$ and has the density*

$$P^0[T_b - T_a \in dt] = \frac{b-a}{\sqrt{2\pi t^3}} e^{-(b-a)^2/2t} dt; \quad 0 < t < \infty.$$

In particular,

$$(8.8)\quad E^0[e^{-\alpha(T_b - T_a)} | \mathscr{F}_{T_a+}] = e^{-(b-a)\sqrt{2\alpha}}; \quad \alpha > 0.$$

PROOF. This is a direct consequence of Theorem 6.16 and the fact that $T_b - T_a = \inf\{t \ge 0; W_{T_a+t} - W_{T_a} = b - a\}$. □

B. Brownian Motion on a Half-Line

When Brownian motion is constrained to have state space $[0, \infty)$, one must specify what happens when the origin is reached. The following problems explore the simplest cases of *absorption* and (instantaneous) *reflection.*

8.6 Problem. Derive the transition density for *Brownian motion absorbed at the origin* $\{W_{t \wedge T_0}, \mathscr{F}_t; 0 \le t < \infty\}$, by verifying that

$$(8.9)\quad P^x[W_t \in dy, T_0 > t] = p_-(t; x, y)\,dy$$
$$\triangleq [p(t; x, y) - p(t; x, -y)]\,dy; \quad t > 0,\ x, y > 0.$$

8.7 Problem. Show that under P^0, *reflected Brownian motion* $|W| \triangleq \{|W_t|, \mathscr{F}_t; 0 \le t < \infty\}$ is a Markov process with transition density

$$(8.10)\quad P^0[|W_{t+s}| \in dy \,|\, |W_t| = x] = p_+(s; x, y)\,dy$$
$$\triangleq [p(s; x, y) + p(s; x, -y)]\,dy; \quad s > 0, t \ge 0 \text{ and } x, y \ge 0.$$

8.8 Problem. Define $Y_t \triangleq M_t - W_t; 0 \le t < \infty$. Show that under P^0, the process $Y = \{Y_t, \mathscr{F}_t; 0 \le t < \infty\}$ is Markov and has transition density

$$(8.11)\quad P^0[Y_{t+s} \in dy \,|\, Y_t = x] = p_+(s; x, y)\,dy; \quad s > 0, t \ge 0 \text{ and } x, y \ge 0.$$

Conclude that under P^0 *the processes* $|W|$ *and* Y *have the same finite-dimensional distributions.*

The surprising equivalence in law of the processes Y and $|W|$ was observed by P. Lévy (1948), who employed it in his deep study of Brownian local time (cf. Chapter 6). The third process M appearing in (8.3) cannot be equivalent in law to Y and $|W|$, since the paths of M are nondecreasing, whereas those of Y and $|W|$ are not. Nonetheless, M will turn out to be of considerable interest in Section 6.2, where we develop a number of deep properties of Brownian local time, using M as the object of study.

C. Brownian Motion on a Finite Interval

In this subsection we consider Brownian motion with state space $[0, a]$, where a is positive and finite. In order to study the case of reflection at both endpoints, consider the function $\varphi\colon \mathbb{R} \to [0, a]$ which satisfies $\varphi(2na) = 0$, $\varphi((2n+1)a) = a$; $n = 0, \pm 1, \pm 2, \ldots$, and is linear between these points.

8.9 Exercise. Show that the *doubly reflected Brownian motion* $\{\varphi(W_t), \mathscr{F}_t; 0 \le t < \infty\}$ satisfies

$$P^x[\varphi(W_t) \in dy] = \sum_{n=-\infty}^{\infty} p_+(t; x, y + 2na)\,dy; \quad 0 < y < a, 0 < x < a, t > 0.$$

The derivation of the transition density for *Brownian motion absorbed at* 0 *and* a i.e., $\{W_{t \wedge T_0 \wedge T_a}, \mathscr{F}_t; 0 \le t < \infty\}$, is the subject of the next proposition.

8.10 Proposition. *Choose* $0 < x < a$. *Then for* $t > 0, 0 < y < a$:

$$P^x[W_t \in dy, T_0 \wedge T_a > t] = \sum_{n=-\infty}^{\infty} p_-(t; x, y + 2na)\, dy. \tag{8.12}$$

PROOF. We follow Dynkin & Yushkevich (1969). Set $\sigma_0 \triangleq 0$, $\tau_0 \triangleq T_0$, and define recursively $\sigma_n \triangleq \inf\{t \geq \tau_{n-1}; W_t = a\}$, $\tau_n = \inf\{t \geq \sigma_n; W_t = 0\}$; $n = 1, 2, \ldots$. We know that $P^x[\tau_0 < \infty] = 1$, and using Theorem 6.16 we can show by induction on n that $\sigma_n - \tau_{n-1}$ is the passage time of the standard Brownian motion $W_{\cdot+\tau_{n-1}} - W_{\tau_{n-1}}$ to a, $\tau_n - \sigma_n$ is the passage time of the standard Brownian moton $W_{\cdot+\sigma_n} - W_{\sigma_n}$ to $-a$, and the sequence of differences $\sigma_1 - \tau_0$, $\tau_1 - \sigma_1$, $\sigma_2 - \tau_1$, $\tau_2 - \sigma_2, \ldots$ consists of independent and identically distributed random variables with moment generating function $e^{-a\sqrt{2\alpha}}$ (c.f. (8.8)). It follows that $\tau_n - \tau_0$, being the sum of $2n$ such differences, has moment generating function $e^{-2na\sqrt{2\alpha}}$, and so

$$P^x[\tau_n - \tau_0 \leq t] = P^0[T_{2na} \leq t].$$

We have then

$$\lim_{n\to\infty} P^x[\tau_n \leq t] = 0; \quad 0 \leq t < \infty. \tag{8.13}$$

For any $y \in (0, \infty)$, we have from Corollary 6.18 and the symmetry of Brownian motion that

$$P^x[W_t \geq y | \mathscr{F}_{\tau_n+}] = P^x[W_t \leq -y | \mathscr{F}_{\tau_n+}] \quad \text{on } \{\tau_n \leq t\},$$

and so for any integer $n \geq 0$,

$$P^x[W_t \geq y, \tau_n \leq t] = P^x[W_t \leq -y, \tau_n \leq t] = P^x[W_t \leq -y, \sigma_n \leq t]. \tag{8.14}$$

Similarly, for $y \in (-\infty, a)$, we have

$$P^x[W_t \leq y | \mathscr{F}_{\sigma_n+}] = P^x[W_t \geq 2a - y | \mathscr{F}_{\sigma_n+}] \quad \text{on } \{\sigma_n \leq t\},$$

whence

$$\begin{aligned} P^x[W_t \leq y, \sigma_n \leq t] &= P^x[W_t \geq 2a - y, \sigma_n \leq t] \\ &= P^x[W_t \geq 2a - y, \tau_{n-1} \leq t]; \quad n \geq 1. \end{aligned} \tag{8.15}$$

We may apply (8.14) and (8.15) alternately and repeatedly to conclude, for $0 < y < a$, $n \geq 0$:

$$P^x[W_t \geq y, \tau_n \leq t] = P^x[W_t \leq -y - 2na],$$

$$P^x[W_t \leq y, \sigma_n \leq t] = P^x[W_t \leq y - 2na],$$

and by differentiating with respect to y, we see that

$$P^x[W_t \in dy, \tau_n \leq t] = p(t; x, -y - 2na)\, dy, \tag{8.16}$$

$$P^x[W_t \in dy, \sigma_n \leq t] = p(t; x, y - 2na)\, dy. \tag{8.17}$$

Now set $\pi_0 = 0$, $\rho_0 = T_a$, and define recursively

$$\pi_n = \inf\{t \ge \rho_{n-1}; W_t = 0\}, \quad \rho_n = \inf\{t \ge \pi_n; W_t = a\}; \quad n = 1, 2, \ldots.$$

We may proceed as previously to obtain the formulas

$$\text{(8.18)} \quad \lim_{n\to\infty} P^x[\rho_n \le t] = 0; \quad 0 \le t < \infty,$$

$$\text{(8.19)} \quad P^x[W_t \in dy, \rho_n \le t] = p(t; x, -y + 2(n+1)a)\,dy; \quad 0 < y < a, n \ge 0,$$

$$\text{(8.20)} \quad P^x[W_t \in dy, \pi_n \le t] = p(t; x, y + 2na)\,dy; \quad 0 < y < a, n \ge 0.$$

It is easily verified by considering the cases $T_0 < T_a$ and $T_0 > T_a$ that $\tau_{n-1} \vee \rho_{n-1} = \sigma_n \wedge \pi_n$ and $\sigma_n \vee \pi_n = \tau_n \wedge \rho_n$; $n \ge 1$. Consequently,

$$\text{(8.21)} \quad \begin{aligned} P^x[W_t \in dy, \tau_{n-1} \wedge \rho_{n-1} \le t] &= P^x[W_t \in dy, \tau_{n-1} \le t] + P^x[W_t \in dy, \rho_{n-1} \le t] \\ &\quad - P^x[W_t \in dy, \sigma_n \wedge \pi_n \le t], \end{aligned}$$

and

$$\text{(8.22)} \quad \begin{aligned} P^x[W_t \in dy, \sigma_n \wedge \pi_n \le t] &= P^x[W_t \in dy, \sigma_n \le t] + P^x[W_t \in dy, \pi_n \le t] \\ &\quad - P^x[W_t \in dy, \tau_n \wedge \rho_n \le t]. \end{aligned}$$

Successive application of (8.21) and (8.22) yields for every integer $k \ge 1$:

(8.23)

$$\begin{aligned} P^x[W_t \in dy, \tau_0 \wedge \rho_0 \le t] &= \sum_{n=1}^{k} \{P^x[W_t \in dy, \tau_{n-1} \le t] + P^x[W_t \in dy, \rho_{n-1} \le t] \\ &\quad - P^x[W_t \in dy, \sigma_n \le t] - P^x[W_t \in dy, \pi_n \le t]\} \\ &\quad + P^x[W_t \in dy, \tau_k \wedge \rho_k \le t]. \end{aligned}$$

Now we let k tend to infinity in (8.23); because of (8.13), (8.18) the last term converges to zero, whereas using (8.16), (8.17) and (8.19), (8.20), we obtain from the remaining terms:

$$\begin{aligned} P^x[W_t \in dy, T_0 \wedge T_a > t] &= P^x[W_t \in dy] - P^x[W_t \in dy, \tau_0 \wedge \rho_0 \le t] \\ &= \sum_{n=-\infty}^{\infty} p_-(t; x, y + 2na)\,dy; \quad 0 < y < a, t > 0. \end{aligned}$$

□

8.11 Exercise. Show that for $t > 0$, $0 \le x \le a$:

$$\text{(8.24)} \quad \begin{aligned} P^x[T_0 \wedge T_a \in dt] = \frac{1}{\sqrt{2\pi t^3}} \sum_{n=-\infty}^{\infty} \Bigg[&(2na + x)\exp\left\{-\frac{(2na+x)^2}{2t}\right\} \\ &+ (2na + a - x)\exp\left\{-\frac{(2na+a-x)^2}{2t}\right\}\Bigg] dt. \end{aligned}$$

It is now tempting to guess the decomposition of (8.24):

(8.25)

$$P^x[T_0 \in dt, T_0 < T_a] = \frac{1}{\sqrt{2\pi t^3}} \sum_{n=-\infty}^{\infty} (2na + x)\exp\left\{-\frac{(2na+x)^2}{2t}\right\} dt,$$

(8.26)

$$P^x[T_a \in dt, T_a < T_0] = \frac{1}{\sqrt{2\pi t^3}} \sum_{n=-\infty}^{\infty} (2na + a - x)\exp\left\{-\frac{(2na + a - x)^2}{2t}\right\} dt.$$

Indeed, one can use the identity (8.6) to compute the Laplace transforms of the right-hand sides; then (8.25), (8.26) are seen to be equivalent to

(8.27) $$E^x[e^{-\alpha T_0} 1_{\{T_0 < T_a\}}] = \frac{\sinh((a - x)\sqrt{2\alpha})}{\sinh(a\sqrt{2\alpha})},$$

(8.28) $$E^x[e^{-\alpha T_a} 1_{\{T_a < T_0\}}] = \frac{\sinh(x\sqrt{2\alpha})}{\sinh(a\sqrt{2\alpha})}.$$

We leave the verification of these identities as a problem. Note that by adding (8.27) and (8.28) we obtain the transform of (8.24):

(8.29) $$E^x[e^{-\alpha(T_0 \wedge T_a)}] = \frac{\cosh\left(\left(x - \frac{a}{2}\right)\sqrt{2\alpha}\right)}{\cosh\left(\frac{a}{2}\sqrt{2\alpha}\right)}.$$

This provides an independent verification of (8.24).

8.12 Problem. Derive the formulas (8.27), (8.28) by applying the optional sampling theorem to the martingale of (8.7).

8.13 Exercise. Show that for $a > 0$, $0 \le x \le a$:

$$P^x[T_0 < T_a] = \frac{a - x}{a}, \quad P^x[T_a < T_0] = \frac{x}{a}.$$

8.14 Problem. Show that $E^x(T_0 \wedge T_a) = x(a - x)$; $\quad 0 \le x \le a$.

D. Distributions Involving Last Exit Times

Proposition 8.1 coupled with the Markov property enables one to compute distributions for a wide variety of Brownian functionals. We illustrate the method by computing some joint distributions involving the last time before t that the reflected Brownian motion Y of Problem 8.8 is at the origin. Note that such *last exit times* are *not* stopping times.

8.15 Proposition. *Define*

(8.30) $$\theta_t \triangleq \sup\{0 \le s \le t;\ W_s = M_t\}.$$

Then for $a \in \mathbb{R}$, $b \ge a^+$, $0 < s < t$, we have:

(8.31)

$$P^0[W_t \in da, M_t \in db, \theta_t \in ds] = \frac{b(b - a)}{\pi\sqrt{s^3(t - s)^3}} \exp\left[-\frac{b^2}{2s} - \frac{(b - a)^2}{2(t - s)}\right] da\, db\, ds.$$

PROOF. For $b \geq 0$, $\varepsilon > \delta > 0$, $x \geq 0$, $a \leq b$ and $0 < s < t$, we have

$$(8.32)\quad P^0\left[b < M_s \leq b + \delta, W_s \in b - dx, \max_{s \leq u \leq t} W_u \leq b, W_t \in da\right]$$
$$\leq P^0[b < M_t \leq b + \delta, \theta_t \leq s, W_s \in b - dx, W_t \in da]$$
$$\leq P^0\left[b < M_s \leq b + \delta, W_s \in b - dx, \max_{s \leq u \leq t} W_u \leq b + \varepsilon, W_t \in da\right].$$

Divide by δ and let $\delta \downarrow 0$, $\varepsilon \downarrow 0$ (in that order). The upper and lower bounds in the preceding inequalities converge to the same limit, which is

$$(8.33)\quad P^0[M_t \in db, \theta_t \leq s, W_s \in b - dx, W_t \in da]$$
$$= P^0\left[M_s \in db, W_s \in b - dx, \max_{s \leq u \leq t} W_u \leq b, W_t \in da\right]$$
$$= P^0[M_s \in db, W_s \in b - dx] \cdot P^{b-x}[M_{t-s} \leq b, W_{t-s} \in da]$$
$$= \frac{b + x}{\pi\sqrt{s^3(t-s)}}\left[\exp\left\{-\frac{(x + \mu_+)^2}{2\sigma^2} - \frac{(2b - a)^2}{2t}\right\}\right.$$
$$\left. - \exp\left\{-\frac{(x + \mu_-)^2}{2\sigma^2} - \frac{a^2}{2t}\right\}\right] dx\, da\, db,$$

where we have used (8.2) and

$$\mu_\pm \triangleq \frac{b(t - s) \pm (a - b)s}{t}, \quad \sigma^2 \triangleq \frac{s(t - s)}{t}.$$

In terms of $\Phi(z) \triangleq (1/\sqrt{2\pi})\int_{-\infty}^z e^{-x^2/2}\,dx$ we may now evaluate the integrals

$$\int_0^\infty (b + x)\exp\left\{-\frac{(x + \mu_\pm)^2}{2\sigma^2}\right\}dx = \sigma^2 e^{-\mu_\pm^2/2\sigma^2}$$
$$+ \frac{s}{t}(b \pm (b - a))\sigma\sqrt{2\pi} \cdot \Phi\left(-\frac{\mu_\pm}{\sigma}\right),$$

and so integrating out x in (8.33) and using the equality

$$(8.34)\quad \frac{\mu_\pm^2}{2\sigma^2} + \frac{(b \pm (b - a))^2}{2t} = \frac{b^2}{2s} + \frac{(b - a)^2}{2(t - s)},$$

we arrive at the formula

$$P^0[M_t \in db, \theta_t \leq s, W_t \in da] = \frac{2}{\sqrt{2\pi t^3}}\left[\Phi\left(-\frac{\mu_+}{\sigma}\right)(2b - a)\exp\left\{-\frac{(2b - a)^2}{2t}\right\}\right.$$
$$\left. - \Phi\left(-\frac{\mu_-}{\sigma}\right)a\exp\left\{-\frac{a^2}{2t}\right\}\right] da\, db.$$

Note that $\partial/\partial s(-\mu_{\pm}/\sigma) = 1/2\sigma((b/s) \pm (b-a)/(t-s))$, and so

$$\frac{\partial}{\partial s} P^0[M_t \in db, \theta_t \le s, W_t \in da]$$

$$= \frac{b(b-a)}{\pi\sqrt{s^3(t-s)^3}} \exp\left\{-\frac{b^2}{2s} - \frac{(b-a)^2}{2(t-s)}\right\} da\, db\, ds. \qquad \square$$

8.16 Remark. If we define $\hat{\theta}_t \triangleq \inf\{0 \le s \le t; W_s = M_t\}$ to be the first time W attains its maximum over $[0, t]$, then (8.32) is still valid when θ_t is replaced by $\hat{\theta}_t$. Thus, θ_t and $\hat{\theta}_t$ have the same distribution, and since $\hat{\theta}_t \le \theta_t$, we see that $P^0[\hat{\theta}_t = \theta_t] = 1$. In other words, the time at which the maximum over $[0, t]$ is attained is almost surely unique.

8.17 Problem. Show that for $b \ge 0$, $0 < s < t$:

$$P^0[M_t \in db, \theta_t \in ds] = \frac{b}{\pi\sqrt{s^3(t-s)}} e^{-b^2/2s}\, db\, ds,$$

whence

$$P^0[\theta_t \in ds] = \frac{ds}{\pi\sqrt{s(t-s)}}, \qquad P^0[M_t \in db \mid \theta_t = s] = \frac{b}{s} e^{-b^2/2s}\, db.$$

In particular, the conditional density of M_t given θ_t does not depend on t. We say that θ_t obeys the *arc-sine law*, since

$$P^0[\theta_t \le s] = \frac{2}{\pi} \arcsin\sqrt{\frac{s}{t}}; \quad 0 \le s \le t, t > 0.$$

8.18 Problem. Define the time of last exit from the origin before t by

$$\gamma_t \triangleq \sup\{0 \le s \le t; W_s = 0\}. \tag{8.35}$$

Show that γ_t obeys the arc-sine law; i.e.,

$$P^0[\gamma_t \in ds] = \frac{ds}{\pi\sqrt{s(t-s)}}; \quad 0 < s < t.$$

(*Hint*: Use Problem 8.8.)

8.19 Exercise. With γ_t defined as in (8.35), derive the quadrivariate density

$$P^0[W_t \in da, M_t \in db, \gamma_t \in ds, \theta_t \in du]$$

$$= \frac{-2ab^2}{(2\pi u(s-u)(t-s))^{3/2}} \exp\left\{-\frac{sb^2}{2u(s-u)} - \frac{a^2}{2(t-s)}\right\} da\, db\, ds\, du;$$

$$0 < u < s < t,\ a < 0 < b.$$

8.20 Exercise (Seshadri, 1988): Deduce from (8.2) that, for fixed $t > 0$, the random variables $M_t(M_t - W_t)$ and W_t are independent, and that $(2/t)M_t(M_t - W_t)$ has a standard exponential distribution: $P^0[M_t(M_t - W_t) \geq tx/2] = e^{-x}$, $x \geq 0$.

2.9. The Brownian Sample Paths

We present in this section a detailed discussion of the basic *absolute properties* of Brownian motion, i.e., those properties which hold with probability one (also called *sample path properties*). These include characterizations of "bad" behavior (*nondifferentiability* and *lack of points of increase*) as well as "good" behavior (*law of the iterated logarithm* and *Lévy modulus of continuity*) of the Brownian paths. We also study the local maxima and the zero sets of these paths. We shall see in Section 3.4 that the sample paths of any continuous martingale can be obtained by running those of a Brownian motion according to a different, path-dependent clock. Thus, this study of Brownian motion has much to say about the sample path properties of much more general classes of processes, including continuous martingales and diffusions.

A. Elementary Properties

We start by collecting together, in Lemma 9.4, the fundamental *equivalence transformations* of Brownian motion. These will prove handy, both in this section and throughout the book; indeed, we made frequent use of symmetry in the previous section.

9.1 Definition. An $\mathbb{R}^d$-valued stochastic process $X = \{X_t; 0 \leq t < \infty\}$ is called *Gaussian* if, for any integer $k \geq 1$ and real numbers $0 \leq t_1 < t_2 < \cdots < t_k < \infty$, the random vector $(X_{t_1}, X_{t_2}, \ldots, X_{t_k})$ has a joint normal distribution. If the distribution of $(X_{t+t_1}, X_{t+t_2}, \ldots, X_{t+t_k})$ does not depend on t, we say that the process is *stationary*.

The finite-dimensional distributions of a Gaussian process X are determined by its expectation vector $m(t) \triangleq EX_t$; $t \geq 0$, and its covariance matrix

$$\rho(s, t) \triangleq E[(X_s - m(s))(X_t - m(t))^T]; \quad s, t \geq 0,$$

where the superscript T indicates transposition. If $m(t) \equiv 0$; $t \geq 0$, we say that X is a *zero-mean Gaussian process*.

9.2 Remark. One-dimensional Brownian motion is a zero-mean Gaussian process with covariance function

$$\rho(s, t) = s \wedge t; \quad s, t \geq 0. \tag{9.1}$$

Conversely, any zero-mean Gaussian process $X = \{X_t, \mathscr{F}_t^X; 0 \le t < \infty\}$ with a.s. continuous paths and covariance function given by (9.1) is a one-dimensional Brownian motion. See Definition 1.1.

Throughout this section, $W = \{W_t, \mathscr{F}_t; 0 \le t < \infty\}$ is a standard, one-dimensional Brownian motion on $(\Omega, \mathscr{F}, P)$. In particular $W_0 = 0$, a.s. P. For fixed $\omega \in \Omega$, we denote by $W_.(\omega)$ the sample path $t \mapsto W_t(\omega)$.

9.3 Problem (Strong Law of Large Numbers). Show that

$$\lim_{t\to\infty} \frac{W_t}{t} = 0, \quad \text{a.s.} \tag{9.2}$$

(*Hint*: Recall the analogous property for the Poisson process, Remark 1.3.10.)

9.4 Lemma. *When $W = \{W_t, \mathscr{F}_t; 0 \le t < \infty\}$ is a standard Brownian motion, so are the processes obtained from the following "equivalence transformations":*

(i) Scaling: $X = \{X_t, \mathscr{F}_{ct}; 0 \le t < \infty\}$ *defined for $c > 0$ by*

$$X_t = \frac{1}{\sqrt{c}} W_{ct} \quad ; \quad 0 \le t < \infty. \tag{9.3}$$

(ii) Time-inversion: $Y = \{Y_t, \mathscr{F}_t^Y; 0 \le t < \infty\}$ *defined by*

$$Y_t = \begin{cases} tW_{1/t} & ; \quad 0 < t < \infty, \\ 0 & ; \quad t = 0. \end{cases} \tag{9.4}$$

(iii) Time-reversal: $Z = \{Z_t, \mathscr{F}_t^Z; 0 \le t < \infty\}$ *defined for $T > 0$ by*

$$Z_t = W_T - W_{T-t} \quad ; \quad 0 \le t \le T. \tag{9.5}$$

(iv) Symmetry: $-W = \{-W_t, \mathscr{F}_t; 0 \le t < \infty\}$.

PROOF. We shall discuss only part (ii), the others being either similar or completely evident. The process Y of (9.4) is easily seen to have a.s. continuous paths; continuity at the origin is a corollary of Problem 9.3. On the other hand, Y is a zero-mean Gaussian process with covariance function

$$E(Y_s Y_t) = st\left(\frac{1}{s} \wedge \frac{1}{t}\right) = s \wedge t; \quad s, t > 0$$

and the conclusion follows from Remark 9.2. □

9.5 Problem. Show that the probability that Brownian motion returns to the origin infinitely often is one.

B. The Zero Set and the Quadratic Variation

We take up now the study of the *zero set of the Brownian path*. Define

$$\mathscr{Z} \triangleq \{(t, \omega) \in [0, \infty) \times \Omega; W_t(\omega) = 0\}, \tag{9.6}$$

and for fixed $\omega \in \Omega$, define the zero set of $W_\cdot(\omega)$:

$$\mathscr{Z}_\omega \triangleq \{0 \le t < \infty; W_t(\omega) = 0\}. \tag{9.7}$$

9.6 Theorem. *For P-a.e.* $\omega \in \Omega$, *the zero set* $\mathscr{Z}_\omega$

(i) *has Lebesgue measure zero,*
(ii) *is closed and unbounded,*
(iii) *has an accumulation point at* $t = 0$,
(iv) *has no isolated point in* $(0, \infty)$, *and therefore*
(v) *is dense in itself.*

PROOF. We start by observing that the set $\mathscr{Z}$ of (9.6) is in $\mathscr{B}([0,\infty)) \otimes \mathscr{F}$, because W is a (progressively) measurable process. By Fubini's theorem,

$$E[\text{meas}(\mathscr{Z}_\omega)] = (\text{meas} \times P)(\mathscr{Z}) = \int_0^\infty P[W_t = 0]\,dt = 0,$$

and therefore $\text{meas}(\mathscr{Z}_\omega) = 0$ for P-a.e. $\omega \in \Omega$, proving (i); here and in the sequel, *meas* means "Lebesgue measure." On the other hand, for P-a.e. $\omega \in \Omega$ the mapping $t \mapsto W_t(\omega)$ is continuous, and $\mathscr{Z}_\omega$ is the inverse image under this mapping of the closed set $\{0\}$. Thus, for every such ω, the set $\mathscr{Z}_\omega$ is closed, is unbounded (Problem 9.5), and has an accumulation point at $t = 0$ (Problem 7.18).

For (iv), let us observe that $\{\omega \in \Omega; \mathscr{Z}_\omega \text{ has an isolated point in } (0,\infty)\}$ can be written as

$$\bigcup_{\substack{a,b \in Q \\ 0 \le a < b < \infty}} \{\omega \in \Omega; \text{there is exactly one } s \in (a,b) \text{ with } W_s(\omega) = 0\} \tag{9.8}$$

where Q is the set of rationals. Let us consider the family of almost surely finite optional times

$$\beta_t \triangleq \inf\{s > t; W_s = 0\}; \quad t \ge 0.$$

According to (iii) we have $\beta_0 = 0$, a.s. P; moreover,

$$\begin{aligned} \beta_{\beta_t(\omega)}(\omega) &= \inf\{s > \beta_t(\omega); W_s(\omega) = 0\} \\ &= \beta_t(\omega) + \inf\{s > 0; W_{s+\beta_t(\omega)}(\omega) - W_{\beta_t(\omega)}(\omega) = 0\} = \beta_t(\omega) \end{aligned}$$

for P-a.e. $\omega \in \Omega$, because $\{W_{s+\beta_t} - W_{\beta_t}; 0 \le s < \infty\}$ is a standard Brownian motion (Theorem 6.16). Therefore, for $0 \le a < b < \infty$,

$$\begin{aligned} &\{\omega \in \Omega; \text{there is exactly one } s \in (a,b) \text{ with } W_s(\omega) = 0\} \\ &\subseteq \{\omega \in \Omega; \beta_a(\omega) < b \text{ and } \beta_{\beta_a(\omega)}(\omega) > b\} \end{aligned}$$

has probability zero, and the same is then true for the union (9.8). □

9.7 Remark. From Theorem 9.6 and the strong Markov property in the form of Theorem 6.16, we see that for every fixed $b \in \mathbb{R}$ and P-a.e. $\omega \in \Omega$, the *level set*

$$\mathscr{Z}_\omega(b) \triangleq \{0 \le t < \infty; W_t(\omega) = b\}$$

is closed, unbounded, of Lebesgue measure zero, and dense in itself.

The following problem strengthens the result of Theorem 1.5.8 in the special case of Brownian motion.

9.8 Problem. Let $\{\Pi_n\}_{n=1}^\infty$ be a sequence of partitions of the interval $[0, t]$ with $\lim_{n\to\infty} \|\Pi_n\| = 0$. Then the quadratic variations

$$V_t^{(2)}(\Pi_n) \triangleq \sum_{k=1}^{m_n} |W_{t_k^{(n)}} - W_{t_{k-1}^{(n)}}|^2$$

of the Brownian motion W over these partitions converge to t in L^2, as $n \to \infty$. If, furthermore, the partitions become so fine that $\sum_{n=1}^\infty \|\Pi_n\| < \infty$ holds, the preceding convergence takes place also with probability one.

C. Local Maxima and Points of Increase

As discussed in Section 1.5, one can easily show by using Problem 9.8 that for almost every $\omega \in \Omega$, *the sample path $W_\cdot(\omega)$ is of unbounded variation on every finite interval* $[0, t]$. In the remainder of this section we describe just how oscillatory the Brownian path is.

9.9 Theorem. *For almost every $\omega \in \Omega$, the sample path $W_\cdot(\omega)$ is monotone in no interval.*

PROOF. If we denote by F the set of $\omega \in \Omega$ with the property that $W_\cdot(\omega)$ is monotone in some interval, we have

$$F = \bigcup_{\substack{s,t \in Q \\ 0 \le s < t < \infty}} \{\omega \in \Omega; W_\cdot(\omega) \text{ is monotone on } [s, t]\},$$

since every nonempty interval includes one with rational endpoints. Therefore, it suffices to show that on any such interval, say on $[0, 1]$, the path $W_\cdot(\omega)$ is monotone for almost no ω. By virtue of the symmetry property (iv) of Lemma 9.4, it suffices then to show that the event

$$A \triangleq \{\omega \in \Omega; W_\cdot(\omega) \text{ is nondecreasing on } [0, 1]\}$$

is in $\mathscr{F}$ and has probability zero. But $A = \bigcap_{n=1}^\infty A_n$, where

$$A_n \triangleq \bigcap_{i=0}^{n-1} \{\omega \in \Omega; W_{(i+1)/n}(\omega) - W_{i/n}(\omega) \ge 0\} \in \mathscr{F}$$

has probability $P(A_n) = \prod_{i=0}^{n-1} P[W_{(i+1)/n} - W_{i/n} \ge 0] = 2^{-n}$. Thus, $P(A) \le \lim_{n\to\infty} P(A_n) = 0$. □

In order to proceed with our study of the Brownian sample paths, we need a few elementary notions and results concerning real-valued functions of one variable.

9.10 Definition. Let $f: [0, \infty) \to \mathbb{R}$ be a given function. A number $t \geq 0$ is called

(i) *a point of increase of size* δ, if for given $\delta > 0$ we have $f(s) \leq f(t) \leq f(u)$ for every $s \in [(t-\delta)^+, t)$ and $u \in (t, t+\delta]$; *a point of strict increase of size* δ, if the preceding inequalities are strict;

(ii) *a point of increase*, if it is a point of increase of size δ for some $\delta > 0$; *a point of strict increase*, if it is a point of strict increase of size δ for some $\delta > 0$;

(iii) *a point of local maximum*, if there exists a number $\delta > 0$ with $f(s) \leq f(t)$ valid for every $s \in [(t-\delta)^+, t+\delta]$; and *a point of strict local maximum*, if there exists a number $\delta > 0$ with $f(s) < f(t)$ valid for every $s \in [(t-\delta)^+, t+\delta] \backslash \{t\}$.

9.11 Problem. Let $f: [0, \infty) \to \mathbb{R}$ be continuous.

(i) Show that the set of points of strict local maximum for f is countable.
(ii) If f is monotone on no interval, then the set of points of local maximum for f is dense in $[0, \infty)$.

9.12 Theorem. *For almost every* $\omega \in \Omega$, *the set of points of local maximum for the Brownian path* $W_.(\omega)$ *is countable and dense in* $[0, \infty)$, *and all local maxima are strict.*

Proof. Thanks to Theorem 9.9 and Problem 9.11, it suffices to show that the set

$$A = \{\omega \in \Omega; \text{ every local maximum of } W_.(\omega) \text{ is strict}\}$$

includes an event of probability one. Indeed, A includes the (countable) intersection of events of the type

$$A_{t_1,\ldots,t_4} \triangleq \left\{\omega \in \Omega; \max_{t_3 \leq t \leq t_4} W_t(\omega) - \max_{t_1 \leq t \leq t_2} W_t(\omega) \neq 0\right\}, \tag{9.9}$$

taken over all quadruples (t_1, t_2, t_3, t_4) of rational numbers satisfying $0 \leq t_1 < t_2 < t_3 < t_4 < \infty$. Therefore, it remains to prove that for every such quadruple, the event in (9.9) has probability one. But the difference of the two random variables in (9.9) can be written as

$$(W_{t_3} - W_{t_2}) + \min_{t_1 \leq t \leq t_2} [W_{t_2}(\omega) - W_t(\omega)] + \max_{t_3 \leq t \leq t_4} [W_t(\omega) - W_{t_3}(\omega)],$$

and the three terms appearing in this sum are independent. Consequently,

$$P[A_{t_1,\ldots,t_4}] = \int_0^\infty \int_{-\infty}^0 P[W_{t_2} - W_{t_3} \neq x + y] \; P\left[\min_{t_1 \le t \le t_2} (W_{t_2} - W_t) \in dx\right]$$
$$\cdot P\left[\max_{t_3 \le t \le t_4} (W_t - W_{t_3}) \in dy\right] = 1$$

because $P[W_{t_2} - W_{t_3} \neq x + y] = 1$. □

Let us now discuss the question of occurrence of points of increase on the Brownian path. We start by observing that the set

$$\Lambda = \{(t,\omega) \in [0,\infty) \times \Omega;\ t \text{ is a point of increase of } W_.(\omega)\}$$

is product-measurable: $\Lambda \in \mathscr{B}([0,\infty)) \otimes \mathscr{F}$. Indeed, Λ can be written as the countable union $\Lambda = \bigcup_{m=1}^\infty \Lambda(m)$, with

$$\Lambda(m) \triangleq \left\{(t,\omega) \in [0,\infty) \times \Omega;\quad \max_{(t-(1/m))^+ \le s \le t} W_s(\omega) = W_t(\omega) = \min_{t \le s \le t+(1/m)} W_s(\omega)\right\},$$

and each $\Lambda(m)$ is in $\mathscr{B}([0,\infty)) \otimes \mathscr{F}$. We denote the sections of Λ by

$$\Lambda_t \triangleq \{\omega \in \Omega; (t,\omega) \in \Lambda\}, \quad \Lambda_\omega \triangleq \{t \in [0,\infty); (t,\omega) \in \Lambda\},$$

and $\Lambda_t(m)$, $\Lambda_\omega(m)$ have a similar meaning. For $0 \le t < \infty$,

$$P[\Lambda_t(m)] \le P[W_{s+t} - W_t \ge 0;\quad \forall s \in [0, 1/m]] = 0$$

because $\{W_{s+t} - W_t; s \ge 0\}$ is a standard Brownian motion (Problem 7.18); now $\Lambda_t = \bigcup_{m=1}^\infty \Lambda_t(m)$ gives also

(9.10) $$P(\Lambda_t) = 0; \quad 0 \le t < \infty$$

as well as

$$\int_\Omega \text{meas}(\Lambda_\omega)\, dP = (\text{meas} \times P)(\Lambda) = \int_0^\infty P(\Lambda_t)\, dt = 0$$

from Fubini's theorem. It follows that $P[\omega \in \Omega; \text{meas}(\Lambda_\omega) = 0] = 1$. The question is whether this assertion can be strengthened to $P[\omega \in \Omega; \Lambda_\omega = \varnothing] = 1$, or equivalently

(9.11) $$P[\omega \in \Omega; \text{the path } W_.(\omega) \text{ has no point of increase}] = 1.$$

That the answer to this question turns out to be *affirmative* is perhaps one of the most surprising aspects of Brownian sample path behavior. We state this result here but defer the proof to Chapter 6.

9.13 Theorem (Dvoretzky, Erdös, & Kakutani (1961)). *Almost every Brownian sample path has no point of increase (or decrease); that is,* (9.11) *holds.*

9.14 Remark. We have already seen that almost every Brownian path has a dense set of local maxima. If $T(\omega)$ is a local maximum for $W_.(\omega)$, then one might imagine that by reflection (replacing $W_t(\omega) - W_{T(\omega)}(\omega)$ by $-(W_t(\omega) - W_{T(\omega)}(\omega))$ for $t \geq T(\omega)$), one could turn the point $T(\omega)$ into a point of increase for a new Brownian motion. Such an approach was used successfully at the beginning of Section 6 to derive the passage time distribution. Here, however, it fails completely. Of course, the results of Section 6 are inappropriate in this context because $T(\omega)$ is *not* a stopping time. Even if the filtration $\{\mathscr{F}_t\}$ is right-continuous, so that $\{\omega \in \Omega; W_\cdot(\omega)$ has a local maximum at $t\}$ is in $\mathscr{F}_t$ for each $t \geq 0$, it is not possible to define a stopping time T for $\{\mathscr{F}_t\}$ such that $W_\cdot(\omega)$ has a local maximum at $T(\omega)$ for all ω in some event of positive probability. In other words, one cannot specify in a "proper way" which of the numerous times of local maximum is to be selected. Indeed, if it were possible to do this, Theorem 9.13 would be violated.

9.15 Remark. It is quite possible that, for each fixed $t \geq 0$, a certain property holds almost surely, but then it fails to hold for all $t \geq 0$ simultaneously on an event whose probability is one (or even positive!). As an extreme and rather trivial example, consider that $P[\omega \in \Omega; W_t(\omega) \neq 1] = 1$ holds for every $0 \leq t < \infty$, but $P[\omega \in \Omega; W_t(\omega) \neq 1$, for every $t \in [0, \infty)] = 0$. The point here is that in order to pass from the consideration of fixed but arbitrary t to the consideration of all t simultaneously, it is usually necessary to reduce the latter consideration to that of a *countable* number of coordinates. This is precisely the problem which must be overcome in the passage from (9.10) to (9.11), and the proof of Theorem 9.13 in Dvoretzky, Erdös & Kakutani (1961) is demanding† because of the difficulty of reducing the property of "being a point of increase" for all $t \geq 0$ to a description involving only countably many coordinates. We choose to give a completely different proof of Theorem 9.13 in Subsection 6.4.B, based on the concept of local time. We do, however, illustrate the technique mentioned previously by taking up a less demanding question, the nondifferentiability of the Brownian path.

D. Nowhere Differentiability

9.16 Definition. For a continuous function $f: [0, \infty) \to \mathbb{R}$, we denote by

(9.12) $$D^{\pm} f(t) = \overline{\lim_{h \to 0\pm}} \frac{f(t+h) - f(t)}{h}$$

the *upper* (*right and left*) *Dini derivates* at t, and by

(9.13) $$D_{\pm} f(t) = \underline{\lim_{h \to 0\pm}} \frac{f(t+h) - f(t)}{h}$$

the *lower* (*right and left*) *Dini derivates* at t. The function f is said to be

† See, however, Adelman (1985) for a simpler argument.

differentiable at t from the right (respectively, *the left*), if $D^+f(t)$ and $D_+f(t)$ (respectively, $D^-f(t)$ and $D_-f(t)$) are finite numbers and equal. The function f is said to be *differentiable at* $t > 0$ if it is differentiable from both the right and the left and the four Dini derivates agree. At $t = 0$, differentiability is defined as differentiability from the right.

9.17 Exercise. Show that for fixed $t \in [0, \infty)$,

$$P[\omega \in \Omega; D^+W_t(\omega) = \infty \text{ and } D_+W_t(\omega) = -\infty] = 1. \tag{9.14}$$

9.18 Theorem (Paley, Wiener & Zygmund (1933)). *For almost every $\omega \in \Omega$, the Brownian sample path $W_.(\omega)$ is nowhere differentiable. More precisely, the set*

$$\{\omega \in \Omega; \textit{for each } t \in [0, \infty), \textit{ either } D^+W_t(\omega) = \infty \textit{ or } D_+W_t(\omega) = -\infty\} \tag{9.15}$$

contains an event $F \in \mathscr{F}$ with $P(F) = 1$.

9.19 Remark. At every point t of local maximum for $W_.(\omega)$ we have $D^+W_t(\omega) \le 0$, and at every point s of local minimum, $D_+W_s(\omega) \ge 0$. Thus, the "or" in (9.15) cannot be replaced by "and." We do not know whether the set of (9.15) belongs to $\mathscr{F}_\infty^W$.

PROOF. It is enough to consider the interval $[0, 1]$. For fixed integers $j \ge 1$, $k \ge 1$, we define the set

$$A_{jk} = \bigcup_{t \in [0,1]} \bigcap_{h \in [0, 1/k]} \{\omega \in \Omega; |W_{t+h}(\omega) - W_t(\omega)| \le jh\}. \tag{9.16}$$

Certainly we have

$$\{\omega \in \Omega; -\infty < D_+W_t(\omega) \le D^+W_t(\omega) < \infty, \text{for some } t \in [0, 1]\} = \bigcup_{j=1}^{\infty} \bigcup_{k=1}^{\infty} A_{jk},$$

and the proof of the theorem will be complete if we find, for each fixed j, k, an event $C \in \mathscr{F}$ with $P(C) = 0$ and $A_{jk} \subseteq C$.

Let us fix a sample path $\omega \in A_{jk}$, i.e., suppose there exists a number $t \in [0, 1]$ with $|W_{t+h}(\omega) - W_t(\omega)| \le jh$ for every $0 \le h \le 1/k$. Take an integer $n \ge 4k$. Then there exists an integer i, $1 \le i \le n$, such that $(i - 1)/n \le t \le i/n$, and it is easily verified that we also have $((i + v)/n)) - t \le (v + 1)/n \le 1/k$, for $v = 1, 2, 3$. It follows that

$$|W_{(i+1)/n}(\omega) - W_{i/n}(\omega)| \le |W_{(i+1)/n}(\omega) - W_t(\omega)| + |W_{i/n}(\omega) - W_t(\omega)|$$
$$\le \frac{2j}{n} + \frac{j}{n} = \frac{3j}{n}.$$

The crucial observation here is that the assumption $\omega \in A_{jk}$ provides information about the size of the Brownian increment, not only over the interval $[i/n, (i + 1)/n]$, but also over the neighboring intervals $[(i + 1)/n, (i + 2)/n]$ and

$[(i+2)/n, (i+3)/n]$. Indeed,

$$|W_{(i+2)/n}(\omega) - W_{(i+1)/n}(\omega)| \le |W_{(i+2)/n} - W_t| + |W_{(i+1)/n} - W_t| \le \frac{3j}{n} + \frac{2j}{n} = \frac{5j}{n},$$

$$|W_{(i+3)/n}(\omega) - W_{(i+2)/n}(\omega)| \le |W_{(i+3)/n} - W_t| + |W_{(i+2)/n} - W_t| \le \frac{4j}{n} + \frac{3j}{n} = \frac{7j}{n}.$$

Therefore, with

$$C_i^{(n)} \triangleq \bigcap_{v=1}^{3} \left\{ \omega \in \Omega; |W_{(i+v)/n}(\omega) - W_{(i+v-1)/n}(\omega)| \le \frac{2v+1}{n} j \right\},$$

we have observed that $A_{jk} \subseteq \bigcup_{i=1}^{n} C_i^{(n)}$ holds for every $n \ge 4k$. But now

$$\sqrt{n}(W_{(i+v)/n} - W_{(i+v-1)/n}) \triangleq Z_v; \quad v = 1, 2, 3$$

are independent, standard normal random variables, and one can easily verify the bound $P[|Z_v| \le \varepsilon] \le \varepsilon$. It develops that

(9.17) $$P(C_i^{(n)}) \le \frac{105j^3}{n^{3/2}}; \quad i = 1, 2, \ldots, n.$$

We have $A_{jk} \subseteq C$ upon taking

(9.18) $$C \triangleq \bigcap_{n=4k}^{\infty} \bigcup_{i=1}^{n} C_i^{(n)} \in \mathscr{F},$$

and (9.17) shows us that $P(C) \le \inf_{n \ge 4k} P(\bigcup_{i=1}^{n} C_i^{(n)}) = 0$. □

9.20 Remark. An alternative approach to Theorem 9.18, based on local time, is indicated in Exercise 3.6.6.

9.21 Exercise. By modifying the preceding proof, establish the following stronger result: *for almost every $\omega \in \Omega$, the Brownian path $W_.(\omega)$ is nowhere Hölder-continuous with exponent $\gamma > \frac{1}{2}$.* (*Hint*: By analogy with (9.16), consider the sets

(9.19)
$$A_{jk} \triangleq \{\omega \in \Omega; |W_{t+h}(\omega) - W_t(\omega)| \le jh^\gamma \text{ for some } t \in [0,1] \text{ and all } h \in [0, 1/k]\}$$

and show that each A_{jk} is included in a P-null event.)

E. Law of the Iterated Logarithm

Our next result is the celebrated *law of the iterated logarithm*, which describes the oscillations of Brownian motion near $t = 0$ and as $t \to \infty$. In preparation for the theorem, we recall the following upper and lower bounds on the tail of the normal distribution.

9.22 Problem. For every $x > 0$, we have

$$\frac{x}{1+x^2}e^{-x^2/2} \le \int_x^\infty e^{-u^2/2}\,du \le \frac{1}{x}e^{-x^2/2}. \tag{9.20}$$

9.23 Theorem. (Law of the Iterated Logarithm (A. Hinčin (1933))). *For almost every $\omega \in \Omega$, we have*

$$\text{(i)}\ \varlimsup_{t\downarrow 0} \frac{W_t(\omega)}{\sqrt{2t\log\log(1/t)}} = 1, \qquad \text{(ii)}\ \varliminf_{t\downarrow 0} \frac{W_t(\omega)}{\sqrt{2t\log\log(1/t)}} = -1,$$

$$\text{(iii)}\ \varlimsup_{t\to\infty} \frac{W_t(\omega)}{\sqrt{2t\log\log t}} = 1, \qquad \text{(iv)}\ \varliminf_{t\to\infty} \frac{W_t(\omega)}{\sqrt{2t\log\log t}} = -1.$$

9.24 Remark. By symmetry, property (ii) follows from (i), and by time-inversion, properties (iii) and (iv) follow from (i) and (ii), respectively (cf. Lemma 9.4). Thus it suffices to establish (i).

PROOF. The submartingale inequality (Theorem 1.3.8 (i)) applied to the exponential martingale $\{X_t, \mathscr{F}_t; 0 \le t < \infty\}$ of (8.7) gives for $\lambda > 0$, $\beta > 0$:

$$P\left[\max_{0\le s\le t}\left(W_s - \frac{\lambda}{2}s\right) \ge \beta\right] = P\left[\max_{0\le s\le t} X_s \ge e^{\lambda\beta}\right] \le e^{-\lambda\beta}. \tag{9.21}$$

With the notation $h(t) \triangleq \sqrt{2t\log\log(1/t)}$ and fixed numbers θ, δ in $(0, 1)$, we choose $\lambda = (1+\delta)\theta^{-n}h(\theta^n)$, $\beta = \frac{1}{2}h(\theta^n)$, and $t = \theta^n$ in (9.21), which becomes

$$P\left[\max_{0\le s\le \theta^n}\left(W_s - \frac{\lambda}{2}s\right) \ge \beta\right] \le \frac{1}{(n\log(1/\theta))^{1+\delta}}; \quad n \ge 1.$$

The last expression is the general term of a convergent series; by the Borel-Cantelli lemma, there exists an event $\Omega_{\theta\delta} \in \mathscr{F}$ of probability one and an integer-valued random variable $N_{\theta\delta}$, so that for every $\omega \in \Omega_{\theta\delta}$ we have

$$\max_{0\le s\le \theta^n}\left[W_s(\omega) - \frac{1+\delta}{2}s\theta^{-n}h(\theta^n)\right] < \frac{1}{2}h(\theta^n); \quad n \ge N_{\theta\delta}(\omega).$$

Thus, for every $t \in (\theta^{n+1}, \theta^n]$:

$$W_t(\omega) \le \max_{0\le s\le \theta^n} W_s(\omega) \le \left(1+\frac{\delta}{2}\right)h(\theta^n) \le \left(1+\frac{\delta}{2}\right)\theta^{-1/2}h(t).$$

Therefore,

$$\sup_{\theta^{n+1}<t\le\theta^n} \frac{W_t(\omega)}{h(t)} \le \left(1+\frac{\delta}{2}\right)\theta^{-1/2}; \quad n \ge N_{\theta\delta}(\omega)$$

holds for every $\omega \in \Omega_{\theta\delta}$, and letting $n \uparrow \infty$ we obtain

$$\varlimsup_{t\downarrow 0} \frac{W_t(\omega)}{h(t)} \le \left(1+\frac{\delta}{2}\right)\theta^{-1/2}, \qquad \text{a.s. } P.$$

By letting $\delta \downarrow 0$, $\theta \uparrow 1$ through the rationals, we deduce

$$\overline{\lim_{t \downarrow 0}} \frac{W_t}{h(t)} \le 1; \qquad \text{a.s. } P. \tag{9.22}$$

In order to obtain an inequality in the opposite direction, we have to employ the second half of the Borel-Cantelli lemma, which relies on independence. We introduce the independent events

$$A_n = \{W_{\theta^n} - W_{\theta^{n+1}} \ge \sqrt{1-\theta}\, h(\theta^n)\}; \quad n = 1, 2, \ldots,$$

again for fixed $0 < \theta < 1$. Inequality (9.20) with $x = \sqrt{2\log n + 2\log\log(1/\theta)}$ provides lower bounds on the probabilities of these events:

$$P(A_n) = P\left[\frac{W_{\theta^n} - W_{\theta^{n+1}}}{\sqrt{\theta^n - \theta^{n+1}}} \ge x\right] \ge \frac{e^{-x^2/2}}{\sqrt{2\pi}(x + 1/x)} \ge \frac{\text{const.}}{n\sqrt{\log n}}; \quad n > \left|\frac{1}{\log\theta}\right|.$$

Now the last expression is the general term of a divergent series, and the second half of the Borel-Cantelli lemma (Chung (1974), p. 76, or Ash (1972), p. 272) guarantees the existence of an event $\Omega_\theta \in \mathscr{F}$ with $P(\Omega_\theta) = 1$ such that, for every $\omega \in \Omega_\theta$ and $k \ge 1$, there exists an integer $m = m(k, \omega) \ge k$ with

$$W_{\theta^m}(\omega) - W_{\theta^{m+1}}(\omega) \ge \sqrt{1-\theta}\, h(\theta^m). \tag{9.23}$$

On the other hand, (9.22) applied to the Brownian motion $-W$ shows that there exist an event $\Omega^* \in \mathscr{F}$ of probability one and an integer-valued random variable N^*, so that for every $\omega \in \Omega^*$

$$-W_{\theta^{n+1}}(\omega) \le 2h(\theta^{n+1}) \le 4\theta^{1/2} h(\theta^n); \quad n \ge N^*(\omega). \tag{9.24}$$

From (9.23) and (9.24) we conclude that, for every $\omega \in \Omega_\theta \cap \Omega^*$ and every integer $k \ge 1$, there exists an integer $m = m(k, \omega) \ge k \vee N^*(\omega)$ such that

$$\frac{W_{\theta^m}(\omega)}{h(\theta^m)} \ge \sqrt{1-\theta} - 4\sqrt{\theta}.$$

By letting $m \to \infty$, we conclude that $\overline{\lim}_{t\downarrow 0}(W_t/h(t)) \ge \sqrt{1-\theta} - \sqrt{4\theta}$ holds a.s. P, and letting $\theta \downarrow 0$ through the rationals we obtain

$$\overline{\lim_{t \downarrow 0}} \frac{W_t}{h(t)} \ge 1; \quad \text{a.s. } P. \qquad \square$$

We observed in Remark 2.12 that almost every Brownian sample path is locally Hölder-continuous with exponent γ for every $\gamma \in (0, \frac{1}{2})$, and we also saw in Exercise 9.21 that Brownian paths are nowhere locally Hölder-continuous for any exponent $\gamma > \frac{1}{2}$. The law of the iterated logarithm applied to $\{W_{t+h} - W_h; 0 \le h < \infty\}$ for fixed $t \ge 0$ gives

$$\overline{\lim_{h \downarrow 0}} \frac{|W_{t+h} - W_t|}{\sqrt{h}} = \infty, \quad P\text{-a.s.} \tag{9.25}$$

Thus a typical Brownian path cannot be "locally Hölder-continuous with exponent $\gamma = \frac{1}{2}$" everywhere on $[0, \infty)$; however, one may not conclude from (9.25) that such a path has this property nowhere on $[0, \infty)$; see Remark 9.15 and the Notes, Section 11.

F. Modulus of Continuity†

Another way to measure the oscillations of the Brownian path is to seek a modulus of continuity. A function $g(\cdot)$ is called a *modulus of continuity for the function* $f\colon [0, T] \to \mathbb{R}$ if $0 \le s < t \le T$ and $|t - s| \le \delta$ imply $|f(t) - f(s)| \le g(\delta)$, for all sufficiently small positive δ. Because of the law of the iterated logarithm, any modulus of continuity for Brownian motion on a bounded interval, say $[0, 1]$, should be at least as large as $\sqrt{2\delta \log\log(1/\delta)}$, but because of the established local Hölder-continuity it need not be any larger than a constant multiple of δ^γ, for any $\gamma \in (0, 1/2)$. A remarkable result by P. Lévy (1937) asserts that with

(9.26) $$g(\delta) \triangleq \sqrt{2\delta \log(1/\delta)}; \quad \delta > 0,$$

$cg(\delta)$ is a modulus of continuity for almost every Brownian path on $[0, 1]$ if $c > 1$, but is a modulus for almost no Brownian path on $[0, 1]$ if $0 < c < 1$. We say that g in (9.26) is the *exact modulus of continuity* of almost every Brownian path. The assertion just made is a straightforward consequence of the following theorem.

9.25 Theorem (Lévy modulus (1937)). *With* $g\colon (0, 1] \to (0, \infty)$ *given by* (9.26), *we have*

(9.27) $$P\left[\overline{\lim_{\delta \downarrow 0}} \frac{1}{g(\delta)} \max_{\substack{0 \le s < t \le 1 \\ t - s \le \delta}} |W_t - W_s| = 1\right] = 1.$$

Proof. With $n \ge 1$, $0 < \theta < 1$, we have by the independence of increments and (9.20):

$$P\left[\max_{1 \le j \le 2^n} |W_{j/2^n} - W_{(j-1)/2^n}| \le (1 - \theta)^{1/2} g(2^{-n})\right] = (1 - \xi)^{2^n} \le \exp(-\xi 2^n),$$

where $\xi \triangleq 2P[2^{n/2} W_{1/2^n} > (1 - \theta)^{1/2} 2^{n/2} g(2^{-n})] \ge 2e^{-x^2/2}/\sqrt{2\pi}(x + 1/x)$ and $x = \sqrt{(1 - \theta)2n \log 2}$. It develops easily that for $n \ge 1$, we have $\xi \ge \alpha 2^{-n(1-\theta)}$ where $\alpha > 0$, and thus

$$P\left[\max_{1 \le j \le 2^n} |W_{j/2^n} - W_{(j-1)/2^n}| \le (1 - \theta)^{1/2} g(2^{-n})\right] \le \exp(-\alpha 2^{n\theta}).$$

By the Borel-Cantelli lemma, there exists an event $\Omega_\theta \in \mathscr{F}$ with $P(\Omega_\theta) = 1$ and

† This subsection may be omitted on first reading; its results will not be used in the sequel.

an integer-valued random variable N_θ such that, for every $\omega \in \Omega_\theta$, we have

$$\frac{1}{g(2^{-n})} \max_{1 \le j \le 2^n} |W_{j/2^n}(\omega) - W_{(j-1)/2^n}(\omega)| > \sqrt{1-\theta}; \quad n \ge N_\theta(\omega).$$

Consequently, we obtain

$$\varlimsup_{\delta \downarrow 0} \frac{1}{g(\delta)} \max_{\substack{0 \le s < t \le 1 \\ t-s \le \delta}} |W_t - W_s| \ge \sqrt{1-\theta},$$

and by letting $\theta \downarrow 0$ along the rationals, we have

$$\varlimsup_{\delta \downarrow 0} \frac{1}{g(\delta)} \max_{\substack{0 \le s < t \le 1 \\ t-s \le \delta}} |W_t - W_s| \ge 1, \quad \text{a.s. } P.$$

For the proof of the opposite inequality, which is much more demanding, we select $\theta \in (0,1)$ and $\varepsilon > ((1+\theta)/(1-\theta)) - 1$, and observe the inequalities

$$(9.28) \quad P\left[\max_{\substack{0 \le i < j \le 2^n \\ k=j-i \le 2^{n\theta}}} \frac{1}{g(k2^{-n})} |W_{j/2^n} - W_{i/2^n}| \ge 1 + \varepsilon\right]$$

$$\le \sum_{k=1}^{[2^{n\theta}]} P\left[\max_{0 \le i < i+k \le 2^n} |W_{(k+i)/2^n} - W_{i/2^n}| \ge (1+\varepsilon) g\left(\frac{k}{2^n}\right)\right]$$

$$\le 2^n \sum_{k=1}^{[2^{n\theta}]} P\left[\frac{|W_{k/2^n}|}{\sqrt{k2^{-n}}} \ge (1+\varepsilon)\sqrt{\log \frac{4^n}{k^2}}\right].$$

The probability in the last summand of (9.28) is bounded above, thanks to (9.20), by a constant multiple of $n^{-1/2}(k2^{-n})^{(1+\varepsilon)^2}$, and

$$\sum_{k=1}^{[2^{n\theta}]} k^{(1+\varepsilon)^2} \le \int_0^{2^{n\theta}+1} x^{(1+\varepsilon)^2}\, dx = \frac{(2^{n\theta}+1)^\nu}{\nu},$$

where $\nu = 1 + (1+\varepsilon)^2$. Therefore,

$$P\left[\max_{\substack{0 \le i < j \le 2^n \\ k=j-i \le 2^{n\theta}}} \frac{|W_{j/2^n} - W_{i/2^n}|}{g(k/2^n)} \ge 1 + \varepsilon\right] \le \frac{\text{const.}}{\sqrt{n}} 2^{-\rho n},$$

with $\rho = (1-\theta)(1+\varepsilon)^2 - (1+\theta)$, a positive constant by choice of ε. Again by the Borel-Cantelli lemma, we have the existence of an event $\Omega_\theta \in \mathscr{F}$ with $P(\Omega_\theta) = 1$, and of an integer-valued random variable N_θ such that

$$(9.29) \qquad 2^{-(1-\theta)N_\theta(\omega)} \le \frac{1}{e}; \quad \forall\, \omega \in \Omega_\theta$$

and

$$(9.30) \qquad \max_{\substack{0 \le i < j \le 2^n \\ k=j-i \le 2^{n\theta}}} \frac{|W_{j/2^n}(\omega) - W_{i/2^n}(\omega)|}{g(k/2^n)} < 1 + \varepsilon; \quad n \ge N_\theta(\omega),\ \omega \in \Omega_\theta.$$

9.26 Problem. Consider the set $D = \bigcup_{n=1}^{\infty} D_n$ of dyadic rationals in $[0, 1]$, with $D_n = \{k2^{-n}; k = 0, 1, \ldots, 2^n\}$. For every $\omega \in \Omega_\theta$ and every $n \geq N_\theta(\omega)$, the inequality

$$|W_t(\omega) - W_s(\omega)| \leq (1+\varepsilon)\left[2 \sum_{j=n+1}^{\infty} g(2^{-j}) + g(t-s)\right] \tag{9.31}$$

is valid for every pair (s, t) of dyadic rationals satisfying $0 < t - s < 2^{-n(1-\theta)}$. (*Hint*: Proceed as in the proof of Theorem 2.8 and use the fact that $g(\cdot)$ is strictly increasing on $(0, 1/e]$.)

Returning to the proof of Theorem 9.25, let us suppose that the dyadic rationals s, t in (9.31) are chosen to satisfy the stronger condition

$$2^{-(n+1)(1-\theta)} \leq \delta \triangleq t - s < 2^{-n(1-\theta)}. \tag{9.32}$$

But then (9.29) implies $2^{-n(1-\theta)} \leq 1/e$; $n \geq N_\theta(\omega)$, and because g is increasing on $(0, 1/e]$ we have

$$\sum_{j=n+1}^{\infty} g(2^{-j}) \leq cg(2^{-n-1}) \leq \frac{c}{\sqrt{1-\theta}} 2^{-\theta(n+1)/2} g(\delta)$$

holds for an appropriate constant $c > 0$. We may conclude from (9.31), (9.32), and the continuity of $W_\cdot(\omega)$ that for every $\omega \in \Omega_\theta$ and $n \geq N_\theta(\omega)$,

$$\frac{1}{g(\delta)} \max_{\substack{0 \leq s < t \leq 1 \\ t-s=\delta}} |W_t(\omega) - W_s(\omega)| \leq (1+\varepsilon)\left[1 + \frac{2c}{1-\theta} 2^{-\theta(n+1)/2}\right]$$

holds for all $\delta \in [2^{-(n+1)(1-\theta)}, 2^{-n(1-\theta)})$. Letting $n \to \infty$, we obtain

$$\overline{\lim_{\delta \downarrow 0}} \frac{1}{g(\delta)} \max_{\substack{0 \leq s < t \leq 1 \\ t-s=\delta}} |W_t(\omega) - W_s(\omega)| \leq 1 + \varepsilon,$$

and because g is strictly increasing on $(0, 1/e]$ we may replace the condition $t - s = \delta$ by $t - s \leq \delta$ in the preceding expression. It remains only to let $\theta \downarrow 0$ (and hence simultaneously $\varepsilon \downarrow 0$) along the rationals, to conclude that

$$\overline{\lim_{\delta \downarrow 0}} \frac{1}{g(\delta)} \max_{\substack{0 \leq s < t \leq 1 \\ t-s\leq\delta}} |W_t(\omega) - W_s(\omega)| \leq 1; \quad \text{a.s. } P.$$

The proof is complete. □

2.10. Solutions to Selected Problems

1.4. For fixed $0 \leq s < t < \infty$, and arbitrary integer $n \geq 1$ and indices $0 = s_0 < s_1 < \cdots < s_n = s$, the σ-field $\sigma(X_0, X_{s_1}, \ldots, X_{s_n}) = \sigma(X_0, X_{s_1} - X_{s_0}, \ldots, X_{s_n} - X_{s_{n-1}})$ is independent of $X_t - X_s$. The union of all σ-fields of this form (over $n \geq 1$,

$(s_1, \ldots, s_{n-1})$ as described) constitutes a collection $\mathscr{C}$ of sets independent of $X_t - X_s$, which is closed under finite intersections. Now $\mathscr{D}$, the collection of all sets in $\mathscr{F}_s^X$ which are independent of $X_t - X_s$, is a Dynkin system containing $\mathscr{C}$. From Theorem 1.3 we conclude that $\mathscr{F}_s^X = \sigma(\mathscr{C})$ is contained in $\mathscr{D}$. In fact, $\mathscr{F}_s^X = \mathscr{D}$.

2.3. Suppose that there exist integers $n \geq 1$, $m \geq 1$, index sequences $\underset{\sim}{t} = (t_1, \ldots, t_n)$, $\underset{\sim}{s} = (s_1, \ldots, s_m)$, and Borel sets $A \in \mathscr{B}(\mathbb{R}^n)$, $B \in \mathscr{B}(\mathbb{R}^m)$ so that $C \in \mathscr{C}$ admits both representations $C = \{\omega \in \mathbb{R}^{[0,\infty)}; (\omega(t_1), \ldots, \omega(t_n)) \in A\} = \{\omega \in \mathbb{R}^{[0,\infty)}; (\omega(s_1), \ldots, \omega(s_m)) \in B\}$. It has to be shown that

$$(2.3)' \qquad Q_{\underset{\sim}{t}}(A) = Q_{\underset{\sim}{s}}(B).$$

Case 1: $m = n$, and there is a permutation $(i_1, \ldots, i_n)$ of $(1, \ldots, n)$ so that $s_j = t_{i_j}$, $1 \leq j \leq n$.

In this case, $A = \{(x_1, \ldots, x_n) \in \mathbb{R}^n; (x_{i_1}, \ldots, x_{i_n}) \in B\}$. Both sides of (2.3)′ are measures on $\mathscr{B}(\mathbb{R}^n)$, and to prove their identity it suffices to verify that they agree on sets of the form $A = A_1 \times \cdots \times A_n$, $A_j \in \mathscr{B}(\mathbb{R})$ (hence $B = A_{i_1} \times \cdots \times A_{i_n}$). But then (2.3)′ is just the first consistency condition (a) in Definition 2.1.

Case 2: $m > n$ and $\{t_1, \ldots, t_n\} \subseteq \{s_1, \ldots, s_m\}$.

Without loss of generality (thanks to Case 1) we may assume that $t_j = s_j$, $1 \leq j \leq n$. Then we have $B = A \times \mathbb{R}^k$ with $k = m - n \geq 1$, and (2.3)′ follows from the second consistency condition (b) of Definition 2.1.

Case 3: None of the above holds.

We enlarge the index set, to wit: $\{q_1, \ldots, q_l\} = \{t_1, \ldots, t_n\} \cup \{s_1, \ldots, s_m\}$, with $m \vee n \leq l \leq m + n$, and obtain a third representation

$$C = \{\omega \in \mathbb{R}^{[0,\infty)}; (\omega(q_1), \ldots, \omega(q_l)) \in E\}, \quad E \in \mathscr{B}(\mathbb{R}^l).$$

By the same reasoning as before, we may assume that $q_j = t_j$, $1 \leq j \leq n$. Then $E = A \times \mathbb{R}^{l-n}$ and, by Case 2, $Q_{\underset{\sim}{t}}(A) = Q_{\underset{\sim}{q}}(E)$ with $\underset{\sim}{q} = (q_1, \ldots, q_l)$. A dual argument shows $Q_{\underset{\sim}{s}}(B) = Q_{\underset{\sim}{q}}(E)$.

The preceding method (adapted from Wentzell (1981)) shows that Q is well defined by (2.3). To prove finite additivity, let us notice that a finite collection $\{C_j\}_{j=1}^m \subseteq \mathscr{C}$ may be represented, by enlargement of the index sequence if necessary, as

$$C_j = \{\omega \in \mathbb{R}^{[0,\infty)}; (\omega(t_1), \ldots, \omega(t_n)) \in A_j\}, \; A_j \in \mathscr{B}(\mathbb{R}^n)$$

for every $1 \leq j \leq m$. The A_j's are pairwise disjoint if the C_j's are, and finite additivity follows easily from (2.3), since

$$\bigcup_{j=1}^m C_j = \left\{\omega \in \mathbb{R}^{[0,\infty)}; (\omega(t_1), \ldots, \omega(t_n)) \in \bigcup_{j=1}^m A_j\right\}.$$

Finally, $\mathbb{R}^{[0,\infty)} = \{\omega \in \mathbb{R}^{[0,\infty)}; \omega(t) \in \mathbb{R}\}$ for every $t \geq 0$, and so $Q(\mathbb{R}^{[0,\infty)}) = Q_t(\mathbb{R}) = 1$.

2.4. Let $\mathscr{F}$ be the collection of all regular sets. We have $\varnothing \in \mathscr{F}$. To show $\mathscr{F}$ is closed under complementation, suppose $A \in \mathscr{F}$, so for each $\varepsilon > 0$, we have a closed set F and an open set G such that $F \subseteq A \subseteq G$ and $Q(G \backslash F) < \varepsilon$. But then F^c is open, G^c is closed, $G^c \subseteq A^c \subseteq F^c$, and $Q(F^c \backslash G^c) = Q(G \backslash F) < \varepsilon$; therefore, $A^c \in \mathscr{F}$. To show $\mathscr{F}$ is closed under countable unions, let $A = \bigcup_{k=1}^\infty A_k$, where $A_k \in \mathscr{F}$ for each k. For each $\varepsilon > 0$, there is a sequence of closed sets $\{F_k\}$ and a sequence of open sets $\{G_k\}$ such that $F_k \subseteq A_k \subseteq G_k$ and $Q(G_k \backslash F_k) < \varepsilon/2^{k+1}$, $k = 1, 2, \ldots$. Let

$G = \bigcup_{k=1}^{\infty} G_k$ and $F = \bigcup_{k=1}^{m} F_k$, where m is chosen so that $Q(\bigcup_{k=1}^{\infty} F_k \backslash \bigcup_{k=1}^{m} F_k) < \varepsilon/2$. Then G is open, F is closed, $F \subseteq A \subseteq G$, and

$$Q(G\backslash F) \le Q\left(G\backslash \bigcup_{k=1}^{\infty} F_k\right) + Q\left(\bigcup_{k=1}^{\infty} F_k\backslash F\right) < \sum_{k=1}^{\infty} \frac{\varepsilon}{2^{k+1}} + \frac{\varepsilon}{2} = \varepsilon.$$

Therefore, $\mathscr{F}$ is a σ-field.

Now choose a closed set F. Let $G_k = \{x \in \mathbb{R}^n;\ \|x - y\| < 1/k, \text{some } y \in F\}$. Then each G_k is open, $G_1 \supseteq G_2 \supseteq \cdots$, and $\bigcap_{k=1}^{\infty} G_k = F$. Thus, for each $\varepsilon > 0$, there exists an m such that $Q(G_m \backslash F) < \varepsilon$. It follows that $F \in \mathscr{F}$.

Since the smallest σ-field containing all closed sets is $\mathscr{B}(\mathbb{R}^n)$, we have $\mathscr{F} = \mathscr{B}(\mathbb{R}^n)$.

2.5. Fix $\underset{\sim}{t} = (t_1, t_2, \ldots, t_n)$, and let $\underset{\sim}{s} = (t_{i_1}, t_{i_2}, \ldots, t_{i_n})$ be a permutation of $\underset{\sim}{t}$. We have constructed a distribution for the random vector $(B_{t_1}, B_{t_2}, \ldots, B_{t_n})$ under which

$$\begin{aligned} Q_{\underset{\sim}{t}}(A_1 \times A_2 \times \cdots \times A_n) &= P[(B_{t_1}, \ldots, B_{t_n}) \in A_1 \times \cdots \times A_n] \\ &= P[(B_{t_{i_1}}, \ldots, B_{t_{i_n}}) \in A_{i_1} \times \cdots \times A_{i_n}] \\ &= Q_{\underset{\sim}{s}}(A_{i_1} \times \cdots \times A_{i_n}). \end{aligned}$$

Furthermore, for $A \in \mathscr{B}(\mathbb{R}^{n-1})$ and $\underset{\sim}{s}' = (t_1, t_2, \ldots, t_{n-1})$,

$$Q_{\underset{\sim}{t}}(A \times \mathbb{R}) = P[(B_{t_1}, \ldots, B_{t_{n-1}}) \in A] = Q_{\underset{\sim}{s}'}(A).$$

2.9. Again we take $T = 1$. It is a bit easier to visualize the proof if we use the maximum norm $|||(t_1, t_2, \ldots, t_d)||| \triangleq \max_{1 \le i \le d} |t_i|$ in $\mathbb{R}^d$ rather than the Euclidean norm. Since all norms are equivalent in $\mathbb{R}^d$, it suffices to prove (2.14) with $\|t - s\|$ replaced by $|||t - s|||$. Introduce the partial ordering $\prec$ in $\mathbb{R}^d$ by $(s_1, s_2, \ldots, s_d) \prec (t_1, t_2, \ldots, t_d)$ if and only if $s_i \le t_i$, $i = 1, \ldots, d$. Define the lattices

$$L_n = \left\{\frac{k}{2^n};\ k = 0, 1, \ldots, 2^n - 1\right\}^d; \quad n \ge 1, \quad L = \bigcup_{n=1}^{\infty} L_n,$$

and for $s \in L_n$, define $N_n(s) = \{t \in L_n;\ s \prec t,\ \|t - s\| = 2^{-n}\}$. Note that L_n has 2nd elements, and for each $s \in L_n$, $N_n(s)$ has d elements. For $s \in L_n$ and $t \in N_n(s)$, Čebyšev's inequality applied to (2.13) gives (with $0 < \gamma < \alpha/\beta$),

$$P[|X_t - X_s| \ge 2^{-\gamma n}] \le C 2^{-n(d + \beta - \alpha\gamma)},$$

and consequently,

$$P\left[\max_{\substack{s \in L_n \\ t \in N_n(s)}} |X_t - X_s| \ge 2^{-\gamma n}\right] \le dC 2^{-n(\beta - \alpha\gamma)}.$$

The Borel-Cantelli lemma gives an event $\Omega^* \in \mathscr{F}$ with $P(\Omega^*) = 1$, and a positive, integer-valued random variable n^*, such that

$$\max_{\substack{s \in L_n \\ t \in N_n(s)}} |X_t(\omega) - X_s(\omega)| < 2^{-\gamma n}, \quad n \ge n^*(\omega), \omega \in \Omega^*.$$

Now let $R_n(s) = \{t \in L_n;\ s \prec t,\ |||t - s||| = 2^{-n}\}$. For $s \in L_n$ and $t \in R_n(s)$, there is a sequence $s = s_0, s_1, \ldots, s_m = t$ with $m \le d$ and $s_i \in N_n(s_{i-1})$, $i = 1, \ldots, m$. Consequently,

(10.1) $$\max_{\substack{s \in L_n \\ t \in R_n(s)}} |X_t(\omega) - X_s(\omega)| \le d2^{-\gamma n}, \quad n \ge n^*(\omega), \omega \in \Omega^*.$$

We now fix $\omega \in \Omega^*$, $n \ge n^*(\omega)$, and show that for every $m > n$, we have

(10.2) $$|X_t(\omega) - X_s(\omega)| \le 2d \sum_{j=n+1}^{m} 2^{-\gamma j}; \quad \forall t, s \in L_m, s \prec t, |||t - s||| < 2^{-n}.$$

For $m = n + 1$, we can only have $t \in R_m(s)$, and (10.2) follows from (10.1). Suppose (10.2) is valid for $m = n + 1, \dots, M - 1$. Take $t, s \in L_M$, $s \prec t$. There is a vector $s^1 \in L_{M-1} \cap R_M(s)$ and a vector $t^1 \in L_{M-1}$ with $t \in R_M(t^1)$ such that $s \prec s^1 \prec t^1 \prec t$. From (10.1) we have,

$$|X_{s^1}(\omega) - X_s(\omega)| \le d2^{-\gamma M}, \quad |X_t(\omega) - X_{t^1}(\omega)| \le d2^{-\gamma M},$$

and from (10.2) with $m = M - 1$, we have

$$|X_{t^1}(\omega) - X_{s^1}(\omega)| \le 2d \sum_{j=n+1}^{M-1} 2^{-\gamma j}.$$

We obtain (10.2) for $m = M$.

For any vectors $s, t \in L$ with $s \prec t$ and $0 < |||t - s||| < h(\omega) \triangleq 2^{-n^*(\omega)}$, we select $n \ge n^*(\omega)$ such that $2^{-(n+1)} \le |||t - s||| < 2^{-n}$. We have from (10.2)

$$|X_t(\omega) - X_s(\omega)| \le 2d \sum_{j=n+1}^{\infty} 2^{-\gamma j} \le \delta |||t - s|||^{\gamma},$$

where $\delta = 2d/(1 - 2^{-\gamma})$. We may now conclude as in the proof of Theorem 2.8.

4.2. The n-dimensional cylinder sets are generated by those among them which are n-fold intersections of one-dimensional cylinder sets; the latter are generated by sets of the form $H = \{\omega \in C[0, \infty); \omega(t_1) \in G\}$, where G is open in $\mathbb{R}$. But H is open in $C[0, \infty)$, because for each $\omega_0 \in H$, this set contains a ball $B(\omega_0, \varepsilon) \triangleq \{\omega \in C[0, \infty); \rho(\omega, \omega_0) < \varepsilon\}$, for suitably small $\varepsilon > 0$. It follows that $\mathscr{G} \subseteq \mathscr{B}(C[0, \infty))$. Because $C[0, \infty)$ is separable, the open sets are countable unions of open balls of the form $B(\omega_0, \varepsilon)$ as previously. Let Q be the set of rationals in $[0, \infty)$. We have

$$B(\omega_0, \varepsilon) = \left\{\omega: \sum_{n=1}^{\infty} \frac{1}{2^n} \sup_{\substack{0 \le t \le n \\ t \in Q}} (|\omega(t) - \omega_0(t)| \wedge 1) < \varepsilon\right\}.$$

The set on the right is $\mathscr{G}$-measurable, so $\mathscr{B}(C[0, \infty)) \subseteq \mathscr{G}$.

For the second claim, notice that with any cylinder set C of the form (2.1)′ we have

$$\begin{aligned} \varphi_t^{-1}(C) &= \{\omega \in C[0, \infty); ((\varphi_t\omega)(t_1), \dots, (\varphi_t\omega)(t_n)) \in A\} \\ &= \{\omega \in C[0, \infty); (\omega(t \wedge t_1), \dots, \omega(t \wedge t_n)) \in A\} \in \mathscr{C}_t, \end{aligned}$$

so $\varphi_t^{-1}(\mathscr{C}) \subseteq \mathscr{C}_t$. On the other hand, for any $C \in \mathscr{C}_t$ we have $t_1, \dots, t_n$ in $[0, t]$ and $A \in \mathscr{B}(\mathbb{R}^n)$ so that

$$\begin{aligned} C &= \{\omega \in C[0, \infty); (\omega(t_1), \dots, \omega(t_n)) \in A\} \\ &= \{\omega \in C[0, \infty); (\omega(t \wedge t_1), \dots \omega(t \wedge t_n)) \in A\} = \varphi_t^{-1}(C), \end{aligned}$$

and thus $\mathscr{C}_t \subseteq \varphi_t^{-1}(\mathscr{C})$. It follows that $\mathscr{C}_t = \varphi_t^{-1}(\mathscr{C})$, which establishes the claim.

4.11. We have to show

(4.6)′ $$\lim_{\lambda\to\infty} \sup_{m\geq 1} P[|X_0^{(m)}| > \lambda] = 0, \quad \text{and}$$

(4.7)′ $$\lim_{\rho\downarrow 0} \sup_{m\geq 1} P\left[\max_{\substack{|t-s|\leq\rho \\ 0\leq t,s\leq T}} |X_t^{(m)} - X_s^{(m)}| > \varepsilon\right] = 0$$

for all positive numbers ε, T. Relation (4.6)′ is an immediate consequence of (i) and of the Čebyšev inequality. It suffices to prove (4.7)′ for $T = 1$. Let $\eta > 0$ and $\varepsilon > 0$ be given. We denote $X^{(m)}$ simply by X, and employ the notation of the proof of Theorem 2.8. With

$$\Omega_l \triangleq \bigcap_{n=l}^{\infty} \left\{ \max_{1\leq k\leq 2^n} |X_{k/2^n} - X_{(k-1)/2^n}| < 2^{-\gamma n} \right\},$$

we have from (2.9): $P(\Omega_l^c) \leqq C_1 \sum_{n=l}^{\infty} 2^{-n(\beta-\alpha\gamma)} \leq \eta$, provided l is a large enough integer. Now for every $\omega\in\Omega_l$ and $n \geq l$, we have from (2.12):

$$\max_{\substack{|t-s|<2^{-n} \\ t,s\in D}} |X_t(\omega) - X_s(\omega)| \leqq \delta 2^{-\gamma n},$$

where $\delta \triangleq 2/(1 - 2^{-\gamma})$. It follows that, given $\varepsilon > 0$, $\eta > 0$, there exists an integer $n = n(\varepsilon, \eta)$ such that for every $\omega\in\Omega_n$:

$$\max_{\substack{|t-s|<2^{-n} \\ 0\leq t,s\leq 1}} |X_t^{(m)}(\omega) - X_s^{(m)}(\omega)| \leq \varepsilon, \qquad \forall m \geq 1$$

(we have used the continuity of the sample path $t \mapsto X_t^{(m)}(\omega)$), and consequently

$$P\left[\max_{\substack{|t-s|<2^{-n} \\ 0\leq t,s\leq 1}} |X_t^{(m)} - X_s^{(m)}| > \varepsilon\right] \leq P(\Omega_n^c) \leq \eta.$$

4.16. Let $(\Omega_n, \mathscr{F}_n, P_n)$ denote the space on which X_n and Y_n are defined, and let E_n denote expectation with respect to P_n. Let X be defined on $(\Omega, \mathscr{F}, P)$. We are given that $\lim_{n\to\infty} E_n f(X^{(n)}) = Ef(X)$ for every bounded, continuous $f: S \to \mathbb{R}$ and that $\rho(X^{(n)}, Y^{(n)}) \to 0$ in probability. To prove $Y^{(n)} \xrightarrow{\mathscr{D}} X$, it suffices to show

$$\lim_{n\to\infty} E_n[f(X^{(n)}) - f(Y^{(n)})] = 0$$

whenever f is bounded and continuous. Let such an f be given, and set $M = \sup_{x\in S} |f(x)| < \infty$. Since $\{X^{(n)}\}_{n=1}^{\infty}$ is relatively compact, it is tight; so for each $\varepsilon > 0$ there exists a compact set $K \subset S$, such that $P_n[X^{(n)}\in K] \geq 1 - \varepsilon/6M$, $\forall n \geq 1$. Choose $0 < \delta < 1$ so $|f(x) - f(y)| < \varepsilon/3$ whenever $x\in K$ and $\rho(x, y) < \delta$. Finally, choose a positive integer N such that $P_n[\rho(X^{(n)}, Y^{(n)}) \geq \delta] \leq \varepsilon/6M$, $\forall n \geq N$. We have

$$\begin{aligned}\left|\int_{\Omega_n} [f(X^{(n)}) - f(Y^{(n)})]\, dP_n\right| &\leq \frac{\varepsilon}{3} P_n[X^{(n)} \in K, \rho(X^{(n)}, Y^{(n)}) < \delta] \\ &\quad + 2M \cdot P_n[X^{(n)} \notin K] \\ &\quad + 2M \cdot P_n[\rho(X^{(n)}, Y^{(n)}) \geq \delta] \leq \varepsilon.\end{aligned}$$

5.2. The collection of sets $F\in\mathscr{B}(C[0, \infty)^d)$ for which $x \mapsto P^x(F)$ is $\mathscr{B}(\mathbb{R}^d)/\mathscr{B}([0, 1])$-measurable forms a Dynkin system, so it suffices to prove this measurability for all finite-dimensional cylinder sets F of the form

$$F = \{\omega \in C[0,\infty)^d;\ \omega(t_0) \in \Gamma_0, \ldots, \omega(t_n) \in \Gamma_n\},$$

where $0 = t_0 < t_1 < \cdots < t_n$, $\Gamma_i \in \mathscr{B}(\mathbb{R}^d)$, $i = 0, 1, \ldots, n$. But

$$P^x(F) = 1_{\Gamma_0}(x) \int_{\Gamma_1} \cdots \int_{\Gamma_n} p_d(t_1; x, y_1) \ldots p_d(t_n - t_{n-1}; y_{n-1}, y_n)\, dy_n \ldots dy_1,$$

where $p_d(t; x, y) \triangleq (2\pi t)^{-d/2} \exp\{-(\|x - y\|^2/2t)\}$. This is a Borel-measurable function of x.

5.7. If for each finite measure μ on $(S, \mathscr{B}(S))$, there exists g_μ as described, then for each $\alpha \in \mathbb{R}$, $\{x \in S; f(x) \le \alpha\} \,\Delta\, \{x \in S; g_\mu(x) \le \alpha\}$ has μ-measure zero. But $\{g_\mu \le \alpha\} \in \mathscr{B}(S)$, so $\{f \le \alpha\} \in \overline{\mathscr{B}(S)}^\mu$. Since this is true for every μ, we have $\{f \le \alpha\} \in \mathscr{U}(S)$.

For the converse, suppose f is universally measurable and let a finite measure μ be given. For $r \in Q$, the set of rationals, let $U(r) = \{x \in S; f(x) \le r\}$. Then $f(x) = \inf\{r \in Q; x \in U(r)\}$. Since $U(r) \in \overline{\mathscr{B}(S)}^\mu$, there exists $B(r) \in \mathscr{B}(S)$ with $\mu[B(r) \,\Delta\, U(r)] = 0$, $r \in Q$. Define

$$g_\mu(x) \triangleq \inf\{r \in Q; x \in B(r)\} = \inf_{r \in Q} \varphi_r(x),$$

where $\varphi_r(x) = r$ if $x \in B(r)$ and $\varphi_r(x) = \infty$ otherwise. Then $g_\mu \colon S \to \mathbb{R}$ is Borel-measurable, and $\{x \in S;\ f(x) \ne g_\mu(x)\} \subseteq \bigcup_{r \in Q} [B(r) \,\Delta\, U(r)]$, which has μ-measure zero.

5.9. We prove (5.5). Let us first show that for $D \in \mathscr{B}(\mathbb{R}^d \times \mathbb{R}^d)$, we have

$$P[(X, Y) \in D \,|\, \mathscr{G}] = P[(X, Y) \in D \,|\, Y]. \tag{10.3}$$

If $D = B \times C$, where $B, C \in \mathscr{B}(\mathbb{R}^d)$, then

$$E[1_{\{X \in B\}} 1_{\{Y \in C\}} | \mathscr{G}] = 1_{\{Y \in C\}} E[1_{\{X \in B\}} | \mathscr{G}] = 1_{\{Y \in C\}} P[X \in B].$$

For the same reasons, $E[1_{\{X \in B\}} 1_{\{Y \in C\}} | Y] = 1_{\{Y \in C\}} P[X \in B]$, so (10.3) holds for this special case. The sets D for which (10.3) holds form a Dynkin system containing all measurable rectangles, so (10.3) holds for every $D \in \mathscr{B}(\mathbb{R}^d \times \mathbb{R}^d)$. To prove (5.5), set $D = \{(x, y);\ x + y \in \Gamma\}$.

A similar proof for (5.6) is possible.

6.9. (ii) By Corollary 1.2.4, S is a stopping time of $\{\mathscr{F}_{t+}\}$. Problem 1.2.17 (i) implies

$$E[Z_2 | \mathscr{F}_{S+}] = E[Z_2 | \mathscr{F}_{(S \wedge s)+}], \quad \text{a.s. on } \{S \le s\}.$$

This equation combined with (i) gives us the desired result.

(iii) Suppose that S is an optional time of $\{\mathscr{F}_t\}$, and that (*e*) holds for every bounded optional time of $\{\mathscr{F}_t\}$. Then for each $s > 0$,

$$P^x[X_{(S \wedge s)+t} \in \Gamma | \mathscr{F}_{(S \wedge s)+}] = (U_t 1_\Gamma)(X_{S \wedge s}), \quad P^x \text{ a.s.}$$

But on $\{S \le s\}$, we have $X_{(S \wedge s)+t} = X_{S+t}$, so (ii) implies

$$\begin{aligned} P^x[X_{S+t} \in \Gamma | \mathscr{F}_{S+}] &= P^x[X_{(S \wedge s)+t} \in \Gamma | \mathscr{F}_{(S \wedge s)+}] \\ &= (U_t 1_\Gamma)(X_{S \wedge s}) = (U_t 1_\Gamma)(X_S), \quad P^x \text{ a.s. on } \{S \le s\}. \end{aligned}$$

Now let $s \uparrow \infty$ to obtain (*e*) for the (possibly unbounded) optional time S. The argument for (*e'*) is the same.

7.1. (i) $\bigcap_{s>t} \mathscr{F}_{s+}^X = \bigcap_{s>t} \bigcap_{\varepsilon>0} \mathscr{F}_{s+\varepsilon}^X = \bigcap_{s>t} \mathscr{F}_s^X = \mathscr{F}_{t+}^X$.

(ii) The σ-field $\mathscr{F}_t^X$ is generated by sets of the form $F = \{(X_{t_1}, \ldots, X_{t_n}) \in \Gamma\}$, where $0 = t_1 < \cdots < t_n = t$ and $\Gamma \in \mathscr{B}(\mathbb{R}^{dn})$. Since $X_t = \lim_{m\to\infty} X_{s_m}$ for any sequence $\{s_m\}_{m=1}^\infty \subseteq [0, t)$ satisfying $s_m \uparrow t$, we have $F \in \mathscr{F}_{t-}^X$.

(iii) Let X be the coordinate mapping process on $C[0, \infty)$. Fix $t > 0$. The nonempty set $F \triangleq \{\omega \in C[0, \infty); \omega \text{ has a local maximum at } t\}$ is in $\mathscr{F}_{t+}^X$, since for each $n \geq 0$,

$$F = \bigcup_{m=n}^{\infty} \bigcap_{\substack{r \in Q \\ |t-r|<1/m}} \{\omega; \omega(t) \geq \omega(r)\} \in \mathscr{F}_{t+1/n}^X.$$

On the other hand, a typical set in $\mathscr{F}_t$ has the form $G = \{\omega \in C[0, \infty); (\omega(t_1), \omega(t_2), \ldots) \in \Gamma\}$, for $\{t_i\}_{i=1}^\infty \subseteq [0, t]$ and $\Gamma \in \mathscr{B}(\mathbb{R} \times \mathbb{R} \times \cdots)$. We claim that $F = G$ cannot hold for any $G \in \mathscr{F}_t^X$. Indeed, suppose we have $F \cap G \neq \varnothing$ for some $G \in \mathscr{F}_t^X$. Then given any $\omega \in F \cap G$, the function

$$\tilde{\omega}(s) \triangleq \begin{cases} \omega(s); \; 0 \leq s \leq t, \\ \omega(t) + s - t; \; s \geq t, \end{cases}$$

is in G but not in F, so F cannot agree with any $G \in \mathscr{F}_t^X$. This shows that the filtration $\{\mathscr{F}_t^X\}$ is not right-continuous. With $\mathscr{G}_t = \mathscr{F}_{t+}^X$, we have

$$\mathscr{G}_{t-} = \sigma\left(\bigcup_{s<t} \bigcap_{\varepsilon>0} \mathscr{F}_{s+\varepsilon}^X\right) \subseteq \sigma\left(\bigcup_{s<t} \mathscr{F}_{(s+t)/2}^X\right) = \mathscr{F}_{t-}^X \subseteq \mathscr{F}_t^X \subset \mathscr{G}_t,$$

so the failure of $\{\mathscr{F}_t^X\}$ to be right-continuous implies the failure of $\{\mathscr{F}_{t+}^X\}$ to be left-continuous.

7.3. Clearly, $\mathscr{H}$ is a σ-field: $\varnothing \in \mathscr{H}$, $F \in \mathscr{H}$ implies that there exists an event $G \in \mathscr{G}$ with $F^c \triangle G^c = F \triangle G \in \mathscr{N}^\mu$, and finally for any sequence $\{F_n\}_{n=1}^\infty \subseteq \mathscr{H}$ we have a companion sequence $\{G_n\}_{n=1}^\infty \subseteq \mathscr{G}$ such that

$$\left(\bigcup_{n=1}^\infty F_n\right) \triangle \left(\bigcup_{n=1}^\infty G_n\right) \subseteq \bigcup_{n=1}^\infty (F_n \triangle G_n) \in \mathscr{N}^\mu,$$

which yields $\bigcup_{n=1}^\infty F_n \in \mathscr{H}$.

Further, the observation $F \triangle G = N \Leftrightarrow F = G \triangle N$ yields the characterization $\mathscr{H} = \{F \subseteq \Omega; \exists\, G \in \mathscr{G}, N \in \mathscr{N}^\mu \text{ such that } F = G \triangle N\}$. It follows that $\mathscr{H} \subseteq \mathscr{G}^\mu$. On the other hand, $\mathscr{H}$ contains both $\mathscr{G}$ and $\mathscr{N}^\mu$, and since it is a σ-field: $\mathscr{G}^\mu = \sigma(\mathscr{G} \cup \mathscr{N}^\mu) \subseteq \mathscr{H}$.

For completeness, let us observe that the requirements $F \in \mathscr{G}^\mu$, $P^\mu(F) = 0$ imply the existence of $G \in \mathscr{G}$ such that $N = F \triangle G \in \mathscr{N}^\mu$ and $P^\mu(G) = 0$. Now F is contained in $N \cup G$ and hence F is in $\mathscr{N}^\mu$, as is any subset of F.

7.4. Let $\{B_t; t \geq 0\}$ be the coordinate mapping process on $C[0, \infty)$. Let P^μ on $(C[0, \infty), \mathscr{B}(C[0, \infty)))$ be the probability measure under which $B = \{B_t, \mathscr{F}_t^B; t \geq 0\}$ is a one-dimensional Brownian motion with initial distribution μ. The set $F = \{\omega; \omega(1) = 0\}$ has P^μ-measure zero, so $F \in \mathscr{N}^\mu \subseteq \mathscr{F}_0^\mu$. If F is also in the completion $\bar{\mathscr{F}}_0^\mu$ of $\mathscr{F}_0^B$ under P^μ, then there must be some $G \in \mathscr{F}_0^B$ with $F \subseteq G$ and $P^\mu(G) = 0$. Such a G must be of the form $G = \{\omega: \omega(0) \in \Gamma\}$ for some $\Gamma \in \mathscr{B}(\mathbb{R})$, and the only way G can contain F is to have $\Gamma = \mathbb{R}$. But then $P^\mu(G) \neq 0$. It follows that F is not in $\bar{\mathscr{F}}_0^\mu$.

7.5. Clearly, $\mathscr{F}_t^\mu \subseteq \mathscr{F}^\mu$ holds for every $0 \le t < \infty$, so $\mathscr{F}_\infty^\mu \subseteq \mathscr{F}^\mu$. For the opposite inclusion, let us take any $F \in \mathscr{F}^\mu$; Problem 7.3 guarantees the existence of an event $G \in \mathscr{F}_\infty^X$ such that $N = F \triangle G \in \mathscr{N}^\mu$. But now $\mathscr{F}_\infty^X \subseteq \mathscr{F}_\infty^\mu$ (we have $\mathscr{F}_t^X \subseteq \mathscr{F}_t^\mu \subseteq \mathscr{F}_\infty^\mu$ for every $0 \le t < \infty$), and thus $G \in \mathscr{F}_\infty^\mu$, $N \in \mathscr{N}^\mu \subseteq \mathscr{F}_\infty^\mu$ imply $F = G \triangle N \in \mathscr{F}_\infty^\mu$.

7.6. Repeat the argument employed in Solution 7.5, replacing $\mathscr{F}^\mu$ by $\mathscr{F}_t^\mu$ and $\mathscr{F}_\infty^\mu$ by $\mathscr{F}_{t-}^X$ and using the left-continuity of the filtration $\{\mathscr{F}_t^X\}$ (Problem 7.1).

7.18. Let $\{B_t, \mathscr{F}_t; t \ge 0\}$, $(\Omega, \mathscr{F})$, $\{P^x\}_{x \in \mathbb{R}}$ be a one-dimensional Brownian family. For $\Gamma \in \mathscr{B}(\mathbb{R})$, define the hitting time $H_\Gamma(\omega) = \inf\{t \ge 0; B_t(\omega) \in \Gamma\}$. According to Problem 1.2.6, $H_{(0,\infty)}$ is optional, so $\{H_{(0,\infty)} = 0\}$ is in $\mathscr{F}_{0+} = \mathscr{F}_0$. Likewise, $H_{(-\infty,0)} \in \mathscr{F}_0$. Because of the symmetry of Brownian motion starting at the origin, $P^0[H_{(0,\infty)} = 0] = P^0[H_{(-\infty,0)} = 0]$. According to the Blumenthal zero-one law (Theorem 7.17), this common value is either zero or one. If it were zero, then $P^0[B_t = 0, \forall 0 \le t \le \varepsilon$ for some $\varepsilon > 0] = 1$, but this contradicts Problem 7.14 (i). Therefore, $P^0[H_{(0,\infty)} = 0] = P^0[H_{(-\infty,0)} = 0] = 1$, and for each $\omega \in \{H_{(0,\infty)} = 0\} \cap \{H_{(-\infty,0)} = 0\}$, there are sequences $s_n \downarrow 0$, $t_n \downarrow 0$ with $B_{s_n}(\omega) > 0$, $B_{t_n}(\omega) < 0$ for every $n \ge 1$.

7.19. For fixed $\omega \in \Omega$, $T_b(\omega)$ is a left-continuous function of b and $S_b(\omega)$ is a right-continuous function of b. For fixed $b \in \mathbb{R}$, T_b is a stopping time and S_b is an optional time (Problem 1.2.6), so both are $\mathscr{F}_\infty^W$-measurable. According to Remark 1.1.14, the set $A = \{(b, \omega) \in [0, \infty) \times \Omega; T_b(\omega) \ne S_b(\omega)\}$ is in $\mathscr{B}([0, \infty)) \otimes \mathscr{F}_\infty^W$. Furthermore, $A_b \triangleq \{\omega \in \Omega; (b, \omega) \in A\}$ is included in the set

$$\{\omega \in \Omega; B_t(\omega) \triangleq W_{T_b(\omega)+t}(\omega) - W_{T_b(\omega)}(\omega) \le 0 \text{ for some } \varepsilon > 0 \text{ and all } t \in [0, \varepsilon]\},$$

which has probability zero because $\{B_t, \mathscr{F}_t^B; 0 \le t < \infty\}$ is a standard Brownian motion (Remark 6.20 and Theorem 6.16), and Problem 7.18 implies that B takes positive values in every interval of the form $[0, \varepsilon]$ with probability one. This establishes (i). For (ii), it suffices to show

$$P[\omega \in \Omega; (L(\omega), \omega) \in A] = \int_0^\infty P(A_b) P[L \in db].$$

If A were a product set $A = C \times D$; $C \in \mathscr{B}([0, \infty))$, $D \in \mathscr{F}_\infty^W$, this would follow from the independence of L and $\mathscr{F}_\infty^W$. The collection of sets $A \in \mathscr{B}([0, \infty)) \otimes \mathscr{F}_\infty^W$ for which this identity holds forms a Dynkin system, so by the Dynkin system Theorem 1.3, this identity holds for every set in $\mathscr{B}([0, \infty)) \otimes \mathscr{F}_\infty^W$.

8.8. We have for $s > 0$, $t \ge 0$, $b \ge a$, $b \ge 0$;

$$P^0[W_{t+s} \le a, M_{t+s} \le b | \mathscr{F}_t] = P^0\left[W_{t+s} \le a, M_t \le b, \max_{0 \le u \le s} W_{t+u} \le b \,\middle|\, \mathscr{F}_t\right]$$

$$= 1_{\{M_t \le b\}} \, P^0\left[W_{t+s} \le a, \max_{0 \le u \le s} W_{t+u} \le b \,\middle|\, W_t\right].$$

The last expression is measurable with respect to the σ-field generated by the pair of random variables (W_t, M_t), and so the process $\{(W_t, M_t); \mathscr{F}_t; 0 \le t < \infty\}$ is Markov. Because Y_t is a function of (W_t, M_t), we have

$$P^0[Y_{t+s} \in \Gamma | \mathscr{F}_t] = P^0[Y_{t+s} \in \Gamma | W_t, M_t]; \quad \Gamma \in \mathscr{B}(\mathbb{R}).$$

In order to prove the Markov property for Y and (8.11), it suffices then to show that

$$P^0[Y_{t+s} \in dy \mid W_t = w, M_t = m] = p_+(s; m - w, y)\,dy.$$

Indeed, for $b > m \geq w$, $b \geq a$, $m \geq 0$ we have

$$P^0[W_{t+s} \in da, M_{t+s} \in db \mid W_t = w, M_t = m]$$

$$= P^0\left[W_{t+s} \in da, \max_{0 \leq u \leq s} W_{t+u} \in db \mid W_t = w, M_t = m\right]$$

$$= P^w[W_s \in da, M_s \in db] = P^0[W_s \in da - w, M_s \in db - w]$$

$$= \frac{2(2b - a - w)}{\sqrt{2\pi s^3}} \exp\left\{-\frac{(2b - a - w)^2}{2s}\right\} da\,db,$$

thanks to (8.2), which also gives

$$P^0[W_{t+s} \in da, M_{t+s} = m \mid W_t = w, M_t = m]$$

$$= P^0\left[W_{t+s} \in da, \max_{0 \leq u \leq s} W_{t+u} \leq m \mid W_t = w, M_t = m\right]$$

$$= P^w[W_s \in da, M_s \leq m] = P^0[W_s \in da - w, M_s \leq m - w]$$

$$= \frac{1}{\sqrt{2\pi s}}\left[\exp\left\{-\frac{(a - w)^2}{2s}\right\} - \exp\left\{-\frac{(2m - a - w)^2}{2s}\right\}\right].$$

Therefore, $P^0[Y_{t+s} \in dy \mid W_t = w, M_t = m]$ is equal to

$$\int_{(m,\infty)} P^0[W_{t+s} \in b - dy, M_{t+s} \in db \mid W_t = w, M_t = m]\,db$$

$$+ P^0[W_{t+s} \in m - dy, M_{t+s} = m \mid W_t = w, M_t = m] = p_+(s; m - w, y)\,dy.$$

Since the finite-dimensional distributions of a Markov process are determined by the initial distribution and the transition density, those of the processes $|W|$ and Y coincide.

8.12. The optional sampling theorem gives

$$e^{\lambda x} = E^x X_0 = E^x X_{t \wedge T_0 \wedge T_a} = E^x[\exp\{\lambda W_{t \wedge T_0 \wedge T_a} - \tfrac{1}{2}\lambda^2(t \wedge T_0 \wedge T_a)\}].$$

Since $W_{t \wedge T_0 \wedge T_a}$ is bounded, we may let $t \to \infty$ to obtain

$$e^{\lambda x} = E^x[\exp\{\lambda W_{T_0 \wedge T_a} - \tfrac{1}{2}\lambda^2(T_0 \wedge T_a)\}]$$

$$= E^x[1_{\{T_0 < T_a\}} e^{-\lambda^2 T_0/2}] + e^{\lambda a} E^x[1_{\{T_a < T_0\}} e^{-\lambda^2 T_a/2}].$$

By choosing $\lambda = \pm\sqrt{2\alpha}$, we obtain two equations which can be solved simultaneously and yield (8.27) and (8.28).

9.3. For every $0 < \sigma < \tau$, Doob's maximal inequality (Theorem 1.3.8 (iv)) gives

$$E\left[\sup_{\sigma \leq t \leq \tau}\left(\frac{W_t}{t}\right)^2\right] \leq \frac{1}{\sigma^2} E\left(\sup_{\sigma \leq t \leq \tau} W_t^2\right) \leq \frac{4}{\sigma^2} E W_\tau^2 = \frac{4\tau}{\sigma^2},$$

and with $\tau = 2^{n+1} = 2\sigma$,

$$P\left[\sup_{2^n \le t \le 2^{n+1}} \frac{|W_t|}{t} > \varepsilon\right] \le \frac{8}{\varepsilon^2} 2^{-n}$$

is seen to hold for every $\varepsilon > 0, n \ge 1$. The result (9.2) follows now from the Borel-Cantelli lemma.

9.5. Use Problem 7.18 and Lemma 9.4 (ii).

9.8. If $\Pi = \{t_0, t_1, \dots, t_m\}$ is a partition of $[0, t]$, we write $V_t^{(2)}(\Pi) - t = \sum_{k=1}^m \{(W_{t_k} - W_{t_{k-1}})^2 - (t_k - t_{k-1})\}$ as a sum of independent, zero-mean random variables. Therefore,

$$E(V_t^{(2)}(\Pi) - t)^2 = \sum_{k=1}^m (t_k - t_{k-1})^2 E\left[\frac{(W_{t_k} - W_{t_{k-1}})^2}{t_k - t_{k-1}} - 1\right]^2 \leqq tE(Z^2 - 1)^2 \|\Pi\|,$$

where Z is a standard normal random variable. The first assertion follows readily. For the second, we observe that

$$\sum_{n=1}^\infty P[|V_t^{(2)}(\Pi_n) - t| > \varepsilon] \leqq \frac{\text{const.}}{\varepsilon^2} \sum_{n=1}^\infty \|\Pi_n\| < \infty$$

holds for every $\varepsilon > 0$, and by the Borel-Cantelli lemma, $P[|V_t^{(2)}(\Pi_n) - t| > \varepsilon$, infinitely often$] = 0$. The conclusion follows.

9.11. (D. Freedman (1971))

(i) Each point of the set

$$M_n = \left\{t \in [0, \infty); f(s) < f(t), \forall s \in \left[\left(t - \frac{1}{n}\right)^+, t + \frac{1}{n}\right] \setminus \{t\}\right\}$$

is isolated, so M_n is countable. But $\bigcup_{n=1}^\infty M_n$ is the set of points of strict local maximum for f.

(ii) It suffices to show that f has a local maximum in an arbitrary, nonempty interval $(a, b) \subseteq [0, \infty)$. Let us begin by assuming $f(a) < f(b)$, and let t be the largest number in $[a, b)$ with $f(t) = f(a)$. Because f is not monotone such a t exists, and in (t, b) there are two numbers r and s with $r < s$ and $f(r) > f(t) \vee f(s)$. Being continuous, f must have a maximum over $[t, s]$, which is a local maximum in the sense of Definition 9.10 (iii).

If $f(a) > f(b)$, we apply the preceding argument in reverse, defining t to be the smallest number in $(a, b]$ with $f(t) = f(b)$. If $f(a) = f(b)$, then f must have a maximum over $[a, b]$ at some $r \in (a, b)$ and this r is also a point of local maximum.

9.22. Use integration by parts (Chung (1974), p. 231; McKean (1969), p. 4).

9.26. It suffices to show that for every $m \ge n \ge N_\theta(\omega)$, we have

$$|W_t(\omega) - W_s(\omega)| \le (1 + \varepsilon)\left[2 \sum_{j=n+1}^m g(2^{-j}) + g(t - s)\right] \tag{10.4}$$

valid for every $s, t \in D_m$ satisfying $0 < t - s < 2^{-n(1-\theta)}$. For $m = n$, (10.4) follows from (9.30). Let us assume that (10.4) holds for $m = n, \dots, M - 1$. With $s, t \in D_M$ and $0 < t - s < 2^{-n(1-\theta)}$, we consider, as in the proof of Theorem 2.8, the numbers $t^1 = \max\{u \in D_{M-1}; u \le t\}$ and $s^1 = \min\{u \in D_{M-1}; u \ge s\}$ and observe the relations $t - t^1 \le 2^{-M}$, $s^1 - s \le 2^{-M}$, and $0 \le t^1 - s^1 \le t - s < 2^{-n(1-\theta)} \le 1/e$, thanks to (9.29). We have

$$|W_{t^1}(\omega) - W_{s^1}(\omega)| \le (1 + \varepsilon)\left[2 \sum_{j=n+1}^{M-1} g(2^{-j}) + g(t^1 - s^1)\right]$$

by the induction assumption, and $|W_t(\omega) - W_{t^1}(\omega)| \le (1 + \varepsilon)g(2^{-M})$ as well as $|W_s(\omega) - W_{s^1}(\omega)| \le (1 + \varepsilon)g(2^{-M})$ because of (9.30). Since $g(t^1 - s^1) \le g(t - s)$, we conclude that (10.4) holds with $m = M$.

2.11. Notes

Section 2.1: The first quantitative work on Brownian motion is due to Bachelier (1900), who was interested in stock price fluctuations. Einstein (1905) derived the transition density for Brownian motion from the molecular-kinetic theory of heat. A rigorous mathematical treatment of Brownian motion began with N. Wiener (1923, 1924a), who provided the first existence proof.

The most profound work in this early period is that of P. Lévy (1939, 1948); he introduced the construction by interpolation expounded in Section 2.3, studied in detail the passage times and other related functionals (Section 2.8), described in detail the so-called fine structure of the typical sample path (Section 2.9), and discovered the notion and properties of the *mesure du voisinage* or "local time" (Section 3.6 and Chapter 6). Most amazingly, he carried out this program without the formal concepts and tools of filtrations, stopping times, or the strong Markov property.

Section 2.2: The construction of a probability measure from a consistent family of finite-dimensional distributions is clearly explained in Kolmogorov (1933); Daniell (1918/1919) had constructed earlier an integral on a space of sequences. The existence of a continuous modification under the conditions of Theorem 2.8 was established by Kolmogorov (published in Slutsky (1937)); Loève ((1978), p. 247) noticed that the same argument also provides local Hölder-continuity with exponent γ for any $0 < \gamma < \beta/\alpha$. For related results, see also Čentsov (1956a). The extension to random fields as in Problem 2.9 was carried out by Čentsov (1956b).

Section 2.3: The Haar function construction of Brownian motion was originally carried out by P. Lévy (1948) and later simplified by Ciesielski (1961). For a similar construction of Brownian motion indexed by directed sets, see Pyke (1983). Yor (1982) shows that the choice of the complete, orthonormal basis in $L^2[0,1]$ is not important for the construction of Brownian motion.

Section 2.4 is adapted from Billingsley (1968). The original proof of Theorem 4.20 is in Donsker (1951), but the one offered here is essentially due to Prohorov (1956). It is also possible to construct a probability space on which all the random walks are defined and converge to Brownian motion almost surely, rather than merely in distribution (Knight (1961)).

Sections 2.5, 2.6: The *Markov property* derives its name from A. A. Markov, whose own work (1906) was in discrete time and state space; in that context, of course, the *usual* and the *strong Markov properties* coincide. It was not immediately realized that the latter is actually stronger than the former; Ray ((1956), pp. 463–464) provides an example of a continuous Markov process which is not strongly Markov. It seems rather amazing today that a complete and rigorous statement about the strongly Markovian character of Brownian motion (Theorem 6.16) was proved only in 1956; see Hunt (1956).

A Markov family for which the function $x \mapsto E^x f(X_t)$ is continuous for any bounded, continuous $f\colon \mathbb{R}^d \to \mathbb{R}$ and $t \in [0, \infty)$ is said to have the *Feller property*, and a right-continuous Markov family with the Feller property is strongly Markovian. Very readable introductions to Markov process theory can be found in Dynkin & Yushkevich (1969), Wentzell ((1981), Chapters 8–13), and Chung (1982); more comprehensive treatments are those by Dynkin (1965), Blumenthal & Getoor (1968), and Ethier & Kurtz (1986). Markov processes with continuous sample paths receive very detailed treatments in the monographs by Itô & McKean (1974), Stroock & Varadhan (1979), and Knight (1981).

Section 2.7: The concept of *enlargement of a filtration* has become very important in Markov process theory and in stochastic integration. There is a substantial body of theory on this topic, which we do not take up here; we instead send the interested reader to the articles in the volume edited by Jeulin & Yor (1985).

Sections 2.8, 2.9: The material here comes mostly from P. Lévy (1939, 1948). Section 1.4 in D. Freedman (1971) was our source for Theorems 9.6, 9.9 and 9.12, and can be consulted for further information on this subject matter. Our discussion of the law of the iterated logarithm and of the Lévy modulus follows McKean (1969) and Williams (1979). Theorem 9.18 was strengthened by Dvoretzky (1963), who showed that there exists a universal constant $c > 0$ such that

$$P\left[\omega \in \Omega;\ \overline{\lim_{h \downarrow 0}} \frac{|W_{t+h}(\omega) - W_t(\omega)|}{\sqrt{h}} \geq c,\ \forall\, t \in [0, \infty)\right] = 1.$$

For every $\omega \in \Omega$, $\mathfrak{S}_\omega \triangleq \{t \in [0, \infty);\ \overline{\lim}_{h \downarrow 0}(|W_{t+h}(\omega) - W_t(\omega)|/\sqrt{h}) < \infty\}$ has been called by Kahane (1976) *the set of slow points from the right for the path* $W_\cdot(\omega)$. Fubini's theorem applied to (9.25) shows that $\text{meas}(\mathfrak{S}_\omega) = 0$ for P a.e. $\omega \in \Omega$, but, for a typical path, $\mathfrak{S}_\omega$ is far from being empty; in fact, we have

$$P\left[\omega \in \Omega;\ \inf_{0 \leq t < \infty} \overline{\lim_{h \downarrow 0}} \frac{|W_{t+h}(\omega) - W_t(\omega)|}{\sqrt{h}} = 1\right] = 1.$$

This is proved in B. Davis (1983), where we send the interested reader for more information and references on this subject.

Chung (1976) and Imhof (1984) offer excellent follow-up reading on the subject matter of Section 2.8.

CHAPTER 3

Stochastic Integration

3.1. Introduction

A tremendous range of problems in the natural, social, and biological sciences came under the dominion of the theory of functions of a real variable when Newton and Leibniz invented the calculus. The primary components of this invention were the use of differentiation to describe rates of change, the use of integration to pass to the limit in approximating sums, and the fundamental theorem of calculus, which relates the two concepts and thereby makes the latter amenable to computation. All of this gave rise to the concept of ordinary differential equations, and it is the application of these equations to the modeling of real-world phenomena which reveals much of the power of calculus.

Stochastic calculus grew out of the need to assign meaning to ordinary differential equations involving continuous stochastic processes. Since the most important such process, Brownian motion, cannot be differentiated, stochastic calculus takes the tack opposite to that of classical calculus: the stochastic integral is defined first, in Section 2, and then the stochastic differential is given meaning through the fundamental "theorem" of calculus. This "theorem" is really a definition in stochastic calculus, because the differential has no meaning apart from that assigned to it when it enters an integral. For this theory to achieve its full potential, it must have some simple rules for computation. These are contained in the change of variable formula (Itô's rule), which is the counterpart of the chain rule from classical calculus. We present it, together with some important applications, in Section 3.

Section 4 advances our recurrent theme that "Brownian motion is the fundamental martingale with continuous paths" by showing how to represent continuous, local martingales in terms of it, either via stochastic integration or via time-change. We also establish the representation of functionals of the Brownian path as stochastic integrals. The important Theorem 4.13 of F. B. Knight is established as an application of these ideas.

Stochastic calculus has a fundamental additional feature not found in its classical counterpart, a feature based on the *Girsanov change of measure* (Theorem 5.1). This result provides a device for solving stochastic differential equations driven by Brownian motion by changing the underlying probability measure, so that the process which was the driving Brownian motion becomes, under the new probability measure, the solution to the differential equation. This profound idea is first presented in Section 5, but does not reach its culmination until the discussion of weak solutions of stochastic differential equations in Chapter 5. In some cases, this device is merely a convenient way of finding the distribution of a functional of an already existent stochastic process; such an example is provided by the computations related to Brownian motion with drift in subsection 5.C. In other cases, the change of measure provides us with a proof of the existence of a solution to a stochastic differential equation, when the more standard methods fail. This is discussed in subsection 5.3.B. A particularly nice application of the Girsanov theorem appears in Section 5.8, where it is used in a model of security markets to remove the differences among the mean rates of return of the securities. This reduction of the model permits a complete analysis by martingale methods. Although "optional" in the sense that stochastic calculus can (and did for 20 years) exist and be useful without it, the Girsanov theorem today plays such a central role in further developments of the subject that the reader would be remiss not to come to grips with this admittedly difficult concept.

In Section 6 we employ the stochastic calculus in the study of P. Lévy's *mesure du voisinage* or *local time*, a device for measuring the "amount of time spent by the Brownian path in the vicinity of a certain point." This concept has become exceedingly important in both theory and applications; we examine its connections with reflected Brownian motion, extend with its help the Itô rule to functions which are not necessarily twice continuously differentiable, and use it as a tool in the study of certain additive functionals of the Brownian path. Finally, Section 7 extends the notion of local time to general, continuous semimartingales.

3.2. Construction of the Stochastic Integral

Let us consider a continuous, square-integrable martingale $M = \{M_t, \mathscr{F}_t; 0 \le t < \infty\}$ on a probability space $(\Omega, \mathscr{F}, P)$ equipped with the filtration $\{\mathscr{F}_t\}$, which will be assumed throughout this chapter to satisfy the usual conditions of Definition 1.2.25. We have shown in Section 2.7 how to obtain such a filtration for standard Brownian motion. We assume $M_0 = 0$ a.s. P. Such a process $M \in \mathscr{M}_2^c$ is of unbounded variation on any finite interval $[0, T]$

(cf. Problems 1.5.11, 1.5.12, and the discussion following them), and consequently integrals of the form

$$I_T(X) = \int_0^T X_t(\omega)\,dM_t(\omega) \tag{2.1}$$

cannot be defined "pathwise" (i.e., for each $\omega \in \Omega$ separately) as ordinary Lebesgue-Stieltjes integrals. Nevertheless, the martingale M has a finite second (or quadratic) variation, given by the continuous, increasing process $\langle M \rangle$; cf. Theorem 1.5.8. It is precisely this fact that allows one to proceed, in a highly nontrivial yet straightforward manner, with the construction of the stochastic integral (2.1) with respect to the continuous, square-integrable martingale M, for an appropriate class of integrands X. The construction is due to Itô (1942a), (1944) for the special case that M is a Brownian motion and to Kunita & Watanabe (1967) for general $M \in \mathscr{M}_2$. We shall first confine ourselves to $M \in \mathscr{M}_2^c$, and denote by $\langle M \rangle$ the unique (up to indistinguishability) adapted, continuous, and increasing process, such that $\{M_t^2 - \langle M \rangle_t, \mathscr{F}_t; 0 \le t < \infty\}$ is a martingale (cf. Definition 1.5.3 and Theorem 1.5.13). The construction will then be extended to general continuous, local martingales M.

We now consider what kinds of integrands are appropriate for (2.1). We first define a measure μ_M on $([0,\infty) \times \Omega, \mathscr{B}([0,\infty)) \otimes \mathscr{F})$ by setting

$$\mu_M(A) = E \int_0^\infty 1_A(t,\omega)\,d\langle M \rangle_t(\omega). \tag{2.2}$$

We shall say that two measurable, adapted processes $X = \{X_t, \mathscr{F}_t; 0 \le t < \infty\}$ and $Y = \{Y_t, \mathscr{F}_t; 0 \le t < \infty\}$ are *equivalent* if

$$X_t(\omega) = Y_t(\omega); \quad \mu_M\text{-a.e. } (t,\omega).$$

This introduces an equivalence relation. For a measurable, $\{\mathscr{F}_t\}$-adapted process X, we define

$$[X]_T^2 \triangleq E \int_0^T X_t^2\,d\langle M \rangle_t, \tag{2.3}$$

provided that the right-hand side is finite. Then $[X]_T$ is the L^2-norm for X, regarded as a function of (t,ω) restricted to the space $[0,T] \times \Omega$, under the measure μ_M. We have $[X - Y]_T = 0$ for all $T > 0$ if and only if X and Y are equivalent. The stochastic integral will be defined in such a manner that whenever X and Y are equivalent, then $I(X)$ and $I(Y)$ will be indistinguishable:

$$P[I_T(X) = I_T(Y), \ \forall 0 \le T < \infty] = 1.$$

2.1 Definition. Let $\mathscr{L}$ denote the set of equivalence classes of all *measurable, $\{\mathscr{F}_t\}$-adapted* processes X, for which $[X]_T < \infty$ for all $T > 0$. We define a metric on $\mathscr{L}$ by $[X - Y]$, where

$$[X] \triangleq \sum_{n=1}^\infty 2^{-n}(1 \wedge [X]_n).$$

Let $\mathscr{L}^*$ denote the set of equivalence classes of *progressively measurable* processes satisfying $[X]_T < \infty$ for all $T > 0$, and define a metric on $\mathscr{L}^*$ in the same way.

We shall follow the usual custom of not being very careful about the distinction between equivalence classes and the processes which are members of those equivalence classes. For example, we will have no qualms about saying, "$\mathscr{L}^*$ consists of those processes in $\mathscr{L}$ which are progressively measurable."

Note that $\mathscr{L}$ (respectively, $\mathscr{L}^*$) contains all bounded, measurable, $\{\mathscr{F}_t\}$-adapted (respectively, bounded, progressively measurable) processes. Both $\mathscr{L}$ and $\mathscr{L}^*$ depend on the martingale $M = \{M_t, \mathscr{F}_t; t \geq 0\}$. When we wish to indicate this dependence explicitly, we write $\mathscr{L}(M)$ and $\mathscr{L}^*(M)$.

If the function $t \mapsto \langle M \rangle_t(\omega)$ is absolutely continuous for P-a.e. ω, we shall be able to construct $\int_0^T X_t \, dM_t$ for all $X \in \mathscr{L}$ and all $T \geq 0$. In the absence of this condition on $\langle M \rangle$, we shall construct the stochastic integral for X in the slightly smaller class $\mathscr{L}^*$. In order to define the stochastic integral with respect to general martingales in $\mathscr{M}_2$ (possibly discontinuous, such as the compensated Poisson process), one has to select an even narrower class of integrands among the so-called *predictable processes*. This notion is a slight extension of left-continuity of the sample paths of the process; since we do not develop stochastic integration with respect to discontinuous martingales, we shall forego further discussion and send the interested reader to the literature: Kunita & Watanabe (1967), Meyer (1976), Liptser & Shiryaev (1977), Ikeda & Watanabe (1981), Elliott (1982), Chung & Williams (1983).

Later in this section, we weaken the conditions that $M \in \mathscr{M}_2^c$ and $[X]_T^2 < \infty$, $\forall\, T \geq 0$, replacing them by $M \in \mathscr{M}^{c,\text{loc}}$ and

$$P\left[\int_0^T X_t^2 \, d\langle M \rangle_t < \infty\right] = 1, \quad \forall\, T \geq 0.$$

This is accomplished by localization.

We pause in our development of the stochastic integral to prove a lemma we will need in Section 4. For $0 < T < \infty$, let $\mathscr{L}_T^*$ denote the class of processes X in $\mathscr{L}^*$ for which $X_t(\omega) = 0$; $\forall\, t > T$, $\omega \in \Omega$. For $T = \infty$, $\mathscr{L}_T^*$ is defined as the class of processes $X \in \mathscr{L}^*$ for which $E\int_0^T X_t^2 \, d\langle M \rangle_t < \infty$ (a condition we already have for $T < \infty$, by virtue of membership in $\mathscr{L}^*$). A process $X \in \mathscr{L}_T^*$ can be identified with one defined only for $(t, \omega) \in [0, T] \times \Omega$, and so we can regard $\mathscr{L}_T^*$ as a subspace of the Hilbert space

$$\mathscr{H}_T \triangleq L^2([0, T] \times \Omega, \mathscr{B}([0, T]) \otimes \mathscr{F}_T, \mu_M). \tag{2.4}$$

More precisely, we regard an equivalence class in $\mathscr{H}_T$ as a member of $\mathscr{L}_T^*$ if it contains a progressively measurable representative. Here and later we replace $[0, T]$ by $[0, \infty)$ when $T = \infty$.

2.2 Lemma. *For $0 < T \leq \infty$, $\mathscr{L}_T^*$ is a closed subspace of $\mathscr{H}_T$. In particular, $\mathscr{L}_T^*$ is complete under the norm $[X]_T$ of (2.3).*

PROOF. Let $\{X^{(n)}\}_{n=1}^{\infty}$ be a convergent sequence in $\mathscr{L}_T^*$ with limit $X \in \mathscr{H}_T$. We may extract a subsequence, also denoted by $\{X^{(n)}\}_{n=1}^{\infty}$, for which

$$\mu_M\{(t,\omega) \in [0,T] \times \Omega; \lim_{n\to\infty} X_t^{(n)}(\omega) \neq X_t(\omega)\} = 0.$$

By virtue of its membership in $\mathscr{H}_T$, X is $\mathscr{B}([0,T]) \otimes \mathscr{F}$-measurable, but may not be progressively measurable. However, with

$$A \triangleq \{(t,\omega) \in [0,T] \times \Omega; \lim_{n\to\infty} X_t^{(n)}(\omega) \text{ exists in } \mathbb{R}\},$$

the process

$$Y_t(\omega) \triangleq \begin{cases} \lim_{n\to\infty} X_t^{(n)}(\omega); & (t,\omega) \in A \\ 0; & (t,\omega) \notin A \end{cases}$$

inherits progressive measurability from $\{X^{(n)}\}_{n=1}^{\infty}$ and is equivalent to X. □

A. Simple Processes and Approximations

2.3 Definition. A process X is called *simple* if there exists a strictly increasing sequence of real numbers $\{t_n\}_{n=0}^{\infty}$ with $t_0 = 0$ and $\lim_{n\to\infty} t_n = \infty$, as well as a sequence of random variables $\{\xi_n\}_{n=0}^{\infty}$ and a nonrandom constant $C < \infty$ with $\sup_{n\geq 0}|\xi_n(\omega)| \leq C$, for every $\omega \in \Omega$, such that ξ_n is $\mathscr{F}_{t_n}$-measurable for every $n \geq 0$ and

$$X_t(\omega) = \xi_0(\omega)1_{\{0\}}(t) + \sum_{i=0}^{\infty} \xi_i(\omega)1_{(t_i,t_{i+1}]}(t); \quad 0 \leq t < \infty, \omega \in \Omega.$$

The class of all simple processes will be denoted by $\mathscr{L}_0$. Note that, because members of $\mathscr{L}_0$ are progressively measurable and bounded, we have $\mathscr{L}_0 \subseteq \mathscr{L}^*(M) \subseteq \mathscr{L}(M)$.

Our program for the construction of the stochastic integral (2.1) can now be outlined as follows: the integral is defined in the obvious way for $X \in \mathscr{L}_0$ as a *martingale transform*:

$$(2.5) \qquad I_t(X) \triangleq \sum_{i=0}^{n-1} \xi_i(M_{t_{i+1}} - M_{t_i}) + \xi_n(M_t - M_{t_n})$$

$$= \sum_{i=0}^{\infty} \xi_i(M_{t\wedge t_{i+1}} - M_{t\wedge t_i}), \quad 0 \leq t < \infty,$$

where $n \geq 0$ is the unique integer for which $t_n \leq t < t_{n+1}$. The definition is then extended to integrands $X \in \mathscr{L}^*$ and $X \in \mathscr{L}$, thanks to the crucial results which show that elements of $\mathscr{L}^*$ and $\mathscr{L}$ can be approximated, in a suitable sense, by simple processes (Propositions 2.6 and 2.8).

2.4 Lemma. *Let X be a bounded, measurable, $\{\mathscr{F}_t\}$-adapted process. Then there exists a sequence $\{X^{(m)}\}_{m=1}^{\infty}$ of simple processes such that*

(2.6) $$\sup_{T>0} \lim_{m\to\infty} E \int_0^T |X_t^{(m)} - X_t|^2\, dt = 0.$$

PROOF. We shall show how to construct, for each fixed $T > 0$, a sequence $\{X^{(n,T)}\}_{n=1}^\infty$ of simple processes so that

$$\lim_{n\to\infty} E \int_0^T |X_t^{(n,T)} - X_t|^2\, dt = 0.$$

Thus, for each positive integer m, there is another integer n_m such that

$$E \int_0^m |X_t^{(n_m, m)} - X_t|^2\, dt \le \frac{1}{m},$$

and the sequence $\{X^{(n_m, m)}\}_{m=1}^\infty$ has the desired properties. Henceforth, T is a fixed, positive number.

We proceed in three steps.

(a) Suppose that X is *continuous*; then the sequence of simple processes

$$X_t^{(n)}(\omega) \triangleq X_0(\omega) 1_{\{0\}}(t) + \sum_{k=0}^{2^n - 1} X_{kT/2^n}(\omega) 1_{(kT/2^n, (k+1)T/2^n]}(t); \quad n \ge 1,$$

satisfies $\lim_{n\to\infty} E \int_0^T |X_t^{(n)} - X_t|^2\, dt = 0$ by the bounded convergence theorem.

(b) Now suppose that X is *progressively measurable*; we consider the continuous, progressively measurable processes

(2.7) $$F_t(\omega) \triangleq \int_0^{t\wedge T} X_s(\omega)\, ds; \quad \tilde{X}_t^{(m)}(\omega) \triangleq m[F_t(\omega) - F_{(t-1/m)\vee 0}(\omega)]; \quad m \ge 1,$$

for $t \ge 0$, $\omega \in \Omega$ (cf. Problem 1.2.19). By virtue of step (a), there exists, for each $m \ge 1$, a sequence of simple processes $\{\tilde{X}^{(m,n)}\}_{n=1}^\infty$ such that $\lim_{n\to\infty} E \int_0^T |\tilde{X}_t^{(m,n)} - \tilde{X}_t^{(m)}|^2\, dt = 0$. Let us consider the $\mathscr{B}([0,T]) \otimes \mathscr{F}_T$-measurable product set

$$A \triangleq \{(t,\omega) \in [0,T] \times \Omega; \lim_{m\to\infty} \tilde{X}_t^{(m)}(\omega) = X_t(\omega)\}^c.$$

For each $\omega \in \Omega$, the cross section $A_\omega \triangleq \{t \in [0,T]; (t,\omega) \in A\}$ is in $\mathscr{B}([0,T])$ and, according to the fundamental theorem of calculus, has Lebesgue measure zero. The bounded convergence theorem now gives $\lim_{m\to\infty} E \int_0^T |\tilde{X}_t^{(m)} - X_t|^2\, dt = 0$, and so a sequence $\{\tilde{X}^{(m, n_m)}\}_{m=1}^\infty$ of bounded, simple processes can be chosen, for which

$$\lim_{m\to\infty} E \int_0^T |\tilde{X}_t^{(m, n_m)} - X_t|^2\, dt = 0.$$

(c) Finally, let X be *measurable and adapted*. We cannot guarantee immediately that the continuous process $F = \{F_t; 0 \le t < \infty\}$ in (2.7) is progressively measurable, because we do not know whether it is adapted. We do know, however, that the process X has a progressively measurable modification Y (Proposition 1.1.12), and we now show that the progressively measurable process $\{G_t \triangleq \int_0^{t\wedge T} Y_s\,ds, \mathscr{F}_t; 0 \le t \le T\}$ is a modification of F. For the measurable process $\eta_t(\omega) = 1_{\{X_t(\omega)\neq Y_t(\omega)\}}$; $0 \le t \le T$, $\omega \in \Omega$, we have from Fubini: $E\int_0^T \eta_t(\omega)\,dt = \int_0^T P[X_t(\omega) \neq Y_t(\omega)]\,dt = 0$. Therefore, $\int_0^T \eta_t(\omega)\,dt = 0$ for P-a.e. $\omega\in\Omega$. Now $\{F_t \neq G_t\}$ is contained in the event $\{\omega; \int_0^T \eta_t(\omega)\,dt > 0\}$, G_t is $\mathscr{F}_t$-measurable, and, by assumption, $\mathscr{F}_t$ contains all subsets of P-null events. Therefore, F_t is also $\mathscr{F}_t$-measurable. Adaptivity and continuity imply progressive measurability, and we may now repeat verbatim the argument in (b). □

2.5 Problem. This problem outlines a method by which the use of Proposition 1.1.12, a result not proved in this text, can be avoided in part (c) of the proof of Lemma 2.4. Let X be a bounded, measurable, $\{\mathscr{F}_t\}$-adapted process. Let $0 < T < \infty$ be fixed. We wish to construct a sequence $\{X^{(k)}\}_{k=1}^\infty$ of simple processes so that

$$\lim_{k\to\infty} E\int_0^T |X_t^{(k)} - X_t|^2\,dt = 0. \tag{2.8}$$

To simplify notation, we set $X_t = 0$ for $t \le 0$. Let $\varphi_n\colon \mathbb{R} \to \{j2^{-n}; j = 0, \pm 1, \pm 2, \ldots\}$ be given by

$$\varphi_n(t) = \frac{j-1}{2^n} \quad \text{for} \quad \frac{j-1}{2^n} < t \le \frac{j}{2^n}.$$

(a) Fix $s \ge 0$. Show that $t - (1/2^n) \le \varphi_n(t-s) + s < t$, and that

$$X_t^{(n,s)} \triangleq X_{\varphi_n(t-s)+s}, \mathscr{F}_t; \quad t \ge 0$$

is a simple, adapted process.

(b) Show that $\lim_{h\downarrow 0} E\int_0^T |X_t - X_{t-h}|^2\,dt = 0$.

(c) Use (a) and (b) to show that

$$\lim_{n\to\infty} E\int_0^T\int_0^1 |X_t^{(n,s)} - X_t|^2\,ds\,dt = 0.$$

(d) Show that for some choice of $s \ge 0$ and some increasing sequence $\{n_k\}_{k=1}^\infty$ of integers, (2.8) holds with $X^{(k)} = X^{(n_k,s)}$.

This argument is adapted from Liptser and Shiryaev (1977).

2.6 Proposition. *If the function $t \mapsto \langle M\rangle_t(\omega)$ is absolutely continuous with respect to Lebesgue measure for P-a.e. $\omega\in\Omega$, then $\mathscr{L}_0$ is dense in $\mathscr{L}$ with respect to the metric of Definition* 2.1.

PROOF.

(a) If $X \in \mathscr{L}$ is bounded, then Lemma 2.4 guarantees the existence of a bounded sequence $\{X^{(m)}\}$ of simple processes satisfying (2.6). From these we extract a subsequence $\{X^{(m_k)}\}$, such that the set

$$\{(t, \omega) \in [0, \infty) \times \Omega; \lim_{k\to\infty} X_t^{(m_k)}(\omega) = X_t(\omega)\}^c$$

has product measure zero. The absolute continuity of $t \mapsto \langle M \rangle_t(\omega)$ and the bounded convergence theorem now imply $[X^{(m_k)} - X] \to 0$ as $k \to \infty$.

(b) If $X \in \mathscr{L}$ is not necessarily bounded, we define

$$X_t^{(n)}(\omega) \triangleq X_t(\omega) 1_{\{|X_t(\omega)| \le n\}}; \quad 0 \le t < \infty, \omega \in \Omega,$$

and thereby obtain a sequence of bounded processes in $\mathscr{L}$. The dominated convergence theorem implies

$$[X^{(n)} - X]_T^2 = E \int_0^T X_t^2 1_{\{|X_t| > n\}} \, d\langle M \rangle_t \xrightarrow[n\to\infty]{} 0$$

for every $T > 0$, whence $\lim_{n\to\infty} [X^{(n)} - X] = 0$. Each $X^{(n)}$ can be approximated by bounded, simple processes, so X can be as well. □

When $t \mapsto \langle M \rangle_t$ is not an absolutely continuous function of the time variable t, we simply choose a more convenient clock. We show how to do this in slightly greater generality than needed for the present application.

2.7 Lemma. *Let $\{A_t; 0 \le t < \infty\}$ be a continuous, increasing (Definition 1.4.4) process adapted to the filtration of the martingale $M = \{M_t, \mathscr{F}_t; 0 \le t < \infty\}$. If $X = \{X_t, \mathscr{F}_t; 0 \le t < \infty\}$ is a progressively measurable process satisfying*

$$E \int_0^T X_t^2 \, dA_t < \infty$$

for each $T > 0$, then there exists a sequence $\{X^{(n)}\}_{n=1}^\infty$ of simple processes such that

$$\sup_{T>0} \lim_{n\to\infty} E \int_0^T |X_t^{(n)} - X_t|^2 \, dA_t = 0.$$

PROOF. We may assume without loss of generality that X is bounded (cf. part (b) in the proof of Proposition 2.6), i.e.,

$$|X_t(\omega)| \le C < \infty; \quad \forall t \ge 0, \omega \in \Omega. \tag{2.9}$$

As in the proof of Lemma 2.4, it suffices to show how to construct, for each fixed $T > 0$, a sequence $\{X^{(n)}\}_{n=1}^\infty$ of simple processes for which

$$\lim_{n\to\infty} E \int_0^T |X_t^{(n)} - X_t|^2 \, dA_t = 0.$$

Henceforth $T > 0$ is fixed, and we assume without loss of generality that

$$X_t(\omega) = 0; \quad \forall t > T, \omega \in \Omega. \tag{2.10}$$

We now describe the time-change. Since $A_t(\omega) + t$ is strictly increasing in $t \geq 0$ for P-a.e. ω, there is a continuous, strictly increasing inverse function $T_s(\omega)$, defined for $s \geq 0$, such that

$$A_{T_s(\omega)}(\omega) + T_s(\omega) = s; \quad \forall s \geq 0.$$

In particular, $T_s \leq s$ and $\{T_s \leq t\} = \{A_t + t \geq s\} \in \mathscr{F}_t$. Thus, for each $s \geq 0$, T_s is a bounded stopping time for $\{\mathscr{F}_t\}$. Taking s as our new time-variable, we define a new filtration $\{\mathscr{G}_s\}$ by

$$\mathscr{G}_s = \mathscr{F}_{T_s}; \quad s \geq 0,$$

and introduce the time-changed process

$$Y_s(\omega) = X_{T_s(\omega)}(\omega); \quad s \geq 0, \omega \in \Omega,$$

which is adapted to $\{\mathscr{G}_s\}$ because of the progressive measurability of X (Proposition 1.2.18). Lemma 2.4 implies that, given any $\varepsilon > 0$ and $R > 0$, there is a simple process $\{Y_s^\varepsilon, \mathscr{G}_s; 0 \leq s < \infty\}$ for which

$$E\int_0^R |Y_s^\varepsilon - Y_s|^2\,ds < \varepsilon/2. \tag{2.11}$$

But from (2.9), (2.10) it develops that

$$\begin{aligned} E\int_0^\infty Y_s^2\,ds &= E\int_0^\infty 1_{\{T_s \leq T\}} X_{T_s}^2\,ds \\ &= E\int_0^{A_T+T} X_{T_s}^2\,ds \leq C^2(EA_T + T) < \infty, \end{aligned}$$

so by choosing R in (2.11) sufficiently large and setting $Y_s^\varepsilon = 0$ for $s > R$, we can obtain

$$E\int_0^\infty |Y_s^\varepsilon - Y_s|^2\,ds < \varepsilon.$$

Now Y_s^ε is simple, and because it vanishes for $s > R$, there is a finite partition $0 = s_0 < s_1 < \cdots < s_n \leq R$ with

$$Y_s^\varepsilon(\omega) = \xi_0(\omega)1_{\{0\}}(s) + \sum_{j=1}^n \xi_{s_{j-1}}(\omega)1_{(s_{j-1},s_j]}(s), \quad 0 \leq s < \infty,$$

where each ξ_{s_j} is measurable with respect to $\mathscr{G}_{s_j} = \mathscr{F}_{T_{s_j}}$ and bounded in absolute value by a constant, say K. Reverting to the original clock, we observe that

$$X_t^\varepsilon \triangleq Y_{t+A_t}^\varepsilon = \xi_0 1_{\{0\}}(t) + \sum_{j=1}^n \xi_{s_{j-1}} 1_{(T_{s_{j-1}},T_{s_j}]}(t), \quad 0 \leq t < \infty,$$

is measurable and adapted, because ξ_{s_j} restricted to $\{T_{s_j} < t\}$ is $\mathscr{F}_t$-measurable (Lemma 1.2.15). We have

$$E\int_0^T |X_t^\varepsilon - X_t|^2\,dA_t \le E\int_0^T |X_t^\varepsilon - X_t|^2(dA_t + dt)$$
$$\le E\int_0^\infty |Y_s^\varepsilon - Y_s|^2\,ds < \varepsilon.$$

The proof is not yet complete because X^ε is not a simple process. To finish it off, we must show how to approximate

$$\eta_t(\omega) \triangleq \xi_{s_{j-1}}(\omega)1_{(T_{s_{j-1}}(\omega),\,T_{s_j}(\omega)]}(t);\quad 0 \le t < \infty,\ \omega\in\Omega,$$

by simple processes. Recall that $T_{s_{j-1}} \le T_{s_j} \le s_j$ and simplify notation by taking $s_{j-1} = 1$, $s_j = 2$. Set

$$T_i^{(m)}(\omega) = \sum_{k=1}^{1+2^{m+1}} \frac{k}{2^m} 1_{[(k-1)/2^m,\,k/2^m)}(T_i(\omega)),\quad i = 1, 2$$

and define

$$\eta_t^{(m)}(\omega) \triangleq \xi_1(\omega)1_{(T_1^{(m)}(\omega),\,T_2^{(m)}(\omega)]}(t)$$
$$= \sum_{k=1}^{2^{m+1}} \xi_1(\omega)1_{\{T_1<(k-1)/2^m\le T_2\}}(\omega)1_{((k-1)/2^m,\,k/2^m]}(t).$$

Because $\{T_1 < (k-1)/2^m \le T_2\} \in \mathscr{F}_{(k-1)/2^m}$ and ξ_1 restricted to $\{T_1 < (k-1)/2^m\}$ is $\mathscr{F}_{(k-1)/2^m}$-measurable, $\eta^{(m)}$ is simple. Furthermore,

$$E\int_0^\infty |\eta_t^{(m)} - \eta_t|^2\,dA_t \le K^2[E(A_{T_2^{(m)}} - A_{T_2}) + E(A_{T_1^{(m)}} - A_{T_1})] \xrightarrow[m\to\infty]{} 0. \qquad \square$$

2.8 Proposition. *The set $\mathscr{L}_0$ of simple processes is dense in $\mathscr{L}^*$ with respect to the metric of Definition* 2.1.

Proof. Take $A = \langle M\rangle$ in Lemma 2.7.

B. Construction and Elementary Properties of the Integral

We have already defined the stochastic integral of a simple process $X \in \mathscr{L}_0$ by the recipe (2.5). Let us list certain properties of this integral: for $X, Y \in \mathscr{L}_0$ and $0 \le s < t < \infty$, we have

(2.12) $$I_0(X) = 0,\quad \text{a.s. } P$$

(2.13) $$E[I_t(X)|\mathscr{F}_s] = I_s(X),\quad \text{a.s. } P$$

(2.14) $$E(I_t(X))^2 = E\int_0^t X_u^2\,d\langle M\rangle_u$$

(2.15) $$\|I(X)\| = [X]$$

(2.16) $$E[(I_t(X) - I_s(X))^2 | \mathscr{F}_s] = E\left[\int_s^t X_u^2 \, d\langle M \rangle_u \middle| \mathscr{F}_s\right], \quad \text{a.s. } P$$

(2.17) $$I(\alpha X + \beta Y) = \alpha I(X) + \beta I(Y); \quad \alpha, \beta \in \mathbb{R}.$$

Properties (2.12) and (2.17) are obvious. Property (2.13) follows from the fact that for any $0 \le s < t < \infty$ and any integer $i \ge 1$, we have, in the notation of (2.5),

$$E[\xi_i(M_{t \wedge t_{i+1}} - M_{t \wedge t_i}) | \mathscr{F}_s] = \xi_i(M_{s \wedge t_{i+1}} - M_{s \wedge t_i}), \quad \text{a.s. } P;$$

this can be verified separately for each of the three cases $s \le t_i$, $t_i < s \le t_{i+1}$, and $t_{i+1} < s$ by using the $\mathscr{F}_{t_i}$-measurability of ξ_i. Thus, we see that $I(X) = \{I_t(X), \mathscr{F}_t; 0 \le t < \infty\}$is a *continuous martingale*. With $0 \le s < t < \infty$ and m and n chosen so that $t_{m-1} \le s < t_m$ and $t_n \le t < t_{n+1}$, we have (cf. the discussion preceding Lemma 1.5.9)

(2.18)

$$\begin{aligned}
&E[(I_t(X) - I_s(X))^2 | \mathscr{F}_s] \\
&= E\left[\left\{\xi_{m-1}(M_{t_m} - M_s) + \sum_{i=m}^{n-1} \xi_i(M_{t_{i+1}} - M_{t_i}) + \xi_n(M_t - M_{t_n})\right\}^2 \middle| \mathscr{F}_s\right] \\
&= E\left[\xi_{m-1}^2(M_{t_m} - M_s)^2 + \sum_{i=m}^{n-1} \xi_i^2(M_{t_{i+1}} - M_{t_i})^2 + \xi_n^2(M_t - M_{t_n})^2 \middle| \mathscr{F}_s\right] \\
&= E\Bigg[\xi_{m-1}^2(\langle M \rangle_{t_m} - \langle M \rangle_s) + \sum_{i=m}^{n-1} \xi_i^2(\langle M \rangle_{t_{i+1}} - \langle M \rangle_{t_i}) \\
&\qquad + \xi_n^2(\langle M \rangle_t - \langle M \rangle_{t_n}) \Bigg| \mathscr{F}_s\Bigg] \\
&= E\left[\int_s^t X_u^2 \, d\langle M \rangle_u \middle| \mathscr{F}_s\right].
\end{aligned}$$

This proves (2.16) and establishes the fact that the continuous martingale $I(X)$ is *square-integrable*: $I(X) \in \mathscr{M}_2^c$, with quadratic variation

(2.19) $$\langle I(X) \rangle_t = \int_0^t X_u^2 \, d\langle M \rangle_u.$$

Setting $s = 0$ and taking expectations in (2.16), we obtain (2.14), and (2.15) follows immediately, upon recalling Definition 1.5.22.

For $X \in \mathscr{L}^*$, Proposition 2.8 implies the existence of a sequence $\{X^{(n)}\}_{n=1}^\infty \subseteq \mathscr{L}_0$ such that $[X^{(n)} - X] \to 0$ as $n \to \infty$. It follows from (2.15) and (2.17) that

$$\|I(X^{(n)}) - I(X^{(m)})\| = \|I(X^{(n)} - X^{(m)})\| = [X^{(n)} - X^{(m)}] \to 0$$

as $n, m \to \infty$. In other words, $\{I(X^{(n)})\}_{n=1}^\infty$ is a Cauchy sequence in $\mathscr{M}_2^c$. By Proposition 1.5.23, there exists a process $I(X) = \{I_t(X); 0 \le t < \infty\}$ in $\mathscr{M}_2^c$, defined modulo indistinguishability, such that $\|I(X^{(n)}) - I(X)\| \to 0$ as $n \to \infty$. Because it belongs to $\mathcal{M}_2^c$, $I(X)$ enjoys properties (2.12) and (2.13). For $0 \le s < t < \infty$, $\{I_s(X^{(n)})\}_{n=1}^\infty$ and $\{I_t(X^{(n)})\}_{n=1}^\infty$ converge in mean-square to $I_s(X)$ and $I_t(X)$, respectively; so for $A \in \mathscr{F}_s$, (2.16) applied to $\{X^{(n)}\}_{n=1}^\infty$ gives

$$E[1_A(I_t(X) - I_s(X))^2] = \lim_{n\to\infty} E[1_A(I_t(X^{(n)}) - I_s(X^{(n)}))^2] \tag{2.20}$$

$$= \lim_{n\to\infty} E\left[1_A \int_s^t (X_u^{(n)})^2 \, d\langle M\rangle_u\right]$$

$$= E\left[1_A \int_s^t X_u^2 \, d\langle M\rangle_u\right],$$

where the last equality follows from $\lim_{n\to\infty} [X^{(n)} - X]_t = 0$. This proves that $I(X)$ also satisfies (2.16) and, consequently, (2.14) and (2.15). Because X and M are progressively measurable, $\int_s^t X_u^2 \, d\langle M\rangle_u$ is $\mathscr{F}_t$-measurable for fixed $0 \le s < t < \infty$, and so (2.16) gives us (2.19). The validity of (2.17) for $X, Y \in \mathscr{L}^*$ also follows from its validity for processes in $\mathscr{L}_0$, upon passage to the limit.

The process $I(X)$ for $X \in \mathscr{L}^*$ is *well defined*; if we have two sequences $\{X^{(n)}\}_{n=1}^\infty$ and $\{Y^{(n)}\}_{n=1}^\infty$ in $\mathscr{L}_0$ with the property $\lim_{n\to\infty} [X^{(n)} - X] = 0$, $\lim_{n\to\infty} [Y^{(n)} - X] = 0$, we can construct a third sequence $\{Z^{(n)}\}_{n=1}^\infty$ with this property, by setting $Z^{(2n-1)} = X^{(n)}$ and $Z^{(2n)} = Y^{(n)}$, for $n \ge 1$. The limit $I(X)$ of the sequence $\{I(Z^{(n)})\}_{n=1}^\infty$ in $\mathcal{M}_2^c$ has to agree with the limits of both sequences, namely $\{I(X^{(n)})\}_{n=1}^\infty$ and $\{I(Y^{(n)})\}_{n=1}^\infty$.

2.9 Definition. For $X \in \mathscr{L}^*$, the *stochastic integral of X with respect to the martingale $M \in \mathcal{M}_2^c$* is the unique, square-integrable martingale $I(X) = \{I_t(X), \mathscr{F}_t; 0 \le t < \infty\}$ which satisfies $\lim_{n\to\infty} \|I(X^{(n)}) - I(X)\| = 0$, for every sequence $\{X^{(n)}\}_{n=1}^\infty \subseteq \mathscr{L}_0$ with $\lim_{n\to\infty} [X^{(n)} - X] = 0$. We write

$$I_t(X) = \int_0^t X_s \, dM_s; \quad 0 \le t < \infty.$$

2.10 Proposition. *For $M \in \mathcal{M}_2^c$ and $X \in \mathscr{L}^*$, the stochastic integral $I(X) = \{I_t(X), \mathscr{F}_t; 0 \le t < \infty\}$ of X with respect to M satisfies* (2.12)–(2.16), *as well as* (2.17) *for every $Y \in \mathscr{L}^*$, and has quadratic variation process given by* (2.19). *Furthermore, for any two stopping times $S \le T$ of the filtration $\{\mathscr{F}_t\}$ and any number $t > 0$, we have*

$$E[I_{t\wedge T}(X) | \mathscr{F}_S] = I_{t\wedge S}(X), \quad a.s.\ P. \tag{2.21}$$

With $X, Y \in \mathscr{L}^*$ we have, a.s. P:

$$E[(I_{t\wedge T}(X) - I_{t\wedge S}(X))(I_{t\wedge T}(Y) - I_{t\wedge S}(Y)) | \mathscr{F}_S] \tag{2.22}$$

$$= E\left[\int_{t\wedge S}^{t\wedge T} X_u Y_u \, d\langle M\rangle_u \,\middle|\, \mathscr{F}_S\right],$$

and in particular, for any number s in $[0, t]$,

(2.23) $E[(I_t(X) - I_s(X))(I_t(Y) - I_s(Y))|\mathscr{F}_s] = E\left[\int_s^t X_u Y_u \, d\langle M\rangle_u \middle| \mathscr{F}_s\right].$

Finally,

(2.24) $$I_{t\wedge T}(X) = I_t(\tilde{X}) \quad a.s.,$$

where $\tilde{X}_t(\omega) \triangleq X_t(\omega)1_{\{t \le T(\omega)\}}$.

PROOF. We have already established (2.12)–(2.17) and (2.19). From (2.13) and the optional sampling theorem (Problem 1.3.24 (ii)), we obtain (2.21). The same result applied to the martingale $\{I_t^2(X) - \int_0^t X_u^2 \, d\langle M\rangle_u, \mathscr{F}_t; t \ge 0\}$ provides the identities

$$\begin{aligned} E[(I_{t\wedge T}(X) - I_{t\wedge S}(X))^2|\mathscr{F}_S] &= E[I_{t\wedge T}^2(X) - I_{t\wedge S}^2(X)|\mathscr{F}_S] \\ &= E\left[\int_{t\wedge S}^{t\wedge T} X_u^2 \, d\langle M\rangle_u \middle| \mathscr{F}_S\right], \quad P\text{-a.s.} \end{aligned}$$

Replacing X in this equation, first by $X + Y$ and then by $X - Y$, and subtracting the resulting equations, we obtain (2.22).

It remains to prove (2.24). We write

$$I_{t\wedge T}(X) - I_t(\tilde{X}) = I_{t\wedge T}(X - \tilde{X}) - (I_t(\tilde{X}) - I_{t\wedge T}(\tilde{X})).$$

Both $\{I_{t\wedge T}(X - \tilde{X}), \mathscr{F}_t; t \ge 0\}$ and $\{I_t(\tilde{X}) - I_{t\wedge T}(\tilde{X}), \mathscr{F}_t; t \ge 0\}$ are in $\mathscr{M}_2^c$; we show that they both have quadratic variation zero, and then appeal to Problem 1.5.12. Now relation (2.22) gives, for the first process,

$$\begin{aligned} &E[(I_{t\wedge T}(X - \tilde{X}) - I_{s\wedge T}(X - \tilde{X}))^2|\mathscr{F}_s] \\ &\quad = E\left[\int_{s\wedge T}^{t\wedge T} (X_u - \tilde{X}_u)^2 \, d\langle M\rangle_u \middle| \mathscr{F}_s\right] = 0 \end{aligned}$$

a.s. P, which gives the desired conclusion. As for the second process, we have

$$E[(I_t(\tilde{X}) - I_{t\wedge T}(\tilde{X}))^2] = E\left[\int_{t\wedge T}^t \tilde{X}_u^2 \, d\langle M\rangle_u\right] = 0,$$

and since this is the expectation of the quadratic variation of this process, we again have the desired result. □

2.11 Remark. If the sample paths $t \mapsto \langle M\rangle_t(\omega)$ of the quadratic variation process $\langle M\rangle$ are absolutely continuous functions of t for P-a.e. ω, then Proposition 2.6 can be used instead of Proposition 2.8 to define $I(X)$ for every $X \in \mathscr{L}$. We have $I(X) \in \mathscr{M}_2^c$ and all the properties of Proposition 2.10 in this case. The only sticking point in the preceding arguments under these conditions is the proof that the measurable process $F_t \triangleq \int_0^t X_s^2 \, d\langle M\rangle_s$ is $\{\mathscr{F}_t\}$-adapted. To see that it is, we can choose Y, a progressively measurable modification of X (Proposition 1.1.12), and define the progressively measurable process $G_t \triangleq \int_0^t Y_s^2\, d\langle M\rangle_s$. Following the proof of Lemma 2.4, step (c), we can then show that $P[F_t = G_t] = 1$ holds for every $t \geq 0$. Because G_t is $\mathscr{F}_t$-measurable, and $\mathscr{F}_t$ contains all P-negligible events in $\mathscr{F}$ (the usual conditions), F is easily seen to be adapted to $\{\mathscr{F}_t\}$.

In the important case that M is standard Brownian motion with $\langle M\rangle_t = t$, the use of the unproven Proposition 1.1.12 can again be avoided. For bounded X, Problem 2.5 shows how to construct a sequence $\{X^{(k)}\}_{k=1}^{\infty}$ of bounded, simple processes so that (2.8) holds; in particular, there is a subsequence, also called $\{X^{(k)}\}_{k=1}^{\infty}$, such that for almost every $t \in [0, T]$ we have

$$F_t \triangleq \int_0^t X_s^2\, ds = \lim_{k\to\infty} \int_0^t (X_s^{(k)})^2\, ds, \quad \text{a.s. } P.$$

Since the right-hand side is $\mathscr{F}_t$-measurable and $\mathscr{F}_t$ contains all null events in $\mathscr{F}$, the left-hand side is also $\mathscr{F}_t$-measurable for a.e. $t \in [0, T]$. The continuity of the samples paths of $\{F_t; t \geq 0\}$ leads to the conclusion that this process is $\mathscr{F}_t$-measurable for *every* t. For unbounded X, we use the localization technique employed in the proof of Proposition 2.6.

We shall not continue to deal explicitly with the case of absolutely continuous $\langle M\rangle$ and $X \in \mathscr{L}$, but all results obtained for $X \in \mathscr{L}^*$ can be modified in the obvious way to account for this case. In later applications involving stochastic integrals with respect to martingales whose quadratic variations are absolutely continuous, we shall require only measurability and adaptivity rather than progressive measurability of integrands. □

2.12 Problem. Let $W = \{W_t, \mathscr{F}_t; 0 \leq t < \infty\}$ be a standard, one-dimensional Brownian motion, and let T be a stopping time of $\{\mathscr{F}_t\}$ with $ET < \infty$. Prove the *Wald identities*

$$E(W_T) = 0, \quad E(W_T^2) = ET.$$

(*Warning*: The optional sampling theorem cannot be applied directly because W does not have a last element and T may not be bounded. The stopping time $t \wedge T$ is bounded for fixed $0 \leq t < \infty$, so $E(W_{t\wedge T}) = 0$, $E(W_{t\wedge T}^2) = E(t \wedge T)$, but it is not a priori evident that

$$\text{(2.25)} \qquad \lim_{t\to\infty} E(W_{t\wedge T}) = EW_T, \quad \lim_{t\to\infty} E(W_{t\wedge T}^2) = E(W_T^2).)$$

2.13 Exercise. Let W be as in Problem 2.12, let b be a real number, and let T_b be the passage time to b of (2.6.1). Use Problem 2.12 to show that for $b \neq 0$, we have $ET_b = \infty$.

C. A Characterization of the Integral

Suppose $M = \{M_t, \mathscr{F}_t; 0 \leq t < \infty\}$ and $N = \{N_t, \mathscr{F}_t; 0 \leq t < \infty\}$ are in $\mathscr{M}_2^c$, and take $X \in \mathscr{L}^*(M)$, $Y \in \mathscr{L}^*(N)$. Then $I_t^M(X) \triangleq \int_0^t X_s\, dM_s$, $I_t^N(Y) \triangleq \int_0^t Y_s\, dN_s$

are also in $\mathcal{M}_2^c$ and, according to (2.19),

$$\langle I^M(X)\rangle_t = \int_0^t X_u^2\,d\langle M\rangle_u, \quad \langle I^N(Y)\rangle_t = \int_0^t Y_u^2\,d\langle N\rangle_u.$$

We propose now to establish the cross-variation formula

$$\langle I^M(X), I^N(Y)\rangle_t = \int_0^t X_u Y_u\,d\langle M, N\rangle_u; \quad t \geq 0,\ P\text{-a.s.} \tag{2.26}$$

If X and Y are simple, then it is straightforward to show by a computation similar to (2.18) that for $0 \leq s < t < \infty$,

$$\begin{aligned} E[(I_t^M(X) - I_s^M(X))(I_t^N(Y) - I_s^N(Y))|\mathscr{F}_s] \\ = E\left[\int_s^t X_u Y_u\,d\langle M, N\rangle_u \middle| \mathscr{F}_s\right]; \quad P\text{-a.s.} \end{aligned} \tag{2.27}$$

This is equivalent to (2.26). It remains to extend this result from simple processes to the case of $X \in \mathscr{L}^*(M)$, $Y \in \mathscr{L}^*(N)$. We carry out this extension in several stages, culminating in Propositions 2.17 and 2.19 with a very useful characterization of the stochastic integral.

2.14 Proposition (An Inequality of Kunita & Watanabe (1967)). *If $M, N \in \mathcal{M}_2^c$, $X \in \mathscr{L}^*(M)$, and $Y \in \mathscr{L}^*(N)$, then a.s.*

$$\int_0^t |X_s Y_s|\,d\check{\xi}_s \leq \left(\int_0^t X_s^2\,d\langle M\rangle_s\right)^{1/2} \left(\int_0^t Y_s^2\,d\langle N\rangle_s\right)^{1/2}; \quad 0 \leq t < \infty,$$

where $\check{\xi}_s$ denotes the total variation of the process $\xi \triangleq \langle M, N\rangle$ on $[0, s]$.

PROOF. According to Problem 1.5.7 (iv), $\check{\xi}(\omega)$ is absolutely continuous with respect to $\varphi(\omega) \triangleq \frac{1}{2}[\langle M\rangle + \langle N\rangle](\omega)$ for every $\omega \in \hat{\Omega}$ with $P(\hat{\Omega}) = 1$, and for every such ω the Radon–Nikodým theorem implies the existence of functions $f_i(\cdot, \omega)\colon [0, \infty) \to \mathbb{R}$; $i = 1, 2, 3$, such that

$$\langle M\rangle_t(\omega) = \int_0^t f_1(s, \omega)\,d\varphi_s(\omega), \quad \langle N\rangle_t(\omega) = \int_0^t f_2(s, \omega)\,d\varphi_s(\omega),$$

$$\xi_t(\omega) = \langle M, N\rangle_t(\omega) = \int_0^t f_3(s, \omega)\,d\varphi_s(\omega); \quad 0 \leq t < \infty.$$

Consequently, for $\alpha, \beta \in \mathbb{R}$ and $\omega \in \tilde{\Omega}_{\alpha\beta} \subseteq \hat{\Omega}$ satisfying $P(\tilde{\Omega}_{\alpha\beta}) = 1$, we have

$$\begin{aligned} 0 &\leq \langle \alpha M + \beta N\rangle_t(\omega) - \langle \alpha M + \beta N\rangle_u(\omega) \\ &= \int_u^t (\alpha^2 f_1(s, \omega) + 2\alpha\beta f_3(s, \omega) + \beta^2 f_2(s, \omega))\,d\varphi_s(\omega); \quad 0 \leq u < t < \infty. \end{aligned}$$

This can happen only if, for every $\omega \in \tilde{\Omega}_{\alpha\beta}$, there exists a set $T_{\alpha\beta}(\omega) \in \mathscr{B}([0, \infty))$ with $\int_{T_{\alpha\beta}(\omega)} d\varphi_t(\omega) = 0$ and such that

$$\alpha^2 f_1(t,\omega) + 2\alpha\beta f_3(t,\omega) + \beta^2 f_2(t,\omega) \ge 0 \tag{2.28}$$

holds for every $t \notin T_{\alpha\beta}(\omega)$. Now let $\tilde{\Omega} \triangleq \bigcap_{\alpha,\beta \in Q} \Omega_{\alpha\beta}$, $T(\omega) \triangleq \bigcup_{\alpha,\beta \in Q} T_{\alpha\beta}(\omega)$ so that $P(\tilde{\Omega}) = 1$, $\int_{T(\omega)} d\varphi_t(\omega) = 0$; $\forall\, \omega \in \tilde{\Omega}$. Fix $\omega \in \tilde{\Omega}$; then (2.28) holds for every $t \notin T(\omega)$ and every pair (α, β) of rational numbers, and thus also for every $t \notin T(\omega)$, $(\alpha, \beta) \in \mathbb{R}^2$; in particular,

$$\alpha^2 |X_t(\omega)|^2 f_1(t,\omega) + 2\alpha |X_t(\omega) Y_t(\omega)| |f_3(t,\omega)| + |Y_t(\omega)|^2 f_2(t,\omega) \ge 0; \quad \forall\, t \notin T(\omega).$$

Integrating with respect to $d\varphi_t$ we obtain

$$\alpha^2 \int_0^t |X_s|^2 \, d\langle M\rangle_s + 2\alpha \int_0^t |X_s Y_s| \, d\check{\xi}_s + \int_0^t |Y_s|^2 \, d\langle N\rangle_s \ge 0; \quad 0 \le t < \infty,$$

almost surely, and the desired result follows by a minimization over α. □

2.15 Lemma. *If $M, N \in \mathcal{M}_2^c$, $X \in \mathcal{L}^*(M)$, and $\{X^{(n)}\}_{n=1}^\infty \subseteq \mathcal{L}^*(M)$ is such that for some $T > 0$,*

$$\lim_{n\to\infty} \int_0^T |X_u^{(n)} - X_u|^2 \, d\langle M\rangle_u = 0; \quad \textit{a.s. } P,$$

then

$$\lim_{n\to\infty} \langle I(X^{(n)}), N\rangle_t = \langle I(X), N\rangle_t; \quad 0 \le t \le T, \quad \textit{a.s. } P.$$

Proof. Problem 1.5.7 (iii) implies for $0 \le t \le T$,

$$|\langle I(X^{(n)}) - I(X), N\rangle_t|^2 \le \langle I(X^{(n)} - X)\rangle_t \, \langle N\rangle_t$$
$$\le \int_0^T |X_u^{(n)} - X_u|^2 \, d\langle M\rangle_u \cdot \langle N\rangle_T. \qquad \square$$

2.16 Lemma. *If $M, N \in \mathcal{M}_2^c$ and $X \in \mathcal{L}^*(M)$, then*

$$\langle I^M(X), N\rangle_t = \int_0^t X_u \, d\langle M, N\rangle_u; \quad \forall\, 0 \le t < \infty, \textit{ a.s.} \tag{2.29}$$

Proof. According to Lemma 2.7, there exists a sequence $\{X^{(n)}\}_{n=1}^\infty$ of simple processes such that

$$\sup_{T>0} \lim_{n\to\infty} E \int_0^T |X_u^{(n)} - X_u|^2 \, d\langle M\rangle_u = 0.$$

Consequently, for each $T > 0$, a subsequence $\{\tilde{X}^{(n)}\}_{n=1}^\infty$ can be extracted for which

$$\lim_{n\to\infty} \int_0^T |\tilde{X}_u^{(n)} - X_u|^2 \, d\langle M\rangle_u = 0, \quad \text{a.s.}$$

But (2.26) holds for simple processes, and so we have

$$\langle I^M(\tilde{X}^{(n)}), N\rangle_t = \int_0^t \tilde{X}_u^{(n)}\, d\langle M, N\rangle_u; \quad 0 \le t \le T$$

almost surely; letting $n \to \infty$ we obtain (2.29) from Lemma 2.15 and the Kunita–Watanabe inequality (Proposition 2.14). □

2.17 Proposition. *If $M, N \in \mathscr{M}_2^c$, $X \in \mathscr{L}^*(M)$, and $Y \in \mathscr{L}^*(N)$, then the equivalent formulas* (2.26) *and* (2.27) *hold.*

PROOF. Lemma 2.16 states that $d\langle M, I^N(Y)\rangle_u = Y_u\, d\langle M, N\rangle_u$. Replacing N in (2.29) by $I^N(Y)$, we have

$$\langle I^M(X), I^N(Y)\rangle_t = \int_0^t X_u\, d\langle M, I^N(Y)\rangle_u$$

$$= \int_0^t X_u Y_u\, d\langle M, N\rangle_u; \quad t \ge 0,\ P\text{-a.s.} \qquad \square$$

2.18 Problem. Let $M = \{M_t, \mathscr{F}_t; 0 \le t < \infty\}$ and $N_t = \{N_t, \mathscr{F}_t; 0 \le t < \infty\}$ be in $\mathscr{M}_2^c$ and suppose $X \in \mathscr{L}_\infty^*(M)$, $Y \in \mathscr{L}_\infty^*(N)$. Then the martingales $I^M(X), I^N(Y)$ are uniformly integrable and have last elements $I_\infty^M(X)$, $I_\infty^N(Y)$, the cross-variation $\langle I^M(X), I^N(Y)\rangle_t$ converges almost surely as $t \to \infty$, and

$$E[I_\infty^M(X) I_\infty^N(Y)] = E\langle I^M(X), I^N(Y)\rangle_\infty = E\int_0^\infty X_t Y_t\, d\langle M, N\rangle_t.$$

In particular,

$$E\left(\int_0^\infty X_t\, dM_t\right)^2 = E\int_0^\infty X_t^2\, d\langle M\rangle_t.$$

2.19 Proposition. *Consider a martingale $M \in \mathscr{M}_2^c$ and a process $X \in \mathscr{L}^*(M)$. The stochastic integral $I^M(X)$ is the unique martingale $\Phi \in \mathscr{M}_2^c$ which satisfies*

$$\langle \Phi, N\rangle_t = \int_0^t X_u\, d\langle M, N\rangle_u; \quad 0 \le t < \infty,\ \text{a.s. } P, \tag{2.30}$$

for every $N \in \mathscr{M}_2^c$.

PROOF. We already know from (2.29) that $\Phi = I^M(X)$ satisfies (2.30). For uniqueness, suppose Φ satisfies (2.30) for every $N \in \mathscr{M}_2^c$. Subtracting (2.29) from (2.30), we have

$$\langle \Phi - I^M(X), N\rangle_t = 0; \quad 0 \le t < \infty,\ \text{a.s. } P.$$

Setting $N = \Phi - I^M(X)$, we see that the continuous martingale $\Phi - I^M(X)$ has quadratic variation zero, so $\Phi = I^M(X)$. □

Proposition 2.19 characterizes the stochastic integral $I^M(X)$ in terms of the more familiar Lebesgue–Stieltjes integral appearing on the right-hand side of (2.30). Such an idea is extremely useful, as the following corollaries illustrate.

In the shorthand "stochastic differential" notation, the first of these states that if $dN = X\,dM$, then $Y\,dN = XY\,dM$.

2.20 Corollary. *Suppose $M \in \mathcal{M}_2^c$, $X \in \mathcal{L}^*(M)$, and $N \triangleq I^M(X)$. Suppose further that $Y \in \mathcal{L}^*(N)$. Then $XY \in \mathcal{L}^*(M)$ and $I^N(Y) = I^M(XY)$.*

PROOF. Because $\langle N\rangle_t = \int_0^t X_s^2\,d\langle M\rangle_s$, we have

$$E\int_0^T X_t^2 Y_t^2\,d\langle M\rangle_t = E\int_0^T Y_t^2\,d\langle N\rangle_t < \infty$$

for all $T > 0$, so $XY \in \mathcal{L}^*(M)$. For any $\tilde{N} \in \mathcal{M}_2^c$, (2.26) gives $d\langle N, \tilde{N}\rangle_s = X_s\,d\langle M, \tilde{N}\rangle_s$, and thus

$$\begin{aligned}\langle I^M(XY), \tilde{N}\rangle_t &= \int_0^t X_s Y_s\,d\langle M, \tilde{N}\rangle_s \\ &= \int_0^t Y_s\,d\langle N, \tilde{N}\rangle_s = \langle I^N(Y), \tilde{N}\rangle_t.\end{aligned}$$

According to Proposition 2.19, $I^M(XY) = I^N(Y)$. □

2.21 Corollary. *Suppose $M, \tilde{M} \in \mathcal{M}_2^c$, $X \in \mathcal{L}^*(M)$, and $\tilde{X} \in \mathcal{L}^*(\tilde{M})$, and there exists a stopping time T of the common filtration for these processes, such that for P-almost every ω,*

$$X_{t\wedge T(\omega)}(\omega) = \tilde{X}_{t\wedge T(\omega)}(\omega), \qquad M_{t\wedge T(\omega)}(\omega) = \tilde{M}_{t\wedge T(\omega)}(\omega); \quad 0 \le t < \infty.$$

Then

$$I^M_{t\wedge T(\omega)}(X)(\omega) = I^{\tilde{M}}_{t\wedge T(\omega)}(\tilde{X})(\omega); \quad 0 \le t < \infty, \textit{ for } P\text{-a.e. } \omega.$$

PROOF. For any $N \in \mathcal{M}_2^c$, we have $\langle M - \tilde{M}, N\rangle_{t\wedge T} = 0$; $0 \le t < \infty$, and so (2.29) implies $\langle I^M(X) - I^{\tilde{M}}(\tilde{X}), N\rangle_{t\wedge T} = 0$; $0 \le t < \infty$. Setting $N = I^M(X) - I^{\tilde{M}}(\tilde{X})$ and using Problem 1.5.12, we obtain the desired result. □

D. Integration with Respect to Continuous, Local Martingales

Corollary 2.21 shows that stochastic integrals are determined locally by the local values of the integrator and integrand. This fact allows us to broaden the classes of both integrators and integrands, a project which we now undertake.

Let $M = \{M_t, \mathscr{F}_t; 0 \le t < \infty\}$ be a continuous, local martingale on a probability space $(\Omega, \mathscr{F}, P)$ with $M_0 = 0$ a.s., i.e., $M \in \mathcal{M}^{c,\text{loc}}$ (Definition 1.5.15). Recall the standing assumption that $\{\mathscr{F}_t\}$ satisfies the usual conditions. We define an equivalence relation on the set of measurable, $\{\mathscr{F}_t\}$-adapted processes just as we did in the paragraph preceding Definition 2.1.

2.22 Definition. We denote by $\mathscr{P}$ the collection of equivalence classes of all *measurable, adapted* processes $X = \{X_t, \mathscr{F}_t; 0 \le t < \infty\}$ satisfying

$$P\left[\int_0^T X_t^2 \, d\langle M\rangle_t < \infty\right] = 1 \quad \text{for every } T \in [0, \infty). \tag{2.31}$$

We denote by $\mathscr{P}^*$ the collection of equivalence classes of all *progressively measurable* processes satisfying this condition.

Again, we shall abuse terminology by speaking of $\mathscr{P}$ and $\mathscr{P}^*$ as if they were classes of processes. As an example of such an abuse, we write $\mathscr{P}^* \subseteq \mathscr{P}$, and if M belongs to $\mathscr{M}_2^c$ (in which case both $\mathscr{L}$ and $\mathscr{L}^*$ are defined) we write $\mathscr{L} \subseteq \mathscr{P}$ and $\mathscr{L}^* \subseteq \mathscr{P}^*$.

We shall continue our development only for integrands in $\mathscr{P}^*$. If a.e. path $t \mapsto \langle M\rangle_t(\omega)$ of the quadratic variation process $\langle M\rangle$ is an absolutely continuous function, we can choose integrands from the wider class $\mathscr{P}$. The reader will see how to accomplish this with the aid of Remark 2.11, once we complete the development for $\mathscr{P}^*$.

Because M is in $\mathscr{M}^{c,\text{loc}}$, there is a nondecreasing sequence $\{S_n\}_{n=1}^\infty$ of stopping times of $\{\mathscr{F}_t\}$, such that $\lim_{n\to\infty} S_n = \infty$ a.s. P, and $\{M_{t\wedge S_n}, \mathscr{F}_t; 0 \le t < \infty\}$ is in $\mathscr{M}_2^c$. For $X \in \mathscr{P}^*$, one constructs another sequence of bounded stopping times by setting

$$R_n(\omega) = n \wedge \inf\left\{0 \le t < \infty; \int_0^t X_s^2(\omega) d\langle M\rangle_s(\omega) \ge n\right\}.$$

This is also a nondecreasing sequence and, because of (2.31), $\lim_{n\to\infty} R_n = \infty$, a.s. P. For $n \ge 1$, $\omega \in \Omega$, set

$$T_n(\omega) = R_n(\omega) \wedge S_n(\omega), \tag{2.32}$$

$$M_t^{(n)}(\omega) \triangleq M_{t\wedge T_n}(\omega), \quad X_t^{(n)}(\omega) = X_t(\omega) 1_{\{T_n(\omega) \ge t\}}; \quad 0 \le t < \infty. \tag{2.33}$$

Then $M^{(n)} \in \mathscr{M}_2^c$ and $X^{(n)} \in \mathscr{L}^*(M^{(n)})$, $n \ge 1$, so $I^{M^{(n)}}(X^{(n)})$ is defined. Corollary 2.21 implies that for $1 \le n \le m$,

$$I_t^{M^{(n)}}(X^{(n)}) = I_t^{M^{(m)}}(X^{(m)}), \quad 0 \le t \le T_n,$$

so we may define the stochastic integral as

$$I_t(X) \triangleq I_t^{M^{(n)}}(X^{(n)}) \quad \text{on } \{0 \le t \le T_n\}. \tag{2.34}$$

This definition is consistent, is independent of the choice of $\{S_n\}_{n=1}^\infty$, and determines a continuous process, which is also a local martingale.

2.23 Definition. For $M \in \mathscr{M}^{c,\text{loc}}$ and $X \in \mathscr{P}^*$, the *stochastic integral of X with respect to M* is the process $I(X) = \{I_t(X), \mathscr{F}_t; 0 \le t < \infty\}$ in $\mathscr{M}^{c,\text{loc}}$ defined by (2.34). As before, we often write $\int_0^t X_s \, dM_s$ instead of $I_t(X)$.

When $M \in \mathcal{M}^{c,\text{loc}}$ and $X \in \mathcal{P}^*$, the integral $I(X)$ will not in general satisfy conditions (2.13)–(2.16), (2.21)–(2.23), or (2.27), which involve expectations at fixed times or unrestricted stopping times. However, *the sample path properties* (2.12), (2.17), (2.19), (2.24), *and* (2.26) *are still valid* and can be easily proved by localization. We also have the following version of Proposition 2.19.

2.24 Proposition. *Consider a local martingale $M \in \mathcal{M}^{c,\text{loc}}$ and a process $X \in \mathcal{P}^*(M)$. The stochastic integral $I^M(X)$ is the unique local martingale $\Phi \in \mathcal{M}^{c,\text{loc}}$ which satisfies* (2.30) *for every $N \in \mathcal{M}_2^c$ (or equivalently, for every $N \in \mathcal{M}^{c,\text{loc}}$).*

2.25 Problem. Suppose $M, N \in \mathcal{M}^{c,\text{loc}}$ and $X \in \mathcal{P}^*(M) \cap \mathcal{P}^*(N)$. Show that for every pair (α, β) of real numbers we have

$$I^{\alpha M + \beta N}(X) = \alpha I^M(X) + \beta I^N(X).$$

2.26 Proposition. *Let $M \in \mathcal{M}^{c,\text{loc}}$, $\{X^{(n)}\}_{n=1}^\infty \subseteq \mathcal{P}^*(M)$, $X \in \mathcal{P}^*(M)$ and suppose that for some stopping time T of $\{\mathcal{F}_t\}$ we have $\lim_{n\to\infty} \int_0^T |X_t^{(n)} - X_t|^2 \, d\langle M \rangle_t = 0$, in probability. Then*

$$\sup_{0 \le t \le T} \left| \int_0^t X_s^{(n)} \, dM_s - \int_0^t X_s \, dM_s \right| \xrightarrow[n\to\infty]{P} 0.$$

PROOF. The proof follows immediately from Problem 1.5.25 and Proposition 2.24.

2.27 Problem. Let $M \in \mathcal{M}^{c,\text{loc}}$ and choose $X \in \mathcal{P}^*$. Show that there exists a sequence of simple processes $\{X^{(n)}\}_{n=1}^\infty$ such that, for every $T > 0$,

$$\lim_{n\to\infty} \int_0^T |X_t^{(n)} - X_t|^2 \, d\langle M \rangle_t = 0$$

and

$$\lim_{n\to\infty} \sup_{0 \le t \le T} |I_t(X^{(n)}) - I_t(X)| = 0$$

hold a.s. P. If M is a standard, one-dimensional Brownian motion, then the preceding also hold with $X \in \mathcal{P}$.

2.28 Problem. Let $M = W$ be standard Brownian motion and $X \in \mathcal{P}$. We define for $0 \le s < t < \infty$

$$\zeta_t^s(X) \triangleq \int_s^t X_u \, dW_u - \frac{1}{2} \int_s^t X_u^2 \, du; \quad \zeta_t(X) \triangleq \zeta_t^0(X). \tag{2.35}$$

The process $\{\exp(\zeta_t(X)), \mathcal{F}_t; 0 \le t < \infty\}$ is a supermartingale; it is a martingale if $X \in \mathcal{L}_0$.

Can one characterize the class of processes $X \in \mathcal{P}^*$, for which the *exponential supermartingale* $\{\exp(\zeta_t(X)), \mathcal{F}_t; 0 \le t < \infty\}$ of Problem 2.28 is in fact a martingale? This question is at the heart of the important result known as the *Girsanov theorem* (Theorem 5.1); we shall try to provide an answer in Section 5.

2.29 Problem. Let W be a standard Brownian motion, ε a number in $[0,1]$, and $\Pi = \{t_0, t_1, \ldots, t_m\}$ a partition of $[0,t]$ with $0 = t_0 < t_1 < \cdots < t_m = t$. Consider the approximating sum

$$S_\varepsilon(\Pi) \triangleq \sum_{i=0}^{m-1} [(1-\varepsilon)W_{t_i} + \varepsilon W_{t_{i+1}}](W_{t_{i+1}} - W_{t_i}) \tag{2.36}$$

for the stochastic integral $\int_0^t W_s\,dW_s$. Show that

$$\lim_{\|\Pi\|\to 0} S_\varepsilon(\Pi) = \frac{1}{2}W_t^2 + \left(\varepsilon - \frac{1}{2}\right)t, \tag{2.37}$$

where the limit is in L^2. The right-hand side of (2.37) is a martingale if and only if $\varepsilon = 0$, so that W is evaluated at the left-hand endpoint of each interval $[t_i, t_{i+1}]$ in the approximating sum (2.36); this corresponds to the Itô integral. With $\varepsilon = \frac{1}{2}$ we obtain the *Fisk-Stratonovich integral*, which obeys the usual rules of calculus such as $\int_0^t W_s \circ dW_s = \frac{1}{2}W_t^2$; we shall have more to say about this in Problems 3.14, 3.15. Finally, $\varepsilon = 1$ leads to the *backward Itô integral* (McKean (1969), p. 35). The sensitivity of the limit in (2.37) to the value of ε is a consequence of the unbounded variation of the Brownian path.

We know all too well that it is one thing to develop a theory of integration in some reasonable generality, and a completely different task to *compute* the integral in any specific case of interest. Indeed, one cannot be expected to repeat the (sometimes arduous) process which fortunately led to an answer in the preceding problem. Just as we develop a calculus for the Riemann integral, which provides us with tools necessary for more or less mechanical computations, we need a *stochastic calculus* for the Itô integral and its extensions. We take up this task in the next section.

2.30 Exercise. For $M \in \mathcal{M}^{c,\mathrm{loc}}$, $X \in \mathcal{P}^*$, and Z an $\mathcal{F}_s$-measurable random variable, show that

$$\int_s^t ZX_u\,dM_u = Z\int_s^t X_u\,dM_u; \quad s \le t < \infty, \text{ a.s.}$$

3.3. The Change-of-Variable Formula

One of the most important tools in the study of stochastic processes of the martingale type is the *change-of-variable formula*, or *Itô's rule*, as it is better known. It provides an integral-differential calculus for the sample paths of such processes.

Let us consider again a basic probability space $(\Omega, \mathscr{F}, P)$ with an associated filtration $\{\mathscr{F}_t\}$ which we always assume to satisfy the usual conditions.

3.1 Definition. *A continuous semimartingale* $X = \{X_t, \mathscr{F}_t; 0 \le t < \infty\}$ is an adapted process which has the decomposition, P a.s.,

$$X_t = X_0 + M_t + B_t; \quad 0 \le t < \infty, \tag{3.1}$$

where $M = \{M_t, \mathscr{F}_t; 0 \le t < \infty\} \in \mathscr{M}^{c,\mathrm{loc}}$ (Definition 1.5.15) and $B = \{B_t, \mathscr{F}_t; 0 \le t < \infty\}$ is the difference of continuous, nondecreasing, adapted processes $\{A_t^{\pm}, \mathscr{F}_t; 0 \le t < \infty\}$:

$$B_t = A_t^+ - A_t^-; \quad 0 \le t < \infty, \tag{3.2}$$

with $A_0^{\pm} = 0$, P a.s. We shall always assume that (3.2) is the minimal decomposition of B; in other words, A_t^+ is the positive variation of B on $[0, t]$ and A_t^- is the negative variation. The total variation of B on $[0, t]$ is then $\check{B}_t \triangleq A_t^+ + A_t^-$.

The following problem discusses the question of uniqueness for the decomposition (3.1) of a continuous semimartingale.

3.2 Problem. Let $X = \{X_t, \mathscr{F}_t; 0 \le t < \infty\}$ be a continuous semimartingale with decomposition (3.1). Suppose that X has another decomposition

$$X_t = X_0 + \tilde{M}_t + \tilde{B}_t; \quad 0 \le t < \infty,$$

where $\tilde{M} \in \mathscr{M}^{c,\mathrm{loc}}$ and $\tilde{B}$ is a continuous, adapted process which has finite total variation on each bounded interval $[0, t]$. Prove that P-a.s.,

$$M_t = \tilde{M}_t, \quad B_t = \tilde{B}_t, \quad 0 \le t < \infty.$$

A. The Itô Rule

Itô's formula states that a "smooth function" of a continuous semimartingale is a continuous semimartingale, and provides its decomposition.

3.3 Theorem. (Itô (1944), Kunita & Watanabe (1967)). *Let $f: \mathbb{R} \to \mathbb{R}$ be a function of class C^2 and let $X = \{X_t, \mathscr{F}_t; 0 \le t < \infty\}$ be a continuous semimartingale with decomposition* (3.1). *Then, P-a.s.,*

$$f(X_t) = f(X_0) + \int_0^t f'(X_s)\,dM_s + \int_0^t f'(X_s)\,dB_s + \frac{1}{2}\int_0^t f''(X_s)\,d\langle M\rangle_s, \quad 0 \le t < \infty. \tag{3.3}$$

3.4 Remark. For fixed ω and $t > 0$, the function $X_s(\omega)$ is bounded for $0 \le s \le t$, so $f'(X_s(\omega))$ is bounded on this interval. It follows that $\int_0^t f'(X_s)\,dM_s$ is defined as in the last section, and this stochastic integral is a continuous, local martingale. The other two integrals in (3.3) are to be understood in the Lebesgue-Stieltjes sense (Remark 1.4.6 (i)), and so, as functions of the upper limit of integration, are of bounded variation. Thus, $\{f(X_t), \mathscr{F}_t; 0 \le t < \infty\}$ is a continuous semimartingale.

3.5 Remark. Equation (3.3) is often written in differential notation:

$$(3.3)' \qquad df(X_t) = f'(X_t)\,dM_t + f'(X_t)\,dB_t + \frac{1}{2} f''(X_t)\,d\langle M\rangle_t$$

$$= f'(X_t)\,dX_t + \frac{1}{2} f''(X_t)\,d\langle M\rangle_t, \quad 0 \le t < \infty.$$

This is the "chain-rule" for stochastic calculus.

PROOF OF THEOREM 3.3. The proof will be accomplished in several steps.

Step 1: Localization. In the notation of Definition 3.1 we introduce, for each $n \ge 1$, the stopping time

$$T_n = \begin{cases} 0; & \text{if } |X_0| \ge n, \\ \inf\{t \ge 0; |M_t| \ge n \text{ or } \check{B}_t \ge n \text{ or } \langle M\rangle_t \ge n\}; & \text{if } |X_0| < n, \\ \infty; & \text{if } |X_0| < n \text{ and } \{\ldots\} = \varnothing. \end{cases}$$

The resulting sequence is nondecreasing with $\lim_{n\to\infty} T_n = \infty$, P-a.s. Thus, if we can establish (3.3) for the stopped processes $X_{t\wedge T_n}$, $M_{t\wedge T_n}$, $t \ge 0$, then we have the desired result upon letting $n \to \infty$. We may assume, therefore, that $X_0(\omega)$ and the random functions $M_t(\omega)$, $\check{B}_t(\omega)$, and $\langle M\rangle_t(\omega)$ on $[0,\infty) \times \Omega$ are all bounded by a common constant K; in particular, M is then a bounded martingale. Under this assumption, we have $|X_t(\omega)| \le 3K$; $0 \le t < \infty$, $\omega \in \Omega$, so the values of f outside $[-3K, 3K]$ are irrelevant. We assume without loss of generality that f has compact support, and so f, f', and f'' are bounded.

Step 2: Taylor expansion. Let us fix $t > 0$ and a partition $\Pi = \{t_0, t_1, \ldots, t_m\}$ of $[0,t]$, with $0 = t_0 < t_1 < \cdots < t_m = t$. A Taylor expansion yields

$$f(X_t) - f(X_0) = \sum_{k=1}^{m} \{f(X_{t_k}) - f(X_{t_{k-1}})\}$$

$$= \sum_{k=1}^{m} f'(X_{t_{k-1}})(X_{t_k} - X_{t_{k-1}}) + \frac{1}{2}\sum_{k=1}^{m} f''(\eta_k)(X_{t_k} - X_{t_{k-1}})^2,$$

where $\eta_k(\omega) = X_{t_{k-1}}(\omega) + \theta_k(\omega)(X_{t_k}(\omega) - X_{t_{k-1}}(\omega))$ for some appropriate $\theta_k(\omega)$ satisfying $0 \le \theta_k(\omega) \le 1$, $\omega \in \Omega$. We may choose θ_k so that $f''(\eta_k)$ is measurable, since we can solve the above equation for $f''(\eta_k)$, unless $X_{t_k} = X_{t_{k-1}}$, in which case we can choose θ_k to be 0. We conclude that

$$(3.4) \qquad f(X_t) - f(X_0) = J_1(\Pi) + J_2(\Pi) + \tfrac{1}{2}J_3(\Pi),$$

where

$$J_1(\Pi) \triangleq \sum_{k=1}^{m} f'(X_{t_{k-1}})(B_{t_k} - B_{t_{k-1}}),$$

$$J_2(\Pi) \triangleq \sum_{k=1}^{m} f'(X_{t_{k-1}})(M_{t_k} - M_{t_{k-1}}),$$

$$J_3(\Pi) \triangleq \sum_{k=1}^{m} f''(\eta_k)(X_{t_k} - X_{t_{k-1}})^2.$$

It is easily seen that $J_1(\Pi)$ converges to the Lebesgue–Stieltjes integral $\int_0^t f'(X_s)\,dB_s$, a.s. P, as the mesh $\|\Pi\| = \max_{1\le k\le m}|t_k - t_{k-1}|$ of the partition decreases to zero. On the other hand, the process

$$Y_s(\omega) \triangleq f'(X_s(\omega)); \quad 0 \le s \le t,\ \omega \in \Omega,$$

is in $\mathscr{L}^*$ (adapted, continuous, and bounded); we intend to approximate it by the simple process

$$Y_s^{\Pi}(\omega) \triangleq f'(X_0(\omega))1_{\{0\}}(s) + \sum_{k=1}^{m} f'(X_{t_{k-1}}(\omega))1_{(t_{k-1},t_k]}(s).$$

Indeed, we have $EI_t^2(Y^{\Pi} - Y) = E\int_0^t |Y_s^{\Pi} - Y_s|^2\,d\langle M\rangle_s \to 0$ as $\|\Pi\| \to 0$, by the bounded convergence theorem, and so

$$J_2(\Pi) = \int_0^t Y_s^{\Pi}\,dM_s \xrightarrow[\|\Pi\|\to 0]{} \int_0^t Y_s\,dM_s$$

in quadratic mean.

Step 3: The quadratic variation term. $J_3(\Pi)$ can be written as

$$J_3(\Pi) = J_4(\Pi) + J_5(\Pi) + J_6(\Pi),$$

where

$$J_4(\Pi) \triangleq \sum_{k=1}^{m} f''(\eta_k)(B_{t_k} - B_{t_{k-1}})^2,$$

$$J_5(\Pi) \triangleq 2\sum_{k=1}^{m} f''(\eta_k)(B_{t_k} - B_{t_{k-1}})(M_{t_k} - M_{t_{k-1}}),$$

$$J_6(\Pi) \triangleq \sum_{k=1}^{m} f''(\eta_k)(M_{t_k} - M_{t_{k-1}})^2.$$

Because B has total variation bounded by K, we have

$$|J_4(\Pi)| + |J_5(\Pi)|$$
$$\le 2K\|f''\|_\infty\left(\max_{1\le k\le m}|B_{t_k} - B_{t_{k-1}}| + \max_{1\le k\le m}|M_{t_k} - M_{t_{k-1}}|\right),$$

and, thanks to the continuity of the processes B and M, this last term converges to zero almost surely as $\|\Pi\| \to 0$ (as well as in $L^1(\Omega, \mathscr{F}, P)$, because of the bounded convergence theorem). As for $J_6(\Pi)$, we define

$$J_6^*(\Pi) \triangleq \sum_{k=1}^{m} f''(X_{t_{k-1}})(M_{t_k} - M_{t_{k-1}})^2$$

and observe

$$|J_6^*(\Pi) - J_6(\Pi)| \le V_t^{(2)}(\Pi) \cdot \max_{1 \le k \le m} |f''(\eta_k) - f''(X_{t_{k-1}})|,$$

where $V_t^{(2)}(\Pi)$ is the quadratic variation of M over the partition Π (c.f. Theorem 1.5.8 and the discussion preceding it). According to Lemma 1.5.9 and the Cauchy-Schwarz inequality,

$$E|J_6^*(\Pi) - J_6(\Pi)| \le \sqrt{6K^4}\sqrt{E\left(\max_{1 \le k \le m} |f''(\eta_k) - f''(X_{t_{k-1}})|\right)^2},$$

and this is seen to converge to zero as $\|\Pi\| \to 0$ because of the continuity of the process X and the bounded convergence theorem. Thus, in order to establish the convergence of the quadratic variation term $J_3(\Pi)$ to the integral $\int_0^t f''(X_s)\,d\langle M\rangle_s$ in $L^1(\Omega, \mathscr{F}, P)$ as $\|\Pi\| \to 0$, it suffices to compare $J_6^*(\Pi)$ to the approximating sum

$$J_7(\Pi) \triangleq \sum_{k=1}^{m} f''(X_{t_{k-1}})(\langle M\rangle_{t_k} - \langle M\rangle_{t_{k-1}}).$$

Recalling the discussion just before Lemma 1.5.9, we obtain

$$\begin{aligned}
&E|J_6^*(\Pi) - J_7(\Pi)|^2 \\
&= E\left|\sum_{k=1}^{m} f''(X_{t_{k-1}})\{(M_{t_k} - M_{t_{k-1}})^2 - (\langle M\rangle_{t_k} - \langle M\rangle_{t_{k-1}})\}\right|^2 \\
&= E\left[\sum_{k=1}^{m} [f''(X_{t_{k-1}})]^2\{(M_{t_k} - M_{t_{k-1}})^2 - (\langle M\rangle_{t_k} - \langle M\rangle_{t_{k-1}})\}^2\right] \\
&\le 2\|f''\|_\infty^2 \cdot E\left[\sum_{k=1}^{m} (M_{t_k} - M_{t_{k-1}})^4 + \sum_{k=1}^{m} (\langle M\rangle_{t_k} - \langle M\rangle_{t_{k-1}})^2\right] \\
&\le 2\|f''\|_\infty^2 \cdot E\left[V_t^{(4)}(\Pi) + \langle M\rangle_t \cdot \max_{1 \le k \le m} (\langle M\rangle_{t_k} - \langle M\rangle_{t_{k-1}})\right],
\end{aligned}$$

and Lemma 1.5.10 together with the bounded convergence theorem shows that the last term in the preceding equations goes to zero as $\|\Pi\| \to 0$. Since convergence in L^2 implies convergence in L^1, we conclude that

$$J_3(\Pi) \xrightarrow[\|\Pi\| \to 0]{} \int_0^t f''(X_s)\,d\langle M\rangle_s \qquad \text{in } L^1(\Omega, \mathscr{F}, P).$$

Step 4: Final Touches. If $\{\Pi^{(n)}\}_{n=1}^\infty$ is a sequence of partitions of $[0, t]$ with $\|\Pi^{(n)}\| \xrightarrow[n \to \infty]{} 0$, then for some subsequence $\{\Pi^{(n_k)}\}_{k=1}^\infty$ we have, P-a.s.,

$$\lim_{k \to \infty} J_1(\Pi^{(n_k)}) = \int_0^t f'(X_s)\,dB_s,$$

$$\lim_{k\to\infty} J_2(\Pi^{(n_k)}) = \int_0^t f'(X_s)\,dM_s,$$

$$\lim_{k\to\infty} J_3(\Pi^{(n_k)}) = \int_0^t f''(X_s)\,d\langle M\rangle_s.$$

Thus, passing to the limit in (3.4), we see that (3.3) holds P-a.s. for each $0 \le t < \infty$. In other words, the processes on the two sides of equality (3.3) are modifications of one another. Since both of them are continuous, they are indistinguishable (Problem 1.1.5). □

We have the following, multidimensional version of Itô's rule.

3.6 Theorem. *Let $\{M_t \triangleq (M_t^{(1)}, \ldots, M_t^{(d)}), \mathscr{F}_t; 0 \le t < \infty\}$ be a vector of local martingales in $\mathscr{M}^{c,\mathrm{loc}}$, $\{B_t \triangleq (B_t^{(1)}, \ldots, B_t^{(d)}), \mathscr{F}_t; 0 \le t < \infty\}$ a vector of adapted processes of bounded variation with $B_0 = 0$, and set $X_t = X_0 + M_t + B_t$; $0 \le t < \infty$, where X_0 is an $\mathscr{F}_0$-measurable random vector in $\mathbb{R}^d$. Let $f(t,x)$: $[0,\infty) \times \mathbb{R}^d \to \mathbb{R}$ be of class $C^{1,2}$. Then, a.s. P,*

$$\begin{aligned}(3.5)\quad f(t,X_t) = f(0,X_0) &+ \int_0^t \frac{\partial}{\partial t} f(s,X_s)\,ds + \sum_{i=1}^d \int_0^t \frac{\partial}{\partial x_i} f(s,X_s)\,dB_s^{(i)} \\ &+ \sum_{i=1}^d \int_0^t \frac{\partial}{\partial x_i} f(s,X_s)\,dM_s^{(i)} \\ &+ \frac{1}{2}\sum_{i=1}^d \sum_{j=1}^d \int_0^t \frac{\partial^2}{\partial x_i \partial x_j} f(s,X_s)\,d\langle M^{(i)}, M^{(j)}\rangle_s, \quad 0 \le t < \infty.\end{aligned}$$

□

3.7 Problem. Prove Theorem 3.6.

3.8 Example. With $M = W =$ Brownian motion, $X_0 = 0$, $B_t \equiv 0$ and $f(x) = x^2$, we deduce from (3.3):

$$W_t^2 = 2\int_0^t W_s\,dW_s + t.$$

Compare this with Problem 2.29.

3.9 Example. Again with $M = W =$ Brownian motion, let us consider $X \in \mathscr{P}$ and recall the exponential supermartingale of Problem 2.28:

$$Z_t = \exp(\zeta_t); \quad 0 \le t < \infty$$

where $\zeta = \zeta(X)$ of (2.35). We now check by application of Itô's rule that this process satisfies the *stochastic integral equation*

$$(3.6)\qquad Z_t = 1 + \int_0^t Z_s X_s\,dW_s; \quad 0 \le t < \infty.$$

Indeed, $\{\zeta_t; \mathscr{F}_t; 0 \le t < \infty\}$ is a semimartingale, with local martingale part $N_t \triangleq \int_0^t X_s\,dW_s$ and bounded variation part $B_t \triangleq -\frac{1}{2}\int_0^t X_s^2\,ds$. With $f(x) = e^x$, we have

$$\begin{aligned} Z_t = f(\zeta_t) &= f(\zeta_0) + \int_0^t f'(\zeta_s)\,dN_s + \int_0^t f'(\zeta_s)\,dB_s + \frac{1}{2}\int_0^t f''(\zeta_s)\,d\langle N\rangle_s \\ &= 1 + \int_0^t Z_s X_s\,dW_s + \int_0^t Z_s\left(-\frac{1}{2}X_s^2\right)ds + \frac{1}{2}\int_0^t Z_s X_s^2\,ds \\ &= 1 + \int_0^t Z_s X_s\,dW_s. \end{aligned}$$

The replacement of dN_s by $X_s\,dW_s$ in this equation is justified by Corollary 2.20 (actually, the extension of Corollary 2.20 to allow for the present case of $M = W$ and $X \in \mathscr{P}$). It is usually more convenient to perform computations like this using *differential notation*. We write

$$d\zeta_t = X_t\,dW_t - \tfrac{1}{2}X_t^2\,dt,$$

and, to reflect the fact that the martingale part of ζ has quadratic variation with differential $X_t^2\,dt$, we let $(d\zeta_t)^2 = X_t^2\,dt$. One may obtain this from the formal computation

$$\begin{aligned} (d\zeta_t)^2 &= (X_t\,dW_t - \tfrac{1}{2}X_t^2\,dt)^2 \\ &= X_t^2(dW_t)^2 - X_t^3\,dW_t\,dt + \tfrac{1}{4}X_t^4(dt)^2 \\ &= X_t^2\,dt, \end{aligned}$$

using the conventional "multiplication table"

	dt	dW_t	$d\tilde{W}_t$
dt	0	0	0
dW_t	0	dt	0
$d\tilde{W}_t$	0	0	dt

where W, $\tilde{W}$ are independent Brownian motions (recall Problem 2.5.5). With this formalism, Itô's rule can be written as

$$df(\zeta_t) = f'(\zeta_t)\,d\zeta_t + \tfrac{1}{2}f''(\zeta_t)(d\zeta_t)^2,$$

and with $f(x) = e^x$, we obtain

$$\begin{aligned} dZ_t &= Z_t X_t\,dW_t - \tfrac{1}{2}Z_t X_t^2\,dt + \tfrac{1}{2}Z_t X_t^2\,dt \\ &= Z_t X_t\,dW_t. \end{aligned}$$

Taking into account the initial condition $Z_0 = 1$, we can then recover (3.6).

3.10 Problem. With $\{Z_t; 0 \le t < \infty\}$ as in Example 3.9, set $Y_t = 1/Z_t$; $0 \le t < \infty$, which is well defined because $P[\inf_{0 \le t \le T} Z_t > 0] = P[\inf_{0 \le t \le T} \zeta_t > -\infty] = 1$. Show that Y satisfies the *stochastic differential equation*

$$dY_t = Y_t X_t^2 \, dt - Y_t X_t \, dW_t, \quad Y_0 = 1.$$

3.11 Example. One of the motivating forces behind the Itô calculus was a desire to understand the effects of *additive noise* on ordinary differential equations. Suppose, for example, that we add a noise term to the linear, ordinary differential equation

$$\dot{\xi}(t) = a(t)\xi(t)$$

to obtain the stochastic differential equation

$$d\xi_t = a(t)\xi_t \, dt + b(t) \, dW_t,$$

where $a(t)$ and $b(t)$ are measurable, nonrandom functions satisfying

$$\int_0^T |a(t)| \, dt + \int_0^T b^2(t) \, dt < \infty; \quad 0 < T < \infty,$$

and W is a Brownian motion. Applying the Itô rule to $X_t^{(1)} X_t^{(2)}$ with $X_t^{(1)} \triangleq \exp[\int_0^t a(s) \, ds]$ and $X_t^{(2)} = \xi_0 + \int_0^t b(s) \exp[-\int_0^s a(u) \, du] \, dW_s$, we see that $\xi_t = X_t^{(1)} X_t^{(2)}$ solves the stochastic equation. Note that ξ_t is well defined because, for $0 < T < \infty$:

$$\int_0^T b^2(s) \exp\left[-2 \int_0^s a(u) \, du\right] ds \le \exp\left[2 \int_0^T |a(u)| \, du\right] \int_0^T b^2(s) \, ds < \infty.$$

A full treatment of linear stochastic differential equations appears in Sec. 5.6.

3.12 Problem. Suppose we have two continuous semimartingales

$$X_t = X_0 + M_t + B_t, \quad Y_t = Y_0 + N_t + C_t; \quad 0 \le t < \infty, \tag{3.7}$$

where M and N are in $\mathcal{M}^{c, \text{loc}}$ and B and C are adapted, continuous processes of bounded variation with $B_0 = C_0 = 0$ a.s. Prove the *integration by parts formula*

$$\int_0^t X_s \, dY_s = X_t Y_t - X_0 Y_0 - \int_0^t Y_s \, dX_s - \langle M, N \rangle_t. \tag{3.8}$$

The Itô calculus differs from ordinary calculus in that familiar formulas, such as the one for integration by parts, now have *correction terms* such as $\langle M, N \rangle_t$ in (3.8). One way to avoid these corrections terms is to absorb them into the definition of the integral, thereby obtaining the *Fisk–Stratonovich integral* of Definition 3.13. Because it obeys the ordinary rules of calculus (Problem 3.14), the Fisk–Stratonovich integral is notationally more convenient than the Itô integral in situations where ordinary and stochastic calculus interact; the primary example of such a situation is the theory of diffusions on differentiable manifolds. The Fisk–Stratonovich integral is also more robust under perturbations of the integrating semimartingale (see subsection 5.2.D), and thus a useful tool in modeling. We note, however, that this integral is defined for a narrower class of integrands than the Itô integral (see Definition 3.13) and requires more smoothness in its chain rule (Problem 3.14). Whenever the Fisk–Stratonovich integral is defined, the Itô integral is also, and the two are related by (3.9).

3.13 Definition. Let X and Y be continuous semimartingales with decompositions given by (3.7). The *Fisk–Stratonovich integral* of Y with respect to X is

$$(3.9)\qquad \int_0^t Y_s \circ dX_s \triangleq \int_0^t Y_s\,dM_s + \int_0^t Y_s\,dB_s + \frac{1}{2}\langle M, N\rangle_t; \quad 0 \le t < \infty,$$

where the first integral on the right-hand side of (3.9) is an Itô integral.

3.14 Problem. Let $X = (X^{(1)}, \ldots, X^{(d)})$ be a vector of continuous semimartingales with decompositions

$$X_t^{(i)} = X_0^{(i)} + M_t^{(i)} + B_t^{(i)}; \quad i = 1, \ldots, d,$$

where each $M^{(i)} \in \mathscr{M}^{c,loc}$ and each $B^{(i)}$ is of the form (3.2). If $f: \mathbb{R}^d \to \mathbb{R}$ is of class C^3, then

$$(3.10)\qquad f(X_t) = f(X_0) + \sum_{i=1}^d \int_0^t \frac{\partial}{\partial x_i} f(X_s) \circ dX_s^{(i)}.$$

3.15 Problem. Let X and Y be continuous semimartingales and $\Pi = \{t_0, t_1, \ldots, t_m\}$ a partition of $[0, t]$ with $0 = t_0 < t_1 < \cdots < t_m = t$. Show that the sum

$$\sum_{i=0}^{m-1} \left(\frac{1}{2} Y_{t_{i+1}} + \frac{1}{2} Y_{t_i}\right)(X_{t_{i+1}} - X_{t_i})$$

converges in probability to $\int_0^t Y_s \circ dX_s$ as $\|\Pi\| \to 0$.

B. Martingale Characterization of Brownian Motion

In the hands of Kunita and Watanabe (1967), the change-of-variable formula (3.5) was shown to be the right tool for providing an elegant proof of P. Lévy's celebrated *martingale characterization of Brownian motion* in $\mathbb{R}^d$. Let us recall here that if $\{B_t = (B_t^{(1)}, \ldots, B_t^{(d)}), \mathscr{F}_t; 0 \le t < \infty\}$ is a d-dimensional Brownian motion on $(\Omega, \mathscr{F}, P)$ with $P[B_0 = 0] = 1$, then $\langle B^{(k)}, B^{(j)}\rangle_t = \delta_{kj} t$; $1 \le k, j \le d$, $0 \le t < \infty$ (Problem 2.5.5). It turns out that this property characterizes Brownian motion among *continuous* local martingales. The compensated Poisson process with intensity $\lambda = 1$ provides an example of a *discontinuous*, square-integrable martingale with $\langle M\rangle_t = t$ (c.f. Example 1.5.4 and Exercise 1.5.20), so the assumption of continuity in the following theorem is essential.

3.16 Theorem (P. Lévy (1948)). *Let* $X = \{X_t = (X_t^{(1)}, \ldots, X_t^{(d)}), \mathscr{F}_t, 0 \le t < \infty\}$ *be a continuous, adapted process in* $\mathbb{R}^d$ *such that, for every component* $1 \le k \le d$, *the process*

$$M_t^{(k)} \triangleq X_t^{(k)} - X_0^{(k)}; \qquad 0 \le t < \infty,$$

is a continuous local martingale relative to $\{\mathscr{F}_t\}$, *and the cross-variations are given by*

$$\langle M^{(k)}, M^{(j)}\rangle_t = \delta_{kj} t; \qquad 1 \le k, j \le d. \tag{3.11}$$

Then $\{X_t, \mathscr{F}_t; 0 \le t < \infty\}$ *is a d-dimensional Brownian motion.*

Proof. We must show that for $0 \le s < t$, the random vector $X_t - X_s$ is independent of $\mathscr{F}_s$ and has the d-variate normal distribution with mean zero and covariance matrix equal to $(t - s)$ times the $(d \times d)$ identity. In light of Lemma 2.6.13, it suffices to prove that for each $u \in \mathbb{R}^d$, with $i = \sqrt{-1}$,

$$E[e^{i(u, X_t - X_s)} | \mathscr{F}_s] = e^{-(1/2)\|u\|^2(t-s)}, \quad \text{a.s. } P. \tag{3.12}$$

For fixed $u = (u_1, \ldots, u_d) \in \mathbb{R}^d$, the function $f(x) = e^{i(u,x)}$ satisfies

$$\frac{\partial}{\partial x_j} f(x) = i u_j f(x), \qquad \frac{\partial^2}{\partial x_j \partial x_k} f(x) = -u_j u_k f(x).$$

Applying Theorem 3.6 to the real and imaginary parts of f, we obtain

$$e^{i(u, X_t)} = e^{i(u, X_s)} + i \sum_{j=1}^d u_j \int_s^t e^{i(u, X_v)}\, dM_v^{(j)} - \frac{1}{2} \sum_{j=1}^d u_j^2 \int_s^t e^{i(u, X_v)}\, dv. \tag{3.13}$$

Now $|f(x)| \le 1$ for all $x \in \mathbb{R}^d$ and, because $\langle M^{(j)}\rangle_t = t$, we have $M^{(j)} \in \mathscr{M}_2^c$. Thus, the real and imaginary parts of $\{\int_0^t e^{i(u, X_v)}\, dM_v^{(j)}, \mathscr{F}_t; 0 \le t < \infty\}$ are not only in $\mathscr{M}^{c,\text{loc}}$, but also in $\mathscr{M}_2^c$. Consequently,

$$E\left[\int_s^t e^{i(u, X_v)}\, dM_v^{(j)} \middle| \mathscr{F}_s\right] = 0, \quad P\text{-a.s.}$$

For $A \in \mathscr{F}_s$, we may multiply (3.13) by $e^{-i(u, X_s)} 1_A$ and take expectations to obtain

$$E[e^{i(u, X_t - X_s)} 1_A] = P(A) - \frac{1}{2}\|u\|^2 \int_s^t E[e^{i(u, X_v - X_s)} 1_A]\, dv.$$

This integral equation for the deterministic function $t \mapsto E[e^{i(u, X_t - X_s)} 1_A]$ is readily solved:

$$E[e^{i(u, X_t - X_s)} 1_A] = P(A) e^{-(1/2)\|u\|^2(t-s)}, \qquad \forall A \in \mathscr{F}_s,$$

and (3.12) follows. □

3.17 Exercise. Let $W_t = (W_t^{(1)}, W_t^{(2)}, W_t^{(3)})$ be a three-dimensional Brownian motion starting at the origin, and define

$$X = \prod_{i=1}^{3} \operatorname{sgn}(W_1^{(i)}),$$

$$M_t^{(1)} = W_t^{(1)}, \quad M_t^{(2)} = W_t^{(2)} \quad \text{and} \quad M_t^{(3)} = XW_t^{(3)}.$$

Show that each of the pairs $(M^{(1)}, M^{(2)})$, $(M^{(1)}, M^{(3)})$ and $(M^{(2)}, M^{(3)})$ is a two-dimensional Brownian motion, but $(M^{(1)}, M^{(2)}, M^{(3)})$ is *not* a three-dimensional Brownian motion. Explain why this does not provide a counter-example to Theorem 3.16, i.e., a three-dimensional process which is not a Brownian motion but which has components in $\mathscr{M}^{c,\text{loc}}$ and satisfies (3.11).

3.18 Problem. Let $W = \{W_t = (W_t^{(1)}, \ldots, W_t^{(d)}), \mathscr{F}_t; 0 \le t < \infty\}$ be a d-dimensional Brownian motion starting at the origin, and let Q be a $d \times d$ orthogonal matrix ($Q^T = Q^{-1}$). Show that $\tilde{W}_t \triangleq QW_t$ is also a d-dimensional Brownian motion. We express this property by saying that "d-dimensional Brownian motion starting at the origin is *rotationally invariant*."

C. Bessel Processes, Questions of Recurrence

Another use of the P. Lévy Theorem 3.16 is to obtain an integral representation for the so-called *Bessel process*. For an integer $d \ge 2$, let $W = \{W_t = (W_t^{(1)}, \ldots, W_t^{(d)}), \mathscr{F}_t; 0 \le t < \infty\}$, $\{P^x\}_{x \in \mathbb{R}^d}$ be a d-dimensional Brownian family on some measurable space $(\Omega, \mathscr{F})$. Consider the distance from the origin

$$(3.14) \qquad R_t \triangleq \|W_t\| = \sqrt{(W_t^{(1)})^2 + \cdots + (W_t^{(d)})^2}; \quad 0 \le t < \infty,$$

so $P^x[R_0 = \|x\|] = 1$. If $x, y \in \mathbb{R}^d$ and $\|x\| = \|y\|$, then there is an orthogonal matrix Q such that $y = Qx$. Under P^x, $\tilde{W} = \{\tilde{W}_t \triangleq QW_t, \mathscr{F}_t; 0 \le t < \infty\}$ is a d-dimensional Brownian motion starting at y, but $\|\tilde{W}_t\| = \|W_t\|$, so for any $F \in \mathscr{B}(C[0,\infty))$, we have

$$(3.15) \qquad P^x[R_. \in F] = P^x[\|\tilde{W}_.\| \in F] = P^y[R_. \in F].$$

In other words, the distribution of the process R under P^x depends on x only through $\|x\|$.

3.19 Definition. Fix an integer $d \ge 2$, and let $W = \{W_t, \mathscr{F}_t; 0 \le t < \infty\}$, $\{P^x\}_{x \in \mathbb{R}^d}$ be a d-dimensional Brownian family on $(\Omega, \mathscr{F})$. The process $R = \{R_t = \|W_t\|, \mathscr{F}_t; 0 \le t < \infty\}$ together with the family of measures $\{\hat{P}^r\}_{r \ge 0} \triangleq \{P^{(r,0,\ldots,0)}\}_{r \ge 0}$ on $(\Omega, \mathscr{F})$ is called a *Bessel family with dimension d*. For fixed $r \ge 0$, we say that R on $(\Omega, \mathscr{F}, \hat{P}^r)$ is a *Bessel process with dimension d starting at r*.

3.20 Problem. Show that for each $d \geq 2$, the Bessel family with dimension d is a strong Markov family (where we modify Definition 2.6.3 to account for the state space $[0, \infty)$).

3.21 Proposition. *Let $d \geq 2$ be an integer and choose $r \geq 0$. The Bessel process R with dimension d starting at r satisfies the integral equation*

$$R_t = r + \int_0^t \frac{d-1}{2R_s} ds + B_t; \qquad 0 \leq t < \infty, \tag{3.16}$$

where $B = \{B_t, \mathscr{F}_t; 0 \leq t < \infty\}$ is the standard, one-dimensional Brownian motion

$$B \triangleq \sum_{i=1}^d B^{(i)} \qquad \text{with} \quad B_t^{(i)} \triangleq \int_0^t \frac{W_s^{(i)}}{R_s} dW_s^{(i)}; \qquad 1 \leq i \leq d. \tag{3.17}$$

PROOF. We use the notation of Definition 3.19, except we write P instead of $\hat{P}^r$. Note first of all that R_t can be at the origin only when $W_t^{(1)}$ is, and so the Lebesgue measure of the set $\{0 \leq s \leq t; R_s = 0\}$ is zero, a.s. P (Theorem 2.9.6). Consequently, the integrand $(d-1)/2R_s$ in (3.16) is defined for Lebesgue almost every s, a.s. P.

Each of the processes $B^{(i)}$ in (3.17) belongs to $\mathscr{M}_2^c$, because

$$E \int_0^t \left(\frac{1}{R_s} W_s^{(i)} \right)^2 ds \leq t; \qquad 0 \leq t < \infty.$$

Moreover,

$$\langle B^{(i)}, B^{(j)} \rangle_t = \int_0^t \frac{1}{R_s^2} W_s^{(i)} W_s^{(j)} d\langle W^{(i)}, W^{(j)} \rangle_s = \delta_{ij} \int_0^t \frac{1}{R_s^2} W_s^{(i)} W_s^{(j)} ds,$$

which implies

$$\langle B \rangle_t = \sum_{i=1}^d \langle B^{(i)} \rangle_t = t,$$

and we conclude from Theorem 3.16 that B is a standard, one-dimensional Brownian motion.

It remains to prove (3.16). A heuristic derivation is to apply Itô's rule (Theorem 3.6) to the function $f(x) \triangleq \|x\| = \sqrt{x_1^2 + \cdots + x_d^2}$: $\mathbb{R}^d \to [0, \infty)$, for which

$$\frac{\partial}{\partial x_i} f(x) = \frac{x_i}{\|x\|}, \quad \frac{\partial^2}{\partial x_i \partial x_j} f(x) = \frac{\delta_{ij}}{\|x\|} - \frac{x_i x_j}{\|x\|^3}; \qquad 1 \leq i, j \leq d,$$

hold on $\mathbb{R}^d \backslash \{0\}$. Then $R_t = f(W_t)$ and (3.16) follows from (3.5). The difficulty here is that f is not differentiable at the origin, and so Theorem 3.6 cannot be applied directly to f. This problem is related to our uneasiness about whether the integral in (3.16) is finite. Here is a resolution. Define

$$Y_t \triangleq \|W_t\|^2 = R_t^2,$$

and use Itô's rule to show that

$$Y_t = r^2 + 2\sum_{i=1}^{d} \int_0^t W_s^{(i)}\, dW_s^{(i)} + td.$$

Let $g(y) = \sqrt{y}$, and for $\varepsilon > 0$, define

$$g_\varepsilon(y) = \begin{cases} \frac{3}{8}\sqrt{\varepsilon} + \frac{3}{4\sqrt{\varepsilon}}y - \frac{1}{8\varepsilon\sqrt{\varepsilon}}y^2; & y < \varepsilon, \\ \sqrt{y}; & y \geq \varepsilon, \end{cases}$$

so g_ε is of class C^2 and $\lim_{\varepsilon \downarrow 0} g_\varepsilon(y) = g(y)$ for all $y \geq 0$. Now apply Itô's rule to obtain

$$g_\varepsilon(Y_t) = g_\varepsilon(r^2) + \sum_{i=1}^{d} I_t^{(i)}(\varepsilon) + J_t(\varepsilon) + K_t(\varepsilon), \tag{3.18}$$

where

$$I_t^{(i)}(\varepsilon) \triangleq \int_0^t \left[1_{\{Y_s \geq \varepsilon\}} \frac{1}{R_s} + 1_{\{Y_s < \varepsilon\}} \frac{1}{2\sqrt{\varepsilon}}\left(3 - \frac{Y_s}{\varepsilon}\right)\right] W_s^{(i)}\, dW_s^{(i)},$$

$$J_t(\varepsilon) \triangleq \int_0^t 1_{\{Y_s \geq \varepsilon\}} \frac{d-1}{2R_s}\, ds,$$

$$K_t(\varepsilon) \triangleq \int_0^t 1_{\{Y_s < \varepsilon\}} \frac{1}{4\sqrt{\varepsilon}} \left[3d - (d+2)\frac{Y_s}{\varepsilon}\right] ds.$$

We now show that, as $\varepsilon \downarrow 0$, (3.18) yields (3.16). From the monotone convergence theorem, we see that

$$\lim_{\varepsilon \downarrow 0} J_t(\varepsilon) = \int_0^t 1_{\{Y_s > 0\}} \frac{d-1}{2R_s}\, ds = \int_0^t \frac{d-1}{2R_s}\, ds, \quad \text{a.s.}$$

We also have $0 \leq EK_t(\varepsilon) \leq (3d/4\sqrt{\varepsilon}) \int_0^t P[Y_s < \varepsilon]\, ds$. The probability in the integrand is bounded above by

$$P[(W_s^{(1)})^2 + (W_s^{(2)})^2 < \varepsilon] = \int_0^{2\pi} \int_0^{\sqrt{\varepsilon}} \frac{1}{2\pi s} e^{-\rho^2/2s} \rho\, d\rho\, d\theta,$$

and so the integral becomes, upon using Fubini's theorem and the change of variable $\xi = \rho/\sqrt{s}$:

$$\begin{aligned} \int_0^t P[Y_s < \varepsilon]\, ds &\leq \int_0^{\sqrt{\varepsilon}} \rho \left(\int_0^t \frac{1}{s} e^{-\rho^2/2s}\, ds \right) d\rho \\ &= 2 \int_0^{\sqrt{\varepsilon}} \rho \left(\int_{\rho/\sqrt{t}}^{\infty} \frac{1}{\xi} e^{-\xi^2/2}\, d\xi \right) d\rho. \end{aligned}$$

But now it is easy to see that this expression is $o(\sqrt{\varepsilon})$ as $\varepsilon \downarrow 0$, using the rule of l'Hôpital. Therefore, $\lim_{\varepsilon \downarrow 0} EK_t(\varepsilon) = 0$. Finally

$$E[B_t^{(i)} - I_t^{(i)}(\varepsilon)]^2 = E\int_0^t 1_{\{Y_s<\varepsilon\}}\left[\frac{1}{R_s} - \frac{1}{2\sqrt{\varepsilon}}\left(3 - \frac{Y_s}{\varepsilon}\right)\right]^2 (W_s^{(i)})^2\,ds$$

$$= E\int_0^t 1_{\{Y_s<\varepsilon\}}\left[1 - \frac{1}{2}\sqrt{\frac{Y_s}{\varepsilon}}\left(3 - \frac{Y_s}{\varepsilon}\right)\right]^2 \left(\frac{W_s^{(i)}}{R_s}\right)^2 ds$$

$$\le \int_0^t P[Y_s < \varepsilon]\,ds = o(\sqrt{\varepsilon}) \qquad \text{as } \varepsilon \downarrow 0.$$

This establishes (3.16). □

Let $\{R_t, \mathscr{F}_t; 0 \le t < \infty\}$ be a Bessel process with dimension $d \ge 2$ starting at $r \ge 0$. Then, for each fixed $t > 0$, it is clear from (3.14) that $P[R_t > 0] = 1$. A more interesting question is whether the origin is nonattainable:

$$P[R_t > 0;\quad \forall\, 0 < t < \infty] = 1. \tag{3.19}$$

The next proposition shows that this is indeed the case. Of course the situation is drastically different in one dimension, since $P[|W_t^{(1)}| > 0; \forall\, 0 < t < \infty] = 0$ (Remark 2.9.7).

3.22 Proposition (Nonattainability of the Origin by the Brownian Path in Dimension $d \ge 2$). *Let $d \ge 2$ be an integer and $r \ge 0$. The Bessel process R with dimension d starting at r satisfies* (3.19).

PROOF. It is sufficient to treat the case $d = 2$, since, for larger d, $(W_t^{(1)})^2 + \cdots + (W_t^{(d)})^2$ can reach zero only if $(W_t^{(1)})^2 + (W_t^{(2)})^2$ does.

We consider first the case $r > 0$. For positive integers k satisfying $(1/k)^k < r < k$ and $n \ge 1$, define stopping times

$$T_k = \inf\left\{t \ge 0; R_t = \left(\frac{1}{k}\right)^k\right\},\quad S_k = \inf\{t \ge 0; R_t = k\},\quad \tau_k = T_k \wedge S_k \wedge n.$$

Because P-almost every Brownian path is unbounded (Theorem 2.9.23), we have

$$P\left[\bigcap_{k=1}^{\infty} \{S_k < \infty\} \cap \left\{\lim_{k\to\infty} S_k = \infty\right\}\right] = 1. \tag{3.20}$$

Using (3.16), apply Itô's rule to $\log(R_t)$ to obtain

$$\log R_{\tau_k} = \log r + \int_0^{\tau_k} \frac{1}{R_s}\,dB_s.$$

This step is permissible because log is of class C^2 in an open interval containing $[(1/k)^k, k]$. For $0 \le s \le \tau_k$, $|1/R_s|$ is bounded, and since τ_k is also bounded, we have $E\int_0^{\tau_k}(1/R_s)\,dB_s = 0$. Therefore,

(3.21) $$\log r = E[\log R_{\tau_k}] = -k(\log k)\, P[T_k \le S_k \wedge n] + (\log k)\, P[S_k \le T_k \wedge n] + E[(\log R_n) 1_{\{n < S_k \wedge T_k\}}].$$

For every $n \geq 1$, $\log R_n$ on $\{n < S_k \wedge T_k\}$ is bounded between $-k(\log k)$ and $\log k$. According to (3.20), as $n \to \infty$ we have $P[n < S_k \wedge T_k] \to 0$. Thus, letting $n \to \infty$ in (3.21), we obtain

$$\log r = -k(\log k)\, P[T_k \le S_k] + (\log k)\, P[S_k \le T_k].$$

If we divide by $k(\log k)$ and let $k \to \infty$, we see that

(3.22) $$\lim_{k\to\infty} P[T_k \le S_k] = 0.$$

Now set $T = \inf\{t > 0; R_t = 0\}$, so that $T_k \le T$ for every $k \geq 1$. From (3.20) and (3.22), we have

(3.23) $$P[T < \infty] = \lim_{k\to\infty} P[T \le S_k] \le \lim_{k\to\infty} P[T_k \le S_k] = 0.$$

It follows that $P[R_t > 0, \forall\, 0 < t < \infty] = 1$.

Finally, we consider the case $r = 0$. Recalling the indexing of probability measures in Definition 3.19, we have from Problem 3.20:

$$\hat{P}^0[R_t > 0; \forall\, \varepsilon < t < \infty] = \hat{E}^0\{\hat{P}^{R_\varepsilon}[R_t > 0; \forall\, 0 < t < \infty]\} = 1$$

for any fixed $\varepsilon > 0$, by what was just proved and the fact that $\hat{P}^0[R_\varepsilon > 0] = 1$. Letting $\varepsilon \downarrow 0$, we obtain the desired result. □

3.23 Problem. Let $R = \{R_t, \mathscr{F}_t; 0 \le t < \infty\}$ be a Bessel process with dimension $d \geq 2$ starting at $r > 0$, and define

$$m = \inf_{0 \le t < \infty} R_t.$$

(i) Show that if $d = 2$, then $m = 0$ a.s. P.
(ii) Show that if $d \geq 3$, then m has the beta distribution

$$P[m \le c] = \left(\frac{c}{r}\right)^{d-2}; \quad 0 \le c \le r.$$

(*Hint*: Adapt the proof of Proposition 3.22. For (ii), an appropriate substitute for the function $f(r) = \log r$ must be used.)

Proposition 3.22 says that, with probability one, a two-dimensional Brownian motion never reaches the origin. Problem 3.23 (i) shows, however, that it comes arbitrarily close. By translation, we can conclude that for any given point $z \in \mathbb{R}^2$, a two-dimensional Brownian path, with any starting position different from z, never reaches the point z, but does reach every disc of positive radius centered at z. In the parlance of Markov chains, one says that "every singleton is *nonrecurrent*," but that "every disc of positive radius is *recurrent*." For a Brownian motion of dimension 3 or greater, Problem 3.23

(ii) shows that, once it gets away from the origin, almost every path of the process remains bounded away from the origin; this lower bound depends, of course, on the particular path. Thus, d-dimensional spheres are nonrecurrent for d-dimensional Brownian motion when $d \geq 3$.

3.24 Problem. Let R be a Bessel process with dimension $d \geq 3$ starting at $r \geq 0$. Show that $P[\lim_{t\to\infty} R_t = \infty] = 1$.

D. Martingale Moment Inequalities

As a final application of Itô's rule in this section, we derive some useful bounds on the moments of martingales. The following exercise illustrates the technique.

3.25 Exercise. With $W = \{W_t, \mathscr{F}_t; 0 \leq t < \infty\}$ a standard, one-dimensional Brownian motion and X a measurable, adapted process satisfying

$$E\int_0^T |X_t|^{2m}\,dt < \infty \tag{3.24}$$

for some real numbers $T > 0$ and $m \geq 1$, show that

$$E\left|\int_0^T X_t\,dW_t\right|^{2m} \leq (m(2m-1))^m T^{m-1} E\int_0^T |X_t|^{2m}\,dt. \tag{3.25}$$

(*Hint*: Consider the martingale $\{M_t = \int_0^t X_s\,dW_s, \mathscr{F}_t; 0 \leq t \leq T\}$, and apply Itô's rule to the submartingale $|M_t|^{2m}$.)

Actually, with a bit of extra effort, we can obtain much stronger results. We shall show, in effect, that for any $M \in \mathscr{M}^{c,\mathrm{loc}}$ the increasing functions $E(|M_t^*|^{2m})$ and $E(\langle M\rangle_t^m)$, with the convention

$$M_t^* \triangleq \max_{0\leq s\leq t} |M_s|, \tag{3.26}$$

have the same growth rate on the entire of $[0, \infty)$, for every $m > 0$. This is the subject of the Burkholder-Davis-Gundy inequalities (Theorem 3.28). We present first some preliminary results.

3.26 Proposition (Martingale Moment Inequalities [Millar (1968), Novikov (1971)]). *Consider a continuous martingale M which, along with its quadratic variation process $\langle M\rangle$, is bounded. For every stopping time T, we have then*

$$E(|M_T|^{2m}) \leq C'_m E(\langle M\rangle_T^m); \qquad m > 0 \tag{3.27}$$

$$B_m E(\langle M\rangle_T^m) \leq E(|M_T|^{2m}); \qquad m > 1/2 \tag{3.28}$$

$$B_m E(\langle M\rangle_T^m) \leq E[(M_T^*)^{2m}] \leq C_m E(\langle M\rangle_T^m); \quad m > 1/2 \tag{3.29}$$

for suitable positive constants B_m, C_m, C'_m which are universal (i.e., depend only on the number m, not on the martingale M nor the stopping time T).

PROOF. We consider the process

$$Y_t \triangleq \delta + \varepsilon\langle M\rangle_t + M_t^2 = \delta + (1+\varepsilon)\langle M\rangle_t + 2\int_0^t M_s\,dM_s, \quad 0 \le t < \infty,$$

where $\delta > 0$ and $\varepsilon \ge 0$ are constants to be chosen later. Applying the change-of-variable formula to $f(x) = x^m$, we obtain

(3.30)
$$Y_t^m = \delta^m + m(1+\varepsilon)\int_0^t Y_s^{m-1}\,d\langle M\rangle_s + 2m(m-1)\int_0^t Y_s^{m-2}M_s^2\,d\langle M\rangle_s + 2m\int_0^t Y_s^{m-1}M_s\,dM_s; \quad 0 \le t < \infty.$$

Because M, Y, and $\langle M\rangle$ are bounded and Y is bounded away from zero, the last integral is a uniformly integrable martingale (Problem 1.5.24). The Optional Sampling Theorem 1.3.22 implies that $E\int_0^T Y_s^{m-1}M_s\,dM_s = 0$, so taking expectations in (3.30), we obtain our basic identity

$$EY_T^m = \delta^m + m(1+\varepsilon)E\int_0^T Y_s^{m-1}\,d\langle M\rangle_s + 2m(m-1)E\int_0^T Y_s^{m-2}M_s^2\,d\langle M\rangle_s. \tag{3.31}$$

Case 1: $0 < m \le 1$, *upper bound*: The last term on the right-hand side of (3.31) is nonpositive; so, letting $\delta \downarrow 0$, we obtain

$$\begin{aligned} E[\varepsilon\langle M\rangle_T + M_T^2]^m &\le m(1+\varepsilon)E\int_0^T (\varepsilon\langle M\rangle_s + M_s^2)^{m-1}\,d\langle M\rangle_s \\ &\le m(1+\varepsilon)\varepsilon^{m-1}E\int_0^T \langle M\rangle_s^{m-1}\,d\langle M\rangle_s \\ &= (1+\varepsilon)\varepsilon^{m-1}E(\langle M\rangle_T^m). \end{aligned} \tag{3.32}$$

The second inequality uses the fact $0 < m \le 1$. But for such m, the function $f(x) = x^m$; $x \ge 0$ is concave, so

$$2^{m-1}(x^m + y^m) \le (x+y)^m; \quad x \ge 0, \quad y \ge 0, \tag{3.33}$$

and (3.32) yields: $\varepsilon^m E(\langle M\rangle_T^m) + E(|M_T|^{2m}) \le (1+\varepsilon)\left(\frac{\varepsilon}{2}\right)^{m-1} E(\langle M\rangle_T^m)$, whence

$$E(|M_T|^{2m}) \le \left[(1+\varepsilon)\left(\frac{2}{\varepsilon}\right)^{1-m} - \varepsilon^m\right]E(\langle M\rangle_T^m). \tag{3.34}$$

Case 2: $m > 1$, *lower bound*: Now the last term in (3.31) is nonnegative, and the direction of all inequalities (3.32)–(3.34) is reversed:

$$E(|M_T|^{2m}) \ge \left[(1+\varepsilon)\left(\frac{\varepsilon}{2}\right)^{m-1} - \varepsilon^m\right] E(\langle M\rangle_T^m).$$

Here, ε has to be chosen in $(0, (2^{m-1}-1)^{-1})$.

Case 3: $\frac{1}{2} < m \le 1$, *lower bound*: Let us evaluate (3.31) with $\varepsilon = 0$ and then let $\delta \downarrow 0$. We obtain

$$E(|M_T|^{2m}) = 2m(m - \tfrac{1}{2})E\int_0^T |M_s|^{2(m-1)}\,d\langle M\rangle_s. \tag{3.35}$$

On the other hand, we have from (3.33), (3.31):

$$\begin{aligned} 2^{m-1}&[\varepsilon^m E(\langle M\rangle_T^m) + E(\delta + M_T^2)^m] \\ &\le E[\varepsilon\langle M\rangle_T + (\delta + M_T^2)]^m \\ &\le \delta^m + m(1+\varepsilon)E\int_0^T (\delta + M_s^2)^{m-1}\,d\langle M\rangle_s. \end{aligned}$$

Letting $\delta \downarrow 0$, we see that

$$2^{m-1}[\varepsilon^m E(\langle M\rangle_T^m) + E(|M_T|^{2m})] \le m(1+\varepsilon)E\int_0^T |M_s|^{2(m-1)}\,d\langle M\rangle_s. \tag{3.36}$$

Relations (3.35) and (3.36) provide us with the lower bound

$$E(|M_T|^{2m}) \ge \varepsilon^m\left(\frac{(1+\varepsilon)2^{1-m}}{2m-1} - 1\right)^{-1} E(\langle M\rangle_T^m)$$

valid for all $\varepsilon > 0$.

Case 4: $m > 1$, *upper bound*: In this case, the inequality (3.36) is reversed, and we obtain

$$E(|M_T|^{2m}) \le \varepsilon^m\left(\frac{(1+\varepsilon)2^{1-m}}{2m-1} - 1\right)^{-1} E(\langle M\rangle_T^m),$$

where now ε has to satisfy $\varepsilon > (2m-1)2^{m-1} - 1$.

This analysis establishes (3.27) and (3.28). From them, and from the Doob maximal inequality (Theorem 1.3.8) applied to the martingale $\{M_{T\wedge t}, \mathscr{F}_t;\ 0 \le t < \infty\}$, we obtain for $m > 1/2$:

$$\begin{aligned} B_m E(\langle M\rangle_{T\wedge t}^m) &\le E(|M_{T\wedge t}|^{2m}) \le E[(M_{T\wedge t}^*)^{2m}] \\ &\le \left(\frac{2m}{2m-1}\right)^{2m} E(|M_{T\wedge t}|^{2m}) \\ &\le C_m'\left(\frac{2m}{2m-1}\right)^{2m} E(\langle M\rangle_{T\wedge t}^m); \quad 0 \le t < \infty, \end{aligned}$$

which is (3.29) with T replaced by $T \wedge t$. Now let $t \to \infty$ in this version of (3.29) and use the monotone convergence theorem. □

3.27 Remark. A straightforward localization argument shows that (3.27), (3.29) are valid for any $M \in \mathcal{M}^{c,\text{loc}}$. The same is true for (3.28), provided that the additional condition $E(\langle M \rangle_T^m) < \infty$ holds.

We can state now the principal result of this subsection.

3.28 Theorem (The Burkholder-Davis-Gundy Inequalities). *Let $M \in \mathcal{M}^{c,\text{loc}}$ and recall the convention* (3.26). *For every $m > 0$ there exist universal positive constants k_m, K_m (depending only on m), such that*

$$k_m E(\langle M \rangle_T^m) \le E[(M_T^*)^{2m}] \le K_m E(\langle M \rangle_T^m) \tag{3.37}$$

holds for every stopping time T. In particular, if we have $E\sqrt{\langle M \rangle_a} < \infty$ for every $0 < a < \infty$, then M is a martingale.

PROOF. From Proposition 3.26 and Remark 3.27, we have the validity of (3.37) for $m > 1/2$. It remains to deal with the case $0 < m \le 1/2$; we assume without loss of generality that M, $\langle M \rangle$ are bounded.

Let us recall now Problem 1.4.15 and its consequence (1.4.17). The right-hand side of (3.29) permits the choice $X = (M^*)^2$, $A = C_1 \langle M \rangle$ in the former, and we obtain from the latter $E[(M_T^*)^{2m}] \le ((2-m)/(1-m))C_1^m E(\langle M \rangle_T^m)$ for every $0 < m < 1$. Similarly, the left-hand side of (3.29) allows us to take $X = B_1 \langle M \rangle$, $A = (M^*)^2$ in Problem 1.4.15, and then (1.4.17) gives for $0 < m < 1$: $((1-m)/(2-m))B_1^m E(\langle M \rangle_T^m)) \le E[(M_T^*)^{2m}]$. The last claim follows from Problem 1.5.19(i), since then (3.37) implies $E(\sup_{0 \le t \le a} |M_t|) < \infty$, $\forall\, 0 < a < \infty$. □

3.29 Problem. Let $M = (M^{(1)}, \ldots, M^{(d)})$ be a vector of continuous, local martingales, i.e., $M^{(i)} \in \mathcal{M}^{c,\text{loc}}$, and denote

$$\|M\|_t^* \triangleq \max_{0 \le s \le t} \|M_s\|, \quad A_t \triangleq \sum_{i=1}^d \langle M^{(i)} \rangle_t; \quad 0 \le t < \infty.$$

Show that for any $m > 0$, there exist (universal) positive constants λ_m, Λ_m such that

$$\lambda_m E(A_T^m) \le E(\|M\|_T^*)^{2m} \le \Lambda_m E(A_T^m) \tag{3.38}$$

holds for every stopping time T.

3.30 Remark. In particular, if the $M^{(i)}$ in Problem 3.29 are given by

$$M_t^{(i)} = \sum_{j=1}^r \int_0^t X_s^{(i,j)}\, dW_s^{(j)},$$

where $\{W_t = (W_t^{(1)}, \ldots, W_t^{(r)}), \mathscr{F}_t; 0 \le t < \infty\}$ is standard, r-dimensional Brownian motion, $\{X_t = (X_t^{(i,j)}); 1 \le i \le d, 1 \le j \le r, 0 \le t < \infty\}$ is a matrix of measurable processes adapted to $\{\mathscr{F}_t\}$, and $\|X_t\|^2 \triangleq \sum_{i=1}^d \sum_{j=1}^r (X_t^{(i,j)})^2$, then (3.38) holds with

$$A_T = \int_0^T \|X_t\|^2 \, dt. \tag{3.39}$$

E. Supplementary Exercises

3.31 Exercise. Define polynomials $H_n(x, y)$; $n = 0, 1, 2, \ldots$ by

$$H_n(x, y) = \frac{\partial^n}{\partial \alpha^n} \exp\left(\alpha x - \frac{1}{2}\alpha^2 y\right)\bigg|_{\alpha=0}; \quad x, y \in \mathbb{R}$$

(e.g., $H_0(x, y) = 1$, $H_1(x, y) = x$, $H_2(x, y) = x^2 - y$, $H_3(x, y) = x^3 - 3xy$, $H_4(x, y) = x^4 - 6x^2y + 3y^2$, etc.). These polynomials satisfy the recursive relations

$$\frac{\partial}{\partial x} H_n(x, y) = nH_{n-1}(x, y); \quad n = 1, 2, \ldots \tag{3.40}$$

as well as the backward heat equation

$$\frac{\partial}{\partial y} H_n(x, y) + \frac{1}{2}\frac{\partial^2}{\partial x^2} H_n(x, y) = 0; \quad n = 0, 1, \ldots. \tag{3.41}$$

For any $M \in \mathscr{M}^{c, \text{loc}}$, verify

(i) the multiple Itô integral computation

$$\int_0^t \int_0^{t_1} \cdots \int_0^{t_{n-1}} dM_{t_n} \ldots dM_{t_2} \, dM_{t_1} = \frac{1}{n!} H_n(M_t, \langle M \rangle_t),$$

(ii) and the expansion

$$\exp\left(\alpha M_t - \frac{\alpha^2}{2}\langle M \rangle_t\right) = \sum_{n=0}^{\infty} \frac{\alpha^n}{n!} H_n(M_t, \langle M \rangle_t).$$

(The polynomials $H_n(x, y)$ are related to the *Hermite polynomials*

$$h_n(x) \triangleq \frac{(-1)^n}{\sqrt{n!}} e^{x^2/2} \frac{d^n}{dx^n} e^{-x^2/2}$$

by the formula $H_n(x, y) = \sqrt{n!}\, y^{n/2} h_n(x/\sqrt{y})$.)

3.32 Exercise. Consider a function $\sigma\colon \mathbb{R} \to (0, \infty)$ which is of class C^1 and such that $1/\sigma$ is not integrable at either $\pm\infty$. Let c, ρ be two real constants, and introduce the (strictly increasing, in x) function $f(t, x) = e^{ct} \int_0^x dy/\sigma(y)$; $0 \le t < \infty$, $x \in \mathbb{R}$ and the continuous, adapted process $\xi_t = \xi_0 + \rho \int_0^t e^{cs} \, ds + \int_0^t e^{cs} \, dW_s$, $\mathscr{F}_t$; $0 \le t < \infty$. Let $g(t, \cdot)$ denote the inverse of $f(t, \cdot)$. Show that the process $X_t = g(t, \xi_t)$ satisfies the stochastic integral equation

$$X_t = X_0 + \int_0^t b(X_s)\,ds + \int_0^t \sigma(X_s)\,dW_s; \quad 0 \le t < \infty \tag{3.42}$$

for an appropriate continuous function $b: \mathbb{R} \to \mathbb{R}$, which you should determine.

3.33 Exercise. Consider two real numbers δ, μ; a standard, one-dimensional Brownian motion W; and let $W_t^{(\mu)} = W_t + \mu t$; $0 \le t < \infty$. Show that the process

$$X_t = \int_0^t \exp[\delta\{W_t^{(\mu)} - W_s^{(\mu)}\} - \tfrac{1}{2}\delta^2(t-s)]\,ds; \quad 0 \le t < \infty$$

satisfies the *Shiryaev-Roberts stochastic integral equation*

$$X_t = \int_0^t (1 + \delta\mu X_s)\,ds + \delta \int_0^t X_s\,dW_s.$$

3.34 Exercise. Let W be a standard, one-dimensional Brownian motion and $0 < T < \infty$. Show that

$$\lim_{\beta\to\infty} \sup_{0\le t\le T} \left| e^{-\beta t} \int_0^t e^{\beta s}\,dW_s \right| = 0, \quad \text{a.s.}$$

3.35 Exercise. In the context of Problem 2.12 but now under the condition $E\sqrt{T} < \infty$, establish the Wald identities

$$E(W_T) = 0, \quad E(W_T^2) = ET.$$

3.36 Exercise (M. Yor). Let R be a Bessel process with dimension $d \ge 3$, starting at $r = 0$. Show that $\{M_t \triangleq (1/R_t^{d-2}); 1 \le t < \infty\}$

(i) is a local martingale,
(ii) satisfies $\sup_{1\le t<\infty} E(M_t^p) < \infty$ for every $0 < p < d/(d-2)$ (and is thus uniformly integrable),
(iii) is *not* a martingale.

3.37 Exercise (M. Yor). Let R be a Bessel process with dimension $d = 2$ starting at $r = 0$. Show that $\{X_t = -\log R_t; 1 \le t < \infty\}$ is a local martingale with $Ee^{\alpha X_t} < \infty$ for $-\infty < \alpha < 2$, $t \ge 1$, but X is *not* a martingale.

3.38 Exercise (Yor, Stricker). Let X be a continuous process and A a continuous, increasing process with $X_0 = A_0 = 0$, a.s.

(i) Suppose that for every $\theta \in \mathbb{R}$, the process

$$Z_t^{(\theta)} \triangleq \exp(\theta X_t - \tfrac{1}{2}\theta^2 A_t); \quad 0 \le t < \infty$$

is a local martingale. Prove that $X \in \mathcal{M}^{c,\text{loc}}$ and $\langle X \rangle = A$.

(ii) Suppose that both X and $Z^{(1)} = \exp(X - \frac{1}{2}A)$ are local martingales. Then again $\langle X \rangle = A$.

3.39 Exercise (Wong & Zakai (1965b)). Let X be a continuous semimartingale of the form (3.1), and $\{B^{(n)}\}_{n=1}^{\infty}$ a sequence of processes of bounded variation, such that $P[\lim_{n\to\infty} B_t^{(n)} = X_t] = 1$ holds for every finite $t > 0$. If the function $f\colon \mathbb{R} \to \mathbb{R}$ is of class $C^1(\mathbb{R})$, show that

$$\lim_{n\to\infty} \int_0^t f(B_s^{(n)})\,dB_s^{(n)} = \int_0^t f(X_s)\,dX_s + \frac{1}{2}\int_0^t f'(X_s)\,d\langle M\rangle_s$$

holds a.s. P, for every fixed $t > 0$.

3.4. Representations of Continuous Martingales in Terms of Brownian Motion

In this section we expound on the theme that Brownian motion is the fundamental continuous martingale, by showing how to represent other continuous martingales in terms of it. We give conditions under which a vector of d continuous local martingales can be represented as *stochastic integrals* with respect to an r-dimensional Brownian motion on a possibly extended probability space. Here we have $r \le d$. We also discuss how a continuous local martingale can be transformed into a Brownian motion by a *random time-change*. In contrast to these representation results, in which one begins with a continuous local martingale, we will also prove a result in which one begins with a Brownian motion $W = \{W_t, \mathscr{F}_t; 0 \le t < \infty\}$ and shows that every continuous local martingale with respect to the Brownian filtration $\{\mathscr{F}_t\}$ is a stochastic integral with respect to W. A related result is that for fixed $0 \le T \le \infty$, every $\mathscr{F}_T$-measurable random variable can be represented as a stochastic integral with respect to W.

We recall our standing assumption that every filtration satisfies the usual conditions, i.e., is right-continuous, and $\mathscr{F}_0$ contains all P-negligible events.

4.1 Remark. Our first representation theorem involves the notion of the *extension of a probability space*. Let $X = \{X_t, \mathscr{F}_t; 0 \le t < \infty\}$ be an adapted process on some $(\Omega, \mathscr{F}, P)$. We may need a d-dimensional Brownian motion independent of X, but because $(\Omega, \mathscr{F}, P)$ may not be rich enough to support this Brownian motion, we must extend the probability space to construct this. Let $(\hat{\Omega}, \hat{\mathscr{F}}, \hat{P})$ be another probability space, on which we consider a d-dimensional Brownian motion $\hat{B} = \{B_t, \hat{\mathscr{F}}_t; 0 \le t < \infty\}$, set $\tilde{\Omega} \triangleq \Omega \times \hat{\Omega}$, $\tilde{\mathscr{G}} \triangleq \mathscr{F} \otimes \hat{\mathscr{F}}$, $\tilde{P} \triangleq P \times \hat{P}$, and define a new filtration by $\tilde{\mathscr{G}}_t \triangleq \mathscr{F}_t \otimes \hat{\mathscr{F}}_t$. The latter may not satisfy the usual conditions, so we augment it and make it right-continuous by defining

$$\mathscr{F}_t \triangleq \bigcap_{s>t} \sigma(\mathscr{G}_s \cup \mathscr{N}),$$

where $\mathscr{N}$ is the collection of $\tilde{P}$-null sets in $\tilde{\mathscr{G}}$. We also complete $\tilde{\mathscr{G}}$ by defining $\tilde{\mathscr{F}} = \sigma(\tilde{\mathscr{G}} \cup \mathscr{N})$. We may extend X and B to $\{\tilde{\mathscr{F}}_t\}$-adapted processes on $(\tilde{\Omega}, \tilde{\mathscr{F}}, \tilde{P})$ by defining for $(\omega, \hat{\omega}) \in \tilde{\Omega}$,

$$\tilde{X}_t(\omega, \hat{\omega}) = X_t(\omega), \quad \tilde{B}_t(\omega, \hat{\omega}) = B_t(\hat{\omega}).$$

Then $\tilde{B} = \{\tilde{B}_t, \tilde{\mathscr{F}}_t; 0 \le t < \infty\}$ is a d-dimensional Brownian motion, independent of $\tilde{X} = \{\tilde{X}_t, \tilde{\mathscr{F}}_t; 0 \le t < \infty\}$. Indeed, $\tilde{B}$ is independent of the extension to $\tilde{\Omega}$ of any $\mathscr{F}$-measurable random variable on Ω. To simplify notation, we henceforth write X and B instead of $\tilde{X}$ and $\tilde{B}$ in the context of extensions.

A. Continuous Local Martingales as Stochastic Integrals with Respect to Brownian Motion

Let us recall (Definition 2.23 and the discussion following it) that if $W = \{W_t, \mathscr{F}_t; 0 \le t < \infty\}$ is a standard Brownian motion and X is a measurable, adapted process with $P[\int_0^t X_s^2\,ds < \infty] = 1$ for every $0 \le t < \infty$, then the stochastic integral $I_t(X) = \int_0^t X_s\,dW_s$ is a continuous local martingale with quadratic variation process $\langle I(X)\rangle_t = \int_0^t X_s^2\,ds$, which is an absolutely continuous function of t, P a.s. Our first representation result provides the converse to this statement; its one-dimensional version is due to Doob (1953).

4.2 Theorem. *Suppose $M = \{M_t = (M_t^{(1)}, \ldots, M_t^{(d)}), \mathscr{F}_t; 0 \le t < \infty\}$ is defined on $(\Omega, \mathscr{F}, P)$ with $M^{(i)} \in \mathscr{M}^{c,\mathrm{loc}}$, $1 \le i \le d$. Suppose also that for $1 \le i,j \le d$, the cross-variation $\langle M^{(i)}, M^{(j)}\rangle_t(\omega)$ is an absolutely continuous function of t for P-almost every ω. Then there is an extension $(\tilde{\Omega}, \tilde{\mathscr{F}}, \tilde{P})$ of $(\Omega, \mathscr{F}, P)$ on which is defined a d-dimensional Brownian motion $W = \{W_t = (W_t^{(1)}, \ldots, W_t^{(d)}), \tilde{\mathscr{F}}_t; 0 \le t < \infty\}$, and a matrix $X = \{(X_t^{(i,k)})_{i,k=1}^d, \tilde{\mathscr{F}}_t; 0 \le t < \infty\}$ of measurable, adapted processes with*

$$\tilde{P}\left[\int_0^t (X_s^{(i,k)})^2\,ds < \infty\right] = 1; \quad 1 \le i,k \le d; \quad 0 \le t < \infty, \tag{4.1}$$

such that we have, $\tilde{P}$-a.s., the representations

$$M_t^{(i)} = \sum_{k=1}^d \int_0^t X_s^{(i,k)}\,dW_s^{(k)}; \quad 1 \le i \le d, \quad 0 \le t < \infty, \tag{4.2}$$

$$\langle M^{(i)}, M^{(j)}\rangle_t = \sum_{k=1}^d \int_0^t X_s^{(i,k)} X_s^{(j,k)}\,ds\,; \quad 1 \le i,j \le d, \quad 0 \le t < \infty. \tag{4.3}$$

PROOF. We prove this theorem by a random, time-dependent rotation of coordinates which reduces it to d separate, one-dimensional cases. We begin by defining

$$(4.4)\qquad z_t^{i,j} = z_t^{j,i} = \frac{d}{dt}\langle M^{(i)}, M^{(j)}\rangle_t = \lim_{n\to\infty} n[\langle M^{(i)}, M^{(j)}\rangle_t - \langle M^{(i)}, M^{(j)}\rangle_{(t-(1/n))^+}],$$

so that the matrix-valued process $Z = \{Z_t = (z_t^{i,j})_{i,j=1}^d, \mathscr{F}_t; 0 \le t < \infty\}$ is symmetric and progressively measurable. For $\alpha = (\alpha_1, \dots, \alpha_d) \in \mathbb{R}^d$, we have

$$\sum_{i=1}^d \sum_{j=1}^d \alpha_i z_t^{i,j} \alpha_j = \frac{d}{dt}\left\langle \sum_{i=1}^d \alpha_i M^{(i)} \right\rangle_t \ge 0,$$

so Z_t is positive-semidefinite for Lebesgue-almost every t, P-a.s.

Any symmetric, positive-semidefinite matrix Z can be diagonalized by an orthogonal matrix Q, i.e., $Q^{-1} = Q^T$, so that $Q^{-1}ZQ = \Lambda$ and Λ is diagonal with the (nonnegative) eigenvalues of Z as its diagonal elements. There are several algorithms which compute Q and Λ from Z, and one can easily verify that these algorithms typically obtain Q and Λ as Borel-measurable functions of Z. In our case, we start with a progressively measurable, symmetric, positive-semidefinite matrix process Z, and so there exist progressively measurable, matrix-valued processes $\{Q_t(\omega) = (q_t^{i,j}(\omega))_{i,j=1}^d; \mathscr{F}_t; 0 \le t < \infty\}$ and $\{\Lambda_t(\omega) = (\delta_{ij}\lambda_t^i(\omega))_{i,j=1}^d, \mathscr{F}_t; 0 \le t < \infty\}$ such that for Lebesgue-almost every t, we have

$$(4.5)\qquad \sum_{k=1}^d q_t^{k,i} q_t^{k,j} = \sum_{k=1}^d q_t^{i,k} q_t^{j,k} = \delta_{ij}\ ; \quad 1 \le i, j \le d,$$

$$(4.6)\qquad \sum_{k=1}^d \sum_{l=1}^d q_t^{k,i} z_t^{k,l} q_t^{l,j} = \delta_{ij}\lambda_t^i \ge 0\ ; \quad 1 \le i, j \le d,$$

a.s. P. From (4.5) with $i = j$ we see that $(q_t^{k,i})^2 \le 1$, so

$$\int_0^t (q_s^{k,i})^2\, d\langle M^{(k)}\rangle_s \le \langle M^{(k)}\rangle_t < \infty,$$

and we can define continuous local martingales by the prescription

$$(4.7)\qquad N_t^{(i)} \triangleq \sum_{k=1}^d \int_0^t q_s^{k,i}\, dM_s^{(k)}; \quad 1 \le i \le d, \quad 0 \le t < \infty.$$

From (4.4) and (4.6) we have, a.s. P,

$$(4.8)\qquad \langle N^{(i)}, N^{(j)}\rangle_t = \sum_{k=1}^d \sum_{l=1}^d \int_0^t q_s^{k,i} q_s^{l,j}\, d\langle M^{(k)}, M^{(l)}\rangle_s = \sum_{k=1}^d \sum_{l=1}^d \int_0^t q_s^{k,i} z_s^{k,l} q_s^{l,j}\, ds = \delta_{ij}\int_0^t \lambda_s^i\, ds.$$

We see, in particular, that

$$\int_0^t \lambda_s^i \, ds = \langle N^{(i)} \rangle_t < \infty. \tag{4.9}$$

We now represent the vector of local martingales $N = \{(N_t^{(1)}, \ldots, N_t^{(d)}), \mathscr{F}_t; 0 \le t < \infty\}$ as a vector of stochastic integrals on an extended probability space $(\tilde{\Omega}, \tilde{\mathscr{F}}, \tilde{P})$, which supports a d-dimensional Brownian motion $B = \{B_t = (B_t^{(1)}, \ldots, B_t^{(d)}), \tilde{\mathscr{F}}_t; 0 \le t < \infty\}$ independent of N (cf. Remark 4.1). Since

$$\int_0^t 1_{\{\lambda_s^i > 0\}} \frac{1}{\lambda_s^i} \, d\langle N^{(i)} \rangle_s = \int_0^t 1_{\{\lambda_s^i > 0\}} \, ds \le t,$$

we can define continuous, local martingales

$$W_t^{(i)} \triangleq \int_0^t 1_{\{\lambda_s^i > 0\}} \frac{1}{\sqrt{\lambda_s^i}} \, dN_s^{(i)} + \int_0^t 1_{\{\lambda_s^i = 0\}} \, dB_s^{(i)}; \quad 1 \le i \le d. \tag{4.10}$$

From (4.8) and Problem 1.5.26 we have

$$\langle W^{(i)}, W^{(j)} \rangle_t = \delta_{ij} t, \quad 1 \le i, j \le d; \quad 0 \le t < \infty,$$

so, according to Theorem 3.16, $W = \{W_t = (W_t^{(1)}, \ldots, W_t^{(d)}), \tilde{\mathscr{F}}_t; 0 \le t < \infty\}$ is a d-dimensional Brownian motion. Moreover,

$$\int_0^t \sqrt{\lambda_s^i} \, dW_s^{(i)} = \int_0^t 1_{\{\lambda_s^i > 0\}} \, dN_s^{(i)} = N_t^{(i)}, \quad 1 \le i \le d; \quad 0 \le t < \infty, \tag{4.11}$$

because the martingale $\int_0^t 1_{\{\lambda_s^i = 0\}} \, dN_s^{(i)}$, having quadratic variation

$$\int_0^t 1_{\{\lambda_s^i = 0\}} \, d\langle N^{(i)} \rangle_s = \int_0^t 1_{\{\lambda_s^i = 0\}} \, \lambda_s^i \, ds = 0,$$

is itself identically zero.

Having thus obtained the stochastic integral representation (4.11) for N in terms of the d-dimensional Brownian motion W, we invert the rotation of coordinates (4.7) to obtain a representation for M. Let us first observe that for $1 \le i, k \le d$,

$$\int_0^t (q_s^{i,k})^2 \lambda_s^k \, ds \le \int_0^t \lambda_s^k \, ds < \infty; \quad 0 \le t < \infty$$

by (4.9), so with $X_t^{(i,k)} \triangleq q_t^{i,k} \sqrt{\lambda_t^k}$, condition (4.1) holds. Furthermore, (4.11), (4.7), and (4.5) imply

$$\begin{aligned} \sum_{k=1}^d \int_0^t X_s^{(i,k)} \, dW_s^{(k)} &= \sum_{k=1}^d \int_0^t q_s^{i,k} \, dN_s^{(k)} \\ &= \sum_{j=1}^d \sum_{k=1}^d \int_0^t q_s^{i,k} q_s^{j,k} \, dM_s^{(j)} \\ &= \sum_{j=1}^d \delta_{ij} \int_0^t dM_s^{(j)} = M_t^{(i)}, \end{aligned} \tag{4.12}$$

which establishes (4.2). Equation (4.3) is an immediate consequence of (4.2). □

4.3 Remark. If for P-a.e. $\omega \in \Omega$, the matrix-valued process $Z_t(\omega) = (z_t^{i,j}(\omega))_{i,j=1}^d$ has constant rank r, $1 \le r \le d$, for Lebesgue-almost every t, then the Brownian motion W used in the representation (4.2) can be chosen to be r-dimensional, and there is no need to introduce the extended probability space $(\tilde{\Omega}, \tilde{\mathscr{F}}, \tilde{P})$. Indeed, we may take $\lambda_t^1, \ldots, \lambda_t^r$ to be the r strictly positive eigenvalues of Z_t, and replace (4.10) by

$$(4.10)' \qquad W_t^{(i)} = \int_0^t \frac{1}{\sqrt{\lambda_s^i}} dN_s^{(i)}; \quad 1 \le i \le r.$$

Since $N_t^{(i)} = 0$; $r + 1 \le i \le d$, $0 \le t < \infty$ (witness (4.9)), (4.12) becomes

$$(4.12)' \quad \sum_{k=1}^{r} \int_0^t X_s^{(i,k)} \, dW_s^{(k)} = \sum_{k=1}^{d} \int_0^t q_s^{i,k} \, dN_s^{(k)} = M_t^{(i)}, \quad 1 \le i \le d.$$

Because (4.10)′ defines $W^{(1)}, \ldots, W^{(r)}$ without reference to the Brownian motion B, there is no need to extend the original probability space.

The following exercise shows that any vector of continuous local martingales can be transformed by a random time-change into a vector of continuous local martingales satisfying the hypotheses of Theorem 4.2.

4.4 Exercise. Let $\{M = (M_t^{(1)}, \ldots, M_t^{(d)}), \mathscr{F}_t; 0 \le t < \infty\}$ be a vector of continuous local martingales on some $(\Omega, \mathscr{F}, P)$, and define

$$A^{(i,j)} \triangleq \langle M^{(i)}, M^{(j)} \rangle, \quad A_t(\omega) \triangleq \sum_{i=1}^{d} \sum_{j=1}^{d} \check{A}_t^{(i,j)}(\omega),$$

where $\check{A}_t^{(i,j)}$ denotes total variation of $A^{(i,j)}$ on $[0, t]$. Let $T_s(\omega)$ be the inverse of the function $A_t(\omega) + t$, i.e., $A_{T_s(\omega)}(\omega) + T_s(\omega) = s$; $0 \le s < \infty$.

(i) Show that for each s, T_s is a stopping time of $\{\mathscr{F}_t\}$.
(ii) Define $\mathscr{G}_s \triangleq \mathscr{F}_{T_s}$; $0 \le s < \infty$. Show that if $\{\mathscr{F}_t\}$ satisfies the usual conditions, then $\{\mathscr{G}_s\}$ does also.
(iii) Define

$$N_s^{(i)} \triangleq M_{T_s}^{(i)}, \quad 1 \le i \le d; \quad 0 \le s < \infty.$$

Show that for each $1 \le i \le d$: $N^{(i)} \in \mathscr{M}^{c,\text{loc}}$, and the cross-variation $\langle N^{(i)}, N^{(j)} \rangle_s$ is an absolutely continuous function of s, a.s. P.

B. Continuous Local Martingales as Time-Changed Brownian Motions

The time-change in Exercise 4.4 is straightforward because the function $A_t + t$ is strictly increasing and continuous in t, and so has a strictly increasing, continuous inverse T_s. Our next representation result requires us to consider the inverse of the quadratic variation of a continuous local martingale; because such a quadratic variation may not be strictly increasing, we begin with a problem describing this situation in some detail.

4.5 Problem. Let $A = \{A(t); 0 \le t < \infty\}$ be a continuous, nondecreasing function with $A(0) = 0$, $S \triangleq A(\infty) \le \infty$, and define for $0 \le s < \infty$:

$$T(s) = \begin{cases} \inf\{t \ge 0; A(t) > s\}; & 0 \le s < S \\ \infty; & s \ge S. \end{cases}$$

The function $T = \{T(s); 0 \le s < \infty\}$ has the following properties:

(i) T is nondecreasing and right-continuous on $[0, S)$, with values in $[0, \infty)$. If $A(t) < S$; $\forall t \ge 0$, then $\lim_{s \uparrow S} T(s) = \infty$.
(ii) $A(T(s)) = s \wedge S$; $0 \le s < \infty$.
(iii) $T(A(t)) = \sup\{\tau \ge t: A(\tau) = A(t)\}$; $0 \le t < \infty$.
(iv) Suppose $\varphi: [0, \infty) \to \mathbb{R}$ is continuous and has the property

$$A(t_1) = A(t) \quad \text{for some } 0 \le t_1 < t \Rightarrow \varphi(t_1) = \varphi(t).$$

Then $\varphi(T(s))$ is continuous for $0 \le s < S$, and

$$\varphi(T(A(t))) = \varphi(t); \quad 0 \le t < \infty. \tag{4.13}$$

(v) For $0 \le t, s < \infty$: $s < A(t) \Leftrightarrow T(s) < t$ and $T(s) \le t \Rightarrow s \le A(t)$.
(vi) If G is a bounded, measurable, real-valued function or a nonnegative, measurable, extended real-valued function defined on $[a, b] \subset [0, \infty)$, then

$$\int_a^b G(t)\,dA(t) = \int_{A(a)}^{A(b)} G(T(s))\,ds. \tag{4.14}$$

4.6 Theorem (Time-Change for Martingales [Dambis (1965), Dubins & Schwarz (1965)]). *Let $M = \{M_t, \mathscr{F}_t; 0 \le t < \infty\} \in \mathscr{M}^{c,\text{loc}}$ satisfy $\lim_{t\to\infty} \langle M \rangle_t = \infty$, a.s. P. Define, for each $0 \le s < \infty$, the stopping time*

$$T(s) = \inf\{t \ge 0; \langle M \rangle_t > s\}. \tag{4.15}$$

Then the time-changed process

$$B_s \triangleq M_{T(s)}, \quad \mathscr{G}_s \triangleq \mathscr{F}_{T(s)}; \quad 0 \le s < \infty \tag{4.16}$$

is a standard one-dimensional Brownian motion. In particular, the filtration $\{\mathscr{G}_s\}$ satisfies the usual conditions and we have, a.s. P:

$$M_t = B_{\langle M \rangle_t}; \quad 0 \le t < \infty. \tag{4.17}$$

Proof. Each $T(s)$ is optional because, by Problem 4.5 (v), $\{T(s) < t\} = \{\langle M \rangle_t > s\} \in \mathscr{F}_t$, and $\{\mathscr{F}_t\}$ satisfies the usual conditions; these are also satisfied by $\{\mathscr{G}_s\}$. Furthermore, for each t, $\langle M \rangle_t$ is a stopping time for the filtration $\{\mathscr{G}_s\}$ because, again by Problem 4.5 (v),

$$\{\langle M \rangle_t \le s\} = \{T(s) \ge t\} \in \mathscr{F}_{T(s)} = \mathscr{G}_s; \quad 0 \le s < \infty.$$

Let us choose $0 \le s_1 < s_2$ and consider the martingale $\{\tilde{M}_t = M_{t \wedge T(s_2)}, \mathscr{F}_t; 0 \le t < \infty\}$, for which we have

$$\langle \tilde{M} \rangle_t = \langle M \rangle_{t \wedge T(s_2)} \le \langle M \rangle_{T(s_2)} = s_2; \quad 0 \le t < \infty$$

by Problem 4.5 (ii). It follows from Problem 1.5.24 that both $\tilde{M}$ and $\tilde{M}^2 - \langle \tilde{M} \rangle$ are uniformly integrable. The Optional Sampling Theorem 1.3.22 implies, a.s. P:

$$E[B_{s_2} - B_{s_1} | \mathscr{G}_{s_1}] = E[\tilde{M}_{T(s_2)} - \tilde{M}_{T(s_1)} | \mathscr{F}_{T(s_1)}] = 0,$$

$$\begin{aligned} E[(B_{s_2} - B_{s_1})^2 | \mathscr{G}_{s_1}] &= E[(\tilde{M}_{T(s_2)} - \tilde{M}_{T(s_1)})^2 | \mathscr{F}_{T(s_1)}] \\ &= E[\langle \tilde{M} \rangle_{T(s_2)} - \langle \tilde{M} \rangle_{T(s_1)} | \mathscr{F}_{T(s_1)}] = s_2 - s_1. \end{aligned}$$

Consequently, $B = \{B_s, \mathscr{G}_s; 0 \le s < \infty\}$ is a square-integrable martingale with quadratic variation $\langle B \rangle_s = s$. We shall know that B is a standard Brownian motion as soon as we establish its continuity (Theorem 3.16). For this we shall use Problem 4.5 (iv).

We must show that for all ω in some $\Omega^* \subseteq \Omega$ with $P(\Omega^*) = 1$, we have:

$$(4.18) \quad \langle M \rangle_{t_1}(\omega) = \langle M \rangle_t(\omega) \quad \text{for some } 0 \le t_1 < t \Rightarrow M_{t_1}(\omega) = M_t(\omega).$$

If the implication (4.18) is valid under the additional assumption that t_1 is rational, then, because of the continuity of $\langle M \rangle$ and M, it is valid even without this assumption. For rational $t_1 \ge 0$, define

$$\sigma = \inf\{t > t_1 : \langle M \rangle_t > \langle M \rangle_{t_1}\},$$

$$N_s = M_{(t_1+s) \wedge \sigma} - M_{t_1}, \quad 0 \le s < \infty,$$

so $\{N_s, \mathscr{F}_{t_1+s}; 0 \le s < \infty\}$ is in $\mathscr{M}^{c,\text{loc}}$ and

$$\langle N \rangle_s = \langle M \rangle_{(t_1+s) \wedge \sigma} - \langle M \rangle_{t_1} = 0, \quad \text{a.s. } P.$$

It follows from Problem 1.5.12 that there is an event $\Omega(t_1) \subseteq \Omega$ with $P(\Omega(t_1)) = 1$ such that for all $\omega \in \Omega(t_1)$,

$$\langle M \rangle_{t_1}(\omega) = \langle M \rangle_t(\omega), \text{ for some } t > t_1 \Rightarrow M_{t_1}(\omega) = M_t(\omega).$$

The intersection of all such events $\Omega(t_1)$ as t_1 range over the nonnegative rationals will serve as Ω^*, so that implication (4.18) is valid for each $\omega \in \Omega^*$. Continuity of B and equality (4.17) now follow from Problem 4.5 (iv). □

4.7 Problem. Show that if $P[S \triangleq \langle M \rangle_\infty < \infty] > 0$, it is still possible to define a Brownian motion B for which (4.17) holds. (*Hint*: The time-change $T(s)$ is now given as in Problem 4.5; assume, as you may, that the probability space has been suitably extended to support an independent Brownian motion (Remark 4.1).)

The proof of the following ramification of Theorem 4.6 is surprisingly technical; the result itself is easily believed. The reader may wish to omit this proof on first reading.

4.8 Proposition. *With the assumptions and the notation of Theorem 4.6, we have the following time-change formula for stochastic integrals. If $X = \{X_t, \mathscr{F}_t; 0 \le t < \infty\}$ is progressively measurable and satisfies*

$$\int_0^\infty X_t^2 \, d\langle M\rangle_t < \infty \quad a.s., \tag{4.19}$$

then the process

$$Y_s \triangleq X_{T(s)}, \mathscr{G}_s; \quad 0 \le s < \infty \tag{4.20}$$

is adapted and satisfies, almost surely:

$$\int_0^\infty Y_s^2 \, ds < \infty \tag{4.21}$$

$$\int_0^t X_v \, dM_v = \int_0^{\langle M\rangle_t} Y_u \, dB_u; \quad 0 \le t < \infty, \tag{4.22}$$

$$\int_0^{T(s)} X_v \, dM_v = \int_0^s Y_u \, dB_u; \quad 0 \le s < \infty. \tag{4.23}$$

PROOF. The process Y is adapted to $\{\mathscr{G}_s\}$ because of Proposition 1.2.18. Relation (4.21) follows from (4.19) and (4.14).

Consider the continuous local martingale $\{J_t \triangleq \int_0^t X_v \, dM_v, \mathscr{F}_t; 0 \le t < \infty\}$. If

$$\langle M\rangle_{t_1}(\omega) = \langle M\rangle_t(\omega) \quad \text{for some } 0 \le t_1 < t, \tag{4.24}$$

then

$$\langle J\rangle_{t_1}(\omega) = \int_0^{t_1} X_v^2(\omega) \, d\langle M\rangle_v(\omega) = \int_0^t X_v^2(\omega) \, d\langle M\rangle_v(\omega) = \langle J\rangle_t(\omega).$$

Applying to J the argument used to obtain (4.18), we conclude that for all ω in some $\Omega^* \subseteq \Omega$ with $P(\Omega^*) = 1$, (4.24) implies the identity $J_{t_1}(\omega) = J_t(\omega)$. According to Problem 4.5 (iv), we have that

$$\tilde{J}_s \triangleq J_{T(s)} = \int_0^{T(s)} X_v \, dM_v; \quad 0 \le s < \infty \tag{4.25}$$

is continuous, and

$$\tilde{J}_{\langle M\rangle_t} = J_{T(\langle M\rangle_t)} = J_t; \quad 0 \le t < \infty, \tag{4.26}$$

almost surely. Let $\tau_n = \inf\{0 \le t < \infty; \langle J\rangle_t \ge n\}$, so $\{J_{t \wedge \tau_n}, \mathscr{F}_t; 0 \le t < \infty\}$ is a martingale. For $0 \le s_1 < s_2$ and $n \ge 1$, we have from the optional sampling theorem:

$$E[\tilde{J}_{s_2 \wedge \langle M\rangle_{n \wedge \tau_n}} | \mathscr{G}_{s_1}] = E[J_{T(s_2) \wedge n \wedge \tau_n} | \mathscr{F}_{T(s_1)}]$$
$$= J_{T(s_1) \wedge n \wedge \tau_n} = \tilde{J}_{s_1 \wedge \langle M\rangle_{n \wedge \tau_n}}, \quad \text{a.s.}$$

Each $\langle M\rangle_{n \wedge \tau_n}$ is a stopping time of the filtration $\{\mathscr{G}_s\}$ because $\{\langle M\rangle_{n \wedge \tau_n} \le s\}$

$= \{n \wedge \tau_n \le T(s)\} = \{n \le T(s),\ \ \langle J\rangle_{T(s)} \ge n\} \in \mathscr{F}_{T(s)}$. By assumption, $\lim_{n\to\infty} \langle M\rangle_{n\wedge\tau_n} = \infty$ a.s.; it follows that $\tilde{J}$ is in $\mathscr{M}^{c,\text{loc}}$ (relative to $\{\mathscr{G}_s\}$). Using this fact, we may repeat the preceding argument to show that $\tilde{J}_{\langle M\rangle_t} = J_t$ is a continuous local martingale *relative to the filtration* $\{\mathscr{G}_{\langle M\rangle_t}\}$, which contains $\{\mathscr{F}_t\}$, and may actually be strictly larger.

In fact, we can choose an arbitrary continuous local martingale $\tilde{N}$ relative to $\{\mathscr{G}_s\}$ and construct $N_t = \tilde{N}_{\langle M\rangle_t}$, a continuous local martingale relative to $\{\mathscr{G}_{\langle M\rangle_t}\}$. If we take $\tilde{N} = B$, then $N = M$ from (4.17) and so *M is in $\mathscr{M}^{c,\text{loc}}$ relative to $\{\mathscr{G}_{\langle M\rangle_t}\}$*. We now establish (4.23) by choosing an arbitrary $\tilde{N} = \{\tilde{N}_s, \mathscr{G}_s; 0 \le s < \infty\} \in \mathscr{M}^{c,\text{loc}}$ and showing that

$$\langle \tilde{J}, \tilde{N}\rangle_s = \int_0^s Y_u\, d\langle B, \tilde{N}\rangle_u; \quad 0 \le s < \infty \tag{4.27}$$

holds a.s. (see Proposition 2.24). Let $N_t = \tilde{N}_{\langle M\rangle_t}$, fix t_1, and set

$$M_t^1 = M_{t+t_1} - M_{t_1}, \quad N_t^1 = N_{t+t_1} - N_{t_1}; \quad 0 < t < \infty.$$

Both M^1, N^1 (respectively; M, N) are local martingales relative to $\{\mathscr{G}_{\langle M\rangle_{t+t_1}}\}$ (resp. $\{\mathscr{G}_{\langle M\rangle_t}\}$). We may compute cross-variations thus:

$$\begin{aligned}|\langle M, N\rangle_{t+t_1} - \langle M, N\rangle_{t_1}| &= |\langle M^1, N^1\rangle_t| \le \sqrt{\langle M^1\rangle_t \langle N^1\rangle_t}\\ &= \sqrt{(\langle M\rangle_{t+t_1} - \langle M\rangle_{t_1})(\langle N\rangle_{t+t_1} - \langle N\rangle_t)},\end{aligned}$$

where Problem 1.5.7 (iii) has been invoked. We see that if $\langle M\rangle$ is constant on an interval, so is $\langle M, N\rangle$. From Problem 4.5 (iv) we conclude that $\langle M, N\rangle_{T(s)}$ is almost surely continuous. Because $\langle M, N\rangle_t$ has finite total variation for t in compact intervals, the composition $\langle M, N\rangle_{T(s)}$ has finite total variation for s in compact intervals. Finally,

$$B_s\tilde{N}_s - \langle M, N\rangle_{T(s)} = M_{T(s)}N_{T(s)} - \langle M, N\rangle_{T(s)}$$

is in $\mathscr{M}^{c,\text{loc}}$ relative to $\{\mathscr{G}_s\}$, so

$$\langle B, \tilde{N}\rangle_s = \langle M, N\rangle_{T(s)}; \quad 0 \le s < \infty, \text{ a.s. } P. \tag{4.28}$$

Setting $s = \langle M\rangle_t$ in (4.28), we obtain

$$\langle B, \tilde{N}\rangle_{\langle M\rangle_t} = \langle M, N\rangle_{T(\langle M\rangle_t)} = \langle M, N\rangle_t; \quad 0 \le t < \infty \tag{4.29}$$

almost surely; we have used Problem 4.5 (iv) with $\varphi = \langle M, N\rangle$. Choose $\omega \in \Omega$ for which (4.29) holds. Then

$$\begin{aligned}\int_0^\infty 1_{[a,b)}(v)\, d\langle M, N\rangle_v(\omega) &= \langle M, N\rangle_b(\omega) - \langle M, N\rangle_a(\omega)\\ &= \langle B, \tilde{N}\rangle_{\langle M\rangle_b(\omega)}(\omega) - \langle B, \tilde{N}\rangle_{\langle M\rangle_a(\omega)}(\omega)\\ &= \int_0^\infty 1_{[\langle M\rangle_a(\omega), \langle M\rangle_b(\omega))}(u)\, d\langle B, \tilde{N}\rangle_u(\omega)\\ &= \int_0^\infty 1_{[a,b)}(T(u,\omega))\, d\langle B, \tilde{N}\rangle_u(\omega),\end{aligned}$$

by virtue of Problem 4.5 (v) in the last step. Thus we have for step functions, and hence for all bounded Borel functions G with compact support on $[0, \infty)$, that

$$\int_0^\infty G(v)\,d\langle M, N\rangle_v = \int_0^\infty G(T(u))\,d\langle B, \tilde{N}\rangle_u \tag{4.30}$$

Indeed, this equation holds for any function $G: [0, \infty) \to \mathbb{R}$ for which the left-hand integral is defined. Returning to (4.27), we obtain from optional sampling and (4.30):

$$\begin{aligned}\langle \tilde{J}, \tilde{N}\rangle_s &= \langle J_{T(\cdot)}, N_{T(\cdot)}\rangle_s = \langle J, N\rangle_{T(s)}\\ &= \int_0^{T(s)} X_v\,d\langle M, N\rangle_v = \int_0^s Y_u\,d\langle B, \tilde{N}\rangle_u; \quad 0 \le s < \infty, \text{ a.s.}\end{aligned}$$

This concludes the proof of (4.23). Replacing s by $\langle M\rangle_t$ in (4.23) and using (4.26), we obtain (4.22). □

4.9 Remark. If, in the context of Theorem 4.6, we have $P[S \triangleq \langle M\rangle_\infty < \infty] > 0$, we may take a Brownian motion B on an extended probability space for which (4.17) holds (Problem 4.7), and the conclusions of Proposition 4.8 are still valid, except now we must define Y by

$$Y_s \triangleq \begin{cases} X_{T(s)}; & 0 \le s < S,\\ 0; & S \le s < \infty.\end{cases} \tag{4.20$'$}$$

The proof is straightforward but tedious, and it is omitted.

4.10 Remark. Lévy's characterization of Brownian motion (Theorem 3.16) permits a bit more generality than expressed in Theorem 4.6. If $X = \{X_t, \mathscr{F}_t: 0 \le t < \infty\}$ is an adapted process with $M \triangleq X - X_0 \in \mathscr{M}^{c,\mathrm{loc}}$ and $\lim_{t\to\infty}\langle M\rangle_t = \infty$ a.s., then the time-changed process

$$X_0 + M_{T(s)}, \quad \mathscr{G}_s \triangleq \mathscr{F}_{T(s)}; \quad 0 \le s < \infty$$

is a one-dimensional Brownian motion with initial distribution PX_0^{-1}. In particular, X_0 is independent of

$$B_s = M_{T(s)}\ , \ \mathscr{F}_s^B\ ; \quad 0 \le s < \infty$$

(Definition 2.5.1). Similar assertions hold in the context of Problem 4.7, Proposition 4.8, and Remark 4.9.

4.11 Problem. We cannot expect to be able to define the stochastic integral $\int_0^1 X_s\,dW_s$ with respect to Brownian motion W for measurable adapted processes X which do not satisfy $\int_0^1 X_s^2\,ds < \infty$ a.s. Indeed, show that if

$$P\left[\int_0^t X_s^2\,ds < \infty\right] = 1, \quad \text{for } 0 \le t < 1 \quad \text{and} \quad E \triangleq \left\{\int_0^1 X_s^2\,ds = \infty\right\},$$

then

$$\overline{\lim_{t\uparrow 1}} \int_0^t X_s\,dW_s = -\underline{\lim_{t\uparrow 1}} \int_0^t X_s\,dW_s = +\infty, \quad \text{a.s. on } E.$$

4.12 Problem. Consider the semimartingale $X_t = x + M_t + C_t$ with $x \in \mathbb{R}$, $M \in \mathscr{M}^{c,\text{loc}}$, C a continuous process of bounded variation, and assume that there exists a constant $\rho > 0$ such that $|C_t| + \langle M \rangle_t \le \rho t$, $\forall\, t \ge 0$ is valid almost surely. Show that for fixed $T > 0$ and sufficiently large $n \ge 1$, we have

$$P\left[\max_{0 \le t \le T} |X_t| \ge n\right] \le \exp\left\{\frac{-n^2}{18\rho T}\right\}.$$

C. A Theorem of F. B. Knight

Let us state and discuss the multivariate extension of Theorem 4.6. The proof will be given in subsection E.

4.13 Theorem (F. B. Knight (1971)). *Let $M = \{M_t = (M_t^{(1)}, \ldots, M_t^{(d)}), \mathscr{F}_t; 0 \le t < \infty\}$ be a continuous, adapted process with $M^{(i)} \in \mathscr{M}^{c,\text{loc}}$, $\lim_{t\to\infty} \langle M^{(i)} \rangle_t = \infty$; a.s. P, and*

$$\langle M^{(i)}, M^{(j)} \rangle_t = 0; \quad 1 \le i \ne j \le d, \quad 0 \le t < \infty. \tag{4.31}$$

Define

$$T_i(s) = \inf\{t \ge 0; \langle M^{(i)} \rangle_t > s\}; \quad 0 \le s < \infty, \quad 1 \le i \le d,$$

so that for each i and s, the random time $T_i(s)$ is a stopping time for the (right-continuous) filtration $\{\mathscr{F}_t\}$. Then the processes

$$B_s^{(i)} \triangleq M_{T_i(s)}^{(i)}; \quad 0 \le s < \infty, \quad 1 \le i \le d,$$

are independent, standard, one-dimensional Brownian motions.

Discussion of Theorem 4.13. The only assertion in Theorem 4.13 which is not already contained in Theorem 4.6 is the independence of the Brownian motions $B^{(i)}$; $1 \le i \le d$. Theorem 4.6 states, in fact, that $B^{(i)}$ is a Brownian motion relative to the filtration $\{\mathscr{G}_s^{(i)} \triangleq \mathscr{F}_{T_i(s)}\}_{s \ge 0}$, but, of course, these filtrations are not independent for different values of i because $\mathscr{G}_\infty^{(i)} = \mathscr{F}_\infty$; $1 \le i \le d$. The additional claim is that the σ-fields $\mathscr{F}_\infty^{B^{(1)}}, \mathscr{F}_\infty^{B^{(2)}}, \ldots, \mathscr{F}_\infty^{B^{(d)}}$ are independent, where $\{\mathscr{F}_s^{B^{(i)}}\}$ is the filtration generated by $B^{(i)}$. This would follow easily if the assumption (4.31) were sufficient to guarantee the independence of $M^{(i)}$, $M^{(j)}$ for $i \ne j$; in general, however, this is not the case. Indeed, if $W = \{W_t, \mathscr{F}_t; 0 \le t < \infty\}$ is a standard Brownian motion, then with

$$M_t^{(1)} \triangleq \int_0^t 1_{\{W_s \ge 0\}}\,dW_s, \quad M_t^{(2)} \triangleq \int_0^t 1_{\{W_s < 0\}}\,dW_s; \quad 0 \le t < \infty,$$

we have $M^{(1)}, M^{(2)} \in \mathscr{M}_2^c$ and

$$\langle M^{(1)}, M^{(2)} \rangle_t = \int_0^t 1_{\{W_s \geq 0\}} 1_{\{W_s < 0\}} \, ds = 0; \quad 0 \leq t < \infty.$$

But $M^{(1)}$ and $M^{(2)}$ are *not* independent, for if they were, $\langle M^{(1)} \rangle$ and $\langle M^{(2)} \rangle$ would also be independent. On the contrary, we have

$$\langle M^{(1)} \rangle_t + \langle M^{(2)} \rangle_t = \int_0^t 1_{\{W_s \geq 0\}} \, ds + \int_0^t 1_{\{W_s < 0\}} \, ds = t \quad ; \quad 0 \leq t < \infty.$$

F. B. Knight's remarkable theorem states that when we apply the proper time-changes to these two intricately connected martingales, and then *forget the time-changes*, independent Brownian motions are obtained. Forgetting the time-changes is accomplished by passing from the filtrations $\{\mathscr{G}_s^{(i)}\}$ to the less informative filtrations $\{\mathscr{F}_s^{B^{(i)}}\}$.

We shall use this example in Section 6.3 to prove the independence of the positive and negative excursion processes associated with a one-dimensional Brownian motion.

D. Brownian Martingales as Stochastic Integrals

In preparation for the proof of Theorem 4.13, we consider a different class of representation results, those for which we begin with a Brownian motion rather than constructing it. We take as the integrator martingale a standard, one-dimensional Brownian motion $W = \{W_t, \mathscr{F}_t; 0 \leq t < \infty\}$ on a probability space $(\Omega, \mathscr{F}, P)$, and we assume $\{\mathscr{F}_t\}$ satisfies the usual conditions. For $0 < T < \infty$, we recall from Lemma 2.2 that $\mathscr{L}_T^*$ is a closed subspace of the Hilbert space $\mathscr{H}_T$. The mapping $X \mapsto I_T(X)$ from $\mathscr{L}_T^*$ to $L^2(\Omega, \mathscr{F}_T, P)$ preserves inner products (see (2.23)):

$$E \int_0^T X_t Y_t \, dt = E[I_T(X) I_T(Y)].$$

Since any convergent sequence in

(4.32) $$\mathscr{R}_T \triangleq \{I_T(X); X \in \mathscr{L}_T^*\}$$

is also Cauchy, its preimage sequence in $\mathscr{L}_T^*$ must have a limit in $\mathscr{L}_T^*$. It follows that $\mathscr{R}_T$ is closed in $L^2(\Omega, \mathscr{F}_T, P)$, a fact we shall need shortly.

Let us denote by $\mathscr{M}_2^*$ the subset of $\mathscr{M}_2^c$ which consists of stochastic integrals

$$I_t(X) = \int_0^t X_s \, dW_s; \quad 0 \leq t < \infty,$$

of processes $X \in \mathscr{L}^*$:

(4.33) $$\mathscr{M}_2^* \triangleq \{I(X); X \in \mathscr{L}^*\} \subseteq \mathscr{M}_2^c \subseteq \mathscr{M}_2.$$

Recall from Definition 1.5.5 the concept of orthogonality in $\mathscr{M}_2$. We have the following fundamental decomposition result of Kunita & Watanabe (1967).

4.14 Proposition. *For every $M \in \mathcal{M}_2$, we have the decomposition $M = N + Z$, where $N \in \mathcal{M}_2^*$, $Z \in \mathcal{M}_2$, and Z is orthogonal to every element of $\mathcal{M}_2^*$.*

PROOF. We have to show the existence of a process $Y \in \mathcal{L}^*$ such that $M = I(Y) + Z$, where $Z \in \mathcal{M}_2$ has the property

$$\langle Z, I(X) \rangle = 0; \quad \forall X \in \mathcal{L}^*. \tag{4.34}$$

Such a decomposition is unique (up to indistinguishability); indeed, if we have $M = I(Y') + Z' = I(Y'') + Z''$ with $Y', Y'' \in \mathcal{L}^*$ and both Z' and Z'' satisfy (4.34), then

$$Z \triangleq Z'' - Z' = I(Y' - Y'')$$

is in $\mathcal{M}_2^c$ and $\langle Z \rangle = \langle Z, I(Y' - Y'') \rangle = 0$. It follows from Problem 1.5.12 that $P[Z_t = 0, \forall 0 \le t < \infty] = 1$.

It suffices, therefore, to establish the decomposition for every finite time-interval $[0, T]$; by uniqueness, we can then extend it to the entire half-line $[0, \infty)$. Let us fix $T > 0$, let $\mathcal{R}_T$ be the closed subspace of $L^2(\Omega, \mathcal{F}_T, P)$ defined by (4.32), and let $\mathcal{R}_T^\perp$ denote its orthogonal complement. The random variable M_T is in $L^2(\Omega, \mathcal{F}_T, P)$, so it admits the decomposition

$$M_T = I_T(Y) + Z_T, \tag{4.35}$$

where $Y \in \mathcal{L}_T^*$ and $Z_T \in L^2(\Omega, \mathcal{F}_T, P)$ satisfies

$$E[Z_T I_T(X)] = 0; \quad \forall X \in \mathcal{L}_T^*. \tag{4.36}$$

Let us denote by $Z = \{Z_t, \mathcal{F}_t; 0 \le t < \infty\}$ a right-continuous version of the martingale $E(Z_T | \mathcal{F}_t)$ (Theorem 1.3.13). Note that $Z_t = Z_T$ for $t \ge T$. Obviously $Z \in \mathcal{M}_2$ and, conditioning (4.35) on $\mathcal{F}_t$, we obtain

$$M_t = I_t(Y) + Z_t; \quad 0 \le t \le T, \text{ a.s. } P. \tag{4.37}$$

It remains to show that Z is orthogonal to every square-integrable martingale of the form $I(X)$; $X \in \mathcal{L}_T^*$, or equivalently, that $\{Z_t I_t(X), \mathcal{F}_t; 0 \le t \le T\}$ is a martingale. But we know from Problem 1.3.26 that this amounts to having $E[Z_S I_S(X)] = 0$ for every stopping time S of the filtration $\{\mathcal{F}_t\}$, with $S \le T$. From (2.24) we have $I_S(X) = I_T(\tilde{X})$, where $\tilde{X}_t(\omega) = X_t(\omega) 1_{\{t \le S(\omega)\}}$ is a process in $\mathcal{L}_T^*$. Therefore,

$$E[Z_S I_S(X)] = E[E(Z_T | \mathcal{F}_S) I_S(X)] = E[Z_T I_T(\tilde{X})] = 0$$

by virtue of (4.36). □

It is useful to have sufficient conditions under which the classes $\mathcal{M}_2^c$ and $\mathcal{M}_2^*$ actually coincide; in other words, the component Z in the decomposition of Proposition 4.14 is actually the trivial martingale $Z \equiv 0$. One such condition is that the filtration $\{\mathcal{F}_t\}$ be the augmentation under P of the filtration $\{\mathcal{F}_t^W\}$ generated by the Brownian motion W. (Recall from Problem 2.7.6 and Proposition 2.7.7 that this augmented filtration is continuous.) We state and prove this result in several dimensions. A martingale relative to this augmented filtration will be called *Brownian*.

4.15 Theorem (Representation of Brownian, Square-Integrable Martingales as Stochastic Integrals). *Let $W = \{W_t = (W_t^{(1)}, \ldots, W_t^{(d)}), \mathscr{F}_t; 0 \le t < \infty\}$ be a d-dimensional Brownian motion on $(\Omega, \mathscr{F}, P)$, and let $\{\mathscr{F}_t\}$ be the augmentation under P of the filtration $\{\mathscr{F}_t^W\}$ generated by W. Then, for any square-integrable martingale $M = \{M_t, \mathscr{F}_t; 0 \le t < \infty\}$ with $M_0 = 0$ and RCLL paths, a.s., there exist progressively measurable processes $Y^{(j)} = \{Y_t^{(j)}, \mathscr{F}_t; 0 \le t < \infty\}$ such that*

$$E \int_0^T (Y_t^{(j)})^2 \, dt < \infty; \quad 1 \le j \le d \tag{4.38}$$

for every $0 < T < \infty$, and

$$M_t = \sum_{j=1}^{d} \int_0^t Y_s^{(j)} \, dW_s^{(j)}; \quad 0 \le t < \infty. \tag{4.39}$$

In particular, M is a.s. continuous. Furthermore, if $\tilde{Y}^{(j)}; 1 \le j \le d$, are any other progressively measurable processes satisfying (4.38), (4.39), *then*

$$\sum_{j=1}^{d} \int_0^\infty |Y_t^{(j)} - \tilde{Y}_t^{(j)}|^2 \, dt = 0, \quad \text{a.s.}$$

PROOF. We first prove by induction on m, where $m = 1, \ldots, d$, that there are processes $Y^{(1)}, \ldots, Y^{(m)}$ in $\mathscr{L}^*$ such that

$$Z_t \triangleq M_t - \sum_{j=1}^{m} \int_0^t Y_s^{(j)} \, dW_s^{(j)}; \quad 0 \le t < \infty \tag{4.40}$$

is orthogonal to every martingale of the form $\sum_{j=1}^m \int_0^t X_s^{(j)} \, dW_s^{(j)}$, where $X^{(j)} \in \mathscr{L}^*$; $1 \le j \le m$. If $m = 1$, this is a direct consequence of Proposition 4.14. Suppose such processes exist for $m - 1$, i.e.,

$$\tilde{Z}_t \triangleq M_t - \sum_{j=1}^{m-1} \int_0^t Y_s^{(j)} \, dW_s^{(j)}; \quad 0 \le t < \infty,$$

is orthogonal to $\sum_{j=1}^{m-1} \int_0^t X_s^{(j)} \, dW_s^{(j)}$ for all $X^{(j)} \in \mathscr{L}^*$; $1 \le j \le m - 1$. Apply Proposition 4.14 to write

$$\tilde{Z}_t = \int_0^t Y_s^{(m)} \, dW_s^{(m)} + Z_t; \quad 0 \le t < \infty,$$

for some $Y^{(m)} \in \mathscr{L}^*$, where Z is orthogonal to $\int_0^\cdot X_s^{(m)} \, dW_s^{(m)}$ for all $X^{(m)} \in \mathscr{L}^*$. For $1 \le j \le m - 1$ and $X^{(j)} \in \mathscr{L}^*$, we have

$$\langle Z, I^{W^{(j)}}(X^{(j)}) \rangle = \langle \tilde{Z}, I^{W^{(j)}}(X^{(j)}) \rangle - \langle I^{W^{(m)}}(Y^{(m)}), I^{W^{(j)}}(X^{(j)}) \rangle = 0.$$

Thus, we have the decomposition (4.40) for M. In particular, with $m = d$,

(4.41) $$\langle M, W^{(j)}\rangle_t = \int_0^t Y_s^{(j)}\,ds;\quad 0 \le t < \infty,\quad 1 \le j \le d.$$

Following Liptser & Shiryaev (1977), pp. 162–163, we now show that, in the notation of (4.40) with $m = d$, we have P-a.s. that

$$Z_t = 0;\quad 0 \le t < \infty.$$

First, we show by induction on n that if $0 = s_0 \le s_1 \le \cdots \le s_n \le t$, and if the functions $f_k\colon \mathbb{R}^d \to \mathbb{C}$, $0 \le k \le n$ are bounded and measurable, then

(4.42) $$E\left[Z_t \cdot \prod_{k=0}^n f_k(W_{s_k})\right] = 0.$$

When $n = 0$, (4.42) can be verified by conditioning on $\mathscr{F}_0$ and using the fact $Z_0 = 0$ a.s. Suppose now that (4.42) holds for some n, and choose $s_n < t$. For $\theta = (\theta_1, \ldots, \theta_d) \in \mathbb{R}^d$ fixed and $s_n \le s \le t$, define with $i = \sqrt{-1}$:

$$\varphi(s) \triangleq E\left[Z_t \cdot \prod_{k=0}^n f_k(W_{s_k}) e^{i(\theta, W_s)}\right] = E\left[Z_s \cdot \prod_{k=0}^n f_k(W_{s_k}) e^{i(\theta, W_s)}\right].$$

Using Itô's rule to justify the identity

$$e^{i(\theta, W_s)} = e^{i(\theta, W_{s_n})} + \sum_{j=1}^d i\theta_j \int_{s_n}^s e^{i(\theta, W_u)}\,dW_u^{(j)} - \frac{\|\theta\|^2}{2}\int_{s_n}^s e^{i(\theta, W_u)}\,du,$$

we may write

(4.43) $$\begin{aligned} E[Z_s e^{i(\theta, W_s)} \mid \mathscr{F}_{s_n}] = Z_{s_n} e^{i(\theta, W_{s_n})} &+ \sum_{j=1}^d i\theta_j E\left[Z_s \int_{s_n}^s e^{i(\theta, W_u)}\,dW_u^{(j)} \,\middle|\, \mathscr{F}_{s_n}\right] \\ &- \frac{\|\theta\|^2}{2} E\left[Z_s \int_{s_n}^s e^{i(\theta, W_u)}\,du \,\middle|\, \mathscr{F}_{s_n}\right]. \end{aligned}$$

But (4.40) and (4.41) imply

$$E\left[Z_s \int_{s_n}^s e^{i(\theta, W_u)}\,dW_u^{(j)} \,\middle|\, \mathscr{F}_{s_n}\right] = E\left[(Z_s - Z_{s_n}) \int_{s_n}^s e^{i(\theta, W_u)}\,dW_u^{(j)} \,\middle|\, \mathscr{F}_{s_n}\right] = 0.$$

Multiplying (4.43) by $\prod_{k=0}^n f_k(W_{s_k})$ and taking expectations, we obtain

(4.44) $$\begin{aligned} \varphi(s) &= \varphi(s_n) - \frac{\|\theta\|^2}{2}\int_{s_n}^s E\left[Z_s \cdot \prod_{k=0}^n f_k(W_{s_k}) e^{i(\theta, W_u)}\right] du \\ &= \varphi(s_n) - \frac{\|\theta\|^2}{2}\int_{s_n}^s \varphi(u)\,du;\quad s_n \le s \le t. \end{aligned}$$

By our induction hypothesis, $\varphi(s_n) = 0$, and the only solution to the integral equation (4.44) satisfying this initial condition is $\varphi(s) = 0$; $s_n \le s \le t$. Thus

$$(4.45) \qquad E\left[Z_t \cdot \prod_{k=0}^{n} f_k(W_{s_k}) e^{i(\theta, W_s)}\right] = 0; \quad \forall \theta \in \mathbb{R}^d.$$

If each f_k is real-valued, we set $D \triangleq Z_t \cdot \prod_{k=0}^{n} f_k(W_{s_k})$ and define measures on $(\mathbb{R}^d, \mathscr{B}(\mathbb{R}^d))$ by $\mu^{\pm}(\Gamma) = E[D^{\pm} 1_\Gamma(W_s)]$; $\Gamma \in \mathscr{B}(\mathbb{R}^d)$. We have shown that

$$\int_{\mathbb{R}^d} e^{i(\theta, x)} \mu^+(dx) = \int_{\mathbb{R}^d} e^{i(\theta, x)} \mu^-(dx) \quad \forall \theta \in \mathbb{R}^d,$$

and by the uniqueness theorem for Fourier transforms, we see that $\mu^+ = \mu^-$. Thus

$$E\left[Z_t \cdot \prod_{k=0}^{n} f_k(W_{s_k}) f(W_s)\right] = 0; \quad s_n \le s_{n+1} \triangleq s \le t,$$

for any bounded, measurable $f\colon \mathbb{R}^d \mapsto \mathbb{C}$. If the functions f_k are complex-valued, then $\prod_{k=0}^{n} f_k(W_{s_k})$ can be written as a finite sum of a product of real-valued functions plus $\sqrt{-1}$ times the finite sum of another such product. Therefore, (4.42) holds for $n+1$ and the induction step is complete.

A standard argument using the Dynkin System Theorem 2.1.3 now shows that we have

$$(4.46) \qquad E[Z_t \xi] = 0$$

for every $\mathscr{F}_t^W$-measurable indicator ξ, and thus, for every $\mathscr{F}_t^W$-measurable, bounded ξ. Since $\mathscr{F}_t$ differs from $\mathscr{F}_t^W$ only by P-null sets, (4.46) also holds for every $\mathscr{F}_t$-measurable, bounded ξ. Setting $\xi = \operatorname{sgn}(Z_t)$, we conclude that $Z_t = 0$ a.s. P for every fixed t, and by right-continuity of Z, for *all* $t \in [0, \infty)$ simultaneously.

The uniqueness assertion in the last sentence of the theorem is proved by observing that the martingale $\sum_{j=1}^{d} \int_0^t (Y_s^{(j)} - \tilde{Y}_s^{(j)})\, dW_s^{(j)}$ is identically zero, and so is its quadratic variation. □

4.16 Problem. Let W and $\{\mathscr{F}_t\}$ be as in Theorem 4.15. Let $M = \{M_t, \mathscr{F}_t; 0 \le t < \infty\}$, satisfying $M_0 = 0$, be an RCLL martingale. Show that M is continuous. (*Hint*: For arbitrary $T \in (0, \infty)$, prove continuity on $[0, T]$ by choosing bounded, $\mathscr{F}_T$-measurable random variables $M_T^{(n)}$ such that $E|M_T^{(n)} - M_T| \le 3^{-n}$, setting $\{M_t^{(n)}; 0 \le t \le T\}$ to be an RCLL modification (Theorem 1.3.13) of $\{E[M_T^{(n)} | \mathscr{F}_t]; 0 \le t \le T\}$, and using Theorem 1.3.8(i) and the Borel-Cantelli Lemma to show that $M^{(n)}$ converges *uniformly* to M on $[0, T]$, almost surely.) Now suppose that M is an RCLL *local* martingale. Show that M is continuous and there exist progressively measurable processes $Y^{(j)} = \{Y_t^{(j)}, \mathscr{F}_t; 0 \le t < \infty\}$ such that

$$\int_0^T (Y_t^{(j)})^2\, dt < \infty; \quad 1 \le j \le d, \quad 0 \le T < \infty,$$

and (4.39) holds.

4.17 Problem. Under the hypotheses of Theorem 4.15 and with $0 < T < \infty$, let ξ be an $\mathscr{F}_T$-measurable random variable with $E\xi^2 < \infty$. Prove that there are progressively measurable processes $Y^{(1)}, \ldots, Y^{(d)}$ satisfying (4.38), and such that

$$\xi = E(\xi) + \sum_{j=1}^{d} \int_0^T Y_s^{(j)}\, dW_s^{(j)}; \quad \text{a.s. } P. \tag{4.47}$$

E. Brownian Functionals as Stochastic Integrals

We extend Problem 4.17 to include the case $T = \infty$. Recall that for $M \in \mathcal{M}_2^c$, we denote by $\mathcal{L}_\infty^*(M)$ the class of processes X which are progressively measurable with respect to the filtration of M and which satisfy $E \int_0^\infty X_t^2\, d\langle M\rangle_t < \infty$. According to Problem 1.5.24, when $X \in \mathcal{L}_\infty^*(M)$, we have $\int_0^\infty X_t\, dM_t$ defined a.s. P. If W is a d-dimensional Brownian motion, we denote by $\mathcal{L}_\infty^*(W)$ the set of processes X which are progressively measurable with respect to the (augmented) filtration of W and which satisfy $E \int_0^\infty X_t^2\, dt < \infty$.

4.18 Proposition. *Under the hypotheses of Theorem 4.15, let ξ be an $\mathscr{F}_\infty$-measurable random variable with $E\xi^2 < \infty$. Then there are processes $Y^{(1)}, \ldots, Y^{(d)}$ in $\mathcal{L}_\infty^*(W)$ such that*

$$\xi = E(\xi) + \sum_{j=1}^{d} \int_0^\infty Y_s^{(j)}\, dW_s^{(j)}; \quad \text{a.s. } P.$$

PROOF. Assume without loss of generality that $E(\xi) = 0$, and let M_t be a right-continuous modification of $E(\xi \mid \mathscr{F}_t)$. According to Theorem 4.15, there exist progressively measurable $Y^{(1)}, \ldots, Y^{(d)}$ satisfying (4.38) and (4.39). Jensen's inequality implies $M_t^2 \le E(\xi^2 \mid \mathscr{F}_t)$, so

$$\sum_{j=1}^{d} E \int_0^t (Y_s^{(j)})^2\, ds = E\langle M\rangle_t = E(M_t^2) \le E(\xi^2) < \infty; \quad 0 \le t < \infty.$$

Hence, $Y^{(j)} \in \mathcal{L}_\infty^*(W)$ for $1 \le j \le d$ and $M_\infty \triangleq \sum_{j=1}^{d} \int_0^\infty Y_s^{(j)}\, dW_s^{(j)}$ is defined. Problem 1.3.20 shows that $M_\infty = E(\xi \mid \mathscr{F}_\infty) = \xi$. □

We leave the proof of *uniqueness* of the representation in Proposition 4.18 as Exercise 4.22 for the reader.

In one dimension, there is a representation result similar to that of Proposition 4.18 in which the Brownian motion is replaced by a continuous local martingale M. This result is instrumental in our eventual proof of Theorem 4.13. Uniqueness is again addressed in Exercise 4.22.

4.19 Proposition. *Let $M = \{M_t, \mathscr{F}_t; 0 \le t < \infty\}$ be in $\mathcal{M}^{c,\text{loc}}$ and assume that $\lim_{t\to\infty} \langle M\rangle_t = \infty$, a.s. P. Define $T(s)$ by (4.15) and let B be the one-dimensional Brownian motion*

$$B_s \triangleq M_{T(s)}, \mathscr{G}_s; \quad 0 \le s < \infty$$

as in Theorem 4.6, except now we take the filtration $\{\mathscr{E}_s\}$ to be the augmentation with respect to P of the filtration $\{\mathscr{F}_s^B\}$ generated by B. Then, for every $\mathscr{E}_\infty$-measurable random variable ξ satisfying $E\xi^2 < \infty$, there is a process $X \in \mathscr{L}_\infty^*(M)$ for which

$$\xi = E(\xi) + \int_0^\infty X_t\,dM_t; \quad \text{a.s. } P. \tag{4.48}$$

PROOF. Let $Y = \{Y_s, \mathscr{E}_s; 0 \le s < \infty\}$ be the progressively measurable process of Proposition 4.18, for which we have

$$E\int_0^\infty Y_s^2\,ds < \infty, \tag{4.49}$$

$$\xi = E(\xi) + \int_0^\infty Y_s\,dB_s. \tag{4.50}$$

Define $\tilde{X}_t = Y_{\langle M\rangle_t}$; $0 \le t < \infty$.

We show how to obtain an $\{\mathscr{F}_t\}$-progressively measurable process X which is equivalent to $\tilde{X}$. Note that because $\{\mathscr{G}_s\} \triangleq \{\mathscr{F}_{T(s)}\}$ contains $\{\mathscr{F}_s^B\}$ and satisfies the usual conditions (Theorem 4.6), we have $\mathscr{E}_s \subseteq \mathscr{G}_s$; $0 \le s \le \infty$. Consequently, Y is progressively measurable relative to $\{\mathscr{G}_s\}$. If Y is a simple process, it is left-continuous (c.f. Definition 2.3), and it is straightforward to show, using Problem 4.5, that $\{Y_{\langle M\rangle_t}; 0 \le t < \infty\}$ is a left-continuous process adapted to $\{\mathscr{F}_t\}$, and hence progressively measurable (Proposition 1.1.13). In the general case, let $\{Y^{(n)}\}_{n=1}^\infty$ be a sequence of progressively measurable (relative to $\{\mathscr{E}_s\}$), simple processes for which

$$\lim_{n\to\infty} E\int_0^\infty |Y_s^{(n)} - Y_s|^2\,ds = 0.$$

(Use Proposition 2.8 and (4.49)). A change of variables (Problem 4.5 (vi)) yields

$$\lim_{n\to\infty} E\int_0^\infty |X_t^{(n)} - \tilde{X}_t|^2\,d\langle M\rangle_t = 0, \tag{4.51}$$

where $X_t^{(n)} \triangleq Y_{\langle M\rangle_t}^{(n)}$. In particular, the sequence $\{X^{(n)}\}_{n=1}^\infty$ is Cauchy in $\mathscr{L}_\infty^*(M)$, and so, by Lemma 2.2, converges to a limit $X \in \mathscr{L}_\infty^*(M)$. From (4.51) we must have

$$E\int_0^\infty |\tilde{X}_t - X_t|^2\,d\langle M\rangle_t = 0,$$

which establishes the desired equivalence of X and $\tilde{X}$.

It remains to prove (4.48), which, in light of (4.50), will follow from

$$\int_0^\infty Y_s\,dB_s = \int_0^\infty X_t\,dM_t; \quad \text{a.s. } P.$$

This equality is a consequence of Proposition 4.8. □

PROOF OF F. B. KNIGHT'S THEOREM 4.13. Our proof is based on that of Meyer (1971). Under the hypotheses of Theorem 4.13, let $\{\mathscr{E}_s^{(i)}\}$ be the augmentation of the filtration $\{\mathscr{F}_s^{B^{(i)}}\}$ generated by $B^{(i)}$; $1 \le i \le d$. All we need to show is that $\mathscr{E}_\infty^{(1)}, \ldots, \mathscr{E}_\infty^{(d)}$ are independent.

For each i, let $\xi^{(i)}$ be a bounded, $\mathscr{E}_\infty^{(i)}$-measurable random variable. According to Proposition 4.19, there is, for each i, a progressively measurable process $X^{(i)} = \{X_t^{(i)}, \mathscr{F}_t; 0 \le t < \infty\}$ which satisfies

$$E\int_0^\infty (X_t^{(i)})^2\, d\langle M^{(i)}\rangle_t < \infty; \quad 1 \le i \le d,$$

and for which

$$\xi^{(i)} = E(\xi^{(i)}) + \int_0^\infty X_t^{(i)}\, dM_t^{(i)}; \quad 1 \le i \le d.$$

Let us assume for the moment that

$$E(\xi^{(i)}) = 0; \quad 1 \le i \le d, \tag{4.52}$$

and define the $\{\mathscr{F}_t\}$-martingale

$$\xi_t^{(i)} \triangleq \int_0^t X_s^{(i)}\, dM_s^{(i)}; \quad 0 \le t < \infty, \quad 1 \le i \le d.$$

Itô's rule and (4.31) imply that

$$\prod_{i=1}^d \xi_t^{(i)} = \sum_{i=1}^d \int_0^t \prod_{j \ne i} \xi_s^{(j)} X_s^{(i)}\, dM_s^{(i)}; \quad 0 \le t < \infty. \tag{4.53}$$

In order to let $t \to \infty$ in (4.53), we must show that

$$E\int_0^\infty \left(\prod_{j \ne i} \xi_s^{(j)} X_s^{(i)}\right)^2 d\langle M^{(i)}\rangle_s < \infty; \quad 1 \le i \le d. \tag{4.54}$$

Because each $\xi^{(i)}$ is assumed to be bounded, there is a finite constant C such that $|\xi^{(i)}| \le C$ almost surely for every i. It follows that

$$\begin{aligned}
&E\int_0^t \left(\prod_{j \ne i} \xi_s^{(j)} X_s^{(i)}\right)^2 d\langle M^{(i)}\rangle_s \\
&\le C^{2(d-1)} E\int_0^t (X_s^{(i)})^2\, d\langle M^{(i)}\rangle_s \\
&= C^{2(d-1)} E\left\langle \xi^{(i)}\right\rangle_t, \quad 0 \le t < \infty \\
&= C^{2(d-1)} E\left[(\xi_t^{(i)})^2\right] \\
&\le C^{2d} < \infty.
\end{aligned}$$

Thus (4.54) holds, and letting $t \to \infty$ in (4.53) we obtain the representation

$$\prod_{i=1}^{d} \xi^{(i)} = \sum_{i=1}^{d} \int_0^\infty \prod_{j \neq i} \xi_s^{(j)} X_s^{(i)} \, dM_s^{(i)}.$$

The right-hand side, being a sum of martingale last elements (Problem 2.18), has expectation zero. Thus, under the assumption (4.52), we have $E\prod_{i=1}^{d} \xi^{(i)} = 0$. Equivalently, we have shown that for any set of bounded random variables $\xi^{(1)}, \ldots, \xi^{(d)}$, where each $\xi^{(i)}$ is $\mathscr{E}_\infty^{(i)}$-measurable, the equality

$$E\prod_{i=1}^{d} [\xi^{(i)} - E(\xi^{(i)})] = 0 \tag{4.55}$$

holds. Using (4.55), one can show by a simple argument of induction on d that

$$E\prod_{i=1}^{d} \xi^{(i)} = \prod_{i=1}^{d} E\xi^{(i)}.$$

Taking $\xi^{(i)} = 1_{A_i}$; $A_i \in \mathscr{E}_\infty^{(i)}$, $1 \leq i \leq d$, we conclude that the σ-fields $\mathscr{E}_\infty^{(1)}, \ldots, \mathscr{E}_\infty^{(d)}$ are independent. □

What happens if the random variable ξ in Problem 4.17 is not square-integrable, but merely a.s. finite? It is reasonable to guess that there is still a representation of the form (4.47), where now the integrands $Y^{(1)}, \ldots, Y^{(d)}$ can only be expected to satisfy

$$\int_0^T (Y_t^{(j)})^2 \, dt < \infty; \quad \text{a.s. } P.$$

In fact, an even stronger result is true. We send the interested reader to Dudley (1977) for the ingenious proof, which uses the representation of stochastic integrals as time-changed Brownian motions. There is, however, no uniqueness in Dudley's theorem; see Exercise 4.22 (iii).

4.20 Theorem (Dudley (1977)). *Let $W = \{W_t, \mathscr{F}_t; 0 \leq t < \infty\}$ be a standard, one-dimensional Brownian motion, where, in addition to satisfying the usual conditions, $\{\mathscr{F}_t\}$ is left-continuous. If $0 < T < \infty$ and ξ is an $\mathscr{F}_T$-measurable, a.s. finite random variable, then there exists a progressively measurable process $Y = \{Y_t, \mathscr{F}_t; 0 \leq t \leq T\}$ satisfying*

$$\int_0^T Y_t^2 \, dt < \infty; \quad \text{a.s. } P,$$

such that

$$\xi = \int_0^T Y_t \, dW_t; \quad \text{a.s. } P.$$

4.21 Remark. Note in Theorem 4.20 that the filtration $\{\mathscr{F}_t\}$ might be generated by a d-dimensional Brownian motion of which W is only one component. For an amplification of this point, see Emery, Stricker & Yan (1983).

4.22 Exercise.

(i) In the setting of Proposition 4.18, show that the processes $Y^{(1)}, \ldots, Y^{(d)}$ are unique in the sense that any other processes $\tilde{Y}^{(1)}, \ldots, \tilde{Y}^{(d)}$ in $\mathscr{L}_\infty^*(W)$ which permit the representation

$$\xi = E(\xi) + \sum_{j=1}^{d} \int_0^\infty \tilde{Y}_t^{(j)}\, dW_t^{(j)}, \quad \text{a.s.}$$

must also satisfy

$$\int_0^\infty \sum_{j=1}^{d} |Y_t^{(j)} - \tilde{Y}_t^{(j)}|^2\, dt = 0, \quad \text{a.s.}$$

(ii) In the setting of Proposition 4.19, show that X is unique in the sense that any other process $\tilde{X} \in \mathscr{L}_\infty^*(M)$ which permits the representation

$$\xi = E(\xi) + \int_0^\infty \tilde{X}_t\, dM_t, \quad \text{a.s.},$$

must also satisfy

$$\int_0^\infty |X_t - \tilde{X}_t|^2\, d\langle M\rangle_t = 0, \quad \text{a.s.}$$

(iii) Find a progressively measurable process Y such that

$$0 < \int_0^1 Y_t^2\, dt < \infty, \quad \text{a.s., but} \quad \int_0^1 Y_t\, dW_t = 0, \quad \text{a.s.}$$

In particular, there can be no assertion of uniqueness of Y in Theorem 4.20.

4.23 Exercise. Is the following assertion true or false? "If $M = \{M_t, \mathscr{F}_t; 0 \le t < \infty\}$ and $N = \{N_t, \mathscr{F}_t; 0 \le t < \infty\}$ are in $\mathscr{M}^{c,\text{loc}}$ and $\langle M\rangle \equiv \langle N\rangle$, then M and N have the same law."

4.24 Exercise (Hajek (1985)). Consider the semimartingales

$$X_t = X_0 + \int_0^t \mu_s\, ds + \int_0^t \sigma_s\, dW_s, \quad Y_t = Y_0 + \int_0^t m(s)\, ds + \int_0^t \rho(s)\, dV_s$$
$$0 \le t < \infty$$

for $0 \le t < \infty$, where W and V are Brownian motions; the progressively measurable processes μ, σ and the Borel-measurable functions $m: [0, \infty) \to \mathbb{R}$, $\rho: [0, \infty) \to [0, \infty)$ are assumed to satisfy

$$\mu_t \le m(t), \quad |\sigma_t| \le \rho(t), \quad \int_0^t \{|\mu_s| + m(s) + \rho^2(s)\}\, ds < \infty, \quad \forall\, 0 \le t < \infty$$

almost surely. If $X_0 \le Y_0$ also holds a.s. and f is a nondecreasing, convex function on $\mathbb{R}$, show that

$$P(X_t \ge c) \le 2P(Y_t \ge c); \quad \forall c \in \mathbb{R},$$
$$Ef(X_t) \le Ef(Y_t)$$

hold for every $t \ge 0$. (*Hint*: By extending the probability space if necessary, take $\hat{W}$ to be a Brownian motion independent of W, and consider the continuous semimartingales

$$Y_t^{(i)} \triangleq Y_0 + \int_0^t m(s)\,ds + \int_0^t \sigma_s\,dW_s + (-1)^i \int_0^t \sqrt{\rho^2(s) - \sigma_s^2}\,d\hat{W}_s; \qquad i = 1, 2.)$$

4.25 Exercise (Elworthy, Li & Yor, 1997): Let M be a continuous, nonnegative local martingale with $M_0 = m > 0$ a real constant and $M_\infty \triangleq \lim_{t\to\infty} M_t = 0$, a.s. Show $l = \sqrt{\pi/2}\sigma = m$, where

$$l \triangleq \lim_{x\to\infty}\left(xP\left[\sup_{t\ge 0} M_t \ge x\right]\right), \quad \sigma \triangleq \lim_{y\to\infty}(yP[\sqrt{\langle M\rangle_\infty} \ge y]).$$

(*Hint*: Use the results of Theorem 4.6 and Problem 4.7, as well as (2.6.2) and Exercise 2.8.13.)

3.5. The Girsanov Theorem

In order to motivate the results of this section, let us consider independent normal random variables $Z_1, \ldots, Z_n$ on $(\Omega, \mathscr{F}, P)$ with $EZ_i = 0$, $EZ_i^2 = 1$. Given a vector $(\mu_1, \ldots, \mu_n) \in \mathbb{R}^n$, we consider the new probability measure $\tilde{P}$ on $(\Omega, \mathscr{F})$ given by

$$\tilde{P}(d\omega) = \exp\left[\sum_{i=1}^n \mu_i Z_i(\omega) - \frac{1}{2}\sum_{i=1}^n \mu_i^2\right] \cdot P(d\omega).$$

Then $\tilde{P}[Z_1 \in dz_1, \ldots, Z_n \in dz_n]$ is given by

$$\exp\left[\sum_{i=1}^n \mu_i z_i - \frac{1}{2}\sum_{i=1}^n \mu_i^2\right] \cdot P[Z_1 \in dz_1, \ldots, Z_n \in dz_n]$$
$$= (2\pi)^{-n/2} \exp\left[-\frac{1}{2}\sum_{i=1}^n (z_i - \mu_i)^2\right] dz_1 \ldots dz_n.$$

Therefore, under $\tilde{P}$ the random variables $Z_1, \ldots, Z_n$ are independent and normal with $\tilde{E}Z_i = \mu_i$ and $\tilde{E}[(Z_i - \mu_i)^2] = 1$. In other words, $\{\tilde{Z}_i = Z_i - \mu_i;\ 1 \le i \le n\}$ are independent, standard normal random variables on $(\Omega, \mathscr{F}, \tilde{P})$. The Girsanov Theorem 5.1 extends this idea of *invariance of Gaussian finite-dimensional distributions* under appropriate translations and changes of the underlying probability measure, from the discrete to the continuous setting. Rather than beginning with an n-dimensional vector $(Z_1, \ldots, Z_n)$ of independent, standard normal random variables, we begin with a d-dimensional Brownian motion under P, and then construct a new measure $\tilde{P}$ under which a "translated" process is a d-dimensional Brownian motion.

A. The Basic Result

Throughout this section, we shall have a probability space $(\Omega, \mathscr{F}, P)$ and a d-dimensional Brownian motion $W = \{W_t = (W_t^{(1)}, \ldots, W_t^{(d)}), \mathscr{F}_t; 0 \le t < \infty\}$ defined on it, with $P[W_0 = 0] = 1$. We assume that the filtration $\{\mathscr{F}_t\}$ satisfies the usual conditions. Let $X = \{X_t = (X_t^{(1)}, \ldots, X_t^{(d)}), \mathscr{F}_t; 0 \le t < \infty\}$ be a vector of measurable, adapted processes satisfying

$$\text{(5.1)} \qquad P\left[\int_0^T (X_t^{(i)})^2\,dt < \infty\right] = 1; \quad 1 \le i \le d, \quad 0 \le T < \infty.$$

Then, for each i, the stochastic integral $I^{W^{(i)}}(X^{(i)})$ is defined and is a member of $\mathscr{M}^{c,\mathrm{loc}}$. We set

$$\text{(5.2)} \qquad Z_t(X) \triangleq \exp\left[\sum_{i=1}^d \int_0^t X_s^{(i)}\,dW_s^{(i)} - \frac{1}{2}\int_0^t \|X_s\|^2\,ds\right].$$

Just as in Example 3.9, we have

$$\text{(5.3)} \qquad Z_t(X) = 1 + \sum_{i=1}^d \int_0^t Z_s(X) X_s^{(i)}\,dW_s^{(i)},$$

which shows that $Z(X)$ is a continuous, local martingale with $Z_0(X) = 1$.

Under certain conditions on X, to be discussed later, $Z(X)$ will in fact be a martingale, and so $EZ_t(X) = 1; 0 \le t < \infty$. In this case we can define, for each $0 \le T < \infty$, a probability measure $\tilde{P}_T$ on $\mathscr{F}_T$ by

$$\text{(5.4)} \qquad \tilde{P}_T(A) \triangleq E[1_A Z_T(X)]; \quad A \in \mathscr{F}_T.$$

The martingale property shows that the family of probability measures $\{\tilde{P}_T; 0 \le T < \infty\}$ satisfies the consistency condition

$$\text{(5.5)} \qquad \tilde{P}_T(A) = \tilde{P}_t(A); \quad A \in \mathscr{F}_t, \quad 0 \le t \le T.$$

5.1 Theorem (Girsanov (1960), Cameron and Martin (1944)). *Assume that $Z(X)$ defined by (5.2) is a martingale. Define a process $\tilde{W} = \{\tilde{W}_t = (\tilde{W}_t^{(1)}, \ldots, \tilde{W}_t^{(d)}), \mathscr{F}_t; 0 \le t < \infty\}$ by*

$$\text{(5.6)} \qquad \tilde{W}_t^{(i)} \triangleq W_t^{(i)} - \int_0^t X_s^{(i)}\,ds; \quad 1 \le i \le d, \quad 0 \le t < \infty.$$

For each fixed $T \in [0, \infty)$, the process $\{\tilde{W}_t, \mathscr{F}_t; 0 \le t \le T\}$ is a d-dimensional Brownian motion on $(\Omega, \mathscr{F}_T, \tilde{P}_T)$.

The preparation for the proof of this result starts with Lemma 5.3; the reader may proceed there directly, skipping the remainder of this subsection on first reading.

Discussion. Occasionally, one wants $\tilde{W}$, as a process defined for *all* $t \in [0, \infty)$, to be a Brownian motion, and for this purpose the measures $\{\tilde{P}_T; 0 \le T < \infty\}$ are inadequate. We would like to have a *single* measure $\tilde{P}$ defined on $\mathscr{F}_\infty$, so that $\tilde{P}$ restricted to any $\mathscr{F}_T$ agrees with $\tilde{P}_T$; however, such a measure does not exist in general.

We thus restrict our attention to d-dimensional Brownian motion W defined on the canonical probability space $\Omega = C[0, \infty)^d$ of continuous, $\mathbb{R}^d$-valued functions, and we content ourselves with a measure $\tilde{P}$ defined only on $\mathscr{F}_\infty^W$, the σ-field generated by W, such that $\tilde{P}$ satisfies

$$\tilde{P}(A) = E[1_A Z_T(X)]; \quad A \in \mathscr{F}_T^W, \quad 0 \le T < \infty. \tag{5.7}$$

If such a $\tilde{P}$ exists, it is clearly unique.

Let P be Wiener measure on the space $(\Omega, \mathscr{F}) \triangleq (C[0, \infty)^d, \mathscr{B}(C[0, \infty)^d))$, under which the coordinate mapping process $W_t(\omega) \triangleq \omega(t), 0 \le t < \infty, \omega \in \Omega$ is a standard Brownian motion. Then $\mathscr{F}_\infty^W = \mathscr{B}(C[0, \infty)^d)$. For every $T \in [0, \infty)$, define the probability measure $\tilde{P}_T$ on $\mathscr{F}_T^W$ by $\tilde{P}_T(A) \triangleq E[1_A Z_T(X)]$ for all $A \in \mathscr{F}_T^W$. The family $\{\tilde{P}\}_{0 \le T < \infty}$ is consistent (cf. (5.5)), and thus (5.7) defines well a finitely additive set function $\tilde{P}$ on the algebra $\bigcup_{0 \le T < \infty} \mathscr{F}_T^W$, which satisfies $\tilde{P}(\emptyset) = 0$, $\tilde{P}(\Omega) = 1$. Once one shows that $\tilde{P}$ is σ-additive on $\bigcup_{0 \le T < \infty} \mathscr{F}_T^W$, then one can use the Carathéodory Extension Theorem to define $\tilde{P}$ on all of $\mathscr{F}_\infty^W$. Rather than providing the details of this argument, we refer the reader to Theorem 4.2 (p. 143) of Parthasarathy (1967).

The process $\tilde{W}$ in Theorem 5.1 is adapted to the filtration $\{\mathscr{F}_t\}$, and so is the process $\{\int_0^t X_s^{(i)} ds; 0 \le t < \infty\}$; this can be seen as in part (c) of the proof of Lemma 2.4, which uses the completeness of $\mathscr{F}_t$. However, when working with the measure $\tilde{P}$ which is defined only on $\mathscr{F}_\infty^W$, we wish $\tilde{W}$ to be adapted to $\{\mathscr{F}_t^W\}$. This filtration does not satisfy the usual conditions, and so we must impose the stronger condition of progressive measurability on X. We have the following corollary to Theorem 5.1.

5.2 Corollary. *Let $W = \{W_t, \mathscr{F}_t; 0 \le t < \infty\}$ be the coordinate mapping process on $\Omega \triangleq C[0, \infty)^d$, so that $\mathscr{F}_\infty^W = \mathscr{B}(C[0, \infty)^d)$. Let P be Wiener measure on $(\Omega, \mathscr{F}_\infty^W)$. Let $X = \{X_t, \mathscr{F}_t^W; 0 \le t < \infty\}$ be a d-dimensional, progressively measurable process satisfying* (5.1). *If $Z(X)$ of* (5.2) *is a martingale, then there is a unique probability measure $\tilde{P}$ satisfying* (5.7), *and $\tilde{W} = \{\tilde{W}_t, \mathscr{F}_t^W; 0 \le t < \infty\}$ defined by* (5.6) *is a d-dimensional Brownian motion on $(\Omega, \mathscr{F}_\infty^W, \tilde{P})$.*

PROOF. We have argued the existence and uniquesness of $\tilde{P}$ above. To see that $\tilde{W}$ is a Brownian motion on $(\Omega, \mathscr{F}_\infty^W, \tilde{P})$, let $0 \le t_1 < \cdots < t_n \le t$ be given. We have

$$\tilde{P}[(\tilde{W}_{t_1},\ldots,\tilde{W}_{t_n})\in\Gamma] = \tilde{P}_t[(\tilde{W}_{t_1},\ldots,\tilde{W}_{t_n})\in\Gamma]; \quad \Gamma\in\mathscr{B}(\mathbb{R}^{dn}).$$

The result now follows from Theorem 5.1. □

Remark. Under the assumptions of Corollary 5.2, the probability measures P and $\tilde{P}$ are mutually absolutely continuous *when restricted to* $\mathscr{F}_T^W$; $0 \le T < \infty$. However, viewed as probability measures on $\mathscr{F}_\infty^W$, *P and $\tilde{P}$ are mutually absolutely continuous if and only if the martingale $Z(X)$ is uniformly integrable.*

For example, when $d = 1$ and $X_t = \mu$, a nonzero constant, the P-martingale

$$Z_t(X) = \exp\left[\mu W_t - \frac{1}{2}\mu^2 t\right]; \quad 0 \le t < \infty$$

is *not* uniformly integrable. Corollary 5.2 and the law of large numbers imply

$$\tilde{P}\left[\lim_{t\to\infty}\frac{1}{t}W_t = \mu\right] = \tilde{P}\left[\lim_{t\to\infty}\frac{1}{t}\tilde{W}_t = 0\right] = 1, \quad P\left[\lim_{t\to\infty}\frac{1}{t}W_t = \mu\right] = 0.$$

In particular, the P-null event $\{\lim_{t\to\infty}(1/t)W_t = \mu\}$ is in $\mathscr{F}_T$ for every $0 \le T < \infty$, so $\tilde{P}$ and $\tilde{P}_T$ cannot agree on $\mathscr{F}_T$. This is the reason we require (5.7) to hold only for $A \in \mathscr{F}_T^W$.

B. Proof and Ramifications

We now proceed with the proof of Theorem 5.1. We denote by $\tilde{E}_T$ ($\tilde{E}$) the expectation operator with respect to $\tilde{P}_T$ ($\tilde{P}$).

5.3 Lemma. *Fix $0 \le T < \infty$ and assume that $Z(X)$ is a martingale. If $0 \le s \le t \le T$ and Y is an $\mathscr{F}_t$-measurable random variable satisfying $\tilde{E}_T|Y| < \infty$, then we have the* Bayes' rule:

$$\tilde{E}_T[Y|\mathscr{F}_s] = \frac{1}{Z_s(X)}E[YZ_t(X)|\mathscr{F}_s], \quad \text{a.s. } P \text{ and } \tilde{P}_T.$$

PROOF. Using the definition of $\tilde{E}_T$, the definition of conditional expectation, and the martingale property, we have for any $A \in \mathscr{F}_s$:

$$\tilde{E}_T\left\{1_A\frac{1}{Z_s(X)}E[YZ_t(X)|\mathscr{F}_s]\right\} = E\{1_A E[YZ_t(X)|\mathscr{F}_s]\}$$

$$= E[1_A YZ_t(X)] = \tilde{E}_T[1_A Y].$$ □

We denote by $\mathscr{M}_T^{c,\mathrm{loc}}$ the class of continuous local martingales $M = \{M_t, \mathscr{F}_t; 0 \le t \le T\}$ on $(\Omega, \mathscr{F}_T, P)$ satisfying $P[M_0 = 0] = 1$, and define $\tilde{\mathscr{M}}_T^{c,\mathrm{loc}}$ similarly, with P replaced by $\tilde{P}_T$.

5.4 Proposition. *Fix $0 \le T < \infty$ and assume that $Z(X)$ is a martingale. If $M \in \mathscr{M}_T^{c,\text{loc}}$, then the process*

$$(5.8) \qquad \tilde{M}_t \triangleq M_t - \sum_{i=1}^d \int_0^t X_s^{(i)}\, d\langle M, W^{(i)}\rangle_s, \mathscr{F}_t; \quad 0 \le t \le T$$

is in $\tilde{\mathscr{M}}_T^{c,\text{loc}}$. If $N \in \mathscr{M}_T^{c,\text{loc}}$ and

$$\tilde{N}_t \triangleq N_t - \sum_{i=1}^d \int_0^t X_s^{(i)}\, d\langle N, W^{(i)}\rangle_s; \quad 0 \le t \le T,$$

then

$$\langle \tilde{M}, \tilde{N}\rangle_t = \langle M, N\rangle_t; \quad 0 \le t \le T, \quad \text{a.s. } P \text{ and } \tilde{P}_T,$$

where the cross-variations are computed under the appropriate measures.

Proof. We consider only the case where M and N are bounded martingales with bounded quadratic variations, and assume also that $Z_t(X)$ and $\sum_{j=1}^d \int_0^t (X_s^{(j)})^2\, ds$ are bounded in t and ω; the general case can be reduced to this one by localization. From Proposition 2.14

$$\left| \int_0^t X_s^{(i)}\, d\langle M, W^{(i)}\rangle_s \right|^2 \le \langle M\rangle_t \cdot \int_0^t (X_s^{(i)})^2\, ds,$$

and thus $\tilde{M}$ is also bounded. The integration-by-parts formula of Problem 3.12 gives

$$Z_t(X)\tilde{M}_t = \int_0^t Z_u(X)\, dM_u + \sum_{i=1}^d \int_0^t \tilde{M}_u X_u^{(i)} Z_u(X)\, dW_u^{(i)},$$

which is a martingale under P. Therefore, for $0 \le s \le t \le T$, we have from Lemma 5.3:

$$\tilde{E}_T[\tilde{M}_t | \mathscr{F}_s] = \frac{1}{Z_s(X)} E[Z_t(X)\tilde{M}_t | \mathscr{F}_s] = \tilde{M}_s, \quad \text{a.s. } P \text{ and } \tilde{P}_T.$$

It follows that $\tilde{M} \in \tilde{\mathscr{M}}^{c,\text{loc}}$.

The change-of-variable formula also implies:

$$\tilde{M}_t\tilde{N}_t - \langle M, N\rangle_t = \int_0^t \tilde{M}_u\, dN_u + \int_0^t \tilde{N}_u\, dM_u - \sum_{i=1}^d \left[\int_0^t \tilde{M}_u X_u^{(i)}\, d\langle N, W^{(i)}\rangle_u + \int_0^t \tilde{N}_u X_u^{(i)}\, d\langle M, W^{(i)}\rangle_u \right]$$

as well as

$$Z_t(X)[\tilde{M}_t\tilde{N}_t - \langle M, N\rangle_t] = \int_0^t Z_u(X)\tilde{M}_u\, dN_u + \int_0^t Z_u(X)\tilde{N}_u\, dM_u + \sum_{i=1}^d \int_0^t [\tilde{M}_u\tilde{N}_u - \langle M, N\rangle_u] X_u^{(i)} Z_u(X)\, dW_u^{(i)}.$$

This last process is consequently a martingale under P, and so Lemma 5.3 implies that for $0 \le s \le t \le T$

$$\tilde{E}_T[\tilde{M}_t\tilde{N}_t - \langle M, N\rangle_t | \mathscr{F}_s] = \tilde{M}_s\tilde{N}_s - \langle M, N\rangle_s; \quad \text{a.s. } P \text{ and } \tilde{P}_T.$$

This proves that $\langle \tilde{M}, \tilde{N}\rangle_t = \langle M, N\rangle_t$; $0 \le t \le T$, a.s. $\tilde{P}_T$ and P. □

Proof of Theorem 5.1. We show that the continuous process $\tilde{W}$ on $(\Omega, \mathscr{F}_T, \tilde{P}_T)$ satisfies the hypotheses of P. Lévy's Theorem 3.16. Setting $M = W^{(j)}$ in Proposition 5.4 we obtain $\tilde{M} = \tilde{W}^{(j)}$ from (5.8), so $\tilde{W}^{(j)} \in \tilde{\mathscr{M}}_T^{c,\text{loc}}$. Setting $N = W^{(k)}$, we obtain

$$\langle \tilde{W}^{(j)}, \tilde{W}^{(k)}\rangle_t = \langle W^{(j)}, W^{(k)}\rangle_t = \delta_{j,k}t; \quad 0 \le t \le T, \text{ a.s. } \tilde{P}_T \text{ and } P. \quad \square$$

Let $\{M_t, \mathscr{F}_t; 0 \le t \le T\}$ be a continuous local martingale under P. With the hypotheses of Theorem 5.1, Proposition 5.4 shows that M is a continuous semimartingale under $\tilde{P}_T$. The converse is also true; if $\{\tilde{M}_t, \mathscr{F}_t; 0 \le t \le T\}$ is a continuous martingale under $\tilde{P}_T$, then Lemma 5.3 implies that for $0 \le s \le t \le T$:

$$E[Z_t(X)\tilde{M}_t | \mathscr{F}_s] = Z_s(X)\tilde{E}_T[\tilde{M}_t | \mathscr{F}_s] = Z_s(X)\tilde{M}_s; \quad \text{a.s. } P \text{ and } \tilde{P}_T,$$

so $Z(X)\tilde{M}$ is a martingale under P. If $\tilde{M} \in \tilde{\mathscr{M}}_T^{c,\text{loc}}$, a localization argument shows that $Z(X)\tilde{M} \in \mathscr{M}_T^{c,\text{loc}}$. But $Z(X) \in \mathscr{M}^c$, and so Itô's rule implies that $\tilde{M} = [Z(X)\tilde{M}]/Z(X)$ is a continuous semimartingale under P (cf. Remark 3.4). Thus, given $\tilde{M} \in \tilde{\mathscr{M}}_T^{c,\text{loc}}$, we have a decomposition

$$\tilde{M}_t = M_t + B_t; \quad 0 \le t \le T,$$

where $M \in \mathscr{M}_T^{c,\text{loc}}$ and B is the difference of two continuous, nondecreasing adapted processes with $B_0 = 0$, P-a.s. According to Proposition 5.4, the process

$$\tilde{M}_t - \left(M_t - \sum_{i=1}^d \int_0^t X_s^{(i)}\, d\langle M, W^{(i)}\rangle_s\right)$$

$$= B_t + \sum_{i=1}^d \int_0^t X_s^{(i)}\, d\langle M, W^{(i)}\rangle_s; \quad 0 \le t \le T,$$

is in $\tilde{\mathscr{M}}_T^{c,\text{loc}}$, and being of bounded variation this process must be indistinguishable from the identically zero process (Problem 3.2). We have proved the following result.

5.5 Proposition. *Under the hypotheses of Theorem* 5.1, *every* $\tilde{M} \in \tilde{\mathscr{M}}_T^{c,\text{loc}}$ *has the representation* (5.8) *for some* $M \in \mathscr{M}_T^{c,\text{loc}}$.

We note now that integrals with respect to $d\tilde{W}_t^{(i)}$ have two possible interpretations. On the one hand, we may interpret them by replacing $d\tilde{W}_t^{(i)}$ by $dW_t^{(i)} - X_t^{(i)}\,dt$ so as to obtain the sum of an Itô integral (under P) and a Lebesgue-Stieltjes integral. On the other hand, $\tilde{W}^{(i)}$ is a Brownian motion under $\tilde{P}_T$, so we may regard integrals with respect to $d\tilde{W}_t^{(i)}$ as Itô integrals under $\tilde{P}_T$. Fortunately, these two interpretations coincide, as the next problem shows.

5.6 Problem. Assume the hypotheses of Theorem 5.1 and suppose $Y = \{Y_t, \mathscr{F}_t; 0 \le t < \infty\}$ is a measurable adapted process satisfying $P[\int_0^T Y_t^2\,dt < \infty] = 1$; $0 \le T < \infty$. Under P we may define the Itô integral $\int_0^t Y_s\,dW_s^{(i)}$, whereas under $\tilde{P}_T$ we may define the Itô integral $\int_0^t Y_s\,d\tilde{W}_s^{(i)}$, $0 \le t \le T$. Show that for $1 \le i \le d$, we have

$$\int_0^t Y_s\,d\tilde{W}_s^{(i)} = \int_0^t Y_s\,dW_s^{(i)} - \int_0^t Y_s X_s^{(i)}\,ds; \quad 0 \le t \le T, \quad \text{a.s. } P \text{ and } \tilde{P}_T.$$

(*Hint*: Use Proposition 2.24.)

C. Brownian Motion with Drift

Let us discuss a rather simple, but interesting, application of the Girsanov theorem: the distribution of *passage times for Brownian motion with drift*. We consider a Brownian motion $W = \{W_t, \mathscr{F}_t; 0 \le t < \infty\}$ and recall from Remark 2.8.3 that the passage time T_b to the level $b \neq 0$ has density and moment generating function, respectively:

$$P[T_b \in dt] = \frac{|b|}{\sqrt{2\pi t^3}} \exp\left[-\frac{b^2}{2t}\right] dt; \quad t > 0, \tag{5.9}$$

$$Ee^{-\alpha T_b} = e^{-|b|\sqrt{2\alpha}}; \quad \alpha > 0. \tag{5.10}$$

For any real number $\mu \neq 0$, the process $\tilde{W} = \{\tilde{W}_t \triangleq W_t - \mu t, \mathscr{F}_t^W; 0 \le t < \infty\}$ is a Brownian motion under the unique measure $P^{(\mu)}$ which satisfies

$$P^{(\mu)}(A) = E[1_A Z_t]; \quad A \in \mathscr{F}_t^W,$$

where $Z_t \triangleq \exp(\mu W_t - \frac{1}{2}\mu^2 t)$ (Corollary 5.2). We say that, *under* $P^{(\mu)}$, $W_t = \mu t + \tilde{W}_t$ *is a Brownian motion with drift* μ. On the set $\{T_b \le t\} \in \mathscr{F}_t^W \cap \mathscr{F}_{T_b}^W = \mathscr{F}_{t \wedge T_b}^W$ we have $Z_{t \wedge T_b} = Z_{T_b}$, so the optional sampling Theorem 1.3.22 and Problem 1.3.23 (i) imply

$$\begin{aligned} P^{(\mu)}[T_b \le t] &= E[1_{\{T_b \le t\}} Z_t] = E[1_{\{T_b \le t\}} E(Z_t | \mathscr{F}_{t \wedge T_b}^W)] \\ &= E[1_{\{T_b \le t\}} Z_{t \wedge T_b}] = E[1_{\{T_b \le t\}} Z_{T_b}] \\ &= E[1_{\{T_b \le t\}} e^{\mu b - (1/2)\mu^2 T_b}] \\ &= \int_0^t \exp\left(\mu b - \frac{1}{2}\mu^2 s\right) P[T_b \in ds]. \end{aligned} \tag{5.11}$$

The relation (5.11) has several consequences. First, together with (5.9) it yields the density of T_b under $P^{(\mu)}$:

(5.12) $$P^{(\mu)}[T_b \in dt] = \frac{|b|}{\sqrt{2\pi t^3}} \exp\left[-\frac{(b-\mu t)^2}{2t}\right] dt, \quad t > 0.$$

Second, letting $t \to \infty$ in (5.11), we see that

$$P^{(\mu)}[T_b < \infty] = e^{\mu b} E[\exp(-\tfrac{1}{2}\mu^2 T_b)],$$

and so we obtain from (5.10):

(5.13) $$P^{(\mu)}[T_b < \infty] = \exp[\mu b - |\mu b|].$$

In particular, *a Brownian motion with drift $\mu \neq 0$ reaches the level $b \neq 0$ with probability one if and only if μ and b have the same sign.* If μ and b have opposite signs, the density in (5.12) is "defective," in the sense that $P^{(\mu)}[T_b < \infty] < 1$.

5.7 Problem. Let T be a stopping time of the filtration $\{\mathscr{F}_t^W\}$ with $P[T < \infty] = 1$. A necessary and sufficient condition for the validity of the *Wald identity*

(5.14) $$E[\exp(\mu W_T - \tfrac{1}{2}\mu^2 T)] = 1,$$

where μ is a given real number, is that

(5.15) $$P^{(\mu)}[T < \infty] = 1.$$

In particular, if $b \in \mathbb{R}$ and $\mu b < 0$, then this condition holds for the stopping time

(5.16) $$S_b \triangleq \inf\{t \geq 0; W_t - \mu t = b\}.$$

5.8 Problem. Denote by

$$h(t; b, \mu) \triangleq \frac{|b|}{\sqrt{2\pi t^3}} \exp\left[-\frac{(b-\mu t)^2}{2t}\right]; \quad t > 0, b \neq 0, \mu \in \mathbb{R},$$

the (possibly defective) density on the right-hand side of (5.12). Use Theorem 2.6.16 to show that

$$h(\,\cdot\,; b_1 + b_2, \mu) = h(\,\cdot\,; b_1, \mu) * h(\,\cdot\,; b_2, \mu); \quad b_1 b_2 > 0, \mu \in \mathbb{R},$$

where * denotes convolution.

5.9 Exercise. With $\mu > 0$ and $W_* \triangleq \inf_{t>0} W_t$, under $P^{(\mu)}$ the random variable $-W_*$ is exponentially distributed with parameter 2μ, i.e.,

$$P^{(\mu)}[-W_* \in db] = 2\mu e^{-2\mu b}\, db, \quad b > 0.$$

5.10 Exercise. Show that

$$E^{(\mu)} e^{-\alpha T_b} = \exp(\mu b - |b|\sqrt{\mu^2 + 2\alpha}), \quad \alpha > 0.$$

5.11 Exercise (Robbins & Siegmund (1973)). Consider, for $\nu > 0$ and $c > 1$, the stopping time of $\{\mathscr{F}_t^W\}$:

$$R_c = \inf\{t \geq 0; \exp(\nu W_t - \tfrac{1}{2}\nu^2 t) = c\}.$$

Show that

$$P[R_c < \infty] = \frac{1}{c}, \quad E^{(\nu)} R_c = \frac{2 \log c}{\nu^2}.$$

D. The Novikov Condition

In order to use the Girsanov theorem effectively, we need some fairly general conditions under which the process $Z(X)$ defined by (5.2) is a martingale. This process is a local martingale because of (5.3). Indeed, with

$$T_n \triangleq \inf\left\{t \geq 0; \max_{1 \leq i \leq d} \int_0^t (Z_s(X) X_s^{(i)})^2 \, ds = n\right\},$$

the "stopped" processes $Z^{(n)} \triangleq \{Z_t^{(n)} \triangleq Z_{t \wedge T_n}(X), \mathscr{F}_t; 0 \leq t < \infty\}$ are martingales. Consequently, we have

$$E[Z_{t \wedge T_n} | \mathscr{F}_s] = Z_{s \wedge T_n}; \quad 0 \leq s \leq t, n \geq 1,$$

and using Fatou's lemma as $n \to \infty$, we obtain $E[Z_t(X) | \mathscr{F}_s] \leq Z_s$; $0 \leq s \leq t$. In other words, $Z(X)$ is always a supermartingale and is a martingale if and only if

(5.17) $$EZ_t(X) = 1; \quad 0 \leq t < \infty$$

(Problem 1.3.25). We provide now sufficient conditions for (5.17).

5.12 Proposition. *Let $M = \{M_t, \mathscr{F}_t; 0 \leq t < \infty\}$ be in $\mathscr{M}^{c,\mathrm{loc}}$ and define*

$$Z_t = \exp[M_t - \tfrac{1}{2}\langle M \rangle_t]; \quad 0 \leq t < \infty.$$

If

(5.18) $$E[\exp\{\tfrac{1}{2}\langle M \rangle_t\}] < \infty; \quad 0 \leq t < \infty,$$

then $EZ_t = 1; 0 \leq t < \infty$.

PROOF. Let $T(s) = \inf\{t \geq 0; \langle M \rangle_t > s\}$, so the time-changed process B of (4.16) is a Brownian motion (Theorem 4.6 and Problem 4.7). For $b < 0$, we define the stopping time for $\{\mathscr{G}_s\}$ as in (5.16):

$$S_b = \inf\{s \geq 0; B_s - s = b\}.$$

Problem 5.7 yields the Wald identity $E[\exp(B_{S_b} - \tfrac{1}{2}S_b)] = 1$, whence $E[\exp(\tfrac{1}{2}S_b)] = e^{-b}$. Consider the exponential martingale $\{Y_s \triangleq \exp(B_s - (s/2)), \mathscr{G}_s; 0 \leq s < \infty\}$ and define $\{N_s \triangleq Y_{s \wedge S_b}, \mathscr{G}_s; 0 \leq s < \infty\}$. According to Problem 1.3.24 (i), N is a martingale, and because $P[S_b < \infty] = 1$ we have

$$N_\infty = \lim_{s \to \infty} N_s = \exp(B_{S_b} - \tfrac{1}{2}S_b).$$

It follows easily from Fatou's lemma that $N = \{N_s, \mathscr{G}_s; 0 \le s \le \infty\}$ is a supermartingale with a last element. However, $EN_\infty = 1 = EN_0$, so $N = \{N_s, \mathscr{G}_s; 0 \le s \le \infty\}$ has constant expectation; thus N is actually a martingale with a last element (Problem 1.3.25). This allows us to use the optional sampling Theorem 1.3.22 to conclude that for any stopping time R of the filtration $\{\mathscr{G}_s\}$:

$$E[\exp\{B_{R \wedge S_b} - \tfrac{1}{2}(R \wedge S_b)\}] = 1.$$

Now let us fix $t \in [0, \infty)$ and recall, from the proof of Theorem 4.6, that $\langle M \rangle_t$ is a stopping time of $\{\mathscr{G}_s\}$. It follows that for $b < 0$:

$$(5.19) \quad E[1_{\{S_b \le \langle M \rangle_t\}} \exp(b + \tfrac{1}{2}S_b)] + E[1_{\{\langle M \rangle_t < S_b\}} \exp(M_t - \tfrac{1}{2}\langle M \rangle_t)] = 1.$$

The first expectation in (5.19) is bounded above by $e^b E[\exp(\frac{1}{2}\langle M \rangle_t)]$, which converges to zero as $b \downarrow -\infty$, thanks to assumption (5.18). As $b \downarrow -\infty$, the second expectation in (5.19) converges to EZ_t because of the monotone convergence theorem. Therefore, $EZ_t = 1; 0 \le t < \infty$. □

5.13 Corollary (Novikov (1972)). *Let* $W = \{W_t = (W_t^{(1)}, \ldots, W_t^{(d)}), \mathscr{F}_t; 0 \le t < \infty\}$ *be a d-dimensional Brownian motion, and let* $X = \{X_t = (X_t^{(1)}, \ldots, X_t^{(d)}), \mathscr{F}_t; 0 \le t < \infty\}$ *be a vector of measurable, adapted processes satisfying* (5.1). *If*

$$(5.20) \qquad E\left[\exp\left(\frac{1}{2}\int_0^T \|X_s\|^2\, ds\right)\right] < \infty; \quad 0 \le T < \infty,$$

then $Z(X)$ *defined by* (5.2) *is a martingale.*

5.14 Corollary. *Corollary* 5.13 *still holds if* (5.20) *is replaced by the following assumption: there exists a sequence* $\{t_n\}_{n=0}^\infty$ *of real numbers with* $0 = t_0 < t_1 < \cdots < t_n \uparrow \infty$, *such that*

$$(5.21) \qquad E\left[\exp\left(\frac{1}{2}\int_{t_{n-1}}^{t_n} \|X_s\|^2\, ds\right)\right] < \infty; \quad \forall n \ge 1.$$

PROOF. Let $X_t(n) = (X_t^{(1)} 1_{[t_{n-1}, t_n)}(t), \ldots, X_t^{(d)} 1_{[t_{n-1}, t_n)}(t))$, so that $Z(X(n))$ is a martingale by Corollary 5.13. In particular,

$$E[Z_{t_n}(X(n)) | \mathscr{F}_{t_{n-1}}] = Z_{t_{n-1}}(X(n)) = 1; \quad n \ge 1.$$

But then,

$$E[Z_{t_n}(X)] = E[Z_{t_{n-1}}(X) E\{Z_{t_n}(X(n)) | \mathscr{F}_{t_{n-1}}\}] = E[Z_{t_{n-1}}(X)],$$

and by induction on n we can show that $E[Z_{t_n}(X)] = 1$ holds for all $n \ge 1$. Since $E[Z_t(X)]$ is nonincreasing in t and $\lim_{n\to\infty} t_n = \infty$, we obtain (5.17). □

5.15 Definition. Let $C[0, \infty)^d$ be the space of continuous functions x: $[0, \infty) \to \mathbb{R}^d$. For $0 \le t < \infty$, define $\mathscr{G}_t \triangleq \sigma(x(s); 0 \le s \le t)$, and set $\mathscr{G} = \mathscr{G}_\infty$ (cf. Problems 2.4.1 and 2.4.2). A *progressively measurable functional* on $C[0, \infty)^d$ is a mapping $\mu\colon [0,\infty) \times C[0,\infty)^d \to \mathbb{R}$ which has the property that for each fixed $0 \le t < \infty$, μ restricted to $[0,t] \times C[0,\infty)^d$ is $\mathscr{B}([0,t]) \otimes \mathscr{G}_t/\mathscr{B}(\mathbb{R})$-measurable.

If $\mu = (\mu^{(1)}, \ldots, \mu^{(d)})$ is a vector of progressively measurable functionals on $C[0,\infty)^d$ and $W = \{W_t = (W_t^{(1)}, \ldots, W_t^{(d)}), \mathscr{F}_t; 0 \le t < \infty\}$ is a d-dimensional Brownian motion on some $(\Omega, \mathscr{F}, P)$, then the processes

$$X_t^{(i)}(\omega) \triangleq \mu^{(i)}(t, W_\cdot(\omega)); \quad 0 \le t < \infty, \quad 1 \le i \le d, \tag{5.22}$$

are progressively measurable relative to $\{\mathscr{F}_t\}$.

5.16 Corollary (Beneš (1971)). *Let the vector $\mu = (\mu^{(1)}, \ldots, \mu^{(d)})$ of progressively measurable functionals on $C[0,\infty)^d$ satisfy, for each $0 \le T < \infty$ and some $K_T > 0$ depending on T, the condition*

$$\|\mu(t,x)\| \le K_T(1 + x^*(t)); \quad 0 \le t \le T, \tag{5.23}$$

where $x^*(t) \triangleq \max_{0 \le s \le t} \|x(s)\|$. Then with $X_t = (X_t^{(1)}, \ldots, X_t^{(d)})$ defined by (5.22), $Z(X)$ of (5.2) is a martingale.

Proof. If, for arbitrary $T > 0$, we can find $\{t_0, \ldots, t_{n(T)}\}$ such that $0 = t_0 < t_1 < \cdots < t_{n(T)} = T$ and (5.21) holds for $1 \le n \le n(T)$, then we can construct a sequence $\{t_n\}_{n=0}^\infty$ satisfying the hypotheses of Corollary 5.14. Thus, fix $T > 0$. We have from (5.22), (5.23) that whenever $0 \le t_{n-1} < t_n \le T$, then

$$\int_{t_{n-1}}^{t_n} \|X_s\|^2\,ds \le (t_n - t_{n-1})K_T^2(1 + W_T^*)^2,$$

where $W_T^* \triangleq \max_{0 \le t \le T} \|W_t\|$. According to Problem 1.3.7, the process $Y_t \triangleq \exp[(t_n - t_{n-1})K_T^2(1 + \|W_t\|)^2/4]$ is a submartingale, and Doob's maximal inequality (Theorem 1.3.8(iv)) yields

$$E\exp[\tfrac{1}{2}(t_n - t_{n-1})K_T^2(1 + W_T^*)^2] = E\left(\max_{0 \le t \le T} Y_t^2\right) \le 4EY_T^2,$$

which is finite provided that $t_n - t_{n-1} \le 1/TK_T^2$. This allows us to construct $\{t_0, \ldots, t_{n(T)}\}$ as described previously. □

5.17 Remark. Liptser & Shiryaev (1977), p. 222, show that with $d = 1$ and $0 < \varepsilon < \frac{1}{2}$, one can construct a process X satisfying the hypotheses of Corollary 5.13 but with (5.20) replaced by the weaker condition

$$E\left[\exp\left\{\left(\frac{1}{2} - \varepsilon\right)\int_0^T X_t^2\,dt\right\}\right] < \infty; \quad 0 \le T < \infty,$$

such that $Z(X)$ is *not* a martingale.

The next exercise, taken from Liptser & Shiryaev (1977), p. 224, provides a simple example in which $Z(X)$ is not a martingale. In particular, it shows that a local martingale (cf. (5.3)) need not be a martingale.

5.18 Exercise. With $W = \{W_t, \mathscr{F}_t; 0 \le t \le 1\}$ a Brownian motion, we define $T = \inf\{0 \le t \le 1; t + W_t^2 = 1\}$, $X_t = -(2/(1-t)^2)W_t 1_{\{t \le T, t<1\}}$; $0 \le t < 1$,

(i) Prove that $P[T < 1] = 1$, and therefore $\int_0^1 X_t^2\,dt < \infty$ a.s.
(ii) Apply Itô's rule to the process $\{(W_t/(1-t))^2; 0 \le t < 1\}$ to conclude that

$$\int_0^1 X_t\,dW_t - \frac{1}{2}\int_0^1 X_t^2\,dt$$
$$= -1 - 2\int_0^T \left[\frac{1}{(1-t)^4} - \frac{1}{(1-t)^3}\right] W_t^2\,dt \le -1.$$

(iii) The exponential supermartingale $\{Z_t(X), \mathscr{F}_t; 0 \le t \le 1\}$ is not a martingale; however, for each $n \ge 1$ and $\sigma_n = 1 - (1/\sqrt{n})$, $\{Z_{t \wedge \sigma_n}(X), \mathscr{F}_t; 0 \le t \le 1\}$ is a martingale.

5.19 Exercise. Let $W = \{W_t, \mathscr{F}_t; 0 \le t < \infty\}$ be a Brownian motion on $(\Omega, \mathscr{F}, P)$ with $P[W_0 = 0] = 1$, and assume that $\{\mathscr{F}_t\}$ is the augmentation under P of the Brownian filtration $\{\mathscr{F}_t^W\}$. Suppose that, for each $0 \le T < \infty$, there is a probability measure $\tilde{P}_T$ on $\mathscr{F}_T$ which is mutually absolutely continuous with respect to P, and that the family of probability measures $\{\tilde{P}_T; 0 \le T < \infty\}$ satisfies the consistency condition (5.5). Show that there exists a measurable, adapted process $X = \{X_t, \mathscr{F}_t; 0 \le t < \infty\}$ satisfying (5.1), such that $Z(X)$ defined by (5.2) is a martingale and (5.4) holds for $0 \le T < \infty$. (*Hint*: Apply Problem 4.16 to the Radon-Nikodým derivative process $d\tilde{P}_t/dP$.)

5.20 Exercise. Suppose that $\{L_t, \mathscr{F}_t; 0 \le t < \infty\} \in \mathscr{M}^{c,\text{loc}}$ is such that $Z_t \triangleq \exp[L_t - \frac{1}{2}\langle L\rangle_t]$ is a martingale under P, and define the new probability measure $\tilde{P}_T(A) \triangleq E(1_A Z_T)$; $A \in \mathscr{F}_T$. Establish the following generalization of Proposition 5.4 and of the Girsanov theorem: if $M \in \mathscr{M}^{c,\text{loc}}$, then

$$\tilde{M}_t \triangleq M_t - \langle L, M\rangle_t = M_t - \int_0^t \frac{1}{Z_s}\,d\langle Z, M\rangle_s, \quad \mathscr{F}_t; \quad 0 \le t \le T$$

is in $\tilde{\mathscr{M}}^{c,\text{loc}}$. (*Hint*: Imitate the proof of Proposition 5.4.)

5.21 Exercise (H. J. Engelbert, R. Höhnle): In the setting of Corollary 5.2, suppose that $\int_0^\infty \|X_t(\omega)\|^2 dt < \infty$ holds for *every* $\omega \in \Omega$. Show that $Z_t(X)$, $0 \le t < \infty$ of (5.2) is then an $\{\mathscr{F}_t^W\}$-martingale.

3.6. Local Time and a Generalized Itô Rule for Brownian Motion

In this section we devise a method for measuring the amount of time spent by the Brownian path in the vicinity of a point $x \in \mathbb{R}$. We saw in Section 2.9 that the Lebesgue measure of the *level set* $\mathscr{L}_\omega(x) = \{0 \le t < \infty; W_t(\omega) = x\}$

turns out to be zero, i.e.,

(6.1) $$\text{meas}\, \mathscr{Z}_\omega(x) = 0, \quad \text{for } P\text{-a.e. } \omega \in \Omega,$$

yielding no information whatsoever about the amount of time spent in the vicinity of the point x (Theorem 2.9.6 and Remark 2.9.7). In search of a nontrivial measure for this amount of time, P. Lévy introduced the two-parameter random field

(6.2) $$L_t(x) = \lim_{\varepsilon \downarrow 0} \frac{1}{4\varepsilon} \text{meas}\{0 \le s \le t; |W_s - x| \le \varepsilon\}; \quad t \in [0, \infty), x \in \mathbb{R}$$

and showed that this limit exists and is finite, but not identically zero. We shall show how $L_t(x)$ can be chosen to be jointly continuous in (t, x) and, for fixed x, nondecreasing in t and constant on each interval in the complement of the closed set $\mathscr{Z}_\omega(x)$. Therefore, $(d/dt)L_t(x)$ exists and is zero for Lebesgue almost every t; i.e., the function $t \mapsto L_t(x)$ is singularly continuous. P. Lévy called $L_t(x)$ the *mesure du voisinage*, or "measure of the time spent by the Brownian path in the vicinity of the point x." We shall refer to $L_t(x)$ as *local time*.

This new concept provides a very powerful tool for the study of Brownian sample paths. In this section, we show how it allows us to generalize Itô's change-of-variable rule to convex but not necessarily differentiable functions, and we use it to study certain additive functionals of the Brownian path. These functionals will be employed in Chapter 5 to provide solutions of stochastic differential equations by the method of random time-change. Local time will be further developed in Chapter 6, where we shall use it to prove that the Brownian path has no point of increase (Theorem 2.9.13). In this section, the reader can appreciate the application of local time to the study of Brownian sample paths by providing a simple proof of their nondifferentiability (Exercise 6.6). This exercise shows that jointly continuous local time cannot exist for processes whose sample paths are of bounded variation on bounded intervals.

Throughout this section, $\{W_t, \mathscr{F}_t; 0 \le t < \infty\}$, $(\Omega, \mathscr{F})$, $\{P^z\}_{z \in \mathbb{R}}$ denotes the one-dimensional Brownian family on the canonical space $\Omega = C[0, \infty)$. This assumption entails no loss of generality, because every standard Brownian motion induces Wiener measure on $C[0, \infty)$ (Remark 2.4.22), and results proved for the latter can be carried back to the original probability space. We take the filtration $\{\mathscr{F}_t\}$ to be $\{\tilde{\mathscr{F}}_t\}$ defined by (2.7.3), and we set $\mathscr{F} = \tilde{\mathscr{F}}_\infty$. This filtration satisfies the usual conditions. In this situation, P^z is just a translate of P^0, i.e.,

(6.3) $$P^z(F) = P^0(F - z); \quad F \in \mathscr{F},$$

(cf. (2.5.1)). We also have at our disposal the shift operators $\{\theta_s\}_{s \ge 0}$ defined by (2.5.15).

6.1 Definition. A measurable, adapted, real-valued process $A = \{A_t, \mathscr{F}_t; 0 \le t < \infty\}$ is called an *additive functional* if, for every $z \in \mathbb{R}$ and P^z-a.e. $\omega \in \Omega$, we have

$$A_{t+s}(\omega) = A_s(\omega) + A_t(\theta_s\omega); \quad 0 \le s, t < \infty. \tag{6.4}$$

6.2 Example. For every fixed Borel set $B \in \mathscr{B}(\mathbb{R})$, we define the *occupation time* of B by the Brownian path up to time t as

$$\Gamma_t(B) \triangleq \int_0^t 1_B(W_s)\,ds = \text{meas}\{0 \le s \le t;\ W_s \in B\}; \quad 0 \le t < \infty, \tag{6.5}$$

where *meas* denotes Lebesgue measure. The resulting process $\Gamma(B) = \{\Gamma_t(B), \mathscr{F}_t; 0 \le t < \infty\}$ is adapted and continuous, and is easily seen to be an additive functional.

A. Definition of Local Time and the Tanaka Formula

Equation (6.2) indicates that doubled local time $2L_t(x)$ should serve as a *density with respect to Lebesgue measure for occupation time.* In other words, we should have

$$\Gamma_t(B, \omega) = \int_B 2L_t(x, \omega)\,dx; \quad 0 \le t < \infty,\ B \in \mathscr{B}(\mathbb{R}). \tag{6.6}$$

We take this property as part of the definition of local time.

6.3 Definition. Let $L = \{L_t(x, \omega); (t, x) \in [0, \infty) \times \mathbb{R}, \omega \in \Omega\}$ be a random field with values in $[0, \infty)$, such that for each fixed value of the parameter pair (t, x) the random variable $L_t(x)$ is $\mathscr{F}_t$-measurable. Suppose that there is an event $\Omega^* \in \mathscr{F}$ with $P^z(\Omega^*) = 1$ for every $z \in \mathbb{R}$ and such that, for each $\omega \in \Omega^*$, the function $(t, x) \mapsto L_t(x, \omega)$ is continuous and (6.6) holds. Then we call L *Brownian local time.*

6.4 Remark. There is no universal agreement in the literature as to whether L in Definition 6.3 or $2L$ is to be called local time. We follow the normalization (6.6), used by Ikeda & Watanabe (1981).

6.5 Remark. With L as in Definition 6.3 and $\omega \in \Omega^*$, one can immediately derive (6.2) from (6.6) and the continuity of $x \mapsto L_t(x, \omega)$. Further, $L(a) = \{L_t(a), \mathscr{F}_t; 0 \le t < \infty\}$ is easily seen to inherit the additive functional property (6.4) from its progenitor, the occupation time Γ (Example 6.2).

6.6 Exercise. Assume that Brownian local time exists and show that for each $\omega \in \Omega^*$ of Definition 6.3, the sample path $t \mapsto W_t(\omega)$ cannot be differentiable anywhere on $(0, \infty)$. (*Hint*: If $t \mapsto W_t(\omega)$ is differentiable at t, then for some sufficiently large C and sufficiently small $\delta > 0$ we must have $|W_{t+h}(\omega) - W_t(\omega)| \le Ch; 0 \le h \le \delta$.)

6.7 Problem.

(i) Show that the validity of (6.6) is equivalent to

$$\int_0^t f(W_s(\omega))\,ds = 2\int_{-\infty}^{\infty} f(x)L_t(x,\omega)\,dx; \quad 0 \le t < \infty, \tag{6.7}$$

for every Borel-measurable function $f: \mathbb{R} \to [0, \infty)$.

(ii) Let $\mathscr{H}$ be the class of continuous functions $h: \mathbb{R} \to [0, 1]$ of the form

$$h(x) = \begin{cases} 0; & x \le q_1, \\ \dfrac{x - q_1}{q_2 - q_1}; & q_1 \le x \le q_2, \\ 1; & q_2 \le x \le q_3, \\ \dfrac{q_4 - x}{q_4 - q_3}; & q_3 \le x \le q_4, \\ 0; & x \ge q_4, \end{cases} \tag{6.8}$$

where $q_1 < q_2 < q_3 < q_4$ are rational numbers.

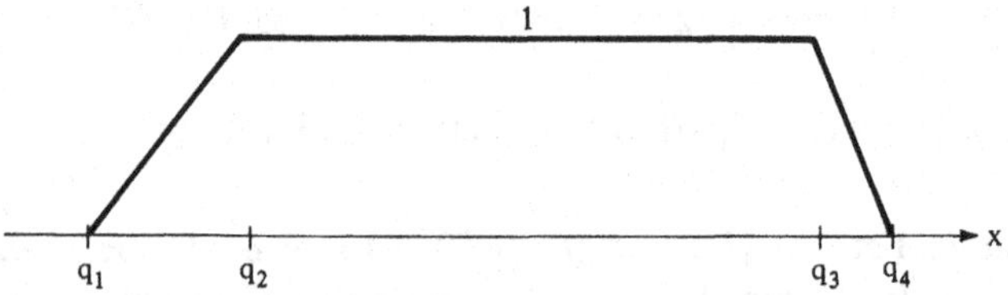

Show that if (6.7) holds for all $h \in \mathscr{H}$, then it holds for every Borel-measurable function $f: \mathbb{R} \to [0, \infty)$. □

Our plan is the following: we shall assume in the present subsection that Brownian local time exists, and we shall derive a convenient representation for it, the *Tanaka formula* (6.11). In the next subsection it will be shown that the right-hand side of (6.11) leads to a random field which satisfies the requirements of Definition 6.3, thus establishing existence.

Let us fix a number $a \in \mathbb{R}$, and take $f(x)$ in (6.7) to be the Dirac delta evaluated at $x - a$, and derive formally the representation

$$L_t(a, \omega) = \frac{1}{2}\int_0^t \delta(W_s(\omega) - a)\,ds. \tag{6.9}$$

But the integral on the right-hand side is only formal. In order to give it meaning, we consider the nondecreasing, convex function $u(x) = (x - a)^+$, which is continuously differentiable on $\mathbb{R}\backslash\{a\}$ and whose second derivative in the distributional sense is $u''(x) = \delta(x - a)$. Bravely assuming that Itô's rule can be applied in this highly irregular situation, we write

$$(6.10)\quad (W_t - a)^+ - (z-a)^+ = \int_0^t 1_{(a,\infty)}(W_s)\,dW_s + \frac{1}{2}\int_0^t \delta(W_s - a)\,ds,$$

and in conjunction with (6.9), we have

$$(6.11)\quad L_t(a) = (W_t - a)^+ - (z-a)^+ - \int_0^t 1_{(a,\infty)}(W_s)\,dW_s;\quad 0 \le t < \infty$$

P^z-a.s. for every $z \in \mathbb{R}$. Despite the heuristic nature of both (6.9) and (6.10), the representation (6.11) for local time is valid and will be established rigorously.

6.8 Proposition. *Let us assume that Brownian local time exists, and fix a number $a \in \mathbb{R}$. Then the process $L(a) = \{L_t(a), \mathscr{F}_t; 0 \le t < \infty\}$ is a nonnegative, continuous additive functional which satisfies* (6.11) *and the companion representations*

$$(6.12)\quad L_t(a) = (W_t - a)^- - (z-a)^- + \int_0^t 1_{(-\infty,a]}(W_s)\,dW_s;\quad 0 \le t < \infty,$$

$$(6.13)\quad 2L_t(a) = |W_t - a| - |z-a| - \int_0^t \operatorname{sgn}(W_s - a)\,dW_s;\quad 0 \le t < \infty$$

a.s. P^z, for every $z \in \mathbb{R}$.

6.9 Remark. Any of the formulas (6.11), (6.12), or (6.13) is referred to as the *Tanaka formula for Brownian local time.* We need establish only (6.11); then (6.12) follows by symmetry and (6.13) by addition, since

$$P^z\left[\int_0^t 1_{\{a\}}(W_s)\,dW_s = 0;\quad \forall\, 0 \le t < \infty\right] = 1;\quad \forall z \in \mathbb{R}.$$

In particular, it does not matter how we define sgn(0) in (6.13); we shall define sgn so as to make it left-continuous, i.e.,

$$(6.14)\qquad \operatorname{sgn}(x) = \begin{cases} 1; & x > 0 \\ -1; & x \leqq 0. \end{cases}$$

6.10 Remark. The process $\{(W_t - a)^+, \mathscr{F}_t; 0 \le t < \infty\}$ is a continuous, nonnegative submartingale (Proposition 1.3.6); it admits, therefore, a unique Doob-Meyer decomposition (under P^z, for any $z \in \mathbb{R}$):

$$(6.15)\qquad (W_t - a)^+ = (z-a)^+ + M_t(a) + A_t(a);\quad 0 \le t < \infty,$$

where $A(a)$ is a continuous, increasing process and $M(a)$ is a martingale (see Section 1.4). The Tanaka formula (6.11) identifies both parts of this decomposition, as $A_t(a) = L_t(a)$ and

$$(6.16)\qquad M_t(a) = \int_0^t 1_{(a,\infty)}(W_s)\,dW_s;\quad 0 \le t < \infty.$$

Similar remarks apply to the representations (6.12) and (6.13).

PROOF OF PROPOSITION 6.8. In order to make rigorous the heuristic discussion which led to (6.11), we must approximate the Dirac delta $\delta(x)$ by a sequence of probability densities with increasing concentration at the origin. More specifically, let us start with the C^∞ function

$$\rho(x) \triangleq \begin{cases} c\exp\left[\dfrac{1}{(x-1)^2-1}\right]; & 0<x<2, \\ 0; & \text{otherwise} \end{cases} \tag{6.17}$$

which satisfies $\int_{-\infty}^{\infty}\rho(x)\,dx = 1$ by appropriate choice of the constant c, and use it to define the probability density functions (called *mollifiers*)

$$\rho_n(x) \triangleq n\rho(nx) \tag{6.18}$$

as well as

$$u_n(x) \triangleq \int_{-\infty}^{x}\int_{-\infty}^{y}\rho_n(z-a)\,dz\,dy; \quad x\in\mathbb{R},\ n\geq 1.$$

We observe that $u_n'(x) = \int_{-\infty}^{x}\rho_n(z-a)\,dz$, and so we have the limiting relations

$$\lim_{n\to\infty} u_n'(x) = 1_{(a,\infty)}(x), \quad \lim_{n\to\infty} u_n(x) = (x-a)^+, \quad x\in\mathbb{R}.$$

We now choose an arbitrary $z\in\mathbb{R}$. According to Itô's rule,

$$u_n(W_t) - u_n(z) = \int_0^t u_n'(W_s)\,dW_s + \frac{1}{2}\int_0^t \rho_n(W_s-a)\,ds; \quad 0\leq t<\infty, \tag{6.19}$$

a.s. P^z. But now from (6.7) and the continuity of local time,

$$\int_0^t \rho_n(W_s-a)\,ds = 2\int_{-\infty}^{\infty}\rho_n(x-a)L_t(x)\,dx \xrightarrow[n\to\infty]{} 2L_t(a); \quad \text{a.s. } P^z.$$

On the other hand,

$$E^z\left|\int_0^t u_n'(W_s)\,dW_s - \int_0^t 1_{(a,\infty)}(W_s)\,dW_s\right|^2$$
$$= E^z\int_0^t |u_n'(W_s) - 1_{(a,\infty)}(W_s)|^2\,ds \leq \int_0^t P^z\left[|W_s-a|\leq\frac{2}{n}\right]ds,$$

which converges to zero as $n\to\infty$. Therefore, for each fixed t, the stochastic integral in (6.19) converges in quadratic mean to the one in (6.16), and (6.11) for each fixed t follows by letting $n\to\infty$ in (6.19). Because of the continuity of the processes in (6.11), we obtain that, except on a P^z-null event, (6.11) holds for $0\leq t<\infty$. □

B. The Trotter Existence Theorem

We can employ now the Tanaka representation (6.11) to settle the question of existence of Brownian local time. This is in fact the only result proved in this subsection.

6.11 Theorem (Trotter (1958)). *Brownian local time exists.*

PROOF. We start by showing that the two-parameter random field, obtained by setting $z = 0$ on the right-hand side of (6.11), admits a continuous modification under P^0. The term $(W_t - a)^+ - (-a)^+$ is obviously jointly continuous in the pair (t, a). For the random field $\{M_t(a); 0 \le t < \infty, a \in \mathbb{R}\}$ in (6.16) we have, with $a < b$, $0 \le s < t \le T$ and any integer $n \ge 1$:

$$
\begin{aligned}
&E^0|M_t(a) - M_s(b)|^{2n} \\
&\quad \le 4^n\left\{E^0\left|\int_s^t 1_{(a,\infty)}(W_u)\,dW_u\right|^{2n} + E^0\left|\int_0^s 1_{(a,b]}(W_u)\,dW_u\right|^{2n}\right\} \\
&\quad \le 4^n C_n\left[E^0\left(\int_s^t 1_{(a,\infty)}(W_u)\,du\right)^n + E^0\left(\int_0^s 1_{(a,b]}(W_u)\,du\right)^n\right],
\end{aligned}
$$

thanks to (3.37). The first expectation is bounded above by $(t-s)^n$, whereas the second is dominated by

$$
\begin{aligned}
&E^0\left[\int_0^T 1_{(a,b]}(W_t)\,dt\right]^n \\
&\quad = \int_0^T \cdots \int_0^T E^0[1_{(a,b]}(W_{t_1})\ldots 1_{(a,b]}(W_{t_n})]\,dt_n \ldots dt_1 \\
&\quad = n! \int_0^T \int_{t_1}^T \cdots \int_{t_{n-1}}^T E^0[1_{(a,b]}(W_{t_1})1_{(a,b]}(W_{t_2})\ldots 1_{(a,b]}(W_{t_n})]\,dt_n \ldots dt_2\,dt_1.
\end{aligned}
$$

With $0 \le t < \theta < T$, we have for every $y \in \mathbb{R}$,

$$
\begin{aligned}
P^0[a < W_\theta \le b \mid W_t = y] &\le P^0\left[a < W_\theta \le b \,\middle|\, W_t = \frac{a+b}{2}\right] \\
&= \sqrt{\frac{2}{\pi}} \int_0^{(b-a)/2\sqrt{\theta - t}} e^{-z^2/2}\,dz \le \frac{b-a}{2\sqrt{\theta - t}},
\end{aligned}
$$

and so

$$
\begin{aligned}
&E^0\left[\int_0^T 1_{(a,b]}(W_t)\,dt\right]^n \\
&\quad \le n!\left(\frac{b-a}{2}\right)^n \int_0^T \int_{t_1}^T \cdots \int_{t_{n-1}}^T [t_1(t_2 - t_1)\ldots(t_n - t_{n-1})]^{-1/2}\,dt_n \ldots dt_2\,dt_1 \\
&\quad \le \hat{C}_{n,T}\,(b-a)^n,
\end{aligned}
$$

where $\hat{C}_{n,T}$ is a constant depending on n and T but not on a or b. Therefore, with $a < b$ and $0 \le s < t \le T$, we have

$$
\begin{aligned}
(6.20) \qquad E^0|M_t(a) - M_s(b)|^{2n} &\le C_{n,T}[(t-s)^n + (b-a)^n] \\
&\le C_{n,T}\|(t,a) - (s,b)\|^n
\end{aligned}
$$

for some constant $C_{n,T}$. By the version of the Kolmogorov-Čentsov theorem for random fields (Problem 2.2.9), there exists a two-parameter random field $\{I_t(a); (t,a) \in [0,\infty) \times \mathbb{R}\}$ such that for P^0-a.e. $\omega \in \Omega$, the mapping $(t,a) \mapsto I_t(a,\omega)$ is locally Hölder-continuous with exponent γ, for any $\gamma \in (0,\frac{1}{2})$, and for each fixed pair (t,a) we have

$$P^0[I_t(a) = M_t(a)] = 1. \tag{6.21}$$

Now we define

$$L_t(a) \triangleq (W_t - a)^+ - (-a)^+ - I_t(a); \quad 0 \le t < \infty, a \in \mathbb{R}.$$

For fixed (t,a), $L_t(a)$ is an $\mathscr{F}_t$-measurable random variable, and the random field L is P^0-a.s. continuous in the pair (t,a). Indeed, because W_t and $I_t(a)$ are both locally Hölder-continuous for any exponent $\gamma \in (0,\frac{1}{2})$, the local time L also has this property: for every $\gamma \in (0,\frac{1}{2})$, $T > 0$, $K > 0$, there exists a P^0-a.s. positive random variable $h(\omega)$ and a constant $\delta > 0$ such that

$$P^0\left[\omega \in \Omega; \sup_{\substack{0 < \|(t,a)-(s,b)\| < h(\omega) \\ 0 \le s,t \le T \\ -K \le a,b \le K}} \frac{|L_t(a,\omega) - L_s(b,\omega)|}{\|(t,a)-(s,b)\|^\gamma} \le \delta\right] = 1. \tag{6.22}$$

Our next task is to show that the random field $L_t(a)$ satisfies the identity (6.6), or equivalently (6.7). For every function h in the class $\mathscr{H}$ of Problem 6.7, define

$$H(x) \triangleq \int_{-\infty}^{\infty} h(u)(x-u)^+\,du = \int_{-\infty}^{x}\int_{-\infty}^{y} h(u)\,du\,dy; \quad x \in \mathbb{R}$$

and observe the identities

$$H'(x) = \int_{-\infty}^{\infty} h(u)1_{(u,\infty)}(x)\,du = \int_{-\infty}^{x} h(u)\,du, \quad H''(x) = h(x).$$

By virtue of Itô's rule and Problem 6.12, we have P^0-a.s. for fixed $t \ge 0$:

$$\begin{aligned}
&\frac{1}{2}\int_0^t h(W_s)\,ds \\
&= H(W_t) - H(0) - \int_0^t H'(W_s)\,dW_s \\
&= \int_{-\infty}^{\infty} h(a)\{(W_t - a)^+ - (-a)^+\}\,da - \int_0^t \left(\int_{-\infty}^{\infty} h(a)1_{(a,\infty)}(W_s)\,da\right)dW_s \\
&= \int_{-\infty}^{\infty} h(a)\left\{(W_t - a)^+ - (-a)^+ - \int_0^t 1_{(a,\infty)}(W_s)\,dW_s\right\}da \\
&= \int_{-\infty}^{\infty} h(a)L_t(a)\,da + \int_{-\infty}^{\infty} h(a)\{I_t(a) - M_t(a)\}\,da.
\end{aligned}$$

But $E^0\int_{-\infty}^{\infty}(I_t(a) - M_t(a))^2\,da = \int_{-\infty}^{\infty} E^0(I_t(a) - M_t(a))^2\,da = 0$ by (6.21).

Thus, for each fixed $t \geq 0$, we have for P^0-a.e. $\omega \in \Omega$

$$\int_0^t h(W_s(\omega))\,ds = 2\int_{-\infty}^{\infty} h(x)L_t(x,\omega)\,dx. \tag{6.23}$$

Since both sides of (6.23) are continuous in t and $\mathscr{H}$ is countable, it is possible to find an event $\Omega_0^* \in \mathscr{F}$ with $P^0(\Omega_0^*) = 1$ such that for each $\omega \in \Omega_0^*$, (6.23) holds for every $h \in \mathscr{H}$ and every $t \geq 0$. Problem 6.7 now implies that for each $\omega \in \Omega_0^*$, (6.7) holds for every Borel function $f\colon \mathbb{R} \to [0, \infty)$.

Recall finally that $\Omega = C[0,\infty)$ and that P^z assigns probability one to the event $\Omega_z \triangleq \{\omega \in \Omega;\, \omega(0) = z\}$. We may assume that $\Omega_0^* \subseteq \Omega_0$, and redefine $L_t(x,\omega)$ for $\omega \notin \Omega_0$ by setting

$$L_t(x,\omega) \triangleq L_t(x - \omega(0), \omega - \omega(0)).$$

We set $\Omega^* = \{\omega \in \Omega;\, \omega - \omega(0) \in \Omega_0^*\}$, so that $P^z(\Omega^*) = 1$ for every $z \in \mathbb{R}$ (c.f. (6.3)). It is easily verified that L and Ω^* have all the properties set forth in Definition 6.3. □

6.12 Problem. For a continuous function $h\colon \mathbb{R} \to [0,\infty)$ with compact support, the following interchange of Lebesgue and Itô integrals is permissible:

$$\int_{-\infty}^{\infty} h(a)\left(\int_0^t 1_{(a,\infty)}(W_s)\,dW_s\right)da = \int_0^t\left(\int_{-\infty}^{\infty} h(a)1_{(a,\infty)}(W_s)\,da\right)dW_s, \quad \text{a.s. } P^0. \tag{6.24}$$

6.13 Problem. We may cast (6.13) in the form

$$|W_t - a| = |z - a| - B_t(a) + 2L_t(a); \quad 0 \leq t < \infty, \tag{6.25}$$

where $B_t(a) \triangleq -\int_0^t \operatorname{sgn}(W_s - a)\,dW_s$, for fixed $a \in \mathbb{R}$.

(i) Show that for any $z \in \mathbb{R}$, the process $B(a) = \{B_t(a), \mathscr{F}_t;\, 0 \leq t < \infty\}$ is a Brownian motion under P^z, with $P^z[B_0(a) = 0] = 1$.

(ii) Using the representation (6.2), show that $L(a) = \{L_t(a), \mathscr{F}_t; 0 \leq t < \infty\}$ is a continuous, increasing process (Definition 1.4.4) which satisfies

$$\int_0^{\infty} 1_{\mathbb{R}\setminus\{a\}}(W_t)\,dL_t(a) = 0; \quad \text{a.s. } P^z. \tag{6.26}$$

In other words, the path $t \mapsto L_t(a,\omega)$ is "flat" off the level set $\mathscr{L}_\omega(a) = \{0 \leq t < \infty;\, W_t(\omega) = a\}$.

(iii) Show that for P^0-a.e. ω, we have $L_t(0,\omega) > 0$ for all $t > 0$.

(iv) Show that for every $z \in \mathbb{R}$ and P^z-a.e. ω, every point t of $\mathscr{L}_\omega(a)$ satisfies $L_q(a,\omega) < L_r(a,\omega)$ for all $q < t < r$.

C. Reflected Brownian Motion and the Skorohod Equation

Our goal in this subsection is to provide a new proof of the celebrated result of P. Lévy (1948) already discussed in Problem 2.8.8, according to which *the processes*

$$(6.27)\quad \{M_t^W - W_t \triangleq \max_{0\le s\le t} W_s - W_t; 0 \le t < \infty\} \quad \text{and} \quad \{|W_t|; 0 \le t < \infty\}$$

have the same laws under P^0. In particular, we shall present the ingenious method of A. V. Skorohod (1961), which provides as a by-product the fact that *the processes*

$$(6.28)\qquad \{M_t^W \triangleq \max_{0\le s\le t} W_s; 0 \le t < \infty\} \quad \text{and} \quad \{2L_t(0); 0 \le t < \infty\}$$

also have the same laws under P^0.

6.14 Lemma (The Skorohod equation (1961)). *Let* $z \ge 0$ *be a given number and* $y(\cdot) = \{y(t); 0 \le t < \infty\}$ *a continuous function with* $y(0) = 0$. *There exists a unique continuous function* $k(\cdot) = \{k(t); 0 \le t < \infty\}$, *such that*

(i) $x(t) \triangleq z + y(t) + k(t) \ge 0;\ 0 \le t < \infty$,
(ii) $k(0) = 0$, $k(\cdot)$ *is nondecreasing, and*
(iii) $k(\cdot)$ *is flat off* $\{t \ge 0; x(t) = 0\}$; *i.e.*, $\int_0^\infty 1_{\{x(s)>0\}}\, dk(s) = 0$.

This function is given by

$$(6.29)\qquad k(t) = \max\left[0, \max_{0\le s\le t}\{-(z + y(s))\}\right], \quad 0 \le t < \infty.$$

Proof. To prove uniqueness, let $k(\cdot)$ and $\tilde{k}(\cdot)$ be continuous functions with properties (i)–(iii), where $x(\cdot)$ and $\tilde{x}(\cdot)$ correspond to $k(\cdot)$ and $\tilde{k}(\cdot)$, respectively. Suppose there exists a number $T > 0$ with $x(T) > \tilde{x}(T)$, and let $\tau \triangleq \max\{0 \le t < T; x(t) - \tilde{x}(t) = 0\}$ so that $x(t) > \tilde{x}(t) \ge 0$, $\forall t \in (\tau, T]$. But $k(\cdot)$ is flat on $\{u \ge 0; x(u) > 0\}$, so $k(\tau) = k(T)$. Therefore,

$$0 < x(T) - \tilde{x}(T) = k(T) - \tilde{k}(T) \le k(\tau) - \tilde{k}(\tau) = x(\tau) - \tilde{x}(\tau),$$

a contradiction. It follows that $x(T) \le \tilde{x}(T)$ for all $T \ge 0$, so $k \le \tilde{k}$. Similarly, $k \ge \tilde{k}$.

We now take $k(\cdot)$ to be defined by (6.29). Conditions (i) and (ii) are obviously satisfied. In order to verify (iii), it suffices to show $\int_0^\infty 1_{\{x(s)>\varepsilon\}}\, dk(s) = 0$ for every $\varepsilon > 0$. Let (t_1, t_2) be a component of the open set $\{s \ge 0; x(s) > \varepsilon\}$ and note that

$$-(z + y(s)) = k(s) - x(s) \le k(t_2) - \varepsilon; \quad t_1 \le s \le t_2.$$

But then

$$k(t_2) = \max\left[k(t_1), \max_{t_1\le s\le t_2}\{-(z + y(s))\}\right] \le \max[k(t_1), k(t_2) - \varepsilon],$$

which shows that $k(t_2) = k(t_1)$. □

6.15 Remark. For every $z \geq 0$ and $y(\cdot) \in C[0, \infty)$ with $y(0) = 0$, we denote by $\mathscr{K}$ the class of functions $k \in C[0, \infty)$ which satisfy conditions (i) and (ii) of Lemma 6.14 and introduce the mappings

$$T_t(z; y) \triangleq \max\left[0, \max_{0 \leq s \leq t} \{-(z + y(s))\}\right]; \quad 0 \leq t < \infty \tag{6.30}$$

$$R_t(z; y) \triangleq z + y(t) + T_t(z; y); \quad 0 \leq t < \infty. \tag{6.31}$$

In terms of these, the solution to the Skorohod equation is given by

$$k(t) = T_t(z; y), \quad x(t) = R_t(z; y), \tag{6.32}$$

and $T(z; y)$ is the minimal element of $\mathscr{K}$, as can be seen from the first part of the proof of Lemma 6.14.

6.16 Proposition. *Let $z \geq 0$ be a given number, and $B = \{B_t, \mathscr{G}_t; 0 \leq t < \infty\}$ a Brownian motion on some probability space $(\Theta, \mathscr{G}, Q)$ with $Q[B_0 = 0] = 1$. We suppose there exists a continuous process $k = \{k_t, \mathscr{G}_t; 0 \leq t < \infty\}$ such that, for Q-a.e. $\theta \in \Theta$, we have*

(i) $X_t(\theta) \triangleq z - B_t(\theta) + k_t(\theta) \geq 0; \quad 0 \leq t < \infty,$
(ii) $k_0(\theta) = 0$, $t \mapsto k_t(\theta)$ *is nondecreasing, and*
(iii) $\int_0^\infty 1_{(0,\infty)}(X_s(\theta))\, dk_s(\theta) = 0.$

Then $X = \{X_t; 0 \leq t < \infty\}$ under Q has the same law as $|W| = \{|W_t|; 0 \leq t < \infty\}$ under P^z.

PROOF. The law of the pair (k, X) is uniquely determined, since by Lemma 6.14 $k_t(\theta) = T_t(z; -B_.(\theta))$, $X_t(\theta) = R_t(z; -B_.(\theta))$; $0 \leq t < \infty$, for Q-a.e. $\theta \in \Theta$. It suffices, therefore, on our given measurable space $(\Omega, \mathscr{F})$ equipped with the Brownian family $\{W_t, \mathscr{F}_t; 0 \leq t < \infty\}$, $\{P^x\}_{x \in \mathbb{R}}$, to exhibit a standard Brownian motion $B = \{B_t, \mathscr{F}_t; 0 \leq t < \infty\}$ and a continuous nondecreasing process $k = \{k_t, \mathscr{F}_t; 0 \leq t < \infty\}$ such that, for P^z-a.e. $\omega \in \Omega$:

$$\begin{gathered} |W_t(\omega)| = z - B_t(\omega) + k_t(\omega); \quad 0 \leq t < \infty, \\ k_0(\omega) = 0, \quad t \mapsto k_t(\omega) \quad \text{is nondecreasing, and} \\ \int_0^\infty 1_{\mathbb{R}\setminus\{0\}}(W_s(\omega))\, dk_s(\omega) = 0. \end{gathered} \tag{6.33}$$

But this has already been accomplished in Problem 6.13 (relations (6.25), (6.26) with $a = 0$), if we make the identifications

$$B_t \equiv -\int_0^t \operatorname{sgn}(W_s)\, dW_s, \quad k_t \equiv 2L_t(0). \qquad \square$$

6.17 Theorem (P. Lévy (1948)). *The pairs of processes $\{(M_t^W - W_t, M_t^W); 0 \leq t < \infty\}$ and $\{(|W_t|, 2L_t(0)); 0 \leq t < \infty\}$ as in (6.27), (6.28) have the same laws under P^0.*

Proof. Because of uniqueness in the Skorohod equation, we have from (6.33)

(6.34) $$2L_t(0,\omega) \equiv M_t^B(\omega), \quad |W_t(\omega)| \equiv M_t^B(\omega) - B_t(\omega); \quad 0 \le t < \infty$$

for P^0-a.e. $\omega \in \Omega$, upon observing that

(6.35) $$M_t^B(\omega) = \max_{0 \le s \le t} B_s(\omega) = T_t(0; -B(\omega))$$

(Remark 6.15). The assertion follows, since both W and B are Brownian motions starting at the origin under P^0. We also notice the useful identity, valid for every fixed $t \in [0, \infty)$:

(6.36) $$M_t^B = \lim_{\varepsilon \downarrow 0} \frac{1}{2\varepsilon} \operatorname{meas}\{0 \le s \le t;\, M_s^B - B_s \le \varepsilon\}, \quad \text{a.s. } P^0. \qquad \square$$

6.18 Problem. Show that for every pair $(a, z) \in \mathbb{R}^2$ we have

$$P^z\left[\omega \in \Omega;\ \lim_{t\to\infty} L_t(a,\omega) = \infty\right] = 1.$$

D. A Generalized Itô Rule for Convex Functions

The functions $f_1(x) = (x-a)^+$, $f_2(x) = (x-a)^-$, and $f_3(x) = |x-a|$ in the Tanaka formulas (6.11)–(6.13) share an important property, namely *convexity*:

(6.37) $$f(\lambda x + (1-\lambda)z) \le \lambda f(x) + (1-\lambda) f(z); \quad x < z, \quad 0 < \lambda < 1,$$

which can be put in the equivalent form

(6.38) $$f(y) \le \frac{z-y}{z-x} f(x) + \frac{y-x}{z-x} f(z); \quad x < y < z$$

upon substituting $y = \lambda x + (1-\lambda)z$. Our success in representing $f(W_t)$ explicitly as a semimartingale, for the particular choices $f(x) = (x-a)^{\pm}$ and $f(x) = |x-a|$, makes us wonder whether it might be possible to obtain a *generalized Itô formula* for convex functions which are not necessarily twice differentiable. This possibility was explored by Meyer (1976) and Wang (1977). We derive the pertinent Itô formula in Theorem 6.22, after a brief digression on the fundamental properties of convex functions on $\mathbb{R}$.

6.19 Problem. Every convex function $f\colon \mathbb{R} \to \mathbb{R}$ is continuous. For fixed $x \in \mathbb{R}$, the difference quotient

(6.39) $$\Delta f(x;h) \triangleq \frac{f(x+h) - f(x)}{h}; \quad h \ne 0$$

is a nondecreasing function of $h \in \mathbb{R}\backslash\{0\}$, and therefore the right- and left-derivatives

(6.40) $$D^{\pm}f(x) \triangleq \lim_{h \to 0\pm} \frac{1}{h}[f(x+h) - f(x)]$$

exist and are finite for every $x \in \mathbb{R}$. Furthermore,

(6.41) $$D^+f(x) \le D^-f(y) \le D^+f(y); \quad x < y,$$

and $D^+f(\cdot)$ (respectively, $D^-f(\cdot)$) is right- (respectively, left-) continuous and nondecreasing on $\mathbb{R}$.

Finally, there exist sequences $\{\alpha_n\}_{n=1}^\infty$ and $\{\beta_n\}_{n=1}^\infty$ of real numbers, such that

(6.42) $$f(x) = \sup_{n \ge 1} (\alpha_n x + \beta_n); \quad x \in \mathbb{R}.$$

(*Hint*: Use (6.38) extensively.)

6.20 Problem. Let the function $\varphi\colon \mathbb{R} \to \mathbb{R}$ be nondecreasing, and define

$$\varphi_{\pm}(x) = \lim_{y \to x\pm} \varphi(y), \quad \Phi(x) = \int_0^x \varphi(u)\,du.$$

(i) The functions φ_+ and φ_- are right- and left-continuous, respectively, with

(6.43) $$\varphi_-(x) \le \varphi(x) \le \varphi_+(x); \quad x \in \mathbb{R}.$$

(ii) The functions $\varphi_{\pm}$ have the same set of continuity points, and equality holds in (6.43) on this set; in particular, except for x in a countable set N, we have $\varphi_{\pm}(x) = \varphi(x)$.

(iii) The function Φ is convex, with

$$D^-\Phi(x) = \varphi_-(x) \le \varphi(x) \le \varphi_+(x) = D^+\Phi(x); \quad x \in \mathbb{R}.$$

(iv) If $f\colon \mathbb{R} \to \mathbb{R}$ is any other convex function for which

(6.44) $$D^-f(x) \le \varphi(x) \le D^+f(x); \quad x \in \mathbb{R},$$

then we have $f(x) = f(0) + \Phi(x)$; $x \in \mathbb{R}$.

6.21 Problem. For any convex function $f\colon \mathbb{R} \to \mathbb{R}$, there is a countable set $N \subset \mathbb{R}$ such that f is differentiable on $\mathbb{R}\backslash N$, and

(6.45) $$f'(x) = D^+f(x) = D^-f(x); \quad x \in \mathbb{R}\backslash N.$$

Moreover

(6.46) $$f(x) - f(0) = \int_0^x f'(u)\,du = \int_0^x D^{\pm}f(u)\,du; \quad x \in \mathbb{R}.$$

The preceding problems show that convex functions are "essentially" differentiable, but Itô's rule requires the existence of a second derivative. For a convex function f, we use instead of its second derivative the *second derivative measure* μ on $(\mathbb{R}, \mathscr{B}(\mathbb{R}))$ defined by

(6.47) $$\mu([a,b)) \triangleq D^-f(b) - D^-f(a); \quad -\infty < a < b < \infty.$$

Of course, if f'' exists, then $\mu(dx) = f''(x)\,dx$. Even without the existence of f'', we may compute Riemann-Stieltjes integrals by parts, to obtain the formula

$$\text{(6.48)} \qquad \int_{-\infty}^{\infty} g(x)\mu(dx) = -\int_{-\infty}^{\infty} g'(x)D^-f(x)\,dx$$

for every function $g: \mathbb{R} \to \mathbb{R}$ which is piecewise C^1 and has compact support.

6.22 Theorem (A Generalized Itô Rule for Convex Functions). *Let $f: \mathbb{R} \to \mathbb{R}$ be a convex function and μ its second derivative measure introduced in* (6.47). *Then, for every $z \in \mathbb{R}$, we have a.s. P^z:*

$$\text{(6.49)} \quad f(W_t) = f(z) + \int_0^t D^-f(W_s)\,dW_s + \int_{-\infty}^{\infty} L_t(x)\mu(dx) \quad ; \quad 0 \le t < \infty.$$

Proof. It suffices to establish (6.49) with t replaced by $t \wedge T_{-n} \wedge T_n$, and by such a localization we may assume without loss of generality that D^-f is uniformly bounded on $\mathbb{R}$. We employ the mollifiers $\{\rho_n\}_{n=1}^{\infty}$ of (6.18) to obtain convex, infinitely differentiable approximations to f by convolution:

$$\text{(6.50)} \qquad f_n(x) \triangleq \int_{-\infty}^{\infty} \rho_n(x-y)f(y)\,dy; \quad n \ge 1.$$

It is not hard to verify that $f_n(x) = \int_{-\infty}^{\infty} \rho(z)f(x-(z/n))\,dz$ and

$$\text{(6.51)} \qquad \lim_{n\to\infty} f_n(x) = f(x), \quad \lim_{n\to\infty} f_n'(x) = D^-f(x)$$

hold for every $x \in \mathbb{R}$. In particular, the nondecreasing functions D^-f and $\{f_n'\}_{n=1}^{\infty}$ are uniformly bounded on compact subsets of $\mathbb{R}$. If $g: \mathbb{R} \to \mathbb{R}$ is of class C^1 and has compact support, then because of (6.48),

$$\begin{aligned}\lim_{n\to\infty} \int_{-\infty}^{\infty} g(x)f_n''(x)\,dx &= -\lim_{n\to\infty} \int_{-\infty}^{\infty} g'(x)f_n'(x)\,dx \\ &= -\int_{-\infty}^{\infty} g'(x)D^-f(x)\,dx = \int_{-\infty}^{\infty} g(x)\mu(dx).\end{aligned}$$

A continuous function g with compact support can be uniformly approximated by functions of class C^1, so that

$$\text{(6.52)} \qquad \lim_{n\to\infty} \int_{-\infty}^{\infty} g(x)f_n''(x)\,dx = \int_{-\infty}^{\infty} g(x)\mu(dx).$$

We can now apply the change-of-variable formula (Theorem 3.3) to $f_n(W_s)$, and obtain, for fixed $t \in (0, \infty)$:

$$f_n(W_t) - f_n(z) = \int_0^t f_n'(W_s)\,dW_s + \frac{1}{2}\int_0^t f_n''(W_s)\,ds, \quad \text{a.s. } P^z.$$

When $n \to \infty$, the left-hand side converges almost surely to $f(W_t) - f(z)$, and the stochastic integral converges in L^2 to $\int_0^t D^-f(W_s)\,dW_s$ because of (6.51) and the uniform boundedness of the functions involved. We also have from (6.7) and (6.52):

$$\lim_{n\to\infty}\int_0^t f_n''(W_s)\,ds = 2\lim_{n\to\infty}\int_{-\infty}^{\infty} f_n''(x)L_t(x)\,dx = 2\int_{-\infty}^{\infty} L_t(x)\mu(dx), \quad \text{a.s. } P^z$$

because, for P^z-a.e. $\omega\in\Omega$, the continuous function $x\mapsto L_t(x,\omega)$ has support on the compact set $[\min_{0\le s\le t} W_s(\omega), \max_{0\le s\le t} W_s(\omega)]$. This proves (6.49) for each fixed t, and because of continuity it is also seen to hold simultaneously for all $t\in[0,\infty)$, a.s. P^z. □

6.23 Corollary. *If $f\colon \mathbb{R}\to\mathbb{R}$ is a linear combination of convex functions, then* (6.49) *holds again for every $z\in\mathbb{R}$; now, μ defined by* (6.47) *is in general a signed measure with finite total variation on each bounded subinterval of* $\mathbb{R}$.

6.24 Problem. Let $a_1 < a_2 < \cdots < a_n$ be real numbers, and denote $D = \{a_1,\ldots,a_n\}$. Suppose that $f\colon \mathbb{R}\to\mathbb{R}$ is continuous and f' and f'' exist and are continuous on $\mathbb{R}\backslash D$, and the limits

$$f'(a_k\pm) \triangleq \lim_{x\to a_k\pm} f'(x), \quad f''(a_k\pm) = \lim_{x\to a_k\pm} f''(x)$$

exist and are finite. Show that f is the difference of two convex functions and, for every $z\in\mathbb{R}$,

$$(6.53)\quad f(W_t) = f(z) + \int_0^t f'(W_s)\,dW_s + \frac{1}{2}\int_0^t f''(W_s)\,ds$$
$$+ \sum_{k=1}^n L_t(a_k)[f'(a_k+) - f'(a_k-)]; \quad 0\le t<\infty, \text{ a.s. } P^z.$$

6.25 Exercise. Obtain the Tanaka formulas (6.11)–(6.13) as corollaries of the generalized Itô rule (6.49).

E. The Engelbert–Schmidt Zero-One Law

Our next application of local time concerns the study of the continuous, nondecreasing additive functional

$$A_t(\omega) = \int_0^t f(W_s(\omega))\,ds; \quad 0\le t<\infty,$$

where $f\colon \mathbb{R}\to[0,\infty)$ is a given Borel-measurable function. We shall be interested in questions of finiteness and asymptotics, but first we need an auxiliary result.

6.26 Lemma. *Let $f\colon \mathbb{R}\to[0,\infty)$ be Borel-measurable; fix $x\in\mathbb{R}$, and suppose there exists a random time T with*

$$P^0[0 < T < \infty] = 1, \quad P^0\left[\int_0^T f(x + W_s)\,ds < \infty\right] > 0.$$

Then, for some $\varepsilon > 0$, we have

$$\int_{-\varepsilon}^{\varepsilon} f(x + y)\,dy < \infty. \tag{6.54}$$

PROOF. From (6.7) and Problem 6.13 (iii), we know there exists an event Ω^* with $P^0(\Omega^*) = 1$, such that for every $\omega \in \Omega^*$:

$$\int_0^{T(\omega)} f(x + W_s(\omega))\,ds = 2\int_{-\infty}^{\infty} f(x + y)L_{T(\omega)}(y, \omega)\,dy$$

and $L_{T(\omega)}(0, \omega) > 0$. By assumption, we may choose $\omega \in \Omega^*$ such that $\int_0^{T(\omega)} f(x + W_s(\omega))\,ds < \infty$ as well. With this choice of ω, we may appeal to the continuity of $L_{T(\omega)}(\cdot, \omega)$ to choose positive numbers ε and c such that $L_{T(\omega)}(y, \omega) \geq c$ whenever $|y| \leq \varepsilon$. Therefore,

$$2c\int_{-\varepsilon}^{\varepsilon} f(x + y)\,dy \leq \int_0^{T(\omega)} f(x + W_s(\omega))\,ds < \infty,$$

which yields (6.54). □

6.27 Proposition (Engelbert–Schmidt (1981) Zero-One Law). *Let $f: \mathbb{R} \to [0, \infty)$ be Borel-measurable. The following three assertions are equivalent:*

(i) $P^0[\int_0^t f(W_s)\,ds < \infty; \quad \forall\, 0 \leq t < \infty] > 0$,
(ii) $P^0[\int_0^t f(W_s)\,ds < \infty; \quad \forall\, 0 \leq t < \infty] = 1$,
(iii) *f is locally integrable; i.e., for every compact set $K \subset \mathbb{R}$, we have $\int_K f(y)\,dy < \infty$.*

PROOF. For the implication *(i) ⇒ (iii)* we fix $b \in \mathbb{R}$ and consider the first passage time T_b. Because $P^0[T_b < \infty] = 1$, (i) gives $P^0[\int_0^{t+T_b} f(W_s)\,ds < \infty; \forall\, 0 \leq t < \infty] > 0$. But then

$$\int_0^{t+T_b(\omega)} f(W_s(\omega))\,ds \geq \int_{T_b(\omega)}^{t+T_b(\omega)} f(W_s(\omega))\,ds = \int_0^t f(b + B_s(\omega))\,ds,$$

where $B_s(\omega) \triangleq W_{s+T_b(\omega)}(\omega) - b;\ 0 \leq s < \infty$ is a new Brownian motion under P^0. It follows that for each $t > 0$, $P^0[\int_0^t f(b + B_s)\,ds < \infty] > 0$, and Lemma 6.26 guarantees the existence of an open neighborhood $U(b)$ of b such that $\int_{U(b)} f(y)\,dy < \infty$. If $K \subset \mathbb{R}$ is compact, the family $\{U(b)\}_{b \in K}$, being an open covering of K, has a finite subcovering. It follows that $\int_K f(y)\,dy < \infty$.

For the implication *(iii) ⇒ (ii)* we have again from (6.7), for P^0-a.e. $\omega \in \Omega$:

$$\begin{aligned}\int_0^t f(W_s(\omega))\,ds &= 2\int_{-\infty}^{\infty} f(y)L_t(y, \omega)\,dy = 2\int_{m_t(\omega)}^{M_t(\omega)} f(y)L_t(y, \omega)\,dy \\ &\leq \left[\max_{m_t(\omega) \leq y \leq M_t(\omega)} 2L_t(y, \omega)\right] \cdot \int_{m_t(\omega)}^{M_t(\omega)} f(y)\,dy; \quad 0 \leq t < \infty,\end{aligned}$$

where $m_t(\omega) = \min_{0\le s\le t} W_s(\omega)$, $M_t(\omega) = \max_{0\le s\le t} W_s(\omega)$. The last integral is finite by assumption, because the set $K = [m_t(\omega), M_t(\omega)]$ is compact. □

6.28 Corollary. *For $0 < t < \infty$, we have the following dichotomy*:

$$P^0\left[\int_0^t \frac{ds}{|W_s|^\alpha} < \infty;\quad \forall\, 0 < t < \infty\right] = \begin{Bmatrix} 1; & \text{if } 0 < \alpha < 1 \\ 0; & \text{if } \alpha \ge 1 \end{Bmatrix}.$$

6.29 Problem. The conditions of Proposition 6.27 are also equivalent to the following assertions:

(iv) $P^0[\int_0^t f(W_s)\,ds < \infty] = 1$, for some $0 < t < \infty$;

(v) $P^x[\int_0^t f(W_s)\,ds < \infty;\ \forall\, 0 \le t < \infty] = 1$, for every $x \in \mathbb{R}$;

(vi) for every $x \in \mathbb{R}$, there exists a Brownian motion $\{B_t, \mathscr{G}_t; 0 \le t < \infty\}$ and a random time S on a suitable probability space $(\Theta, \mathscr{G}, Q)$, such that $Q[B_0 = 0, 0 < S < \infty] = 1$ and

$$Q\left[\int_0^S f(x + B_s)\,ds < \infty\right] > 0.$$

(*Hint*: It suffices to justify the implications (ii) ⇒ (iv) ⇒ (vi) ⇒ (iii) ⇒ (v) ⇒ (vi), the first and last of which are obvious.)

6.30 Problem. Suppose that the Borel-measurable function $f: \mathbb{R} \to [0, \infty)$ satisfies: $\text{meas}\{y \in \mathbb{R}; f(y) > 0\} > 0$. Show that

$$P^x\left[\omega \in \Omega;\ \int_0^\infty f(W_s(\omega))\,ds = \infty\right] = 1 \tag{6.55}$$

holds for every $x \in \mathbb{R}$. Assume further that f has compact support, and consider the sequence of continuous processes

$$X_t^{(n)} \triangleq \frac{1}{\sqrt{n}} \int_0^{nt} f(W_s)\,ds;\quad 0 \le t < \infty, n \ge 1.$$

Establish then, under P^0, the convergence

$$X^{(n)} \xrightarrow[n\to\infty]{\mathscr{D}} X \tag{6.56}$$

in the sense of Definition 2.4.4, where $X_t \triangleq 2\|f\|_1 L_t(0)$ and $\|f\|_1 \triangleq \int_{-\infty}^\infty f(y)\,dy > 0$.

3.7. Local Time for Continuous Semimartingales†

The concept of local time and its application to obtain a generalized Itô rule can be extended from the case of Brownian motion in the previous section to that of continuous semimartingales. The significant differences are that

† This section may be omitted on first reading; its results will be used only in Section 5.5.

time-integrals such as in formula (6.7) now become integrals with respect to quadratic variation, and that the local time is not necessarily jointly continuous in the time and space variables. We shall use the generalized Itô rule developed in this section as a very important tool in the treatment of existence and uniqueness questions for one-dimensional stochastic differential equations, presented in Section 5.5.

Let

$$X_t = X_0 + M_t + V_t; \quad 0 \le t < \infty \tag{7.1}$$

be a continuous semimartingale, where $M = \{M_t, \mathscr{F}_t; 0 \le t < \infty\}$ is in $\mathscr{M}^{c,\text{loc}}$, $V = \{V_t, \mathscr{F}_t; 0 \le t < \infty\}$ is the difference of continuous, nondecreasing, adapted processes with $V_0 = 0$ a.s., and $\{\mathscr{F}_t\}$ satisfies the usual conditions. The results of this section are contained in the following theorem and are inspired by a more general treatment in Meyer (1976); they say in particular that *convex functions of continuous semimartingales are themselves continuous semimartingales*, and they provide the requisite decomposition.

7.1 Theorem. *Let X be a continuous semimartingale of the form* (7.1) *on some probability space $(\Omega, \mathscr{F}, P)$. There exists then a* semimartingale local time *for X, i.e., a nonnegative random field $\Lambda = \{\Lambda_t(a, \omega); (t, a) \in [0, \infty) \times \mathbb{R}, \omega \in \Omega\}$ such that the following hold:*

(i) *The mapping $(t, a, \omega) \mapsto \Lambda_t(a, \omega)$ is measurable and, for each fixed (t, a), the random variable $\Lambda_t(a)$ is $\mathscr{F}_t$-measurable.*

(ii) *For every fixed $a \in \mathbb{R}$, the mapping $t \mapsto \Lambda_t(a, \omega)$ is continuous and nondecreasing with $\Lambda_0(a, \omega) = 0$, and*

$$\int_0^\infty 1_{\mathbb{R}\setminus\{a\}}(X_t(\omega))\, d\Lambda_t(a, \omega) = 0, \quad \textit{for } P\textit{-a.e. } \omega \in \Omega. \tag{7.2}$$

(iii) *For every Borel-measurable $k: \mathbb{R} \to [0, \infty)$, the identity*

$$\int_0^t k(X_s(\omega))\, d\langle M\rangle_s(\omega) = 2\int_{-\infty}^\infty k(a)\Lambda_t(a, \omega)\, da; \quad 0 \le t < \infty \tag{7.3}$$

holds for P-a.e. $\omega \in \Omega$.

(iv) *For P-a.e. $\omega \in \Omega$, the limits*

$$\lim_{\substack{\tau \to t \\ b \downarrow a}} \Lambda_\tau(b, \omega) = \Lambda_t(a, \omega) \quad \textit{and} \quad \Lambda_t(a-, \omega) \triangleq \lim_{\substack{\tau \to t \\ b \uparrow a}} \Lambda_\tau(b, \omega)$$

exist for all $(t, a) \in [0, \infty) \times \mathbb{R}$. We express this property by saying that Λ is a.s. jointly continuous in t and RCLL in a.

(v) *For every convex function $f: \mathbb{R} \to \mathbb{R}$, we have the* generalized change of variable formula

$$f(X_t) = f(X_0) + \int_0^t D^- f(X_s)\, dM_s + \int_0^t D^- f(X_s)\, dV_s + \int_{-\infty}^\infty \Lambda_t(a)\mu(da); \quad 0 \le t < \infty, \quad \textit{a.s. } P, \tag{7.4}$$

where D^-f is the left-hand derivative in (6.40) *and μ is the second derivative measure* (6.47).

7.2 Corollary. *If $f: \mathbb{R} \to \mathbb{R}$ is a linear combination of convex functions,* (7.4) *still holds. Now μ defined by* (6.47) *is a signed measure, finite on each bounded subinterval of $\mathbb{R}$.*

7.3 Problem. Let X be a continuous semimartingale with decomposition (7.1) and let $f: \mathbb{R} \to \mathbb{R}$ be a function whose derivative is absolutely continuous. Then f'' exists Lebesgue-almost everywhere, and we have the Itô formula:

$$f(X_t) = f(X_0) + \int_0^t f'(X_s)\,dM_s + \int_0^t f'(X_s)\,dV_s + \frac{1}{2}\int_0^t f''(X_s)\,d\langle M\rangle_s; \quad 0 \le t < \infty, \text{ a.s. } P.$$

7.4 Remark. In the setting of Theorem 6.17, we observe that the reflected Brownian motion

$$X_t \triangleq |W_t| = -B_t + 2L_t(0); \quad 0 \le t < \infty$$

is a semimartingale with $X_0 \equiv 0$, $M \equiv -B$, $V \equiv 2L(0)$ under P^0. The generalized Itô rule (7.4) applied to this semimartingale with $f(x) = |x|$ gives in conjunction with (6.26): $X_t = -B_t - 2L_t(0) + 2\Lambda_t(0)$, and therefore $\Lambda_t(0) = 2L_t(0)$; $0 \le t < \infty$, a.s. P^0. In other words, the *semimartingale local time of reflected Brownian motion at the origin is twice the Brownian local time at the origin*, as one would expect intuitively.

By way of preparation for the proof of Theorem 7.1, we provide a construction similar to that used in the proof of Theorem 6.22. Let $f: \mathbb{R} \to \mathbb{R}$ be convex. Thanks to the usual localization argument, we may assume that D^-f is uniformly bounded on $\mathbb{R}$ and $\langle M\rangle_t$ and V_t are uniformly bounded in $0 \le t < \infty$ and $\omega \in \Omega$. Applying Itô's rule to the smooth function f_n of (6.50), we obtain

$$(7.5) \quad f_n(X_t) = f_n(X_0) + \int_0^t f_n'(X_s)\,dM_s + \int_0^t f_n'(X_s)\,dV_s + C_t^{(n)}; \quad 0 \le t < \infty,$$

where

$$C_t^{(n)} = \frac{1}{2}\int_0^t f_n''(X_s)\,d\langle M\rangle_s; \quad 0 \le t < \infty$$

is a continuous, nondecreasing (by the convexity of f_n), and $\{\mathscr{F}_t\}$-adapted process. As in Theorem 6.22, we have as $n \to \infty$:

$$f_n(X_t) \to f(X_t) \quad \text{and} \quad \int_0^t f_n'(X_s)\,dV_s \to \int_0^t D^-f(X_s)\,dV_s, \quad \text{a.s.}$$

$$\int_0^t f_n'(X_s)\,dM_s \to \int_0^t D^-f(X_s)\,dM_s, \quad \text{in probability}$$

for every fixed t. It follows that the remaining term $C_t^{(n)}$ in (7.5) must also converge in probability to a limit $C_t(f)$, and

$$(7.6) \quad f(X_t) = f(X_0) + \int_0^t D^-f(X_s)\,dM_s + \int_0^t D^-f(X_s)\,dV_s + C_t(f); \quad 0 \le t < \infty.$$

Now $f(X_t)$ is continuous in t and both integrals have continuous modifications, so we may and do choose a continuous modification of $C_t(f)$. Each $C^{(n)}$ is nondecreasing and adapted to $\{\mathscr{F}_t\}$; the limit $C(f)$ inherits both these properties.

With $a \in \mathbb{R}$ fixed, we apply (7.6) to the functions $f_1(x) \triangleq (x-a)^+$, $f_2(x) = (x-a)^-$, and $f_3(x) = |x-a|$ to obtain the *Tanaka-Meyer formulas*

$$(7.7) \quad (X_t - a)^+ = (X_0 - a)^+ + \int_0^t 1_{(a,\infty)}(X_s)\,dM_s + \int_0^t 1_{(a,\infty)}(X_s)\,dV_s + C_t^+(a)$$

$$(7.8) \quad (X_t - a)^- = (X_0 - a)^- - \int_0^t 1_{(-\infty,a]}(X_s)\,dM_s - \int_0^t 1_{(-\infty,a]}(X_s)\,dV_s + C_t^-(a)$$

$$(7.9) \quad |X_t - a| = |X_0 - a| + \int_0^t \operatorname{sgn}(X_s - a)\,dM_s + \int_0^t \operatorname{sgn}(X_s - a)\,dV_s + 2\Lambda_t(a),$$

with the conventions (6.14), $C_t^+(a) \triangleq C_t(f_1)$, $C_t^-(a) \triangleq C_t(f_2)$, and $2\Lambda_t(a) \triangleq C_t(f_3)$. The processes $C_t^\pm(a)$, $\Lambda_t(a)$ are adapted, continuous, and nondecreasing in t, and the random field $\Lambda_t(a,\omega)$ will be our candidate for the local time of X. Now (7.7) and (7.8) yield $C_t^+(a) = C_t^-(a)$ (upon subtraction), as well as

$$(7.10) \qquad \Lambda_t(a) = C_t^\pm(a); \quad 0 \le t < \infty,\ a \in \mathbb{R}$$

(upon addition and comparison with (7.9)).

Although the process $\{\Lambda_t(a); 0 \le t < \infty\}$ is continuous for every fixed a, we do not yet have any information about the regularity of $\Lambda_t(a)$ in the pair (t,a). We approach this issue by studying the regularity of the other terms appearing in (7.7).

7.5. Lemma. *Let X be a continuous semimartingale with decomposition* (7.1). *Define*

$$I_t(a) \triangleq \int_0^t 1_{(a,\infty)}(X_s)\,dM_s; \quad 0 \le t < \infty,\ a \in \mathbb{R}.$$

The random field $I = \{I_t(a), \mathscr{F}_t; 0 \le t < \infty, a \in \mathbb{R}\}$ has a continuous modification. In other words, there exists a random field $\hat{I} = \{\hat{I}_t(a), \mathscr{F}_t; 0 \le t < \infty, a \in \mathbb{R}\}$ such that:

(i) *For P-a.e.* $\omega \in \Omega$, *the mapping* $(t, a) \mapsto \hat{I}_t(a, \omega)$ *is continuous on* $[0, \infty) \times \mathbb{R}$.
(ii) *For every* $t \in [0, \infty)$ *and* $a \in \mathbb{R}$, *we have* $I_t(a) = \hat{I}_t(a)$, *a.s.* P.

PROOF. By the usual localization argument, we may assume without loss of generality that there is a constant K for which

$$(7.11) \qquad \sup_{0 \le t < \infty} |X_t| \le K, \quad \langle M \rangle_\infty \le K, \quad \check{V}_\infty \le K,$$

where $\check{V}_\infty$ is the total variation of V on $[0, \infty)$. According to Remark 4.9, we may choose a Brownian motion B for which we have the equations

$$(7.12) \quad I_t(a) \triangleq \int_0^t 1_{(a,\infty)}(X_v)\,dM_v = \int_0^{\langle M \rangle_t} 1_{(a,\infty)}(Y_u)\,dB_u; \quad 0 \le t < \infty,$$

$$(7.13) \quad H_s(a) \triangleq \int_0^s 1_{(a,\infty)}(Y_u)\,dB_u = \int_0^{T(s)} 1_{(a,\infty)}(X_v)\,dM_v; \quad 0 \le s < \infty,$$

where $T(\cdot)$ is given by (4.15), $Y_s \triangleq X_{T(s)}$ for $0 \le s < \langle M \rangle_\infty$, and Y_s is chosen so that $1_{(b,\infty)}(Y_s) = 0$ for $s \ge \langle M \rangle_\infty$ and b in some neighborhood of a (cf. (4.20)′, Remark 4.9). We shall prove the existence of a continuous modification of H by using the extension of the Kolmogorov-Čentsov theorem (Problem 2.2.9). According to the latter, it suffices to show

$$(7.14) \qquad E|H_{s_2}(a) - H_{s_1}(b)|^\alpha \le C[(s_2 - s_1)^2 + (b - a)^2]^{1+\beta};$$
$$0 < s_1 < s_2 < \infty, \quad a, b \in \mathbb{R}$$

for suitable positive constants α, β, and C. Note that

$$(7.15) \quad E|H_{s_2}(a) - H_{s_1}(b)|^\alpha \le 2^\alpha E|H_{s_2}(a) - H_{s_2}(b)|^\alpha + 2^\alpha E|H_{s_2}(b) - H_{s_1}(b)|^\alpha,$$

and, according to the martingale moment inequality (3.37), we may bound the latter expectation by

$$E|H_{s_2}(b) - H_{s_1}(b)|^\alpha \le C_\alpha E[\langle H(b) \rangle_{s_2} - \langle H(b) \rangle_{s_1}]^{\alpha/2} \le C_\alpha (s_2 - s_1)^{\alpha/2}.$$

When $\alpha > 4$, this bound is of the type required by (7.14). Thus, it remains only to deal with the first expectation on the right-hand side of (7.15), i.e., to show

$$(7.16) \quad E|H_s(a) - H_s(b)|^\alpha \le C(b - a)^{2(1+\beta)}; \quad 0 \le s < \infty, \quad -\infty < a < b < \infty,$$

where $\alpha > 4$, $\beta > 0$, and C is a positive constant. We fix $a < b$ and introduce the convex function

$$f(x) = \int_0^x \int_0^y 1_{(a,b]}(z)\,dz\,dy,$$

for which $|f'|$ is bounded by $b - a$ and $D^- f' = 1_{(a,b]}$. In particular, passage to the limit in (7.5) yields

$$f(X_t) = f(X_0) + \int_0^t f'(X_\theta)\,dM_\theta + \int_0^t f'(X_\theta)\,dV_\theta + \frac{1}{2}\int_0^t 1_{(a,b]}(X_\theta)\,d\langle M \rangle_\theta.$$

Assumption (7.11) and Problem 1.5.24 show that X_t and all the preceding integrals have limits as $t \to \infty$, so we may replace t by $T(s)$ (which may be infinite) to obtain, for every $k \geq 1$, the bound

(7.17)
$$\left| \int_0^{T(s)} 1_{(a,b]}(X_\theta)\, d\langle M\rangle_\theta \right|^k \leq 6^k |f(X_{T(s)}) - f(X_0)|^k + 6^k \left| \int_0^{T(s)} f'(X_\theta)\, dM_\theta \right|^k$$
$$+ 6^k \left| \int_0^{T(s)} f'(X_\theta)\, dV_\theta \right|^k.$$

We bound the terms on the right-hand side of (7.17). The mean-value theorem implies

$$|f(X_{T(s)}) - f(X_0)|^k \leq (b-a)^k |X_{T(s)} - X_0|^k \leq 2^k K^k (b-a)^k,$$

and it is also clear that

$$\left| \int_0^{T(s)} f'(X_\theta)\, dV_\theta \right|^k \leq K^k (b-a)^k.$$

Applying the martingale moment inequality (3.37) to the stochastic integral in (7.17), we obtain the bound

$$E\left| \int_0^{T(s)} f'(X_v)\, dM_v \right|^k \leq \varliminf_{t\to\infty} E\left| \int_0^{T(s)\wedge t} f'(X_v)\, dM_v \right|^k$$
$$\leq C_k E\left[\int_0^{\infty} |f'(X_v)|^2\, d\langle M\rangle_v \right]^{k/2}$$
$$\leq C_k K^{k/2} (b-a)^k.$$

We conclude from these considerations that there exists a constant C, depending only on k and on the bound K in (7.11), such that

$$E\left| \int_0^{T(s)} 1_{(a,b]}(X_v)\, d\langle M\rangle_v \right|^k \leq C(b-a)^k; \quad 0 < s < \infty, \quad -\infty < a < b < \infty.$$

Now (3.37) can be invoked again to establish (7.16) with $\alpha = 2k$ and $2(1+\beta) = k$, provided $k > 2$. From (7.12), (7.13) we see that $I_t(a) = H_{\langle M\rangle_t}(a)$, and since $\langle M\rangle$ is continuous, the existence of a continuous modification for H implies the same for I. □

The Lebesgue-Stieltjes integral

$$J_t(a) \triangleq \int_0^t 1_{(a,\infty)}(X_s)\, dV_s \tag{7.18}$$

appearing in (7.7) can fail to be jointly continuous in (t, a); see Exercise 7.9. However, it is *jointly continuous in t and RCLL in a*; the proof is left as a problem for the reader.

7.6 Problem. Let X be a continuous semimartingale with decomposition (7.1). For P-almost every $\omega \in \Omega$, we have for all $(t, a) \in [0, \infty) \times \mathbb{R}$:

$$\lim_{\substack{\tau \to t \\ b \downarrow a}} J_\tau(b, \omega) = J_t(a, \omega), \tag{7.19}$$

$$\lim_{\substack{\tau \to t \\ b \uparrow a}} J_\tau(b, \omega) = \int_0^t 1_{[a,\infty)}(X_s(\omega))\, dV_s(\omega). \tag{7.20}$$

Proof of Theorem 7.1. Using Lemma 7.5 and Problem 7.6, we choose a modification of $C_t^+(a)$ in (7.7) which is jointly continuous in t and RCLL in a. We take $\Lambda_t(a)$ and $C_t^{\pm}(a)$ to satisfy (7.10) for every t, a, and ω. In particular, $\Lambda_t(a, \omega)$ is jointly $\mathscr{B}([0, \infty)) \otimes \mathscr{B}(\mathbb{R}) \otimes \mathscr{F}$-measurable and satisfies (iv), and the other measurability claims of (i) and (ii) hold. In particular, it follows from (7.19), (7.20) that

$$\Lambda_t(a) - \Lambda_t(a-) = \textstyle\int_0^t 1_{\{a\}}(X_s)\, dV_s.$$

For the *proof of (7.2)*, consider any two rational numbers $0 \le u < v < \infty$ and the event

$$H_{uv} \triangleq \{\omega \in \Omega;\ X_s(\omega) < a, \quad \forall s \in [u, v]\}.$$

From relation (7.7) we have on H_{uv}: $\Lambda_u(a, \omega) = \Lambda_v(a, \omega)$, except for ω in a null set N_{uv}. Let $N \triangleq \bigcup_{\substack{0 \le u < v < \infty \\ u, v \in Q}} N_{uv}$ and fix $\omega \in \Omega \backslash N$. The set $S(a, \omega) \triangleq \{0 < t < \infty;\ X_t(\omega) < a\}$ is open and, as such, is the countable union of disjoint open intervals. Let (α, β) be such an interval. If $\alpha < u < v < \beta$, where u and v are rational, then $\Lambda_u(a, \omega) = \Lambda_v(a, \omega)$. It follows that $\int_{(\alpha,\beta)} d\Lambda_t(a, \omega) = 0$, and thus

$$\int_0^\infty 1_{(-\infty,a)}(X_t(\omega))\, d\Lambda_t(a, \omega) = 0. \tag{7.21}$$

A similar argument based on (7.8) shows that

$$\int_0^\infty 1_{(a,\infty)}(X_t(\omega))\, d\Lambda_t(a, \omega) = 0, \quad \text{a.s.} \tag{7.22}$$

which establishes (7.2).

For the *proof of (7.4)*, we may assume, by the usual localization argument, that there exists a constant $K > 0$ such that (7.11) holds. Consequently, we may also assume without loss of generality that $D^- f$ is constant outside $(-K, K)$, so the second derivative measure μ has support on $[-K, K)$. Let $x \in [-K, K]$ be fixed and introduce the function

$$g(a) = \begin{cases} 0; & a \le -K - 1, \\ (x + K)(a + K + 1); & -K - 1 \le a \le -K, \\ x - a; & -K \le a \le x, \\ 0; & x \le a. \end{cases}$$

According to (6.48), (6.46), we have for $-K \le x \le K$,

$$\begin{aligned}(7.23)\quad \int_{-\infty}^{\infty} (x-a)^+ \mu(da) &= -\int_{-\infty}^{\infty} g'(a) D^- f(a)\, da \\ &= -(x+K)\int_{-K-1}^{-K} D^- f(a)\, da + \int_{-K}^{x} D^- f(a)\, da \\ &= -(x+K)\, D^- f(-K) + f(x) - f(-K),\end{aligned}$$

and from the definition of μ:

$$(7.24)\qquad \int_{-\infty}^{\infty} 1_{(a,\infty)}(x)\mu(da) = \mu([-K,x)) = D^- f(x) - D^- f(-K).$$

We may now integrate with respect to μ in the Tanaka-Meyer formula (7.7) and use (7.23) to obtain

$$\begin{aligned}(7.25)\quad f(X_t) = (X_t - X_0)\, D^- f(-K) + f(X_0) &+ \int_{-\infty}^{\infty}\int_0^t 1_{(a,\infty)}(X_s)\, dM_s\, \mu(da) \\ &+ \int_{-\infty}^{\infty}\int_0^t 1_{(a,\infty)}(X_s)\, dV_s\, \mu(da) + \int_{-\infty}^{\infty} \Lambda_t(a)\mu(da).\end{aligned}$$

Fubini's theorem and (7.24) allow us to write

$$(7.26)\qquad \int_{-\infty}^{\infty}\int_0^t 1_{(a,\infty)}(X_s)\, dV_s\, \mu(da) = \int_0^t D^- f(X_s)\, dV_s - V_t\, D^- f(-K).$$

A similar interchange of the order of integration in the integral

$$(7.27)\qquad \int_{-\infty}^{\infty}\int_0^t 1_{(a,\infty)}(X_s)\, dM_s\, \mu(da) = \int_0^t D^- f(X_s)\, dM_s - M_t\, D^- f(-K)$$

is justified by Problem 7.7 following this proof. Substitution of (7.26), (7.27) into (7.25) results in (7.4).

Finally, let us consider (7.4) in the special case $f(x) = x^2$. Then $\mu(da) = 2\, da$, and comparison of (7.4) with the result from the usual Itô rule reveals that

$$\langle M\rangle_t = 2\int_{-\infty}^{\infty} \Lambda_t(a)\, da; \quad 0 \le t < \infty, \text{ a.s.}$$

Thus, for any measurable function $h(s,\omega)\colon [0,\infty)\times\Omega \to [0,\infty)$,

$$\int_0^{\infty} h(s,\omega)\, d\langle M\rangle_s(\omega) = 2\int_{-\infty}^{\infty}\int_0^{\infty} h(s,\omega)\, d\Lambda_s(a,\omega)\, da$$

holds for P-a.e. $\omega\in\Omega$. Now if $k\colon \mathbb{R}\to[0,\infty)$ is measurable, we may take $h(s,\omega) = 1_{[0,t]}(s)\cdot k(X_s(\omega))$ and obtain for P-a.e. ω:

$$\int_0^t k(X_s(\omega))\,d\langle M\rangle_s(\omega) = 2\int_{-\infty}^{\infty}\int_0^t k(X_s(\omega))\,d\Lambda_s(a,\omega)\,da$$
$$= 2\int_{-\infty}^{\infty} k(a)\Lambda_t(a,\omega)\,da; \quad 0 \le t < \infty,$$

thanks to (7.2). This completes the proof. □

7.7 Problem. Let X be a continuous semimartingale with decomposition (7.1), μ be a σ-finite measure on $(\mathbb{R}, \mathscr{B}(\mathbb{R}))$, and $h: \mathbb{R} \to [0, \infty)$ be a continuous function with compact support. Then

$$\int_{-\infty}^{\infty} h(a)\left(\int_0^t 1_{(a,\infty)}(X_s)\,dM_s\right)\mu(da) = \int_0^t \left(\int_{-\infty}^{\infty} h(a)1_{(a,\infty)}(X_s)\mu(da)\right)dM_s.$$

7.8 Remark. The proof of Theorem 7.1 shows that the semimartingale local time $\Lambda_t^M(a)$ for a continuous local martingale M is jointly continuous in (t, a), because the possibly discontinuous term $J_t(a)$ of (7.18) is not present. In particular, (7.9) becomes then a.s. P:

$$(7.28) \quad |M_t - a| = |M_0 - a| + \int_0^t \operatorname{sgn}(M_s - a)\,dM_s + 2\Lambda_t^M(a); \quad 0 \le t < \infty.$$

Comparison of (7.28) with the Tanaka formula (6.13) shows that *the semimartingale local time $\Lambda^W(a)$ for Brownian motion W coincides with the local time $L(a)$ of the previous section.* If $M \in \mathscr{M}_2^c$, then for any stopping time τ,

$$(7.29) \qquad E|M_{t\wedge\tau}| = 2E[\Lambda_{t\wedge\tau}^M(0)]; \quad 0 \le t < \infty.$$

7.9 Exercise. Show by example that $J_t(a)$ defined by (7.18) can fail to be continuous in a.

7.10 Exercise. Let X be a continuous semimartingale with decomposition (7.1). Show that for every $a \in \mathbb{R}$,

$$\int_0^\infty 1_{\{a\}}(X_s)\,d\langle M\rangle_s = 0, \quad \text{a.s. } P.$$

7.11 Exercise. Show that the semimartingale local time of a continuous process of bounded variation is identically zero.

7.12 Exercise (LeGall (1983)). Let X be a continuous semimartingale with decomposition (7.1), and suppose that there exists a Borel-measurable function $k: (0, \infty) \to (0, \infty)$ such that $\int_{(0,\varepsilon)} (du/k(u)) = \infty, \forall\, \varepsilon > 0$, but for every $t \in (0, \infty)$ we have

$$\int_0^t \frac{d\langle M\rangle_s}{k(X_s)} 1_{\{X_s>0\}} < \infty, \quad \text{a.s. } P.$$

Then the local time $\Lambda(0)$ of X at the origin is identically zero, almost surely.

7.13 Exercise. Consider a continuous local martingale M and denote $S_t \triangleq \max_{0\le s\le t} M_s$, $L_t \triangleq 2\Lambda_t(0)$. Suppose now that $f(t,x,y)\colon \mathbb{R}^3 \to \mathbb{R}$ is a function of class $C^2(\mathbb{R}^3)$ which satisfies

$$\frac{\partial f}{\partial t} + \frac{1}{2}\frac{\partial^2 f}{\partial x^2} = 0$$

in $\mathbb{R}^3$ and

$$\frac{\partial f}{\partial x}(t,0,y) + \frac{\partial f}{\partial y}(t,0,y) = 0$$

for every $(t,y) \in \mathbb{R}^2$. Show then that the processes $f(\langle M\rangle_t, |M_t|, L_t)$ and $f(\langle M\rangle_t, S_t - M_t, S_t)$ are local martingales.

Deduce also the following:

(i) The process $(S_t - M_t)^2 - \langle M\rangle_t$ is a local martingale.
(ii) For every real-valued function g of class $C^1(\mathbb{R})$, the processes

$$g(L_t) - |M_t|\, g'(L_t) \quad \text{and} \quad g(S_t) - (S_t - M_t)\, g'(S_t)$$

are local martingales.

7.14 Exercise. For a nonnegative, continuous semimartingale X of the form (7.1) with $X_0 \equiv 0$, the following conditions are equivalent:

(i) V is flat off $\{t \ge 0;\, X_t = 0\}$.
(ii) The process $\int_0^t 1_{\{X_s \neq 0\}}\, dX_s;\ 0 \le t < \infty$ belongs to $\mathcal{M}^{c,\,\mathrm{loc}}$.
(iii) There exists $N \in \mathcal{M}^{c,\,\mathrm{loc}}$ such that $X_t = \max_{0\le s\le t} N_s - N_t$.

3.8. Solutions to Selected Problems

2.5. (a) It is easily verified that $u - (1/2^n) \le \varphi_n(u) < u$. Consequently, $X_t^{(n,s)}$ is $\mathscr{F}_t$-measurable, and since φ_n takes only discrete values, $X^{(n,s)}$ is simple.

(b) The procedure (2.7) results in measurable (but perhaps not adapted) processes $\{\tilde{X}^{(m)}\}_{m=1}^\infty$, such that

$$\lim_{m\to\infty} E\int_0^T |\tilde{X}_t^{(m)} - X_t|^2\, dt = 0.$$

Thus, for given $\varepsilon > 0$, we can find $m \ge 1$ so that with $X^\varepsilon \triangleq \tilde{X}^{(m)}$ we have $E\int_0^T |X_t^\varepsilon - X_t|^2\, dt \le \varepsilon^2$. The Minkowski inequality leads then to

$$\left(E\int_0^T |X_t - X_{t-h}|^2\,dt\right)^{1/2}$$
$$\leq \left(E\int_0^T |X_t - X_t^\varepsilon|^2\,dt\right)^{1/2} + \left(E\int_0^T |X_t^\varepsilon - X_{t-h}^\varepsilon|^2\,dt\right)^{1/2}$$
$$+ \left(E\int_0^T |X_{t-h}^\varepsilon - X_{t-h}|^2\,dt\right)^{1/2}$$
$$\leq 2\varepsilon + \left(E\int_0^T |X_t^\varepsilon - X_{t-h}^\varepsilon|^2\,dt\right)^{1/2}.$$

We can now let $h \downarrow 0$ and conclude, from the continuity of X^ε and the bounded convergence theorem, that

$$\overline{\lim_{h\downarrow 0}}\, E\int_0^T |X_t - X_{t-h}|^2\,dt \leq 4\varepsilon^2.$$

(c) Let i be any nonnegative integer. As s ranges over $[i/2^n, (i+1)/2^n)$, $\varphi_n(t-s)+s$ ranges over $[t-(1/2^n), t)$. Therefore,

$$E\int_0^T\int_0^1 |X_t^{(n,s)} - X_t|^2\,ds\,dt = 2^n E\int_0^T\int_0^{2^{-n}} |X_t - X_{t-h}|^2\,dh\,dt$$
$$= 2^n\int_0^{2^{-n}}\left[E\int_0^T |X_t - X_{t-h}|^2\,dt\right]dh$$
$$\leq \max_{0\leq h\leq 2^{-n}} E\int_0^T |X_t - X_{t-h}|^2\,dt,$$

which converges to zero as $n\to\infty$ because of (b).

(d) From (c) there exists a sequence $\{n_k\}_{k=1}^\infty$ of integers, increasing to infinity as $k\to\infty$, such that for meas $\times$ meas $\times$ P-a.e. triple (s,t,ω) in $[0,1]\times[0,T]\times\Omega$, where *meas* means "Lebesgue measure," we have

(8.1) $$\lim_{k\to\infty} |X_t^{(n_k,s)}(\omega) - X_t(\omega)|^2 = 0.$$

Therefore, we can select $s\in[0,1]$ such that for meas $\times$ P-a.e. pair (t,ω) in $[0,T]\times\Omega$, we have (8.1). Setting $X^{(k)} \triangleq X^{(n_k,s)}$, we obtain (2.8) from the bounded convergence theorem.

2.18. By assumption, we have

$$E\langle I^M(X)\rangle_\infty = E\int_0^\infty X_s^2\,d\langle M\rangle_s < \infty.$$

Uniform integrability and the existence of a last element for $I^M(X)$ follow from Problem 1.5.24, as does uniform integrability of $(I^M(X))^2$; similarly for $I^N(Y)$.

Applying Proposition 2.14 with X_u, Y_u replaced by $X_u 1_{\{u\geq T\}}$, $Y_u 1_{\{u\geq T\}}$, respectively, we obtain

$$\left|\int_T^{T+t} X_s Y_s\,d\langle M,N\rangle_s\right| \leq \left(\int_T^{T+t} X_s^2\,d\langle M\rangle_s\cdot\int_T^{T+t} Y_s^2\,d\langle N\rangle_s\right)^{1/2},$$

whence

$$\left|\int_T^\infty X_s Y_s \, d\langle M,N\rangle_s\right| \le \left(\int_T^\infty X_s^2 \, d\langle M\rangle_s \cdot \int_T^\infty Y_s^2 \, d\langle N\rangle_s\right)^{1/2}$$

a.s. P. As $T \to \infty$, the right-hand side of this inequality converges to zero; therefore,

$$\langle I^M(X), I^N(Y)\rangle_t = \int_0^t X_s Y_s \, d\langle M,N\rangle_s$$

converges as $t \to \infty$ and is bounded by the integrable random variable

$$\left(\int_0^\infty X_s^2 \, d\langle M\rangle_s\right)^{1/2} \cdot \left(\int_0^\infty Y_s^2 \, d\langle N\rangle_s\right)^{1/2},$$

a.s. P. The dominated convergence theorem gives then

$$\begin{aligned}\lim_{t\to\infty} E[I_t^M(X) I_t^N(Y)] &= \lim_{t\to\infty} E\langle I^M(X), I^N(Y)\rangle_t = E\langle I^M(X), I^N(Y)\rangle_\infty \\ &= E\int_0^\infty X_s Y_s \, d\langle M,N\rangle_s.\end{aligned}$$

We also have

$$\begin{aligned}E[I_\infty^M(X) I_\infty^N(Y)] &= E[(I_\infty^M(X) - I_t^M(X))(I_\infty^N(Y) - I_t^N(Y))] \\ &\quad + E[I_t^M(X)(I_\infty^N(Y) - I_t^N(Y))] \\ &\quad + E[I_t^N(Y)(I_\infty^M(X) - I_t^M(X))] \\ &\quad + E[I_t^M(X) I_t^N(Y)].\end{aligned}$$

We have just shown that the fourth term on the right-hand side converges to $E\langle I^M(X), I^N(Y)\rangle_\infty$ as $t \to \infty$. The other three terms converge to zero because of Hölder's inequality and the uniform integrability of $(I^M(X))^2$ and $(I^N(Y))^2$.

2.27. (S. Dayanik): With $X \in \mathscr{P}^*(M)$, we construct the sequence of bounded stopping times $\{T_n\}_{n=1}^\infty$ in (2.32). In the notation of (2.33), each $X^{(n)}$ is in $\mathscr{L}^*(M^{(n)})$ and therefore can be approximated by a sequence of simple processes $\{X^{(n,k)}\}_{k=1}^\infty \subset \mathscr{L}_0$ in the sense

$$\lim_{k\to\infty} E\int_0^T |X_t^{(n,k)} - X_t^{(n)}|^2 d\langle M^{(n)}\rangle_t = 0 \quad \forall\, T < \infty \tag{1}$$

(Proposition 2.8). Let us fix a positive $T < \infty$. By (1), for every n, we can find some m_n such that

$$E\int_0^T |X_t^{(n,m_n)} - X_t^{(n)}|^2 d\langle M^{(n)}\rangle_t < \frac{1}{n}.$$

We claim that

$$\int_0^T |X_t^{(n,m_n)} - X_t|^2 d\langle M\rangle_t \xrightarrow{P} 0 \quad \text{as } n \to \infty. \tag{2}$$

To show this, we first observe that, for every n, $X_t^{(n)} = X_t$ and $\langle M^{(n)}\rangle_t = \langle M\rangle_t$ for every $0 \le t \le T$ on $\{T \le T_n\}$. Therefore, for every fixed $\varepsilon > 0$,

$$P\left\{\int_0^T |X_t^{(n,m_n)} - X_t|^2 d\langle M\rangle_t > \varepsilon\right\}$$
$$\le P\left[\left\{\int_0^T |X_t^{(n,m_n)} - X_t|^2 d\langle M\rangle_t > \varepsilon\right\} \cap \{T \le T_n\}\right] + P\{T > T_n\}$$
$$= P\left[\left\{\int_0^T |X_t^{(n,m_n)} - X_t^{(n)}|^2 d\langle M^{(n)}\rangle_t > \varepsilon\right\} \cap \{T \le T_n\}\right] + P\{T > T_n\}$$
$$\le P\left\{\int_0^T |X_t^{(n,m_n)} - X_t^{(n)}|^2 d\langle M^{(n)}\rangle_t > \varepsilon\right\} + P\{T > T_n\}$$
$$\le \frac{1}{\varepsilon} E\int_0^T |X_t^{(n,m_n)} - X_t^{(n)}|^2 d\langle M^{(n)}\rangle_t + P\{T_n < T\}$$
(by the Markov Inequality)
$$\le \frac{1}{n\varepsilon} + P\{T_n < T\}$$

for every n. Since $\lim_{n\to\infty} P\{T_n < T\} = 0$, for every large enough n we have $P\{\int_0^T |X_t^{(n,m_n)} - X_t|^2 d\langle M\rangle_t > \varepsilon\} < \varepsilon$. This proves (2).

Denote every simple process $X^{(n,m_n)}$ by $Y^{(T,n)}$ to emphasize its dependence on T. By (2) and Proposition 2.26, both sequences of random variables

$$\int_0^T |Y_t^{(T,n)} - X_t|^2 d\langle M\rangle_t, \quad \sup_{0\le t\le T} |I_t(Y^{(T,n)}) - I_t(X)|$$

converge to zero in probability, as $n \to \infty$, and there exists a subsequence for which the convergence takes place almost surely. Having done this construction for T fixed, we use a diagonalization argument, as in the first paragraph of the proof of Lemma 2.4, to obtain a sequence which works for all T.

In the case that M is Brownian Motion, we use Proposition 2.6 rather than Proposition 2.8 in this construction.

2.29. We have

$$S_\varepsilon(\Pi) = \frac{1}{2}\sum_{i=0}^{m-1}(W_{t_{i+1}}^2 - W_{t_i}^2) + \left(\varepsilon - \frac{1}{2}\right)\sum_{i=0}^{m-1}(W_{t_{i+1}} - W_{t_i})^2$$
$$= \frac{1}{2}W_t^2 + \left(\varepsilon - \frac{1}{2}\right)\sum_{i=0}^{m-1}(W_{t_{i+1}} - W_{t_i})^2.$$

Recalling the discussion preceding Lemma 1.5.9, we may write

$$E\left[\sum_{i=0}^{m-1}(W_{t_{i+1}} - W_{t_i})^2 - t\right]^2 = E\left\{\sum_{i=0}^{m-1}[(W_{t_{i+1}} - W_{t_i})^2 - (t_{i+1} - t_i)]\right\}^2$$
$$= \sum_{i=0}^{m-1} E[(W_{t_{i+1}} - W_{t_i})^2 - (t_{i+1} - t_i)]^2$$
$$= \sum_{i=0}^{m-1} E(W_{t_{i+1}} - W_{t_i})^4 - \sum_{i=0}^{m-1}(t_{i+1} - t_i)^2$$

$$\leq C_2 \sum_{i=0}^{m-1} (t_{i+1} - t_i)^2$$
$$\leq C_2 t \|\Pi\|,$$

where we have used Problem 2.2.10. This proves (2.37). To see that $\varepsilon = 0$ corresponds to the Itô integral, consider the (piecewise constant) process

$$X_s^\Pi \triangleq \sum_{i=0}^{m-1} W_{t_i} 1_{(t_i, t_{i+1}]}(s); \quad 0 \leq s \leq t$$

in $\mathscr{L}_t^*(W)$, for which

$$E \int_0^t |X_s^\Pi - W_s|^2 \, ds = \sum_{i=0}^{m-1} \int_{t_i}^{t_{i+1}} E|W_{t_i} - W_t|^2 \, dt$$
$$= \sum_{i=0}^{m-1} \int_{t_i}^{t_{i+1}} (t - t_i) \, dt = \frac{1}{2} \sum_{i=0}^{m-1} (t_{i+1} - t_i)^2 \to 0$$

as $\|\Pi\| \to 0$. By definition, the Itô integral $\int_0^t W_s \, dW_s$ is the L^2-limit of $S_\varepsilon(\Pi) = \int_0^t X_s^\Pi \, dW_s$.

3.7. The proof is much like that of Theorem 3.3. The Taylor expansion in Step 2 of that proof is replaced by

$$f(t_k, X_{t_k}) - f(t_{k-1}, X_{t_{k-1}})$$
$$= [f(t_k, X_{t_k}) - f(t_{k-1}, X_{t_k})] + [f(t_{k-1}, X_{t_k}) - f(t_{k-1}, X_{t_{k-1}})]$$
$$= \frac{\partial}{\partial t} f(\tau_k, X_{t_k})(t_k - t_{k-1}) + \sum_{i=1}^{d} \frac{\partial}{\partial x_i} f(t_{k-1}, X_{t_{k-1}})(X_{t_k}^{(i)} - X_{t_{k-1}}^{(i)})$$
$$+ \frac{1}{2} \sum_{i=1}^{d} \sum_{j=1}^{d} \frac{\partial^2}{\partial x_i \partial x_j} f(t_{k-1}, \eta_k)(X_{t_k}^{(i)} - X_{t_{k-1}}^{(i)})(X_{t_k}^{(j)} - X_{t_{k-1}}^{(j)}),$$

where $t_{k-1} \leq \tau_k \leq t_k$ and η_k is as before.

3.14. Let $Y_t^{(i)} = \partial f(X_t)/\partial x_i$, so according to Itô's rule (Theorem 3.6),

$$Y_t^{(i)} = Y_0^{(i)} + \sum_{j=1}^{d} \int_0^t \frac{\partial^2}{\partial x_i \partial x_j} f(X_s) \, dM_s^{(j)} + \sum_{j=1}^{d} \int_0^t \frac{\partial^2}{\partial x_i \partial x_j} f(X_s) \, dB_s^{(j)}$$
$$+ \frac{1}{2} \sum_{k=1}^{d} \sum_{j=1}^{d} \int_0^t \frac{\partial^3}{\partial x_i \partial x_j \partial x_k} f(X_s) d\langle M^{(j)}, M^{(k)} \rangle_s.$$

It follows from (3.9) that

$$\int_0^t Y_s^{(i)} \circ dX_s^{(i)} = \int_0^t Y_s^{(i)} \, dX_s^{(i)} + \frac{1}{2} \sum_{j=1}^{d} \int_0^t \frac{\partial^2}{\partial x_i \partial x_j} f(X_s) \, d\langle M^{(i)}, M^{(j)} \rangle_s,$$

and now (3.10) reduces to the Itô rule applied to $f(X_t)$.

3.15. Let X and Y have the decomposition (3.7). The sum in question is

$$\sum_{i=0}^{m-1} Y_{t_i}(X_{t_{i+1}} - X_{t_i}) + \frac{1}{2} \sum_{i=0}^{m-1} (Y_{t_{i+1}} - Y_{t_i})(X_{t_{i+1}} - X_{t_i}).$$

The first term is $\int_0^t Y_s^\Pi \, dM_s + \int_0^t Y_s^\Pi \, dB_s$, where

$$Y_s^\Pi \triangleq \sum_{i=0}^{m-1} Y_{t_i} 1_{(t_i, t_{i+1}]}(s); \quad 0 \le s \le t,$$

and the continuity of Y implies the convergence in probability of $\int_0^t (Y_s^\Pi - Y_s)^2 \, d\langle M \rangle_s$ and $\int_0^t [Y_s^\Pi - Y_s] \, dB_s$ to zero. It follows from Proposition 2.26 that

$$\sum_{i=0}^{m-1} Y_{t_i}(X_{t_{i+1}} - X_{t_i}) \xrightarrow[\|\Pi\| \to 0]{P} \int_0^t Y_s \, dX_s.$$

The other term is

$$\frac{1}{2} \sum_{i=0}^{m-1} (M_{t_{i+1}} - M_{t_i})(N_{t_{i+1}} - N_{t_i}) + \frac{1}{2} \sum_{i=0}^{m-1} (M_{t_{i+1}} - M_{t_i})(C_{t_{i+1}} - C_{t_i})$$
$$+ \frac{1}{2} \sum_{i=0}^{m-1} (N_{t_{i+1}} - N_{t_i})(B_{t_{i+1}} - B_{t_i}) + \frac{1}{2} \sum_{i=0}^{m-1} (B_{t_{i+1}} - B_{t_i})(C_{t_{i+1}} - C_{t_i}),$$

which converges in probability to $\frac{1}{2}\langle M, N \rangle_t$ because of Problem 1.5.14 and the bounded variation of B and C on $[0, t]$.

3.29. For $x_i \ge 0$, $i = 1, \ldots, d$, we have

$$x_1^m + \cdots + x_d^m \le d(x_1 + \cdots + x_d)^m \le d^{m+1}(x_1^m + \cdots + x_d^m).$$

Therefore

$$\|M_t\|^{2m} = \left[\sum_{i=1}^d (M_t^{(i)})^2 \right]^m \le d^m \sum_{i=1}^d |M_t^{(i)}|^{2m} \tag{8.2}$$

and

$$\sum_{i=1}^d \langle M^{(i)} \rangle_T^m \le d \left(\sum_{i=1}^d \langle M^{(i)} \rangle_T \right)^m = dA_T^m. \tag{8.3}$$

Taking maxima in (8.2), expectations in the resulting inequality and in (8.3), and applying the right-hand side of (3.37) to each $M^{(i)}$, we obtain

$$E(\|M\|_T^*)^{2m} \le d^m \sum_{i=1}^d E[(M^{(i)})_T^*]^{2m} \le d^m \sum_{i=1}^d K_m E[\langle M^{(i)} \rangle_T^m] \le K_m d^{m+1} E(A_T^m).$$

A similar proof can be given for the lower bound on $E(\|M\|_T^*)^{2m}$.

4.5. (i) The nondecreasing character of T is obvious. Thus, for right-continuity, we need only show that $\lim_{\theta \downarrow s} T(\theta) \le T(s)$, for $0 \le s < S$. Set $t = T(s)$. The definition of $T(s)$ implies that for each $\varepsilon > 0$, we have $A(t + \varepsilon) > s$, and for $s < \theta < A(t + \varepsilon)$, we have $T(\theta) \le t + \varepsilon$. Therefore, $\lim_{\theta \downarrow s} T(\theta) \le t$.

(ii) The identity is trivial for $s \ge S$; if $s < S$, set $t = T(s)$ and choose $\varepsilon > 0$. We have $A(t + \varepsilon) > s$, and letting $\varepsilon \downarrow 0$, we see from the continuity of A that $A(T(s)) \ge s$. If $t = T(s) = 0$, we are done. If $t > 0$, then for $0 < \varepsilon < t$, the definition of $T(s)$ implies $A(t - \varepsilon) \le s$. Letting $\varepsilon \downarrow 0$, we obtain $A(T(s)) \le s$.

(iii) This follows immediately from the definition of $T(\cdot)$.

(iv) Since, by (i), T is right-continuous, so is $\varphi(T(\cdot))$. To show the left-continuity, take any $s \in [0, S)$ and any increasing sequence, $\{s_n\}$, such that $s_n \to s$. Since T is nondecreasing, $\{T(s_n)\}$ is a nondecreasing sequence of real numbers bounded from above by $T(s)$. Therefore $\lim_{n \to \infty} T(s_n)$ exists.

Now, we claim that $\varphi(\lim_{n\to\infty} T(s_n)) = \varphi(T(s))$. To see this, note that, by continuity of A and (ii), we have $A(\lim_{n\to\infty} T(s_n)) = \lim_{n\to\infty} A(T(s_n)) = \lim_{n\to\infty} s_n = s$. This, together with propery (3), proves our claim. Finally, by continuity of φ, it follows that $\lim_{n\to\infty} \varphi(T(s_n)) = \varphi(\lim_{n\to\infty} T(s_n)) = \varphi(T(s))$. Hence, $\varphi(T(\cdot))$ is continuous.

Finally, to prove (4), note that, by (ii), we have $A(T(A(t))) = A(t) \wedge S = A(t)$. Now (4) follows from property (3) of φ.

(v) This is a direct consequence of the definition of T and the continuity of A.

(vi) For $a \le t_1 < t_2 \le b$, let $G(t) = 1_{[t_1,t_2)}(t)$. According to (v), $t_1 \le T(s) < t_2$ if and only if $A(t_1) \le s < A(t_2)$, so

$$\int_a^b G(t)\,dA(t) = A(t_2) - A(t_1) = \int_{A(a)}^{A(b)} G(T(s))\,ds.$$

The linearity of the integral and the monotone convergence theorem imply that the collection of sets $C \in \mathscr{B}([a,b])$ for which

$$\int_a^b 1_C(t)\,dA(t) = \int_{A(a)}^{A(b)} 1_C(T(s))\,ds \tag{8.4}$$

forms a Dynkin system. Since it contains all intervals of the form $[t_1, t_2) \subset [a,b]$, and these are closed under finite intersection and generate $\mathscr{B}([a,b])$, we have (8.4) for every $C \in \mathscr{B}([a,b])$ (Dynkin System Theorem 2.1.3). The proof of (vi) is now straightforward.

4.7. Again as before, every $\langle M\rangle_t$ (resp., $T(s)$) is a stopping time of $\{\mathscr{G}_s\}$ (resp., $\{\mathscr{F}_t\}$), and the same is true of $S \triangleq \lim_{t\to\infty} \langle M\rangle_t$ (Lemma 1.2.11). The local martingale $\tilde{M}$ has quadratic variation $\langle \tilde{M}\rangle_t \le \langle M\rangle_{T(s_2)} = S \wedge s_2 \le s_2 < \infty$ (Problem 4.5 (ii)), so again both $\tilde{M}$, $\tilde{M}^2 - \langle \tilde{M}\rangle$ are uniformly integrable martingales, and by optional sampling:

$$E[\tilde{M}_{T(s_2)} - \tilde{M}_{T(s_1)} | \mathscr{F}_{T(s_1)}] = 0,$$

$$E[(\tilde{M}_{T(s_2)} - \tilde{M}_{T(s_1)})^2 | \mathscr{F}_{T(s_1)}] = E[\langle M\rangle_{T(s_2)} - \langle M\rangle_{T(s_1)} | \mathscr{F}_{T(s_1)}]; \quad \text{a.s. } P.$$

It follows that $M \circ T \triangleq \{M_{T(s)}, \mathscr{G}_s; 0 \le s < \infty\}$ is a martingale with $\langle M \circ T\rangle_s = \langle M\rangle_{T(s)}$, and by Problem 4.5 (iv), $M \circ T$ has continuous paths. Now if $\{W_s, \mathscr{G}_s; 0 \le s < \infty\}$ is an independent Brownian motion, the process

$$B_s \triangleq W_s - W_{s\wedge S} + M_{T(s)}, \mathscr{G}_s; \quad 0 \le s < \infty$$

is a continuous martingale with quadratic variation $\langle B\rangle_s = s - (s \wedge S) + \langle M\rangle_{T(s)} = s$, i.e., a Brownian motion. For this process, (4.17) is established by using Problem 4.5 (iv).

4.11. Let φ be a deterministic, strictly increasing function mapping $[0,\infty)$ onto $[0,1)$, and define $M \in \mathscr{M}^{c,\mathrm{loc}}$ by

$$M_t \triangleq \int_0^{\varphi(t)} X_s\,dW_s, \quad \text{so} \quad \langle M\rangle_t = \int_0^{\varphi(t)} X_s^2\,ds; \quad 0 \le t < \infty,$$

and $\lim_{t\to\infty} \langle M\rangle_t = \infty$, a.s. on E. According to Problem 4.7, there is a Brownian motion B such that

$$\int_0^t X_s\,dW_s = B_{p(t)}, \quad \text{where} \quad p(t) \triangleq \langle M\rangle_{\varphi^{-1}(t)}.$$

We have $\lim_{t\uparrow 1} p(t) = \infty$ a.s. on E, and so, by the law of the iterated logarithm for Brownian motion,

$$\overline{\lim_{t\uparrow 1}}\, B_{p(t)} = -\underline{\lim_{t\uparrow 1}}\, B_{p(t)} = +\infty, \quad \text{a.s. on } E.$$

4.12. (Adapted from Watanabe (1984).) From Problem 4.7 we have the representation (4.17) for a suitable Brownian motion B. Taking $n > 3\max(|x|, \rho T)$ and denoting $R_n = \inf\{t \geq 0; |B_t| \geq n/3\}$, we have the inclusions

$$\left\{\max_{0\leq t\leq T} |X_t| \geq n\right\} \subseteq \left\{\max_{0\leq t\leq T} |M_t| \geq \frac{n}{3}\right\} = \{\langle M\rangle_T \geq R_n\} \subseteq \{\rho T \geq R_n\},$$

which lead, via (2.6.1), (2.6.2), and (2.9.20), to

$$P\left[\max_{0\leq t\leq T} |X_t| \geq n\right] \leq P[R_n \leq \rho T] \leq 2P[T_{n/3} \leq \rho T] = 4P\left[B_{\rho T} \geq \frac{n}{3}\right]$$

$$\leq \frac{4}{\sqrt{2\pi}} \int_{n/3\sqrt{\rho T}}^{\infty} e^{-z^2/2}\,dz \leq \frac{12}{n}\sqrt{\frac{\rho T}{2\pi}} \exp\left\{-\frac{n^2}{18\rho T}\right\}$$

The conclusion follows.

6.12. Let h have support in $[0, b]$, and consider the sequence of partitions

$$D_n = \left\{b_k^{(n)} = \frac{k}{2^n} b;\ k = 0, 1, \ldots, 2^n\right\}; \quad n \geq 1$$

of this interval. Choosing a modification of $\int_0^t 1_{(a,\infty)}(W_s)\,dW_s$ which is continuous in a (cf. (6.21)), we see that the Lebesgue (and Riemann) integral on the left-hand side of (6.24) is approximated by the sum

$$\sum_{k=0}^{2^n-1} \frac{b}{2^n} h(b_k^{(n)}) \left(\int_0^t 1_{(b_k^{(n)},\infty)}(W_s)\,dW_s\right) = \int_0^t F_n(W_s)\,dW_s,$$

where the uniformly bounded sequence of functions

$$F_n(x) \triangleq \sum_{k=0}^{2^n-1} \frac{b}{2^n} h(b_k^{(n)}) 1_{(b_k^{(n)},\infty)}(x); \quad n \geq 1$$

converges uniformly, as $n \to \infty$, to the Lebesgue (and Riemann) integral

$$F(x) \triangleq \int_{-\infty}^{\infty} h(a) 1_{(a,\infty)}(x)\,da.$$

Therefore, the sequence of stochastic integrals $\{\int_0^t F_n(W_s)\,dW_s\}_{n=1}^{\infty}$ converges in L^2 to the stochastic integral $\int_0^t F(W_s)\,dW_s$, which is the right-hand side of (6.24).

6.13. (i) Under any P^z, $B(a)$ is a continuous, square-integrable martingale with quadratic variation process

$$\langle B(a)\rangle_t = \int_0^t [\operatorname{sgn}(W_s - a)]^2\,ds = t; \quad 0 \leq t < \infty, \text{ a.s. } P^z.$$

According to Theorem 3.16, $B(a)$ is a Brownian motion.

(ii) For ω in the set Ω^* of Definition 6.3, we have (6.2) (Remark 6.5), and from this we see immediately that $L_0(a, \omega) = 0$ and $L_t(a, \omega)$ is nondecreasing in t.

For each $z \in \mathbb{R}$, there is a set $\tilde{\Omega} \in \mathscr{F}$ with $P^z(\tilde{\Omega}) = 1$ such that $\mathscr{Z}_\omega(a)$ is closed for all $\omega \in \tilde{\Omega}$. For $\omega \in \tilde{\Omega} \cap \Omega^*$, the complement of $\mathscr{Z}_\omega(a)$ is the countable union of open intervals $\bigcup_{\alpha \in \mathbb{N}} I_\alpha$. To prove (6.26), it suffices to show that $\int_{I_\alpha} dL_t(\omega) = 0$ for each $\alpha \in \mathbb{N}$. Fix an index α and let $I_\alpha = (u, v)$. Since $W_.(\omega) - a$ has no zero in (u, v), we know that $|W_.(\omega) - a|$ is bounded away from the origin on $[u + (1/n), v - (1/n)]$, where $n > 2/(v - u)$. Thus, for all sufficiently small $\varepsilon > 0$,

$$\text{meas}\left\{0 \le s \le u + \frac{1}{n}; |W_s - a| \le \varepsilon\right\} = \text{meas}\left\{0 \le s \le v - \frac{1}{n}; |W_s - a| \le \varepsilon\right\},$$

whence $L_{u+(1/n)}(a, \omega) = L_{v-(1/n)}(a, \omega)$. It follows that $\int_{[u+(1/n), v-(1/n)]} dL_t(a, \omega) = 0$, and letting $n \to \infty$ we obtain the desired result.

(iii) Set $z = a = 0$ in (6.25) to obtain $|W_t| = -B_t(0) + 2L_t(0)$; $0 \le t < \infty$, a.s. P^0. The left-hand side of this relation is nonnegative; $B_t(0)$ changes sign infinitely often in any interval $[0, \varepsilon]$, $\varepsilon > 0$ (Problem 2.7.18). It follows that $L_t(0)$ cannot remain zero in any such interval.

(iv) It suffices to show that for any two rational numbers $0 \le q < r < \infty$, if $W_t(\omega) = a$ for some $t \in (q, r)$ then $L_q(a, \omega) < L_r(a, \omega)$, P^z-a.e. ω. Let $T(\omega) \triangleq \inf\{t \ge q; W_t(\omega) = a\}$. Applying (iii) to the Brownian motion $\{W_{s+T} - a; 0 \le s < \infty\}$ we conclude that

$$L_{T(\omega)}(a, \omega) < L_{T(\omega)+s}(a, \omega) \quad \text{for all } s > 0,\ P^z\text{-a.e. } \omega,$$

by the additive functional property of local time (Definition 6.1 and Remark 6.5). For every $\omega \in \{T < r\}$ we may take $s = r - T(\omega)$ above, and this yields $L_q(a, \omega) < L_r(a, \omega)$.

6.19. From (6.38) we obtain $\overline{\lim}_{y \downarrow x} f(y) \le f(x)$, $\overline{\lim}_{y \uparrow z} f(y) \le f(z)$ and $f(y) \le \underline{\lim}_{x \uparrow y} f(x)$, $f(y) \le \underline{\lim}_{z \downarrow y} f(z)$. This establishes the continuity of f on $\mathbb{R}$.

For $\xi \in \mathbb{R}$ fixed and $0 < h_1 < h_2$, we have from (6.38), with $x = \xi$, $y = \xi + h_1$, $z = \xi + h_2$:

$$\Delta f(\xi; h_1) \le \Delta f(\xi; h_2). \tag{8.5}$$

On the other hand, applying (6.38) with $x = \xi - h_2$, $y = \xi - h_1$, and $z = \xi$ yields

$$\Delta f(\xi; -h_2) \le \Delta f(\xi; -h_1). \tag{8.6}$$

Finally, with $x = \xi - \varepsilon$, $y = \xi$, $z = \xi + \delta$, we have

$$\Delta f(\xi; -\varepsilon) \le \Delta f(\xi; \delta); \quad \varepsilon, \delta > 0. \tag{8.7}$$

Relations (8.5)–(8.7) establish the requisite monotonicity in h of the difference quotient (6.39), and hence the existence and finiteness of the limits in (6.40).

In particular, (8.7) gives $D^- f(x) \le D^+ f(x)$ upon letting $\varepsilon \downarrow 0$, $\delta \downarrow 0$, which establishes the second inequality in (6.41). On the other hand, we obtain easily from (8.5) and (8.6) the bounds

$$(y - x)\, D^+ f(x) \le f(y) - f(x) \le (y - x)\, D^- f(y); \quad x < y, \tag{8.8}$$

which establish (6.41).

For the right-continuity of the function $D^+ f(\cdot)$, we begin by observing the inequality $D^+ f(x) \le \lim_{y \downarrow x} D^+ f(y)$; $x \in \mathbb{R}$, which is a consequence of (6.41). In the opposite direction, we employ the continuity of f, as well as (8.8), to obtain for $x < z$:

$$\frac{f(z) - f(x)}{z - x} = \lim_{y \downarrow x} \frac{f(z) - f(y)}{z - y} \geq \lim_{y \downarrow x} D^+ f(y).$$

Upon letting $z \downarrow x$, we obtain $D^+f(x) \geq \lim_{y \downarrow x} D^+f(y)$. The left-continuity of $D^-f(\cdot)$ is proved similarly.

From (8.8) we observe that, for any function $\varphi\colon \mathbb{R} \to \mathbb{R}$ satisfying

(8.9) $$D^-f(x) \leq \varphi(x) \leq D^+f(x); \quad x \in \mathbb{R},$$

we have for fixed $y \in \mathbb{R}$,

(8.10) $$f(x) \geq G_y(x) \triangleq f(y) + (x - y)\varphi(y); \quad x \in \mathbb{R}.$$

The function $G_y(\cdot)$ is called a *line of support* for the convex function $f(\cdot)$. It is immediate from (8.10) that $f(x) = \sup_{y \in \mathbb{R}} G_y(x)$; the point of (6.42) is that $f(\cdot)$ can be expressed as the supremum of *countably many* lines of support. Indeed, let E be a countable, dense subset of $\mathbb{R}$. For any $x \in \mathbb{R}$, take a sequence $\{y_n\}_{n=1}^\infty$ of numbers in E, converging to x. Because this sequence is bounded, so are the sequences $\{D^\pm f(y_n)\}_{n=1}^\infty$ (by monotonicity and finiteness of the functions $D^\pm f(\cdot)$) and $\{\varphi(y_n)\}_{n=1}^\infty$ (by (8.9)). Therefore, $\lim_{n\to\infty} G_{y_n}(x) = f(x)$, which implies that $f(x) = \sup_{y \in E} G_y(x)$.

6.20. (iii) For any $x < y < z$, we have

(8.11) $$\varphi(x) \leq \frac{\Phi(y) - \Phi(x)}{y - x} = \frac{1}{y - x}\int_x^y \varphi(u)\,du \leq \varphi(y)$$

$$\leq \frac{1}{z - y}\int_y^z \varphi(u)\,du = \frac{\Phi(z) - \Phi(y)}{z - y} \leq \varphi(z).$$

This gives

$$\Phi(y) \leq \frac{z - y}{z - x}\Phi(x) + \frac{y - x}{z - x}\Phi(z),$$

which verifies convexity in the form (6.38). Now let $x \uparrow y$, $z \downarrow y$ in (8.11), to obtain

$$\varphi_-(y) \leq D^-\Phi(y) \leq \varphi(y) \leq D^+\Phi(y) \leq \varphi_+(y); \quad y \in \mathbb{R}.$$

At every continuity point x of φ, we have $\varphi_\pm(x) = \varphi(x) = D^\pm\Phi(x)$. The left- (respectively, right-) continuity of φ_- and $D^-\varphi$ (respectively, φ_+ and $D^+\Phi$) implies $\varphi_-(y) = D^-\Phi(y)$ (respectively, $\varphi_+(y) = D^+\Phi(y)$) for all $y \in \mathbb{R}$.

(iv) Letting $x \downarrow y$ (respectively, $x \uparrow y$) in (6.44), we obtain

$$D^-f(y) \leq \varphi_-(y) \leq \varphi(y) \leq \varphi_+(y) \leq D^+f(y); \quad y \in \mathbb{R}.$$

But now from (6.41) one gets

$$\varphi_+(x) \leq D^+f(x) \leq D^-f(y) \leq \varphi_-(y) \leq \varphi(y); \quad x < y,$$

and letting $y \downarrow x$ we conclude $\varphi_+(x) = D^+f(x)$; $x \in \mathbb{R}$. Similarly, we conclude $\varphi_-(x) = D^-f(x)$; $x \in \mathbb{R}$. Now consider the function $G \triangleq f - \Phi$, and simply notice the consequences $D^\pm G(x) = D^\pm f(x) - D^\pm\Phi(x) = 0$; $x \in \mathbb{R}$, of the preceding discussion; in other words, G is differentiable on $\mathbb{R}$ with derivative which is identically zero. It follows that G is identically constant.

6.29. *(iv)* ⇒ *(vi)*: Let $t \in (0, \infty)$ be such that $P^0[\int_0^t f(W_s)\,ds < \infty] = 1$. For $x = 0$, just take $S = t$. For $x \neq 0$, consider the first passage time T_x and notice that $P^0[0 < T_x < \infty] = 1$, $P^0[2T_x \le t] > 0$, and that $\{B_s \triangleq W_{s+T_x} - x, 0 \le s < \infty\}$ is a Brownian motion under P^0. Now, for every $\omega \in \{2T_x \le t\}$:

$$\int_0^{T_x(\omega)} f(x + B_s(\omega))\,ds = \int_{T_x(\omega)}^{2T_x(\omega)} f(W_u(\omega))\,du \le \int_0^t f(W_u(\omega))\,du < \infty,$$

whence $\{2T_x \le t\} \subseteq \{\int_0^{T_x} f(x + B_s)\,ds < \infty\}$, a.s. P^0. We conclude that this latter event has positive probability under P^0, and (vi) follows upon taking $S = T_x$.

(vi) ⇒ *(iii)*: Lemma 6.26 gives, for each $x \in K$, the existence of an open neighborhood $U(x)$ of x with $\int_{U(x)} f(y)\,dy < \infty$. Now (iii) follows from the compactness of K.

(iii) ⇒ *(v)*: For fixed $x \in \mathbb{R}$, define $g_x(y) = f(x + y)$ and apply the known implication (iii) ⇒ (ii) to the function g_x.

7.3. We may write f as the difference of the convex functions

$$f_1(x) \triangleq f(0) + xf'(0) + \int_0^x \int_0^y [f''(z)]^+\,dz\,dy, \quad f_2(x) \triangleq \int_0^x \int_0^y [f''(z)]^-\,dz\,dy,$$

and apply (7.4). In this case, $\mu(dx) = f''(x)\,dx$, and (7.3) shows that $\int_{-\infty}^{\infty} \Lambda_t(a)\mu(da) = \frac{1}{2}\int_0^t f''(X_s)\,d\langle M\rangle_s$.

7.6. Let $\check{V}_t(\omega)$ denote the total variation of $V_.(\omega)$ on $[0, t]$. For P-a.e. $\omega \in \Omega$, we have $\check{V}_t(\omega) < \infty$; $0 \le t < \infty$. Consequently, for $a < b$,

$$|J_t(a) - J_\tau(b)| \le |J_t(a) - J_\tau(a)| + |J_\tau(a) - J_\tau(b)| \le |\check{V}_t - \check{V}_\tau| + \int_0^\tau 1_{(a,b]}(X_s)\,d\check{V}_s,$$

and these last expressions converge to zero a.s. as $\tau \to t$ and $b \downarrow a$. Furthermore, the exceptional set of $\omega \in \Omega$ for which convergence fails does not depend on t or a. Relation (7.20) is proved similarly.

7.7. The solution is a slight modification of Solution 6.12, where now we use Lemma 7.5 to establish the continuity in a of the integrand on the left-hand side.

3.9. Notes

Section 3.2: The concept of the stochastic integral with respect to Brownian motion was introduced by Paley, Wiener & Zygmund (1933) for nonrandom integrands, and by K. Itô (1942a, 1944) in the generality of the present section. Itô's motivation was to achieve a rigorous treatment of the stochastic differential equation which governs the diffusion processes of A. N. Kolmogorov

(1931). Doob (1953) was the first to study the stochastic integral as a martingale, and to suggest a unified treatment of stochastic integration as a chapter of martingale theory. This task was accomplished by Courrège (1962/1963), Fisk (1963), Kunita & Watanabe (1967), Meyer (1967), Millar (1968), Doléans-Dade & Meyer (1970). Much of this theory has become standard and has received monograph treatment; we mention in this respect the books by McKean (1969), Gihman & Skorohod (1972), Arnold (1973), Friedman (1975), Liptser & Shiryaev (1977), Stroock & Varadhan (1979), and Ikeda & Watanabe (1981) and the monographs by Skorohod (1965) and Chung & Williams (1983). Our presentation draws on most of these sources, but is closest in spirit to Ikeda & Watanabe (1981) and Liptser & Shiryaev (1977). The approach suggested by Lemma 2.4 and Problem 2.5 is due to Doob (1953). A major recent development has been the extension of this theory by the "French school" to include integration of left-continuous, or more generally, "predictable," processes with respect to discontinuous martingales. The fundamental reference for this material is Meyer (1976), supplemented by Dellacherie & Meyer (1975/1980); other accounts can be found in Métivier & Pellaumail (1980), Métivier (1982), Kopp (1984), Kussmaul (1977), and Elliott (1982).

Section 3.3: Theorem 3.16 was discovered by P. Lévy (1948: p. 78); a different proof appears on p. 384 of Doob (1953). Theorem 3.28 extends the Burkholder-Davis-Gundy inequalities of discrete-parameter martingale theory; see the excellent expository article by Burkholder (1973). The approach that we follow was suggested by M. Yor (personal communication). For more information on the approximations of stochastic integrals as in Problem 3.15, see Yor (1977).

Section 3.4: The idea of extending the probability space in order to accommodate the Brownian motion W in the representation of Theorem 4.2 is due to Doob (1953; pp. 449–451) for the case $d = 1$. Problem 4.11 is essentially from McKean (1969; p. 31). Chapters II of Ikeda & Watanabe (1981) and XII of Elliott (1982) are good sources for further reading on the subject matter of Sections 3.3 and 3.4. For a different proof and further extensions of the F. B. Knight theorem, see Cocozza & Yor (1980) and Pitman & Yor (1986) (Theorems B.2, B.4), respectively. Our solution of Problem 4.16 is taken from Rogers & Williams (1987).

Section 3.5: The celebrated Theorem 5.1 was proved by Cameron & Martin (1944) for nonrandom integrands X, and by Girsanov (1960) in the present generality. Our treatment was inspired by the lecture notes of S. Orey (1974). Girsanov's work was presaged by that of Maruyama (1954), (1955). Kazamaki (1977) (see also Kazamaki & Sekiguchi (1979)) provides a condition different from the Novikov condition (5.18): if $\exp(\frac{1}{2}M_t)$ is a submartingale, then $Z_t = \exp(M_t - \frac{1}{2}\langle M\rangle_t)$ is a martingale. The same is true if $E[\exp(\frac{1}{2}M_t)] < \infty$ (Kazamaki (1978)). Proposition 5.4 is due to Van Schuppen & Wong (1974).

Section 3.6: Brownian local time is the creation of P. Lévy (1948), although the first rigorous proof of its existence was given by Trotter (1958). Our approach to Theorem 6.11 follows that of Ikeda & Watanabe (1981) and

McKean (1969). One can study the local time of a nonrandom function divorced from probability theory, and the general pattern that develops is that *regular local times correspond to irregular functions*; for instance, for the highly irregular Brownian paths we obtained Hölder-continuous local times (relation (6.22)). See Geman & Horowitz (1980) for more information on this topic. On the other hand, Yor (1986) shows directly that the occupation time $B \mapsto \Gamma_t(B, \omega)$ of (6.6) has a density.

The Skorohod problem of Lemma 6.14, for RCLL trajectories y, was treated by Chaleyat-Maurel, El Karoui & Marchal (1980).

The generalized Itô rule (Theorem 6.22) is due to Meyer (1976) and Wang (1977). There is a converse to Corollary 6.23: if $f(W_t)$ is a continuous semimartingale, then f is the difference of convex functions (Wang (1977), Çinlar, Jacod, Protter & Sharpe (1980)). A multidimensional version of Theorem 6.22, in which convex functions are replaced by potentials, has been proved by Brosamler (1970).

Tanaka's formula (6.11) provides a representation of the form $f(W_t) - f(W_0) + \int_0^t g(W_s)\,dW_s$ for the continuous additive functional $L_t(a)$, with $a \in \mathbb{R}$ fixed. In fact, any continuous additive functional has such a representation, where f may be chosen to be continuous; see Ventsel (1962), Tanaka (1963).

We follow Ikeda & Watanabe (1981) in our exposition of Theorem 6.17. For more information on the subject matter of Problem 6.30, the reader is referred to Papanicolaou, Stroock & Varadhan (1977).

Section 3.7: Local time for semimartingales is discussed in the volume edited by Azéma & Yor (1978); see in particular the articles by Azéma & Yor (pp. 3–16) and Yor (pp. 23–36). Local time for Markov processes is treated by Blumenthal & Getoor (1968). Yor (1978) proved that local time $\Lambda_t(a)$ for a continuous semimartingale is jointly continuous in t and RCLL in a. His proof assumes the existence of local time, whereas ours is a step in the proof of existence.

Exercise 7.13 comes from Azéma & Yor (1979); see also Jeulin & Yor (1980) for applications of these martingales in the study of distributions of random variables associated with local time. Exercise 7.14 is taken from Yor (1979).

CHAPTER 4

Brownian Motion and Partial Differential Equations

4.1. Introduction

There is a rich interplay between probability theory and analysis, the study of which goes back at least to Kolmogorov (1931). It is not possible in a few sections to develop this subject systematically; we instead confine our attention to a few illustrative cases of this interplay. Recent monographs on this subject are those of Doob (1984) and Durrett (1984).

The solutions to many problems of elliptic and parabolic partial differential equations can be represented as expectations of stochastic functionals. Such representations allow one to infer properties of these solutions and, conversely, to determine the distributions of various functionals of stochastic processes by solving related partial differential equation problems.

In the next section, we treat the Dirichlet problem of finding a function which is harmonic in a given region and assumes specified boundary values. One can use Brownian motion to characterize those Dirichlet problems for which a solution exists, to construct a solution, and to prove uniqueness. We shall also derive Poisson integral formulas and see how they are related to exit distributions for Brownian motion.

The Laplacian appearing in the Dirichlet problem is the simplest elliptic operator; the simplest parabolic operator is that appearing in the heat equation. Section 3 is devoted to a study of the connections between Brownian motion and the one-dimensional heat equation, and, again, we give probabilistic proofs of existence and uniqueness theorems and probabilistic interpretations of solutions. Exploiting the connections in the opposite direction, we show how solutions to the heat equation enable us to compute boundary crossing probabilities for Brownian motion.

Section 4 takes up the study of more complicated elliptic and parabolic equations based on the Laplacian. Here we develop formulas necessary for the treatment of Brownian functionals which are more complex than those appearing in Section 2.8.

The connections established in this chapter between Brownian motion and elliptic and parabolic differential equations based on the Laplacian are a foreshadowing of a more general relationship between diffusion processes and second-order elliptic and parabolic differential equations. A good deal of the more general theory appears in Section 5.7, but it is never so elegant and surprisingly powerful as in the simple cases of the Laplace and heat equations developed here. In particular, in the more general setting, one must rely on existence theorems from the theory of partial differential equations, whereas in this chapter we can give probabilistic proofs of the existence of solutions to the relevant partial differential equations.

4.2. Harmonic Functions and the Dirichlet Problem

The connection between Brownian motion and harmonic functions is profound, yet simply explained. For this reason, we take this connection as our first illustration of the interplay between probability theory and analysis.

Recall that a function u mapping an open subset D of $\mathbb{R}^d$ into $\mathbb{R}$ is called *harmonic in D* if u is of class C^2 and $\Delta u \triangleq \sum_{i=1}^d (\partial^2 u/\partial x_i^2) = 0$ in D. As we shall prove shortly, a harmonic function is necessarily of class C^∞ and has the mean-value property. It is this mean-value property which introduces Brownian motion in a natural way into the study of harmonic functions.

Throughout this section, $\{W_t, \mathscr{F}_t; 0 \le t < \infty\}$, $(\Omega, \mathscr{F})$, $\{P^x\}_{x \in \mathbb{R}^d}$ is a d-dimensional Brownian family and $\{\mathscr{F}_t\}$ satisfies the usual conditions. We denote by D an open set in $\mathbb{R}^d$ and introduce the stopping time (Problem 1.2.7)

$$\tau_D \triangleq \inf\{t \ge 0; W_t \in D^c\}, \tag{2.1}$$

the time of first exit from D. The boundary of D will be denoted by ∂D, and $\bar{D} = D \cup \partial D$ is the closure of D. Recall (Theorem 2.9.23) that each component of W is almost surely unbounded, so

$$P^x[\tau_D < \infty] = 1; \quad \forall x \in D \subset \mathbb{R}^d, D \text{ bounded}. \tag{2.2}$$

Let $B_r \triangleq \{x \in \mathbb{R}^d; \|x\| < r\}$ be the open ball of radius r centered at the origin. The volume of this ball is

$$V_r \triangleq \frac{2r^d \pi^{d/2}}{d\,\Gamma\left(\frac{d}{2}\right)}, \tag{2.3}$$

and its surface area is

$$S_r \triangleq \frac{2r^{d-1}\pi^{d/2}}{\Gamma(d/2)} = \frac{d}{r} V_r. \tag{2.4}$$

We define a probability measure μ_r on ∂B_r by

$$\mu_r(dx) = P^0[W_{\tau_{B_r}} \in dx]; \quad r > 0. \tag{2.5}$$

A. The Mean-Value Property

Because of the rotational invariance of Brownian motion (Problem 3.3.18), the measure μ_r is also rotationally invariant and thus proportional to surface measure on ∂B_r. In particular, the Lebesgue integral of a function f over B_r can be written in iterated form as

$$\int_{B_r} f(x)\,dx = \int_0^r S_\rho \int_{\partial B_\rho} f(x)\mu_\rho(dx)\,d\rho. \tag{2.6}$$

2.1 Definition. We say that the function $u: D \to \mathbb{R}$ has the *mean-value property* if, for every $a \in D$ and $0 < r < \infty$ such that $a + \bar{B}_r \subset D$, we have

$$u(a) = \int_{\partial B_r} u(a + x)\mu_r(dx).$$

With the help of (2.6) one can derive the consequence

$$u(a) = \frac{1}{V_r}\int_{B_r} u(a + x)\,dx$$

of the mean-value property, which asserts that the mean integral value of u over a ball is equal to the value at the center. Using the divergence theorem one can prove analytically (cf. Gilbarg & Trudinger (1977), p. 14) that a harmonic function possesses the mean-value property. A very simple probabilistic proof can be based on Itô's rule.

2.2 Proposition. *If u is harmonic in D, then it has the mean-value property there.*

PROOF. With $a \in D$ and $0 < r < \infty$ such that $a + \bar{B}_r \subset D$, we have from Itô's rule

$$u(W_{t \wedge \tau_{a+B_r}}) = u(W_0) + \sum_{i=1}^d \int_0^{t \wedge \tau_{a+B_r}} \frac{\partial u}{\partial x_i}(W_s)\,dW_s^{(i)}$$
$$+ \frac{1}{2}\int_0^{t \wedge \tau_{a+B_r}} \Delta u(W_s)\,ds; \quad 0 \le t < \infty.$$

Because u is harmonic, the last (Lebesgue) integral vanishes, and since $(\partial u/\partial x_i)$; $1 \le i \le d$, are bounded functions on $a + B_r$, the expectations under P^a of the stochastic integrals are all equal to zero. After taking these expectations on both sides and letting $t \to \infty$, we use (2.2) to obtain

$$u(a) = E^a u(W_{\tau_{a+B_r}}) = \int_{\partial B_r} u(a + x)\mu_r(dx). \qquad \square$$

2.3 Corollary (Maximum Principle). *Suppose that u is harmonic in the open, connected domain D. If u achieves its supremum over D at some point in D, then u is identically constant.*

PROOF. Let $M = \sup_{x \in D} u(x)$, and let $D_M = \{x \in D; u(x) = M\}$. We assume that D_M is nonempty and show that $D_M = D$. Since u is continuous, D_M is closed relative to D. But for $a \in D_M$ and $0 < r < \infty$ such that $a + \bar{B}_r \subset D$, we have the mean value property:

$$M = u(a) = \frac{1}{V_r} \int_{B_r} u(a + x)\, dx,$$

which shows that $u = M$ on $a + B_r$. Therefore, D_M is open. Because D is connected, either D_M or $D \setminus D_M$ must be empty. □

2.4 Exercise. Suppose D is bounded and connected, u is defined and continuous on $\bar{D}$, and u is harmonic in D. Then u attains its maximum over $\bar{D}$ on ∂D. If v is another function, harmonic in D and continuous on $\bar{D}$, and $v = u$ on ∂D, then $v = u$ on D as well.

For the sake of completeness, we state and prove the converse of Proposition 2.2. Our proof, which uses no probability, is taken from Dynkin & Yushkevich (1969).

2.5 Proposition. *If u maps D into $\mathbb{R}$ and has the mean value property, then u is of class C^∞ and harmonic.*

PROOF. We first prove that u is of class C^∞. For $\varepsilon > 0$, let $g_\varepsilon: \mathbb{R}^d \Rightarrow [0, \infty)$ be the C^∞ function

$$g_\varepsilon(x) = \begin{cases} c(\varepsilon) \exp\left[\dfrac{1}{\|x\|^2 - \varepsilon^2}\right]; & \|x\| < \varepsilon, \\ 0; & \|x\| \geq \varepsilon, \end{cases}$$

where $c(\varepsilon)$ is chosen so that (because of (2.6))

$$\int_{B_\varepsilon} g_\varepsilon(x)\, dx = c(\varepsilon) \int_0^\varepsilon S_\rho \exp\left(\frac{1}{\rho^2 - \varepsilon^2}\right) d\rho = 1. \tag{2.7}$$

For $\varepsilon > 0$ and $a \in D$ such that $a + \bar{B}_\varepsilon \subset D$, define

$$u_\varepsilon(a) \triangleq \int_{B_\varepsilon} u(a + x) g_\varepsilon(x)\, dx = \int_{\mathbb{R}^d} u(y) g_\varepsilon(y - a)\, dy.$$

From the second representation, it is clear that u_ε is of class C^∞ on the open subset of D where it is defined. Furthermore, for every $a \in D$ there exists $\varepsilon > 0$ so that $a + \bar{B}_\varepsilon \subset D$; from (2.6), (2.7), and the mean-value property of u, we may then write

$$\begin{aligned}u_\varepsilon(a) &= \int_{B_\varepsilon} u(a+x)g_\varepsilon(x)\,dx \\ &= c(\varepsilon)\int_0^\varepsilon S_\rho \int_{\partial B_\rho} u(a+x)\exp\left(\frac{1}{\rho^2-\varepsilon^2}\right)\mu_\rho(dx)\,d\rho \\ &= c(\varepsilon)\int_0^\varepsilon S_\rho u(a)\exp\left(\frac{1}{\rho^2-\varepsilon^2}\right)d\rho = u(a),\end{aligned}$$

and conclude that u is also of class C^∞.

In order to show that $\Delta u = 0$ in D, we choose $a \in D$ and expand à la Taylor in the neighborhood $a + \bar{B}_\varepsilon$,

$$(2.8)\qquad u(a+y) = u(a) + \sum_{i=1}^d y_i \frac{\partial u}{\partial x_i}(a) + \frac{1}{2}\sum_{i=1}^d \sum_{j=1}^d y_i y_j \frac{\partial^2 u}{\partial x_i \partial x_j}(a) + o(\|y\|^2);\quad y \in \bar{B}_\varepsilon,$$

where again $\varepsilon > 0$ is chosen so that $a + \bar{B}_\varepsilon \subset D$. Odd symmetry gives us

$$\int_{\partial B_\varepsilon} y_i \mu_\varepsilon(dy) = 0,\quad \int_{\partial B_\varepsilon} y_i y_j \mu_\varepsilon(dy) = 0;\quad i \neq j,$$

so upon integrating in (2.8) over ∂B_ε and using the mean-value property we obtain

$$(2.9)\qquad \begin{aligned}u(a) &= \int_{\partial B_\varepsilon} u(a+y)\mu_\varepsilon(dy) \\ &= u(a) + \frac{1}{2}\sum_{i=1}^d \frac{\partial^2 u}{\partial x_i^2}(a)\int_{\partial B_\varepsilon} y_i^2 \mu_\varepsilon(dy) + o(\varepsilon^2).\end{aligned}$$

But

$$\int_{\partial B_\varepsilon} y_i^2 \mu_\varepsilon(dy) = \frac{1}{d}\sum_{i=1}^d \int_{\partial B_\varepsilon} y_i^2 \mu_\varepsilon(dy) = \frac{\varepsilon^2}{d},$$

and so (2.9) becomes

$$\frac{\varepsilon^2}{2d}\Delta u(a) + o(\varepsilon^2) = 0.$$

Dividing by ε^2 and letting $\varepsilon \downarrow 0$, we see that $\Delta u(a) = 0$. □

B. The Dirichlet Problem

We take up now the *Dirichlet problem* (D, f): with D an open subset of $\mathbb{R}^d$ and $f: \partial D \to \mathbb{R}$ a given continuous function, find a continuous function $u: \bar{D} \to \mathbb{R}$ such that u is harmonic in D and takes on boundary values specified by f; i.e., u is of class $C^2(D)$ and

(2.10) $$\Delta u = 0; \quad \text{in } D,$$

(2.11) $$u = f; \quad \text{on } \partial D.$$

Such a function, when it exists, will be called a *solution to the Dirichlet problem* (D,f). One may interpret $u(x)$ as the steady-state temperature at $x \in D$ when the boundary temperatures of D are specified by f.

The power of the probabilistic method is demonstrated by the fact that we can immediately write down a very likely solution to (D,f), namely

(2.12) $$u(x) \triangleq E^x f(W_{\tau_D}); \quad x \in \bar{D},$$

provided of course that

(2.13) $$E^x|f(W_{\tau_D})| < \infty; \quad \forall x \in D.$$

By the definition of τ_D, u satisfies (2.11). Furthermore, for $a \in D$ and B_r chosen so that $a + \bar{B}_r \subset D$, we have from the strong Markov property:

$$\begin{aligned} u(a) &= E^a f(W_{\tau_D}) = E^a\{E^a[f(W_{\tau_D})|\mathscr{F}_{\tau_{a+B_r}}]\} \\ &= E^a\{u(W_{\tau_{a+B_r}})\} = \int_{\partial B_r} u(a+x)\mu_r(dx). \end{aligned}$$

Therefore, u has the mean-value property, and so it must satisfy (2.10). The only unresolved issue is whether u is continuous up to and including ∂D. It turns out that this depends on the *regularity of* ∂D, as we shall see later. We summarize our discussion so far and establish a uniqueness result for (D,f) which strengthens Exercise 2.4.

2.6 Proposition. *If* (2.13) *holds, then u defined by* (2.12) *is harmonic in D.*

2.7 Proposition. *If f is bounded and*

(2.14) $$P^a[\tau_D < \infty] = 1; \quad \forall a \in D,$$

then any bounded solution to (D,f) has the representation (2.12).

PROOF. Let u be any bounded solution to (D,f), and let $D_n \triangleq \{x \in D; \inf_{y \in \partial D} \|x - y\| > 1/n\}$. From Itô's rule we have

$$u(W_{t \wedge \tau_{B_n} \wedge \tau_{D_n}}) = u(W_0) + \sum_{i=1}^d \int_0^{t \wedge \tau_{B_n} \wedge \tau_{D_n}} \frac{\partial u}{\partial x_i}(W_s)\,dW_s^{(i)}; \quad 0 \le t < \infty, \quad n \ge 1.$$

Since $(\partial u/\partial x_i)$ is bounded in $\overline{B_n \cap D_n}$, we may take expectations and conclude that

$$u(a) = E^a u(W_{t \wedge \tau_{B_n} \wedge \tau_{D_n}}); \quad 0 \le t < \infty, \quad n \ge 1, \quad a \in D_n.$$

As $t \to \infty$, $n \to \infty$, (2.14) implies that $u(W_{t \wedge \tau_{B_n} \wedge \tau_{D_n}})$ converges to $f(W_{\tau_D})$, a.s. P^a. The representation (2.12) follows from the bounded convergence theorem. □

2.8 Exercise. With $D = \{(x_1, x_2); x_2 > 0\}$ and $f(x_1, 0) = 0$; $x_1 \in \mathbb{R}$, show by example that (D, f) can have unbounded solutions not given by (2.12).

In the light of Propositions 2.6 and 2.7, the existence of a solution to the Dirichlet problem boils down to the question of the continuity of u defined by (2.12) at the boundary of D. We therefore undertake to characterize those points $a \in \partial D$ for which

$$\lim_{\substack{x \to a \\ x \in D}} E^x f(W_{\tau_D}) = f(a) \tag{2.15}$$

holds for every bounded, measurable function $f: \partial D \to \mathbb{R}$ which is continuous at the point a.

2.9 Definition. Consider the stopping time of the right-continuous filtration $\{\mathscr{F}_t\}$ given by $\sigma_D \triangleq \inf\{t > 0; W_t \in D^c\}$ (contrast with the definition of τ_D in (2.1)). We say that a point $a \in \partial D$ is *regular* for D if $P^a[\sigma_D = 0] = 1$; i.e., a Brownian path started at a does not immediately return to D and remain there for a nonempty time interval.

2.10 Remark. A point $a \in \partial D$ is called *irregular* if $P^a[\sigma_D = 0] < 1$; however, the event $\{\sigma_D = 0\}$ belongs to $\mathscr{F}^W_{0+}$, and so the Blumenthal zero-one law (Theorem 2.7.17) gives for an irregular point a: $P^a[\sigma_D = 0] = 0$.

2.11 Remark. It is evident that regularity is a local condition; i.e., $a \in \partial D$ is regular for D if and only if a is regular for $(a + B_r) \cap D$, for some $r > 0$.

In the one-dimensional case every point of ∂D is regular (Problem 2.7.18) and the Dirichlet problem is always solvable, the solution being piecewise-linear. When $d \geq 2$, more interesting behavior can occur. In particular, if $D = \{x \in \mathbb{R}^d; 0 < \|x\| < 1\}$ is a *punctured ball*, then for any $x \in D$ the Brownian motion starting at x exits from D on its outer boundary, not at the origin (Proposition 3.3.22). This means that u defined by (2.12) is determined solely by the values of f along the outer boundary of D and, except at the origin, this u will agree with the harmonic function

$$\tilde{u}(x) \triangleq E^x f(W_{\tau_{B_1}}) = E^x f(W_{\sigma_D}); \quad x \in B_1.$$

Now $u(0) \triangleq f(0)$, so u is continuous at the origin if and only if $f(0) = \tilde{u}(0)$. When $d \geq 3$, it is even possible for ∂D to be connected but contain irregular points (Example 2.17).

2.12 Theorem. *Assume that $d \geq 2$ and fix $a \in \partial D$. The following are equivalent:*

(i) *equation* (2.15) *holds for every bounded, measurable function $f: \partial D \to \mathbb{R}$ which is continuous at a;*
(ii) *a is regular for D;*

(iii) *for all* $\varepsilon > 0$, *we have*

$$\lim_{\substack{x \to a \\ x \in D}} P^x[\tau_D > \varepsilon] = 0. \tag{2.16}$$

PROOF. We assume without loss of generality that $a = 0$, and begin by proving the implication *(i)* ⇒ *(ii)* by contradiction. If the origin is irregular, then $P^0[\sigma_D = 0] = 0$ (Remark 2.10). Since a Brownian motion of dimension $d \geq 2$ never returns to its starting point (Proposition 3.3.22), we have

$$\lim_{r \downarrow 0} P^0[W_{\sigma_D} \in B_r] = P^0[W_{\sigma_D} = 0] = 0.$$

Fix $r > 0$ for which $P^0[W_{\sigma_D} \in B_r] < (1/4)$, and choose a sequence $\{\delta_n\}_{n=1}^{\infty}$ for which $0 < \delta_n < r$ for all n and $\delta_n \downarrow 0$. With $\tau_n \triangleq \inf\{t \geq 0; \|W_t\| \geq \delta_n\}$, we have $P^0[\tau_n \downarrow 0] = 1$, and thus, $\lim_{n\to\infty} P^0[\tau_n < \sigma_D] = 1$. Furthermore, on the event $\{\tau_n < \sigma_D\}$ we have $W_{\tau_n} \in D$. For n large enough so that $P^0[\tau_n < \sigma_D] \geq (1/2)$, we may write

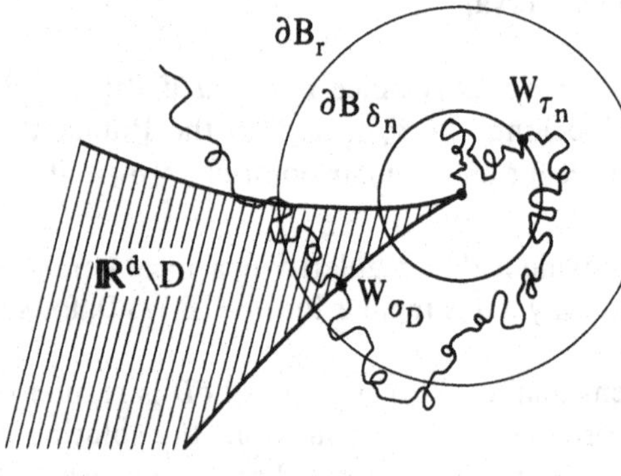

$$\begin{aligned}
\frac{1}{4} &> P^0[W_{\sigma_D} \in B_r] \geq P^0[W_{\sigma_D} \in B_r, \tau_n < \sigma_D] \\
&= E^0(1_{\{\tau_n < \sigma_D\}} P^0[W_{\sigma_D} \in B_r \mid \mathscr{F}_{\tau_n}]) \\
&= \int_{D \cap B_{\delta_n}} P^x[W_{\tau_D} \in B_r] P^0[\tau_n < \sigma_D, W_{\tau_n} \in dx] \\
&\geq \frac{1}{2} \inf_{x \in D \cap B_{\delta_n}} P^x[W_{\tau_D} \in B_r],
\end{aligned}$$

from which we conclude that $P^{x_n}[W_{\tau_D} \in B_r] \leq (1/2)$ for some $x_n \in D \cap B_{\delta_n}$. Now choose a bounded, continuous function $f: \partial D \to \mathbb{R}$ such that $f = 0$ outside B_r, $f \leq 1$ inside B_r, and $f(0) = 1$. For such a function we have

$$\overline{\lim_{n\to\infty}} E^{x_n} f(W_{\tau_D}) \leq \overline{\lim_{n\to\infty}} P^{x_n}[W_{\tau_D} \in B_r] \leq \frac{1}{2} < f(0),$$

and (i) fails.

We next show that *(ii)* ⇒ *(iii)*. Observe first of all that for $0 < \delta < \varepsilon$, the function

$$g_\delta(x) \triangleq P^x[W_s \in D; \delta \le s \le \varepsilon] = E^x(P^{W_\delta}[\tau_D > \varepsilon - \delta])$$
$$= \int_{\mathbb{R}^d} P^y[\tau_D > \varepsilon - \delta]\, P^x[W_\delta \in dy]$$

is continuous in x. But

$$g_\delta(x) \downarrow g(x) \triangleq P^x[W_s \in D; 0 < s \le \varepsilon] = P^x[\sigma_D > \varepsilon]$$

as $\delta \downarrow 0$, so g is upper semicontinuous. From this fact and the inequality $\tau_D \le \sigma_D$, we conclude that $\overline{\lim}_{x \to 0, x \in D} P^x[\tau_D > \varepsilon] \le \overline{\lim}_{x \to 0} g(x) \le g(0) = 0$, by (ii).

Finally, we establish *(iii)* ⇒ *(i)*. We know that for each $r > 0$, $P^x[\max_{0 \le t \le \varepsilon} \|W_t - W_0\| < r]$ does not depend on x and approaches one as $\varepsilon \downarrow 0$. But then

$$P^x[\|W_{\tau_D} - W_0\| < r] \ge P^x\left[\left\{\max_{0 \le t \le \varepsilon} \|W_t - W_0\| < r\right\} \cap \{\tau_D \le \varepsilon\}\right]$$
$$\ge P^0\left[\max_{0 \le t \le \varepsilon} \|W_t\| < r\right] - P^x[\tau_D > \varepsilon].$$

Letting $x \to 0$ ($x \in D$) and $\varepsilon \downarrow 0$, successively, we obtain from (iii)

$$\lim_{\substack{x \to 0 \\ x \in D}} P^x[\|W_{\tau_D} - x\| < r] = 1; \quad 0 < r < \infty.$$

The continuity of f at the origin and its boundedness on ∂D give us (2.15). □

C. Conditions for Regularity

For many open sets D and boundary points $a \in \partial D$, we can convince ourselves intuitively that a Brownian motion originating at a will exit from $\bar{D}$ immediately; i.e., a is regular. We formalize this intuition with a careful discussion of regularity.

We have already seen that when $d = 2$, the center of a punctured disc is an irregular boundary point. The following development, culminating with Problem 2.16, shows that, in $\mathbb{R}^2$, any irregular boundary point of D must be "isolated" in the sense that it cannot be connected to any other point outside D by a simple arc lying outside D.

2.13 Definition. Let $D \subset \mathbb{R}^d$ be open and $a \in \partial D$. A *barrier at a* is a continuous function $v: \bar{D} \to \mathbb{R}$ which is harmonic in D, positive on $\bar{D} \backslash \{a\}$, and equal to zero at a.

2.14 Example. Let $D \subset B_r \subset \mathbb{R}^2$ be open, where $0 < r < 1$, and assume $(0,0) \in \partial D$. If a single-valued, analytic branch of $\log(x_1 + ix_2)$ can be defined in $\bar{D}\setminus(0,0)$, then

$$v(x_1,x_2) \triangleq \begin{cases} -\mathrm{Re}\dfrac{1}{\log(x_1+ix_2)} = -\dfrac{\log\sqrt{x_1^2+x_2^2}}{|\log(x_1+ix_2)|^2}; & (x_1,x_2)\in \bar{D}\setminus(0,0), \\ 0; & (x_1,x_2)=(0,0), \end{cases}$$

is a barrier at $(0,0)$. Indeed, being the real part of an analytic function, v is harmonic in D, and because $0 < \sqrt{x_1^2+x_2^2} \le r < 1$ in $\bar{D}\setminus(0,0)$, v is positive on this set.

2.15 Proposition. *Let D be bounded and $a \in \partial D$. If there exists a barrier at a, then a is regular.*

PROOF. Let v be a barrier at a. We establish condition (i) of Theorem 2.12. With $f: \partial D \to \mathbb{R}$ bounded and continuous at a, define $M = \sup_{x\in\partial D}|f(x)|$. Choose $\varepsilon > 0$ and let $\delta > 0$ be such that $|f(x) - f(a)| \le \varepsilon$ if $x \in \partial D$ and $\|x - a\| \le \delta$. Choose k so that $kv(x) \ge 2M$ for $x \in \bar{D}$ and $\|x - a\| \ge \delta$. We then have $|f(x) - f(a)| \le \varepsilon + kv(x)$; $x \in \partial D$, so

$$|E^x f(W_{\tau_D}) - f(a)| \le \varepsilon + k\cdot E^x v(W_{\tau_D}) = \varepsilon + k\cdot v(x); \quad x \in D$$

by Proposition 2.7. But v is continuous and $v(a) = 0$, so

$$\varlimsup_{\substack{x\to a\\ x\in D}} |E^x f(W_{\tau_D}) - f(a)| \le \varepsilon.$$

Finally, we let $\varepsilon \downarrow 0$ to obtain (2.15). □

2.16 Problem. Let $D \subset \mathbb{R}^2$ be open, and suppose that $a \in \partial D$ has the property that there exists a point $b \ne a$ in $\mathbb{R}^2\setminus D$, and a simple arc in $\mathbb{R}^2\setminus D$ connecting a to b. Show that a is regular.

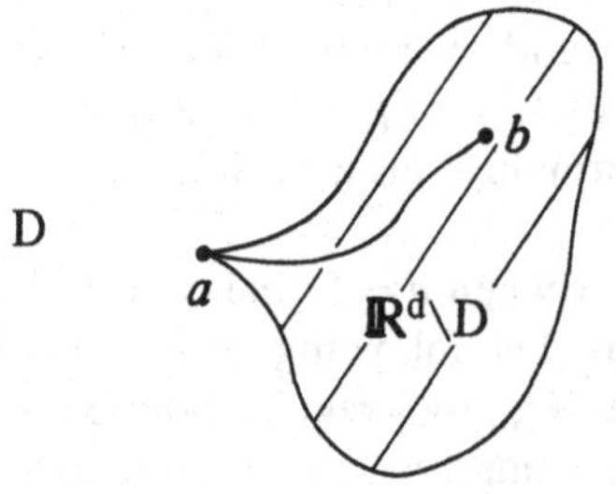

In three or more dimensions, it is possible to create a cusped region D so that the boundary point at the end of the cusp is irregular. We illustrate this situation in $\mathbb{R}^3$. In particular, we will construct a region as demarcated in the following figure by the broken line, so that, when this region is rotated about the x_1-axis, the resulting solid has an irregular boundary point at the origin. It is a simple matter to replace this solid by an even larger one, having a smooth boundary except for a cusp at the origin (dotted curve). It is a direct consequence of Definition 2.9 that the origin is also an irregular boundary point for this larger solid.

2.17 Example (Lebesgue's Thorn). With $d = 3$ and $\{\varepsilon_n\}_{n=1}^{\infty}$ a sequence of positive numbers decreasing to zero, define

$$E = \{(x_1, x_2, x_3);\ -1 < x_1 < 1, x_2^2 + x_3^2 < 1\},$$

$$F_n = \left\{(x_1, x_2, x_3);\ 2^{-n} \le x_1 \le 2^{-n+1}, x_2^2 + x_3^2 \le \varepsilon_n\right\},$$

$$D = E \Big\backslash \left(\bigcup_{n=1}^{\infty} F_n\right).$$

Now $P^0[(W_t^{(2)}, W_t^{(3)}) = (0,0), \text{ for some } t > 0] = 0$ (Proposition 3.3.22), so the P^0-probability that $W = (W^{(1)}, W^{(2)}, W^3)$ ever hits the compact set $K_n \triangleq \{(x_1, x_2, x_3);\ 2^{-n} \le x_1 \le 2^{-n+1}, x_2 = x_3 = 0\}$ is zero. According to Problem 3.3.24, $\lim_{t\to\infty} \|W_t\| = \infty$ a.s. P^0, so for P^0-a.e. $\omega \in \Omega$, the path $t \mapsto W_t(\omega)$ remains bounded away from K_n. Thus, if ε_n is chosen sufficiently small, we can ensure that $P^0[W_t \in F_n, \text{ for some } t \ge 0] \le 3^{-n}$. If W, beginning at the origin, does not return to D immediately, it must avoid D by entering $\bigcup_{n=1}^{\infty} F_n$. In other words,

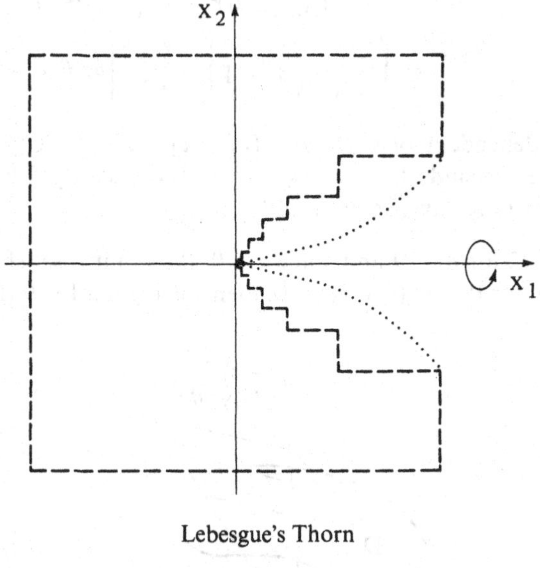

Lebesgue's Thorn

$$P^0[\sigma_D = 0] \le P^0[W_t \in F_n, \text{ for some } t \ge 0 \text{ and } n \ge 1] \le \sum_{n=1}^{\infty} 3^{-n} < 1. \qquad \square$$

If cusplike behavior is avoided, then the boundary points of D are regular, regardless of dimension. To make this statement precise, let us define for

$y \in \mathbb{R}^d \setminus \{0\}$ and $0 \le \theta \le \pi$, the *cone* $C(y, \theta)$ *with direction* y *and aperture* θ by

$$C(y, \theta) = \{x \in \mathbb{R}^d; (x, y) \ge \|x\| \cdot \|y\| \cdot \cos \theta\}.$$

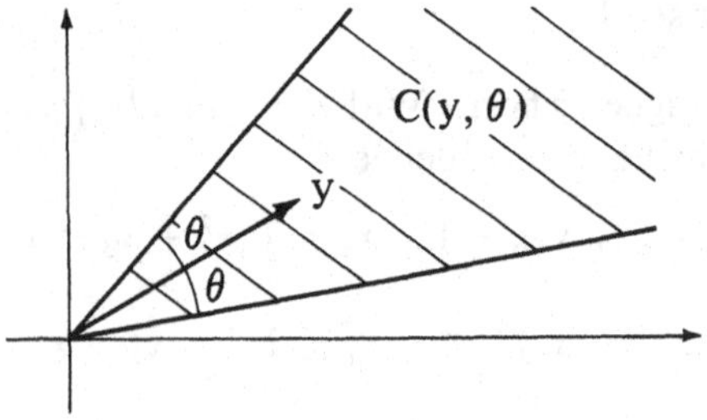

2.18 Definition. We say that the point $a \in \partial D$ satisfies *Zaremba's cone condition* if there exists $y \ne 0$ and $0 < \theta < \pi$ such that the translated cone $a + C(y, \theta)$ is contained in $\mathbb{R}^d \setminus D$.

2.19 Theorem. *If a point* $a \in \partial D$ *satisfies Zaremba's cone condition, then it is regular.*

PROOF. We assume without loss of generality that a is the origin and $C(y, \theta) \subseteq \mathbb{R}^d \setminus D$, where $y \ne 0$ and $0 < \theta < \pi$. Because the change of variable $z = (x/\sqrt{t})$ maps $C(y, \theta)$ onto itself, we have for any $t > 0$,

$$P^0[W_t \in C(y, \theta)] = \int_{C(y,\theta)} \frac{1}{(2\pi t)^{d/2}} \exp\left[-\frac{\|x\|^2}{2t}\right] dx$$

$$= \int_{C(y,\theta)} \frac{1}{(2\pi)^{d/2}} \exp\left[-\frac{\|z\|^2}{2}\right] dz \triangleq q > 0,$$

where q is independent of t. Now $P^0[\sigma_D \le t] \ge P^0[W_t \in C(y, \theta)] = q$, and letting $t \downarrow 0$ we conclude that $P^0[\sigma_D = 0] > 0$. Regularity follows from the Blumenthal zero-one law (Remark 2.10). $\square$

2.20 Remark. If, for $a \in \partial D$ and some $r > 0$, the point a satisfies Zaremba's cone condition for the set $(a + B_r) \cap D$, then a is regular for D (Remark 2.11).

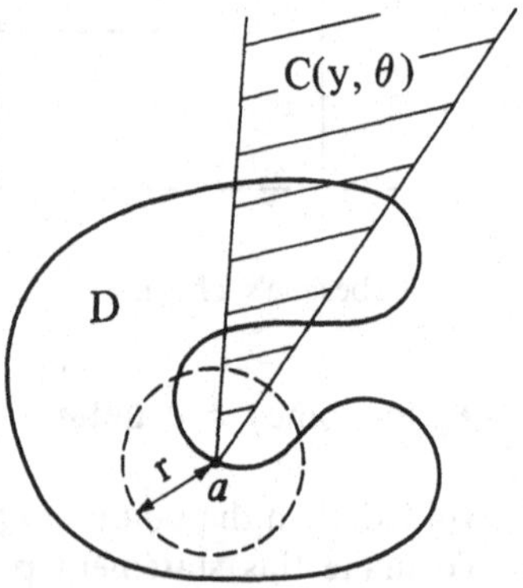

D. Integral Formulas of Poisson†

We now have a complete solution to the Dirichlet problem for a large class of open sets D and bounded, continuous boundary data functions $f: \partial D \to \mathbb{R}$. Indeed, *if every boundary point of D is regular and D satisfies* (2.14), *then the unique bounded solution to (D, f) is given by* (2.12) (Propositions 2.6, 2.7 and Theorem 2.12). In some cases, we can actually compute the right-hand side of (2.12) and thereby obtain Poisson integral formulas.

2.21 Theorem (Poisson Integral Formula for a Half-Space). *With $d \geq 2$, $D = \{(x_1, \ldots, x_d); x_d > 0\}$ and $f: \partial D \to \mathbb{R}$ bounded and continuous, the unique bounded solution to the Dirichlet problem (D, f) is given by*

$$u(x) = \frac{\Gamma(d/2)}{\pi^{d/2}} \int_{\partial D} \frac{x_d f(y)}{\|y - x\|^d} dy; \quad x \in D. \tag{2.17}$$

2.22 Problem. Prove Theorem 2.21.

The Poisson integral formula for a d-dimensional sphere can be obtained from Theorem 2.21 via the *Kelvin transformation.* Let $\varphi: \mathbb{R}^d \backslash \{0\} \to \mathbb{R}^d \backslash \{0\}$ be defined by $\varphi(x) = (x/\|x\|^2)$. Note that φ is its own inverse. We simplify notation by writing x^* instead of $\varphi(x)$.

For $r > 0$, let $B = \{x \in \mathbb{R}^d: \|x - c\| < r\}$, where $c = re_d$ and e_i is the unit vector with a one in the i-th position. Suppose $f: \partial B \to \mathbb{R}$ is continuous (and hence bounded), so there exists a unique function u which solves the Dirichlet problem (B, f). The reader may easily verify that

$$\varphi(B) = H \triangleq \left\{x^* \in \mathbb{R}^d; (x^*, c) > \frac{1}{2}\right\}$$

and $\varphi(\partial B \backslash \{0\}) = \partial H = \{x^* \in \mathbb{R}^d; (x^*, c) = \frac{1}{2}\}$. We define $u^*: \bar{H} \to \mathbb{R}$, the Kelvin transform of u, by

$$u^*(x^*) = \frac{1}{\|x^*\|^{d-2}} u(x). \tag{2.18}$$

A tedious but straightforward calculation shows that $\Delta u^*(x^*) = \|x\|^{-d-2} \Delta u(x)$, so u^* is a bounded solution to the Dirichlet problem (H, f^*) where

$$f^*(x^*) = \frac{1}{\|x^*\|^{d-2}} f(x); \quad x^* \in \partial H. \tag{2.19}$$

Because $H = (1/2r)e_d + D$, where D is as in Theorem 2.21, we may apply (2.17) to obtain

† The results of this subsection will not be used later in the text.

(2.20) $$u^*(x^*) = \frac{\Gamma(d/2)}{\pi^{d/2}} \int_{\partial H} \frac{\left(x_d^* - \frac{1}{2r}\right) f^*(y^*)}{\|y^* - x^*\|^d} dy^*; \quad x^* \in H.$$

Formulas (2.18)–(2.20) provide us with the unique solution to the Dirichlet problem (B, f). These formulas are, however, a bit unwieldy, a problem which can be remedied by the change of variable $y = \varphi(y^*)$ in the integral of (2.20). This change maps the hyperplane ∂H into the sphere ∂B. The surface element on ∂B is $S_r \mu_r(dy - c)$ (recall (2.4), (2.5)). A little bit of algebra and calculus on manifolds (Spivak (1965), p. 126) shows that the proposed change of variable in (2.20) involves

(2.21) $$dy^* = \frac{S_r \mu_r(dy - c)}{\|y\|^{2(d-1)}}.$$

(The reader familiar with calculus on manifolds may wish to verify (2.21) first for the case $y^* = y_1^* e_1 + (1/2r)e_d$ and then observe that the general case may be reduced to this one by a rotation. The reader unfamiliar with calculus on manifolds can content himself with the verification when $d = 2$, or can refer to Gilbarg & Trudinger (1977), p. 20, for a proof of Theorem 2.23 which uses the divergence theorem but avoids formula (2.21).)

On the other hand,

(2.22) $$\|y^* - x^*\|^2 = \frac{\|x - y\|^2}{\|x\|^2 \|y\|^2},$$

(2.23) $$r^2 - \|x - c\|^2 = \|x\|^2[2(c, x^*) - 1] = 2r\|x\|^2\left(x_d^* - \frac{1}{2r}\right).$$

Using (2.18), (2.19), and (2.21)–(2.23) to change the variable of integration in (2.20), we obtain

(2.24) $$u(x) = r^{d-2}(r - \|x - c\|^2) \int_{\partial B} \frac{f(y)\mu_r(dy - c)}{\|y - x\|^d}; \quad x \in B.$$

Translating this formula to a sphere centered at the origin, we obtain the following classical result.

2.23 Theorem (Poisson Integral Formula for a Sphere). *With $d \geq 2$, $B_r = \{x \in \mathbb{R}^d; \|x\| < r\}$, and $f\colon \partial B_r \to \mathbb{R}$ continuous, the unique solution to the Dirichlet problem (B_r, f) is given by*

(2.25) $$u(x) = r^{d-2}(r - \|x\|^2) \int_{\partial B_r} \frac{f(y)\mu_r(dy)}{\|y - x\|^d}; \quad x \in B_r.$$

2.24 Exercise. Show that for $x \in B_r$, we have the *exit distribution*

(2.26) $$P^x[W_{\tau_{B_r}} \in dy] = \frac{r^{d-2}(r^2 - \|x\|^2)\mu_r(dy)}{\|x - y\|^d}; \quad \|y\| = r.$$

E. Supplementary Exercises

2.25 Problem. Consider as given an open, bounded subset D of $\mathbb{R}^d$ and the bounded, continuous functions $g: D \to \mathbb{R}$ and $f: \partial D \to \mathbb{R}$. Assume that $u: \bar{D} \to \mathbb{R}$ is continuous, of class $C^2(D)$, and solves the *Poisson equation*

$$\frac{1}{2}\Delta u = -g; \quad \text{in } D$$

subject to the boundary condition

$$u = f; \quad \text{on } \partial D.$$

Then establish the representation

$$u(x) = E^x\left[f(W_{\tau_D}) + \int_0^{\tau_D} g(W_t)\,dt\right]; \quad x \in \bar{D}. \tag{2.27}$$

In particular, the *expected exit time from a ball* is given by

$$E^x \tau_{B_r} = \frac{r^2 - \|x\|^2}{d}; \quad x \in B_r. \tag{2.28}$$

(*Hint*: Show that the process $\{M_t \triangleq u(W_{t \wedge \tau_D}) + \int_0^{t \wedge \tau_D} g(W_s)\,ds, \mathscr{F}_t; 0 \le t < \infty\}$ is a uniformly integrable martingale.)

2.26 Exercise. Suppose we remove condition (2.14) in Proposition 2.7. Show that $v(x) \triangleq P^x[\tau_D = \infty]$ is harmonic in D, and if $a \in \partial D$ is regular, then $\lim_{\substack{x \to a \\ x \in D}} v(x) = 0$. In particular, if every point of ∂D is regular, then with $u(x) = E^x[f(W_{\tau_D}) 1_{\{\tau_D < \infty\}}]$, the function $u + \lambda v$ is a bounded solution to the Dirichlet problem (D, f) for any $\lambda \in \mathbb{R}$. (It is possible to show that every bounded solution to (D, f) is of this form; see Port & Stone (1978), Theorem 4.2.12.)

2.27 Exercise. Let D be bounded with every boundary point regular. Prove that every boundary point has a barrier.

2.28 Exercise. *A complex-valued Brownian motion* is defined to be a process $\overline{W} = \{W_t^{(1)} + iW_t^{(2)}, \mathscr{F}_t; 0 \le t < \infty\}$, where $W = \{(W_t^{(1)}, W_t^{(2)}), \mathscr{F}_t; 0 \le t < \infty\}$ is a two-dimensional Brownian motion and $i = \sqrt{-1}$:

(i) Use Theorem 3.4.13 to show that if $\overline{W}$ is a complex-valued Brownian motion and $f: \mathbb{C} \to \mathbb{C}$ is analytic and nonconstant, then (under an appropriate condition) $f(\overline{W})$ is a complex-valued Brownian motion with a random time-change (P. Lévy (1948)).
(ii) With $\xi \in \mathbb{C} \backslash \{0\}$, show that $M_t \triangleq \xi e^{\overline{W}_t}$, $0 \le t < \infty$ is a time-changed, complex-valued Brownian motion. (*Hint:* Use Problem 3.6.30.)
(iii) Use the result in (ii) to provide a new proof of Proposition 3.3.22.

For additional information see B. Davis (1979).

4.3. The One-Dimensional Heat Equation

In this section we establish stochastic representations for the temperatures in infinite, semi-infinite, and finite rods. We then show how such representations allow one to compute boundary-crossing probabilities for Brownian motion.

Consider an infinite rod, insulated and extended along the x-axis of the (t, x) plane, and let $f(x)$ denote the temperature of the rod at time $t = 0$ and location x. If $u(t, x)$ is the temperature of the rod at time $t \geq 0$ and position $x \in \mathbb{R}$, then, with appropriate choice of units, u will satisfy the *heat equation*

$$\frac{\partial u}{\partial t} = \frac{1}{2}\frac{\partial^2 u}{\partial x^2}, \tag{3.1}$$

with initial condition $u(0, x) = f(x)$; $x \in \mathbb{R}$. The starting point of our probabilistic treatment of (3.1) is furnished by the observation that the transition density

$$p(t; x, y) \triangleq \frac{1}{dy} P^x[W_t \in dy] = \frac{1}{\sqrt{2\pi t}} e^{-(x-y)^2/2t}; \quad t > 0, \quad x, y \in \mathbb{R},$$

of the one-dimensional Brownian family satisfies the partial differential equation

$$\frac{\partial p}{\partial t} = \frac{1}{2}\frac{\partial^2 p}{\partial x^2}. \tag{3.2}$$

Suppose then that $f: \mathbb{R} \to \mathbb{R}$ is a Borel-measurable function satisfying the condition

$$\int_{-\infty}^{\infty} e^{-ax^2}|f(x)|\,dx < \infty \tag{3.3}$$

for some $a > 0$. It is well known (see Problem 3.1) that

$$u(t, x) \triangleq E^x f(W_t) = \int_{-\infty}^{\infty} f(y) p(t; x, y)\,dy \tag{3.4}$$

is defined for $0 < t < (1/2a)$ and $x \in \mathbb{R}$, has derivatives of all orders, and satisfies the heat equation (3.1).

3.1 Problem. Show that for any nonnegative integers n and m, under the assumption (3.3), we have

$$\frac{\partial^{n+m}}{\partial t^n \partial x^m} u(t, x) = \int_{-\infty}^{\infty} f(y) \frac{\partial^{n+m}}{\partial t^n \partial x^m} p(t; x, y)\,dy; \quad 0 < t < \frac{1}{2a}, \quad x \in \mathbb{R}. \tag{3.5}$$

If f is bounded and continuous, then rewriting (3.4) as $u(t, x) = E^0 f(x + W_t)$, we can use the bounded convergence theorem to conclude

(3.6) $$f(x) = \lim_{\substack{t \downarrow 0 \\ y \to x}} u(t, y), \quad \forall x \in \mathbb{R}.$$

In fact, we have the stronger result contained in the following problem.

3.2 Problem. If $f: \mathbb{R} \to \mathbb{R}$ is a Borel-measurable function satisfying (3.3) and f is continuous at x, then (3.6) holds.

A. The Tychonoff Uniqueness Theorem

We shall call $p(t; x, y)$ a *fundamental solution* to the problem of finding a function u which satisfies (3.1) and agrees with the specified function f at time $t = 0$.

We shall say that a function $u: \mathbb{R}^m \to \mathbb{R}$ has continuous derivatives up to a certain order on a set G, if these derivatives exist and are continuous in the interior of G, and have continuous extensions to that part of the boundary ∂G which is included in G. With this convention, we can state the following uniqueness theorem. For nonnegative functions, a substantially stronger result is given in Exercise 3.8.

3.3 Theorem (Tychonoff (1935)). *Suppose that the function u is $C^{1,2}$ on the strip $(0, T] \times \mathbb{R}$ and satisfies (3.1) there, as well as the conditions*

(3.7) $$\lim_{\substack{t \downarrow 0 \\ y \to x}} u(t, y) = 0; \quad x \in \mathbb{R},$$

(3.8) $$\sup_{0 < t \le T} |u(t, x)| \le K e^{ax^2}; \quad x \in \mathbb{R},$$

for some positive constants K and a. Then $u = 0$ on $(0, T] \times \mathbb{R}$.

3.4 Remark. If u_1 and u_2 satisfy (3.1), (3.8) and

$$\lim_{\substack{t \downarrow 0 \\ y \to x}} u_1(t, y) = \lim_{\substack{t \downarrow 0 \\ y \to x}} u_2(t, y),$$

then Theorem 3.3 applied to $u_1 - u_2$ asserts that $u_1 = u_2$ on $(0, T) \times \mathbb{R}$.

3.5 Remark. Any probabilistic treatment of the heat equation involves a time-reversal. This is already suggested by the representation (3.4), in which the initial temperature function f is evaluated at W_t rather than W_0. We shall see this time-reversal many times in this section, beginning with the following probabilistic proof of Theorem 3.3.

PROOF OF THEOREM 3.3. Let T_y be the passage time of W to y as in (2.6.1). Fix $x \in \mathbb{R}$, choose $n > |x|$, and let $R_n = T_n \wedge T_{-n}$. With $t \in [0, T)$ fixed and

$$v(\theta, x) \triangleq u(T - t - \theta, x); \quad 0 \le \theta < T - t,$$

we have from Itô's rule, for $0 \le s < T - t$,

$$u(T-t,x) = v(0,x) = E^x v(s \wedge R_n, W_{s \wedge R_n}) \tag{3.9}$$
$$= E^x[v(s, W_s)1_{\{s<R_n\}}] + E^x[v(R_n, W_{R_n})1_{\{s \ge R_n\}}].$$

Now $|v(s, W_s)|1_{\{s<R_n\}}$ is dominated by

$$\max_{\substack{0 \le s < T-t \\ |y| \le n}} |u(T-t-s, y)| \le K e^{an^2},$$

and $v(s, W_s)$ converges P^x-a.s. to zero as $s \uparrow T - t$, thanks to (3.7). Likewise, $|v(R_n, W_{R_n})|1_{\{s \ge R_n\}}$ is dominated by Ke^{an^2}. Letting $s \uparrow T - t$ in (3.9), we obtain from the bounded convergence theorem:

$$u(T-t,x) = E^x[v(R_n, W_{R_n})1_{\{R_n < T-t\}}].$$

Therefore, with $0 \le t < T$, $|x| < n$,

$$|u(T-t,x)| \le Ke^{an^2} P^x[R_n < T-t]$$
$$\le Ke^{an^2}(P^0[T_{n-x} < T] + P^0[T_{n+x} < T])$$
$$\le Ke^{an^2}\sqrt{\frac{2}{\pi}}\left(\int_{(n-x)/\sqrt{T}}^{\infty} e^{-z^2/2}\,dz + \int_{(n+x)/\sqrt{T}}^{\infty} e^{-z^2/2}\,dz\right),$$

where we have used (2.6.2). But from (2.9.20) it is evident that $\lim_{n\to\infty} e^{an^2}\int_{(n \pm x)/\sqrt{T}}^{\infty} e^{-z^2/2}\,dz = 0$, provided $a < 1/2T$.

Having proved the theorem for $a < (1/2T)$, we can easily extend it to the case where this inequality does not hold by choosing $T_0 = 0 < T_1 < \cdots < T_n = T$ such that $a < (1/2(T_i - T_{i-1}))$; $i = 1, \ldots, n$, and then showing successively that $u = 0$ in each of the strips $(T_{i-1}, T_i]$; $i = 1, \ldots, n$. □

It is instructive to note that the function

$$h(t,x) \triangleq \frac{x}{t} p(t; x, 0) = -\frac{\partial}{\partial x} p(t; x, 0); \quad t > 0, \quad x \in \mathbb{R}, \tag{3.10}$$

solves the heat equation (3.1) on every strip of the form $(0, T] \times \mathbb{R}$; furthermore, it satisfies condition (3.8) for every $0 < a < (1/2T)$, as well as (3.7) for every $x \ne 0$. However, the limit in (3.7) fails to exist for $x = 0$, although we do have $\lim_{t \downarrow 0} h(t, 0) = 0$.

B. Nonnegative Solutions of the Heat Equation

If the initial temperature f is nonnegative, as it always is if measured on the absolute scale, then the temperature should remain nonnegative for all $t > 0$; this is evident from the representation (3.4). Is it possible to characterize the nonnegative solutions of the heat equation? This was done by Widder (1944), who showed that such functions u have a representation

$$u(t,x) = \int_{-\infty}^{\infty} p(t;x,y)\,dF(y); \quad x \in \mathbb{R},$$

where $F\colon \mathbb{R} \to \mathbb{R}$ is nondecreasing. Corollary 3.7 (i)′, (ii)′ is a precise statement of Widder's result. We extend Widder's work by providing *probabilistic* characterizations of nonnegative solutions to the heat equation; these appear as Corollary 3.7 (iii)′, (iv)′.

3.6 Theorem. *Let $v(t,x)$ be a nonnegative function defined on a strip $(0,T) \times \mathbb{R}$, where $0 < T < \infty$. The following four conditions are equivalent:*

(i) *for some nondecreasing function $F\colon \mathbb{R} \to \mathbb{R}$,*

$$v(t,x) = \int_{-\infty}^{\infty} p(T-t;x,y)\,dF(y); \quad 0 < t < T, \quad x \in \mathbb{R}; \tag{3.11}$$

(ii) *v is of class $C^{1,2}$ on $(0,T) \times \mathbb{R}$ and satisfies the "backward" heat equation*

$$\frac{\partial v}{\partial t} + \frac{1}{2}\frac{\partial^2 v}{\partial x^2} = 0 \tag{3.12}$$

on this strip;

(iii) *for a Brownian family $\{W_s, \mathscr{F}_s; 0 \le s < \infty\}$, $(\Omega, \mathscr{F})$, $\{P^x\}_{x\in\mathbb{R}}$ and each fixed $t \in (0,T)$, $x \in \mathbb{R}$, the process $\{v(t+s, W_s), \mathscr{F}_s; 0 \le s < T-t\}$ is a martingale on $(\Omega, \mathscr{F}, P^x)$;*

(iv) *for a Brownian family $\{W_s, \mathscr{F}_s; 0 \le s < \infty\}$, $(\Omega, \mathscr{F})$, $\{P^x\}_{x\in\mathbb{R}}$ we have*

$$v(t,x) = E^x v(t+s, W_s); \quad 0 < t \le t+s < T, \quad x \in \mathbb{R}. \tag{3.13}$$

Proof. Since $(\partial/\partial t)p(T-t;x,y) + (1/2)(\partial^2/\partial x^2)p(T-t;x,y) = 0$, the implication *(i) ⇒ (ii)* can be proved by showing that the partial derivatives of v can be computed by differentiating under the integral in (3.11). For $a > 1/2T$ we have

$$\int_{-\infty}^{\infty} e^{-ay^2}\,dF(y) = \sqrt{\frac{\pi}{a}}\, v\left(T - \frac{1}{2a}, 0\right) < \infty.$$

This condition is analogous to (3.3) and allows us to proceed as in Solution 3.1.

For the implications *(ii)⇒(iii)* and *(ii) ⇒ (iv)*, we begin by applying Itô's rule to $v(t+s, W_s)$; $0 \le s < T-t$. With $a < x < b$, we consider the passage times T_a and T_b as in (2.6.1) and obtain:

$$v(t + (s \wedge T_a \wedge T_b), W_{s\wedge T_a \wedge T_b}) = v(t, W_0) + \int_0^{s\wedge T_a\wedge T_b} \frac{\partial}{\partial x} v(t+\sigma, W_\sigma)\,dW_\sigma$$
$$+ \int_0^{s\wedge T_a\wedge T_b} \left(\frac{\partial}{\partial t} + \frac{1}{2}\frac{\partial^2}{\partial x^2}\right) v(t+\sigma, W_\sigma)\,d\sigma.$$

Under assumption (ii) the Lebesgue integral vanishes, as does the expectation of the stochastic integral because of the boundedness of $(\partial/\partial x)v(t + \sigma, y)$ when $a \leq y \leq b$ and $0 \leq \sigma \leq s < T - t$. Therefore,

$$v(t, x) = E^x v(t + (s \wedge T_a \wedge T_b), W_{s \wedge T_a \wedge T_b}). \tag{3.14}$$

Now let $a \downarrow -\infty$, $b \uparrow \infty$ and rely on the nonnegativity of v and Fatou's lemma to obtain

$$v(t, x) \geq E^x v(t + s, W_s); \quad 0 < t \leq t + s < T, \quad x \in \mathbb{R}. \tag{3.15}$$

Inequality (3.15) implies that for fixed $t \in (0, T)$ and $x \in \mathbb{R}$, the process $\{v(t + s, W_s), \mathscr{F}_s; 0 \leq s < T - t\}$ is a supermartingale on $(\Omega, \mathscr{F}, P^x)$. Indeed, for $0 \leq s_1 \leq s_2 < T - t$, the Markov property yields

$$E^x[v(t + s_2, W_{s_2}) | \mathscr{F}_{s_1}](\omega) = f(W_{s_1}(\omega)), \quad \text{for } P^x\text{-a.e. } \omega \in \Omega, \tag{3.16}$$

where

$$f(y) \triangleq E^y v(t + s_2, W_{s_2 - s_1}) \tag{3.17}$$

(see Proposition 2.5.13). From (3.15), we have

$$E^y v(t + s_2, W_{s_2 - s_1}) \leq v(t + s_1, y),$$

and so for $0 < t \leq t + s_1 \leq t + s_2 < T, \quad x \in \mathbb{R}$:

$$v(t + s_1, W_{s_1}) \geq E^x[v(t + s_2, W_{s_2}) | \mathscr{F}_{s_1}], \quad \text{a.s. } P^x. \tag{3.18}$$

It is clear from this argument that if equality holds in (3.15), then $\{v(t+s, W_s), \mathscr{F}_s; 0 \leq s < T - t\}$ is a martingale. To complete our proof of *(ii)* ⇒ *(iii)* and *(ii)* ⇒ *(iv)*, we must establish the reverse of inequality (3.15).

Returning to (3.14), we may write

$$\begin{aligned} v(t, x) &= E^x[v(t + s, W_s) 1_{\{s \leq T_a \wedge T_b\}}] + E^x[v(t + T_a, a) 1_{\{T_a < s \wedge T_b\}}] \\ &\quad + E^x[v(t + T_b, b) 1_{\{T_b < s \wedge T_a\}}] \\ &\leq E^x v(t + s, W_s) + E^x[v(t + T_a, a) 1_{\{T_a < s\}}] \\ &\quad + E^x[v(t + T_b, b) 1_{\{T_b < s\}}]. \end{aligned}$$

We will have established (3.13) as soon as we prove

$$\varliminf_{b \to \infty} E^x[v(t + T_b, b) 1_{\{T_b < s\}}] = 0 \tag{3.19}$$

(a dual argument then shows that $\varliminf_{a \to -\infty} E^x[v(t + T_a, a) 1_{T_a < s\}}] = 0$). For (3.19), it suffices to show that with $B > 0$ large enough, we have

$$\int_B^\infty E^x[v(t + T_b, b) 1_{\{T_b < s\}}] \, db < \infty.$$

We choose $x \in \mathbb{R}$, $0 < t < T$, and $0 \leq s < t$ so that $s + t < T$. From (2.6.3) and (3.10) we have

$$P^x[T_b \in d\sigma] = h(\sigma; b - x) \, d\sigma; \quad b > x, \sigma > 0.$$

For $B \geq x$ sufficiently large, $h(\sigma; b - x)$ is an increasing function of $\sigma \in (0, s)$, provided $b \geq B$. Furthermore, for $r \in (s, t)$ and B perhaps larger, we have

$$h(s, b - x) \leq \sqrt{\frac{r}{s^3}}\; p(r; x, b); \quad b \geq B.$$

It follows that

$$\begin{aligned}\int_B^\infty E^x[v(t + T_b, b)1_{\{T_b < s\}}]\,db &= \int_B^\infty \int_0^s v(t + \sigma, b)h(\sigma, b - x)\,d\sigma\,db \\ &\leq \sqrt{\frac{r}{s^3}} \int_0^s \int_B^\infty v(t + \sigma, b)p(r; x, b)\,db\,d\sigma \\ &\leq \sqrt{\frac{r}{s^3}} \int_0^s E^x v(t + \sigma, W_r)\,d\sigma \\ &\leq \sqrt{\frac{r}{s^3}} \int_0^s v(t + \sigma - r, x)\,d\sigma < \infty,\end{aligned}$$

where the next to last inequality is a consequence of (3.15). This proves (3.13) for $x \in \mathbb{R}$, $0 < t \leq t + s < T$, as long as $s < t$.

We now remove the unwanted restriction $s < t$. We show by induction on the positive integers k that if

(3.20) $$0 < t \leq t + s < T, \quad s < kt,$$

then

(3.21) $$v(t, x) = E^x v(t + s, W_s); \quad x \in \mathbb{R}.$$

This will yield (3.13) for the range of values indicated there. We have just established that (3.20) implies (3.21) when $k = 1$. Assume this implication for some $k \geq 1$, so $\{v(t + s, W_s), \mathscr{F}_s; 0 \leq s < kt\}$ is a martingale. Choose $s_2 \in [kt, (k + 1)t)$ and $s_1 \in [0, kt)$ so that $0 < s_2 - s_1 < t$. Then

$$\begin{aligned}E^x v(t + s_2, W_{s_2}) &= E^x\{E^x[v(t + s_2, W_{s_2})|\mathscr{F}_{s_1}]\} \\ &= E^x v(t + s_1, W_{s_1}) = v(t, x),\end{aligned}$$

where we have used (3.16), (3.17), and the induction hypothesis in the form

$$E^y v(t + s_2, W_{s_2 - s_1}) = v(t + s_1, y).$$

Finally, we take up the implication *(iv)*⇒*(i)*. For $0 < \varepsilon < (T/4)$, $(T/2) < t < T$, (3.13) gives

$$v(t - \varepsilon, x) = E^x v(T - \varepsilon, W_{T-t}) = \int_{-\infty}^\infty \frac{p(T - t; x, y)}{p\left(\frac{T}{2}; 0, y\right)}\,dF_\varepsilon(y),$$

where F_ε is the nondecreasing function

$$F_\varepsilon(x) \triangleq \int_{-\infty}^x p\left(\frac{T}{2}; 0, y\right) v(T - \varepsilon, y)\,dy; \quad x \in \mathbb{R}.$$

Again from (3.13), $F_\varepsilon(\infty) = E^0 v(T-\varepsilon, W_{T/2}) = v((T/2)-\varepsilon, 0)$, and thus

$$\sup_{0<\varepsilon<(T/4)} F_\varepsilon(\infty) \le \max_{(T/4)\le s\le(T/2)} v(s,0) < \infty.$$

By Helly's theorem (Ash (1972), p. 329), there exists a sequence $\varepsilon_1 > \varepsilon_2 > \cdots > \varepsilon_k \downarrow 0$ and a nondecreasing function $F^*\colon \mathbb{R} \to [0,\infty)$, such that $\lim_{k\to\infty} F_{\varepsilon_k}(x) = F^*(x)$ for every x at which F^* is continuous. Because for fixed $x \in \mathbb{R}$ and $t \in ((T/2), T)$ the ratio $(p(T-t; x, y)/p((T/2); 0, y))$ is a bounded, continuous function of y, converging to zero as $|y| \to \infty$, we have

$$v(t,x) = \lim_{k\to\infty} v(t-\varepsilon_k, x) = \int_{-\infty}^{\infty} \frac{p(T-t; x, y)}{p\left(\frac{T}{2}; 0, y\right)}\, dF^*(y)$$

by the extended Helly-Bray lemma (Loève (1977), p. 183). Defining $F(x) = \int_0^x (dF^*(y)/p((T/2); 0, y))$, we have (3.11) for $(T/2) < t < T$, $x \in \mathbb{R}$.

If $0 < t \le (T/2)$, we choose $t_1 \in ((T/2), T)$ and use (3.13) to write

$$\begin{aligned} v(t,x) &= \int_{-\infty}^{\infty} p(t_1 - t; x, y) v(t_1, y)\, dy \\ &= \int_{-\infty}^{\infty}\int_{-\infty}^{\infty} p(t_1 - t; x, y) p(T - t_1; y, z)\, dy\, dF(z) \\ &= \int_{-\infty}^{\infty} p(T-t; x, z)\, dF(z). \end{aligned}$$

□

3.7 Corollary. *Let $u(t,x)$ be a nonnegative function defined on a strip $(0,T)\times\mathbb{R}$, where $0 < T \le \infty$. The following four conditions are equivalent:*

(i)′ *for some nondecreasing function $F\colon \mathbb{R}\to\mathbb{R}$,*

$$u(t,x) = \int_{-\infty}^{\infty} p(t; x, y)\, dF(y); \quad 0 < t < T, x \in \mathbb{R}; \tag{3.22}$$

(ii)′ *u is of class $C^{1,2}$ on $(0,T)\times\mathbb{R}$ and satisfies the heat equation (3.1) there;*

(iii)′ *for a Brownian family $\{W_s, \mathscr{F}_s; 0 \le s < \infty\}$, $(\Omega, \mathscr{F})$, $\{P^x\}_{x\in\mathbb{R}}$ and each fixed $t \in (0,T)$, $x \in \mathbb{R}$, the process $\{u(t-s, W_s), \mathscr{F}_s; 0 \le s < t\}$ is a martingale on $(\Omega, \mathscr{F}, P^x)$;*

(iv)′ *for a Brownian family $\{W_s, \mathscr{F}_s; 0 \le s < \infty\}$, $(\Omega, \mathscr{F})$, $\{P^x\}_{x\in\mathbb{R}}$ we have*

$$u(t,x) = E^x u(t-s, W_s); \quad 0 \le s < t < T, x \in \mathbb{R}. \tag{3.23}$$

Proof. If T is finite, we obtain this corollary by defining $v(t,x) = u(T-t,x)$ and appealing to Theorem 3.6. If $T = \infty$, then for each integer $n \ge 1$ we set $v_n(t,x) = u(n-t,x)$; $0 < t < n$, $x \in \mathbb{R}$. Applying Theorem 3.6 to each v_n we see that conditions (ii)′, (iii)′, and (iv)′ are equivalent, they are implied by (i)′, and they imply the existence, for any fixed $n \ge 1$, of a nondecreasing function $F\colon \mathbb{R}\to\mathbb{R}$ such that (3.22) holds on $(0,n)\times\mathbb{R}$. For $t \ge n$, we have from (3.23):

$$u(t,x) = E^x u\left(\frac{n}{2}, W_{t-(n/2)}\right) = \int_{-\infty}^{\infty} u\left(\frac{n}{2}, z\right) p\left(t - \frac{n}{2}; x, z\right) dz$$
$$= \int_{-\infty}^{\infty} \int_{-\infty}^{\infty} p\left(\frac{n}{2}; z, y\right) p\left(t - \frac{n}{2}; x, z\right) dz\, dF(y)$$
$$= \int_{-\infty}^{\infty} p(t; x, y)\, dF(y). \qquad \square$$

3.8 Exercise (Widder's Uniqueness Theorem).

(i) Let $u(t,x)$ be a nonnegative function of class $C^{1,2}$ defined on the strip $(0,T) \times \mathbb{R}$, where $0 < T \le \infty$, and assume that u satisfies (3.1) on this strip and

$$\lim_{\substack{t \downarrow 0 \\ y \to x}} u(t,y) = 0; \quad x \in \mathbb{R}.$$

Show that $u = 0$ on $(0,T) \times \mathbb{R}$. (*Hint*: Establish the uniform integrability of the martingale $u(t-s, W_s)$; $0 \le s < t$.)

(ii) Let u be as in (i), except now assume that $\lim_{t \downarrow 0, y \to x} u(t,y) = f(x)$; $x \in \mathbb{R}$. Assuming that $f(\cdot)$ is continuous, show that

$$u(t,x) = \int_{-\infty}^{\infty} p(t; x, y) f(y)\, dy; \quad 0 < t < T, x \in \mathbb{R}.$$

Can we represent nonnegative solutions $v(t,x)$ of the backward heat equation (3.12) on the entire half-plane $(0,\infty) \times \mathbb{R}$, just as we did in Corollary 3.7 for nonnegative solutions $u(t,x)$ of the heat equation (3.1)? Certainly this cannot be achieved by a simple time-reversal on the results of Corollary 3.7. Instead, we can relate the functions u and v by the formula

$$\text{(3.24)} \qquad v(t,x) = \sqrt{\frac{2\pi}{t}} \exp\left(\frac{x^2}{2t}\right) u\left(\frac{1}{t}, \frac{x}{t}\right); \quad 0 < t < \infty, x \in \mathbb{R}.$$

The reader can readily verify that v satisfies (3.12) on $(0,\infty) \times \mathbb{R}$ if and only if u satisfies (3.1) there. The change of variables implicit in (3.24) allows us to deduce the following proposition from Corollary 3.7.

3.9 Proposition (Robbins & Siegmund (1973)). *Let $v(t,x)$ be a nonnegative function defined on the half-plane $(0,\infty) \times \mathbb{R}$. With $T = \infty$, conditions* (ii), (iii), *and* (iv) *of Theorem* 3.6 *are equivalent to one another, and to* (i)″:

(i)″ *for some nondecreasing function $F: \mathbb{R} \to \mathbb{R}$,*

$$\text{(3.25)} \qquad v(t,x) = \int_{-\infty}^{\infty} \exp\left(yx - \frac{1}{2} y^2 t\right) dF(y); \quad 0 < t < \infty, x \in \mathbb{R}.$$

PROOF. The equivalence of (ii), (iii), and (iv) for $T = \infty$ follows from their equivalence for all finite T. If v is given by (3.25), then differentiation under the integral can be justified as in Theorem 3.6, and it results in (3.12). If v satisfies (ii), then u given by (3.24) satisfies (ii)′, and hence (i)′, of Corollary 3.7. But (3.24) and (3.22) reduce to (3.25). □

C. Boundary-Crossing Probabilities for Brownian Motion

The representation (3.25) has rather unexpected consequences in the computation of *boundary-crossing probabilities* for Brownian motion. Let us consider a positive function $v(t, x)$ which is defined and of class $C^{1,2}$ on $(0, \infty) \times \mathbb{R}$, and satisfies the backward heat equation. Then v admits the representation (3.25) for some F, and differentiating under the integral we see that

$$\text{(3.26)} \qquad \frac{\partial}{\partial t} v(t, x) < 0; \quad 0 < t < \infty, x \in \mathbb{R}$$

and that $v(t, \cdot)$ is convex for each $t > 0$. In particular, $\lim_{t \downarrow 0} v(t, 0)$ exists. We assume that this limit is finite, and, without loss of generality (by scaling, if necessary), that

$$\text{(3.27)} \qquad \lim_{t \downarrow 0} v(t, 0) = 1.$$

We also assume that

$$\text{(3.28)} \qquad \lim_{t \to \infty} v(t, 0) = 0,$$

$$\text{(3.29)} \qquad \lim_{x \to \infty} v(t, x) = \infty; \quad 0 < t < \infty,$$

$$\text{(3.30)} \qquad \lim_{x \to -\infty} v(t, x) = 0, \quad 0 < t < \infty.$$

It is easily seen that (3.27)–(3.30) are satisfied if and only if F *is a probability distribution function with* $F(0+) = 0$. We impose this condition, so that (3.25) becomes

$$\text{(3.31)} \qquad v(t, x) = \int_{0+}^{\infty} \exp\left(yx - \frac{1}{2}y^2 t\right) dF(y); \quad 0 < t < \infty, x \in \mathbb{R},$$

where $F(\infty) = 1$, $F(0+) = 0$. This representation shows that $v(t, \cdot)$ is strictly increasing, so for each $t > 0$ and $b > 0$ there is a unique number $A(t, b)$ such that

$$\text{(3.32)} \qquad v(t, A(t, b)) = b.$$

It is not hard to verify that the function $A(\cdot, b)$ is continuous and strictly increasing (cf. (3.26)). We may define $A(0, b) = \lim_{t \downarrow 0} A(t, b)$.

We shall show how *one can compute the probability that a Brownian path* W, *starting at the origin, will eventually cross the curve* $A(\cdot, b)$. The problem of computing the probability that a Brownian motion crosses a given, time-dependent continuous boundary $\{\psi(t); 0 \le t < \infty\}$ is thereby reduced to finding a solution v to the backward heat equation which also satisfies (3.27)–(3.30) and $v(t, \psi(t)) = b; 0 \le t < \infty$, for some $b > 0$. In this generality both problems are quite difficult; our point is that the probabilistic problem can be traded for a partial differential equation problem. We shall provide an explicit solution to both of them when the boundary is linear.

Let $\{W_t, \mathscr{F}_t; 0 \le t < \infty\}$, $(\Omega, \mathscr{F})$, $\{P^x\}_{x \in \mathbb{R}}$ be a Brownian family, and define

$$Z_t = v(t, W_t); \quad 0 < t < \infty.$$

For $0 < s < t$, we have from the Markov property and condition (iv) of Proposition 3.9:

$$E^0[Z_t | \mathscr{F}_s] = f(W_s) = v(s, W_s) = Z_s, \quad \text{a.s. } P^0,$$

where $f(y) \triangleq E^y v(t, W_{t-s})$. In other words, $\{Z_t, \mathscr{F}_t; 0 < t < \infty\}$ is a continuous, nonnegative martingale on $(\Omega, \mathscr{F}, P^0)$. Let $\{t_n\}$ be a sequence of positive numbers with $t_n \downarrow 0$, and set $Z_0 = \lim_{n \to \infty} Z_{t_n}$. This limit exists, P^0-a.s., and is independent of the particular sequence $\{t_n\}$ chosen; see the proof of Proposition 1.3.14(i). Being $\mathscr{F}^W_{0+}$-measurable, Z_0 must be a.s. constant (Theorem 2.7.17).

3.10 Lemma. *The extended process $Z \triangleq \{Z_t, \mathscr{F}_t; 0 \le t < \infty\}$ is a continuous, nonnegative martingale under P^0 and satisfies $Z_0 = 1$, $Z_\infty = 0$, P^0-a.s.*

Proof. Let $\{t_n\}$ be a sequence of positive numbers with $t_n \downarrow 0$. The sequence $\{Z_{t_n}\}_{n=1}^\infty$ is uniformly integrable (Problem 1.3.11, Remark 1.3.12), so by the Markov property for W, we have for all $t > 0$:

$$E^0[Z_t | \mathscr{F}_0] = E^0 Z_t = \lim_{n \to \infty} E^0 Z_{t_n} = E^0 Z_0 = Z_0.$$

This establishes that $\{Z_t, \mathscr{F}_t; 0 \le t < \infty\}$ is a martingale.

Since $Z_\infty \triangleq \lim_{t \to \infty} Z_t$ exists P^0-a.s. (Problem 1.3.16), as does $Z_0 \triangleq \lim_{t \downarrow 0} Z_t$, it suffices to show that $\lim_{t \downarrow 0} Z_t = 1$ and $\lim_{t \to \infty} Z_t = 0$ in P^0-probability. For every finite $c > 0$, we shall show that

$$\lim_{t \downarrow 0} \sup_{|x| \le c\sqrt{t}} |v(t, x) - 1| = 0. \tag{3.33}$$

Indeed, for $t > 0$, $|x| \le c\sqrt{t}$:

$$\int_{0+}^\infty \exp\left(-yc\sqrt{t} - \frac{1}{2}y^2 t\right) dF(y) \le v(t, x) \le \int_{0+}^\infty \exp\left(yc\sqrt{t} - \frac{1}{2}y^2 t\right) dF(y). \tag{3.34}$$

Because $\pm yc\sqrt{t} - y^2 t/2 \le c^2/2$; $\forall\, y > 0$, the bounded convergence theorem implies that both integrals in (3.34) converge to 1, as $t \downarrow 0$, and (3.33) follows. Thus, for any $\varepsilon > 0$, we can find $t_{c,\varepsilon}$, depending on c and ε, such that

$$1 - \varepsilon < v(t,x) < 1 + \varepsilon; \quad |x| \le c\sqrt{t}, \quad 0 < t < t_{c,\varepsilon}.$$

Consequently, for $0 < t < t_{c,\varepsilon}$,

$$P^0[|Z_t - 1| > \varepsilon] = P^0[|v(t, W_t) - 1| > \varepsilon] \le P^0[|W_t| > c\sqrt{t}] = 2[1 - \Phi(c)],$$

where

$$\Phi(x) \triangleq \frac{1}{\sqrt{2\pi}} \int_{-\infty}^{x} e^{-z^2/2}\, dz.$$

Letting first $t \downarrow 0$ and then $c \to \infty$, we conclude that $Z_t \to 1$ in probability as $t \downarrow 0$. A similar argument shows that

$$\lim_{t\to\infty} \sup_{|x| \le c\sqrt{t}} v(t,x) = 0, \tag{3.35}$$

and, using (3.35) instead of (3.33), one can also show that $Z_t \to 0$ in probability as $t \to \infty$. □

It is now a fairly straightforward matter to apply Problem 1.3.28 to the martingale Z and obtain the probability that the Brownian path $\{W_t(\omega); 0 \le t < \infty\}$ ever crosses the boundary $\{A(t,b); 0 \le t < \infty\}$.

3.11 Problem. Suppose that $v: (0,\infty) \times \mathbb{R} \to (0,\infty)$ is of class $C^{1,2}$ and satisfies (3.12) and (3.27)–(3.30). For fixed $b > 0$, let $A(\cdot\,, b): [0,\infty) \to \mathbb{R}$ be the continuous function satisfying (3.32). Then, for any $s > 0$ and Lebesgue-almost every $a \in \mathbb{R}$ with $v(s,a) < b$, we have

$$P^0[W_t \ge A(t,b), \text{ for some } t \ge s \mid W_s = a] = \frac{v(s,a)}{b}, \tag{3.36}$$

$$\begin{aligned} &P^0[W_t \ge A(t,b), \text{ for some } t \ge s] \\ &\quad = 1 - \Phi\left(\frac{A(s,b)}{\sqrt{s}}\right) + \frac{1}{b}\int_{0+}^{\infty} \Phi\left(\frac{A(s,b)}{\sqrt{s}} - y\sqrt{s}\right) dF(y), \end{aligned} \tag{3.37}$$

where F is the probability distribution function in (3.31).

3.12 Example. With $\mu > 0$, let $v(t,x) = \exp(\mu x - \mu^2 t/2)$, so $A(t,b) = \beta t + \gamma$, where $\beta = (\mu/2)$, $\gamma = (1/\mu)\log b$. Then $F(y) = 1_{[\mu,\infty)}(y)$, and so for any $s > 0$, $\beta > 0$, $\gamma \in \mathbb{R}$, and Lebesgue-almost every $a < \gamma + \beta s$:

$$P^0[W_t \ge \beta t + \gamma, \text{ for some } t \ge s \mid W_s = a] = e^{-2\beta(\gamma - a + \beta s)}, \tag{3.38}$$

and for any $s > 0$, $\beta > 0$, and $\gamma \in \mathbb{R}$:

$$\begin{aligned} P^0[W_t \ge \beta t + \gamma, \text{ for some } t \ge s] &= 1 - \Phi\left(\frac{\gamma}{\sqrt{s}} + \beta\sqrt{s}\right) \\ &\quad + e^{-2\beta\gamma}\Phi\left(\frac{\gamma}{\sqrt{s}} - \beta\sqrt{s}\right). \end{aligned} \tag{3.39}$$

The observation that the time-inverted process Y of Lemma 2.9.4 is a Brownian motion allows one to cast (3.38) with $\gamma = 0$ into the following formula for the maximum of the so-called "tied-down" Brownian motion or "Brownian bridge":

(3.40)
$$P^0\left[\max_{0 \le t \le T} W_t \ge \beta \,\middle|\, W_T = a\right] = e^{-2\beta(\beta-a)/T}$$

for $T > 0$, $\beta > 0$, a.e. $a \le \beta$, and (3.39) into a boundary-crossing probability on the bounded interval $[0, T]$:

(3.41) $P^0[W_t \ge \beta + \gamma t, \text{ for some } t \in [0, T]]$
$$= 1 - \Phi\left(\gamma\sqrt{T} + \frac{\beta}{\sqrt{T}}\right) + e^{-2\beta\gamma}\Phi\left(\gamma\sqrt{T} - \frac{\beta}{\sqrt{T}}\right); \quad \beta > 0, \gamma \in \mathbb{R}.$$

3.13 Exercise. Show that $P^0[W_t \ge \beta t + \gamma, \text{ for some } t \ge 0] = e^{-2\beta\gamma}$, for $\beta > 0$ and $\gamma > 0$ (recall Exercise 3.5.9).

D. Mixed Initial/Boundary Value Problems

We now discuss briefly the concept of temperatures in a semi-infinite rod and the relation of this concept to *Brownian motion absorbed at the origin.* Suppose that $f: (0, \infty) \to \mathbb{R}$ is a Borel-measurable function satisfying

(3.42)
$$\int_0^\infty e^{-ax^2}|f(x)|\,dx < \infty$$

for some $a > 0$. We define

(3.43)
$$u_1(t, x) \triangleq E^x[f(W_t)1_{\{T_0 > t\}}]; \quad 0 < t < \frac{1}{2a}, \quad x > 0.$$

The reflection principle gives us the formula (2.8.9)

$$P^x[W_t \in dy, T_0 > t] = p_-(t; x, y)\,dy \triangleq [p(t; x, y) - p(t; x, -y)]\,dy$$

for $t > 0$, $x > 0$, $y > 0$, and so

(3.44)
$$u_1(t, x) = \int_0^\infty f(y)p(t; x, y)\,dy - \int_{-\infty}^0 f(-y)p(t; x, y)\,dy,$$

which gives us a definition for u_1 valid on the whole strip $(0, 1/2a) \times \mathbb{R}$. This representation is of the form (3.4), where the initial datum f satisfies $f(y) = -f(-y)$; $y > 0$. It is clear then that u_1 has derivatives of all orders, satisfies the heat equation (3.1), satisfies (3.6) at all continuity points of f, and

$$\lim_{\substack{s \to t \\ x \downarrow 0}} u_1(s, x) = 0; \quad 0 < t < \frac{1}{2a}.$$

We may regard $u_1(t,x)$; $0 < t < (1/2a)$, $x \geq 0$, as the temperature in a semi-infinite rod along the nonnegative x-axis, when the end $x = 0$ is held at a constant temperature (equal to zero) and the initial temperature at $y > 0$ is $f(y)$.

Suppose now that the initial temperature in a semi-infinite rod is identically zero, but the temperature at the endpoint $x = 0$ at time t is $g(t)$, where $g\colon (0, 1/2a) \to \mathbb{R}$ is bounded and continuous. The *Abel transform* of g, namely

$$
\begin{aligned}
(3.45)\qquad u_2(t,x) &\triangleq E^x[g(t - T_0)1_{\{T_0 \leq t\}}] \\
&= \int_0^t g(t-\tau)h(\tau,x)\,d\tau \\
&= \int_0^t g(s)h(t-s,x)\,ds; \quad 0 < t < \frac{1}{2a},\ x > 0
\end{aligned}
$$

with h given by (3.10), is a solution to (3.1) because h is, and $h(0,x) = 0$ for $x > 0$. We may rewrite this formula as

$$u_2(t,x) = E^0[g(t - T_x)1_{\{T_x \leq t\}}]; \quad 0 < t < \frac{1}{2a},\ x > 0,$$

and then the bounded convergence theorem shows that

$$\lim_{\substack{s \to t \\ x \downarrow 0}} u_2(s,x) = g(t); \quad 0 < t < \frac{1}{2a},$$

$$\lim_{\substack{t \downarrow 0 \\ y \to x}} u_2(t,y) = 0; \quad 0 < x < \infty.$$

We may add u_1 and u_2 to obtain a solution to the problem with initial datum f and time-dependent boundary condition $g(t)$ at $x = 0$.

3.14 Exercise (Neumann Boundary Condition). Suppose that $f\colon (0,\infty) \to \mathbb{R}$ is a Borel-measurable function satisfying (3.42), and define

$$u(t,x) \triangleq E^x f(|W_t|); \quad 0 < t < \frac{1}{2a},\ x > 0.$$

Show that u is of class $C^{1,2}$, satisfies (3.1) on $(0, 1/2a) \times (0,\infty)$ and (3.6) at all continuity points of f, as well as

$$\lim_{\substack{s \to t \\ x \downarrow 0}} \frac{\partial}{\partial x} u(s,x) = 0; \quad 0 < t < \frac{1}{2a}.$$

3.15 Exercise (Finite Rod). Suppose that g, k are bounded, continuous functions from $(0,\infty)$ into $\mathbb{R}$, and f is a bounded, continuous function from $(0,b)$ into $\mathbb{R}$. We seek a function u which is of class $C^{1,2}$ on $(0,\infty) \times (0,b)$ and which has a continuous extension to the boundaries $\{0\} \times (0,b)$, $(0,\infty) \times \{0\}$, and

$(0, \infty) \times \{b\}$, such that

$$\frac{\partial u}{\partial t} = \frac{1}{2}\frac{\partial^2 u}{\partial x^2} \quad \text{on } (0,\infty)\times(0,b),$$

$$u(0,x) = f(x); \quad 0 < x < b,$$

$$u(t,0) = g(t); \quad 0 < t < \infty,$$

$$u(t,b) = k(t); \quad 0 < t < \infty.$$

Show that the unique bounded solution to this problem is given by the expression

(3.46) $$u(t,x) = E^x[f(W_t)1_{\{t < T_0 \wedge T_b\}} + g(t - T_0)1_{\{T_0 < t \wedge T_b\}} + k(t - T_b)1_{\{T_b < t \wedge T_0\}}]; \quad 0 < t < \infty, 0 < x < b.$$

(*Hint:* Use Proposition 2.8.10 and Formulas (2.8.25), (2.8.26).)

4.4. The Formulas of Feynman and Kac

We continue our program of obtaining stochastic representations for solutions of partial differential equations. In the first subsection, we introduce the *Feynman-Kac formula*, which provides such a representation for the solution of the *parabolic equation*

(4.1) $$\frac{\partial u}{\partial t} + ku = \frac{1}{2}\Delta u + g; \quad (t,x) \in (0,\infty)\times \mathbb{R}^d$$

subject to the initial condition

(4.2) $$u(0,x) = f(x); \quad x \in \mathbb{R}^d$$

for suitable functions $k: \mathbb{R}^d \to [0,\infty)$, $g: (0,\infty)\times\mathbb{R}^d \to \mathbb{R}$ and $f: \mathbb{R}^d \to \mathbb{R}$.

In the special case of $g = 0$, we may define the Laplace transform

$$z_\alpha(x) \triangleq \int_0^\infty e^{-\alpha t} u(t,x)\,dt; \quad x \in \mathbb{R}^d,$$

and using (4.1), (4.2), integration by parts, and the assumption that $\lim_{t\to\infty} e^{-\alpha t}u(t,x) = 0$; $\alpha > 0$, $x \in \mathbb{R}^d$, we may compute formally

(4.3) $$\frac{1}{2}\Delta z_\alpha = \frac{1}{2}\int_0^\infty e^{-\alpha t}\Delta u\,dt = (\alpha + k)z_\alpha - f.$$

The stochastic representation for the solution z_α of the *elliptic equation* (4.3) is known as the *Kac formula*; in the second subsection we illustrate its use when $d = 1$ by computing the distributions of occupation times for Brownian motion. The second subsection may be read independently of the first one.

Throughout this section, $\{W_t, \mathscr{F}_t; 0 \le t < \infty\}$, $(\Omega, \mathscr{F})$, $\{P^x\}_{x \in \mathbb{R}^d}$ is a d-dimensional Brownian family.

A. The Multidimensional Formula

4.1 Definition. Consider continuous functions $f: \mathbb{R}^d \to \mathbb{R}$, $k: \mathbb{R}^d \to [0, \infty)$, and $g: [0, T] \times \mathbb{R}^d \to \mathbb{R}$. Suppose that v is a continuous, real-valued function on $[0, T] \times \mathbb{R}^d$, of class $C^{1,2}$ on $[0, T) \times \mathbb{R}^d$ (see the explanation preceding Theorem 3.3), and satisfies

$$-\frac{\partial v}{\partial t} + kv = \frac{1}{2}\Delta v + g; \quad \text{on } [0, T) \times \mathbb{R}^d, \tag{4.4}$$

$$v(T, x) = f(x); \quad x \in \mathbb{R}^d. \tag{4.5}$$

Then the function v is said to be a solution of the *Cauchy problem* for the backward heat equation (4.4) with *potential k* and *Lagrangian g*, subject to the terminal condition (4.5).

4.2 Theorem (Feynman (1948), Kac (1949)). *Let v be as in Definition* 4.1 *and assume that*

$$\max_{0 \le t \le T} |v(t, x)| + \max_{0 \le t \le T} |g(t, x)| \le K e^{a\|x\|^2}; \quad \forall x \in \mathbb{R}^d, \tag{4.6}$$

for some constants $K > 0$ and $0 < a < 1/(2Td)$. Then v admits the stochastic representation

$$v(t, x) = E^x\left[f(W_{T-t})\exp\left\{-\int_0^{T-t} k(W_s)\,ds\right\} + \int_0^{T-t} g(t + \theta, W_\theta)\exp\left\{-\int_0^\theta k(W_s)\,ds\right\}d\theta\right]; \quad 0 \le t \le T, x \in \mathbb{R}^d. \tag{4.7}$$

In particular, such a solution is unique.

4.3 Remark. If $g \ge 0$ on $[0, T] \times \mathbb{R}^d$, then condition (4.6) may be replaced by

$$\max_{0 \le t \le T} |v(t, x)| \le K e^{a\|x\|^2}; \quad \forall x \in \mathbb{R}^d. \tag{4.8}$$

This leads to the following *maximum principle* for the Cauchy problem: if the continuous function v: $[0, T] \times \mathbb{R}^d \to \mathbb{R}$ is of class $C^{1,2}$ on $[0, T) \times \mathbb{R}^d$ and satisfies the growth condition (4.8), as well as the differential inequality

$$-\frac{\partial v}{\partial t} + kv \ge \frac{1}{2}\Delta v \quad \text{on } [0, T) \times \mathbb{R}^d$$

with a continuous potential $k: \mathbb{R}^d \to [0, \infty)$, then $v \ge 0$ on $\{T\} \times \mathbb{R}^d$ implies $v \ge 0$ on $[0, T] \times \mathbb{R}^d$.

4.4 Remark. If we do not assume the existence of a $C^{1,2}$ solution to the Cauchy problem (4.4), (4.5), then the function defined by the right-hand side of (4.7) need not be $C^{1,2}$. The reader is referred to Friedman (1964), Chapter 1, Friedman (1975), p. 147, or Dynkin (1965), Theorem 13.16, for conditions under which the Cauchy problem of Definition 4.1 admits a solution. This is the case, for example, if k is bounded and uniformly Hölder-continuous on compact subsets of $\mathbb{R}^d$, g is continuous on $[0, T] \times \mathbb{R}^d$ and Hölder-continuous in x uniformly with respect to $(t, x) \in [0, T] \times \mathbb{R}^d$, f is continuous, and for some constants L and $\nu > 0$ we have

$$\max_{0 \le t \le T} |g(t,x)| + |f(x)| \le L(1 + \|x\|^\nu); \quad x \in \mathbb{R}^d.$$

Proof of Theorem 4.2. We obtain from Itô's rule, in conjunction with (4.4):

$$d\left[v(t+\theta, W_\theta)\exp\left\{-\int_0^\theta k(W_s)\,ds\right\}\right]$$
$$= \exp\left\{-\int_0^\theta k(W_s)\,ds\right\}\left[-g(t+\theta, W_\theta)\,d\theta + \sum_{i=1}^d \frac{\partial}{\partial x_i} v(t+\theta, W_\theta)\,dW_\theta^{(i)}\right].$$

Let $S_n = \inf\{t \ge 0;\ \|W_t\| \ge n\sqrt{d}\};\ n \ge 1$. We choose $0 < r < T - t$ and integrate on $[0, r \wedge S_n]$; the resulting stochastic integrals have expectation zero, so

$$v(t,x) = E^x \int_0^{r \wedge S_n} g(t+\theta, W_\theta)\exp\left\{-\int_0^\theta k(W_s)\,ds\right\} d\theta$$
$$+ E^x\left[v(t+S_n, W_{S_n})\exp\left\{-\int_0^{S_n} k(W_s)\,ds\right\} 1_{\{S_n \le r\}}\right]$$
$$+ E^x\left[v(t+r, W_r)\exp\left\{-\int_0^r k(W_s)\,ds\right\} 1_{\{S_n > r\}}\right].$$

The first term on the right-hand side converges to

$$E^x \int_0^{T-t} g(t+\theta, W_\theta)\exp\left\{-\int_0^\theta k(W_s)\,ds\right\} d\theta$$

as $n \to \infty$ and $r \uparrow T - t$, either by monotone convergence (if $g \ge 0$) or by dominated convergence (it is bounded in absolute value by $\int_0^{T-t} |g(t+\theta, W_\theta)|\,d\theta$, which has finite expectation by virtue of (4.6)). The second term is dominated by

$$\begin{aligned} E^x[|v(t+S_n, W_{S_n})| 1_{\{S_n \le T-t\}}] &\le K e^{adn^2} P^x[S_n \le T] \\ &\le K e^{adn^2} \sum_{j=1}^d P^x\left[\max_{0 \le t \le T} |W_t^{(j)}| \ge n\right] \\ &\le 2K e^{adn^2} \sum_{j=1}^d \{P^x[W_T^{(j)} \ge n] \\ &\qquad + P^x[-W_T^{(j)} \ge n]\}, \end{aligned}$$

where we have used (2.6.2). But by (2.9.20),

$$e^{adn^2}P^x[\pm W_T^{(j)} \geq n] \leq e^{adn^2}\sqrt{\frac{T}{2\pi}}\frac{1}{n \mp x^{(j)}}e^{-(n\mp x^{(j)})^2/2T}$$

which converges to zero as $n \to \infty$, because $0 < a < 1/(2Td)$. Again by the dominated convergence theorem, the third term is shown to converge to $E^x[v(T, W_{T-t})\exp\{-\int_0^{T-t}k(W_s)\,ds\}]$ as $n \to \infty$ and $r \uparrow T - t$. The Feynman-Kac formula (4.7) follows. □

4.5 Corollary. *Assume that $f\colon \mathbb{R}^d \to \mathbb{R}$, $k\colon \mathbb{R}^d \to [0, \infty)$, and $g\colon [0, \infty) \times \mathbb{R}^d \to \mathbb{R}$ are continuous, and that the continuous function $u\colon [0, \infty) \times \mathbb{R}^d \to \mathbb{R}$ is of class $C^{1,2}$ on $(0, \infty) \times \mathbb{R}^d$ and satisfies (4.1) and (4.2). If for each finite $T > 0$ there exist constants $K > 0$ and $0 < a < 1/(2Td)$ such that*

$$\max_{0 \leq t \leq T}|u(t,x)| + \max_{0 \leq t \leq T}|g(t,x)| \leq Ke^{a\|x\|^2}; \quad \forall x \in \mathbb{R}^d,$$

then u admits the stochastic representation

(4.9)
$$u(t,x) = E^x\left[f(W_t)\exp\left\{-\int_0^t k(W_s)\,ds\right\} + \int_0^t g(t-\theta, W_\theta)\exp\left\{-\int_0^\theta k(W_s)\,ds\right\}d\theta\right]; \quad 0 \leq t < \infty, x \in \mathbb{R}^d.$$

In the case $g = 0$ we can think of $u(t,x)$ in (4.1) as the temperature at time $t \geq 0$ at the point $x \in \mathbb{R}^d$ of a medium which is not a perfect heat conductor, but instead dissipates heat locally at the rate k (*heat flow with cooling*). The Feynman-Kac formula (4.9) suggests that this situation is equivalent to *Brownian motion with annihilation* (killing) of particles at the same rate k: the probability that the particle survives up to time t, conditional on the path $\{W_s; 0 \leq s \leq t\}$, is then $\exp\{-\int_0^t k(W_s)\,ds\}$.

4.6 Exercise. Consider the Cauchy problem for the "quasilinear" parabolic equation

(4.10)
$$\frac{\partial V}{\partial t} = \frac{1}{2}\Delta V - \frac{1}{2}\|\nabla V\|^2 + k; \quad \text{in } (0, \infty) \times \mathbb{R}^d,$$

(4.11)
$$V(0,x) = 0; \quad x \in \mathbb{R}^d$$

(linear in $(\partial V/\partial t)$ and the Laplacian ΔV, nonlinear in the gradient ∇V), where $k\colon \mathbb{R}^d \to [0, \infty)$ is a continuous function. Show that if $V\colon [0, \infty) \times \mathbb{R}^d \to \mathbb{R}$ is a solution which is continuous on its domain, of class $C^{1,2}$ on $(0, \infty) \times \mathbb{R}^d$, and satisfies the quadratic growth condition for every $T > 0$:

$$-V(t,x) \leq C + a\|x\|^2; \quad (t,x) \in [0, T] \times \mathbb{R}^d$$

where $T > 0$ is arbitrary and $0 < a < 1/(2Td)$, then V is given by

$$(4.12)\quad V(t,x) = -\log E^x\left[\exp\left\{-\int_0^t k(W_s)\,ds\right\}\right]; \quad 0 \le t < \infty, x \in \mathbb{R}^d.$$

We turn now our attention to equation (4.3). The discussion at the beginning of this section and equation (4.9) suggest that any solution z to the equation

$$(4.13)\qquad (\alpha + k)z = \frac{1}{2}\Delta z + f; \quad \text{on } \mathbb{R}^d$$

should be represented as

$$(4.14)\qquad z(x) = E^x \int_0^\infty f(W_t)\exp\left\{-\alpha t - \int_0^t k(W_s)\,ds\right\}dt.$$

4.7 Exercise. Let $f\colon \mathbb{R}^d \to \mathbb{R}$ and $k\colon \mathbb{R}^d \to [0, \infty)$ be continuous, with

$$(4.15)\qquad E^x \int_0^\infty |f(W_t)| \exp\left\{-\alpha t - \int_0^t k(W_s)\,ds\right\}dt < \infty; \quad \forall\, x \in \mathbb{R}^d,$$

for some constant $\alpha > 0$. Suppose that $\psi\colon \mathbb{R}^d \to \mathbb{R}$ is a solution of class C^2 to (4.13), and let z be defined by (4.14). If ψ is bounded, then $\psi = z$; if ψ is nonnegative, then $\psi \ge z$. *(Hint:* Use Problem 2.25).

B. The One-Dimensional Formula

In the one-dimensional case, the stochastic representation (4.14) has the remarkable feature that it defines a function of class C^2 when f and k are continuous. Contrast this, for example, to Remark 4.4. We prove here a slightly more general result.

4.8 Definition. A Borel-measurable function $f\colon \mathbb{R} \to \mathbb{R}$ is called *piecewise-continuous* if it admits left- and right-hand limits everywhere on $\mathbb{R}$ and it has only finitely many points of discontinuity in every bounded interval. We denote by D_f the set of discontinuity points of f. A continuous function $f\colon \mathbb{R} \to \mathbb{R}$ is called *piecewise* C^j, $j \ge 1$, if its derivatives $f^{(i)}$, $1 \le i \le j-1$ are continuous, and the derivative $f^{(j)}$ is piecewise-continuous.

4.9 Theorem (Kac (1951)). *Let $f\colon \mathbb{R} \to \mathbb{R}$ and $k\colon \mathbb{R} \to [0, \infty)$ be piecewise-continuous functions with*

$$(4.16)\qquad \int_{-\infty}^\infty |f(x+y)| e^{-|y|\sqrt{2\alpha}}\,dy < \infty; \quad \forall\, x \in \mathbb{R},$$

for some fixed constant $\alpha > 0$. Then the function z defined by (4.14) *is piecewise C^2 and satisfies*

(4.17) $$(\alpha + k)z = \frac{1}{2}z'' + f; \quad on\ \mathbb{R}\backslash(D_f \cup D_k).$$

4.10 Remark. The Laplace transform computation

$$\int_0^\infty e^{-\alpha t}\frac{1}{\sqrt{2\pi t}}e^{-\xi^2/2t}\,dt = \frac{1}{\sqrt{2\alpha}}e^{-|\xi|\sqrt{2\alpha}}; \quad \alpha > 0,\ \xi \in \mathbb{R}$$

enables us to replace (4.16) by the equivalent condition

(4.16)′ $$E^x \int_0^\infty e^{-\alpha t}|f(W_t)|\,dt < \infty, \quad \forall x \in \mathbb{R}.$$

PROOF OF THEOREM 4.9. For piecewise-continuous functions g which satisfy condition (4.16), we introduce the *resolvent operator* G_α given by

$$(G_\alpha g)(x) \triangleq E^x \int_0^\infty e^{-\alpha t} g(W_t)\,dt = \frac{1}{\sqrt{2\alpha}}\int_{-\infty}^\infty e^{-|y-x|\sqrt{2\alpha}} g(y)\,dy$$

$$= \frac{1}{\sqrt{2\alpha}}\left[\int_{-\infty}^x e^{(y-x)\sqrt{2\alpha}} g(y)\,dy + \int_x^\infty e^{(x-y)\sqrt{2\alpha}} g(y)\,dy\right]; \quad x \in \mathbb{R}.$$

Differentiating, we obtain

$$(G_\alpha g)'(x) = \int_x^\infty e^{(x-y)\sqrt{2\alpha}} g(y)\,dy - \int_{-\infty}^x e^{(y-x)\sqrt{2\alpha}} g(y)\,dy; \quad x \in \mathbb{R},$$

(4.18) $$(G_\alpha g)''(x) = -2g(x) + 2\alpha(G_\alpha g)(x); \quad x \in \mathbb{R}\backslash D_g.$$

It will be shown later that

(4.19) $$G_\alpha(kz) = G_\alpha f - z$$

and

(4.20) $$G_\alpha(|kz|)(x) < \infty; \quad \forall x \in \mathbb{R}.$$

If we then write (4.18) successively with $g = f$ and $g = kz$ and subtract, we obtain the desired equation (4.17) for $x \in \mathbb{R}\backslash(D_f \cup D_{kz})$, thanks to (4.19). One can easily check via the dominated convergence theorem that z is continuous, so $D_{kz} \subseteq D_k$. Integration of (4.17) yields the continuity of z'.

In order to verify (4.19), we start with the observation

$$0 \le \int_0^t k(W_s)\exp\left\{-\int_s^t k(W_u)\,du\right\}ds = 1 - \exp\left\{-\int_0^t k(W_u)\,du\right\} \le 1; \quad t \ge 0,$$

and so by Fubini's theorem and the Markov property:

$$(G_\alpha f - z)(x) = E^x \int_0^\infty e^{-\alpha t}\left(1 - e^{-\int_0^t k(W_s)\,ds}\right) f(W_t)\,dt$$

$$= E^x \int_0^\infty e^{-\alpha t} f(W_t) \int_0^t k(W_s)\exp\left\{-\int_s^t k(W_u)\,du\right\}ds\,dt$$

$$= \int_0^\infty E^x\left[k(W_s)\int_s^\infty \exp\left\{-\alpha t - \int_s^t k(W_u)\,du\right\} f(W_t)\,dt\right] ds$$

$$= \int_0^\infty e^{-\alpha s} E^x\left[k(W_s)\int_0^\infty \exp\left\{-\alpha t - \int_0^t k(W_{s+u})\,du\right\} f(W_{s+t})\,dt\right] ds$$

$$= E^x \int_0^\infty e^{-\alpha s} k(W_s)\cdot E^x\left[\left[\int_0^\infty \exp\left\{-\alpha t - \int_0^t k(W_{s+u})\,du\right\} f(W_{s+t})\,dt \,\middle|\, \mathscr{F}_s\right]\right] ds$$

$$= E^x \int_0^\infty e^{-\alpha s} k(W_s) z(W_s)\,ds = (G_\alpha(kz))(x); \quad x \in \mathbb{R},$$

which gives us (4.19). We may replace f in (4.14) by $|f|$ to obtain a nonnegative function $\hat{z} \geq |z|$, and just as earlier we have

$$G_\alpha(|kz|)(x) \leq (G_\alpha(k\hat{z}))(x) = (G_\alpha(|f|) - \hat{z})(x) < \infty; \quad x \in \mathbb{R}.$$

Relation (4.20) follows. □

Here are some applications of Theorem 4.9.

4.11 Proposition (P. Lévy's Arc-Sine Law for the Occupation Time of $(0, \infty)$). *Let $\Gamma_+(t) \triangleq \int_0^t 1_{(0,\infty)}(W_s)\,ds$. Then*

$$(4.21) \quad P^0[\Gamma_+(t) \leq \theta] = \int_0^{\theta/t} \frac{ds}{\pi\sqrt{s(1-s)}} = \frac{2}{\pi}\cdot \arcsin\sqrt{\frac{\theta}{t}}; \quad 0 \leq \theta \leq t.$$

Proof. For $\alpha > 0$, $\beta > 0$ the function

$$z(x) = E^x \int_0^\infty \exp\left(-\alpha t - \beta\int_0^t 1_{(0,\infty)}(W_s)\,ds\right) dt$$

(with potential $k = \beta \cdot 1_{(0,\infty)}$ and Lagrangian $f = 1$) satisfies, according to Theorem 4.9, the equation

$$\alpha z(x) = \tfrac{1}{2}z''(x) - \beta z(x) + 1; \quad x > 0,$$

$$\alpha z(x) = \tfrac{1}{2}z''(x) + 1; \quad x < 0,$$

and the conditions

$$z(0+) = z(0-); \quad z'(0+) = z'(0-).$$

The unique bounded solution to the preceding equation has the form

$$z(x) = \begin{cases} Ae^{-x\sqrt{2(\alpha+\beta)}} + \dfrac{1}{\alpha+\beta}; & x > 0 \\[2ex] Be^{x\sqrt{2\alpha}} + \dfrac{1}{\alpha}; & x < 0. \end{cases}$$

The continuity of $z(\cdot)$ and $z'(\cdot)$ at $x = 0$ allows us to solve for $A = (\sqrt{\alpha+\beta} - \sqrt{\alpha})/(\alpha+\beta)\sqrt{\alpha}$, so

$$\int_0^\infty e^{-\alpha t} E^0 e^{-\beta\Gamma_+(t)}\,dt = z(0) = \frac{1}{\sqrt{\alpha(\alpha+\beta)}}; \quad \alpha > 0, \beta > 0.$$

We have the related computation

$$\int_0^\infty e^{-\alpha t} \int_0^t \frac{e^{-\beta\theta}}{\pi\sqrt{\theta(t-\theta)}}\,d\theta\,dt = \int_0^\infty \frac{e^{-\beta\theta}}{\pi\sqrt{\theta}} \int_\theta^\infty \frac{e^{-\alpha t}}{\sqrt{t-\theta}}\,dt\,d\theta$$

$$= \frac{1}{\pi}\int_0^\infty \frac{e^{-(\alpha+\beta)\theta}}{\sqrt{\theta}} \int_0^\infty \frac{e^{-\alpha s}}{\sqrt{s}}\,ds\,d\theta$$

$$= \frac{1}{\sqrt{\alpha(\alpha+\beta)}},$$

since

$$\int_0^\infty \frac{e^{-\gamma t}}{\sqrt{t}}\,dt = \sqrt{\frac{\pi}{\gamma}}; \quad \gamma > 0. \tag{4.22}$$

The uniqueness of Laplace transforms implies (4.21). □

4.12 Proposition (Occupation Time of $(0, \infty)$ until First Hitting $b > 0$). *For $\beta > 0$, $b > 0$, we have*

$$E^0 \exp[-\beta\Gamma_+(T_b)] \triangleq E^0 \exp\left[-\beta\int_0^{T_b} 1_{(0,\infty)}(W_s)\,ds\right] \tag{4.23}$$

$$= \frac{1}{\cosh b\sqrt{2\beta}}.$$

Proof. With $\Gamma_b(t) \triangleq \int_0^t 1_{(b,\infty)}(W_s)\,ds$, positive numbers α, β, γ and

$$z(x) \triangleq E^x \int_0^\infty 1_{(0,\infty)}(W_t)\exp(-\alpha t - \beta\Gamma_+(t) - \gamma\Gamma_b(t))\,dt,$$

we have

$$z(0) = E^0 \int_0^{T_b} \exp(-\alpha t - \beta\Gamma_+(t))\,d\Gamma_+(t)$$

$$+ E^0 \int_{T_b}^\infty \exp(-\alpha t - \beta\Gamma_+(t) - \gamma\Gamma_b(t))\,d\Gamma_+(t).$$

Since $\Gamma_b(t) > 0$ a.s. on $\{T_b < t\}$ (Problem 2.7.19), we have

$$\lim_{\gamma\uparrow\infty} z(0) = E^0 \int_0^{T_b} \exp(-\alpha t - \beta\Gamma_+(t))\,d\Gamma_+(t)$$

$$\lim_{\alpha\downarrow 0}\lim_{\gamma\uparrow\infty} z(0) = E^0 \int_0^{T_b} \exp(-\beta\Gamma_+(t))\,d\Gamma_+(t) \tag{4.24}$$

$$= \frac{1}{\beta}[1 - E^0\exp(-\beta\Gamma_+(T_b))].$$

According to Theorem 4.9, the function $z(\cdot)$ is piecewise C^2 on $\mathbb{R}$ and satisfies the equation (with $\sigma = \alpha + \beta$):

$$\alpha z(x) = \tfrac{1}{2}z''(x); \qquad x < 0,$$
$$\sigma z(x) = \tfrac{1}{2}z''(x) + 1; \quad 0 < x < b,$$
$$(\sigma + \gamma)z(x) = \tfrac{1}{2}z''(x) + 1; \quad x > b.$$

The unique bounded solution is of the form

$$z(x) = \begin{cases} Ae^{x\sqrt{2\alpha}}; & x < 0, \\ Be^{x\sqrt{2\sigma}} + Ce^{-x\sqrt{2\sigma}} + \dfrac{1}{\sigma}; & 0 < x < b, \\ De^{-x\sqrt{2(\sigma+\gamma)}} + \dfrac{1}{\sigma + \gamma}; & x > b. \end{cases}$$

Matching the values of $z(\cdot)$ and $z'(\cdot)$ across the points $x = 0$ and $x = b$, we obtain the values of the four constants A, B, C, and D. In particular, $z(0) = A$ is given by

$$2\frac{\dfrac{\sinh b\sqrt{2\sigma}}{\sqrt{2\sigma}} + \sqrt{\dfrac{\sigma+\gamma}{\sigma}}\left[\dfrac{\cosh b\sqrt{2\sigma} - 1}{\sqrt{2\sigma}}\right] + \dfrac{1}{\sqrt{2(\sigma+\gamma)}}}{(\sqrt{2\alpha} + \sqrt{2(\sigma+\gamma)})\cosh b\sqrt{2\sigma} + \left(\sqrt{2\alpha\left(\dfrac{\sigma+\gamma}{\sigma}\right)} + \sqrt{2\sigma}\right)\sinh b\sqrt{2\sigma}},$$

whence

$$\lim_{\gamma\uparrow\infty} z(0) = \frac{\sqrt{\dfrac{2}{\alpha+\beta}}(\cosh b\sqrt{2(\alpha+\beta)} - 1)}{\sqrt{2(\alpha+\beta)}\cosh b\sqrt{2(\alpha+\beta)} + \sqrt{2\alpha}\sinh b\sqrt{2(\alpha+\beta)}}$$

and

$$\lim_{\alpha\downarrow 0}\lim_{\gamma\uparrow\infty} z(0) = \frac{1}{\beta}\left[1 - \frac{1}{\cosh b\sqrt{2\beta}}\right].$$

The result (4.23) now follows from (4.24). □

4.13 Exercise. (D. Ocone): If W is Brownian motion in $\mathbb{R}^d$, show that for every $x \in \mathbb{R}^d$, $t > 0$ and $0 < a < 1/(2t)$ we have

$$E^x\left[\exp\left\{a.\sup_{0\le s\le t}\|W_s\|^2\right\}\right] < \infty.$$

(*Hint*: Use the estimate (2.8.3)′ of Problem 2.8.2.)

4.14 Exercise. (D. Ocone, H. Wang): Show that Theorem 4.2 and Corollary 4.5 remain valid, under the assumption $0 < a < 1/(2T)$. (Hint: Use Exercise 4.13.)

4.5. Solutions to Selected Problems

2.16. Assume without loss of generality that $a = 0$. Choose $0 < r < \|b\| \wedge 1$. It suffices to show that a is regular for $B_r \cap D$ (Remark 2.11). But there is a simple arc C in $\mathbb{R}^d \backslash D$ connecting $a = 0$ to b, and in $B_r \backslash C$, a single-valued, analytic branch of $\log(x_1 + ix_2)$ can be defined because winding about the origin is not possible. Regularity of $a = 0$ is an immediate consequence of Example 2.14 and Proposition 2.15.

2.22. Every boundary point of D satisfies Zaremba's cone condition, and so is regular. It remains only to evaluate (2.12).

Whenever Y and Z are independent random variables taking values in measurable spaces $(G, \mathscr{G})$ and $(H, \mathscr{H})$, respectively, and $f: G \times H \to \mathbb{R}$ is bounded and measurable, then

$$Ef(Y, Z) = \int_H \int_G f(y, z) P[Y \in dy] P[Z \in dz].$$

We apply this identity to the independent random variables $\tau_D = \inf\{t \geq 0; W_t^{(d)} = 0\}$ and $\{(W_t^{(1)}, \ldots, W_t^{(d-1)}); 0 \leq t < \infty\}$, the latter taking values in $C[0, \infty)^{d-1}$. This results in the evaluation (see (2.8.5))

$$E^x f(W_{\tau_D}) = \int_0^\infty \int_{\mathbb{R}^{d-1}} f(y_1, \ldots, y_{d-1}, 0) P^x[(W_t^{(1)}, \ldots, W_t^{(d-1)}) \in (dy_1, \ldots, dy_{d-1})] \cdot P^x[\tau_D \in dt]$$

$$= \int_{\partial D} x_d f(y) \int_0^\infty \frac{1}{t(2\pi t)^{d/2}} \exp\left[-\frac{\|y - x\|^2}{2t}\right] dt\, dy,$$

and it remains only to verify that

$$\int_0^\infty \frac{1}{t^{(d+2)/2}} \exp\left[-\frac{\lambda^2}{2t}\right] dt = \frac{2^{d/2}\Gamma(d/2)}{\lambda^d}, \quad \lambda > 0. \tag{5.1}$$

For $d = 1$, equation (5.1) follows from Remark 2.8.3. For $d = 2$, (5.1) can be verified by direct integration. The cases of $d \geq 3$ can be reduced to one of these two cases by successive application of the integration by parts identity

$$\int_0^\infty \frac{1}{t^{\alpha+1}} e^{-\lambda^2/2t}\, dt = \frac{2(\alpha - 1)}{\lambda^2} \int_0^\infty \frac{1}{t^\alpha} e^{-\lambda^2/2t}\, dt, \quad \alpha > 1.$$

2.25. Consider an increasing sequence $\{D_n\}_{n=1}^\infty$ of open sets with $\bar{D}_n \subset D$; $\forall\, n \geq 1$ and $\bigcup_{n=1}^\infty D_n = D$, so that the stopping times $\tau_n = \inf\{t \geq 0; W_t \notin D_n\}$ satisfy $\lim_{n\to\infty} \tau_n = \tau_D$, a.s. P^x. It is seen from Itô's rule that

$$M_t^{(n)} \triangleq u(W_{t \wedge \tau_n}) + \int_0^{t \wedge \tau_n} g(W_s)\, ds, \mathscr{F}_t; \quad 0 \leq t < \infty$$

is a P^x-martingale for every $n \geq 1, x \in D$; also, both $|M_t(\omega)|, |M_t^{(n)}(\omega)|$ are bounded above by

$$\max_{\bar{D}} |u| + (t \wedge \tau_D(\omega)) \cdot \max_{\bar{D}} |g|, \quad \text{for } P^x\text{-a.e. } \omega \in \Omega.$$

By letting $n \to \infty$ and using the bounded convergence theorem, we obtain the martingale property of the process M; its uniform integrability will follow as soon as we establish

$$E^x \tau_D < \infty, \quad \forall x \in D. \tag{5.2}$$

Then the limiting random variable $M_\infty = \lim_{t\to\infty} M_t$ of the uniformly integrable martingale M is identified as

$$M_\infty = u(W_{\tau_D}) + \int_0^{\tau_D} g(W_t)\,dt, \quad \text{a.s. } P^x \text{ (Problem 1.3.20)}$$

and the identity $E^x M_0 = E^x M_\infty$ yields the representation (2.27).

As for (5.2), it only has to be verified for $D = B_r$. But then the function $v(x) = (r^2 - \|x\|^2)/d$ of (2.28) satisfies $(1/2)\Delta v = -1$ in B_r and $v = 0$ on ∂B_r, and by what we have already shown, $\{v(W_{t\wedge\tau_{B_r}}) + (t \wedge \tau_{B_r}), \mathscr{F}_t, 0 \le t < \infty\}$ is a P^x-martingale. So $v(x) = E^x[v(W_{t\wedge\tau_{B_r}}) + (t \wedge \tau_{B_r})] \ge E^x(t \wedge \tau_{B_r})$, and upon letting $t \to \infty$ we obtain $E^x \tau_{B_r} \le v(x) < \infty$.

3.1. (Copson (1975)): Fix $\beta > 0$, $\varepsilon > 0$, $0 < t_0 < t_1 \triangleq (1/2(a+\varepsilon))$, and set $B = \{(t,x); t_0 < t < t_1, |x| < \beta\}$. For $(t,x) \in B$, $y \in \mathbb{R}$, we have

$$\begin{aligned} ay^2 - \frac{(x-y)^2}{2t} &\le ay^2 - \frac{1}{2t_1}(|x| - |y|)^2 \le ay^2 - \frac{1}{2t_1}y^2 + \frac{1}{t_1}|x||y| \\ &\le -\frac{\varepsilon}{2}y^2 - \frac{\varepsilon}{2}y^2 + \frac{\beta}{t_1}|y| \le -\frac{\varepsilon}{2}y^2 + \frac{\beta^2}{2\varepsilon t_1^2}. \end{aligned}$$

For any nonnegative integers n and m, there is a constant $C(n,m)$ such that

$$\left|\frac{\partial^{n+m}}{\partial t^n \partial x^m} p(t;x,y)\right| \le C(n,m)(1 + |y|^{2n+m}) \exp\left\{-\frac{(x-y)^2}{2t}\right\}; \quad (t,x) \in B,\ y \in \mathbb{R},$$

and so

$$\begin{aligned} &\left| f(y) \frac{\partial^{n+m}}{\partial t^n \partial x^m} p(t;x,y)\right| \\ &\quad \le |f(y)|\, C(n,m)(1 + |y|^{2n+m}) \exp\left\{-\left(a + \frac{\varepsilon}{2}\right)y^2 + \frac{\beta^2}{2\varepsilon t_1^2}\right\} \\ &\quad \le D(n,m)|f(y)| e^{-ay^2}; \quad (t,x) \in B,\ y \in \mathbb{R}, \end{aligned} \tag{5.3}$$

where $D(n,m)$ is a constant independent of t, x, and y. It follows from (3.3) that the integral in (3.5) converges uniformly for $(t,x) \in B$, and is thus a continuous function of (t,x) on $(0, 1/2a) \times \mathbb{R}$.

We prove (3.5) for the case $n = 0$, $m = 1$; the general case is easily established by induction. For $(t,x) \in B$, $(t, x+h) \in B$, we have

$$\begin{aligned} \frac{1}{h}[u(t,x+h) - u(t,x)] &= \int_{-\infty}^{\infty} \frac{1}{h}[p(t;x+h,y) - p(t,x,y)] f(y)\,dy \\ &= \int_{-\infty}^{\infty} \frac{\partial}{\partial x} p(t;\theta_h(t,y),y) f(y)\,dy, \end{aligned}$$

where, according to the mean-value theorem, $\theta_h(t, y)$ lies between x and $x + h$. We now let $h \to 0$, using the bound (5.3) and the dominated convergence theorem, to obtain (3.5).

3.2. (Widder (1944)): We suppose f is continuous at x_0 and assume without loss of generality that $f(x_0) = 0$. For each $\varepsilon > 0$, there exists $\delta > 0$ such that $|f(y)| \le \varepsilon$ for $|y - x_0| \le \delta$. We have for $x \in [x_0 - (\delta/2), x_0 + (\delta/2)]$,

$$(5.4) \qquad |u(t,x)| \le \int_{-\infty}^{x_0-\delta} |f(y)| p(t; x, y)\, dy + \int_{x_0-\delta}^{x_0+\delta} |f(y)| p(t; x, y)\, dy + \int_{x_0+\delta}^{\infty} |f(y)| p(t; x, y)\, dy.$$

The middle integral is bounded above by ε; we show that the other two converge to zero, as $t \downarrow 0$. For the third integral, we have the upper bound

$$(5.5) \qquad \frac{1}{\sqrt{2\pi t}} \int_{x_0+\delta}^{\infty} e^{-ay^2} |f(y)| \exp\left[ay^2 - \frac{\left(y - x_0 - \frac{\delta}{2}\right)^2}{2t} \right] dy.$$

For t sufficiently small, $\exp[ay^2 - (y - x_0 - \delta/2)^2/2t]$ is a decreasing function of y for $y \ge x_0 + \delta$ (it has its maximum at $y = (x_0 + \delta/2)/(1 - 2at)$). Therefore, the expression in (5.5) is bounded above by

$$p\left(t; 0, \frac{\delta}{2}\right) \exp[a(x_0 + \delta)^2] \int_{x_0+\delta}^{\infty} e^{-ay^2} |f(y)|\, dy$$

which approaches zero as $t \downarrow 0$. The first integral in (5.4) is treated similarly.

4.6. Notes

Section 4.2: The Dirichlet problem has a long and venerable history (see, e.g., Poincaré (1899) and Kellogg (1929)). Zaremba (1911) was the first to observe that the problem was not always solvable, citing the example of a punctured region. Lebesgue (1924) subsequently pointed out that in three or more dimensions, if D has a sufficiently sharp, inward-pointing cusp, then the problem can fail to have a solution (our Example 2.17). Poincaré (1899) used barriers to show that if every point on ∂D lies on the surface of a sphere which does not otherwise intersect $\bar{D}$, then the Dirichlet problem can be solved in D. Zaremba (1909) replaced the sphere in Poincaré's sufficient condition by a cone. Wiener (1924b) has given a necessary and sufficient condition involving the capacity of a set.

The beautiful connection between the Dirichlet problem and Brownian motion was made by Kakutani (1944a, b), and his pioneering work laid the foundation for the probabilistic exposition we have given here. Hunt (1957/1958) studied the links between potential theory and a large class of transient Markov processes. These matters are explored in greater depth in Itô & McKean (1974), Sections 7.10–7.12; Port & Stone (1978); and Doob (1984).

Section 4.3: The representation (3.4) for the solution of the heat equation is usually attributed to Poisson (1835, p. 140), although it was known to both Fourier (1822, p. 454) and Laplace (1809, p. 241). The heat equation for the semi-infinite rod was studied by Widder (1953), who established uniqueness and representation results similar to Theorems 3.3 and 3.6. Hartman & Wintner (1950) considered the rod of finite length. For more examples and further information on the subject matter of Subsection C, including applications to the theory of statistical tests of power one, the reader is referred to Robbins & Siegmund (1973), Novikov (1981), and the references therein.

Section 4.4: Theorem 4.2 was first established by M. Kac (1949) for $d = 1$; his work was influenced by the derivation of the Schrödinger equation achieved by R. P. Feynman in his doctoral dissertation. Kac's results were strengthened and extended to the multidimensional case by M. Rosenblatt (1951), who also provided Hölder continuity conditions on the potential k in order to guarantee a $C^{1,2}$ solution. Proposition 4.12 is taken from Itô & McKean (1974).

Let $k: \mathbb{R} \to [0, \infty)$ be continuous and satisfy $\lim_{x \to \pm\infty} k(x) = \infty$; then the eigenvalue problem

$$(k(x) - \lambda)\psi(x) = \frac{1}{2}\psi''(x); \quad x \in \mathbb{R},$$

with $\psi \in L^2(\mathbb{R})$, has a discrete spectrum $\lambda_1 < \lambda_2 < \ldots$ and corresponding eigenfunctions $\{\psi_j\}_{j=1}^{\infty} \subseteq L^2(\mathbb{R})$. Kac (1951) derived the stochastic representation

$$\lambda_1 = -\lim_{t \to \infty} \frac{1}{t} \log E^x\left[\exp\left\{-\int_0^t k(W_s)\,ds\right\}\right] \tag{6.1}$$

for the principal eigenvalue, by combining the Feynman-Kac expression

$$u(t,x) = E^x\left[\exp\left\{-\int_0^t k(W_s)\,ds\right\}\right]$$

for the solution of the Cauchy problem

$$\begin{aligned} \frac{\partial u}{\partial t} + k \cdot u &= \frac{1}{2}\Delta u; \quad (0,\infty) \times \mathbb{R} \\ u(0,x) &= 1; \qquad x \in \mathbb{R} \end{aligned} \tag{6.2}$$

(Corollary 4.5), with the formal eigenfunction expansion

$$u(t,x) = \sum_{j=1}^{\infty} c_j e^{-\lambda_j t}\psi_j(x)$$

for the solution of (6.2). Recall also Exercise 4.6, and see Karatzas (1980) for a control-theoretic interpretation (and derivation) of this result.

Sweeping generalizations of (6.1), as well as an explanation of its connection with the classical variational expression

$$(6.3) \qquad \lambda_1 = \inf_{\substack{\psi \in L^2 \\ \int_{-\infty}^{\infty} \psi^2(x)\,dx = 1,}} \left\{ \int_{-\infty}^{\infty} k(x)\psi^2(x)\,dx + \frac{1}{2}\int_{-\infty}^{\infty} (\psi'(x))^2\,dx \right\},$$

are provided in the context of the *theory of large deviations* of Donsker & Varadhan (1975), (1976). This theory constitutes an important recent development in probability theory, and is overviewed succinctly in the monographs by Stroock (1984) and Varadhan (1984).

The reader interested in the relations of the results in this section with quantum physics is referred to Simon (1979).

Alternative approaches to the arc-sine law for $\Gamma_+(t)$ can be found in Exercise 6.3.8 and Remark 6.3.12.

CHAPTER 5

Stochastic Differential Equations

5.1. Introduction

We explore in this chapter questions of existence and uniqueness for solutions to stochastic differential equations and offer a study of their properties. This endeavor is really a study of *diffusion processes.* Loosely speaking, the term *diffusion* is attributed to a Markov process which has continuous sample paths and can be characterized in terms of its infinitesimal generator.

In order to fix ideas, let us consider a d-dimensional Markov family $X = \{X_t, \mathscr{F}_t; 0 \le t < \infty\}$, $(\Omega, \mathscr{F})$, $\{P^x\}_{x \in \mathbb{R}^d}$, and assume that X has continuous paths. We suppose, further, that the relation

$$\lim_{t \downarrow 0} \frac{1}{t}[E^x f(X_t) - f(x)] = (\mathscr{A}f)(x); \quad \forall x \in \mathbb{R}^d \tag{1.1}$$

holds for every f in a suitable subclass of the space $C^2(\mathbb{R}^d)$ of real-valued, twice continuously differentiable functions on $\mathbb{R}^d$; the operator $\mathscr{A}f$ in (1.1) is given by

$$(\mathscr{A}f)(x) \triangleq \frac{1}{2}\sum_{i=1}^d \sum_{k=1}^d a_{ik}(x)\frac{\partial^2 f(x)}{\partial x_i \partial x_k} + \sum_{i=1}^d b_i(x)\frac{\partial f(x)}{\partial x_i} \tag{1.2}$$

for suitable Borel-measurable functions $b_i, a_{ik}: \mathbb{R}^d \to \mathbb{R}$, $1 \le i, k \le d$. The left-hand side of (1.1) is the *infinitesimal generator* of the Markov family, applied to the test function f. On the other hand, the operator in (1.2) is called the *second-order differential operator* associated with the *drift vector* $b = (b_1, \ldots, b_d)$ and the *diffusion matrix* $a = \{a_{ik}\}_{1 \le i,k \le d}$, which is assumed to be symmetric and nonnegative-definite for every $x \in \mathbb{R}^d$.

The drift and diffusion coefficients can be interpreted heuristically in the following manner: fix $x \in \mathbb{R}^d$ and let $f_i(y) \triangleq y_i$, $f_{ik}(y) \triangleq (y_i - x_i)(y_k - x_k)$;

$y \in \mathbb{R}^d$. Assuming that (1.1) holds for these test functions, we obtain

$$E^x[X_t^{(i)} - x_i] = tb_i(x) + o(t) \tag{1.3}$$

$$E^x[(X_t^{(i)} - x_i)(X_t^{(k)} - x_k)] = ta_{ik}(x) + o(t) \tag{1.4}$$

as $t \downarrow 0$, for $1 \le i, k \le d$. In other words, the drift vector $b(x)$ measures locally the mean velocity of the random motion modeled by X, and $a(x)$ approximates the rate of change in the covariance matrix of the vector $X_t - x$, for small values of $t > 0$. The monograph by Nelson (1967) can be consulted for a detailed study of the kinematics and dynamics of such random motions.

1.1 Definition. Let $X = \{X_t, \mathscr{F}_t; 0 \le t < \infty\}$, $(\Omega, \mathscr{F})$, $\{P^x\}_{x \in \mathbb{R}^d}$ be a d-dimensional Markov family, such that

(i) X has continuous sample paths;
(ii) relation (1.1) holds for every $f \in C^2(\mathbb{R}^d)$ which is bounded and has bounded first- and second-order derivatives;
(iii) relations (1.3), (1.4) hold for every $x \in \mathbb{R}^d$; and
(iv) the tenets (a)–(d) of Definition 2.6.3 are satisfied, but only for stopping times S.

Then X is called a (Kolmogorov-Feller) *diffusion process.*

There are several approaches to the study of diffusions, ranging from the purely analytical to the purely probabilistic. In order to illustrate the traditional *analytical approach,* let us suppose that the Markov family of Definition 1.1 has a transition probability density function

$$P^x[X_t \in dy] = \Gamma(t; x, y)\, dy; \quad \forall\, x \in \mathbb{R}^d,\ t > 0. \tag{1.5}$$

Various heuristic arguments, with (1.1) as their starting point, can then be employed to suggest that $\Gamma(t; x, y)$ should satisfy the *forward Kolmogorov equation,* for every fixed $x \in \mathbb{R}^d$:

$$\frac{\partial}{\partial t}\Gamma(t; x, y) = \mathscr{A}^*\Gamma(t; x, y); \quad (t, y) \in (0, \infty) \times \mathbb{R}^d, \tag{1.6}$$

and the *backward Kolmogorov equation,* for every fixed $y \in \mathbb{R}^d$:

$$\frac{\partial}{\partial t}\Gamma(t; x, y) = \mathscr{A}\Gamma(t; x, y); \quad (t, x) \in (0, \infty) \times \mathbb{R}^d. \tag{1.7}$$

The operator $\mathscr{A}^*$ in (1.6) is given by

$$(\mathscr{A}^* f)(y) \triangleq \frac{1}{2}\sum_{i=1}^d \sum_{k=1}^d \frac{\partial^2}{\partial y_i \partial y_k}[a_{ik}(y) f(y)] - \sum_{i=1}^d \frac{\partial}{\partial y_i}[b_i(y) f(y)], \tag{1.8}$$

the formal adjoint of $\mathscr{A}$ in (1.2), provided of course that the coefficients b_i, a_{ik} possess the smoothness requisite in (1.8). The early work of Kolmogorov (1931) and Feller (1936) used tools from the theory of partial differential equations to establish, under suitable and rather restrictive conditions, the existence of a solution $\Gamma(t; x, y)$ to (1.6), (1.7). Existence of a continuous Markov process X satisfying (1.5) can then be shown via the consistency Theorem 2.2.2 and the Čentsov-Kolmogorov Theorem 2.2.8, very much in the spirit of our approach in Section 2.2. A modern account of this methodology is contained in Chapters 2 and 3 of Stroock & Varadhan (1979).

The methodology of *stochastic differential equations* was suggested by P. Lévy as an "alternative," probabilistic approach to diffusions and was carried out in a masterly way by K. Itô (1942a, 1946, 1951). Suppose that we have a continuous, adapted d-dimensional process $X = \{X_t, \mathscr{F}_t; 0 \le t < \infty\}$ which satisfies, for every $x \in \mathbb{R}^d$, the stochastic integral equation

$$(1.9) \quad X_t^{(i)} = x_i + \int_0^t b_i(X_s)\,ds + \sum_{j=1}^r \int_0^t \sigma_{ij}(X_s)\,dW_s^{(j)}; \quad 0 \le t < \infty, 1 \le i \le d$$

on a probability space $(\Omega, \mathscr{F}, P^x)$, where $W = \{W_t, \mathscr{F}_t; 0 \le t < \infty\}$ is a Brownian motion in $\mathbb{R}^r$ and the coefficients $b_i, \sigma_{ij}: \mathbb{R}^d \to \mathbb{R}; 1 \le i \le d, 1 \le j \le r$ are Borel-measurable. Then it is reasonable to expect that, under certain conditions, (1.1)–(1.4) will indeed be valid, with

$$(1.10) \qquad a_{ik}(x) \triangleq \sum_{j=1}^r \sigma_{ij}(x)\sigma_{kj}(x).$$

We leave the verification of this fact as an exercise for the reader.

1.2 Problem. Assume that the coefficients b_i, σ_{ij} are bounded and continuous, and the $\mathbb{R}^d$-valued process X satisfies (1.9). Show that (1.3), (1.4) hold for every $x \in \mathbb{R}^d$, and that (1.1) holds for every $f \in C^2(\mathbb{R}^d)$ which is bounded and has bounded first- and second-order derivatives.

Itô's theory is developed in Section 2 under the rubric of *strong solutions*. A strong solution of (1.9) is constructed on a given probability space, with respect to a given filtration and a given Brownian motion W. In Section 3 we take up the idea of *weak solutions*, a notion in which the probability space, the filtration, and the driving Brownian motion are part of the solution rather than the statement of the problem. The reformulation of a stochastic differential equation as a *martingale problem* is presented in Section 4. The solution of this problem is equivalent to constructing a weak solution. Employing martingale methods, we establish a version of the strong Markov property—corresponding to (iv) of Definition 1.1—for these solutions; they thereby earn the right to be called diffusions.

The stochastic differential equation approach to diffusions provides a powerful methodology and the useful representation (1.9) for a very large class of such processes. Indeed, the only important strong Markov processes with continuous sample paths which are not directly included in such a development are those which exhibit "anomalous" boundary behavior (e.g., reflection, absorption, or killing on a boundary).

Certain aspects of the one-dimensional case are discussed at some length in Section 5; a state-space transformation leads from the general equation to one without drift, and the latter is studied by the method of *random time-change*. The notion and properties of local time from Sections 3.6, 3.7 play an important role here, as do the new concepts of scale function, speed measure, and explosions.

Section 6 studies linear equations; Section 7 takes up the connections with *partial differential equations*, in the spirit of Chapter 4 but not in the same detail.

We devote Section 8 to applications of stochastic calculus and differential equations in mathematical economics. The related *option pricing* and *consumption/investment* problems are discussed in some detail, providing concrete illustrations of the power and usefulness of our methodology. In particular, the second of these problems echoes the more general themes of *stochastic control theory*.

The field of stochastic differential equations is now vast, both in theory and in applications; we attempt in the notes (Section 10) a brief survey, but we abandon any claim to completeness.

5.2. Strong Solutions

In this section we introduce the concept of a stochastic differential equation with respect to Brownian motion and its solution in the so-called strong sense. We discuss the questions of existence and uniqueness of such solutions, as well as some of their elementary properties.

Let us start with Borel-measurable functions $b_i(t,x)$, $\sigma_{ij}(t,x)$; $1 \le i \le d$, $1 \le j \le r$, from $[0,\infty) \times \mathbb{R}^d$ into $\mathbb{R}$, and define the $(d \times 1)$ *drift vector* $b(t,x) = \{b_i(t,x)\}_{1\le i\le d}$ and the $(d \times r)$ *dispersion matrix* $\sigma(t,x) = \{\sigma_{ij}(t,x)\}_{\substack{1\le i\le d\\1\le j\le r}}$. The intent is to assign a meaning to the *stochastic differential equation*

$$dX_t = b(t,X_t)\,dt + \sigma(t,X_t)\,dW_t, \tag{2.1}$$

written componentwise as

$$dX_t^{(i)} = b_i(t,X_t)\,dt + \sum_{j=1}^{r} \sigma_{ij}(t,X_t)\,dW_t^{(j)}; \quad 1 \le i \le d, \tag{2.1$'$}$$

where $W = \{W_t; 0 \le t < \infty\}$ is an r-dimensional Brownian motion and $X = \{X_t; 0 \le t < \infty\}$ is a suitable stochastic process with continuous sample paths and values in $\mathbb{R}^d$, the "solution" of the equation. The drift vector $b(t,x)$ and the dispersion matrix $\sigma(t,x)$ are the *coefficients* of this equation; the $(d \times d)$ matrix $a(t,x) \triangleq \sigma(t,x)\sigma^T(t,x)$ with elements

$$a_{ik}(t,x) \triangleq \sum_{j=1}^{r} \sigma_{ij}(t,x)\sigma_{kj}(t,x); \quad 1 \le i, k \le d \tag{2.2}$$

will be called the *diffusion matrix*.

A. Definitions

In order to develop the concept of *strong solution*, we choose a probability space $(\Omega, \mathscr{F}, P)$ as well as an r-dimensional Brownian motion $W = \{W_t, \mathscr{F}_t^W; 0 \le t < \infty\}$ on it. We assume also that this space is rich enough to accommodate a random vector ξ taking values in $\mathbb{R}^d$, independent of $\mathscr{F}_\infty^W$, and with given distribution

$$\mu(\Gamma) = P[\xi \in \Gamma]; \quad \Gamma \in \mathscr{B}(\mathbb{R}^d).$$

We consider the left-continuous filtration

$$\mathscr{G}_t \triangleq \sigma(\xi) \vee \mathscr{F}_t^W = \sigma(\xi, W_s; 0 \le s \le t); \quad 0 \le t < \infty,$$

as well as the collection of null sets

$$\mathscr{N} \triangleq \{N \subseteq \Omega; \exists\, G \in \mathscr{G}_\infty \text{ with } N \subseteq G \text{ and } P(G) = 0\},$$

and create the *augmented filtration*

$$\mathscr{F}_t \triangleq \sigma(\mathscr{G}_t \cup \mathscr{N}), \quad 0 \le t < \infty; \qquad \mathscr{F}_\infty \triangleq \sigma\left(\bigcup_{t \ge 0} \mathscr{F}_t\right), \tag{2.3}$$

by analogy with the construction of Definition 2.7.2. Obviously, $\{W_t, \mathscr{G}_t; 0 \le t < \infty\}$ is an r-dimensional Brownian motion, and then so is $\{W_t, \mathscr{F}_t; 0 \le t < \infty\}$ (cf. Theorem 2.7.9). It follows also, just as in the proof of Proposition 2.7.7, that the filtration $\{\mathscr{F}_t\}$ satisfies the usual conditions.

2.1 Definition. A *strong solution* of the stochastic differential equation (2.1), on the given probability space $(\Omega, \mathscr{F}, P)$ and with respect to the fixed Brownian motion W and initial condition ξ, is a process $X = \{X_t; 0 \le t < \infty\}$ with continuous sample paths and with the following properties:

(i) X is adapted to the filtration $\{\mathscr{F}_t\}$ of (2.3),

(ii) $P[X_0 = \xi] = 1$,

(iii) $P[\int_0^t \{|b_i(s, X_s)| + \sigma_{ij}^2(s, X_s)\}\, ds < \infty] = 1$ holds for every $1 \le i \le d$, $1 \le j \le r$ and $0 \le t < \infty$, and

(iv) the integral version of (2.1)

$$X_t = X_0 + \int_0^t b(s, X_s)\, ds + \int_0^t \sigma(s, X_s)\, dW_s; \quad 0 \le t < \infty, \tag{2.4}$$

or equivalently,

$$X_t^{(i)} = X_0^{(i)} + \int_0^t b_i(s, X_s)\, ds + \sum_{j=1}^r \int_0^t \sigma_{ij}(s, X_s)\, dW_s^{(j)}; \tag{2.4'}$$

$$0 \le t < \infty, 1 \le i \le d,$$

holds almost surely.

2.2 Remark. The crucial requirement of this definition is captured in condition (i); it corresponds to our intuitive understanding of X as the "output" of a dynamical system described by the pair of coefficients (b, σ), whose "input" is W and which is also fed by the initial datum ξ.

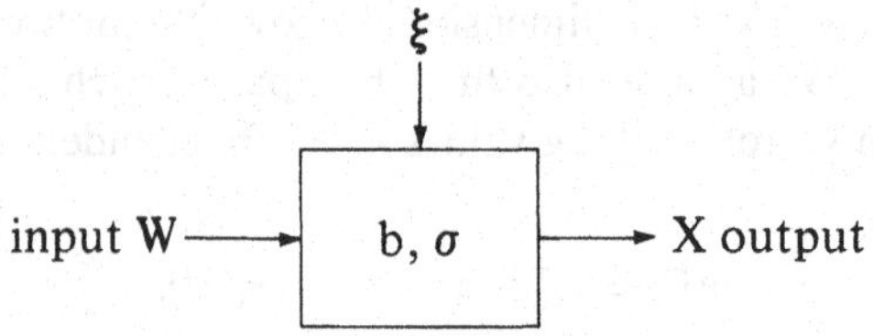

The *principle of causality* for dynamical systems requires that the output X_t at time t depend only on ξ and the values of the input $\{W_s; 0 \le s \le t\}$ up to that time. This principle finds its mathematical expression in (i).

Furthermore, when both ξ and $\{W_t; 0 \le t < \infty\}$ are given, their specification should determine the output $\{X_t; 0 \le t < \infty\}$ in an unambiguous way. We are thus led to expect the following form of uniqueness.

2.3 Definition. Let the drift vector $b(t,x)$ and dispersion matrix $\sigma(t,x)$ be given. Suppose that, whenever W is an r-dimensional Brownian motion on some $(\Omega, \mathscr{F}, P)$, ξ is an independent, d-dimensional random vector, $\{\mathscr{F}_t\}$ is given by (2.3), and X, $\tilde{X}$ are two strong solutions of (2.1) relative to W with initial condition ξ, then $P[X_t = \tilde{X}_t; 0 \le t < \infty] = 1$. Under these conditions, we say that *strong uniqueness holds for the pair* (b, σ).

We sometimes abuse the terminology by saying that *strong uniqueness holds for equation* (2.1), even though the condition of strong uniqueness requires us to consider every r-dimensional Brownian motion, not just a particular one.

2.4 Example. Consider the one-dimensional equation

$$dX_t = b(t, X_t)\,dt + dW_t,$$

where $b: [0,\infty) \times \mathbb{R} \to \mathbb{R}$ is bounded, Borel-measurable, and nonincreasing in the space variable; i.e., $b(t,x) \le b(t,y)$ for all $0 \le t < \infty$, $-\infty < y \le x < \infty$. For this equation, strong uniqueness holds. Indeed, for any two processes $X^{(1)}$, $X^{(2)}$ satisfying P-a.s.

$$X_t^{(i)} = X_0 + \int_0^t b(s, X_s^{(i)})\,ds + W_t; \quad 0 \le t < \infty \text{ and } i = 1, 2,$$

we may define the continuous process $\Delta_t = X_t^{(1)} - X_t^{(2)}$ and observe that

$$\Delta_t^2 = 2\int_0^t (X_s^{(1)} - X_s^{(2)})[b(s, X_s^{(1)}) - b(s, X_s^{(2)})]\,ds \le 0; \quad 0 \le t < \infty, \text{ a.s. } P.$$

B. The Itô Theory

If the dispersion matrix $\sigma(t,x)$ is identically equal to zero, (2.4) reduces to the *ordinary* (nonstochastic, except possibly in the initial condition) *integral equation*

$$X_t = X_0 + \int_0^t b(s, X_s)\,ds. \tag{2.5}$$

In the theory for such equations (e.g., Hale (1969), Theorem I.5.3), it is customary to impose the assumption that the vector field $b(t, x)$ satisfies a local Lipschitz condition in the space variable x and is bounded on compact subsets of $[0, \infty) \times \mathbb{R}^d$. These conditions ensure that for sufficiently small $t > 0$, the *Picard-Lindelöf iterations*

$$X_t^{(0)} \equiv X_0; \quad X_t^{(n+1)} = X_0 + \int_0^t b(s, X_s^{(n)})\,ds, \quad n \geq 0, \tag{2.6}$$

converge to a solution of (2.5), and that this solution is unique. In the absence of such conditions the equation might fail to be solvable or might have a continuum of solutions. For instance, the one-dimensional equation

$$X_t = \int_0^t |X_s|^\alpha\,ds \tag{2.7}$$

has only one solution for $\alpha \geq 1$, namely $X_t \equiv 0$; however, for $0 < \alpha < 1$, all functions of the form

$$X_t = \begin{cases} 0; & 0 \leq t \leq s, \\ \left(\dfrac{t-s}{\beta}\right)^\beta; & s \leq t < \infty, \end{cases}$$

with $\beta = 1/(1-\alpha)$ and arbitrary $0 \leq s \leq \infty$, solve (2.7).

It seems then reasonable to attempt developing a theory for stochastic differential equations by imposing Lipschitz-type conditions, and investigating what kind of existence and/or uniqueness results one can obtain this way. Such a program was first carried out by K. Itô (1942a, 1946).

2.5 Theorem. *Suppose that the coefficients $b(t,x)$, $\sigma(t,x)$ are locally Lipschitz-continuous in the space variable; i.e., for every integer $n \geq 1$ there exists a constant $K_n > 0$ such that for every $t \geq 0$, $\|x\| \leq n$ and $\|y\| \leq n$:*

$$\|b(t,x) - b(t,y)\| + \|\sigma(t,x) - \sigma(t,y)\| \leq K_n\|x - y\|. \tag{2.8}$$

Then strong uniqueness holds for equation (2.1).

2.6 Remark on Notation. For every $(d \times r)$ matrix σ, we write

$$\|\sigma\|^2 \triangleq \sum_{i=1}^d \sum_{j=1}^r \sigma_{ij}^2. \tag{2.9}$$

Before proceeding with the proof, let us recall the useful *Gronwall inequality*.

2.7 Problem. Suppose that the continuous function $g(t)$ satisfies

$$(2.10) \qquad 0 \le g(t) \le \alpha(t) + \beta \int_0^t g(s)\,ds; \quad 0 \le t \le T,$$

with $\beta \ge 0$ and $\alpha\colon [0, T] \to \mathbb{R}$ integrable. Then

$$(2.11) \qquad g(t) \le \alpha(t) + \beta \int_0^t \alpha(s)e^{\beta(t-s)}\,ds; \quad 0 \le t \le T.$$

PROOF OF THEOREM 2.5. Let us suppose that X and $\tilde{X}$ are both strong solutions, defined for all $t \ge 0$, of (2.1) relative to the same Brownian motion W and the same initial condition ξ, on some $(\Omega, \mathscr{F}, P)$. We define the stopping times $\tau_n = \inf\{t \ge 0;\ \|X_t\| \ge n\}$ for $n \ge 1$, as well as their tilded counterparts, and we set $S_n \triangleq \tau_n \wedge \tilde{\tau}_n$. Clearly $\lim_{n\to\infty} S_n = \infty$, a.s. P, and

$$X_{t\wedge S_n} - \tilde{X}_{t\wedge S_n} = \int_0^{t\wedge S_n} \{b(u, X_u) - b(u, \tilde{X}_u)\}\,du + \int_0^{t\wedge S_n} \{\sigma(u, X_u) - \sigma(u, \tilde{X}_u)\}\,dW_u.$$

Using the vector inequality $\|v_1 + \cdots + v_k\|^2 \le k^2(\|v_1\|^2 + \cdots + \|v_k\|^2)$, the Hölder inequality for Lebesgue integrals, the basic property (3.2.27) of stochastic integrals, and (2.8), we may write for $0 \le t \le T$:

$$\begin{aligned} E\|X_{t\wedge S_n} - \tilde{X}_{t\wedge S_n}\|^2 &\le 4E\left[\int_0^{t\wedge S_n} \|b(u, X_u) - b(u, \tilde{X}_u)\|\,du\right]^2 \\ &\quad + 4E\sum_{i=1}^d \left[\sum_{j=1}^r \int_0^{t\wedge S_n} (\sigma_{ij}(u, X_u) - \sigma_{ij}(u, \tilde{X}_u))\,dW_u^{(j)}\right]^2 \\ &\le 4tE\int_0^{t\wedge S_n} \|b(u, X_u) - b(u, \tilde{X}_u)\|^2\,du \\ &\quad + 4E\int_0^{t\wedge S_n} \|\sigma(u, X_u) - \sigma(u, \tilde{X}_u)\|^2\,du \\ &\le 4(T+1)K_n^2 \int_0^t E\|X_{u\wedge S_n} - \tilde{X}_{u\wedge S_n}\|^2\,du. \end{aligned}$$

We now apply Problem 2.7 with $g(t) \triangleq E\|X_{t\wedge S_n} - \tilde{X}_{t\wedge S_n}\|^2$ to conclude that $\{X_{t\wedge S_n}; 0 \le t < \infty\}$ and $\{\tilde{X}_{t\wedge S_n}; 0 \le t < \infty\}$ are modifications of one another, and thus are indistinguishable. Letting $n \to \infty$, we see that the same is true for $\{X_t; 0 \le t < \infty\}$ and $\{\tilde{X}_t; 0 \le t < \infty\}$. □

2.8 Remark. It is worth noting that even for ordinary differential equations, a local Lipschitz condition is not sufficient to guarantee global existence of a solution. For example, the unique (because of Theorem 2.5) solution to the equation

$$X_t = 1 + \int_0^t X_s^2\,ds$$

is $X_t = 1/(1-t)$, which "explodes" as $t \uparrow 1$. We thus impose stronger conditions in order to obtain an existence result.

2.9 Theorem. *Suppose that the coefficients $b(t,x)$, $\sigma(t,x)$ satisfy the global Lipschitz and linear growth conditions*

(2.12) $$\|b(t,x) - b(t,y)\| + \|\sigma(t,x) - \sigma(t,y)\| \le K\|x - y\|,$$

(2.13) $$\|b(t,x)\|^2 + \|\sigma(t,x)\|^2 \le K^2(1 + \|x\|^2),$$

for every $0 \le t < \infty$, $x \in \mathbb{R}^d$, $y \in \mathbb{R}^d$, where K is a positive constant. On some probability space $(\Omega, \mathscr{F}, P)$, let ξ be an $\mathbb{R}^d$-valued random vector, independent of the r-dimensional Brownian motion $W = \{W_t, \mathscr{F}_t^W; 0 \le t < \infty\}$, and with finite second moment:

(2.14) $$E\|\xi\|^2 < \infty.$$

Let $\{\mathscr{F}_t\}$ be as in (2.3). Then there exists a continuous, adapted process $X = \{X_t, \mathscr{F}_t; 0 \le t < \infty\}$ which is a strong solution of equation (2.1) relative to W, with initial condition ξ. Moreover, this process is square-integrable: for every $T > 0$, there exists a constant C, depending only on K and T, such that

(2.15) $$E\|X_t\|^2 \le C(1 + E\|\xi\|^2)e^{Ct}; \quad 0 \le t \le T.$$

The idea of the proof is to mimic the deterministic situation and to construct recursively, by analogy with (2.6), a sequence of successive approximations by setting $X_t^{(0)} \equiv \xi$ and

(2.16) $$X_t^{(k+1)} \triangleq \xi + \int_0^t b(s, X_s^{(k)})\,ds + \int_0^t \sigma(s, X_s^{(k)})\,dW_s; \quad 0 \le t < \infty,$$

for $k \ge 0$. These processes are obviously continuous and adapted to the filtration $\{\mathscr{F}_t\}$. The hope is that the sequence $\{X^{(k)}\}_{k=1}^\infty$ will converge to a solution of equation (2.1).

Let us start with the observation which will ultimately lead to (2.15).

2.10 Problem. For every $T > 0$, there exists a positive constant C depending only on K and T, such that for the iterations in (2.16) we have

(2.17) $$E\|X_t^{(k)}\|^2 \le C(1 + E\|\xi\|^2)e^{Ct}; \quad 0 \le t \le T, k \ge 0.$$

PROOF OF THEOREM 2.9. We have $X_t^{(k+1)} - X_t^{(k)} = B_t + M_t$ from (2.16), where

$$B_t \triangleq \int_0^t \{b(s, X_s^{(k)}) - b(s, X_s^{(k-1)})\}\,ds, \quad M_t \triangleq \int_0^t \{\sigma(s, X_s^{(k)}) - \sigma(s, X_s^{(k-1)})\}\,dW_s.$$

Thanks to the inequalities (2.13) and (2.17), the process $\{M_t = (M_t^{(1)}, \ldots, M_t^{(d)}),$

$\mathscr{F}_t; 0 \le t < \infty\}$ is seen to be a vector of square-integrable martingales, for which Problem 3.3.29 and Remark 3.3.30 give

$$E\left[\max_{0\le s\le t} \|M_s\|^2\right] \le \Lambda_1 E\int_0^t \|\sigma(s, X_s^{(k)}) - \sigma(s, X_s^{(k-1)})\|^2\,ds$$
$$\le \Lambda_1 K^2 E\int_0^t \|X_s^{(k)} - X_s^{(k-1)}\|^2\,ds.$$

On the other hand, we have $E\|B_t\|^2 \le K^2 t \int_0^t E\|X_s^{(k)} - X_s^{(k-1)}\|^2\,ds$, and therefore, with $L = 4K^2(\Lambda_1 + T)$,

$$\text{(2.18)}\quad E\left[\max_{0\le s\le t} \|X_s^{(k+1)} - X_s^{(k)}\|^2\right] \le L\int_0^t E\|X_s^{(k)} - X_s^{(k-1)}\|^2\,ds;\quad 0 \le t \le T.$$

Inequality (2.18) can be iterated to yield the successive upper bounds

$$\text{(2.19)}\quad E\left[\max_{0\le s\le t} \|X_s^{(k+1)} - X_s^{(k)}\|^2\right] \le C^*\frac{(Lt)^k}{k!};\quad 0 \le t \le T,$$

where $C^* = \max_{0\le t\le T} E\|X_t^{(1)} - \xi\|^2$, a finite quantity because of (2.17). Relation (2.19) and the Čebyšev inequality now give

$$\text{(2.20)}\quad P\left[\max_{0\le t\le T} \|X_t^{(k+1)} - X_t^{(k)}\| > \frac{1}{2^{k+1}}\right] \le 4C^*\frac{(4LT)^k}{k!};\quad k = 1, 2, \ldots,$$

and this upper bound is the general term in a convergent series. From the Borel-Cantelli lemma, we conclude that there exists an event $\Omega^* \in \mathscr{F}$ with $P(\Omega^*) = 1$ and an integer-valued random variable $N(\omega)$ such that for every $\omega \in \Omega^*$: $\max_{0\le t\le T} \|X_t^{(k+1)}(\omega) - X_t^{(k)}(\omega)\| \le 2^{-(k+1)}$, $\forall\, k \ge N(\omega)$. Consequently,

$$\text{(2.21)}\quad \max_{0\le t\le T} \|X_t^{(k+m)}(\omega) - X_t^{(k)}(\omega)\| \le 2^{-k},\quad \forall\, m \ge 1,\ k \ge N(\omega).$$

We see then that the sequence of sample paths $\{X_t^{(k)}(\omega); 0 \le t \le T\}_{k=1}^\infty$ is convergent in the supremum norm on continuous functions, from which follows the existence of a continuous limit $\{X_t(\omega); 0 \le t \le T\}$ for all $\omega \in \Omega^*$. Since T is arbitrary, we have the existence of a continuous process $X = \{X_t; 0 \le t < \infty\}$ with the property that for P-a.e. ω, the sample paths $\{X_\cdot^{(k)}(\omega)\}_{k=1}^\infty$ converge to $X_\cdot(\omega)$, uniformly on compact subsets of $[0, \infty)$. Inequality (2.15) is a consequence of (2.17) and Fatou's lemma. From (2.15) and (2.13) we have condition (iii) of Definition 2.1. Conditions (i) and (ii) are also clearly satisfied by X. The following problem concludes the proof. □

2.11 Problem. Show that the just constructed process

$$\text{(2.22)}\quad X_t \triangleq \lim_{k\to\infty} X_t^{(k)};\quad 0 \le t < \infty$$

satisfies requirement (iv) of Definition 2.1.

2.12 Problem. With the exception of (2.15) and the square-integrability of X, the assertions of Theorem 2.9 remain valid if the assumption (2.14) is removed.

C. Comparison Results and Other Refinements

In the one-dimensional case, the Lipschitz condition on the dispersion coefficient can be relaxed considerably.

2.13 Proposition (Yamada & Watanabe (1971)). *Let us suppose that the coefficients of the one-dimensional equation* ($d = r = 1$)

$$dX_t = b(t, X_t)\,dt + \sigma(t, X_t)\,dW_t \tag{2.1}$$

satisfy the conditions

$$|b(t,x) - b(t,y)| \le K|x - y|, \tag{2.23}$$

$$|\sigma(t,x) - \sigma(t,y)| \le h(|x - y|), \tag{2.24}$$

for every $0 \le t < \infty$ *and* $x \in \mathbb{R}$, $y \in \mathbb{R}$, *where* K *is a positive constant and* $h\colon [0, \infty) \to [0, \infty)$ *is a strictly increasing function with* $h(0) = 0$ *and*

$$\int_{(0,\varepsilon)} h^{-2}(u)\,du = \infty; \quad \forall\, \varepsilon > 0. \tag{2.25}$$

Then strong uniqueness holds for the equation (2.1).

2.14 Example. One can take the function h in this proposition to be $h(u) = u^\alpha$; $\alpha \ge (1/2)$.

Proof of Proposition 2.13. Because of the conditions imposed on the function h, there exists a strictly decreasing sequence $\{a_n\}_{n=0}^\infty \subseteq (0, 1]$ with $a_0 = 1$, $\lim_{n\to\infty} a_n = 0$ and $\int_{a_n}^{a_{n-1}} h^{-2}(u)\,du = n$, for every $n \ge 1$. For each $n \ge 1$, there exists a continuous function ρ_n on $\mathbb{R}$ with support in (a_n, a_{n-1}) so that $0 \le \rho_n(x) \le (2/nh^2(x))$ holds for every $x > 0$, and $\int_{a_n}^{a_{n-1}} \rho_n(x)\,dx = 1$. Then the function

$$\psi_n(x) \triangleq \int_0^{|x|} \int_0^y \rho_n(u)\,du\,dy; \quad x \in \mathbb{R} \tag{2.26}$$

is even and twice continuously differentiable, with $|\psi_n'(x)| \le 1$ and $\lim_{n\to\infty} \psi_n(x) = |x|$ for $x \in \mathbb{R}$. Furthermore, the sequence $\{\psi_n\}_{n=1}^\infty$ is nondecreasing.

Now let us suppose that there are two strong solutions $X^{(1)}$ and $X^{(2)}$ of (2.1) with $X_0^{(1)} = X_0^{(2)}$ a.s. It suffices to prove the indistinguishability of $X^{(1)}$ and $X^{(2)}$ under the assumption

$$E\int_0^t |\sigma(s, X_s^{(i)})|^2\,ds < \infty; \quad 0 \le t < \infty,\ i = 1, 2; \tag{2.27}$$

otherwise, we may use condition (iii) of Definition 2.1 and a localization argument to reduce the situation to one in which (2.27) holds. We have

$$\Delta_t \triangleq X_t^{(1)} - X_t^{(2)} = \int_0^t \{b(s, X_s^{(1)}) - b(s, X_s^{(2)})\}\, ds + \int_0^t \{\sigma(s, X_s^{(1)}) - \sigma(s, X_s^{(2)})\}\, dW_s,$$

and by the Itô rule,

$$\begin{aligned}\text{(2.28)}\qquad \psi_n(\Delta_t) = &\int_0^t \psi_n'(\Delta_s)[b(s, X_s^{(1)}) - b(s, X_s^{(2)})]\, ds \\ &+ \frac{1}{2}\int_0^t \psi_n''(\Delta_s)[\sigma(s, X_s^{(1)}) - \sigma(s, X_s^{(2)})]^2\, ds \\ &+ \int_0^t \psi_n'(\Delta_s)[\sigma(s, X_s^{(1)}) - \sigma(s, X_s^{(2)})]\, dW_s.\end{aligned}$$

The expectation of the stochastic integral in (2.28) is zero because of assumption (2.27), whereas the expectation of the second integral in (2.28) is bounded above by $E\int_0^t \psi_n''(\Delta_s)h^2(|\Delta_s|)\, ds \le 2t/n$. We conclude that

$$\begin{aligned}\text{(2.29)}\qquad E\psi_n(\Delta_t) &\le E\int_0^t \psi_n'(\Delta_s)[b(s, X_s^{(1)}) - b(s, X_s^{(2)})]\, ds + \frac{t}{n} \\ &\le K\int_0^t E|\Delta_s|\, ds + \frac{t}{n}; \quad t \ge 0,\ n \ge 1.\end{aligned}$$

A passage to the limit as $n \to \infty$ yields $E|\Delta_t| \le K\int_0^t E|\Delta_s|\, ds$; $t \ge 0$, and the conclusion now follows from the Gronwall inequality and sample path continuity. □

2.15 Example (Girsanov (1962)). From what we have just proved, it follows that strong uniqueness holds for the one-dimensional stochastic equation

$$\text{(2.30)}\qquad X_t = \int_0^t |X_s|^\alpha\, dW_s; \quad 0 \le t < \infty,$$

as long as $\alpha \ge (1/2)$, and it is obvious that the unique solution is the trivial one $X_t \equiv 0$. This is also a solution when $0 < \alpha < (1/2)$, but it is no longer the only solution. We shall in fact see in Remark 5.6 that not only does strong uniqueness fail when $0 < \alpha < (1/2)$, but we do not even have uniqueness in the weaker sense developed in the next section.

2.16 Remark. Yamada & Watanabe (1971) actually establish Proposition 2.13 under a condition on $b(t, x)$ weaker than (2.23), namely,

$$\text{(2.23)}'\qquad |b(t,x) - b(t,y)| \le \kappa(|x - y|); \quad 0 \le t < \infty,\ x \in \mathbb{R},\ y \in \mathbb{R},$$

where $\kappa: [0, \infty) \to [0, \infty)$ is strictly increasing and concave with $\kappa(0) = 0$ and $\int_{(0,\varepsilon)} (du/\kappa(u)) = \infty$ for every $\varepsilon > 0$.

2.17 Exercise (Itô & Watanabe (1978)). The stochastic equation

$$X_t = 3\int_0^t X_s^{1/3}\,ds + 3\int_0^t X_s^{2/3}\,dW_s$$

has uncountably many strong solutions of the form

$$X_t^{(\theta)} = \begin{cases} 0; & 0 \le t < \beta_\theta, \\ W_t^3; & \beta_\theta \le t < \infty, \end{cases}$$

where $0 \le \theta \le \infty$ and $\beta_\theta \triangleq \inf\{s \ge \theta; W_s = 0\}$. Note that the function $\sigma(x) = 3x^{2/3}$ satisfies condition (2.24), but the function $b(x) = 3x^{1/3}$ fails to satisfy the condition of Remark 2.16.

The methodology employed in the proof of Proposition 2.13 can be used to great advantage in establishing *comparison results* for solutions of one-dimensional stochastic differential equations. Such results amount to a certain kind of "monotonicity" of the solution process X with respect to the drift coefficient $b(t, x)$, and they are useful in a variety of situations, including the study of certain simple stochastic control problems. We develop some comparison results in the following proposition and problem.

2.18 Proposition. *Suppose that on a certain probability space $(\Omega, \mathscr{F}, P)$ equipped with a filtration $\{\mathscr{F}_t\}$ which satisfies the usual conditions, we have a standard, one-dimensional Brownian motion $\{W_t, \mathscr{F}_t; 0 \le t < \infty\}$ and two continuous, adapted processes $X^{(j)}; j = 1, 2$, such that*

$$(2.31)\quad X_t^{(j)} = X_0^{(j)} + \int_0^t b_j(s, X_s^{(j)})\,ds + \int_0^t \sigma(s, X_s^{(j)})\,dW_s; \quad 0 \le t < \infty$$

holds a.s. for $j = 1, 2$. We assume that

(i) *the coefficients $\sigma(t, x)$, $b_j(t, x)$ are continuous, real-valued functions on $[0, \infty) \times \mathbb{R}$,*
(ii) *the dispersion matrix $\sigma(t, x)$ satisfies condition (2.24), where h is as described in Proposition 2.13,*
(iii) $X_0^{(1)} \le X_0^{(2)}$ *a.s.,*
(iv) $b_1(t, x) \le b_2(t, x)$, $\forall\, 0 \le t < \infty$, $x \in \mathbb{R}$, *and*
(v) *either $b_1(t, x)$ or $b_2(t, x)$ satisfies condition (2.23).*

Then

$$(2.32)\qquad P[X_t^{(1)} \le X_t^{(2)}, \forall\, 0 \le t < \infty] = 1.$$

PROOF. For concreteness, let us suppose that (2.23) is satisfied by $b_1(t, x)$. Proceeding as in the proof of Proposition 2.13, we assume without loss of generality that (2.27) holds. We recall the functions $\psi_n(x)$ of (2.26) and create a new sequence of auxiliary functions by setting $\varphi_n(x) = \psi_n(x) \cdot 1_{(0,\infty)}(x)$; $x \in \mathbb{R}$, $n \geq 1$. With $\Delta_t = X_t^{(1)} - X_t^{(2)}$, the analogue of relation (2.29) is

$$E\varphi_n(\Delta_t) - \frac{t}{n} \leq E \int_0^t \varphi_n'(\Delta_s)[b_1(s, X_s^{(1)}) - b_2(s, X_s^{(2)})]\, ds$$

$$= E \int_0^t \varphi_n'(\Delta_s)[b_1(s, X_s^{(1)}) - b_1(s, X_s^{(2)})]\, ds$$

$$+ E \int_0^t \varphi_n'(\Delta_s)[b_1(s, X_s^{(2)}) - b_2(s, X_s^{(2)})]\, ds \leq K \int_0^t E(\Delta_s^+)\, ds,$$

by virtue of (iv) and (2.23) for $b_1(t,x)$. Now we can let $n \to \infty$ to obtain $E(\Delta_t^+) \leq K \int_0^t E(\Delta_s^+)\, ds$; $0 \leq t < \infty$, and by the Gronwall inequality (Problem 2.7), we have $E(\Delta_t^+) = 0$; i.e., $X_t^{(1)} \leq X_t^{(2)}$ a.s. P. □

2.19 Exercise. Suppose that in Proposition 2.18 we drop condition (v) but strengthen condition (iv) to

(iv)′ $$b_1(t,x) < b_2(t,x); \quad 0 \leq t < \infty,\ x \in \mathbb{R}.$$

Then the conclusion (2.32) still holds. (*Hint*: For each integer $m \geq 3$, construct a Lipschitz-continuous function $b_m(t,x)$ such that

(2.33) $$b_1(t,x) \leq b_m(t,x) \leq b_2(t,x); \quad 0 \leq t \leq m,\ |x| \leq m).$$

It should be noted that for the equation

(2.34) $$X_t = \xi + \int_0^t b(s, X_s)\, ds + W_t; \quad 0 \leq t < \infty,$$

with unit dispersion coefficient and drift $b(t,x)$ satisfying the conditions of Theorem 2.9, the proof of that theorem can be simplified considerably. Indeed, since there is no stochastic integral in (2.34), we may fix an arbitrary $\omega \in \Omega$ and regard (2.34) as a deterministic integral equation with forcing function $\{W_t(\omega); 0 \leq t < \infty\}$. For the iterations defined by (2.16), and with

$$D_t^{(k)}(\omega) \triangleq \max_{0 \leq s \leq t} \|X_s^{(k)}(\omega) - X_s^{(k-1)}(\omega)\|; \quad k = 1, 2, \ldots,$$

we have the bound $D_t^{(k)}(\omega) \leq K \int_0^t D_s^{(k-1)}(\omega)\, ds$; $0 \leq t < \infty$, valid for *every* $\omega \in \Omega$. The latter can be iterated to prove convergence of the scheme (2.16), path by path, to a continuous, adapted process X which obeys (2.34) surely. This is the standard Picard-Lindelöf proof from ordinary differential equations and makes no use of probabilistic tools such as the martingale inequality or the Borel-Cantelli lemma.

Lamperti (1964) has observed that, under appropriate conditions on the coefficients b and σ, the general, one-dimensional integral equation

(2.4)″
$$X_t = \xi + \int_0^t b(X_s)\,ds + \int_0^t \sigma(X_s)\,dW_s; \quad 0 \le t < \infty,$$

can be reduced by a change of scale to one of the form (2.34); see the following exercise.

2.20 Exercise. Suppose that the coefficients $\sigma: \mathbb{R} \to (0, \infty)$ and $b: \mathbb{R} \to \mathbb{R}$ are of class C^2 and C^1, respectively; that $b' - (1/2)\sigma\sigma'' - (b\sigma'/\sigma)$ is bounded; and that $(1/\sigma)$ is not integrable at either $\pm\infty$. Then (2.4)″ has a unique, strong solution X. (*Hint*: Consider the function $f(x) = \int_0^x (du/\sigma(u))$ and apply Itô's rule to $f(X_t)$).

A second important class of equations that can be solved by first fixing the Brownian path and then solving a deterministic differential equation was discovered by Doss (1977); see Proposition 2.21.

D. Approximations of Stochastic Differential Equations

Stochastic differential equations have been widely applied to the study of the effect of adding random perturbations (noise) to deterministic differential systems. Brownian motion offers an idealized model for this noise, but in many applications the actual noise process is of bounded variation and non-Markov. Then the following modeling issue arises.

Suppose that $\{V_n\}_{n=1}^{\infty}$ is such a sequence of stochastic processes which converges, in an appropriate strong sense, to the Brownian motion $W = \{W_t, \mathscr{F}_t; 0 \le t < \infty\}$. Suppose, furthermore, that $\{X_n\}_{n=1}^{\infty}$ is a corresponding sequence of solutions to the stochastic integral equations

(2.35)
$$X_t^{(n)} = \xi + \int_0^t b(X_s^{(n)})\,ds + \int_0^t \sigma(X_s^{(n)})\,dV_s^{(n)}; \quad n \ge 1,$$

where the second integral is to be understood in the Lebesgue-Stieltjes sense. As $n \to \infty$, will $\{X^{(n)}\}_{n=1}^{\infty}$ converge to a process X, and if so, what kind of integral equation will X satisfy? It turns out that under fairly general conditions, the proper equation for X is

(2.36)
$$X_t = \xi + \int_0^t b(X_s)\,ds + \int_0^t \sigma(X_s) \circ dW_s,$$

where the second integral is in the *Fisk-Stratonovich* sense.

Our proof of this depends on the following result by Doss (1977).

2.21 Proposition. *Suppose that σ is of class $C^2(\mathbb{R})$ with bounded first and second derivatives, and that b is Lipschitz-continuous. Then the one-dimensional stochastic differential equation*

$$(2.36)' \qquad X_t = \xi + \int_0^t \left\{ b(X_s) + \frac{1}{2}\sigma(X_s)\sigma'(X_s) \right\} ds + \int_0^t \sigma(X_s)\,dW_s$$

has a unique, strong solution; this can be written in the form

$$X_t(\omega) = u(W_t(\omega), Y_t(\omega)); \quad 0 \le t < \infty, \omega \in \Omega$$

for a suitable, continuous function $u: \mathbb{R}^2 \to \mathbb{R}$ *and a process* Y *which solves an ordinary differential equation, for every* $\omega \in \Omega$.

2.22 Remark. Under the conditions of Proposition 2.21, the process $\{\sigma(X_t), \mathscr{F}_t; 0 \le t < \infty\}$ is a continuous semimartingale with decomposition

$$\sigma(X_t) = \sigma(\xi) + \int_0^t \left[b(X_s)\sigma'(X_s) + \frac{1}{2}\sigma(X_s)(\sigma'(X_s))^2 + \frac{1}{2}\sigma''(X_s)\sigma^2(X_s) \right] ds$$
$$+ \int_0^t \sigma(X_s)\sigma'(X_s)\,dW_s,$$

and so, according to Definition 3.3.13,

$$\int_0^t \sigma(X_s) \circ dW_s = \frac{1}{2}\int_0^t \sigma(X_s)\sigma'(X_s)\,ds + \int_0^t \sigma(X_s)\,dW_s.$$

In other words, equations (2.36) and (2.36)′ are equivalent.

PROOF OF PROPOSITION 2.21. Let $u(x, y): \mathbb{R}^2 \to \mathbb{R}$ be the solution of the ordinary differential equation

$$(2.37) \qquad \frac{\partial u}{\partial x} = \sigma(u), \quad u(0, y) = y;$$

such a solution exists globally, thanks to our assumptions. We have then

$$(2.38) \qquad \frac{\partial^2 u}{\partial x^2} = \sigma(u)\sigma'(u), \quad \frac{\partial^2 u}{\partial x \partial y} = \sigma'(u)\frac{\partial u}{\partial y}, \quad \frac{\partial}{\partial y}u(0, y) = 1,$$

which give

$$(2.39) \qquad \frac{\partial}{\partial y}u(x, y) = \exp\left\{ \int_0^x \sigma'(u(z, y))\,dz \right\} \triangleq \frac{1}{\rho(x, y)}.$$

Let $A > 0$ be a bound on σ' and σ''. Then $e^{-A|x|} \le \rho(x, y) \le e^{A|x|}$, and (2.39) implies the Lipschitz condition

$$|u(x, y_1) - u(x, y_2)| \le e^{A|x|}|y_1 - y_2|.$$

If L is a Lipschitz constant for b, then

$$|b(u(x, y_1)) - b(u(x, y_2))| \le Le^{A|x|}|y_1 - y_2|$$

and consequently, for fixed x, $b(u(x, y))$ is Lipschitz-continuous and exhibits linear growth in y. Using the inequality $|e^{z_1} - e^{z_2}| \le (e^{z_1} \vee e^{z_2})|z_1 - z_2|$, we may write

$$|\rho(x, y_1) - \rho(x, y_2)| \le [\rho(x, y_1) \vee \rho(x, y_2)] \cdot \int_0^{|x|} |\sigma'(u(z, y_1)) - \sigma'(u(z, y_2))| \, dz$$

$$\le e^{A|x|} \int_0^{|x|} A|u(z, y_1) - u(z, y_2)| \, dz$$

$$\le A|x|e^{2A|x|}|y_1 - y_2|.$$

For fixed x, $\rho(x, y)$ is thus Lipschitz-continuous and bounded in y. It follows that the product $f(x, y) \triangleq \rho(x, y) \cdot b(u(x, y))$ satisfies Lipschitz and growth conditions of the form

$$(2.40) \qquad |f(x, y_1) - f(x, y_2)| \le L_k |y_1 - y_2|; \quad -k \le x, y_1, y_2 \le k$$

$$(2.41) \qquad |f(x, y)| \le K_1 + K_k|y|; \quad |x| \le k, y \in \mathbb{R},$$

where the constants L_k and K_k depend on k.

We may fix $\omega \in \Omega$ and let $Y_t(\omega)$ be the solution to the ordinary differential equation

$$(2.42) \qquad \frac{d}{dt} Y_t(\omega) = f(W_t(\omega), Y_t(\omega)) \quad \text{with } Y_0(\omega) = \xi(\omega).$$

Such a solution exists globally and is unique because of (2.40) and (2.41). The resulting process Y is adapted to $\{\mathscr{F}_t\}$, and the same is true of $X_t(\omega) \triangleq u(W_t(\omega), Y_t(\omega))$. An application of Itô's rule shows that X satisfies (2.36)′. □

We turn our attention to the integral equation (2.35).

2.23 Lemma. *Let P-a.e. path $V_.(\omega)$ of the process $V = \{V_t; 0 \le t < \infty\}$ be continuous and have finite total variation $\check{V}_t(\omega)$ on compact intervals of the form $[0, t]$. If $b, \sigma: \mathbb{R} \to \mathbb{R}$ are Lipschitz-continuous, then the equation* (2.35) *with $V^n \equiv V$ possesses a unique solution.*

PROOF. Set $X_t^{(0)} = \xi$; $0 \le t < \infty$ and define recursively for $k \ge 1$:

$$(2.43) \qquad X_t^{(k+1)} \triangleq \xi + \int_0^t b(X_s^{(k)}) \, ds + \int_0^t \sigma(X_s^{(k)}) \, dV_s; \quad 0 \le t < \infty.$$

Then $D_t^{(k+1)} \triangleq \max_{0 \le s \le t} |X_s^{(k+1)} - X_s^{(k)}|$ satisfies

$$D_t^{(k+1)} \le L \int_0^t D_s^{(k)} (ds + d\check{V}_s),$$

where L is a Lipschitz constant for b and σ. Iteration of this inequality leads to

$$D_t^{(k+1)} \le D_t^{(1)} \frac{L^k (t + \check{V}_t)^k}{k!}; \quad 0 \le t < \infty, k \ge 0.$$

For $0 \le t < \infty$ fixed, the right-hand side of this last inequality is summable over k, so $\{X_s^{(k)}; 0 \le s \le t\}_{k=0}^{\infty}$ is Cauchy in the supremum norm on $C[0,t]$. Since t is arbitrary, $X^{(k)}$ must converge to a continuous process X, and the convergence is uniform on bounded intervals. Letting $k \to \infty$ in (2.43), we see that X solves (2.35).

If Y is another solution to (2.35), then $D_t \triangleq \max_{0 \le s \le t} |X_s - Y_s|$ must satisfy

$$D_t \le D_t \frac{L^k(t + \check{V}_t)^k}{k!}, \quad 0 \le t < \infty, k \ge 0,$$

which implies that $D \equiv 0$. □

2.24 Proposition. *Suppose that σ is of class $C^2(\mathbb{R})$ with bounded first and second derivatives, and that b is Lipschitz-continuous. Let $\{V_t^{(n)}; 0 \le t < \infty\}_{n=1}^{\infty}$ be a sequence of processes satisfying the same conditions as V in Lemma 2.23, and let $\{W_t, \mathscr{F}_t; 0 \le t < \infty\}$ be a Brownian motion with*

$$\lim_{n\to\infty} \sup_{0 \le s \le t} |V_s^{(n)} - W_s| = 0, \quad a.s.,$$

for every $0 \le t < \infty$. Then the sequence of (unique) solutions to the integral equation (2.35) converges almost surely, uniformly on bounded intervals, to the unique solution X of equation (2.36).

PROOF. Let u and f be as in the proof of Proposition 2.21; let $Y_t^{(n)}(\omega)$ be the solution to the ordinary differential equation (cf. (2.42)):

$$\frac{d}{dt} Y_t^{(n)}(\omega) = f(V_t^{(n)}(\omega), Y_t^{(n)}(\omega)); \quad Y_0^{(n)}(\omega) = \xi(\omega),$$

and define $X_t^{(n)}(\omega) \triangleq u(V_t^{(n)}(\omega), Y_t^{(n)}(\omega))$. Ordinary calculus shows that $X^{(n)}$ is the (unique) solution to (2.35). It remains only to show that with Y defined by (2.42), we have for every $0 \le t < \infty$:

$$\lim_{n\to\infty} \sup_{0 \le s \le t} |Y_s^{(n)} - Y_s| = 0, \quad \text{a.s.} \tag{2.44}$$

Let us fix $\omega \in \Omega$, $0 \le t < \infty$, and a positive integer k. With L_k as in (2.40), we may choose $\varepsilon > 0$ satisfying $\varepsilon < e^{-L_k t}$. Let $\tau_k(\omega) = t \wedge \inf\{0 \le s \le t; |Y_s(\omega)| \ge k - 1$ or $|W_s(\omega)| \ge k - 1\}$, $\tau_k^{(n)}(\omega) = t \wedge \inf\{0 \le s \le t; |Y_s^{(n)}(\omega)| \ge k\}$. For fixed $\omega \in \mathbb{R}$, we may choose n sufficiently large, so that $|f(V_s^{(n)}(\omega), Y_s(\omega)) - f(W_s(\omega), Y_s(\omega))| \le \varepsilon^2$ and $|V_s^{(n)}(\omega)| \le k$ hold for every $s \in [0, \tau_k(\omega) \wedge \tau_k^{(n)}(\omega)]$, and thus

$$\begin{aligned}\left|\frac{d}{ds}(Y_s^{(n)}(\omega) - Y_s(\omega))\right| &\le |f(V_s^{(n)}(\omega), Y_s^{(n)}(\omega)) - f(V_s^{(n)}(\omega), Y_s(\omega))| \\ &\quad + |f(V_s^{(n)}(\omega), Y_s(\omega)) - f(W_s(\omega), Y_s(\omega))| \\ &\le L_k |Y_s^{(n)}(\omega) - Y_s(\omega)| + \varepsilon^2.\end{aligned}$$

An application of Gronwall's inequality (Problem 2.7) yields

$$|Y_s^{(n)}(\omega) - Y_s(\omega)| \le \varepsilon^2 e^{L_k t} < \varepsilon; \quad 0 \le s \le \tau_k(\omega) \wedge \tau_k^{(n)}(\omega).$$

This last inequality shows that $\tau_k(\omega) \le \tau_k^{(n)}(\omega)$, so we may let first $n \to \infty$ and then $\varepsilon \downarrow 0$ to conclude that

$$\lim_{n\to\infty} \sup_{0\le s\le \tau_k(\omega)} |Y_s^{(n)}(\omega) - Y_s(\omega)| = 0.$$

For k sufficiently large we have $\tau_k(\omega) = t$, and (2.44) follows. □

2.25 Remark. The proof of Proposition 2.24 works only for one-dimensional stochastic differential equations with time-independent coefficients. In higher dimensions there may not be a counterpart of the function u satisfying (2.37) (see, however, Exercise 2.28 (i)), and consequently Proposition 2.24 does not hold in the same generality as in one dimension. However, if the continuous processes $\{V^{(n)}\}_{n=1}^{\infty}$ of bounded variation are obtained by mollification or piecewise-linear interpolation of the paths of a Brownian motion W, then the multidimensional version of the Fisk-Stratonovich stochastic differential equation (2.36) is still the correct limit of the Lebesgue-Stieltjes differential equations (2.35). The reader is referred to Ikeda & Watanabe (1981), Chapter VI, Section 7, for a full discussion.

E. Supplementary Exercises

2.26 Exercise (The Kramers-Smoluchowski Approximation; Nelson (1967)). Let $b(t,x)$: $[0,\infty) \times \mathbb{R} \to \mathbb{R}$ be a continuous, bounded function which satisfies the Lipschitz condition (2.23), and for every finite $\alpha > 0$ consider the stochastic differential system

$$dX_t^{(\alpha)} = Y_t^{(\alpha)}\,dt; \qquad X_0^{(\alpha)} = \xi$$
$$dY_t^{(\alpha)} = \alpha b(t, X_t^{(\alpha)})\,dt - \alpha Y_t^{(\alpha)}\,dt + \alpha\,dW_t; \quad Y_0^{(\alpha)} = \eta,$$

where ξ, η are a.s. finite random variables, jointly independent of the Brownian motion W.

(i) This system admits a unique, strong solution for every value of $\alpha \in (0,\infty)$.
(ii) For every fixed, finite $T > 0$, we have

$$\lim_{\alpha\to\infty} \sup_{0\le t\le T} |X_t^{(\alpha)} - X_t| = 0, \quad \text{a.s.},$$

where X is the unique, strong solution to (2.34).

2.27 Exercise. Solve explicitly the one-dimensional equation

$$dX_t = \left(\sqrt{1 + X_t^2} + \tfrac{1}{2}X_t\right)dt + \sqrt{1 + X_t^2}\,dW_t.$$

2.28 Exercise.

(i) Suppose that there exists an $\mathbb{R}^d$-valued function $u(t,y) = (u_1(t,y),\dots,u_d(t,y))$ of class $C^{1,2}([0,\infty) \times \mathbb{R}^d)$, such that

$$\frac{\partial u_i}{\partial t}(t,y) = b_i(t,u(t,y)), \quad \frac{\partial u_i}{\partial y_j}(t,y) = \sigma_{ij}(t,u(t,y)); \quad 1 \le i,j \le d$$

hold on $[0,\infty) \times \mathbb{R}^d$, where each $b_i(t,x)$ is continuous and each $\sigma_{ij}(t,x)$ is of class $C^{1,2}$ on $[0,\infty) \times \mathbb{R}^d$. Show then that the process

$$X_t \triangleq u(t, W_t); \quad 0 \le t < \infty,$$

where W is a d-dimensional Brownian motion, solves the Fisk-Stratonovich equation

$$(2.36)'' \qquad dX_t = b(t,X_t)\,dt + \sigma(t,X_t) \circ dW_t.$$

(ii) Use the above result to find the unique, strong solution of the one-dimensional Itô equation

$$dX_t = \left[\frac{2}{1+t}X_t - a(1+t)^2\right]dt + a(1+t)^2\,dW_t; \quad 0 \le t < \infty.$$

5.3. Weak Solutions

Our intent in this section is to discuss a notion of solvability for the stochastic differential equation (2.1) which, although weaker than the one introduced in the preceding section, is yet extremely useful and fruitful in both theory and applications. In particular, one can prove existence of solutions under assumptions on the drift term $b(t,x)$ much weaker than those of the previous section, and the notion of uniqueness attached to this new mode of solvability will lead naturally to the *strong Markov property* of the solution process (Theorem 4.20).

3.1 Definition. A *weak solution* of equation (2.1) is a triple (X, W), $(\Omega, \mathscr{F}, P)$, $\{\mathscr{F}_t\}$, where

(i) $(\Omega, \mathscr{F}, P)$ is a probability space, and $\{\mathscr{F}_t\}$ is a filtration of sub-σ-fields of $\mathscr{F}$ satisfying the usual conditions,

(ii) $X = \{X_t, \mathscr{F}_t; 0 \le t < \infty\}$ is a continuous, adapted $\mathbb{R}^d$-valued process, $W = \{W_t, \mathscr{F}_t; 0 \le t < \infty\}$ is an r-dimensional Brownian motion, and

(iii), (iv) of Definition 2.1 are satisfied.

The probability measure $\mu(\Gamma) \triangleq P[X_0 \in \Gamma]$, $\Gamma \in \mathscr{B}(\mathbb{R}^d)$ is called the *initial distribution* of the solution.

The filtration $\{\mathscr{F}_t\}$ in Definition 3.1 is not necessarily the augmentation of the filtration $\mathscr{G}_t = \sigma(\xi) \vee \mathscr{F}_t^W, 0 \le t < \infty$, generated by the "driving Brownian motion" and by the "initial condition" $\xi = X_0$. Thus, the value of the solution $X_t(\omega)$ at time t is not necessarily given by a measurable functional of the Brownian path $\{W_s(\omega); 0 \le s \le t\}$ and the initial condition $\xi(\omega)$. On the other hand, because W is a Brownian motion relative to $\{\mathscr{F}_t\}$, the solution $X_t(\omega)$ at time t cannot anticipate the future of the Brownian motion; besides $\{W_s(\omega); 0 \le s \le t\}$ and $\xi(\omega)$, whatever extra information is required to compute $X_t(\omega)$ must be independent of $\{W_\theta(\omega) - W_t(\omega); t \le \theta < \infty\}$.

One consequence of this arrangement is that the existence of a weak solution (X, W), $(\Omega, \mathscr{F}, P)$, $\{\mathscr{F}_t\}$ does *not* guarantee, for a given Brownian motion $\{\tilde{W}_t, \tilde{\mathscr{F}}_t; 0 \le t < \infty\}$ on a (possibly different) probability space $(\tilde{\Omega}, \tilde{\mathscr{F}}, \tilde{P})$, the existence of a process $\tilde{X}$ such that the triple $(\tilde{X}, \tilde{W})$, $(\tilde{\Omega}, \tilde{\mathscr{F}}, \tilde{P})$, $\{\tilde{\mathscr{F}}_t\}$ is again a weak solution. It is clear, however, that strong solvability implies weak solvability.

A. Two Notions of Uniqueness

There are two reasonable concepts of uniqueness which can be associated with weak solutions. The first is a straightforward generalization of strong uniqueness as set forth in Definition 2.3; the second, uniqueness in distribution, is better suited to the concept of weak solutions.

3.2 Definition. Suppose that whenever (X, W), $(\Omega, \mathscr{F}, P)$, $\{\mathscr{F}_t\}$, and $(\tilde{X}, W)$, $(\Omega, \mathscr{F}, P)$, $\{\tilde{\mathscr{F}}_t\}$, are weak solutions to (2.1) with common Brownian motion W (relative to possibly different filtrations) on a common probability space $(\Omega, \mathscr{F}, P)$ and with common initial value, i.e., $P[X_0 = \tilde{X}_0] = 1$, the two processes X and $\tilde{X}$ are indistinguishable: $P[X_t = \tilde{X}_t; \forall\, 0 \le t < \infty] = 1$. We say then that *pathwise uniqueness holds for equation* (2.1).

3.3 Remark. All the strong uniqueness results of Section 2 are also valid for pathwise uniqueness; indeed, none of the proofs given there takes advantage of the special form of the filtration for a strong solution.

3.4 Definition. We say that *uniqueness in the sense of probability law* holds for equation (2.1) if, for any two weak solutions (X, W), $(\Omega, \mathscr{F}, P)$, $\{\mathscr{F}_t\}$, and $(\tilde{X}, \tilde{W})$, $(\tilde{\Omega}, \tilde{\mathscr{F}}, \tilde{P})$, $\{\tilde{\mathscr{F}}_t\}$, with the same initial distribution, i.e.,

$$P[X_0 \in \Gamma] = \tilde{P}[\tilde{X}_0 \in \Gamma]; \quad \forall\, \Gamma \in \mathscr{B}(\mathbb{R}^d),$$

the two processes $X, \tilde{X}$ have the same law.

Existence of a weak solution does not imply that of a strong solution, and uniqueness in the sense of probability law does not imply pathwise uniqueness. The following example illustrates these points amply. However, pathwise uniqueness does imply uniqueness in the sense of probability law (see Proposition 3.20).

3.5 Example. (H. Tanaka (e.g., Zvonkin (1974))). Consider the one-dimensional equation

(3.1) $$X_t = \int_0^t \operatorname{sgn}(X_s)\,dW_s; \quad 0 \le t < \infty,$$

where

$$\operatorname{sgn}(x) = \begin{cases} 1; & x > 0, \\ -1; & x \le 0. \end{cases}$$

If (X, W), $(\Omega, \mathscr{F}, P)$, $\{\mathscr{F}_t\}$ is a weak solution, then the process $X = \{X_t, \mathscr{F}_t; 0 \le t < \infty\}$ is a continuous, square-integrable martingale with quadratic variation process $\langle X \rangle_t = \int_0^t \operatorname{sgn}^2(X_s)\,ds = t$. Therefore, X is a Brownian motion (Theorem 3.3.16), and *uniqueness in the sense of probability law holds.* On the other hand, $(-X, W)$, $(\Omega, \mathscr{F}, P)$, $\{\mathscr{F}_t\}$ is also a weak solution, so once we establish existence of a weak solution, we shall also have shown that *pathwise uniqueness cannot hold for equation* (3.1).

Now start with a probability space $(\Omega, \mathscr{F}, P)$ and a one-dimensional Brownian motion $X = \{X_t, \tilde{\mathscr{F}}_t^X; 0 \le t < \infty\}$ on it; we assume $P[X_0 = 0] = 1$ and denote by $\{\tilde{\mathscr{F}}_t^X\}$ the augmentation of the filtration $\{\mathscr{F}_t^X\}$ under P. The same argument as before shows that

$$W_t \triangleq \int_0^t \operatorname{sgn}(X_s)\,dX_s; \quad 0 \le t < \infty$$

is a Brownian motion adapted to $\{\tilde{\mathscr{F}}_t^X\}$. Corollary 3.2.20 shows that (X, W), $(\Omega, \mathscr{F}, P)$, $\{\tilde{\mathscr{F}}_t^X\}$ is a weak solution to (3.1). With $\{\tilde{\mathscr{F}}_t^W\}$ denoting the augmentation of $\{\mathscr{F}_t^W\}$, this construction gives $\tilde{\mathscr{F}}_t^W \subseteq \tilde{\mathscr{F}}_t^X$, which is the opposite inclusion from that required for a strong solution!

Let us now show that equation (3.1) *does not admit a strong solution.* Assume the contrary; i.e., let X satisfy (3.1) on a given $(\Omega, \mathscr{F}, P)$ with respect to a given Brownian motion W, and assume $\tilde{\mathscr{F}}_t^X \subseteq \tilde{\mathscr{F}}_t^W$ for every $t \ge 0$. Then X is necessarily a Brownian motion, and from Tanaka's formula (3.6.13) with $a = 0$, we have

$$\begin{aligned} W_t &= \int_0^t \operatorname{sgn}(X_s)\,dX_s = |X_t| - 2L_t^X(0) \\ &= |X_t| - \lim_{\varepsilon \downarrow 0} \frac{1}{2\varepsilon} \operatorname{meas}\{0 \le s \le t; |X_s| \le \varepsilon\}, \quad 0 \le t < \infty, \text{ a.s. } P, \end{aligned}$$

where $L_t^X(0)$ is the local time at the origin for X. Consequently, $\tilde{\mathscr{F}}_t^W \subseteq \tilde{\mathscr{F}}_t^{|X|}$, and thus also $\tilde{\mathscr{F}}_t^X \subseteq \tilde{\mathscr{F}}_t^{|X|}$ holds for every $t \ge 0$. But this last inclusion is absurd.

B. Weak Solutions by Means of the Girsanov Theorem

The principal method for creating weak solutions to stochastic differential equations is *transformation of drift* via the Girsanov theorem. The proof of the next proposition illustrates this approach.

3.6 Proposition. *Consider the stochastic differential equation*

$$dX_t = b(t, X_t)\,dt + dW_t; \quad 0 \le t \le T, \tag{3.2}$$

where T is a fixed positive number, W is a d-dimensional Brownian motion, and $b(t, x)$ *is a Borel-measurable,* $\mathbb{R}^d$*-valued function on* $[0, T] \times \mathbb{R}^d$ *which satisfies*

$$\|b(t,x)\| \le K(1 + \|x\|); \quad 0 \le t \le T, x \in \mathbb{R}^d \tag{3.3}$$

for some positive constant K. For any probability measure μ *on* $(\mathbb{R}^d, \mathscr{B}(\mathbb{R}^d))$*, equation* (3.2) *has a weak solution with initial distribution* μ.

PROOF. We begin with a d-dimensional Brownian family $X = \{X_t, \mathscr{F}_t; 0 \le t \le T\}$, $(\Omega, \mathscr{F})$, $\{P^x\}_{x \in \mathbb{R}^d}$. According to Corollary 3.5.16,

$$Z_t \triangleq \exp\left\{\sum_{j=1}^{d} \int_0^t b_j(s, X_s)\,dX_s^{(j)} - \frac{1}{2}\int_0^t \|b(s, X_s)\|^2\,ds\right\}$$

is a martingale under each measure P^x, so the Girsanov Theorem 3.5.1 implies that, under Q^x given by $(dQ^x/dP^x) = Z_T$, the process

$$W_t \triangleq X_t - X_0 - \int_0^t b(s, X_s)\,ds; \quad 0 \le t \le T \tag{3.4}$$

is a Brownian with $Q^x[W_0 = 0] = 1$, $\forall x \in \mathbb{R}^d$. Simply rewriting (3.4) as

$$X_t = X_0 + \int_0^t b(s, X_s)\,ds + W_t; \quad 0 \le t \le T,$$

we see that, with $Q^\mu(A) \triangleq \int_{\mathbb{R}^d} Q^x(A)\mu(dx)$, the triple (X, W), $(\Omega, \mathscr{F}, Q^\mu)$, $\{\mathscr{F}_t\}$ is a weak solution to (3.2). □

3.7 Remark. If we seek a solution to (3.2) defined for all $0 \le t < \infty$, we can repeat the preceding argument using the filtration $\{\mathscr{F}_t^X\}$ instead of $\{\mathscr{F}_t\}$ and citing Corollary 3.5.2 rather than Theorem 3.5.1. Whereas $\{\mathscr{F}_t\}$ in Proposition 3.6 can be chosen to satisfy the usual conditions, $\{\mathscr{F}_t^X\}$ does not have this property. Thus, as a last step in this construction, we take $\mathscr{N}$ to be the collection of null sets of $(\Omega, \mathscr{F}_\infty^X, Q^\mu)$, set $\tilde{\mathscr{G}}_t = \sigma(\mathscr{F}_t^X \cup \mathscr{N})$ and $\mathscr{G}_t = \tilde{\mathscr{G}}_{t+}$. The filtration $\{\mathscr{G}_t\}$ satisfies the usual conditions.

3.8 Remark. It is apparent from Corollary 3.5.16 that Proposition 3.6 can be extended to include the case

$$X_t = X_0 + \int_0^t b(s, X)\,ds + W_t; \quad 0 \le t \le T, \tag{3.5}$$

where $b(t, x)$ is a vector of progressively measurable functionals on $C[0, \infty)^d$; see Definition 3.14.

3.9 Remark. Even when the initial distribution μ degenerates to unit point mass at some $x \in \mathbb{R}^d$, the filtration $\{\mathscr{F}_t^W\}$ of the driving Brownian motion in

(3.5) may be strictly smaller than the filtration $\{\mathscr{F}_t^X\}$ of the solution process (see the discussion following Definition 3.1). This is shown by the celebrated example of Cirel'son (1975). Nice accounts of Cirel'son's result appear in Liptser & Shiryaev (1977), pp. 150–151, and Kallianpur (1980), pp. 189–191.

The Girsanov theorem is also helpful in the study of uniqueness in law of weak solutions. We use it to establish a companion to Proposition 3.6.

3.10 Proposition. *Assume that* $(X^{(i)}, W^{(i)})$, $(\Omega^{(i)}, \mathscr{F}^{(i)}, P^{(i)})$, $\{\mathscr{F}_t^{(i)}\}$; $i = 1, 2$, *are weak solutions to* (3.2) *with the same initial distribution. If*

$$(3.6) \qquad P^{(i)}\left[\int_0^T \|b(t, X_t^{(i)})\|^2\, dt < \infty\right] = 1; \quad i = 1, 2,$$

then $(X^{(1)}, W^{(1)})$ *and* $(X^{(2)}, W^{(2)})$ *have the same law under their respective probability measures.*

PROOF. For each $k \geq 1$, let

$$(3.7) \qquad \tau_k^{(i)} \triangleq T \wedge \inf\left\{0 \leq t \leq T; \int_0^t \|b(s, X_s^{(i)})\|^2\, ds = k\right\}.$$

According to Novikov's condition (Corollary 3.5.13),

$$(3.8) \quad \xi_t^{(k)}(X^{(i)}) \triangleq \exp\left\{-\int_0^{t\wedge\tau_k^{(i)}} (b(s, X_s^{(i)}), dW_s^{(i)}) - \frac{1}{2}\int_0^{t\wedge\tau_k^{(i)}} \|b(s, X_s^{(i)})\|^2\, ds\right\}$$

is a martingale, so we may define probability measures $\tilde{P}_k^{(i)}$ on $\mathscr{F}_T^{(i)}$, $i = 1, 2$, according to the prescription $(d\tilde{P}_k^{(i)}/dP^{(i)}) = \xi_T^{(k)}(X^{(i)})$. The Girsanov Theorem 3.5.1 states that, under $\tilde{P}_k^{(i)}$, the process

$$(3.9) \quad X_{t\wedge\tau_k^{(i)}}^{(i)} = X_0^{(i)} + \int_0^{t\wedge\tau_k^{(i)}} b(s, X_s^{(i)})\, ds + W_{t\wedge\tau_k^{(i)}}^{(i)}; \quad 0 \leq t \leq T$$

is a d-dimensional Brownian motion with initial distribution μ, stopped at time $\tau_k^{(i)}$. But $\tau_k^{(i)}$, $\{W_t^{(i)}; 0 \leq t \leq \tau_k^{(i)}\}$, and $\xi_T^{(k)}(X^{(i)})$ can all be defined in terms of the process in (3.9) (see Problem 3.5.6). Therefore, for $0 = t_0 < t_1 < \cdots < t_n = T$ and $\Gamma \in \mathscr{B}(\mathbb{R}^{2d(n+1)})$, we have

$$(3.10) \quad P^{(1)}[(X_{t_0}^{(1)}, W_{t_0}^{(1)}, \ldots, X_{t_n}^{(1)}, W_{t_n}^{(1)}) \in \Gamma; \tau_k^{(1)} = T]$$

$$= \int_{\Omega^{(1)}} \frac{1}{\xi_T^{(k)}(X^{(1)})} 1_{\{(X_{t_0}^{(1)}, W_{t_0}^{(1)}, \ldots, X_{t_n}^{(1)}, W_{t_n}^{(1)}) \in \Gamma; \tau_k^{(1)} = T\}}\, d\tilde{P}_k^{(1)}$$

$$= \int_{\Omega^{(2)}} \frac{1}{\xi_T^{(k)}(X^{(2)})} 1_{\{(X_{t_0}^{(2)}, W_{t_0}^{(2)}, \ldots, X_{t_n}^{(2)}, W_{t_n}^{(2)}) \in \Gamma; \tau_k^{(2)} = T\}}\, d\tilde{P}_k^{(2)}$$

$$= P^{(2)}[(X_{t_0}^{(2)}, W_{t_0}^{(2)}, \ldots, X_{t_n}^{(2)}, W_{t_n}^{(2)}) \in \Gamma; \tau_k^{(2)} = T].$$

By assumption (3.6), $\lim_{k\to\infty} P^{(i)}[\tau_k^{(i)} = T] = 1$; $i = 1, 2$, so passage to the limit as $k \to \infty$ in (3.10) gives the desired conclusion. □

3.11 Corollary. *If the drift term $b(t, x)$ in (3.2) is uniformly bounded, then uniqueness in the sense of probability law holds for equation (3.2). Furthermore, with $0 \le t_1 < t_2 < \cdots < t_n \le T$ and with the notation developed in the proof of Proposition 3.6, we have then*

$$(3.11)\quad Q^{\mu}[(X_{t_1}, \ldots, X_{t_n}) \in \Gamma] = \int_{\mathbb{R}^d} E^x[1_{\{(X_{t_1}, \ldots, X_{t_n}) \in \Gamma\}} Z_T]\mu(dx); \quad \Gamma \in \mathscr{B}(\mathbb{R}^{dn}).$$

3.12 Exercise. According to Proposition 3.6 and Corollary 3.11, the one-dimensional stochastic differential equation

$$(3.12)\qquad dX_t = -\operatorname{sgn}(X_t)\,dt + dW_t, \quad X_0 = 0$$

possesses a weak solution which is unique in the sense of probability law. Show that if (X, W), $(\Omega, \mathscr{F}, Q)$, $\{\mathscr{F}_t\}$ is such a solution and $L_t(0)$ is the local time at the origin for the Brownian family $\{X_t, \mathscr{F}_t\}$, $(\Omega, \mathscr{F})$, $\{P^x\}_{x\in\mathbb{R}}$, then

$$(3.13)\quad Q[X_t \in \Gamma] = e^{-t/2} E^0[1_{\{X_t \in \Gamma\}} \exp(-|X_t| + 2L_t(0))]; \quad \Gamma \in \mathscr{B}(\mathbb{R}).$$

(An explicit formula for the right-hand side of (3.13) can be derived from Theorem 3.6.17 and relation (2.8.2). See Problem 6.3.4 and Exercise 6.3.5.)

3.13 Problem. Consider the stochastic differential equation (2.1) with $\sigma(t, x)$ a $(d \times d)$ nonsingular matrix for every $t \ge 0$ and $x \in \mathbb{R}^d$. Assume that $b(t, x)$ is uniformly bounded, the smallest eigenvalue of $\sigma(t, x)\sigma^{tr}(t, x)$ is uniformly bounded away from zero, and the equation

$$(3.14)\qquad dX_t = \sigma(t, X_t)\,dW_t; \quad 0 \le t \le T$$

has a weak solution with initial distribution μ. Show that (2.1) also has a weak solution for $0 \le t \le T$ with initial distribution μ. (We shall have more to say about existence and uniqueness of solutions to (3.14) in Sections 4 and 5.)

3.14 Definition. Let $b_i(t, y)$ and $\sigma_{ij}(t, y)$; $1 \le i \le d$, $1 \le j \le r$, be progressively measurable functionals from $[0, \infty) \times C[0, \infty)^d$ into $\mathbb{R}$ (Definition 3.5.15). A *weak solution to the functional stochastic differential equation*

$$(3.15)\qquad dX_t = b(t, X)\,dt + \sigma(t, X)\,dW_t; \quad 0 \le t < \infty,$$

is a triple (X, W), $(\Omega, \mathscr{F}, P)$, $\{\mathscr{F}_t\}$ satisfying (i), (ii) of Definition 3.1, as well as

$$\text{(iii)}\qquad \int_0^t \{|b_i(s, X)| + \sigma_{ij}^2(s, X)\}\,ds < \infty; \quad 1 \le i \le d, \quad 1 \le j \le r, \quad t \ge 0,$$

$$\text{(iv)}\qquad X_t = X_0 + \int_0^t b(s, X)\,ds + \int_0^t \sigma(s, X)\,dW_s; \quad 0 \le t < \infty,$$

almost surely.

3.15 Problem. Suppose $b_i(t, y)$ and $\sigma_{ij}(t, y)$; $1 \le i \le d$, $1 \le j \le r$, are progressively measurable functionals from $[0, \infty) \times C[0, \infty)^d$ into $\mathbb{R}$ satisfying

$$\|b(t,y)\|^2 + \|\sigma(t,y)\|^2 \le K\left(1 + \max_{0 \le s \le t} \|y(s)\|^2\right); \quad \forall\, 0 \le t < \infty, \quad y \in C[0,\infty)^d, \tag{3.16}$$

where K is a positive constant. If (X, W), $(\Omega, \mathscr{F}, P)$, $\{\mathscr{F}_t\}$ is a weak solution to (3.15) with $E\|X_0\|^{2m} < \infty$ for some $m \ge 1$, then for any finite $T > 0$, we have

$$E\left(\max_{0 \le s \le t} \|X_s\|^{2m}\right) \le C(1 + E\|X_0\|^{2m})e^{Ct}; \quad 0 \le t \le T, \tag{3.17}$$

$$E\|X_t - X_s\|^{2m} \le C(1 + E\|X_0\|^{2m})(t-s)^m; \quad 0 \le s < t \le T, \tag{3.18}$$

where C is a positive constant depending only on m, T, K, and d.

C. A Digression on Regular Conditional Probabilities

We know that indistinguishable processes have the same finite-dimensional distributions, and this causes us to suspect that pathwise uniqueness implies uniqueness in the sense of probability law. The remainder of this section is devoted to the confirmation of this conjecture. As preparation, we need to state certain results about regular conditional probabilities, in the spirit of Definition 2.6.12, but in a form better suited to our present needs. We refer the reader to Parthasarathy (1967), pp. 131–150, and Ikeda & Watanabe (1981), pp. 12–16, for proofs and further information.

3.16 Definition. Let $(\Omega, \mathscr{F}, P)$ be a probability space and $\mathscr{G}$ a sub-σ-field of $\mathscr{F}$. A function $Q(\omega; A)\colon \Omega \times \mathscr{F} \to [0, 1]$ is called a *regular conditional probability for $\mathscr{F}$ given $\mathscr{G}$* if

(i) for each $\omega \in \Omega$, $Q(\omega; \cdot)$ is a probability measure on $(\Omega, \mathscr{F})$,
(ii) for each $A \in \mathscr{F}$, the mapping $\omega \mapsto Q(\omega; A)$ is $\mathscr{G}$-measurable, and
(iii) for each $A \in \mathscr{F}$, $Q(\omega; A) = P[A|\mathscr{G}](\omega)$; P-a.e. $\omega \in \Omega$.

Suppose that, whenever $Q'(\omega; A)$ is another function with these properties, there exists a null set $N \in \mathscr{F}$ such that $Q(\omega; A) = Q'(\omega; A)$ for all $A \in \mathscr{F}$ and $\omega \in \Omega \backslash N$. We then say that the regular conditional probability for $\mathscr{F}$ given $\mathscr{G}$ is *unique*.

Note that if X in Definition 2.6.12 is the identity mapping, then conditions (i)–(iii) of that definition coincide with those of Definition 3.16.

3.17 Definition. Let $(\Omega, \mathscr{F})$ be a measurable space. We say that $\mathscr{F}$ is *countably determined* if there exists a countable collection of sets $\mathscr{M} \subseteq \mathscr{F}$ such that, whenever two probability measures agree on $\mathscr{M}$, they also agree on $\mathscr{F}$. We say that $\mathscr{F}$ is *countably generated* if there exists a countable collection of sets $\mathscr{C} \subseteq \mathscr{F}$ such that $\mathscr{F} = \sigma(\mathscr{C})$.

In the space $C[0, \infty)^m$, for an arbitrary integer $m \geq 1$, let us introduce the σ-fields

$$(3.19) \qquad \mathscr{B}_t(C[0, \infty)^m) \triangleq \sigma(z(s); 0 \leq s \leq t) = \varphi_t^{-1}(\mathscr{B}(C[0, \infty)^m))$$

for $0 \leq t < \infty$, where φ_t: $C[0, \infty)^m \to C[0, \infty)^m$ is the truncation mapping $(\varphi_t z)(s) \triangleq z(t \wedge s)$; $z \in C[0, \infty)^m$, $0 \leq s < \infty$. As in Problem 2.4.2, it is shown that $\mathscr{B}_t(C[0, \infty)^m) = \sigma(\mathscr{C}_t)$, where $\mathscr{C}_t$ is the countable collection of all finite-dimensional cylinder sets of the form

$$C = \{z \in C[0, \infty)^m; (z(t_1), \ldots, z(t_n)) \in A\}$$

with $n \geq 1$, $t_i \in [0, t] \cap Q$, for $1 \leq i \leq n$, and $A \in \mathscr{B}(\mathbb{R}^{mn})$ equal to the product of intervals with rational endpoints. The continuity of $z \in C[0, \infty)^m$ allows us to conclude that a set of the form C is in $\sigma(\mathscr{C}_t)$, even if the points t_i are not necessarily rational.

It follows that $\mathscr{B}_t(C[0, \infty)^m)$ is *countably generated.* On the other hand, the generating class $\mathscr{C}_t$ is closed under finite intersections, so any two probability measures on $\mathscr{B}_t(C[0, \infty)^m)$ which agree on $\mathscr{C}_t$ must also agree on $\mathscr{B}_t(C[0, \infty)^m)$, by the Dynkin System Theorem 2.1.3. It follows that $\mathscr{B}_t(C[0, \infty)^m)$ is also *countably determined.*

More generally, Theorem 2.1.3 shows that if a σ-field $\mathscr{F}$ is generated by a countable collection of sets $\mathscr{C}$ which happens to be closed under pairwise intersection, then $\mathscr{F}$ is also countably determined. In a topological space with a countable base (e.g., a separable metric space), we may take this $\mathscr{C}$ to be the collection of all finite intersections of complements of these basic open sets.

3.18 Theorem. *Suppose that Ω is a complete, separable metric space, and denote the Borel σ-field $\mathscr{F} = \mathscr{B}(\Omega)$. Let P be a probability measure on $(\Omega, \mathscr{F})$, and let $\mathscr{G}$ be a sub-σ-field of $\mathscr{F}$. Then a regular conditional probability Q for $\mathscr{F}$ given $\mathscr{G}$ exists and is unique. Furthermore, if $\mathscr{H}$ is a countably determined sub-σ-field of $\mathscr{G}$, then there exists a null set $N \in \mathscr{G}$ such that*

$$\text{(iv)} \qquad Q(\omega; A) = 1_A(\omega); \quad A \in \mathscr{H}, \omega \in \Omega \backslash N.$$

In particular, if X is a $\mathscr{G}$-measurable random variable taking values in another complete, separable metric space, then with $\mathscr{H}$ denoting the σ-field generated by X, (iv) *implies*

$$\text{(iv)}' \qquad Q(\omega; \{\omega' \in \Omega; X(\omega') = X(\omega)\}) = 1; \quad P\text{-a.e. } \omega \in \Omega.$$

When the σ-field $\mathscr{G}$ is generated by a random variable, we may recast the assertions of Theorem 3.18 as follows.

3.19 Theorem. *Let $(\Omega, \mathscr{F}, P)$ be as in Theorem 3.18, and let X be a measurable mapping from this space into a measurable space $(S, \mathscr{S})$, on which it induces the* *distribution* $PX^{-1}(B) \triangleq P[\omega \in \Omega; X(\omega) \in B], B \in \mathscr{S}$. *There exists then a function* $Q(x; A): S \times \mathscr{F} \to [0, 1]$, *called a* regular conditional probability for $\mathscr{F}$ given X, *such that*

(i) *for each* $x \in S$, $Q(x; \cdot)$ *is a probability measure on* $(\Omega, \mathscr{F})$,
(ii) *for each* $A \in \mathscr{F}$, *the mapping* $x \mapsto Q(x; A)$ *is* $\mathscr{S}$*-measurable, and*
(iii) *for each* $A \in \mathscr{F}$, $Q(x; A) = P[A | X = x]$, PX^{-1}*-a.e.* $x \in S$.

If $Q'(x; A)$ *is another function with these properties, then there exists a set* $N \in \mathscr{S}$ *with* $PX^{-1}(N) = 0$ *such that* $Q(x; A) = Q'(x; A)$ *for all* $A \in \mathscr{F}$ *and* $x \in S \backslash N$. *Furthermore, if S is also a complete, separable metric space and* $\mathscr{S} = \mathscr{B}(S)$, *then N can be chosen so that we have the additional property*:

(iv) $$Q(x; \{\omega \in \Omega; X(\omega) \in B\}) = 1_B(x); \quad B \in \mathscr{S}, x \in S \backslash N.$$

In particular,

(iv)′ $$Q(x; \{\omega \in \Omega; X(\omega) = x\}) = 1; \quad PX^{-1}\text{-a.e. } x \in S.$$

D. Results of Yamada and Watanabe on Weak and Strong Solutions

Returning to our initial question about the relation between pathwise uniqueness and uniqueness in the sense of probability law, let us consider two weak solutions $(X^{(j)}, W^{(j)})$, $(\Omega_j, \mathscr{F}_j, \nu_j)$, $\{\mathscr{F}_t^{(j)}\}$; $j = 1, 2$, of equation (2.1) with

(3.20) $$\mu(B) \triangleq \nu_1[X_0^{(1)} \in B] = \nu_2[X_0^{(2)} \in B]; \quad B \in \mathscr{B}(\mathbb{R}^d).$$

We set $Y_t^{(j)} = X_t^{(j)} - X_0^{(j)}$; $0 \le t < \infty$, and we regard the j-th solution as consisting of three parts: $X_0^{(j)}$, $W^{(j)}$, and $Y^{(j)}$. This triple induces a measure P_j on

$$(\Theta, \mathscr{B}(\Theta)) \triangleq (\mathbb{R}^d \times C[0, \infty)^r \times C[0, \infty)^d,$$
$$\mathscr{B}(\mathbb{R}^d) \otimes \mathscr{B}(C[0, \infty)^r) \otimes \mathscr{B}(C[0, \infty)^d))$$

according to the prescription

(3.21) $$P_j(A) \triangleq \nu_j[(X_0^{(j)}, W^{(j)}, Y^{(j)}) \in A]; \quad A \in \mathscr{B}(\Theta), j = 1, 2.$$

We denote by $\theta = (x, w, y)$ the generic element of Θ. The marginal of each P_j on the x-coordinate of θ is μ, the marginal on the w-coordinate is Wiener measure P_*, and the distribution of the (x, w) pair is the product measure $\mu \times P_*$ because $X_0^{(j)}$ is $\mathscr{F}_0^{(j)}$-measurable and $W^{(j)}$ is independent of $\mathscr{F}_0^{(j)}$ (Problem 2.5.5). Furthermore, under P_j, the initial value of the y-coordinate is zero, almost surely.

The two weak solutions $(X^{(1)}, W^{(1)})$ and $(X^{(2)}, W^{(2)})$ are defined on (possibly) different sample spaces. Our first task is to bring them together on the same, canonical space, while preserving their joint distributions. Toward this end, we note that on $(\Theta, \mathscr{B}(\Theta), P_j)$ there exists a regular conditional probability for $\mathscr{B}(\Theta)$ given (x, w). We shall be interested only in conditional probabilities of sets in $\mathscr{B}(\Theta)$ of the form $\mathbb{R}^d \times C[0,\infty)^r \times F$, where $F \in \mathscr{B}(C[0,\infty)^d)$. Thus, with a slight abuse of terminology, we speak of

$$Q_j(x, w; F): \mathbb{R}^d \times C[0,\infty)^r \times \mathscr{B}(C[0,\infty)^d) \to [0,1]$$

as the *regular conditional probability for* $\mathscr{B}(C[0,\infty)^d)$ *given* (x, w). According to Theorem 3.19, this regular conditional probability enjoys the following properties:

(3.22) (i) for each $x \in \mathbb{R}^d$, $w \in C[0,\infty)^r$, $Q_j(x, w; \cdot)$ is a probability measure on $(C[0,\infty)^d, \mathscr{B}(C[0,\infty)^d))$,

(3.22) (ii) for each $F \in \mathscr{B}(C[0,\infty)^d)$, the mapping $(x, w) \mapsto Q_j(x, w; F)$ is $\mathscr{B}(\mathbb{R}^d) \otimes \mathscr{B}(C[0,\infty)^r)$-measurable, and

(3.22) (iii) $P_j(G \times F) = \int_G Q_j(x, w; F)\mu(dx)P_*(dw)$; $F \in \mathscr{B}(C[0,\infty)^d)$,
$G \in \mathscr{B}(\mathbb{R}^d) \otimes \mathscr{B}(C[0,\infty)^r)$.

Finally, we consider the measurable space $(\Omega, \mathscr{F})$, where $\Omega = \Theta \times C[0,\infty)^d$ and $\mathscr{F}$ is the completion of the σ-field $\mathscr{B}(\Theta) \otimes \mathscr{B}(C[0,\infty)^d)$ by the collection $\mathscr{N}$ of null sets under the probability measure

$$P(d\omega) \triangleq Q_1(x, w; dy_1)Q_2(x, w; dy_2)\mu(dx)P_*(dw). \tag{3.23}$$

We have denoted by $\omega = (x, w, y_1, y_2)$ a generic element of Ω. In order to endow $(\Omega, \mathscr{F}, P)$ with a filtration that satisfies the usual conditions, we take

$$\mathscr{G}_t \triangleq \sigma\{(x, w(s), y_1(s), y_2(s)); 0 \le s \le t\}, \quad \tilde{\mathscr{G}}_t \triangleq \sigma(\mathscr{G}_t \cup \mathscr{N}), \quad \mathscr{F}_t \triangleq \tilde{\mathscr{G}}_{t+},$$

for $0 \le t < \infty$. It is evident from (3.21), (3.22) (iii), and (3.23) that

$$P[\omega \in \Omega; (x, w, y_j) \in A] = \nu_j[(X_0^{(j)}, W^{(j)}, Y^{(j)}) \in A]; \quad A \in \mathscr{B}(\Theta), j = 1, 2, \tag{3.21$'$}$$

and so *the distribution of* $(x + y_j, w)$ *under* P *is the same as the distribution of* $(X^{(j)}, W^{(j)})$ *under* ν_j. In particular, the w-coordinate process $\{w(t), \mathscr{G}_t; 0 \le t < \infty\}$ is an r-dimensional Brownian motion on $(\Omega, \mathscr{F}, P)$, and it is then not difficult to see (cf. Ikeda & Watanabe (1981), Lemma IV.1.2) that the same is true for $\{w(t), \mathscr{F}_t; 0 \le t < \infty\}$.

3.20 Proposition (Yamada & Watanabe (1971)). *Pathwise uniqueness implies uniqueness in the sense of probability law.*

Proof. We started with two weak solutions $(X^{(j)}, W^{(j)})$, $(\Omega_j, \mathscr{F}_j, \nu_j)$, $\{\mathscr{F}_t^{(j)}\}$; $j = 1, 2$, of equation (2.1), with (3.20) satisfied. We have created two weak solutions $(x + y_j, w)$, $j = 1, 2$, on a single probability space $(\Omega, \mathscr{F}, P)$, $\{\mathscr{F}_t\}$, such that $(X^{(j)}, W^{(j)})$ under ν_j has the same law as $(x + y_j, w)$ under P. Pathwise uniqueness implies $P[x + y_1(t) = x + y_2(t), \forall\, 0 \le t < \infty] = 1$, or equivalently,

$$P[\omega = (x, w, y_1, y_2) \in \Omega; y_1 = y_2] = 1. \tag{3.24}$$

It develops from (3.21)$'$, (3.24) that

$$\begin{aligned}\nu_1[(X_0^{(1)}, W^{(1)}, Y^{(1)}) \in A] &= P[\omega \in \Omega; (x, w, y_1) \in A]\\ &= P[\omega \in \Omega; (x, w, y_2) \in A]\\ &= \nu_2[(X_0^{(2)}, W^{(2)}, Y^{(2)}) \in A]; \quad A \in \mathscr{B}(\Theta),\end{aligned}$$

and this is uniqueness in the sense of probability law. □

Proposition 3.20 has the remarkable corollary that *weak existence and pathwise uniqueness imply strong existence.* We develop this result.

3.21 Problem. For every fixed $t \geq 0$ and $F \in \mathscr{B}_t(C[0, \infty)^d)$, the mapping $(x, w) \mapsto Q_j(x, w; F)$ is $\hat{\mathscr{B}}_t$-measurable, where $\{\hat{\mathscr{B}}_t\}$ is the augmentation of the filtration $\{\mathscr{B}(\mathbb{R}^d) \otimes \mathscr{B}_t(C[0, \infty)^r)\}$ by the null sets of $\mu(dx)P_*(dw)$.

(*Hint*: Consider the regular conditional probabilities $Q_j^t(x, w; F)$: $\mathbb{R}^d \times C[0, \infty)^r \times \mathscr{B}_t(C[0, \infty)^d) \rightarrow [0, 1]$ for $\mathscr{B}_t(C[0, \infty)^d)$, given $(x, \varphi_t w)$. These enjoy properties analogous to those of $Q_j(x, w; F)$; in particular, for every $F \in \mathscr{B}_t(C[0, \infty)^d)$, the mapping $(x, w) \mapsto Q_j^t(x, w; F)$ is $\mathscr{B}(\mathbb{R}^d) \otimes \mathscr{B}_t(C[0, \infty)^r)$-measurable, and

$$P_j(G \times F) = \int_G Q_j^t(x, w; F)\mu(dx)P_*(dw) \tag{3.25}$$

for every $G \in \mathscr{B}(\mathbb{R}^d) \otimes \mathscr{B}_t(C[0, \infty)^r)$. If (3.25) is shown to be valid for every $G \in \mathscr{B}(\mathbb{R}^d) \otimes \mathscr{B}(C[0, \infty)^r)$, then comparison of (3.25) with (3.22) (iii) shows that $Q_j(x, w; F) = Q_j^t(x, w; F)$ for $\mu \times P_*$-a.e. (x, w), and the conclusion follows. Establish (3.25), first for sets of the form

$$G = G_1 \times (\varphi_t^{-1} G_2 \cap \sigma_t^{-1} G_3); \quad G_1 \in \mathscr{B}(\mathbb{R}^d), \quad G_2, G_3 \in \mathscr{B}(C[0, \infty)^r), \tag{3.26}$$

where $(\sigma_t w)(s) \triangleq w(t + s) - w(t)$; $s \geq 0$, and then in the generality required.)

3.22 Problem. In the context of Proposition 3.20, there exists a function k: $\mathbb{R}^d \times C[0, \infty)^r \rightarrow C[0, \infty)^d$ such that, for $\mu \times P_*$-a.e. $(x, w) \in \mathbb{R}^d \times C[0, \infty)^r$, we have

$$Q_1(x, w; \{k(x, w)\}) = Q_2(x, w; \{k(x, w)\}) = 1. \tag{3.27}$$

This function k is $\mathscr{B}(\mathbb{R}^d) \otimes \mathscr{B}(C[0, \infty)^r)/\mathscr{B}(C[0, \infty)^d)$-measurable and, for each $0 \leq t < \infty$, it is also $\hat{\mathscr{B}}_t/\mathscr{B}_t(C[0, \infty)^d)$-measurable (see Problem 3.21 for the definition of $\hat{\mathscr{B}}_t$). We have, in addition,

$$P[\omega = (x, w, y_1, y_2) \in \Omega; y_1 = y_2 = k(x, w)] = 1. \tag{3.28}$$

3.23 Corollary. *Suppose that the stochastic differential equation* (2.1) *has a weak solution* (X, W), $(\Omega, \mathscr{F}, P)$, $\{\mathscr{F}_t\}$ *with initial distribution* μ, *and suppose that pathwise uniqueness holds for* (2.1). *Then there exists a* $\mathscr{B}(\mathbb{R}^d) \otimes \mathscr{B}(C[0, \infty)^r)/\mathscr{B}(C[0, \infty)^d)$-*measurable function* h: $\mathbb{R}^d \times C[0, \infty)^r \rightarrow C[0, \infty)^d$, *which is also* $\hat{\mathscr{B}}_t/\mathscr{B}_t(C[0, \infty)^d)$-*measurable for every fixed* $0 \leq t < \infty$, *such that*

$$X_{\cdot} = h(X_0, W_{\cdot}), \quad \text{a.s. } P. \tag{3.29}$$

Moreover, given any probability space $(\tilde{\Omega}, \tilde{\mathscr{F}}, \tilde{P})$ *rich enough to support an* $\mathbb{R}^d$*-valued random variable* ξ *with distribution* μ *and an independent Brownian motion* $\{\tilde{W}_t, \mathscr{F}_t^{\tilde{W}}; 0 \le t < \infty\}$, *the process*

$$\tilde{X}_{.} \triangleq h(\xi, \tilde{W}_{.}) \tag{3.30}$$

is a strong solution of equation (2.1) *with initial condition* ξ.

PROOF. Let $h(x, w) = x + k(x, w)$, where k is as in Problem 3.22. From (3.28) and (3.21)′ we see that (3.29) holds. For ξ and $\tilde{W}$ as described, both $(X_0, W_.)$ and $(\xi, \tilde{W}_.)$ induce the same measure $\mu \times P_*$ on $\mathbb{R}^d \times \mathscr{B}(C[0, \infty)^r)$, and since $(X_. = h(X_0, W_.), W_.)$ satisfies (2.1), so does $(\tilde{X}_. = h(\xi, \tilde{W}_.), \tilde{W}_.)$. The process $\tilde{X}$ is adapted to $\{\mathscr{F}_t\}$ given by (2.3), because h is $\hat{\mathscr{B}}_t/\mathscr{B}_t(C[0, \infty)^d)$-measurable. □

The functional relations (3.29), (3.30) provide a very satisfactory formulation of the *principle of causality* articulated in Remark 2.2.

5.4. The Martingale Problem of Stroock and Varadhan

We have seen that when the drift and dispersion coefficients of a stochastic differential equation satisfy the Lipschitz and linear growth conditions of Theorem 2.9, then the equation possesses a unique strong solution. For more general coefficients, though, a strong solution to the stochastic differential equation might not exist (Example 3.5); then the questions of existence and uniqueness, as well as the properties of a solution, have to be discussed in a different setting. One possibility is indicated by Definitions 3.1 and 3.4: one attempts to solve the stochastic differential equation in the "weak" sense of finding a process with the right law (finite-dimensional distributions), and to do so uniquely. A variation on this approach, developed by Stroock & Varadhan (1969), formulates the search for the law of a diffusion process with given drift and dispersion coefficients in terms of a *martingale problem*. The latter is equivalent to solving the related stochastic differential equation in the weak sense, but does not involve the equation explicitly. This formulation has the advantage of being particularly well suited for the continuity and weak convergence arguments which yield existence results (Theorem 4.22) and "invariance principles", i.e., the convergence of Markov chains to diffusion processes (Stroock & Varadhan (1969), Section 10). Furthermore, it casts the question of uniqueness in terms of the solvability of a certain parabolic equation (Theorem 4.28), for which sufficient conditions are well known.

This section is organized as follows. First, the martingale problem is formulated and its equivalence with the problem of finding a weak solution to the corresponding stochastic differential equation is established. Using this martingale formulation and the optional sampling theorem, we next establish the strong Markov property for these solution processes. Finally, conditions for existence and uniqueness of solutions to the martingale problem are provided. These conditions are different from, and not comparable to, those given in the previous section for existence and uniqueness of weak solutions to stochastic differential equations.

4.1 Remark on Notation. We shall follow the accepted practice of denoting by $C^k(E)$ the collection of all continuous functions $f: E \to \mathbb{R}$ which have continuous derivatives of every order up to k; here, E is an open subset of some Euclidean space $\mathbb{R}^d$. If $f(t, x): [0, T) \times E \to \mathbb{R}$ is a continuous function, we write $f \in C([0, T) \times E)$, and if the partial derivatives $(\partial f/\partial t)$, $(\partial f/\partial x_i)$, $(\partial^2 f/\partial x_i \partial x_j)$; $1 \le i, j \le d$, exist and are continuous on $(0, T) \times E$, we write $f \in C^{1,2}((0, T) \times E)$. The notation $f \in C^{1,2}([0, T) \times E)$ means that $f \in C^{1,2}((0, T) \times E)$ and the indicated partial derivatives have continuous extensions to $[0, T) \times E$. We shall denote by $C_b^k(E)$, $C_0^k(E)$, the subsets of $C^k(E)$ of bounded functions and functions having compact support, respectively. In particular, a function in $C_0^k(E)$ has *bounded* partial derivatives up to order k; this might not be true for a function in $C_b^k(E)$.

A. Some Fundamental Martingales

In order to provide motivation for the martingale problem, let us suppose that (X, W), $(\Omega, \mathscr{F}, P)$, $\{\mathscr{F}_t\}$ is a weak solution to the stochastic differential equation (2.1). For every $t \ge 0$, we introduce the second-order differential operator

$$(4.1) \quad (\mathscr{A}_t f)(x) \triangleq \frac{1}{2} \sum_{i=1}^d \sum_{k=1}^d a_{ik}(t, x) \frac{\partial^2 f(x)}{\partial x_i \partial x_k} + \sum_{i=1}^d b_i(t, x) \frac{\partial f(x)}{\partial x_i}; \quad f \in C^2(\mathbb{R}^d),$$

where $a_{ik}(t, x)$ are the components of the diffusion matrix (2.2). If, as in the next proposition, f is a function of $t \in [0, \infty)$ and $x \in \mathbb{R}^d$, then $(\mathscr{A}_t f)(t, x)$, is obtained by applying $\mathscr{A}_t$ to $f(t, \cdot)$.

4.2 Proposition. *For every continuous function $f(t, x): [0, \infty) \times \mathbb{R}^d \to \mathbb{R}$ which belongs to $C^{1,2}([0, \infty) \times \mathbb{R}^d)$, the process $M^f = \{M_t^f, \mathscr{F}_t; 0 \le t < \infty\}$ given by*

$$(4.2) \qquad M_t^f \triangleq f(t, X_t) - f(0, X_0) - \int_0^t \left(\frac{\partial f}{\partial s} + \mathscr{A}_s f\right)(s, X_s)\, ds$$

is a continuous, local martingale; i.e., $M^f \in \mathscr{M}^{c,\text{loc}}$. If g is another member of $C^{1,2}([0, \infty) \times \mathbb{R}^d)$, then $M^g \in \mathscr{M}^{c,\text{loc}}$ and

$$(4.3) \qquad \langle M^f, M^g \rangle_t = \sum_{i=1}^d \sum_{k=1}^d \int_0^t a_{ik}(s, X_s) \frac{\partial}{\partial x_i} f(s, X_s) \frac{\partial}{\partial x_k} g(s, X_s)\, ds.$$

Furthermore, if $f \in C_0([0, \infty) \times \mathbb{R}^d)$ and the coefficients σ_{ij}; $1 \le i \le d$, $1 \le j \le r$, are bounded on the support of f, then $M^f \in \mathscr{M}_2^c$.

PROOF. The Itô rule expresses M^f as a sum of stochastic integrals:

$$(4.4)\quad M_t^f = \sum_{i=1}^{d}\sum_{j=1}^{r} M_t^{(i,j)}, \quad \text{with} \quad M_t^{(i,j)} \triangleq \int_0^t \sigma_{ij}(s,X_s)\frac{\partial}{\partial x_i} f(s,X_s)\,dW_s^{(j)}.$$

Introducing the stopping times

$$S_n \triangleq \inf\left\{t \geq 0;\ \|X_t\| \geq n \text{ or } \int_0^t \sigma_{ij}^2(s,X_s)\,ds \geq n \text{ for some } (i,j)\right\}$$

and recalling that a weak solution must satisfy condition (iii) of Definition 2.1, we see that $\lim_{n\to\infty} S_n = \infty$ a.s. The processes

$$(4.5)\quad M_t^f(n) \triangleq M_{t\wedge S_n}^f = \sum_{i=1}^{d}\sum_{j=1}^{r}\int_0^{t\wedge S_n} \sigma_{ij}(s,X_s)\frac{\partial}{\partial x_i} f(s,X_s)\,dW_s^{(j)}; \quad n \geq 1,$$

are continuous martingales, and so $M^f \in \mathscr{M}^{c,\text{loc}}$. The cross-variation in (4.3) follows readily from (4.5). If f has compact support on which each σ_{ij} is bounded, then the integrand in the expression for $M^{(i,j)}$ in (4.4) is bounded, so $M^f \in \mathscr{M}_2^c$. □

With the exception of the last assertion, a completely analogous result is valid for functional stochastic differential equations (Definition 3.14). We elaborate in the following problem.

4.3 Problem. Let $b_i(t,y)$ and $\sigma_{ij}(t,y)$; $1 \leq i \leq d$, $1 \leq j \leq r$, be progressively measurable functionals from $[0,\infty) \times C[0,\infty)^d$ into $\mathbb{R}$. By analogy with (2.2), we define the diffusion matrix $a(t,y)$ with components

$$(4.6)\qquad a_{ik}(t,y) \triangleq \sum_{j=1}^{r} \sigma_{ij}(t,y)\sigma_{kj}(t,y); \quad 0 \leq t < \infty,\ y \in C[0,\infty)^d.$$

Suppose that (X,W), $(\Omega,\mathscr{F},P)$, $\{\mathscr{F}_t\}$, is a weak solution to the functional stochastic differential equation (3.15), and set

$$(4.1)'\qquad (\mathscr{A}_t'u)(y) = \frac{1}{2}\sum_{i=1}^{d}\sum_{k=1}^{d} a_{ik}(t,y)\frac{\partial^2 u(y(t))}{\partial x_i\,\partial x_k} + \sum_{i=1}^{d} b_i(t,y)\frac{\partial u(y(t))}{\partial x_i};$$
$$0 \leq t < \infty,\ u \in C^2(\mathbb{R}^d),\ y \in C[0,\infty)^d.$$

Then, for any functions $f, g \in C^{1,2}([0,\infty) \times \mathbb{R}^d)$, the process

$$(4.2)'\quad M_t^f \triangleq f(t,X_t) - f(0,X_0) - \int_0^t \left[\frac{\partial f}{\partial s} + \mathscr{A}_s'f\right](s,X_s)\,ds,\ \mathscr{F}_t; \quad 0 \leq t < \infty$$

is in $\mathscr{M}^{c,\text{loc}}$, and

$$(4.3)'\qquad \langle M^f, M^g\rangle_t = \sum_{i=1}^{d}\sum_{k=1}^{d}\int_0^t a_{ik}(s,X)\frac{\partial}{\partial x_i} f(s,X_s)\frac{\partial}{\partial x_k} g(s,X_s)\,ds.$$

Furthermore, if the first derivatives of f are bounded, and for each $0 < T < \infty$ we have

(4.7) $$\|\sigma(t,y)\| \le K_T; \quad 0 \le t \le T, \quad y \in C[0,\infty)^d,$$

where K_T is a constant depending on T, then $M^f \in \mathscr{M}_2^c$.

The simplest case in Proposition 4.2 is that of a d-dimensional Brownian motion, which corresponds to $b_i(t,x) \equiv 0$ and $\sigma_{ij}(t,x) \equiv \delta_{ij}$; $1 \le i, j \le d$. Then the operator in (4.1) becomes

$$\mathscr{A}f = \frac{1}{2}\Delta f = \frac{1}{2}\sum_{i=1}^{d} \frac{\partial^2 f}{\partial x_i^2}; \quad f \in C^2(\mathbb{R}^d).$$

4.4 Problem. A continuous, adapted process $W = \{W_t, \mathscr{F}_t; 0 \le t < \infty\}$ is a d-dimensional Brownian motion if and only if

$$f(W_t) - f(W_0) - \frac{1}{2}\int_0^t \Delta f(W_s)\,ds, \mathscr{F}_t; \quad 0 \le t < \infty,$$

is in $\mathscr{M}^{c,\text{loc}}$ for every $f \in C^2(\mathbb{R}^d)$.

B. Weak Solutions and Martingale Problems

Problem 4.4 provides a novel *martingale characterization* of Brownian motion. The basic idea in the theory of Stroock & Varadhan is to employ M^f of (4.2) in a similar fashion to characterize diffusions with general drift and dispersion coefficients.

To explain how this characterization works, we shall find it convenient to deal temporarily with progressively measurable functionals $b_i(t,y)$, $\sigma_{ij}(t,y)$: $[0,\infty) \times C[0,\infty)^d \to \mathbb{R}$, $1 \le i \le d$, $1 \le j \le r$. We recall the family of operators $\{\mathscr{A}_t'\}$ of (4.1)'.

4.5 Definition. A probability measure P on $(C[0,\infty)^d, \mathscr{B}(C[0,\infty)^d))$, under which

(4.8) $$M_t^f = f(y(t)) - f(y(0)) - \int_0^t (\mathscr{A}_s'f)(y)\,ds, \mathscr{F}_t; \quad 0 \le t < \infty,$$

is a continuous, local martingale for every $f \in C^2(\mathbb{R}^d)$, is called a *solution to the local martingale problem* associated with $\{\mathscr{A}_t'\}$. Here $\mathscr{F}_t = \mathscr{G}_{t+}$, and $\{\mathscr{G}_t\}$ is the augmentation under P of the canonical filtration $\mathscr{B}_t \triangleq \mathscr{B}_t(C[0,\infty)^d)$ as in (3.19).

According to Problem 4.3, a weak solution to the functional stochastic differential equation (3.15) induces on $(C[0,\infty)^d, \mathscr{B}(C[0,\infty)^d))$ a probability measure P which solves the local martingale problem associated with $\{\mathscr{A}_t'\}$. The converse of this assertion is also true, as we now show.

4.6 Proposition. *Let P be a probability measure on $(C[0,\infty)^d, \mathscr{B}(C[0,\infty)^d))$ under which the process M^f of (4.8) is a continuous, local martingale for the choices $f(x) = x_i$ and $f(x) = x_i x_k$; $1 \le i, k \le d$. Then there is an r-dimensional Brownian motion $W = \{W_t, \tilde{\mathscr{F}}_t; 0 \le t < \infty\}$ defined on an extension $(\tilde{\Omega}, \tilde{\mathscr{F}}, \tilde{P})$ of $(C[0,\infty)^d, \mathscr{B}(C[0,\infty)^d), P)$, such that $(X_t \triangleq y(t), W_t), (\tilde{\Omega}, \tilde{\mathscr{F}}, \tilde{P}), \{\tilde{\mathscr{F}}_t\}$, is a weak solution to equation* (3.15).

PROOF. By assumption,

$$M_t^{(i)} \triangleq X_t^{(i)} - X_0^{(i)} - \int_0^t b_i(s, X)\,ds, \mathscr{F}_t; \quad 0 \le t < \infty$$

is a continuous, local martingale under P. In particular,

$$P\left[\int_0^t |b_i(s, X)|\,ds < \infty; 0 \le t < \infty\right] = 1; \quad 1 \le i \le d. \tag{4.9}$$

With $f(x) = x_i x_k$, we see that

$$M_t^{(i,k)} \triangleq X_t^{(i)} X_t^{(k)} - X_0^{(i)} X_0^{(k)} - \int_0^t [X_s^{(i)} b_k(s, X) + X_s^{(k)} b_i(s, X) + a_{ik}(s, X)]\,ds$$

is also a continuous, local martingale. But one can express

$$M_t^{(i)} M_t^{(k)} - \int_0^t a_{ik}(s, X)\,ds \tag{4.10}$$

as the sum of the continuous local martingale $M_t^{(i,k)} - X_0^{(i)} M_t^{(k)} - X_0^{(k)} M_t^{(i)}$ and the process

$$\begin{aligned}
&\int_0^t (X_s^{(i)} - X_t^{(i)}) b_k(s, X)\,ds + \int_0^t (X_s^{(k)} - X_t^{(k)}) b_i(s, X)\,ds \\
&\qquad + \int_0^t b_i(s, X)\,ds \int_0^t b_k(s, X)\,ds \\
&= \int_0^t (M_s^{(i)} - M_t^{(i)}) b_k(s, X)\,ds + \int_0^t (M_s^{(k)} - M_t^{(k)}) b_i(s, X)\,ds \\
&= -\int_0^t \left[\int_0^s b_k(u, X)\,du\right] dM_s^{(i)} - \int_0^t \left[\int_0^s b_i(u, X)\,du\right] dM_s^{(k)}.
\end{aligned} \tag{4.11}$$

The last identity may be verified by applying Itô's rule to both processes claimed to be equal. We see then that the process of (4.11) is a continuous local martingale. Therefore, the process of (4.10) is in $\mathscr{M}^{c,\text{loc}}$, and

$$\langle M^{(i)}, M^{(k)} \rangle_t = \int_0^t a_{ik}(s, X)\,ds; \quad 0 \le t < \infty, \text{ a.s.} \tag{4.12}$$

We may now invoke Theorem 3.4.2 to conclude the existence of a d-dimensional Brownian motion $\{\tilde{W}_t, \tilde{\mathscr{F}}_t; 0 \le t < \infty\}$ on an extension $(\tilde{\Omega}, \tilde{\mathscr{F}}, \tilde{P})$ of $(C[0,\infty)^d, \mathscr{B}(C[0,\infty)^d), P)$ endowed with a filtration $\{\tilde{\mathscr{F}}_t\}$ which satisfies the usual conditions, as well as the existence of a matrix $\rho = \{\rho_{ij}(t), \tilde{\mathscr{F}}_t; 0 \le t < \infty\}_{1 \le i,j \le d}$ of measurable, adapted processes with

$$\tilde{P}\left[\int_0^t \rho_{ij}^2(s)\,ds < \infty\right] = 1; \quad 1 \le i, j \le d, 0 \le t < \infty, \tag{4.13}$$

such that

$$M_t^{(i)} = \sum_{j=1}^d \int_0^t \rho_{ij}(s)\,d\tilde{W}_s^{(j)}; \quad 1 \le i \le d, 0 \le t < \infty \tag{4.14}$$

holds a.s. $\tilde{P}$. This last equation can be rewritten as

$$X_t = X_0 + \int_0^t b(s, X)\,ds + \int_0^t \rho(s)\,d\tilde{W}_s; \quad 0 \le t < \infty. \tag{4.15}$$

In order to complete the proof, we need to establish the existence of an r-dimensional Brownian motion $W = \{W_t, \tilde{\mathscr{F}}_t; 0 \le t < \infty\}$ on $(\tilde{\Omega}, \tilde{\mathscr{F}}, \tilde{P})$, such that

$$\int_0^t \rho(s)\,d\tilde{W}_s = \int_0^t \sigma(s, X)\,dW_s; \quad 0 \le t < \infty \tag{4.16}$$

holds $\tilde{P}$-almost surely. From (4.12), (4.14) and with the notation (4.6), it will then be clear that

$$\tilde{P}\left[\sum_{j=1}^d \rho_{ij}(t)\rho_{kj}(t) = a_{ik}(t, X), \text{ for a.e. } t \ge 0\right] = 1; \quad 1 \le i, k \le d$$

and (4.13) will imply

$$\tilde{P}\left[\int_0^t \sigma_{ij}^2(s, X)\,ds < \infty\right] = 1; \quad 1 \le i \le d, 1 \le j \le r, 0 \le t < \infty. \tag{4.17}$$

The relations (4.9), (4.15)–(4.17) will then yield (X, W), $(\tilde{\Omega}, \tilde{\mathscr{F}}, \tilde{P})$, $\{\tilde{\mathscr{F}}_t\}$ as a weak solution to (3.15).

It suffices to construct W satisfying (4.16) under the assumption $r = d$. Indeed, if $r > d$, we may augment X, b, and σ by setting $X_t^{(i)} = b_i(t, y) = \sigma_{ij}(t, y) = 0$; $d + 1 \le i \le r$, $1 \le j \le r$. This r-dimensional process X satisfies an appropriately modified version of (4.8), and we may proceed as before except now we shall obtain a matrix ρ which, like σ, will be of dimension $(r \times r)$. On the other hand, if $r < d$, we need only augment σ by setting $\sigma_{ij}(t, y) = 0$; $1 \le i \le d$, $r + 1 \le j \le d$, and nothing else is affected. Both ρ and σ are then $(d \times d)$ matrices.

According to Problem 4.7 following this proof, there exists a Borel-measurable, $(d \times d)$-matrix-valued function $R(\rho, \sigma)$ defined on the set

$$D \triangleq \{(\rho, \sigma); \rho \text{ and } \sigma \text{ are } (d \times d) \text{ matrices with } \rho\rho^T = \sigma\sigma^T\} \tag{4.18}$$

such that $\sigma = \rho R(\rho, \sigma)$ and $R(\rho, \sigma) R^T(\rho, \sigma) = I$, the $(d \times d)$ identity matrix. We set

$$W_t \triangleq \int_0^t R^T(\rho_s, \sigma(s, X))\, d\tilde{W}_s; \quad 0 \le t < \infty.$$

Then $W^{(i)} \in \mathcal{M}^{c,\mathrm{loc}}$; $1 \le i \le d$, and

$$\langle W^{(i)}, W^{(j)} \rangle_t = t\delta_{ij}; \quad 1 \le i, j \le d, 0 \le t < \infty.$$

It follows from Lévy's Theorem 3.3.16 that $\{W_t, \tilde{\mathscr{F}}_t; 0 \le t < \infty\}$ is a d-dimensional Brownian motion. Relation (4.16) is apparent. □

4.7 Problem. Show that there exists a Borel-measurable, $(d \times d)$-matrix-valued function $R(\rho, \sigma)$ defined on the set D of (4.18) and such that

$$\sigma = \rho R(\rho, \sigma), \quad R(\rho, \sigma) R^T(\rho, \sigma) = I; \quad (\rho, \sigma) \in D.$$

(*Hint*: Diagonalize $\rho\rho^T = \sigma\sigma^T$ and study the effect of the diagonalization transformation on ρ and σ.)

4.8 Corollary. *The existence of a solution P to the local martingale problem associated with $\{\mathscr{A}_t'\}$ is equivalent to the existence of a weak solution (X, W), $(\tilde{\Omega}, \tilde{\mathscr{F}}, \tilde{P})$, $\{\tilde{\mathscr{F}}_t\}$ to the equation* (3.15). *The two solutions are related by $P = \tilde{P}X^{-1}$; i.e., X induces the measure P on $(C[0,\infty)^d, \mathscr{B}(C[0,\infty)^d))$.*

4.9 Corollary. *The uniqueness of the solution P to the local martingale problem with fixed but arbitrary initial distribution*

$$P[y \in C[0,\infty)^d;\, y(0) \in \Gamma] = \mu(\Gamma); \quad \Gamma \in \mathscr{B}(\mathbb{R}^d)$$

is equivalent to uniqueness in the sense of probability law for the equation (3.15).

Because of the difficulty in computing expectations for local martingales, it is helpful to introduce the following modification of Definition 4.5.

4.10 Definition (Martingale Problem of Stroock & Varadhan (1969)). A probability measure P on $(C[0,\infty)^d, \mathscr{B}(C[0,\infty)^d))$ under which M^f in (4.8) is a continuous martingale for every $f \in C_0^2(\mathbb{R}^d)$ is called a *solution to the martingale problem* associated with $\{\mathscr{A}_t'\}$.

Given progressively measurable functionals $b_i(t, y)$, $\sigma_{ij}(t, y)$: $[0, \infty) \times C[0,\infty)^d \to \mathbb{R}$; $1 \le i \le d$, $1 \le j \le r$, the associated family of operators $\{\mathscr{A}_t'\}$, and a probability measure μ on $\mathscr{B}(\mathbb{R}^d)$, we can consider the following three conditions:

(A) There exists a *weak solution* to the functional stochastic differential equation (3.15) with initial distribution μ.
(B) There exists a solution P to the *local martingale problem* associated with $\{\mathscr{A}_t'\}$ with $P[y(0) \in \Gamma] = \mu(\Gamma)$; $\Gamma \in \mathscr{B}(\mathbb{R}^d)$.

(C) There exists a solution P to the *martingale problem* associated with $\{\mathscr{A}_t'\}$ with $P[y(0) \in \Gamma] = \mu(\Gamma)$; $\Gamma \in \mathscr{B}(\mathbb{R}^d)$.

4.11 Proposition. *Conditions* (A) *and* (B) *are equivalent and are implied by* (C). *Furthermore,* (A) *implies* (C) *under either of the additional assumptions:*

(A.1) *For each* $0 < T < \infty$, *condition* (4.7) *holds.*

(A.2) *Each* $\sigma_{ij}(t, y)$ *is of the form* $\sigma_{ij}(t, y) = \tilde{\sigma}_{ij}(t, y(t))$, *where the Borel-measurable functions* $\tilde{\sigma}_{ij}: [0, \infty) \times \mathbb{R}^d \to \mathbb{R}$ *are bounded on compact sets.*

PROOF. We have already established the equivalence of (A) and (B.) If P is a solution to the martingale problem and $f \in C^2(\mathbb{R}^d)$ does not necessarily have compact support, we can define for every integer $k \geq 1$ the stopping time

$$S_k \triangleq \inf\{t \geq 0: \|y(t)\| \geq k\}. \tag{4.19}$$

Let $g_k \in C_0^2(\mathbb{R}^d)$ agree with f on $\{x \in \mathbb{R}^d; \|x\| \leq k\}$. Under P, each $M_t^{g_k}$ is a martingale which agrees with M_t^f for $t \leq S_k$. It follows that $M^f \in \mathscr{M}^{c,\mathrm{loc}}$; thus (C) $\Rightarrow$ (B). Under (A.1), Problem 4.3 shows (A) $\Rightarrow$ (C); under (A.2), the argument for this implication is given in Proposition 4.2.

4.12 Remark. It is not always necessary to verify the martingale property of M^f under P for every $f \in C_0^2(\mathbb{R}^d)$, in order to conclude that P solves the martingale problem. To wit, consider $f_i(x) \triangleq x_i$ and $f_{ij}(x) \triangleq x_i x_j$ for $1 \leq i$, $j \leq d$, and choose sequences $\{g_i^{(k)}\}_{k=1}^\infty$, $\{g_{ij}^{(k)}\}_{k=1}^\infty$ of functions in $C_0^2(\mathbb{R}^d)$ such that $g_i^{(k)}(x) = f_i(x)$, $g_{ij}^{(k)}(x) = f_{ij}(x)$ for $\|x\| \leq k$. If $M^{g_i^{(k)}}$ and $M^{g_{ij}^{(k)}}$ are martingales for $1 \leq i, j \leq d$ and $k \geq 1$, then M^{f_i} and $M^{f_{ij}}$ are local martingales. According to Proposition 4.6, there is a weak solution to (3.15), and Corollary 4.8 now implies that P solves the local martingale problem. Under either of the assumptions (A.1) or (A.2), P must also solve the martingale problem.

It is also instructive to note that in the Definitions 3.1 and 3.14 of weak solution to a stochastic differential equation, the Brownian motion W appears only as a "nuisance parameter." The Stroock & Varadhan formulation eliminates this "parametric" process completely. Indeed, the essence of the implication (B) $\Rightarrow$ (A) proved in Proposition 4.6 is the construction of this process.

In keeping with our practice of working with filtrations which satisfy the usual conditions, we have constructed from $\mathscr{B}_t \triangleq \mathscr{B}_t(C[0,\infty)^d)$ such a filtration $\{\mathscr{F}_t\}$ for the Definitions 4.5 and 4.10. Later in this section, we shall instead want to deal with $\{\mathscr{B}_t\}$ itself, because this filtration does not depend on the probability measure under consideration and each $\mathscr{B}_t$ is countably determined. Toward this end, we need the following result.

4.13 Problem. Assume either

(A.1)′ $\|b(t, y)\| + \|\sigma_{ij}(t, y)\| \leq K_T$; $0 \leq t \leq T$, $y \in C[0, \infty)^d$, for every $0 < T < \infty$, where K_T is a constant depending on T, or else

(A.2)′ $b_i(t, y)$ and $\sigma_{ij}(t, y)$ are of the form $b_i(t, y) = \tilde{b}_i(t, y(t))$, $\sigma_{ij}(t, y) = \tilde{\sigma}_{ij}(t, y(t))$, where the Borel-measurable functions $\tilde{b}_i$, $\tilde{\sigma}_{ij}$: $[0, \infty) \times \mathbb{R}^d \to \mathbb{R}$ are bounded on compact sets.

Let P be a probability measure on $((C[0,\infty)^d, \mathscr{B}(C[0,\infty)^d))$; $\{\mathscr{F}_t\}$ be as in Definition 4.5; and $f \in C_0^2(\mathbb{R}^d)$. Show that if $\{M_t^f, \mathscr{B}_t; 0 \le t < \infty\}$ is a martingale, then so is $\{M_t^f, \mathscr{F}_t; 0 \le t < \infty\}$.

C. Well-Posedness and the Strong Markov Property

We pause here in our development of the martingale problem to discuss the *strong Markov property* for solutions of stochastic differential equations. Consistent with our discussion of Markov families in Chapter 2, we shed the trappings of time-dependence. Thus, we have time-homogeneous and Borel-measurable drift and dispersion coefficients b_i: $\mathbb{R}^d \to \mathbb{R}$, σ_{ij}: $\mathbb{R}^d \to \mathbb{R}$; $1 \le i \le d$, $1 \le j \le r$, and we shall study the time-homogeneous version of (2.1), written here in integral form as

$$X_t = x + \int_0^t b(X_s)\,ds + \int_0^t \sigma(X_s)\,dW_s; \quad 0 \le t < \infty. \tag{4.20}$$

This model actually does allow for time-dependence because time can be appended to the state variable; i.e., we may take $X_t^{(d+1)} = t$, $b_{d+1}(x) = 1$, $\sigma_{d+1,j}(x) = 0$; $1 \le j \le r$. We adopt the time-homogeneous assumption primarily for simplicity of exposition. Note, however, that some results (e.g., Remark 4.30 and Refinements 4.32) require the nondegeneracy of the diffusion coefficient, which is not valid in such an augmented model.

4.14 Definition. The stochastic integral equation (4.20) is said to be *well posed* if, for every initial condition $x \in \mathbb{R}^d$, it admits a weak solution which is unique in the sense of probability law.

We know, for instance, that (4.20) is well posed if b and σ satisfy Lipschitz and linear growth conditions (Theorems 2.5, 2.9). If σ is the $(d \times d)$ identity matrix and b is uniformly bounded, (4.20) is again well posed (Proposition 3.6 and Corollary 3.11). Later in this section, we shall obtain well-posedness under even less restrictive conditions on σ (Theorems 4.22 and 4.28, Corollary 4.29).

If (4.20) is well posed, then the solution X with initial condition $X_0 = x$ induces a measure P^x on $(C[0,\infty)^d, \mathscr{B}(C[0,\infty)^d))$. One can then ask whether the coordinate mapping process on this canonical space, the filtration $\{\mathscr{B}_t\}$, and the family of probability measures $\{P^x\}_{x \in \mathbb{R}^d}$ constitute a strong Markov family. We shall see that if b and σ are bounded on compact subsets of $\mathbb{R}^d$, the answer to this question is essentially affirmative. Our analysis proceeds via the martingale problem, which we now specialize to the case at hand. We denote by

(1.2) $$(\mathscr{A}f)(x) = \frac{1}{2}\sum_{i=1}^{d}\sum_{k=1}^{d} a_{ik}(x)\frac{\partial^2 f(x)}{\partial x_i \partial x_k} + \sum_{i=1}^{d} b_i(x)\frac{\partial f(x)}{\partial x_i}$$

the time-homogeneous version of the operator in (4.1).

4.15 Definition. Assume that b_i: $\mathbb{R}^d \to \mathbb{R}$ and σ_{ij}: $\mathbb{R}^d \to \mathbb{R}$; $1 \le i \le d$, $1 \le j \le r$ are bounded on compact subsets of $\mathbb{R}^d$. A probability measure P on $(\Omega, \mathscr{B}) \triangleq (C[0,\infty)^d, \mathscr{B}(C[0,\infty)^d))$ under which

(4.21) $$E\left[f(y(t)) - f(y(s)) - \int_s^t (\mathscr{A}f)(y(u))\,du \,\middle|\, \mathscr{B}_s\right] = 0, \quad \text{a.s. } P$$

holds for every $0 \le s < t < \infty$, $f \in C_0^2(\mathbb{R}^d)$, is called a *solution to the time-homogeneous martingale problem*. We denote by P^x any solution for which

(4.22) $$P^x[y \in C[0,\infty)^d;\ y(0) = x] = 1.$$

We say that the time-homogeneous martingale problem is *well posed* if, for every $x \in \mathbb{R}^d$, there is exactly one such measure P^x.

4.16 Remark. The replacement of $\mathscr{F}_s$ by $\mathscr{B}_s$ in (4.21) is justified by Problem 4.13.

4.17 Remark. Under the conditions of Definition 4.15, well-posedness of the time-homogeneous martingale problem is equivalent to well-posedness of the stochastic integral equation (4.20) (Corollaries 4.8 and 4.9 and Proposition 4.11).

4.18 Lemma. *For every bounded stopping time T of the filtration $\{\mathscr{B}_t\}$, we have*

$$\mathscr{B}_T = \sigma(y(t \wedge T);\ 0 \le t < \infty).$$

PROOF. The $\mathscr{B}_T$-measurability of each $y(t \wedge T)$ follows directly from Definition 1.2.12 and Proposition 1.2.18. It remains to show $\mathscr{B}_T \subseteq \sigma(y(t \wedge T); 0 \le t < \infty)$.

Let φ_t: $C[0,\infty)^d \to C[0,\infty)^d$ be given by $(\varphi_t y)(s) = y(t \wedge s)$; $0 \le s < \infty$, for arbitrary $t \ge 0$. Problem 1.2.2 shows that $T(y) = T(\varphi_{T(y)}(y))$ holds for every $y \in C[0,\infty)^d$ and so, with $A \in \mathscr{B}_T$ and $t \triangleq T(y)$, we have

$$y \in A \Leftrightarrow y \in [A \cap \{T \le t\}] \Leftrightarrow \varphi_t y \in [A \cap \{T \le t\}] \Leftrightarrow \varphi_t y \in A.$$

The second of these equivalences is a consequence of the facts $A \cap \{T \le t\} \in \mathscr{B}_t$ and $y(s) = (\varphi_t(y))(s)$; $0 \le s \le t$. We conclude that

$$A = \{y \in C[0,\infty)^d;\ \varphi_{T(y)}(y) \in A\} = \{y \in C[0,\infty)^d;\ y(\cdot \wedge T) \in A\}. \qquad \square$$

For the next lemma, we recall the discussion of regular conditional probabilities in Subsection 3.C, as well as the formula (2.5.15) for the shift operators: $(\theta_s y)(t) = y(s+t)$; $0 \le t < \infty$ for $s \ge 0$ and $y \in C[0,\infty)^d$.

4.19 Lemma. *Let T be a bounded stopping time of $\{\mathscr{B}_t\}$ and $\mathscr{G}$ a countably determined sub-σ-field of $\mathscr{B}_T$ such that $y(T)$ is $\mathscr{G}$-measurable. Suppose that b and σ are bounded on compact subsets of $\mathbb{R}^d$, and that the probability measure P on $(\Omega, \mathscr{B}) = (C[0,\infty)^d, \mathscr{B}(C[0,\infty)^d))$ solves the time-homogeneous martingale problem of Definition 4.15. We denote by $Q_\omega(F) = Q(\omega; F): \Omega \times \mathscr{B} \to [0,1]$ the regular conditional probability for $\mathscr{B}$ given $\mathscr{G}$.*

There exists then a P-null event $N \in \mathscr{G}$ such that, for every $\omega \notin N$, the probability measure

$$P_\omega \triangleq Q_\omega \circ \theta_T^{-1}$$

solves the martingale problem (4.21), (4.22) with $x = \omega(T(\omega))$.

PROOF. We notice first that, thanks to the assumptions imposed on $\mathscr{G}$, Theorem 3.18 (iv)′ implies the existence of a P-null event $N \in \mathscr{G}$, such that

$$Q(\omega; \{y \in \Omega;\ y(T(y)) = \omega(T(\omega))\}) = 1,$$

and therefore also

$$P_\omega[y \in \Omega;\ y(0) = \omega(T(\omega))] = Q_\omega[y \in \Omega;\ y(T(y)) = \omega(T(\omega))] = 1$$

hold for every $\omega \notin N$. Thus (4.22) is satisfied with $x = \omega(T(\omega))$.

In order to establish (4.21), we choose $0 \le s < t < \infty$, $G \in \mathscr{G}$, $F \in \mathscr{B}_s$, $f \in C_0^2(\mathbb{R}^d)$; define

$$Z(y) \triangleq f(y(t)) - f(y(s)) - \int_s^t (\mathscr{A}f)(y(u))\,du; \quad y \in C[0,\infty)^d,$$

and observe that

$$\begin{aligned}(4.23) \qquad \int_\Omega Z(y)1_F(y)P_\omega(dy) &= \int_\Omega Z(\theta_{T(y)}y)1_F(\theta_{T(y)}y)Q(\omega; dy)\\ &= E[1_{\theta_T^{-1}F}(Z \circ \theta_T)|\mathscr{G}](\omega)\\ &= E[E(Z \circ \theta_T|\mathscr{B}_{T+s}) \cdot 1_{\theta_T^{-1}F}|\mathscr{G}](\omega)\\ &= 0, \quad P\text{-a.e. } \omega.\end{aligned}$$

We have used in the last step the martingale property (4.21) for P and the optional sampling theorem (Problem 1.3.23 (i)).

Let us observe that, because of our assumptions, the random variable Z is bounded; relation (4.23) shows that the $\mathscr{G}$-measurable random variable $\omega \mapsto \int_F Z(y)P_\omega(dy)$ is zero except on a P-null event depending on s, t, f, and F. Consider a countable subcollection $\mathscr{E}$ of $\mathscr{B}_s$ and a P-null event $N(s,t,f) \in \mathscr{G}$, such that

$$\int_F Z(y)P_\omega(dy) = 0; \quad \forall\, \omega \notin N(s,t,f), \quad \forall\, F \in \mathscr{E}.$$

Then the finite measures $\nu_\omega^\pm(F) \triangleq \int_F Z^\pm(y)P_\omega(dy)$; $F \in \mathscr{B}_s$ agree on $\mathscr{E}$, and since $\mathscr{B}_s$ is countably determined, the subcollection $\mathscr{E}$ can be chosen so as to permit the conclusion

$$\int_F Z(y)P_\omega(dy) = 0; \quad \forall\, \omega \notin N(s,t,f), \quad \forall\, F \in \mathscr{B}_s.$$

We may set now $N(f) = \bigcup_{\substack{s,t \in Q\\ 0 \le s < t < \infty}} N(s,t,f)$, and use the boundedness and continuity (in s, t) of Z to conclude that

$$M_t^f \triangleq f(y(t)) - f(y(0)) - \int_0^t (\mathscr{A}f)(y(u))\,du, \ \mathscr{B}_t; \quad 0 \le t < \infty$$

is a martingale under P_ω, for every $\omega \notin N(f)$. Finally, we see that there exists a P-null event $N \in \mathscr{G}$ under which M^f is a P_ω-martingale for all $\omega \notin N$ and countably many $f \in C_0^2(\mathbb{R}^d)$; because of Remark 4.12, P_ω solves the time-homogeneous martingale problem for all $\omega \notin N$. □

4.20 Theorem. *Suppose that the coefficients b, σ are bounded on compact subsets of $\mathbb{R}^d$, and that the time-homogeneous martingale problem of Definition* 4.15 *(or equivalently, the stochastic integral equation* (4.20)*) is well posed. Then for every stopping time T of $\{\mathscr{B}_t\}$, $F \in \mathscr{B}(C[0,\infty)^d)$ and $x \in \mathbb{R}^d$, we have the* strong Markov property

$$P^x[\theta_T^{-1}F \mid \mathscr{B}_T](\omega) = P^{\omega(T)}[F], \quad P^x\text{-a.s. on } \{T < \infty\}. \tag{4.24}$$

Proof. If the stopping time T is bounded, we let $\mathscr{G}$ in Lemma 4.19 be $\mathscr{B}_T$, which is countably determined by Lemma 4.18. Using the notation of Lemma 4.19, we may write then, for every $F \in \mathscr{B}(C[0,\infty)^d)$:

$$P^x[\theta_T^{-1}F \mid \mathscr{B}_T](\omega) = Q(\omega; \theta_T^{-1}F) = P_\omega(F) = P^{\omega(T(\omega))}[F],$$

for P^x-a.e. $\omega \in \Omega$. The last identity is a consequence of the uniqueness of solution to the time-homogeneous martingale problem with initial condition $x = \omega(T(\omega))$.

Unbounded stopping times are handled as in Problem 2.6.9 (iii). □

4.21 Remark. The strong Markov property of (4.24) is the same as condition (e'') of Theorem 2.6.10, except that we have succeeded in proving it only for stopping (rather than optional) times.

Condition (a) of Definition 2.6.3 is satisfied under the assumptions of Theorem 4.20. Indeed, *well-posedness implies that the mapping $x \mapsto P^x(F)$ is Borel-measurable for every $F \in \mathscr{B}(C[0,\infty)^d)$*, but the proof of this statement requires a rather extensive set-theoretic development (see Stroock & Varadhan (1979), Exercise 6.7.4, and Parthasarathy (1967), Corollary 3.3, p. 22). This result is of rather limited interest, however, because when a proof of well-posedness is given, it typically provides a constructive demonstration of the measurability of the mapping $x \mapsto P^x(F)$.

D. Questions of Existence

It is time now to use the martingale problem in order to establish the fundamental existence result for weak solutions of stochastic differential equations with bounded, continuous coefficients.

4.22 Theorem (Skorohod (1965), Stroock & Varadhan (1969)). *Consider the stochastic differential equation*

$$dX_t = b(X_t)\,dt + \sigma(X_t)\,dW_t, \tag{4.25}$$

where the coefficients $b_i, \sigma_{ij}: \mathbb{R}^d \to \mathbb{R}$ *are bounded and continuous functions. Corresponding to every initial distribution* μ *on* $\mathscr{B}(\mathbb{R}^d)$ *with*

$$\int_{\mathbb{R}^d} \|x\|^{2m} \mu(dx) < \infty, \quad \textit{for some } m > 1,$$

there exists a weak solution of (4.25).

PROOF. For integers $j \geq 0$, $n \geq 1$ we consider the dyadic rationals $t_j^{(n)} = j2^{-n}$ and introduce the functions $\psi_n(t) = t_j^{(n)}$; $t \in [t_j^{(n)}, t_{j+1}^{(n)})$. We define the new coefficients

$$b^{(n)}(t, y) \triangleq b(y(\psi_n(t))), \quad \sigma^{(n)}(t, y) \triangleq \sigma(y(\psi_n(t))); \tag{4.26}$$
$$0 \leq t < \infty,\ y \in C[0, \infty)^d,$$

which are progressively measurable functionals.

Now let us consider on some probability space $(\Omega, \mathscr{F}, P)$ an r-dimensional Brownian motion $W = \{W_t, \mathscr{F}_t^W; 0 \leq t < \infty\}$ and an independent random vector ξ with the given initial distribution μ, and let us construct the filtration $\{\mathscr{F}_t\}$ as in (2.3). For each $n \geq 1$, we define the continuous process $X^{(n)} = \{X_t^{(n)}, \mathscr{F}_t; 0 \leq t < \infty\}$ by setting $X_0^{(n)} = \xi$ and then recursively:

$$X_t^{(n)} = X_{t_j^{(n)}}^{(n)} + b(X_{t_j^{(n)}}^{(n)})(t - t_j^{(n)}) + \sigma(X_{t_j^{(n)}}^{(n)})(W_t - W_{t_j^{(n)}}); \quad j \geq 0,\ t_j^{(n)} < t \leq t_{j+1}^{(n)}.$$

Then $X^{(n)}$ solves the functional stochastic integral equation

$$X_t^{(n)} = \xi + \int_0^t b^{(n)}(s, X^{(n)})\,ds + \int_0^t \sigma^{(n)}(s, X^{(n)})\,dW_s; \quad 0 \leq t < \infty. \tag{4.27}$$

Fix $0 < T < \infty$. From Problem 3.15 we obtain

$$\sup_{n \geq 1} E\|X_t^{(n)} - X_s^{(n)}\|^{2m} \leq C(1 + E\|\xi\|^{2m})(t - s)^m; \quad 0 \leq s < t \leq T,$$

where C is a constant depending only on m, T, the dimension d, and the bound on $\|b\|^2 + \|\sigma\|^2$. Let $P^{(n)} \triangleq P(X^{(n)})^{-1}$; $n \geq 1$ be the sequence of probability measures induced on $(C[0, \infty)^d, \mathscr{B}(C[0, \infty)^d))$ by these processes; it follows from Problem 2.4.11 and Remark 2.4.13 that this sequence is tight. We may then assert by the Prohorov Theorem 2.4.7, relabeling indices if necessary, that the sequence $\{P^{(n)}\}_{n=1}^{\infty}$ converges weakly to a probability measure P^* on this canonical space.

According to Proposition 4.11 and Problem 4.13, it suffices to show

$$P^*[y \in C[0,\infty)^d;\ y(0) \in \Gamma] = \mu(\Gamma);\quad \Gamma \in \mathscr{B}(\mathbb{R}^d), \tag{4.28}$$

$$E^*\left[f(y(t)) - f(y(s)) - \int_s^t (\mathscr{A}f)(y(u))\,du \,\middle|\, \mathscr{B}_s\right] = 0, \quad \text{a.s. } P^*, \tag{4.29}$$

for every $0 \le s < t < \infty$, $f \in C_0^2(\mathbb{R}^d)$. For every $f \in C_b(\mathbb{R}^d)$, the weak convergence of $\{P^{(n)}\}_{n=1}^{\infty}$ to P^* gives

$$E^* f(y(0)) = \lim_{n\to\infty} E^{(n)} f(y(0)) = \int_{\mathbb{R}^d} f(x)\mu(dx),$$

and (4.28) follows. In order to establish (4.29), let us recall from (4.27) and Proposition 4.11 that for every $f \in C_0^2(\mathbb{R}^d)$ and $n \ge 1$,

$$f(y(t)) - f(y(0)) - \int_0^t (\mathscr{A}_u^{(n)} f)(y)\,du,\ \mathscr{B}_t;\quad 0 \le t < \infty$$

is a martingale under $P^{(n)}$, where

$$(\mathscr{A}_t^{(n)} f)(y) \triangleq \sum_{i=1}^d b_i^{(n)}(t,y)\frac{\partial f(y(t))}{\partial x_i} + \frac{1}{2}\sum_{i=1}^d\sum_{k=1}^d\sum_{j=1}^r \sigma_{ij}^{(n)}(t,y)\sigma_{kj}^{(n)}(t,y)\frac{\partial^2 f(y(t))}{\partial x_i \partial x_k}.$$

Therefore, with $0 \le s < t < \infty$ and $g\colon C[0,\infty)^d \to \mathbb{R}$ a bounded, continuous, $\mathscr{B}_s$-measurable function, we have

$$E^{(n)}\left[\left\{f(y(t)) - f(y(s)) - \int_s^t (\mathscr{A}_u^{(n)} f)(y)\,du\right\} g(y)\right] = 0. \tag{4.30}$$

We shall show that for fixed $0 \le s < t < \infty$, the expression

$$F_n(y) \triangleq f(y(t)) - f(y(s)) - \int_s^t (\mathscr{A}_u^{(n)} f)(y)\,du$$

converges uniformly on compact subsets of $C[0,\infty)^d$ to

$$F(y) \triangleq f(y(t)) - f(y(s)) - \int_s^t (\mathscr{A}f)(y(u))\,du.$$

Then Problem 2.4.12 and Remark 2.4.13 will imply that we may let $n \to \infty$ in (4.30) to obtain $E^*[F(y)g(y)] = 0$ for any function g as described, and (4.29) will follow. Let $K \subseteq C[0,\infty)^d$ be compact, so that (Theorem 2.4.9 and (2.4.3)):

$$M \triangleq \sup_{\substack{y \in K\\ 0 \le u \le t}} \|y(u)\| < \infty, \quad \lim_{n\to\infty}\sup_{y \in K} m^t(y, 2^{-n}) = 0.$$

Because b and σ are uniformly continuous on $\{x \in \mathbb{R}^d;\ \|x\| \le M\}$, we can find for every $\varepsilon > 0$ an integer $n(\varepsilon)$ such that

$$\sup_{\substack{0 \le s \le t \\ y \in K}} (\|b^{(n)}(s, y) - b(y(s))\| + \|\sigma^{(n)}(s, y) - \sigma(y(s))\|) \le \varepsilon; \quad n \ge n(\varepsilon).$$

The uniform convergence on K of F_n to F follows. □

4.23 Remark. It is not difficult to modify the proof of Theorem 4.22 to allow b and σ to be bounded, continuous functions of $(t, x) \in [0, \infty) \times \mathbb{R}^d$, or even to allow them to be bounded, continuous, progressively measurable functionals.

E. Questions of Uniqueness

Finally, we take up the issue of uniqueness in the martingale problem.

4.24 Definition. A collection $\mathscr{D}$ of Borel-measurable functions $\varphi: \mathbb{R}^d \to \mathbb{R}$ is called a *determining class* on $\mathbb{R}^d$ if, for any two finite measures μ_1 and μ_2 on $\mathscr{B}(\mathbb{R}^d)$, the identity

$$\int_{\mathbb{R}^d} \varphi(x)\mu_1(dx) = \int_{\mathbb{R}^d} \varphi(x)\mu_2(dx), \quad \forall \varphi \in \mathscr{D}$$

implies $\mu_1 = \mu_2$.

4.25 Problem. The collection $C_0^\infty(\mathbb{R}^d)$ is a determining class on $\mathbb{R}^d$.

4.26 Lemma. *Suppose that for every $f \in C_0^\infty(\mathbb{R}^d)$, the Cauchy problem*

$$\frac{\partial u}{\partial t} = \mathscr{A}u; \quad \text{in } (0, \infty) \times \mathbb{R}^d, \tag{4.31}$$

$$u(0, \cdot) = f; \quad \text{in } \mathbb{R}^d, \tag{4.32}$$

has a solution $u_f \in C([0, \infty) \times \mathbb{R}^d) \cap C^{1,2}((0, \infty) \times \mathbb{R}^d)$ which is bounded on each strip of the form $[0, T] \times \mathbb{R}^d$. Then, if P^x and $\tilde{P}^x$ are any two solutions of the time-homogeneous martingale problem with initial condition $x \in \mathbb{R}^d$, their one-dimensional marginal distributions agree; i.e., for every $0 \le t < \infty$:

$$P^x[y(t) \in \Gamma] = \tilde{P}^x[y(t) \in \Gamma]; \quad \forall \Gamma \in \mathscr{B}(\mathbb{R}^d). \tag{4.33}$$

PROOF. For a fixed finite $T > 0$, the function $g(t, x) \triangleq u_f(T - t, x)$; $0 \le t \le T$, $x \in \mathbb{R}^d$ is of class $C_b([0, T] \times \mathbb{R}^d) \cap C^{1,2}((0, T) \times \mathbb{R}^d)$ and satisfies

$$\frac{\partial g}{\partial t} + \mathscr{A}g = 0; \quad \text{in } (0, T) \times \mathbb{R}^d,$$

$$g(T, \cdot) = f; \quad \text{in } \mathbb{R}^d.$$

Under either P^x or $\tilde{P}^x$, the coordinate mapping process $X_t = y(t)$ on $C[0, \infty)^d$ is a solution to the stochastic integral equation (4.20) (with respect to some

Brownian motion, on some possibly extended probability space; see Proposition 4.11). According to Proposition 4.2, the process $\{g(t, y(t)), \mathscr{B}_t; 0 \le t \le T\}$ is a local martingale under both P^x and $\tilde{P}^x$; being bounded and continuous, the process is in fact a martingale. Therefore,

$$E^x f(y(T)) = E^x g(T, y(T)) = g(0, x) = \tilde{E}^x g(T, y(T)) = \tilde{E}^x f(y(T)).$$

Since f can be any function in the determining class $C_0^\infty(\mathbb{R}^d)$, we conclude that (4.33) holds. □

We are witnessing here a remarkable duality: the *existence* of a solution to the Cauchy problem (4.31), (4.32) implies the *uniqueness*, at least for one-dimensional marginal distributions, of solutions to the martingale problem. In order to proceed beyond the uniqueness of the one-dimensional marginals to that of all finite-dimensional distributions, we utilize the Markov-like property obtained in Lemma 4.19. (We cannot, of course, use the Markov property contained in Theorem 4.20, since uniqueness is assumed in that result.)

4.27 Proposition. *Suppose that for every $x \in \mathbb{R}^d$, any two solutions P^x and $\tilde{P}^x$ of the time-homogeneous martingale problem with initial condition x have the same one-dimensional marginal distributions; i.e.,* (4.33) *holds. Then, for every $x \in \mathbb{R}^d$, the solution to the time-homogeneous martingale problem with initial condition x is unique.*

PROOF. Since a measure on $C[0, \infty)^d$ is determined by its finite-dimensional distributions, it suffices to fix $0 \le t_1 < t_2 < \cdots < t_n < \infty$ and show that P^x and $\tilde{P}^x$ agree on $\mathscr{G}(t_1, \ldots, t_n) \triangleq \sigma(y(t_1), \ldots, y(t_n))$. We proceed by induction on n. We have assumed the truth of this assertion for $n = 1$. Suppose now that P^x and $\tilde{P}^x$ agree on $\mathscr{G}(t_1, \ldots, t_{n-1})$, and let Q_y be a regular conditional probability for $\mathscr{B}(C[0, \infty)^d)$ given $\mathscr{G}(t_1, \ldots, t_{n-1})$, corresponding to P^x. According to Lemma 4.19, there exists a P^x-null set $N \in \mathscr{G}(t_1, \ldots, t_{n-1})$ such that for $y \notin N$, the measure $P_y \triangleq Q_y \circ \theta_{t_{n-1}}^{-1}$ solves the time-homogeneous martingale problem with initial condition $y(t_{n-1})$. Likewise, there is a regular conditional probability $\tilde{Q}_y$ corresponding to $\tilde{P}^x$ and a $\tilde{P}^x$-null set $\tilde{N} \in \mathscr{G}(t_1, \ldots, t_{n-1})$, such that $\tilde{P}_y \triangleq \tilde{Q}_y \circ \theta_{t_{n-1}}^{-1}$ solves this martingale problem whenever $y \notin \tilde{N}$. For y not in the null (under both P^x, $\tilde{P}^x$) set $N \cup \tilde{N}$, we know that P_y and $\tilde{P}_y$ have the same one-dimensional marginals. Thus, with $A \in \mathscr{B}(\mathbb{R}^{d(n-1)})$ and $B \in \mathscr{B}(\mathbb{R}^d)$, we have

$$\begin{aligned}
&P^x[(y(t_1), \ldots, y(t_{n-1})) \in A, y(t_n) \in B] \\
&= E^x[1_{\{(y(t_1), \ldots, y(t_{n-1})) \in A\}} P_y\{\omega \in \Omega; \omega(t_n - t_{n-1}) \in B\}] \\
&= E^x[1_{\{(y(t_1), \ldots, y(t_{n-1})) \in A\}} \tilde{P}_y\{\omega \in \Omega; \omega(t_n - t_{n-1}) \in B\}] \\
&= \tilde{E}^x[1_{\{(y(t_1), \ldots, y(t_{n-1})) \in A\}} \tilde{P}_y\{\omega \in \Omega; \omega(t_n - t_{n-1}) \in B\}] \\
&= \tilde{P}^x[(y(t_1), \ldots, y(t_{n-1})) \in A, y(t_n) \in B],
\end{aligned}$$

where we have used not only the equality of $P_y\{\omega\in\Omega;\ \omega(t_n - t_{n-1})\in B\}$ and $\tilde{P}_y\{\omega\in\Omega;\ \omega(t_n - t_{n-1})\in B\}$, but also their $\mathscr{G}(t_1,\ldots,t_{n-1})$-measurability. It is now clear that P^x and $\tilde{P}^x$ agree on $\mathscr{G}(t_1,\ldots,t_n)$. □

We can now put the various results together.

4.28 Theorem (Stroock & Varadhan (1969)). *Suppose that the coefficients $b(x)$ and $\sigma(x)$ in Definition* 4.15 *are such that, for every $f\in C_0^\infty(\mathbb{R}^d)$, the Cauchy problem* (4.31), (4.32) *has a solution $u_f\in C([0,\infty)\times\mathbb{R}^d)\cap C^{1,2}((0,\infty)\times\mathbb{R}^d)$ which is bounded on each strip of the form $[0,T]\times\mathbb{R}^d$. Then, for every $x\in\mathbb{R}^d$, there exists at most one solution to the time-homogeneous martingale problem.*

4.29 Corollary. *Let the coefficients $b(x)$, $\sigma(x)$ be bounded and continuous and satisfy the assumptions of Theorem* 4.28. *Then the time-homogeneous martingale problem is well posed.*

4.30 Remark. A sufficient condition for the solvability of the Cauchy problem (4.31), (4.32) in the way required by Theorem 4.28 is that the coefficients $b_i(x)$, $a_{ik}(x)$; $1\le j,k\le d$ be bounded and Hölder-continuous on $\mathbb{R}^d$, and the matrix $a(x)$ be *uniformly positive definite*; i.e.,

$$\text{(4.34)}\qquad \sum_{i=1}^{d}\sum_{k=1}^{d} a_{ik}(x)\xi_i\xi_k \ge \lambda\|\xi\|^2;\quad \forall x,\xi\in\mathbb{R}^d \quad\text{and some } \lambda>0.$$

We refer to Friedman (1964), Chapter 1, and Friedman (1975), §6.4, §6.5, or Stroock & Varadhan (1979), Theorem 3.2.1, for such results; see also Remark 7.8 later in this chapter.

4.31 Remark. If $a_{ik}\in C^2(\mathbb{R}^d)$ for $1\le i,k\le d$, then $a(x)$ has a locally Lipschitz-continuous square root, i.e., a $(d\times d)$-matrix-valued function $\tilde{\sigma}(x) = \{\tilde{\sigma}_{ij}(x)\}_{1\le i,j\le d}$ such that $a_{ik}(x) = \sum_{j=1}^d \tilde{\sigma}_{ij}(x)\tilde{\sigma}_{kj}(x)$; $x\in\mathbb{R}^d$ (Friedman (1975), Theorem 6.1.2). We do not necessarily have $\sigma(x) = \tilde{\sigma}(x)$ (indeed, one interesting case is the one-dimensional problem with $\sigma(x) = \operatorname{sgn}(x)$ and $\tilde{\sigma}(x) = 1$; $x\in\mathbb{R}$). If, in addition, $b_i\in C^1(\mathbb{R}^d)$; $1\le i\le d$, then according to Theorem 2.5, Remark 3.3, Proposition 3.20, Corollary 4.9, and Proposition 4.11, for every $x\in\mathbb{R}^d$ there exists *at most* one solution to the time-homogeneous martingale problem. This result imposes no condition analogous to (4.34) and is thus especially helpful in the study of degenerate diffusions.

4.32 Refinements. It can be shown that if the coefficients $b(x)$, $\sigma(x)$ are bounded and Borel-measurable, and the matrix $a(x)$ is uniformly positive definite on compact subsets of $\mathbb{R}^d$, then the time-homogeneous martingale problem admits a solution. In the cases $d=1$ and $d=2$, this solution is unique. See Stroock & Varadhan (1979), Exercises 7.3.2–7.3.4, and Krylov (1969), (1974).

F. Supplementary Exercises

4.33 Exercise. Assume that the coefficients $b_i: \mathbb{R}^d \to \mathbb{R}$, $\sigma_{ij}: \mathbb{R}^d \to \mathbb{R}$; $1 \le i \le d$, $1 \le j \le r$ are measurable and bounded on compact subsets of $\mathbb{R}^d$, and let $\mathscr{A}$ be the associated operator (1.2). Let $X = \{X_t, \mathscr{F}_t; 0 \le t < \infty\}$ be a continuous process on some probability space $(\Omega, \mathscr{F}, P)$ and assume that $\{\mathscr{F}_t\}$ satisfies the usual conditions. With $f \in C^2(\mathbb{R}^d)$ and $\alpha \in \mathbb{R}$, introduce the processes

$$M_t \triangleq f(X_t) - f(X_0) - \int_0^t \mathscr{A}f(X_s)\,ds, \mathscr{F}_t; \quad 0 \le t < \infty,$$

$$\Lambda_t \triangleq e^{-\alpha t} f(X_t) - f(X_0) + \int_0^t e^{-\alpha s}(\alpha f(X_s) - \mathscr{A}f(X_s))\,ds, \mathscr{F}_t; \quad 0 \le t < \infty,$$

and show $M \in \mathscr{M}^{c,\mathrm{loc}} \Leftrightarrow \Lambda \in \mathscr{M}^{c,\mathrm{loc}}$. If f is bounded away from zero on compact sets and

$$N_t \triangleq f(X_t) \exp\left\{-\int_0^t \frac{\mathscr{A}f(X_s)}{f(X_s)}\,ds\right\} - f(X_0), \mathscr{F}_t; \quad 0 \le t < \infty,$$

then these two conditions are also equivalent to: $N \in \mathscr{M}^{c,\mathrm{loc}}$. (*Hint*: Recall from Problem 3.3.12 that if $M \in \mathscr{M}^{c,\mathrm{loc}}$ and C is a continuous process of bounded variation, then $C_t M_t - \int_0^t M_s\,dC_s = \int_0^t C_s\,dM_s$ is in $\mathscr{M}^{c,\mathrm{loc}}$.)

4.34 Exercise. Let (X, W), $(\Omega, \mathscr{F}, P)$, $\{\mathscr{F}_t\}$ be a weak solution to the functional stochastic differential equation (3.15), where condition (A.1)′ of Problem 4.13 holds. For any continuous function $f: [0,\infty) \times \mathbb{R}^d \to \mathbb{R}$ of class $C^{1,2}([0,\infty) \times \mathbb{R}^d)$ and any progressively measurable, locally integrable process $\{k_t, \mathscr{F}_t; 0 \le t < \infty\}$, show that

$$\Lambda_t \triangleq f(t, X_t)e^{-\int_0^t k_u\,du} - f(0, X_0) - \int_0^t \left(\frac{\partial f}{\partial s} + \mathscr{A}'_s f - k_s f\right) e^{-\int_0^s k_u\,du}\,ds, \mathscr{F}_t; \quad 0 \le t < \infty$$

is in $\mathscr{M}^{c,\mathrm{loc}}$. If, furthermore, f and its indicated derivatives are bounded and k is bounded from below, then Λ is a martingale.

4.35 Exercise. Let the coefficients b, σ be bounded on compact subsets of $\mathbb{R}^d$, and assume that for each $x \in \mathbb{R}^d$, the time-homogeneous martingale problem of Definition 4.15 has a solution P^x satisfying (4.22). Suppose that there exists a function $f: \mathbb{R}^d \to [0, \infty)$ of class $C^2(\mathbb{R}^d)$ such that

$$\mathscr{A}f(x) + \lambda f(x) \le c, \quad \forall x \in \mathbb{R}^d$$

holds for some $\lambda > 0$, $c \ge 0$. Then

$$E^x f(y(t)) \le f(x)e^{-\lambda t} + \frac{c}{\lambda}(1 - e^{-\lambda t}); \quad 0 \le t < \infty, x \in \mathbb{R}^d.$$

5.5. A Study of the One-Dimensional Case

This section presents the definitive results of Engelbert and Schmidt concerning weak solutions of the one-dimensional, time-homogeneous stochastic differential equation

$$dX_t = b(X_t)\,dt + \sigma(X_t)\,dW_t \tag{5.1}$$

with Borel-measurable coefficients $b: \mathbb{R} \to \mathbb{R}$ and $\sigma: \mathbb{R} \to \mathbb{R}$. These authors provide simple necessary and sufficient conditions for existence and uniqueness when $b \equiv 0$, and sharp sufficient conditions in the case of general drift coefficients. The principal tools required for this analysis are local time, the generalized Itô rule, the Engelbert-Schmidt zero-one law (see Subsection 3.6.E), and the notion of time-change.

Solutions of equation (5.1) may not exist globally, but rather only up to an "explosion time" S; see, for example, Remark 2.8. We formalize the idea of explosion.

5.1 Definition. *A weak solution up to an explosion time* of equation (5.1) is a triple (X, W), $(\Omega, \mathscr{F}, P)$, $\{\mathscr{F}_t\}$, where

(i) $(\Omega, \mathscr{F}, P)$ is a probability space, and $\{\mathscr{F}_t\}$ is a filtration of sub-σ-fields of $\mathscr{F}$ satisfying the usual conditions,

(ii) $X = \{X_t, \mathscr{F}_t; 0 \le t < \infty\}$ is a continuous, adapted, $\mathbb{R} \cup \{\pm\infty\}$-valued process with $|X_0| < \infty$ a.s., and $\{W_t, \mathscr{F}_t; 0 \le t < \infty\}$ is a standard, one-dimensional Brownian motion,

(iii) with

$$S_n \triangleq \inf\{t \ge 0; |X_t| \ge n\}, \tag{5.2}$$

we have

$$P\left[\int_0^{t \wedge S_n} \{|b(X_s)| + \sigma^2(X_s)\}\,ds < \infty\right] = 1; \quad \forall\, 0 \le t < \infty \tag{5.3}$$

and

(iv)

$$P\left[X_{t \wedge S_n} = X_0 + \int_0^t b(X_s) 1_{\{s \le S_n\}}\,ds + \int_0^t \sigma(X_s) 1_{\{s \le S_n\}}\,dW_s;\ \forall\, 0 \le t < \infty\right] = 1 \tag{5.4}$$

valid for every $n \ge 1$.

We refer to

$$S = \lim_{n\to\infty} S_n \tag{5.5}$$

as the *explosion time* for X. The assumption of continuity of X in the extended real numbers implies that

(5.6) $S = \inf\{t \geq 0: X_t \notin \mathbb{R}\}$ and $X_S = \pm\infty$ a.s. on $\{S < \infty\}$.

We stipulate that $X_t = X_S$; $S \leq t < \infty$.

The assumption of finiteness of X_0 gives $P[S > 0] = 1$. We do not assume that $\lim_{t\to\infty} X_t$ exists on $\{S = \infty\}$, so X_S may not be defined on this event. If $P[S = \infty] = 1$, then Definition 5.1 reduces to Definition 3.1 for a weak solution to (5.1).

We begin with a discussion of the time-change which will be employed in both the existence and uniqueness proofs.

A. The Method of Time-Change

Suppose we have defined on a probability space a standard, one-dimensional Brownian motion $B = \{B_s, \mathscr{F}_s^B; 0 \leq s < \infty\}$ and an independent random variable ξ with distribution μ. Let $\{\mathscr{G}_s\}$ be a filtration satisfying the usual conditions, relative to which B is still a Brownian motion, and such that ξ is $\mathscr{G}_0$-measurable (a filtration with these properties was constructed for Definition 2.1). We introduce

(5.7) $$T_s \triangleq \int_0^{s+} \frac{du}{\sigma^2(\xi + B_u)}; \quad 0 \leq s < \infty,$$

a nondecreasing, extended real-valued process which is continuous in the topology of $[0, \infty]$ except for a possible jump to infinity at a finite time s. From Problem 3.6.30 we have

(5.8) $$T_\infty \triangleq \lim_{s\uparrow\infty} T_s = \infty \quad \text{a.s.}$$

We define the "inverse" of T_s by

(5.9) $$A_t \triangleq \inf\{s \geq 0; T_s > t\}; \quad 0 \leq t < \infty \quad \text{and} \quad A_\infty \triangleq \lim_{t\to\infty} A_t.$$

Whether or not T_s reaches infinity in finite time, we have a.s.

(5.10) $$A_0 = 0, \quad A_t < \infty; \quad 0 \leq t < \infty \quad \text{and} \quad A_\infty = \inf\{s \geq 0: T_s = \infty\}.$$

Because T_s is continuous and strictly increasing on $[0, A_\infty)$, A_t is also continuous on $[0, \infty)$ and strictly increasing on $[0, T_{A_\infty}-)$. Note, however, that if T_s jumps to infinity (at $s = A_\infty$), then

(5.11) $$A_t = A_\infty; \forall t \geq T_{A_\infty}-;$$

if not, then $T_{A_\infty}- = T_{A_\infty} = \infty$, and (5.11) is vacuously valid. The identities

(5.12) $$T_{A_t} = t; \quad 0 \leq t < T_{A_\infty}-,$$

$$A_{T_s} = s; \quad 0 \le s < A_\infty, \tag{5.13}$$

hold almost surely. From these considerations we deduce that

$$T_s = \inf\{t \ge 0; A_t > s\}; \quad 0 \le s < \infty, \text{ a.s.} \tag{5.14}$$

In other words, A_t and T_s are related as in Problem 3.4.5.

Let us consider now the closed set

$$I(\sigma) = \left\{x \in \mathbb{R}; \int_{-\varepsilon}^{\varepsilon} \frac{dy}{\sigma^2(x+y)} = \infty, \forall\, \varepsilon > 0\right\}, \tag{5.15}$$

and define

$$R \triangleq \inf\{s \ge 0; \xi + B_s \in I(\sigma)\}. \tag{5.16}$$

5.2 Lemma. *We have $R = A_\infty$, a.s. In particular,*

$$\int_0^{R+} \frac{du}{\sigma^2(\xi + B_u)} = \infty, \quad a.s. \tag{5.17}$$

Proof. Define a sequence of stopping times

$$R_n \triangleq \inf\left\{s \ge 0; \rho(\xi + B_s, I(\sigma)) \le \frac{1}{n}\right\}; \quad n \ge 1,$$

where

$$\rho(x, I(\sigma)) \triangleq \inf\{|x - y|; y \in I(\sigma)\}. \tag{5.18}$$

Because $I(\sigma)$ is closed, we have $\lim_{n\to\infty} R_n = R$, a.s. (recall Solution 1.2.7). For $n \ge 1$, set

$$\sigma_n(x) = \begin{cases} \sigma(x); & \rho(x, I(\sigma)) \ge \dfrac{1}{n}, \\[2ex] 1; & \rho(x, I(\sigma)) < \dfrac{1}{n}. \end{cases}$$

We have $\int_K \sigma_n^{-2}(x)\,dx < \infty$ for each compact set $K \subset \mathbb{R}$, and the Engelbert-Schmidt zero-one law (Proposition 3.6.27 (iii) $\Rightarrow$ Problem 3.6.29 (v)) gives a.s.

$$\int_0^{R_n \wedge s} \frac{du}{\sigma^2(\xi + B_u)} = \int_0^{R_n \wedge s} \frac{du}{\sigma_n^2(\xi + B_u)} \le \int_0^s \frac{du}{\sigma_n^2(\xi + B_u)} < \infty; \quad 0 \le s < \infty.$$

For P-a.e. $\omega \in \Omega$ and s chosen to satisfy $s < R(\omega)$, we can let $n \to \infty$ to conclude

$$\int_0^s \frac{du}{\sigma^2(\xi(\omega) + B_u(\omega))} < \infty.$$

It follows that $T_s < \infty$ on $\{s < R\}$, and thus $A_\infty \ge R$ a.s. (see (5.10)).

For the reverse inequality, observe first that $I(\sigma) = \varnothing$ implies $R = \infty$, and in this case there is nothing to prove. If $I(\sigma) \ne \varnothing$ then $R < \infty$ a.s. and for every $s > 0$,

$$\int_0^{R+s} \frac{du}{\sigma^2(\xi + B_u)} \geq \int_0^s \frac{du}{\sigma^2(\xi + B_R + W_u)},$$

where $W_u \triangleq B_{R+u} - B_R$ is a standard Brownian motion, independent of B_R (Theorem 2.6.16). Because $\xi + B_R \in I(\sigma)$, Lemma 3.6.26 shows that the latter integral is infinite. It follows that $T_R = \infty$, so from (5.10), $A_\infty \leq R$ a.s. □

We take up first the stochastic differential equation (5.1) when the drift is identically zero. One important feature of the resulting equation

$$dX_t = \sigma(X_t)\,dW_t \tag{5.19}$$

is that *solutions cannot explode*. We leave the verification of this claim to the reader.

5.3 Problem. Suppose (X, W), $(\Omega, \mathscr{F}, P)$, $\{\mathscr{F}_t\}$ is a weak solution of equation (5.19) up to an explosion time S. Show that $S = \infty$, a.s. (*Hint*: Recall Problem 3.4.11.)

We also need to introduce the set

$$Z(\sigma) = \{x \in \mathbb{R};\ \sigma(x) = 0\}. \tag{5.20}$$

The fundamental existence result for the stochastic differential equation (5.19) is the following.

5.4 Theorem (Engelbert & Schmidt (1984)). *Equation* (5.19) *has a non-exploding weak solution for every initial distribution* μ *if and only if*

$$I(\sigma) \subseteq Z(\sigma); \tag{E}$$

i.e., if

$$\int_{-\varepsilon}^{\varepsilon} \frac{dy}{\sigma^2(x + y)} = \infty, \quad \forall \varepsilon > 0 \Rightarrow \sigma(x) = 0.$$

5.5 Remark. Every continuous function σ satisfies (E), but so do many discontinuous functions, e.g., $\sigma(x) = \operatorname{sgn}(x)$. The function $\sigma(x) = 1_{\{0\}}(x)$ does not satisfy (E).

PROOF OF THEOREM 5.4. Let us assume first that (E) holds and let $\{\xi + B_s, \mathscr{G}_s;\ 0 \leq s < \infty\}$ be a Brownian motion with initial distribution μ, as described at the beginning of this subsection. Using the notation of (5.7)–(5.16), we can verify from Problem 3.4.5 (v) that each A_t is a stopping time for $\{\mathscr{G}_s\}$. We set

$$M_t = B_{A_t}, \quad X_t = \xi + M_t, \quad \mathscr{F}_t = \mathscr{G}_{A_t}; \quad 0 \leq t < \infty. \tag{5.21}$$

Because A is continuous, $\{\mathscr{F}_t\}$ inherits the usual conditions from $\{\mathscr{G}_s\}$ (cf. Problem 1.2.23). From the optional sampling theorem (Problem 1.3.23 (i)) and the identity $A_{t \wedge T_s} = A_t \wedge s$ (Problem 3.4.5 (ii), (v)), we have for $0 \le t_1 \le t_2 < \infty$ and $n \ge 1$:

$$E[M_{t_2 \wedge T_n} | \mathscr{F}_{t_1}] = E[B_{A_{t_2} \wedge n} | \mathscr{G}_{A_{t_1}}] = B_{A_{t_1} \wedge n} = M_{t_1 \wedge T_n}, \quad \text{a.s.}$$

Since $\lim_{n\to\infty} T_n = \infty$ a.s., we conclude that $M \in \mathscr{M}^{c,\text{loc}}$. Furthermore, $M^2_{t \wedge T_n} - A_{t \wedge T_n} = B^2_{A_t \wedge n} - (A_t \wedge n)$ is in $\mathscr{M}^c$ for each $n \ge 1$, so

$$\langle M \rangle_t = A_t; \quad 0 \le t < \infty, \text{ a.s.} \tag{5.22}$$

As the next step, we show that the process of (5.9) is given by

$$A_t = \int_0^t \sigma^2(X_v)\,dv; \quad 0 \le t < \infty, \text{ a.s.} \tag{5.23}$$

Toward this end, fix $\omega \in \{R = A_\infty\}$. For $s < A_\infty(\omega)$, (5.7) and (5.10) show that the function $u \mapsto T_u(\omega)$ restricted to $u \in [0, s]$ is absolutely continuous. The change of variable $v = T_u(\omega)$ is equivalent to $A_v(\omega) = u$ (see (5.12), (5.13)) and leads to the formula

$$A_t(\omega) = \int_0^{A_t(\omega)} \sigma^2(\xi(\omega) + B_u(\omega))\,dT_u(\omega) = \int_0^t \sigma^2(X_v(\omega))\,dv, \tag{5.24}$$

valid as long as $A_t(\omega) < A_\infty(\omega)$, i.e., $t < \tau(\omega)$, where

$$\tau \triangleq T_{A_\infty -} = \inf\{u \ge 0;\ A_u = A_\infty\}. \tag{5.25}$$

If $\tau(\omega) = \infty$, we are done. If not, letting $t \uparrow \tau(\omega)$ in (5.24) and using the continuity of $A(\omega)$, we obtain (5.23) for $0 \le t \le \tau(\omega)$. On the interval $[\tau(\omega), \infty]$, $A_t(\omega) \equiv A_{\tau(\omega)}(\omega) = A_\infty(\omega) = R(\omega)$. If $R(\omega) < \infty$, then

$$X_{\tau(\omega)}(\omega) = \xi(\omega) + B_{R(\omega)}(\omega) \in I(\sigma) \subseteq Z(\sigma),$$

and so

$$\sigma(X_t(\omega)) = \sigma(X_{\tau(\omega)}(\omega)) = 0; \quad \tau(\omega) \le t < \infty.$$

Thus, for $t \ge \tau(\omega)$, equation (5.23) holds with both sides equal to $A_{\tau(\omega)}(\omega)$.

From (5.22), (5.23), and the finiteness of A_t (see (5.10)), we conclude that $\langle M \rangle$ is a.s. absolutely continuous. Theorem 3.4.2 asserts the existence of a Brownian motion $\tilde{W} = \{\tilde{W}_t, \tilde{\mathscr{F}}_t; 0 \le t < \infty\}$ and a measurable, adapted process $\rho = \{\rho_t, \tilde{\mathscr{F}}_t; 0 \le t < \infty\}$ on a possibly extended probability space $(\tilde{\Omega}, \tilde{\mathscr{F}}, \tilde{P})$, such that

$$M_t = \int_0^t \rho_v\,d\tilde{W}_v, \quad \langle M \rangle_t = \int_0^t \rho_v^2\,dv; \quad 0 \le t < \infty, \tilde{P} \text{ a.s.}$$

In particular,

$$\tilde{P}[\rho_t^2 = \sigma^2(X_t) \text{ for Lebesgue a.e. } t \ge 0] = 1.$$

We may set

$$W_t = \int_0^t \operatorname{sgn}(\rho_v \sigma(X_v))\,d\tilde{W}_v; \quad 0 \le t < \infty;$$

observe that W is itself a Brownian motion (Theorem 3.3.16); and write

$$X_t = \xi + M_t = \xi + \int_0^t \sigma(X_v)\,dW_v; \quad 0 \le t < \infty, \tilde{P}\text{-a.s.}$$

Thus, (X, W) is a weak solution to (5.19) with initial distribution μ.

To prove the *necessity* of (E), we suppose that for every $x \in \mathbb{R}$, (5.19) has a nonexploding weak solution (X, W) with $X_0 = x$ a.s. Here $W = \{W_t, \mathscr{F}_t; 0 \le t < \infty\}$ is a Brownian motion with $W_0 = 0$ a.s. Then

$$M_t = X_t - X_0, \mathscr{F}_t; \quad 0 \le t < \infty \tag{5.26}$$

is in $\mathscr{M}^{c,\text{loc}}$ and

$$\langle M \rangle_t = \int_0^t \sigma^2(X_v)\,dv < \infty; \quad 0 \le t < \infty, \text{ a.s.} \tag{5.27}$$

According to Problem 3.4.7, there is a Brownian motion $B = \{B_s, \mathscr{G}_s; 0 \le s < \infty\}$ on a possibly extended probability space, such that

$$M_t = B_{\langle M \rangle_t}; \quad 0 \le t < \infty, \text{ a.s.} \tag{5.28}$$

Let $T_s = \inf\{t \ge 0; \langle M \rangle_t > s\}$. Then $s \wedge \langle M \rangle_\infty = \langle M \rangle_{T_s}$ (Problem 3.4.5 (ii)), so using the change of variable $u = \langle M \rangle_v$ (ibid. (vi)) and the fact that $d\langle M \rangle$ assigns zero measure to the set $\{v \ge 0; \sigma^2(X_v) = 0\}$, we may write

$$\begin{aligned} \int_0^{s \wedge \langle M \rangle_\infty} \frac{du}{\sigma^2(X_0 + B_u)} &= \int_0^{\langle M \rangle_{T_s}} \frac{du}{\sigma^2(X_0 + B_u)} = \int_0^{T_s} \frac{d\langle M \rangle_v}{\sigma^2(X_v)} \\ &= \int_0^{T_s} 1_{\{\sigma^2(X_v) > 0\}} \frac{d\langle M \rangle_v}{\sigma^2(X_v)} \\ &= \int_0^{T_s} 1_{\{\sigma^2(X_v) > 0\}}\,dv \le T_s. \end{aligned} \tag{5.29}$$

Let us choose the initial condition $X_0 = x$ in $Z(\sigma)^c$. Then $\sigma(x) \ne 0$, so a solution to (5.19) with such an initial condition cannot be almost surely constant. Moreover $P[\langle M \rangle_\infty > 0] > 0$, and thus for sufficiently small positive s, $P[T_s < \infty, \langle M \rangle_\infty > 0] > 0$. Now apply Lemma 3.6.26 with

$$T(\omega) \triangleq \begin{cases} s \wedge \langle M \rangle_\infty(\omega); & \text{if } T_s(\omega) < \infty, \langle M \rangle_\infty(\omega) > 0, \\ s; & \text{otherwise.} \end{cases}$$

We conclude from this lemma and (5.29) that $x \in I(\sigma)^c$. It follows that $I(\sigma) \subseteq Z(\sigma)$. □

5.6 Remark. The solution (5.21) constructed in the proof of Theorem 5.4 is nonconstant up until the time

$$\tau \triangleq \inf\{t \ge 0; X_t \in I(\sigma)\}. \tag{5.30}$$

(This definition agrees with (5.25).) In particular, if $X_0 = x \in I(\sigma)^c$, then X is not identically constant. On the other hand, if $x \in Z(\sigma)$, then $Y_t \equiv x$ also solves (5.19). Thus *there can be no uniqueness in the sense of probability law if $Z(\sigma)$ is strictly larger than $I(\sigma)$.* This is the case in the Girsanov Example 2.15; if $\sigma(x) = |x|^\alpha$ and $0 < \alpha < (1/2)$, then $I(\sigma) = \emptyset$ and $Z(\sigma) = \{0\}$.

Engelbert & Schmidt (1985) show that if $\sigma: \mathbb{R} \to \mathbb{R}$ is any Borel-measurable function satisfying $I(\sigma) = \emptyset$, then all solutions of (5.19) can be obtained from the one constructed in Theorem 5.4 by "delaying" it when it is in $Z(\sigma)$. An identically constant solution corresponds to infinite delay, but other, less drastic, delays are also possible.

5.7 Theorem. (Engelbert & Schmidt (1984)). *For every initial distribution μ, the stochastic differential equation* (5.19) *has a solution which is unique in the sense of probability law if and only if*

$$I(\sigma) = Z(\sigma). \tag{E + U}$$

Proof. The inclusion $I(\sigma) \subseteq Z(\sigma)$, i.e., condition (E), is necessary for existence (Theorem 5.4); in the presence of (E), the reverse inclusion is necessary for uniqueness (Remark 5.6). Condition (E) is also sufficient for existence (Theorem 5.4), so it remains only to show that the equality $I(\sigma) = Z(\sigma)$ is sufficient for uniqueness. We shall show, in fact, that the inclusion $I(\sigma) \supseteq Z(\sigma)$ is sufficient for uniqueness.

Assume $I(\sigma) \supseteq Z(\sigma)$ and let (X, W), $(\Omega, \mathscr{F}, P)$, $\{\mathscr{F}_t\}$ be a weak solution of (5.19) with initial distribution μ. We define M as in (5.26) and obtain (5.27)–(5.29), where B is a standard Brownian motion. We also introduce τ via (5.30). By assumption,

$$\sigma^2(X_t) > 0; \quad 0 \le t < \tau, \text{ a.s.,} \tag{5.31}$$

so $\langle M \rangle$ is strictly increasing on $[0, \tau]$. In particular,

$$R \triangleq \inf\{s \ge 0; X_0 + B_s \in I(\sigma)\} = \langle M \rangle_\tau, \quad \text{a.s.} \tag{5.32}$$

We set

$$T_s \triangleq \inf\{t \ge 0; \langle M \rangle_t > s\}, \tag{5.33}$$

so T is nondecreasing, right-continuous, and

$$\{T_s < t\} = \{s < \langle M \rangle_t\}. \tag{5.34}$$

We claim that

$$\int_0^s \frac{du}{\sigma^2(X_0 + B_u)} = T_s; \quad 0 \le s < \langle M \rangle_\infty, \text{ a.s.} \tag{5.35}$$

In light of (5.29), (5.31), to verify this claim it suffices to show

$$R \ge \langle M \rangle_\infty, \quad \text{a.s.} \tag{5.36}$$

Indeed, on the set $\{R < \langle M \rangle_\infty\}$ we have by the definition (5.33) that $T_R < \infty$;

letting $s \downarrow R$ in (5.29), we obtain then

$$\int_0^{R+} \frac{du}{\sigma^2(X_0 + B_u)} \le T_R < \infty \quad \text{on } \{R < \langle M\rangle_\infty\}.$$

Comparing this with (5.17), we see that $P[R < \langle M\rangle_\infty] = 0$, and (5.35) is established.

We next argue that

$$T_s = \int_0^{s+} \frac{du}{\sigma^2(X_0 + B_u)}; \quad 0 \le s < \infty, \text{ a.s.} \tag{5.37}$$

For $s < \langle M\rangle_\infty$, we have $T_s < \infty$ and so (5.37) follows immediately from (5.35). For $s = \langle M\rangle_\infty = R$, it is clear from (5.33) that $T_s = \infty$, and (5.37) follows from (5.17). It is then apparent that (5.37) also holds for $s \ge \langle M\rangle_\infty$ with both sides equal to infinity.

Comparing (5.37) and (5.7), we conclude that the discussion preceding Lemma 5.2 can be brought to bear. The process A defined in (5.9) coincides then with $\langle M\rangle$, which is thus seen to be continuous and real-valued and to satisfy

$$\langle M\rangle_t = \inf\left\{s \ge 0; \int_0^{s+} \frac{du}{\sigma^2(X_0 + B_u)} > t\right\}; \quad 0 \le t < \infty. \tag{5.38}$$

We now prove uniqueness in the sense of probability law by showing that the distribution induced by the solution process X on the canonical space $(C[0,\infty), \mathscr{B}(C[0,\infty)))$ is completely determined by the distribution μ of X_0. The distribution induced on this space by the process $X_0 + B$ is completely determined by μ because B is a standard Brownian motion independent of X_0 (Remark 3.4.10). For $\omega \in C[0,\infty)$, define $\varphi_\cdot(\omega)$ to be the right-continuous process

$$\varphi_t(\omega) \triangleq \inf\left\{s \ge 0; \int_0^{s+} \frac{du}{\sigma^2(\omega(u))} > t\right\}; \quad 0 \le t < \infty$$

with values in $[0,\infty]$. Because of its right-continuity, φ is $\mathscr{B}([0,\infty)) \otimes \mathscr{B}(C[0,\infty))$-measurable (Remark 1.1.14). Let $\pi: [0,\infty) \times C[0,\infty) \to \mathbb{R}$ be the measurable projection mapping $\pi_t(\omega) = \omega(t)$. Consider the $\mathscr{B}([0,\infty)) \otimes \mathscr{B}(C[0,\infty))$-measurable process $\psi_t(\omega) \triangleq \omega(0) + \pi_{\varphi_t(\omega)}(\omega)$. We have

$$X_t = X_0 + B_{\langle M\rangle_t} = \psi_t(X_0 + B_\cdot), \quad 0 \le t < \infty, \tag{5.39}$$

and so the law of X is completely determined by that of $X_0 + B_\cdot$. □

5.8 Remark. It is apparent from the proof of Theorem 5.7 that under the assumption $I(\sigma) \supseteq Z(\sigma)$, any solution to (5.19) remains constant after arriving in $I(\sigma)$. See, in particular, the representation (5.39) and recall that $\langle M\rangle_t$ is constant for $\tau \le t < \infty$.

Using local time for continuous semimartingales, it is possible to give a simple but powerful *sufficient condition for pathwise uniqueness* of the solution to equation (5.19).

5.9 Theorem. *Suppose that there exist functions* $f: \mathbb{R} \to [0, \infty)$ *and* $h: [0, \infty) \to [0, \infty)$ *such that:*

(i) *at every* $x \in I(\sigma)^c$, *the quotient* $(f/\sigma)^2$ *is locally integrable; i.e., there exists* $\varepsilon > 0$ *(depending on* x*) such that*

$$\int_{x-\varepsilon}^{x+\varepsilon} \left(\frac{f(y)}{\sigma(y)}\right)^2 dy < \infty;$$

(ii) *the function* h *is strictly increasing and satisfies* $h(0) = 0$ *and* (2.25);

(iii) *there exists a constant* $a > 0$ *such that*

$$|\sigma(x + y) - \sigma(x)| \le f(x)h(|y|); \quad \forall\, x \in \mathbb{R},\ y \in [-a, a].$$

Then pathwise uniqueness holds for the equation (5.19).

PROOF. Observe first of all that if $\sigma(x) = 0$, then (ii) and (iii) imply that

$$\int_{-\varepsilon}^{\varepsilon} \frac{dy}{\sigma^2(x+y)} \ge 2f^{-2}(x) \int_0^{\varepsilon} h^{-2}(y)\,dy = \infty, \quad \forall\, \varepsilon > 0.$$

Thus, $Z(\sigma) \subseteqq I(\sigma)$.

Suppose now that $X^{(i)} = \{X_t^{(i)}, \mathscr{F}_t; 0 \le t < \infty\}$; $i = 1, 2$, are solutions to (5.19) relative to the same Brownian motion $W = \{W_t, \mathscr{F}_t; 0 \le t < \infty\}$ on some probability space $(\Omega, \mathscr{F}, P)$, with $P[X_0^{(1)} = X_0^{(2)}] = 1$. Defining $\tau^{(i)} = \inf\{t \ge 0;\ X_t^{(i)} \in I(\sigma)\}$, we recall from Remark 5.8 that $X_t^{(i)} = X_{t \wedge \tau^{(i)}}^{(i)}$; $0 \le t < \infty$, a.s., so it suffices to prove that

$$X_t^{(1)} = X_t^{(2)}; \quad 0 \le t \le \tau^{(1)}, \text{ a.s.} \tag{5.40}$$

We set $\Delta_t = X_t^{(2)} - X_t^{(1)}$. For each integer $k \ge (1/a)$, there exists $\varepsilon_k > 0$ for which $\int_{\varepsilon_k}^{1/k} h^{-2}(y)\,dy = k$. Using the continuity of the local time for Δ (Remark 3.7.8), (3.7.3), and assumption (iii), we may write for every random time τ:

$$\begin{aligned} \Lambda_\tau^\Delta(0) &= \lim_{k\to\infty} \frac{1}{k} \int_{\varepsilon_k}^{1/k} h^{-2}(x)\Lambda_\tau^\Delta(x)\,dx \\ &= \lim_{k\to\infty} \frac{1}{k} \int_0^\tau h^{-2}(\Delta_s) 1_{[\varepsilon_k, (1/k)]}(\Delta_s)(\sigma(X_s^{(2)}) - \sigma(X_s^{(1)}))^2\,ds \\ &\le \overline{\lim_{k\to\infty}} \frac{1}{k} \int_0^\tau f^2(X_s^{(1)})\,ds, \quad \text{a.s. } P. \end{aligned} \tag{5.41}$$

Now $M_t \triangleq X_t^{(1)} - X_0^{(1)}$ is in $\mathscr{M}^{c,\mathrm{loc}}$, and so M admits the representation $M_t = B_{\langle M \rangle_t}$, where B is a Brownian motion (Problem 3.4.7). Furthermore,

$$\langle M \rangle_t = \int_0^t \sigma^2(X_v^{(1)})\,dv,$$

and $\sigma^2(X_v^{(1)}) > 0$ for $0 \le v < \tau^{(1)}$, so a change of variable results in the relation

$$\int_0^t f^2(X_v^{(1)})\,dv = \int_0^{\langle M \rangle_t} f^2(X_0^{(1)} + B_u)\sigma^{-2}(X_0^{(1)} + B_u)\,du; \quad 0 \le t < \tau^{(1)}.$$

We take

$$\tau_n \triangleq \inf\left\{t \ge 0;\ \rho(X_t^{(1)}, I(\sigma)) \le \frac{1}{n}\right\},$$

$$R_n \triangleq \langle M \rangle_{\tau_n} = \inf\left\{s \ge 0;\ \rho(X_0 + B_s, I(\sigma)) \le \frac{1}{n}\right\},$$

where ρ is given by (5.18). We alter f and σ near $I(\sigma)$ by setting

$$f_n(x) = \sigma_n(x) = 1; \qquad \text{if } \rho(x, I(\sigma)) < \frac{1}{n},$$

$$f_n(x) = f(x),\ \sigma_n(x) = \sigma(x); \quad \text{if } \rho(x, I(\sigma)) \ge \frac{1}{n},$$

so $f_n^2\sigma_n^{-2}$ is locally integrable at every point in $\mathbb{R}$. From the implication (iii) $\Rightarrow$ (v) in Proposition 3.6.27 and Problem 3.6.29, it follows that

$$\int_0^{\tau_n} f^2(X_v^{(1)})\,dv = \int_0^{R_n} f^2(X_0^{(1)} + B_u)\sigma^{-2}(X_0^{(1)} + B_u)\,du$$

$$= \int_0^{R_n} f_n^2(X_0^{(1)} + B_u)\sigma_n^{-2}(X_0^{(1)} + B_u)\,du < \infty, \quad \text{a.s.}$$

From (5.41) we see now that $\Lambda_{\tau_n}^{\Delta}(0) = 0;\ n \ge 1$, a.s. Since $\tau_n \uparrow \tau^{(1)}$ as $n \to \infty$, we conclude from the continuity of $\Lambda_{\cdot}^{\Delta}(0)$ that $\Lambda_{\tau^{(1)}}^{\Delta}(0) = 0$, a.s. P. Remark 3.7.8 shows that $E|X_{t\wedge\tau^{(1)}}^{(2)} - X_{t\wedge\tau^{(1)}}^{(1)}| = 0;\ 0 \le t < \infty$, and (5.40) follows. □

We recall from Corollary 3.23 that the existence of a weak solution and pathwise uniqueness imply strong existence. This leads to the following result.

5.10 Corollary. *Under condition* (E) *of Theorem* 5.4 *and conditions* (i)–(iii) *of Theorem* 5.9, *the equation* (5.19) *possesses a unique strong solution for every initial distribution* μ.

5.11 Remark. If the function f is locally bounded, condition (i) of Theorem 5.9 follows directly from the definition of $I(\sigma)$. We may take $h(y) = y^\alpha;\ \alpha \ge (1/2)$, and (iii) becomes the condition that σ be locally Hölder-continuous with exponent at least (1/2); see Examples 2.14 and 2.15.

B. The Method of Removal of Drift

It is time to return to the stochastic differential equation (5.1), in which a drift term appears. We transform this equation so as to remove the drift and thereby reduce it to the case already studied. This reduction requires assumptions of *nondegeneracy* and *local integrability*:

(ND) $$\sigma^2(x) > 0; \quad \forall\, x \in \mathbb{R},$$

(LI) $$\forall\, x \in \mathbb{R},\ \exists\, \varepsilon > 0 \text{ such that } \int_{x-\varepsilon}^{x+\varepsilon} \frac{|b(y)|\,dy}{\sigma^2(y)} < \infty.$$

Under these assumptions, we fix a number $c \in \mathbb{R}$ and define the *scale function*

(5.42) $$p(x) \triangleq \int_c^x \exp\left\{-2\int_c^\xi \frac{b(\zeta)\,d\zeta}{\sigma^2(\zeta)}\right\} d\xi; \quad x \in \mathbb{R}.$$

The function p has a continuous, strictly positive derivative, and p'' exists almost everywhere and satisfies

(5.43) $$p''(x) = -\frac{2b(x)}{\sigma^2(x)}p'(x).$$

Henceforth, whenever we write p'', we shall mean the (locally integrable) function defined by (5.43) *on the entire of* $\mathbb{R}$; this definition is possible because of (ND).

The function p maps $\mathbb{R}$ onto $(p(-\infty), p(\infty))$ and has a continuously differentiable inverse $q\colon (p(-\infty), p(\infty)) \to \mathbb{R}$. This latter function has derivative $q'(y) = (1/p'(q(y)))$, which is actually absolutely continuous, with

(5.44) $$\begin{aligned} q''(y) &= -\frac{p''(q(y))}{p'(q(y))}(q'(y))^2 \\ &= \frac{2b(q(y))}{\sigma^2(q(y))}\frac{1}{(p'(q(y)))^2}; \quad \text{a.e. in } (p(-\infty), p(\infty)). \end{aligned}$$

We extend p to $[-\infty, \infty]$ and q to $[p(-\infty), p(\infty)]$ so that the resulting functions are continuous in the topology on the extended real number system.

5.12 Problem. Although not explicitly indicated by the notation, $p(x)$ defined by (5.42) depends on the number $c \in \mathbb{R}$. Let us for the moment display this dependence by writing $p_c(x)$. Show that

(5.45) $$p_a(x) = p_a(c) + p_a'(c)p_c(x).$$

In particular, the finiteness (or nonfiniteness) of $p_c(\pm\infty)$ does not depend on the choice of c.

5.13 Proposition. *Assume* (ND) *and* (LI). *A process* $X = \{X_t, \mathscr{F}_t; 0 \le t < \infty\}$ *is a weak (or strong) solution of equation* (5.1) *if and only if the process*

$Y = \{Y_t \triangleq p(X_t), \mathscr{F}_t; 0 \le t < \infty\}$ *is a weak (or strong) solution of*

$$Y_t = Y_0 + \int_0^t \tilde{\sigma}(Y_s)\,dW_s; \quad 0 \le t < \infty, \tag{5.46}$$

where

$$p(-\infty) < Y_0 < p(\infty) \quad a.s., \tag{5.47}$$

$$\tilde{\sigma}(y) = \begin{cases} p'(q(y))\sigma(q(y)); & p(-\infty) < y < p(\infty), \\ 0; & otherwise. \end{cases} \tag{5.48}$$

The process X may explode in finite time, but the process Y does not.

PROOF. Let X satisfy (5.1) up to the explosion time S, define $Y_t = p(X_t)$, and recall S_n from (5.2); we obtain from the generalized Itô rule (Problem 3.7.3) and (5.43):

$$Y_{t\wedge S_n} = p(X_0) + \int_0^{t\wedge S_n} \tilde{\sigma}(Y_s)\,dW_s; \quad 0 \le t < \infty, n \ge 1. \tag{5.49}$$

Because X is a continuous, extended-real-valued process defined for all $0 \le t < \infty$ (Definition 5.1) and $p\colon [-\infty, \infty] \to [p(-\infty), p(\infty)]$ is continuous, Y is also continuous for all $0 \le t < \infty$. Hence, as $n \to \infty$, the left-hand side of (5.49) converges to $Y_{t\wedge S}$, and the right-hand side must also have a limit. In light of Problem 3.4.11, this means that $\int_0^{t\wedge S} \tilde{\sigma}^2(Y_s)\,ds < \infty$; $0 \le t < \infty$, a.s. Because X_s, and hence Y_s, is constant for $S \le s < \infty$, we must in fact have $\int_0^t \tilde{\sigma}^2(Y_s)\,ds < \infty$; $0 \le t < \infty$, a.s. It follows that $\int_0^t \tilde{\sigma}(Y_s)\,dW_s$ is defined for all $0 \le t < \infty$ and is a continuous process. From (5.49) we see that (5.46) holds for $0 \le t < S$. This equality must also hold for $0 \le t \le S$ on the set $\{S < \infty\}$, because of continuity. The integrand $\tilde{\sigma}(Y_s)$ vanishes for $S \le s < \infty$, and so (5.46) in fact holds for all $0 \le t < \infty$. The solution of this equation cannot explode in finite time; see Problem 5.3.

For the converse, let us begin with a process Y satisfying (5.46), (5.47) which takes values in $[p(-\infty), p(\infty)]$. We can always arrange for Y to be constant after reaching one of the endpoints of this interval, because the integrand in (5.46) vanishes when $Y_s \notin (p(-\infty), p(\infty))$. Define $X_t \triangleq q(Y_t)$; $0 \le t < \infty$, and let S_n be given by (5.2). Then the generalized Itô rule of Problem 3.7.3 gives, in conjunction with the properties of the function q (in particular, (5.44)) and (5.48),

$$\begin{aligned} X_{t\wedge S_n} &= X_0 + \int_0^{t\wedge S_n} q'(Y_s)\,dY_s + \frac{1}{2}\int_0^{t\wedge S_n} q''(Y_s)\,d\langle Y\rangle_s \\ &= X_0 + \int_0^{t\wedge S_n} \frac{1}{p'(X_s)}\tilde{\sigma}(Y_s)\,dW_s + \int_0^{t\wedge S_n} \frac{b(X_s)}{\sigma^2(X_s)}\frac{1}{(p'(X_s))^2}\tilde{\sigma}^2(Y_s)\,ds \\ &= X_0 + \int_0^{t\wedge S_n} b(X_s)\,ds + \int_0^{t\wedge S_n} \sigma(X_s)\,dW_s; \quad 0 \le t < \infty \end{aligned}$$

almost surely, for every $n \ge 1$. □

5.14 Exercise. Show by example that if (LI) holds but (ND) fails, then (5.46) can have a solution Y which is *not* of the form $p(X)$, for some solution X of (5.1).

5.15 Theorem. *Assume that* σ^{-2} *is locally integrable at every point in* $\mathbb{R}$, *and conditions* (ND) *and* (LI) *hold. Then for every initial distribution* μ, *the equation* (5.1) *has a weak solution up to an explosion time, and this solution is unique in the sense of probability law.*

PROOF. Let $\tilde{\sigma}$ be defined by (5.48). According to Theorem 5.7 and Proposition 5.13, it suffices to prove that $I(\tilde{\sigma}) = Z(\tilde{\sigma})$. Now $Z(\tilde{\sigma}) = (p(-\infty), p(+\infty))^c$, and $I(\tilde{\sigma})$ contains this set. We must show that $\tilde{\sigma}^{-2}$ is locally integrable at every point $y_0 \in (p(-\infty), p(\infty))$. At such a point, choose $\varepsilon > 0$ so that

$$p(-\infty) < y_0 - \varepsilon < y_0 + \varepsilon < p(\infty),$$

and write

$$\int_{y_0-\varepsilon}^{y_0+\varepsilon} \frac{dy}{\tilde{\sigma}^2(y)} = \int_{q(y_0-\varepsilon)}^{q(y_0+\varepsilon)} \frac{dx}{p'(x)\sigma^2(x)}.$$

The second integral is finite, because p' is bounded away from zero on finite intervals and σ^{-2} is locally integrable. □

5.16 Corollary. *Assume that* σ^{-2} *is locally integrable at every point in* $\mathbb{R}$, *and that conditions* (ND), (LI), *and*

$$\text{(5.50)} \qquad \begin{aligned} |b(x) - b(y)| &\le K|x - y|; \quad (x, y) \in \mathbb{R}^2 \\ |\sigma(x) - \sigma(y)| &\le h(|x - y|); \quad (x, y) \in \mathbb{R}^2 \end{aligned}$$

hold, where K *is a positive constant and* $h: [0, \infty) \to [0, \infty)$ *is a strictly increasing function for which* $h(0) = 0$ *and* (2.25) *hold. Then, for every initial condition* ξ *independent of the driving Brownian motion* $W = \{W_t, \mathscr{F}_t; 0 \le t < \infty\}$, *the equation* (5.1) *has a unique strong solution* (*possibly up to an explosion time*).

PROOF. Weak existence (Theorem 5.15) and pathwise uniqueness (Proposition 2.13 and Remark 3.3) imply strong existence (Corollary 3.23). These results can easily be localized to deal with the case of possible explosion of the solution. □

5.17 Proposition. *Assume that* $b: \mathbb{R} \to \mathbb{R}$ *is bounded and* $\sigma: \mathbb{R} \to \mathbb{R}$ *is Lipschitz-continuous with* σ^2 *bounded away from zero on every compact subset of* $\mathbb{R}$. *Then, for every initial condition* ξ *independent of the driving Brownian motion* $W = \{W_t, \mathscr{F}_t; 0 \le t < \infty\}$, *equation* (5.1) *has a nonexploding, unique strong solution.*

PROOF. We show first that *the boundedness of* b *prevents the explosion of any solution* X to (5.1). We fix $t \in (0, \infty)$ and let $n \to \infty$ in (5.4); the Lebesgue integral

$\int_0^t b(X_s)1_{\{s\le S_n\}}\,ds$ converges to $\int_0^{t\wedge S} b(X_s)\,ds$, a finite expression because b is bounded. On the other hand, the stochastic integral $\int_0^t \sigma(X_s)1_{\{s\le S_n\}}\,dW_s$ converges to $\int_0^{t\wedge S}\sigma(X_s)\,dW_s$ on the event $A \triangleq \{\int_0^{t\wedge S}\sigma^2(X_s)\,ds < \infty\}$, and does not have a limit on A^c; cf. Problem 3.4.11. It develops that the limit of the right-hand side of (5.4) exists and is finite a.s. on A, and does not exist on A^c. On the left-hand side of (5.4) we have $\lim_{n\to\infty} X_{t\wedge S_n} = X_{t\wedge S}$, which is defined a.s. and is equal to $\pm\infty$ on $\{S < t\}$. It follows that $P[S < t] = 0$ holds for every $t > 0$, so $P[S = \infty] = 1$.

We turn now to the questions of *existence of a weak solution*, and of *pathwise uniqueness*, for equation (5.1). The assumptions on b and σ imply (ND), (LI), and the local integrability of σ^{-2}, so weak existence follows from Theorem 5.15. According to Proposition 5.13, the pathwise uniqueness of (5.1) is equivalent to that of (5.46). Because $p'' \triangleq -(2b/\sigma^2)p'$ is locally bounded, p' is locally Lipschitz. It follows that $\tilde{\sigma}(y)$ defined by (5.48) is locally Lipschitz at every point $y \in (p(-\infty), p(\infty))$. We have shown that any solution X to (5.1) does not explode, so any solution Y to (5.46), (5.47) must remain in the interval $(p(-\infty), p(\infty))$. Under these conditions, the proof given for Theorem 2.5 shows that pathwise uniqueness holds for (5.46). We appeal to Corollary 3.23 in order to conclude the argument. □

Proposition 5.13 raises the interesting issue of determining necessary and sufficient conditions for explosion of the solution X to (5.1). Since Y given by (5.46) does not explode, and $Y_t = p(X_t)$, it is clear that the condition $p(\pm\infty) = \pm\infty$ guarantees that X is also nonexploding; however, this sufficient condition is unfortunately not necessary (see Remark 5.18). We develop the necessary and sufficient condition known as *Feller's test for explosions* in Theorem 5.29.

5.18 Remark. Consider the case of $b(x) = \mathrm{sgn}(x)$, $\sigma(x) \equiv \sigma > 0$. The scale function p of (5.42) is bounded, and according to Proposition 5.17 the equation (5.1) has a nonexploding, unique strong solution for any initial distribution.

5.19 Remark. The linear growth condition

$$|b(x)| + |\sigma(x)| \le K(1 + |x|); \quad \forall x \in \mathbb{R}$$

is sufficient for $P[S = \infty] = 1$; cf. Problem 3.15.

C. Feller's Test for Explosions

We begin here a systematic discussion of explosions. Rather than working exclusively with processes taking values on the entire real line, we start with an interval

$$I = (\ell, r); \quad -\infty \le \ell < r \le \infty$$

and assume that the coefficients $\sigma: I \to \mathbb{R}$, $b: I \to \mathbb{R}$ satisfy

(ND)′ $$\sigma^2(x) > 0; \quad \forall x \in I,$$

(LI)′ $$\forall x \in I, \exists \varepsilon > 0 \quad \text{such that} \quad \int_{x-\varepsilon}^{x+\varepsilon} \frac{1 + |b(y)|}{\sigma^2(y)} dy < \infty.$$

We define the *scale function* p by (5.42), where now the number c must be in I. We also introduce the *speed measure*

(5.51) $$m(dx) \triangleq \frac{2\,dx}{p'(x)\sigma^2(x)}; \quad x \in I$$

and the *Green's function*

(5.52) $$G_{a,b}(x,y) \triangleq \frac{(p(x \wedge y) - p(a))(p(b) - p(x \vee y))}{p(b) - p(a)}; \quad x, y \in [a,b] \subseteq I.$$

In terms of these two objects, a solution to the equation

(5.53) $$b(x)M'(x) + \tfrac{1}{2}\sigma^2(x)M''(x) = -1; \quad a < x < b,$$

(5.54) $$M(a) = M(b) = 0$$

is given by

(5.55) $$M_{a,b}(x) \triangleq \int_a^b G_{a,b}(x,y)m(dy)$$
$$= -\int_a^x (p(x) - p(y))m(dy) + \frac{p(x) - p(a)}{p(b) - p(a)} \int_a^b (p(b) - p(y))m(dy).$$

As was the case with p'', the second derivative $M''_{a,b}$ exists except possibly on a set of Lebesgue measure zero; we define $M''_{a,b}$ at every point in (a,b) by using (5.53). Note that $G_{a,b}(\cdot,\cdot)$, and hence $M_{a,b}(\cdot)$, is nonnegative.

5.20 Definition. A *weak solution in the interval* $I = (\ell, r)$ of equation (5.1) is a triple (X, W), $(\Omega, \mathscr{F}, P)$, $\{\mathscr{F}_t\}$, where

(i)′ condition (i) of Definition 5.1 holds,

(ii)′ $X = \{X_t, \mathscr{F}_t; 0 \le t < \infty\}$ is a continuous, adapted, $[\ell, r]$-valued process with $X_0 \in I$ a.s., and $\{W_t, \mathscr{F}_t; 0 \le t < \infty\}$ is a standard, one-dimensional Brownian motion,

(iii)′ with $\{\ell_n\}_{n=1}^\infty$ and $\{r_n\}_{n=1}^\infty$ strictly monotone sequences satisfying $\ell < \ell_n < r_n < r$, $\lim_{n\to\infty} \ell_n = \ell$, $\lim_{n\to\infty} r_n = r$, and

(5.56) $$S_n \triangleq \inf\{t \ge 0: X_t \notin (\ell_n, r_n)\}; \quad n \ge 1,$$

the equations (5.3) and (5.4) hold.

We refer to

(5.57) $$S = \inf\{t \ge 0: X_t \notin (\ell, r)\} = \lim_{n\to\infty} S_n$$

as the *exit time* from I. The assumption $X_0 \in I$ a.s. guarantees that $P[S > 0] = 1$. If $\ell = -\infty$ and $r = +\infty$, Definition 5.20 reduces to Definition 5.1, once we stipulate that $X_t = X_S$; $S \le t < \infty$.

Let (X, W) be a weak solution in I of equation (5.1) with $X_0 = x \in (a, b) \subseteq I$, and set

$$\tau_n = \inf\left\{t \ge 0: \int_0^t \sigma^2(X_s)\,ds \ge n\right\}; \quad n = 1, 2, \ldots,$$

$$T_{a,b} = \inf\{t \ge 0; X_t \notin (a, b)\}; \quad \ell < a < b < r.$$

We may apply the generalized Itô rule (Problem 3.7.3) to $M_{a,b}(X_t)$ and obtain

$$M_{a,b}(X_{t \wedge \tau_n \wedge T_{a,b}}) = M_{a,b}(x) - (t \wedge \tau_n \wedge T_{a,b}) + \int_0^{t \wedge \tau_n \wedge T_{a,b}} M'_{a,b}(X_s)\sigma(X_s)\,dW_s.$$

Taking expectations and then letting $n \to \infty$, we see that

$$E(t \wedge T_{a,b}) = M_{a,b}(x) - EM_{a,b}(X_{t \wedge T_{a,b}}) \le M_{a,b}(x) < \infty, \tag{5.58}$$

and then letting $t \to \infty$ we obtain $ET_{a,b} \le M_{a,b}(x) < \infty$.

In other words, *X exits from every compact subinterval of (ℓ, r) in finite expected time.* Armed with this observation, we may return to (5.58), observe from (5.54) that $\lim_{t\to\infty} EM_{a,b}(X_{t \wedge T_{a,b}}) = 0$, and conclude

$$ET_{a,b} = M_{a,b}(x); \quad a < x < b. \tag{5.59}$$

On the other hand, the generalized Itô rule applied in the same way to $p(X_t)$ gives $p(x) = Ep(X_{t \wedge T_{a,b}})$, whence

$$p(x) = Ep(X_{T_{a,b}}) = p(a)P[X_{T_{a,b}} = a] + p(b)P[X_{T_{a,b}} = b], \tag{5.60}$$

upon letting $t \to \infty$. The two probabilities in (5.60) add up to one, and thus

$$P[X_{T_{a,b}} = a] = \frac{p(b) - p(x)}{p(b) - p(a)}, \quad P[X_{T_{a,b}} = b] = \frac{p(x) - p(a)}{p(b) - p(a)}. \tag{5.61}$$

These expressions will help us obtain information about the behavior of X near the endpoints of the interval (ℓ, r) from the corresponding behavior of the scale function. Problem 5.12 shows that the expressions on the right-hand sides of the relations in (5.61) do not depend on the choice of c in the definition of p.

5.21 Remark. For Brownian motion W on $I = (-\infty, \infty)$, we have (with $c = 0$ in (5.42)) $p(x) = x$, $m(dx) = 2\,dx$. For a process Y satisfying

$$Y_t = \tilde{x} + \int_0^t \tilde{\sigma}(Y_s)\,dW_s,$$

we again have $p(\tilde{x}) = \tilde{x}$, but $m(dx) = (2\,d\tilde{x}/\tilde{\sigma}^2(\tilde{x}))$. Now Y is a Brownian motion run "according to a different clock" (Theorem 3.4.6), and the speed measure simply records how this change of clock affects the expected value of exit times:

$$ET_{\tilde{a},\tilde{b}}(\tilde{x}) = \int_{\tilde{a}}^{\tilde{b}} \frac{((\tilde{x} \wedge \tilde{y}) - \tilde{a})(\tilde{b} - (\tilde{x} \vee \tilde{y}))}{\tilde{b} - \tilde{a}} \cdot \frac{2\,d\tilde{y}}{\tilde{\sigma}^2(\tilde{y})} \tag{5.62}$$

Once drift is introduced, the formulas become a bit more complicated, but the idea remains the same. Indeed, suppose that we begin with X satisfying (5.1), compute the scale function p and speed measure m for X by (5.42) and (5.51), and adopt the notation $\tilde{x} = p(x)$, $\tilde{y} = p(y)$, $\tilde{a} = p(a)$, $\tilde{b} = p(b)$; then (5.55), (5.59) show that $ET_{a,b}(x)$ is still given by the right-hand side of (5.62), where now $\tilde{\sigma}$ is the dispersion coefficient (5.48) of the process $Y_t \triangleq p(X_t)$. We say $Y_t = p(X_t)$ is in the *natural scale* because it satisfies a stochastic differential equation without drift and thus has the identity as its scale function.

5.22 Proposition. *Assume that* (ND)′, (LI)′ *hold, and let X be a weak solution of* (5.1) *in I, with nonrandom initial condition $X_0 = x \in I$. Let p be given by* (5.42) *and S by* (5.57). *We distinguish four cases:*

(a) $p(\ell+) = -\infty$, $p(r-) = \infty$. *Then*

$$P[S = \infty] = P\left[\sup_{0 \le t < \infty} X_t = r\right] = P\left[\inf_{0 \le t < \infty} X_t = \ell\right] = 1.$$

In particular, the process X is recurrent: for every $y \in I$, we have

$$P[X_t = y;\ \text{some}\ 0 \le t < \infty] = 1.$$

(b) $p(\ell+) > -\infty$, $p(r-) = \infty$. *Then*

$$P\left[\lim_{t \uparrow S} X_t = \ell\right] = P\left[\sup_{0 \le t < S} X_t < r\right] = 1.$$

(c) $p(\ell+) = -\infty$, $p(r-) < \infty$. *Then*

$$P\left[\lim_{t \uparrow S} X_t = r\right] = P\left[\inf_{0 \le t < S} X_t > \ell\right] = 1.$$

(d) $p(\ell+) > -\infty$, $p(r-) < \infty$. *Then*

$$P\left[\lim_{t \uparrow S} X_t = \ell\right] = 1 - P\left[\lim_{t \uparrow S} X_t = r\right] = \frac{p(r-) - p(x)}{p(r-) - p(\ell+)}.$$

5.23 Remark. In cases (b), (c), and (d), we make no claim concerning the finiteness of S. Even in case (d), we may have $P[S = \infty] = 1$, as demonstrated in Remark 5.18.

Proof of Proposition 5.22. For case (a), we have from (5.61) for $\ell < a < x < b < r$:

$$P\left[\inf_{0 \le t < S} X_t \le a\right] \ge P[X_{T_{a,b}} = a] = \frac{1 - (p(x)/p(b))}{1 - (p(a)/p(b))}. \tag{5.63}$$

Letting $b \uparrow r$, we obtain $P[\inf_{0\le t<S} X_t \le a] = 1$ for every $a \in I$. Now we let $a \downarrow \ell$ to get $P[\inf_{0\le t<S} X_t = \ell] = 1$. A dual argument shows that $P[\sup_{0\le t<S} X_t = r] = 1$. Suppose now that $P[S<\infty] > 0$; then the event $\{\lim_{t\uparrow S} X_t$ exists and is equal to ℓ or $r\}$ has positive probability, and so $\{\sup_{0\le t<S} X_t = r\}$ and $\{\inf_{0\le t<S} X_t = \ell\}$ cannot both have probability one. This contradiction shows that $P[S<\infty] = 0$.

For case (b), we first observe that (5.63) still implies $P[\inf_{0\le t<S} X_t = \ell] = 1$. If, however, we recall $P[X_{T_{a,b}} = b] = (p(x) - p(a))/(p(b) - p(a))$ from (5.61) and let $a \downarrow \ell$, we see that

$$P[X_t = b; \text{ some } 0 \le t < S] = \frac{p(x) - p(\ell+)}{p(b) - p(\ell+)}.$$

Letting now $b \uparrow r$, we conclude that $P[\sup_{0\le t<S} X_t = r] = 0$. We have thus shown

$$P\left[\inf_{0\le t<S} X_r = \ell\right] = P\left[\sup_{0\le t<S} X_t < r\right] = 1.$$

It remains only to show that $\lim_{t\uparrow S} X_t = \inf_{0\le t<S} X_t$, and for this it suffices to establish that the limit exists, almost surely. With S_n as in (5.56), the process $Y_t^{(n)} \triangleq p(X_{t\wedge S_n}) - p(\ell+)$; $0 \le t < \infty$ is for each $n \ge 1$ a nonnegative local martingale (see (5.49)); letting $n\to\infty$ and using Fatou's lemma, we see that $Y_t \triangleq p(X_{t\wedge S}) - p(\ell+)$; $0\le t<\infty$ is a nonnegative supermartingale. As such, it converges almost surely as $t\to\infty$ (Problem 1.3.16). Because $p\colon [\ell, r) \to \mathbb{R}$ has a continuous inverse, $\lim_{t\to\infty} X_{t\wedge S}$ must exist.

Case (c) is dual to (b), and case (d) is obtained easily, by taking limits in (5.61). □

5.24 Example. For the *Brownian motion* $X_t = \mu t + \sigma W_t$ *on* $I = (-\infty, \infty)$ *with drift* $\mu > 0$ *and variance* $\sigma^2 > 0$, we have (setting $c = 0$ in (5.42)) that $p(x) = (1 - e^{-\beta x})/\beta$ and $m(dx) = (2e^{\beta x}/\sigma^2)\,dx$, where $\beta = 2\mu/\sigma^2$. We are in case (c). Compare this result with Exercise 3.5.9.

5.25 Example. For the *Bessel process* with dimension $d \ge 2$ (Proposition 3.3.21), we have $I = (0,\infty)$, $b(x) = (d-1)/2x$, and $\sigma^2(x) \equiv 1$. With $c = 1$, we obtain

(i) for $d = 2$: $p(x) = \log x$, $m(dx) = 2x\,dx$ (case (a)),
(ii) for $d \ge 3$: $p(x) = (1 - x^{2-d})/(d-2)$, $m(dx) = 2x^{d-1}\,dx$ (case (c)).

Compare these results with Problem 3.3.23.

Proposition 5.17, Remark 5.19, and part (a) of Proposition 5.22 provide sufficient conditions for nonexplosion of the process X in (5.1), i.e., for $P[S=\infty] = 1$. In our search for conditions which are both necessary and sufficient, we shall need the following result about an ordinary differential equation.

We define, by recursion, the sequence $\{u_n\}_{n=0}^{\infty}$ of real-valued functions on I, by setting $u_0 \equiv 1$ and

$$u_n(x) = \int_c^x p'(y) \int_c^y u_{n-1}(z)m(dz)\,dy; \quad x \in I, n \geq 1 \tag{5.64}$$

where, as before, c is a fixed number in I. In particular we set for $x \in I$:

$$v(x) \triangleq u_1(x) = \int_c^x p'(y) \int_c^y \frac{2\,dz}{p'(z)\sigma^2(z)}\,dy = \int_c^x (p(x) - p(y))m(dy). \tag{5.65}$$

5.26 Lemma. *Assume that* (ND)′ *and* (LI)′ *hold. The series*

$$u(x) = \sum_{n=0}^{\infty} u_n(x); \quad x \in I \tag{5.66}$$

converges uniformly on compact subsets of I and defines a differentiable function with absolutely continuous derivative on I. Furthermore, u is strictly increasing (decreasing) in the interval (c, r) (respectively, (ℓ, c)) and satisfies

$$\tfrac{1}{2}\sigma^2(x)u''(x) + b(x)u'(x) = u(x); \quad \text{a.e. } x \in I, \tag{5.67}$$

$$u(c) = 1, \quad u'(c) = 0, \tag{5.68}$$

as well as

$$1 + v(x) \leq u(x) \leq e^{v(x)}; \quad x \in I. \tag{5.69}$$

PROOF. It is verified easily that the functions $\{u_n\}_{n=1}^{\infty}$ in (5.64) are nonnegative, are strictly increasing (decreasing) on (c, r) (respectively, (ℓ, c)), and satisfy

$$\tfrac{1}{2}\sigma^2(x)u_n''(x) + b(x)u_n'(x) = u_{n-1}(x), \quad \text{a.e. } x \in I. \tag{5.70}$$

We show by induction that

$$u_n(x) \leq \frac{v^n(x)}{n!}; \quad n = 0, 1, 2, \ldots. \tag{5.71}$$

Indeed, (5.71) is valid for $n = 0$; assuming it is true for $n = k - 1$ and noting that

$$u_k'(x) = p'(x) \int_c^x u_{k-1}(z)m(dz); \quad x \in I, \tag{5.72}$$

we obtain for $c \leq x < r$:

$$u_k(x) \leq \int_c^x p'(y) \int_c^y \frac{v^{k-1}(z)}{(k-1)!} m(dz)\,dy \leq \frac{1}{(k-1)!} \int_c^x p'(y)v^{k-1}(y)m((c, y])\,dy$$
$$= \frac{1}{(k-1)!} \int_c^x v^{k-1}(y)\,dv(y) = \frac{v^k(x)}{k!}.$$

A similar inequality holds for $\ell < x \leq c$. This proves (5.71), and from (5.72) we have also

$$|u_n'(x)| \le |v'(x)| \frac{v^{n-1}(x)}{(n-1)!}; \quad n = 1, 2, \ldots.$$

It follows that the series in (5.66), as well as $\sum_{n=0}^{\infty} u_n'(x)$, converges absolutely on I, uniformly on compact subsets. Solving (5.70) for $u_n''(x)$, we see that $\sum_{n=1}^{\infty} u_n''(x)$ also converges absolutely, at each point $x \in I$, to an integrable function. Term-by-term integration of this sum shows that $\sum_{n=1}^{\infty} u_n''(x)$ is almost everywhere the second derivative of u in (5.66), and that $u'(x) = \sum_{n=0}^{\infty} u_n'(x)$ holds for every $x \in I$. The other claims follow readily. □

5.27 Problem. Prove the implications

(5.73) $$p(r-) = \infty \Rightarrow v(r-) = \infty,$$

(5.74) $$p(\ell+) = -\infty \Rightarrow v(\ell+) = \infty.$$

5.28 Problem. In the spirit of Problem 5.12, we could display the dependence of $v(x)$ on c by writing

(5.75) $$v_c(x) \triangleq \int_c^x p_c'(y) \int_c^y \frac{2\,dz}{p_c'(z)\sigma^2(z)}\,dy.$$

Show that for $a, c \in I$:

(5.76) $$v_a(x) = v_a(c) + v_a'(c) p_c(x) + v_c(x); \quad x \in I.$$

In particular, the finiteness or nonfiniteness of $v_c(r-)$, $v_c(\ell+)$ does not depend on c.

5.29 Theorem (Feller's (1952) Test for Explosions). *Assume that* (ND)′ *and* (LI)′ *hold, and let* (X, W), $(\Omega, \mathscr{F}, P)$, $\{\mathscr{F}_t\}$ *be a weak solution in* $I = (\ell, r)$ *of* (5.1) *with nonrandom initial condition* $X_0 = x \in I$. *Then* $P[S = \infty] = 1$ *or* $P[S = \infty] < 1$, *according to whether*

(5.77) $$v(\ell+) = v(r-) = \infty$$

or not.

Proof. Set $\tau_n \triangleq \inf\{t \ge 0: \int_0^t \sigma^2(X_s)\,ds \ge n\}$ and $Z_t^{(n)} \triangleq u(X_{t \wedge S_n \wedge \tau_n})$. According to the generalized Itô rule (Problem 3.7.3) and relation (5.67), $Z^{(n)}$ has the representation

$$Z_t^{(n)} = Z_0^{(n)} + \int_0^{t \wedge S_n \wedge \tau_n} u(X_s)\,ds + \int_0^{t \wedge S_n \wedge \tau_n} u'(X_s)\sigma(X_s)\,dW_s.$$

Consequently, $M_t^{(n)} \triangleq e^{-t \wedge S_n \wedge \tau_n} Z_t^{(n)}$ has the representation

$$M_t^{(n)} = M_0^{(n)} + \int_0^{t \wedge S_n \wedge \tau_n} e^{-s} u'(X_s)\sigma(X_s)\,dW_s$$

as a nonnegative local martingale. Fatou's lemma shows that any nonnegative local martingale is a supermartingale, and that this property is also enjoyed by the process $M_t \triangleq \lim_{n\to\infty} M_t^{(n)} = e^{-t\wedge S}u(X_{t\wedge S})$; $0 \le t < \infty$. Therefore,

$$(5.78) \qquad M_\infty = \lim_{t\to\infty} e^{-t\wedge S}u(X_{t\wedge S})$$

exists and is finite, almost surely (Problem 1.3.16).

Let us now suppose that (5.77) holds. From (5.69) we see that $u(\ell+) = u(r-) = \infty$, and (5.78) shows that $M_\infty = \infty$ a.s. on the event $\{S < \infty\}$. It follows that $P[S < \infty] = 0$.

For the converse, assume that (5.77) fails; for instance, suppose that $v(r-) < \infty$. Then (5.69) yields $u(r-) < \infty$. In light of Problem 5.28, we may assume without loss of generality that $c < x < r$, and set $T_c = \inf\{t \ge 0; X_t = c\}$. The continuous process

$$M_{t\wedge T_c} = e^{-(t\wedge S\wedge T_c)}u(X_{t\wedge S\wedge T_c}); \quad 0 \le t < \infty$$

is a bounded local martingale, hence a bounded martingale, which therefore converges almost surely (and in L^1) as $t \to \infty$; cf. Problem 1.3.20. It develops that

$$u(x) = Ee^{-S\wedge T_c}u(X_{S\wedge T_c}) = u(r-)Ee^{-S}1_{\{S<T_c\}} + u(c)Ee^{-T_c}1_{\{T_c<S\}}.$$

If $P[S = \infty] = 1$, the preceding identity gives $u(x) = u(c)Ee^{-T_c} \le u(c)$, contradicting the fact that u is strictly increasing on $[c, x]$. It follows that $P[S = \infty] < 1$. □

5.30 Example. For Brownian motion W on $I = (-\infty, \infty)$, we have already computed $p(x) = x$, $m(dx) = 2\,dx$ (with $c = 0$). Consequently, $v(x) = x^2$ and $v(\pm\infty) = \infty$.

5.24 Example (continued). For Brownian motion with drift $\mu > 0$ and variance $\sigma^2 > 0$, it turns out that $v(x) = 2(\beta x - 1 + e^{-\beta x})/\beta^2\sigma^2$, with $\beta = (2\mu/\sigma^2)$ and $c = 0$. Again, $v(\pm\infty) = \infty$.

5.25 Example (continued). For the Bessel process with dimension $d \ge 2$ and $c = 1$, we have $v(x) = [x^2 - 1 - 2p(x)]/d$; $0 < x < \infty$, and $v(0+) = v(\infty) = \infty$.

5.31 Exercise. Consider the *geometric Brownian motion* X, which satisfies the stochastic integral equation

$$(5.79) \qquad X_t = x + \mu\int_0^t X_s\,ds + \nu\int_0^t X_s\,dW_s,$$

where $x > 0$. Use Theorem 5.29 and Proposition 5.22 to show that $X_t \in (0, \infty)$ for all t, and

(i) if $\mu < \nu^2/2$, then $\lim_{t\to\infty} X_t = 0$, $\sup_{0\le t<\infty} X_t < \infty$, a.s.;
(ii) if $\mu > \nu^2/2$, then $\inf_{0\le t<\infty} X_t > 0$, $\lim_{t\to\infty} X_t = \infty$, a.s.;
(iii) if $\mu = \nu^2/2$, then $\inf_{0\le t<\infty} X_t = 0$, $\sup_{0\le t<\infty} X_t = \infty$, a.s.

(The solution to (5.79) is given by

$$X_t = x\exp\{(\mu - \tfrac{1}{2}v^2)t + vW_t\},$$

and so (i)–(iii) can also be deduced from the properties of Brownian motion with drift (see, e.g., Problem 2.9.3 and Proposition 2.9.23).)

Let us suppose now that (5.77) is violated, so that $P[S < \infty]$ is positive. Under what additional conditions can we guarantee that this probability is actually equal to one, i.e., that explosion occurs almost surely?

5.32 Proposition. *Assume that* (ND)′ *and* (LI)′ *hold. We have* $P[S < \infty] = 1$ *if and only if one of the following conditions holds*:

(i) $v(r-) < \infty$ *and* $v(\ell+) < \infty$,
(ii) $v(r-) < \infty$ *and* $p(\ell+) = -\infty$, *or*
(iii) $v(\ell+) < \infty$ *and* $p(r-) = \infty$.

In the first case, we actually have $ES < \infty$.

5.33 Remark. If (i) prevails, then we also have finiteness of $p(r-)$ and $p(\ell+)$ (Problem 5.27), and we can define by analogy with (5.52) the Green's function

$$G(x, y) = \frac{[p(x \wedge y) - p(\ell+)][p(r-) - p(x \vee y)]}{p(r-) - p(\ell+)}; \quad (x, y) \in I^2$$

for the entire interval $I = (\ell, r)$. We also define the counterpart of (5.55):

$$M(x) = \int_\ell^r G(x, y)m(dy); \quad x \in (\ell, r).$$

This function satisfies $M(\ell+) = M(r-) = 0$, has an absolutely continuous derivative on I, and satisfies the equation (5.53) there. The same procedure that led to (5.59) now gives $ES = M(x) < \infty$, justifying the last claim in Proposition 5.32.

Proof of Sufficiency in Proposition 5.32. We just dealt with (i). Suppose that (ii) holds; then with $\{\ell_n\}_{n=1}^\infty$ and $\{r_n\}_{n=1}^\infty$ as in Definition 5.20 (iii)′, we introduce the stopping times

$$R_n \triangleq \inf\{t \geq 0; X_t = \ell_n\}; \quad n \geq 1,$$

$$T_r \triangleq \inf\{t \geq 0; X_t = r\}, \quad T_\ell \triangleq \inf\{t \geq 0; X_t = \ell\} = \lim_{n\to\infty} R_n.$$

Because $v(r-) < \infty$, $v(\ell_n) < \infty$, we obtain as in Remark 5.33 that $E(R_n \wedge T_r) < \infty$; $n \geq 1$. On the other hand, (5.73) gives $p(r-) < \infty$, so we are in the case (c) of Proposition 5.22. Therefore, for P-a.e. $\omega \in \Omega$, $R_n(\omega) = \infty$ for sufficiently large n (depending on ω), and thus $S(\omega) = T_r(\omega) < \infty$.

Condition (iii) is dual to (ii). □

PROOF OF NECESSITY IN PROPOSITION 5.32. Assume that $P[S < \infty] = 1$; from Theorem 5.29 we conclude that either $v(\ell+) < \infty$ or $v(r-) < \infty$. Let us suppose $v(\ell+) < \infty$, and that none of (i), (ii), (iii) holds. Then necessarily $p(r-) < \infty = v(r-)$ and $p(\ell+) > -\infty$ (remember (5.74)), so we are in case (d) of Proposition 5.22, and thus $A_r \triangleq \{\lim_{t\uparrow S} X_t = r\}$ has positive probability. But now we recall from the proof of Theorem 5.29 that $M_t = e^{-t\wedge S}u(X_{t\wedge S})$ is a nonnegative supermartingale with M_∞ of (5.78) an almost surely finite random variable. According to (5.69), $u(r-) = \infty$, and so $S = \infty$ on A_r. This shows that $P[S < \infty] < 1$, contradicting our initial assumption. It follows that at least one of (i), (ii), (iii) must hold, if $P[S < \infty] = 1$ does. □

D. Supplementary Exercises

5.34 Exercise. Take $\sigma(x) \equiv 1$ and $I = (-\infty, \infty)$.

(a) If $b(x) = \frac{3}{2}x^2$, show that $P[S < \infty] = 1$.
(b) If $b(x) = 2x^3$, show that $ES < \infty$.

5.35 Exercise. Take $I = (0, \infty)$ and $b(x) \equiv k$, $\sigma(x) = \ell\sqrt{x}$, where k and ℓ are positive constants.

(i) Show that we are in case (a), (b), or (c) of Proposition 5.22 according as $\mu \triangleq (2k/\ell^2)$ is equal to, less than, or greater than one, respectively.
(ii) In the first and third cases, the solution is nonexploding; in the second case the origin is reached in finite time with probability one.
(iii) The cases $\ell = 2$, $k = 2, 3, \ldots$ should be familiar; can you relate the solution to a known process?
(iv) Solve the stochastic differential equation explicitly in the case $\ell^2 = 4k$. (*Hint*: Recall Proposition 2.21.)

5.36 Exercise. Show that the solution of the equation

$$dX_t = (1 + X_t)(1 + X_t^2)\,dt + (1 + X_t^2)\,dW_t; \quad X_0 \in \left(-\frac{\pi}{2}, \frac{\pi}{2}\right),$$

explodes (to $\pm\infty$) in finite expected time.

5.37 Exercise. Discuss the possibility of explosion in the cases:

(i) $b(x) = -x$, $\sigma(x) = \sqrt{2(1 + x^2)}$, $I = \mathbb{R}$;
(ii) $b(x) = \mu x$, $\sigma(x) = \delta x$, $I = (0, \infty)$, and $\mu \in \mathbb{R}$, $\delta > 0$;
(iii) $b(x) = 1 + \delta\mu x$, $\sigma(x) = \delta x$, $I = (0, \infty)$, and $\mu \in \mathbb{R}$, $\delta > 0$.

(*Hint for (iii)*: Recall Exercise 3.3.33.)

5.38 Exercise. Let the function $b: \mathbb{R} \to \mathbb{R}$ be locally square-integrable, i.e., $\forall\, x \in \mathbb{R}$, $\exists\, \varepsilon > 0$ such that $\int_{x-\varepsilon}^{x+\varepsilon} b^2(y)\,dy < \infty$.

(i) Show that, corresponding to every initial distribution μ on $(\mathbb{R}, \mathscr{B}(\mathbb{R}))$, the equation

$$dX_t = b(X_t)\,dt + dW_t$$

has a weak solution up to an explosion time S, and this solution is unique in the sense of probability law.

(ii) Prove that for every finite $T > 0$, this solution obeys the *generalized Girsanov formula*:

$$P[X_{.} \in \Gamma, S > T] = E\left[\exp\left(\int_0^T b(W_s)\,dW_s - \frac{1}{2}\int_0^T b^2(W_s)\,ds\right)\cdot 1_{\{W_{.} \in \Gamma\}}\right]$$

for every $\Gamma \in \mathscr{B}_T(C[0,\infty))$, the σ-field of (3.19) with $m = 1$. Here, W is a one-dimensional Brownian motion with $P(W_0 \in B) = \mu(B)$; $B \in \mathscr{B}(\mathbb{R})$. (*Hint*: Try to emulate the proof of Proposition 3.10, and recall the Engelbert-Schmidt zero-one law of Section 3.6.)

(iii) In particular, conclude that with W as in (ii), the nonnegative supermartingale

$$Z_t = \exp\left\{\int_0^t b(W_s)\,dW_s - \frac{1}{2}\int_0^t b^2(W_s)\,ds\right\}; \quad 0 \le t < \infty,$$

is a martingale if and only if $P[S = \infty] = 1$; this is equivalent to $\varphi(\pm\infty) = \infty$, where

$$\varphi(x) = \int_0^x \int_0^y \exp\left\{-2\int_z^y b(u)\,du\right\} dz\,dy.$$

5.39 Exercise. In the setting of Subsection C, show that for every bounded, piecewise continuous function $h: I \to [0,\infty)$ and $x \in (a,b) \subsetneqq I$, we have

$$E^x \int_0^{T_{a,b}} h(X_t)\,dt = \int_a^b G_{a,b}(x,y)h(y)m(dy).$$

Here and in the following exercises we denote by a superscript x on probabilities and/or expectations the initial condition $X_0 = x$. (*Hint*: Proceed by analogy with (5.55), (5.59).)

5.40 Exercise (Pollack & Siegmund (1985)). In the setting of Subsection C with b and σ bounded on compact subintervals of I and with the scale function $p(x)$ and the speed measure $m(dx)$ satisfying

$$p(\ell+) = -\infty, \quad p(r-) = \infty, \quad m(I) < \infty,$$

the solution X to (5.1) starting at $x \in I$ never exits I (Proposition 5.22). We assume that

$$P^x(X_t = z) = 0; \quad \forall\, x, z \in I, t > 0. \tag{5.80}$$

Show in the following steps that *the normalized speed measure* $(m(dx)/m(I))$ *is the limiting distribution of* X:

(5.81) $$\lim_{t\to\infty} P^x(X_t < z) = \frac{m((\ell, z))}{m(I)}; \quad x, z \in I.$$

(i) Introduce the stopping times $T_a = \inf\{t \geq 0; X_t = a\}$, $S_{a;b} = \inf\{t \geq T_a; X_t = b\}$ for $a, b \in I$. Deduce from the assumptions and (5.59) the *positive recurrence* properties, with $\ell < a < x < u < b < r$:

$$E^x T_a = -\int_a^x (p(x) - p(y))m(dy) + (p(x) - p(a)) \cdot m((a, r)) < \infty,$$

$$E^x T_b = -\int_x^b (p(y) - p(x))m(dy) + (p(b) - p(x)) \cdot m((\ell, b)) < \infty,$$

$$E^x S_{u;x} = E^x T_u + E^u T_x = (p(u) - p(x)) \cdot m(I) < \infty.$$

(ii) Show that the same methodology as in (i) allows us to conclude, with the help of Exercise 5.39:

$$E^x \int_0^{S_{u;x}} 1_{(\ell,z)}(X_t)\, dt = (p(u) - p(x)) \cdot m((\ell, z)); \quad \forall z \in I.$$

(iii) With the aid of assumption (5.80), show that for $\ell < x < u < r$ and $z \in I$, $z \neq x$, the function

$$a(t) \triangleq P^x(X_t < z, S_{u;x} > t); \quad 0 \leq t < \infty,$$

is continuous and $\sum_{n=1}^{\infty} \max_{n-1 \leq t \leq n} a(t) < \infty$. This last condition implies direct Riemann integrability of $a(\cdot)$ (see Feller (1971), Chapter XI, Section 1).

(iv) Establish a renewal-type equation

$$F(t) = a(t) + \int_0^t F(t - s)\nu(ds); \quad 0 \leq t < \infty$$

for the function $F(t) \triangleq P^x(X_t < z)$, with appropriate measure $\nu(dt)$ on $[0, \infty)$. Show that the renewal theorem (Feller (1971), Chapter XI) implies (5.81).

5.41 Exercise (Le Gall (1983)). Provide a proof of Proposition 2.13 based on semimartingale local time.

(*Hint*: Apply the Tanaka-Meyer formula (3.7.9) to the difference $X \triangleq X^{(1)} - X^{(2)}$ between two solutions $X^{(1)}, X^{(2)}$ of (2.1). Use Exercise 3.7.12 to show that the local time at the origin for X is identically zero, almost surely.)

5.42 Exercise (Le Gall (1983)). Suppose that uniqueness in the sense of probability law holds for (2.1), and that for any two solutions $X^{(1)}, X^{(2)}$ on the same probability space and with respect to the same Brownian motion and initial condition, the local time of $X = X^{(1)} - X^{(2)}$ at the origin is identically equal to zero, almost surely. Then pathwise uniqueness holds for (2.1).

(*Hint*: Use a Tanaka-Meyer formula to show that $X^{(1)} \vee X^{(2)}$ also solves (2.1).)

5.6. Linear Equations

In this section we consider d-dimensional stochastic differential equations in which the solution process enters linearly. Such processes arise in estimation and control of linear systems, in economics (see Section 5.8), and in various other fields. As we shall see, one can provide a fairly explicit representation for the solution of a linear stochastic differential equation.

For most of this section, we study the equation

$$dX_t = [A(t)X_t + a(t)]\,dt + \sigma(t)\,dW_t, \quad 0 \le t < \infty, \qquad X_0 = \xi, \tag{6.1}$$

where W is an r-dimensional Brownian motion independent of the d-dimensional initial vector ξ, and the $(d \times d)$, $(d \times 1)$ and $(d \times r)$ matrices $A(t)$, $a(t)$, and $\sigma(t)$ are nonrandom, measurable, and locally bounded. In Problem 6.15 we generalize (6.1) for one-dimensional equations by allowing the solution X also to appear in the dispersion coefficient.

The deterministic equation corresponding to (6.1) is

$$\dot{\xi}(t) = A(t)\xi(t) + a(t); \quad \xi(0) = \xi. \tag{6.2}$$

Standard existence and uniqueness results (Hale (1969), Section I.5) imply that for every initial condition $\xi \in \mathbb{R}^d$, (6.2) has an absolutely continuous solution $\xi(t)$ defined for $0 \le t < \infty$. Likewise, the matrix differential equation

$$\dot{\Phi}(t) = A(t)\Phi(t), \quad \Phi(0) = I \tag{6.3}$$

has a unique (absolutely continuous) solution defined for $0 \le t < \infty$. (Here I is the $(d \times d)$ identity matrix.) This matrix function Φ is called the *fundamental solution* to the homogeneous equation

$$\dot{\xi}(t) = A(t)\xi(t). \tag{6.4}$$

For each $t \ge 0$, the matrix $\Phi(t)$ is *nonsingular*, for otherwise there would be a $t_0 \ge 0$ and a nonzero vector $\lambda \in \mathbb{R}^d$ such that $\Phi(t_0)\lambda = 0$. But $\Phi(t)\lambda$ is a solution to (6.4), and since the identically zero function is the unique solution which vanishes at t_0, we must have $\Phi(t)\lambda = 0$ for all t. This would contradict the initial condition $\Phi(0) = I$.

In terms of Φ, the solution of the deterministic equation (6.2) is simply

$$\xi(t) = \Phi(t)\Big[\xi(0) + \int_0^t \Phi^{-1}(s)a(s)\,ds\Big]. \tag{6.5}$$

(See Hale (1969), Chapter 3, for additional information.)

A pleasant fact is that the solution of (6.1) has a representation similar to (6.5). Indeed, it is easily verified by Itô's rule that

$$X_t \triangleq \Phi(t)\left[X_0 + \int_0^t \Phi^{-1}(s)a(s)\,ds + \int_0^t \Phi^{-1}(s)\sigma(s)\,dW_s\right]; \quad 0 \le t < \infty \tag{6.6}$$

solves (6.1). Pathwise uniqueness for equation (6.1) follows from Theorem 2.5.

6.1 Problem. Suppose that $E\|X_0\|^2 < \infty$, and introduce the *mean vector* and *covariance matrix* functions

$$m(t) \triangleq EX_t, \tag{6.7}$$

$$\rho(s,t) \triangleq E[(X_s - m(s))(X_t - m(t))^T], \tag{6.8}$$

$$V(t) \triangleq \rho(t,t). \tag{6.9}$$

Show that

$$m(t) = \Phi(t)\left[m(0) + \int_0^t \Phi^{-1}(s)a(s)\,ds\right], \tag{6.10}$$

$$\rho(s,t) = \Phi(s)\left[V(0) + \int_0^{s\wedge t} \Phi^{-1}(u)\sigma(u)(\Phi^{-1}(u)\sigma(u))^T\,du\right]\Phi^T(t), \tag{6.11}$$

hold for every $0 \le s,\ t < \infty$. In particular, $m(t)$ and $V(t)$ solve the linear equations

$$\dot{m}(t) = A(t)m(t) + a(t), \tag{6.12}$$

$$\dot{V}(t) = A(t)V(t) + V(t)A^T(t) + \sigma(t)\sigma^T(t). \tag{6.13}$$

6.2 Problem. Show that if X_0 has a d-variate normal distribution, then X is a Gaussian process (Definition 2.9.1).

A. Gauss-Markov Processes

If X_0 is normally distributed, then the finite-dimensional distributions of the Gaussian process X in (6.6) are completely determined by the mean and covariance functions. In this case, we would like to know under what additional conditions we can guarantee the nondegeneracy of the distribution of X_t, i.e., the positive definiteness of the matrix

$$V(t) = \Phi(t)\left[V(0) + \int_0^t \Phi^{-1}(u)\sigma(u)(\Phi^{-1}(u)\sigma(u))^T\,du\right]\Phi^T(t), \tag{6.14}$$

for every $t \ge 0$. In order to settle this question, we shall introduce the concept of *controllability* from linear system theory.

6.3 Definition. The pair of matrix-valued, measurable, locally bounded functions (A, σ) is called *controllable on* $[0, T]$ if for every pair $x, y \in \mathbb{R}^d$, there exists a measurable, bounded function $v\colon [0,T] \to \mathbb{R}^r$, such that

$$Y(t) = x + \int_0^t A(s)Y(s)\,ds + \int_0^t \sigma(s)v(s)\,ds; \quad 0 \le t \le T, \tag{6.15}$$

satisfies $Y(T) = y$. In other words, for every pair $x, y \in \mathbb{R}^d$, there exists a *control function* $v(\cdot)$ which steers the linear system (6.15) from $Y(0) = x$ to $Y(T) = y$.

6.4 Proposition. *The pair of functions* (A, σ) *is controllable on* $[0, T]$ *if and only if the matrix*

$$M(T) \triangleq \int_0^T \Phi^{-1}(t)\sigma(t)(\Phi^{-1}(t)\sigma(t))^T\,dt \tag{6.16}$$

is nonsingular.

PROOF. Let $G(t) = \Phi^{-1}(t)\sigma(t)$. From (6.5), the solution to (6.15) is

$$Y(t) = \Phi(t)\left[x + \int_0^t G(s)v(s)\,ds\right],$$

and because of the nonsingularity of $\Phi(T)$, controllability is equivalent to the condition that $\int_0^T G(s)v(s)\,ds$ range over all of $\mathbb{R}^d$ as v ranges over the bounded, measurable functions from $[0, T]$ to $\mathbb{R}^r$.

Choose an arbitrary $z \in \mathbb{R}^d$. Under the assumption of nonsingularity of $M(T)$, we may set $v(s) = G^T(s)M^{-1}(T)z$, and then we have $z = \int_0^T G(s)v(s)\,ds$. On the other hand, if $M(T)$ is singular, then there exists a nonzero $z \in \mathbb{R}^d$ such that $z^T M(T)z = 0$, i.e., $\int_0^T z^T G(s)G^T(s)z\,ds = 0$, which shows that $z^T G(s) = 0$ for Lebesgue-almost every $s \in [0, T]$. Consequently, $z^T \int_0^T G(s)v(s)\,ds = 0$ for any bounded, measurable v, which contradicts controllability. □

We see from its definition that $V(0)$ is positive semidefinite, so the nonsingularity (and hence the positive-definiteness) of $M(T)$ implies the same property for $V(T)$. We obtain thereby the following result: *if the pair* (A, σ) *is controllable on* $[0, T]$, *then the matrix* $V(T)$ *of* (6.14) *is nonsingular*. A bit of reflection shows that the two conditions are actually equivalent, provided $V(0) = 0$, i.e., X_0 in (6.1) is nonrandom.

When the matrices A and σ appearing in (6.1) are constant, this result takes a more explicit form. In this case, the fundamental solution to (6.4) is

$$\Phi(t) = e^{tA} \triangleq \sum_{n=0}^{\infty} \frac{t^n}{n!} A^n, \tag{6.17}$$

and controllability reduces to the following rank condition.

6.5 Proposition. *The pair of constant matrices* (A, σ) *is controllable* (*on any interval* $[0, T]$) *if and only if the* $(d \times d)$ *controllability matrix* $C \triangleq [\sigma, A\sigma, A^2\sigma, \ldots, A^{d-1}\sigma]$ *has rank d.*

PROOF. Let us first assume that $\text{rank}(C) < d$, and show that $M(T)$ given by (6.16) is singular for every $0 < T < \infty$. The assumption $\text{rank}(C) < d$ is equivalent to the existence of a nonzero vector $z \in \mathbb{R}^d$ for which

$$z^T\sigma = z^T A\sigma = z^T A^2\sigma = \cdots = z^T A^{d-1}\sigma = 0. \tag{6.18}$$

By the Hamilton-Cayley theorem, A satisfies its characteristic equation; thus, every positive, integral power of A can be written as a linear combination of $I, A, A^2, \ldots, A^{d-1}$. It follows from (6.18) that $z^T A^n \sigma = 0$; $n \geq 0$, and thereby

$$z^T \Phi^{-1}(t)\sigma = z^T e^{-tA}\sigma = \sum_{n=0}^{\infty} \frac{(-t)^n}{n!} z^T A^n \sigma = 0; \quad t \geq 0.$$

The singularity of $M(T)$ follows from $z^T M(T) z = 0$.

Let us now assume that for some $T > 0$, $M(T)$ is singular and show that $\operatorname{rank}(C) < d$. The singularity of $M(T)$ enables us to find a nonzero vector $z \in \mathbb{R}^d$ such that

$$0 = z^T M(T) z = \int_0^T \|z^T e^{-At}\sigma\|^2 \, dt.$$

From this we see that $f(t) \triangleq z^T e^{-At}\sigma$ is identically zero on $[0, T]$, and evaluating $f(0), f'(0), \ldots, f^{(d-1)}(0)$ we obtain (6.18). □

When A and σ are constant, equations (6.13) and (6.14) take the simplified form

$$(6.13)' \qquad \dot{V}(t) = AV(t) + V(t)A^T + \sigma\sigma^T,$$

$$(6.14)' \qquad V(t) = e^{tA}\left[V(0) + \int_0^t e^{-sA}\sigma\sigma^T e^{-sA^T} \, ds\right] e^{tA^T}.$$

One could hope that by proper choice of $V(0)$, it would be possible to obtain a constant solution to these equations. Under the assumption that all the eigenvalues of A have negative real parts, so that the integral

$$(6.19) \qquad V \triangleq \int_0^\infty e^{sA}\sigma\sigma^T e^{sA^T} \, ds$$

converges, one can verify that $V(t) \equiv V$ does indeed solve (6.13)′, (6.14)′. We leave this verification as a problem for the reader.

6.6 Problem. Show that if $V(0)$ in (6.14)′ is given by (6.19), then $V(t) \equiv V(0)$. In particular, V of (6.19) satisfies the algebraic matrix equation

$$(6.20) \qquad AV + VA^T = -\sigma\sigma^T.$$

We have established the following result.

6.7 Theorem. *Suppose in the stochastic differential equation* (6.1) *that* $\sigma(t) \equiv \sigma$, $a(t) \equiv 0$, *all the eigenvalues of* $A(t) \equiv A$ *have negative real parts, and the initial random vector* $\xi = X_0$ *has a d-variate normal distribution with mean* $m(0) = 0$ *and covariance* $V = E(X_0 X_0^T)$ *as in* (6.19). *Then the solution X is a stationary, zero-mean Gaussian process, with covariance function*

$$(6.21) \qquad \rho(s,t) = \begin{cases} e^{(s-t)A}V; & 0 \leq t \leq s < \infty \\ Ve^{(t-s)A^T}; & 0 \leq s \leq t < \infty. \end{cases}$$

PROOF. We have already seen that $V(t) \equiv V$; (6.21) follows from (6.14)′ and (6.11). □

6.8 Example (The Ornstein-Uhlenbeck Process). In the case $d = r = 1$, $a(t) \equiv 0$, $A(t) = -\alpha < 0$, and $\sigma(t) \equiv \sigma > 0$, (6.1) gives the oldest example

$$dX_t = -\alpha X_t\,dt + \sigma\,dW_t \tag{6.22}$$

of a stochastic differential equation (Uhlenbeck & Ornstein (1930), Doob (1942), Wang & Uhlenbeck (1945)). This corresponds to the Langevin (1908) equation for the Brownian motion of a particle with friction. According to (6.6), the solution of this equation is

$$X_t = X_0 e^{-\alpha t} + \sigma \int_0^t e^{-\alpha(t-s)}\,dW_s; \quad 0 \le t < \infty.$$

If $EX_0^2 < \infty$, the expectation, variance, and covariance functions in (6.7)–(6.9) become

$$m(t) \triangleq EX_t = m(0)e^{-\alpha t},$$

$$V(t) \triangleq \operatorname{Var}(X_t) = \frac{\sigma^2}{2\alpha} + \left(V(0) - \frac{\sigma^2}{2\alpha}\right)e^{-2\alpha t},$$

$$\rho(s,t) \triangleq \operatorname{Cov}(X_s, X_t) = \left[V(0) + \frac{\sigma^2}{2\alpha}(e^{2\alpha(t \wedge s)} - 1)\right]e^{-\alpha(t+s)}.$$

If the initial random variable X_0 has a normal distribution with mean zero and variance $(\sigma^2/2\alpha)$, then X is a stationary, zero-mean Gaussian process with covariance function $\rho(s,t) = (\sigma^2/2\alpha)e^{-\alpha|t-s|}$.

B. Brownian Bridge

Let us consider now the one-dimensional equation

$$dX_t = \frac{b - X_t}{T - t}\,dt + dW_t; \quad 0 \le t < T, \quad \text{and} \quad X_0 = a, \tag{6.23}$$

for given real numbers a, b and $T > 0$. This is of the form (6.1) with $A(t) = -1/(T-t)$, $a(t) = b/(T-t)$ and $\sigma(t) \equiv 1$, whence $\Phi(t) = 1 - (t/T)$. From (6.6) we have

$$X_t = a\left(1 - \frac{t}{T}\right) + b\frac{t}{T} + (T - t)\int_0^t \frac{dW_s}{T - s}; \quad 0 \le t < T.$$

6.9 Lemma. *The process*

$$Y_t = \begin{cases} (T - t)\displaystyle\int_0^t \frac{dW_s}{T - s}; & 0 \le t < T, \\ 0; & t = T, \end{cases} \tag{6.24}$$

is continuous, zero-mean, and Gaussian, with covariance function

$$\rho(s,t) = (s \wedge t) - \frac{st}{T}; \quad 0 \le s, t \le T. \tag{6.25}$$

PROOF. The process $M_t \triangleq \int_0^t (T-s)^{-1}\,dW_s$; $0 \le t < T$ is a continuous, square-integrable martingale with quadratic variation

$$\langle M \rangle_t \triangleq \int_0^t \frac{ds}{(T-s)^2} = \frac{1}{T-t} - \frac{1}{T}.$$

According to Theorem 3.4.6, there exists a standard, one-dimensional Brownian motion B such that $M_t = B_{\langle M \rangle_t}$; $0 \le t < T$. It follows from the strong law of large numbers for Brownian motion (Problem 2.9.3) that $Y_t \triangleq B_{\langle M \rangle_t}/(\langle M \rangle_t + T^{-1})$ converges almost surely to zero as $t \uparrow T$. The process Y of (6.24) is thus seen to be almost surely continuous on $[0, T]$, and to be a zero-mean Gaussian process with

$$E(Y_s Y_t) = (T-t)(T-s)\int_0^{t \wedge s} \frac{du}{(T-u)^2} = (s \wedge t) - \frac{st}{T},$$

provided $0 \le s, t < T$. If $s \vee t = T$, the preceding expectation is trivially zero, in agreement with (6.25). □

6.10 Corollary. *The process*

$$X_t = \begin{cases} a\left(1 - \dfrac{t}{T}\right) + b\dfrac{t}{T} + (T-t)\displaystyle\int_0^t \frac{dW_s}{T-s}; & 0 \le t < T, \\ b; & t = T, \end{cases} \tag{6.26}$$

is Gaussian with a.s. continuous paths, expectation function

$$m(t) \triangleq E(X_t) = a\left(1 - \frac{t}{T}\right) + b\frac{t}{T}; \quad 0 \le t \le T, \tag{6.27}$$

and covariance function $\rho(s,t)$ given by (6.25). *This process is the pathwise unique solution of equation* (6.23) *on* $[0, T)$.

6.11 Problem. Show that the finite-dimensional distributions for the process X in (6.26) are given by

$$P[X_{t_1} \in dx_1, \ldots, X_{t_n} \in dx_n] = \prod_{i=1}^n p(t_i - t_{i-1}; x_{i-1}, x_i) \cdot \frac{p(T - t_n; x_n, b)}{p(T; a, b)} dx_1 \ldots dx_n, \tag{6.28}$$

where $0 = t_0 < t_1 < \cdots < t_n < T$, $x_0 = a$, $(x_1, \ldots, x_n) \in \mathbb{R}^n$, and $p(t; x, y)$ is the Gaussian kernel (2.2.6). (*Hint*: The normal random variables $Z_i \triangleq X_{t_i}/(T - t_i) - X_{t_{i-1}}/(T - t_{i-1})$; $i = 1, \ldots, n$, are independent.)

6.12 Definition. A *Brownian bridge from a to b on* $[0, T]$ is any almost surely continuous process defined on $[0, T]$, with finite dimensional distributions specified by (6.28).

It is apparent that any continuous, Gaussian process on $[0, T]$ with mean and covariance functions specified by (6.27) and (6.25), respectively, is a Brownian bridge from a to b. Besides the representation of Brownian bridge as the solution to (6.23), there are two other ways of thinking about this process; they appear in the next two problems.

6.13 Problem. Let $\{W_t, \mathscr{F}_t; 0 \le t < \infty\}, (\Omega, \mathscr{F}), \{P^a\}_{a \in \mathbb{R}}$ be a one-dimensional Brownian family. Show that for $0 = t_0 < t_1 < \cdots < t_n < T$, $x_0 = a$, and $(x_1, \ldots, x_n) \in \mathbb{R}^n$, the conditional finite-dimensional distributions $P^a[W_{t_1} \in dx_1, \ldots, W_{t_n} \in dx_n | W_T = b]$ are given by the right-hand side of (6.28), for Lebesgue-almost every $b \in \mathbb{R}$. In other words, *Brownian bridge from a to b on* $[0, T]$ *is Brownian motion started at a and conditioned to arrive at b at time T.*

6.14 Problem. Let W be a standard, one-dimensional Brownian motion and define

$$B_t^{a \to b} \triangleq a\left(1 - \frac{t}{T}\right) + b\frac{t}{T} + \left(W_t - \frac{t}{T}W_T\right); \quad 0 \le t \le T. \tag{6.29}$$

Then $B^{a \to b}$ is a Brownian bridge from a to b on $[0, T]$.

C. The General, One-Dimensional Linear Equation

Let us consider the one-dimensional ($d = 1, r \ge 1$) stochastic differential equation

$$dX_t = [A(t)X_t + a(t)]\,dt + \sum_{j=1}^{r} [S_j(t)X_t + \sigma_j(t)]\,dW_t^{(j)}, \tag{6.30}$$

where $W = \{W_t = (W_t^{(1)}, \ldots, W_t^{(r)}), \mathscr{F}_t;\ 0 \le t < \infty\}$ is an r-dimensional Brownian motion, and the coefficients A, a, S_j, σ_j are measurable, $\{\mathscr{F}_t\}$-adapted, almost surely locally bounded processes. We set

$$\begin{aligned} \zeta_t &\triangleq \sum_{j=1}^{r} \int_0^t S_j(u)\,dW_u^{(j)} - \frac{1}{2}\sum_{j=1}^{r} \int_0^t S_j^2(u)\,du, \\ Z_t &\triangleq \exp\left[\int_0^t A(u)\,du + \zeta_t\right]. \end{aligned} \tag{6.31}$$

6.15 Problem. Show that the unique solution of equation (6.30) is

$$(6.32)\quad X_t = Z_t\left[X_0 + \int_0^t \frac{1}{Z_u}\{a(u) - \sum_{j=1}^r S_j(u)\sigma_j(u)\}\,du + \sum_{j=1}^r \int_0^t \frac{\sigma_j(u)}{Z_u}\,dW_u^{(j)}\right].$$

In particular, the solution of the equation

$$(6.33)\qquad dX_t = A(t)X_t\,dt + \sum_{j=1}^r S_j(t)X_t\,dW_t^{(j)}$$

is given by

$$(6.34)\quad X_t = X_0 \exp\left[\int_0^t \{A(u) - \frac{1}{2}\sum_{j=1}^r S_j^2(u)\}\,du + \sum_{j=1}^r \int_0^t S_j(u)\,dW_u^{(j)}\right].$$

In the case of constant coefficients $A(t) \equiv A$, $S_j(t) \equiv S_j$ with $2A < \sum_{j=1}^r S_j^2$ in (6.34), show that $\lim_{t\to\infty} X_t = 0$ a.s., for arbitrary initial condition X_0. □

D. Supplementary Exercises

6.16 Exercise. Write down the stochastic differential equation satisfied by $Y_t = X_t^k$; $0 \le t < \infty$, with $k > 1$ arbitrary but fixed and X the solution of equation (5.79). Use your equation to compute $E(X_t^k)$.

6.17 Exercise. Define the *d-dimensional Brownian bridge from a to b on* $[0, T]$ $(a, b \in \mathbb{R}^d)$ to be any almost surely continuous process defined on $[0, T]$, with finite dimensional distributions specified by (6.28), where now

$$p(t; x, y) = (2\pi t)^{-(d/2)} \exp\left\{-\frac{\|x - y\|^2}{2t}\right\}; \quad x, y \in \mathbb{R}^d,\ t > 0.$$

(i) Prove that the processes X given by (6.26) and $B^{a\to b}$ given by (6.29), where W is a d-dimensional Brownian motion with $W_0 = 0$ a.s., are d-dimensional Brownian bridges from a to b on $[0, T]$.
(ii) Prove that the d-dimensional processes $\{B_t^{a\to b}; 0 \le t \le T\}$ and $\{B_{T-t}^{b\to a}; 0 \le t \le T\}$ have the same law.
(iii) Show that for any bounded, measurable function $F: C[0, T]^d \to \mathbb{R}$, we have

$$(6.35)\qquad EF(a + W_\cdot) = \int_{\mathbb{R}^d} EF(B_\cdot^{a\to b})p(T; a, b)\,db.$$

6.18 Exercise. Let $\Phi: \mathbb{R}^d \to \mathbb{R}$ be of class C^2 with bounded second partial derivatives and bounded gradient $\nabla\Phi$, and consider the *Smoluchowski equation*

$$(6.36)\qquad dX_t = \nabla\Phi(X_t)\,dt + dW_t; \quad 0 \le t < \infty,$$

where W is a standard, $\mathbb{R}^d$-valued Brownian motion. According to Theorems 2.5, 2.9 and Problem 2.12, this equation admits a unique strong solution for every initial distribution on X_0. Show that the measure

$$\mu(dx) = e^{2\Phi(x)}\,dx \quad \text{on } \mathscr{B}(\mathbb{R}^d) \tag{6.37}$$

is *invariant* for (6.36); i.e., if $X^{(a)}$ is the unique strong solution of (6.36) with initial condition $X_0^{(a)} = a \in \mathbb{R}^d$, then

$$\mu(A) = \int_{\mathbb{R}^d} P(X_t^{(a)} \in A)\mu(da); \quad \forall\, A \in \mathscr{B}(\mathbb{R}^d) \tag{6.38}$$

holds for every $0 \le t < \infty$.

(*Hint*: From Corollary 3.11 and the Itô rule, we have

$$Ef(X_t^{(a)}) = E\left[f(a + W_t)\exp\left\{\Phi(a + W_t) - \Phi(a) - \frac{1}{2}\int_0^t (\Delta\Phi + \|\nabla\Phi\|^2)(a + W_s)\,ds\right\}\right] \tag{6.39}$$

for every $f \in C_0^\infty(\mathbb{R}^d)$. Now use Exercise 6.17 (ii), (iii) and Problem 4.25.)

6.19 Remark. If $d = 1$ in Exercise 6.18, the *speed measure* of the process X is given by $m(dx) = 2\exp\{-2\Phi(c)\}\mu(dx)$ and is therefore invariant. Recall Exercise 5.40.

6.20 Exercise (The Brownian Oscillator). Consider the Langevin system

$$dX_t = Y_t\,dt$$
$$dY_t = -\beta X_t\,dt - \alpha Y_t\,dt + \sigma\,dW_t,$$

where W is a standard, one-dimensional Brownian motion and β, σ, and α are positive constants.

(i) Solve this system explicitly.
(ii) Show that if (X_0, Y_0) has an appropriate Gaussian distribution, then (X_t, Y_t) is a stationary Gaussian process.
(iii) Compute the covariance function of this stationary Gaussian process.

6.21 Exercise. Consider the one-dimensional equation (6.1) with $a(t) \equiv 0$, $\sigma(t) \equiv \sigma > 0$, $A(t) \le -\alpha < 0$, $\forall\, 0 \le t < \infty$, and $\xi = x \in \mathbb{R}$. Show that

$$EX_t^2 \le \frac{\sigma^2}{2\alpha} + \left(x^2 - \frac{\sigma^2}{2\alpha}\right)e^{-2\alpha t}; \quad \forall t \ge 0.$$

6.22 Exercise. Let $W = \{W_t = (W_t^{(1)}, \ldots, W_t^{(r)}), \mathscr{F}_t;\ 0 \le t < \infty\}$ be an r-dimensional Brownian motion, and let $A(t)$, $S^{(p)}(t)$; $p = 1, \ldots, r$, be adapted, bounded, $(d \times d)$ matrix-valued processes on $[0, T]$. Then the matrix stochastic integral equation

$$X(t) = I + \int_0^t A(s)X(s)\,ds + \sum_{p=1}^{r} \int_0^t S^{(p)}(s)X(s)\,dW_s^{(p)} \tag{6.40}$$

has a unique, strong solution (Theorems 2.5, 2.9). The componentwise formulation of (6.40) is

$$X_{kj}(t) = \delta_{kj} + \sum_{\ell=1}^{d} \int_0^t A_{k\ell}(s)X_{\ell j}(s)\,ds + \sum_{p=1}^{r} \sum_{\ell=1}^{d} \int_0^t S_{k\ell}^{(p)}(s)X_{\ell j}(s)\,dW_s^{(p)}.$$

Show that $X(t)$ has an inverse, which satisfies

$$X^{-1}(t) = I + \int_0^t X^{-1}(s)\left[\sum_{p=1}^{r} (S^{(p)}(s))^2 - A(s)\right] ds - \sum_{p=1}^{r} \int_0^t X^{-1}(s)S^{(p)}(s)\,dW_s^{(p)}. \tag{6.41}$$

5.7. Connections with Partial Differential Equations

The connections between Brownian motion on one hand, and the Dirichlet and Cauchy problems (for the Poisson and heat equations, respectively) on the other, were explored at some length in Chapter 4. In this section we document analogous connections between solutions of stochastic differential equations, and the Dirichlet and Cauchy problems for the associated, more general elliptic and parabolic equations. Such connections have already been presaged in Section 4 of this chapter, in the prominent role played there by the differential operators $\mathscr{A}_t$ and $(\partial/\partial t) + \mathscr{A}_t$, as well as in the relevance of the Cauchy problem to the question of uniqueness in the martingale problem (Theorem 4.28).

In Chapter 4 we employed probabilistic arguments to establish the existence and uniqueness of solutions to the Dirichlet and Cauchy problems considered there. The stochastic representations of solutions, which were so useful for uniqueness, will carry over to the generality of this section. As far as existence is concerned, however, the mean-value property for harmonic functions and the explicit form of the fundamental solution for the heat equation will no longer be available to us. We shall content ourselves, therefore, with representation and uniqueness results, and fall back on standard references in the theory of partial differential equations when an existence result is needed. The reader is referred to the notes for a brief discussion of probabilistic methods for proving existence.

Throughout this section, we shall be considering a solution to the stochastic integral equation

$$(7.1)\quad X_s^{(t,x)} = x + \int_t^s b(\theta, X_\theta^{(t,x)})\,d\theta + \int_t^s \sigma(\theta, X_\theta^{(t,x)})\,dW_\theta; \quad t \le s < \infty$$

under the standing assumptions that

(7.2) $\begin{cases} \text{the coefficients } b_i(t,x), \sigma_{ij}(t,x)\colon [0,\infty) \times \mathbb{R}^d \to \mathbb{R} \text{ are} \\ \textit{continuous} \text{ and satisfy the } \textit{linear growth} \text{ condition (2.13);} \end{cases}$

(7.3) $\begin{cases} \text{the equation (7.1) has a weak solution } (X^{(t,x)}, W), \\ (\Omega, \mathscr{F}, P), \{\mathscr{F}_s\} \text{ for every pair } (t,x); \text{ and} \end{cases}$

(7.4) this solution is unique in the sense of probability law.

We frequently suppress the superscripts (t, x) in $X^{(t,x)}$, and write $E^{t,x}$ to indicate the expectation computed under these initial conditions. Associated with (7.1) is the second-order differential operator $\mathscr{A}_t$ of (4.1). When b and σ do not depend on t, we write $\mathscr{A}$ (as in (1.2)) instead of $\mathscr{A}_t$ and E^x instead of $E^{t,x}$.

A. The Dirichlet Problem

Let D be an open subset of $\mathbb{R}^d$, and assume that b and σ do not depend on t.

7.1 Definition. The operator $\mathscr{A}$ of (1.2) is called *elliptic at the point* $x \in \mathbb{R}^d$ if

$$\sum_{i=1}^d \sum_{k=1}^d a_{ik}(x)\xi_i\xi_k > 0; \quad \forall \xi \in \mathbb{R}^d \backslash \{0\}.$$

If $\mathscr{A}$ is elliptic at every point of D, we say that $\mathscr{A}$ is *elliptic in* D; if there exists a number $\delta > 0$ such that

$$\sum_{i=1}^d \sum_{k=1}^d a_{ik}(x)\xi_i\xi_k \ge \delta\|\xi\|^2; \quad \forall x \in D,\ \xi \in \mathbb{R}^d,$$

we say that $\mathscr{A}$ is *uniformly elliptic in* D.

Let $\mathscr{A}$ be elliptic in the open, bounded domain D, and consider the continuous functions $k\colon \bar{D} \to [0, \infty)$, $g\colon \bar{D} \to \mathbb{R}$, and $f\colon \partial D \to \mathbb{R}$. The *Dirichlet problem* is to find a continuous function $u\colon \bar{D} \to \mathbb{R}$ such that u is of class $C^2(D)$ and satisfies the elliptic equation

$$(7.5)\qquad \mathscr{A}u - ku = -g; \quad \text{in } D$$

as well as the boundary condition

$$(7.6)\qquad u = f; \quad \text{on } \partial D.$$

7.2 Proposition. *Let u be a solution of the Dirichlet problem* (7.5), (7.6) *in the open, bounded domain D, and let* $\tau_D \triangleq \inf\{t \ge 0; X_t \notin D\}$. *If*

$$(7.7)\qquad E^x\tau_D < \infty; \quad \forall x \in D,$$

then under the assumptions (7.2)–(7.4) *we have*

(7.8)

$$u(x) = E^x\left[f(X_{\tau_D})\exp\left\{-\int_0^{\tau_D} k(X_s)\,ds\right\} + \int_0^{\tau_D} g(X_t)\exp\left\{-\int_0^t k(X_s)\,ds\right\}dt\right]$$

for every $x \in \bar{D}$.

7.3 Problem. Prove Proposition 7.2.

When should the condition (7.7) be expected to hold? Intuitively speaking, if there is enough "diffusion" to guarantee that, in at least one component, X behaves like a Brownian motion, then (7.7) is valid. We render this idea precise in the following lemma.

7.4 Lemma. *Suppose that for the open, bounded domain* D, *we have for some* $1 \le \ell \le d$:

$$\min_{x \in \bar{D}} a_{\ell\ell}(x) > 0. \tag{7.9}$$

Then (7.7) *holds.*

Proof (Friedman (1975), p. 145). With $b \triangleq \max_{x \in \bar{D}} |b(x)|$, $a \triangleq \min_{x \in \bar{D}} a_{\ell\ell}(x)$, $q \triangleq \min_{x \in \bar{D}} x_\ell$, and $\nu > (2b/a)$, we consider the function $h(x) = -\mu \exp(\nu x_\ell)$; $x = (x_1, \ldots, x_d) \in D$, where the constant $\mu > 0$ will be determined later. This function is of class $C^\infty(D)$ and satisfies

$$-(\mathscr{A}h)(x) = \mu e^{\nu x_\ell}\{\tfrac{1}{2}\nu^2 a_{\ell\ell}(x) + \nu b_\ell(x)\} \ge \frac{1}{2}\mu\nu a\, e^{\nu q}\left(\nu - \frac{2b}{a}\right); \quad x \in D.$$

Choosing μ sufficiently large, we can guarantee that $\mathscr{A}h \le -1$ holds in D. Now the function h and its derivatives are bounded on $\bar{D}$, so by Itô's rule we have for every $x \in D$, $t \ge 0$:

$$E^x(t \wedge \tau_D) \le h(x) - E^x h(X_{t \wedge \tau_D}) \le 2 \max_{y \in \bar{D}} |h(y)| < \infty.$$

Let $t \to \infty$ to obtain (7.7). □

7.5 Remark. Condition (7.9) is stronger than ellipticity but weaker than uniform ellipticity in D. Now suppose that in the open bounded domain D, we have that

(i) $\mathscr{A}$ is uniformly elliptic,
(ii) the coefficients a_{ij}, b_i, k, g are Hölder-continuous, and
(iii) every point $a \in \partial D$ has the *exterior sphere property*; i.e., there exists a ball $B(a)$ such that $\bar{B}(a) \cap D = \varnothing$, $\bar{B}(a) \cap \partial D = \{a\}$.

We also retain the assumption that f is continuous on ∂D. Then there exists a function u of class $C(\bar{D}) \cap C^2(D)$ (in fact, with Hölder-continuous second partial derivatives in D), which solves the Dirichlet problem (7.5), (7.6); see Gilbarg & Trudinger (1977), p. 101, Friedman (1964), p. 87, or Friedman (1975), p. 134. By virtue of Proposition 7.2, such a function is unique and is given by (7.8).

B. The Cauchy Problem and a Feynman-Kac Representation

With an arbitrary but fixed $T > 0$ and appropriate constants $L > 0$, $\lambda \geq 1$, we consider functions $f(x)\colon \mathbb{R}^d \to \mathbb{R}$, $g(t,x)\colon [0,T] \times \mathbb{R}^d \to \mathbb{R}$ and $k(t,x)\colon [0,T] \times \mathbb{R}^d \to [0,\infty)$ which are *continuous* and satisfy

$$\text{(7.10)} \qquad \text{(i) } |f(x)| \leq L(1 + \|x\|^{2\lambda}) \quad \text{or} \quad \text{(ii) } f(x) \geq 0; \quad \forall\, x \in \mathbb{R}^d$$

as well as

$$\text{(7.11)} \quad \text{(i) } |g(t,x)| \leq L(1 + \|x\|^{2\lambda}) \quad \text{or} \quad \text{(ii) } g(t,x) \geq 0; \quad \forall\, 0 \leq t \leq T,\, x \in \mathbb{R}^d.$$

We recall also the operator $\mathscr{A}_t$ of (4.1), and formulate the analogue of the *Feynman-Kac Theorem* 4.4.2:

7.6 Theorem. *Under the preceding assumptions and* (7.2)–(7.4), *suppose that* $v(t,x)\colon [0,T] \times \mathbb{R}^d \to \mathbb{R}^d$ *is continuous, is of class* $C^{1,2}([0,T) \times \mathbb{R}^d)$ *(Remark* 4.1), *and satisfies the Cauchy problem*

$$\text{(7.12)} \qquad -\frac{\partial v}{\partial t} + kv = \mathscr{A}_t v + g; \quad \text{in } [0,T) \times \mathbb{R}^d,$$

$$\text{(7.13)} \qquad v(T,x) = f(x); \quad x \in \mathbb{R}^d,$$

as well as the polynomial growth condition

$$\text{(7.14)} \qquad \max_{0 \leq t \leq T} |v(t,x)| \leq M(1 + \|x\|^{2\mu}); \quad x \in \mathbb{R}^d,$$

for some $M > 0$, $\mu \geq 1$. *Then* $v(t,x)$ *admits the stochastic representation*

$$\text{(7.15)} \qquad v(t,x) = E^{t,x}\left[f(X_T)\exp\left\{ -\int_t^T k(\theta, X_\theta)\,d\theta \right\} + \int_t^T g(s, X_s)\exp\left\{ -\int_t^s k(\theta, X_\theta)\,d\theta \right\} ds \right]$$

on $[0,T] \times \mathbb{R}^d$; *in particular, such a solution is unique.*

PROOF. Proceeding as in the proof of Theorem 4.4.2, we apply the Itô rule to the process $v(s, X_s)\exp\{-\int_t^s k(\theta, X_\theta)\,d\theta\}$; $s \in [t,T]$, and obtain, with $\tau_n \triangleq \inf\{s \geq t;\ \|X_s\| \geq n\}$,

$$(7.16)\quad v(t,x) = E^{t,x}\left[\int_t^{T\wedge\tau_n} g(s,X_s)\exp\left\{-\int_t^s k(\theta,X_\theta)\,d\theta\right\}ds\right]$$
$$+ E^{t,x}\left[v(\tau_n,X_{\tau_n})\exp\left\{-\int_t^{\tau_n} k(\theta,X_\theta)\,d\theta\right\}1_{\{\tau_n\le T\}}\right]$$
$$+ E^{t,x}\left[f(X_T)\exp\left\{-\int_t^{T} k(\theta,X_\theta)\,d\theta\right\}1_{\{\tau_n> T\}}\right].$$

Let us recall from (3.17) the estimate

$$(7.17)\quad E^{t,x}\left[\max_{t\le\theta\le s}\|X_\theta\|^{2m}\right]\le C(1+\|x\|^{2m})e^{C(s-t)};\quad t\le s\le T,$$

valid for every $m \ge 1$ and some $C = C(m, K, T, d) > 0$. Now the first term on the right-hand side of (7.16) converges as $n \to \infty$ to

$$E^{t,x}\int_t^T g(s,X_s)\exp\left\{-\int_t^s k(\theta,X_\theta)\,d\theta\right\}ds,$$

by either the dominated convergence theorem (thanks to (7.11) (i) and (7.17)) or the monotone convergence theorem (if (7.11) (ii) prevails). The second term is bounded in absolute value by

$$(7.18)\quad E^{t,x}[|v(\tau_n,X_{\tau_n})|\,1_{\{\tau_n\le T\}}]\le M(1+n^{2\mu})P^{t,x}[\tau_n\le T].$$

However, this last probability can be written as

$$P^{t,x}[\tau_n\le T] = P^{t,x}\left[\max_{t\le\theta\le T}\|X_\theta\|\ge n\right]\le n^{-2m}E^{t,x}\left[\max_{t\le\theta\le T}\|X_\theta\|^{2m}\right]$$
$$\le Cn^{-2m}(1+\|x\|^{2m})e^{CT},$$

by virtue of (7.17) and the Čebyšev inequality. Simply selecting $m > \mu$, we see that the right-hand side of (7.18) converges to zero as $n \to \infty$. Finally, the last term in (7.16) converges to

$$E^{t,x}\left[f(X_T)\exp\left\{-\int_t^{T} k(\theta,X_\theta)\,d\theta\right\}\right],$$

either by the dominated or by the monotone convergence theorem. □

7.7 Problem. In the case of bounded coefficients, i.e.,

$$(7.19)\quad |b_i(t,x)| + \sum_{j=1}^r \sigma_{ij}^2(t,x)\le\rho;\quad 0\le t<\infty,\quad x\in\mathbb{R}^d,\quad 1\le i\le d,$$

the polynomial growth condition (7.14) in Theorem 7.6 may be replaced by

$$(7.20)\quad \max_{0\le t\le T}|v(t,x)|\le Me^{\mu\|x\|^2};\quad x\in\mathbb{R}^d$$

for some $M > 0$ and $0 < \mu < (1/18\rho Td)$. (*Hint*: Use Problem 3.4.12.)

7.8 Remark. For conditions under which the Cauchy problem (7.12), (7.13) has a solution satisfying the exponential growth condition (7.20), one should consult Friedman (1964), Chapter I. A set of conditions *sufficient for the existence* of a solution v satisfying the polynomial growth condition (7.14) is:

(i) *Uniform ellipticity*: There exists a positive constant δ such that

$$\sum_{i=1}^{d} \sum_{k=1}^{d} a_{ik}(t,x)\xi_i\xi_k \geq \delta\|\xi\|^2$$

holds for every $\xi \in \mathbb{R}^d$ and $(t,x) \in [0,\infty) \times \mathbb{R}^d$;

(ii) *Boundedness*: The functions $a_{ik}(t,x)$, $b_i(t,x)$, $k(t,x)$ are bounded in $[0,T] \times \mathbb{R}^d$;

(iii) *Hölder continuity*: The functions $a_{ik}(t,x)$, $b_i(t,x)$, $k(t,x)$, and $g(t,x)$ are uniformly Hölder-continuous in $[0,T] \times \mathbb{R}^d$;

(iv) *Polynomial growth*: The functions $f(x)$ and $g(t,x)$ satisfy (7.10) (i) and (7.11) (i), respectively.

Conditions (i), (ii), and (iii) can be relaxed somewhat by making them local requirements. We refer the reader to Friedman (1975), p. 147, for precise formulations.

7.9 Definition. A *fundamental solution* of the second-order partial differential equation

$$-\frac{\partial u}{\partial t} + ku = \mathscr{A}_t u \tag{7.21}$$

is a nonnegative function $G(t,x;\tau,\xi)$ defined for $0 \leq t < \tau \leq T$, $x \in \mathbb{R}^d$, $\xi \in \mathbb{R}^d$, with the property that for every $f \in C_0(\mathbb{R}^d)$, $\tau \in (0,T]$, the function

$$u(t,x) \triangleq \int_{\mathbb{R}^d} G(t,x;\tau,\xi)f(\xi)\,d\xi; \quad 0 \leq t < \tau,\ x \in \mathbb{R}^d \tag{7.22}$$

is bounded, of class $C^{1,2}$, satisfies (7.21), and

$$\lim_{t \uparrow \tau} u(t,x) = f(x); \quad x \in \mathbb{R}^d. \tag{7.23}$$

Under conditions (i)–(iii) of Remark 7.8 imposed on the coefficients $b_i(t,x)$, $a_{ik}(t,x)$, and $k(t,x)$, a fundamental solution of (7.21) exists (see Friedman (1975), pp. 141, 148 and Friedman (1964), Chapter I). For fixed $(\tau,\xi) \in (0,T] \times \mathbb{R}^d$, the function

$$\varphi(t,x) \triangleq G(t,x;\tau,\xi)$$

is of class $C^{1,2}([0,\tau) \times \mathbb{R}^d)$ and satisfies the *backward Kolmogorov equation* (7.21) in the *backward variables* (t,x). If, in addition, the functions $(\partial/\partial x_i)b_i(t,x)$, $(\partial/\partial x_i)a_{ik}(t,x)$, $(\partial^2/\partial x_i\,\partial x_k)a_{ik}(t,x)$ are bounded and Hölder-continuous, then for fixed $(t,x) \in [0,T) \times \mathbb{R}^d$ the function

$$\psi(\tau,\xi) \triangleq G(t,x;\tau,\xi)$$

is of class $C^{1,2}((t, T] \times \mathbb{R}^d)$ and satisfies the *forward Kolmogorov equation*

$$(7.24)\qquad \frac{\partial}{\partial \tau}\psi(\tau, \xi) = \frac{1}{2}\sum_{i=1}^{d}\sum_{k=1}^{d}\frac{\partial^2}{\partial \xi_i \partial \xi_k}[a_{ik}(\tau, \xi)\psi(\tau, \xi)] - \sum_{i=1}^{d}\frac{\partial}{\partial \xi_i}[b_i(\tau, \xi)\psi(\tau, \xi)] - k(\tau, \xi)\psi(\tau, \xi)$$

in the *forward variables* (τ, ξ).

Returning to the Cauchy problem (7.21), (7.23) with $k \equiv 0$, we recall from Theorem 7.6 that its solution is given by

$$(7.25)\qquad u(t, x) = Ef(X_\tau^{(t,x)}); \quad f \in C_0(\mathbb{R}^d).$$

A comparison of (7.22), (7.25), in conjunction with Problem 4.25, leads to the conclusion that *any fundamental solution* $G(t, x; \tau, \xi)$ *is also the transition probability density for the process* $X^{(t,x)}$ *determined by* (7.1); i.e.,

$$(7.26)\qquad P[X_\tau^{(t,x)} \in A] = \int_A G(t, x; \tau, \xi)\, d\xi; \quad A \in \mathscr{B}(\mathbb{R}^d), \quad 0 \le t < \tau \le T.$$

In particular, under the conditions (7.2)–(7.4), this fundamental solution is *unique*, and the representation (7.15) of the solution to the Cauchy problem

$$(7.27)\qquad -\frac{\partial v}{\partial t} = \mathscr{A}_t v + g; \quad \text{in } [0, T) \times \mathbb{R}^d,$$

$$(7.28)\qquad v(T, x) = f(x); \quad x \in \mathbb{R}^d,$$

now takes the form

$$(7.29)\qquad v(t, x) = \int_{\mathbb{R}^d} G(t, x; T, \xi) f(\xi)\, d\xi + \int_t^T \int_{\mathbb{R}^d} G(t, x; \tau, \xi) g(\tau, \xi)\, d\xi\, d\tau.$$

C. Supplementary Exercises

7.10 Exercise. The Cauchy problem (7.27), (7.28) does not include the *potential* term kv appearing in (7.12). The case of nonzero k corresponds to a *diffusion with killing*. In particular, let $X^{(t,x)}$ be the solution to (7.1); let Y be an independent, exponentially distributed random variable with mean 1; and define the *lifetime*

$$\rho^{(t,x)} \triangleq \inf\left\{s \ge t; \int_t^s k(\theta, X_\theta^{(t,x)})\, d\theta \ge Y\right\}.$$

The killed diffusion process is

$$\tilde{X}_s^{(t,x)} \triangleq \begin{cases} X_s^{(t,x)}; & t \le s < \rho^{(t,x)}, \\ \Delta; & s \ge \rho^{(t,x)}, \end{cases}$$

where Δ is a *cemetery state* isolated from $\mathbb{R}^d$. Assume the conditions of Theorem 7.6 and let $G(t, x; \tau, \xi)$ be a fundamental solution of (7.21). Then we have

$$P[\tilde{X}_\tau^{(t,x)} \in A] = \int_A G(t, x; \tau, \xi)\, d\xi; \quad A \in \mathscr{B}(\mathbb{R}^d), \quad 0 \le t < \tau \le T, \tag{7.30}$$

and the solution (7.15) of the Cauchy problem (7.12), (7.13) takes the form (7.29).

7.11 Exercise. Suppose that $b_i(t, x)$; $1 \le i \le d$, are uniformly bounded on $[0, T] \times \mathbb{R}^d$ and that $f(x)$ and $g(t, x)$ satisfy (7.10) and (7.11), respectively. If $v(t, x)$ is a solution to the Cauchy problem

$$-\frac{\partial v}{\partial t} = \frac{1}{2}\Delta v + (b, \nabla v) + g; \quad \text{in } [0, T) \times \mathbb{R}^d$$

$$v(T, x) = f(x); \quad x \in \mathbb{R}^d$$

and (7.20) holds, then

$$\begin{aligned} v(t, x) = E^x \Bigg[& f(W_{T-t}) \exp\left\{ \int_0^{T-t} (b(t + \theta, W_\theta), dW_\theta) - \frac{1}{2}\int_0^{T-t} \|b(t + \theta, W_\theta)\|^2\, d\theta \right\} \\ & + \int_0^{T-t} g(t + s, W_s) \exp\left\{ \int_0^s (b(t + \theta, W_\theta), dW_\theta) - \frac{1}{2}\int_0^s \|b(t + \theta, W_\theta)\|^2\, d\theta \right\} ds \Bigg], \end{aligned}$$

where $\{W_t, \mathscr{F}_t; 0 \le t \le T\}$, $(\Omega, \mathscr{F})$, $\{P^x\}_{x \in \mathbb{R}^d}$ is a d-dimensional Brownian family.

7.12 Exercise. Write down the Kolmogorov forward and backward equations with $k \equiv 0$ for one-dimensional Brownian motion with constant drift μ, and verify that the transition probability density of this process satisfies these equations in the appropriate variables.

7.13 Exercise. Let the coefficients b, σ in (7.1) be independent of t, and assume that condition (7.9) holds for every open, bounded domain $D \subset \mathbb{R}^d$. Suppose also that there exists a function $f: \mathbb{R}^d \backslash \{0\} \to \mathbb{R}$ of class C^2, which satisfies

$$\mathscr{A}f(x) \le 0 \quad \text{on } \mathbb{R}^d \backslash \{0\} \tag{7.31}$$

and is such that $F(r) \triangleq \min_{\|x\|=r} f(x)$ is strictly increasing with $\lim_{r\to\infty} F(r) = \infty$.

(i) Show that we have the *recurrence property*

(7.32) $$P^x(\tau_r < \infty) = 1; \quad \forall x \in \mathbb{R}^d \backslash \bar{B}_r$$

for every $r > 0$, where $B_r = \{x \in \mathbb{R}^d; \|x\| < r\}$ and $\tau_r = \inf\{t \geq 0; X_t \in \bar{B}_r\}$.

(ii) Verify that (7.32) holds in the case

(7.33) $$(x, b(x)) + \frac{1}{2}\sum_{i=1}^{d} a_{ii}(x) \leq \frac{(x, a(x)x)}{\|x\|^2}; \quad \forall x \in \mathbb{R}^d \backslash \{0\}.$$

(iii) If (7.31) is strengthened to $\mathscr{A}f(x) \leq -1$; $\forall x \in \mathbb{R}^d \backslash \{0\}$, then we have the *positive recurrence property*

(7.34) $$E^x \tau_r < \infty \quad \forall x \in \mathbb{R}^d \backslash \bar{B}_r.$$

5.8. Applications to Economics

In this section we apply the theory of stochastic calculus and differential equations to two related problems in financial economics. The first of these is *option pricing*, where we derive the celebrated Black & Scholes (1973) option pricing formula. The second application is the *optimal consumption/investment problem* formulated by Merton (1971). These problems are unified by their reliance on the theory of stochastic differential equations to model the trading of risky securities in continuous time. In the second problem, this theory allows us to characterize the value function and optimal consumption process in a context more general than considered heretofore. We subsequently specialize the model to the case of constant coefficients, so as to illustrate the use of the Hamilton-Jacobi-Bellman equation in stochastic control.

A. Portfolio and Consumption Processes

Let us consider a market in which $d + 1$ assets (or "securities") are traded continuously. We assume throughout this section that there is a fixed time horizon $0 \leq T < \infty$. One of the assets, called the *bond*, has a price $P_0(t)$ which evolves according to the differential equation

(8.1) $$dP_0(t) = r(t)P_0(t)\,dt, \quad P_0(0) = p_0; \quad 0 \leq t \leq T.$$

The remaining d assets, called *stocks*, are "risky"; their prices are modeled by the linear stochastic differential equation for $i = 1, \ldots, d$:

(8.2) $$dP_i(t) = b_i(t)P_i(t)\,dt + P_i(t)\sum_{j=1}^{d} \sigma_{ij}(t)\,dW_t^{(j)}, \quad P_i(0) = p_i; \quad 0 \leq t \leq T.$$

The process $W = \{W_t = (W_t^{(1)}, \ldots, W_t^{(d)})^T, \mathscr{F}_t; 0 \leq t \leq T\}$ is a d-dimensional Brownian motion on a probability space $(\Omega, \mathscr{F}, P)$, and the filtration $\{\mathscr{F}_t\}$ is the augmentation under P of the filtration $\{\mathscr{F}_t^W\}$ generated by W. The *interest rate* process $\{r(t), \mathscr{F}_t; 0 \leq t \leq T\}$, as well as the vector of *mean rates of return*

$\{b(t) = (b_1(t),\ldots,b_d(t))^T, \mathscr{F}_t: 0 \le t \le T\}$ and the *dispersion* matrix $\{\sigma(t) = (\sigma_{ij}(t))_{1 \le i,j \le d}, \mathscr{F}_t; 0 \le t \le T\}$, are assumed to be measurable, adapted, and bounded uniformly in $(t,\omega) \in [0,T] \times \Omega$. We set $a(t) \triangleq \sigma(t)\sigma^T(t)$ and assume that for some number $\varepsilon > 0$,

$$\xi^T a(t)\xi \ge \varepsilon\|\xi\|^2; \quad \forall\, \xi \in \mathbb{R}^d, \quad 0 \le t \le T, \text{ a.s.} \tag{8.3}$$

8.1 Problem. Under assumption (8.3), $\sigma^T(t)$ has an inverse and

$$\|(\sigma^T(t))^{-1}\xi\| \le \frac{1}{\sqrt{\varepsilon}}\|\xi\|; \quad \forall\, \xi \in \mathbb{R}^d, 0 \le t \le T, \quad \text{a.s.} \tag{8.4}$$

Moreover, with $\hat{a}(t) \triangleq \sigma^T(t)\sigma(t)$, we have

$$\xi^T \hat{a}(t)\xi \ge \varepsilon\|\xi\|^2; \quad \forall\, \xi \in \mathbb{R}^d, \quad 0 \le t \le T, \quad \text{a.s.} \tag{8.5}$$

so $\sigma(t)$ also has an inverse and

$$\|(\sigma(t))^{-1}\xi\| \le \frac{1}{\sqrt{\varepsilon}}\|\xi\|; \quad \forall\, \xi \in \mathbb{R}^d, \quad 0 \le t \le T, \quad \text{a.s.} \tag{8.6}$$

We imagine now an investor who starts with some initial endowment $x \ge 0$ and invests it in the $d+1$ assets described previously. Let $N_i(t)$ denote the number of shares of asset i owned by the investor at time t. Then $X_0 \equiv x = \sum_{i=0}^d N_i(0)p_i$, and the investor's *wealth* at time t is

$$X_t = \sum_{i=0}^d N_i(t)P_i(t). \tag{8.7}$$

If trading of shares takes place at discrete time points, say at t and $t+h$, and there is no infusion or withdrawal of funds, then

$$X_{t+h} - X_t = \sum_{i=0}^d N_i(t)[P_i(t+h) - P_i(t)]. \tag{8.8}$$

If, on the other hand, the investor chooses at time $t+h$ to consume an amount hC_{t+h} and reduce the wealth accordingly, then (8.8) should be replaced by

$$X_{t+h} - X_t = \sum_{i=0}^d N_i(t)[P_i(t+h) - P_i(t)] - hC_{t+h}. \tag{8.9}$$

The continuous-time analogue of (8.9) is

$$dX_t = \sum_{i=0}^d N_i(t)\,dP_i(t) - C_t\,dt.$$

Taking (8.1), (8.2), (8.7) into account and denoting by $\pi_i(t) \triangleq N_i(t)P_i(t)$ the amount invested in the i-th stock, $1 \le i \le d$, we may write this as

(8.10)

$$dX_t = (r(t)X_t - C_t)\,dt + \sum_{i=1}^d (b_i(t) - r(t))\pi_i(t)\,dt + \sum_{i=1}^d \sum_{j=1}^d \pi_i(t)\sigma_{ij}(t)\,dW_t^{(j)}.$$

8.2 Definition. A *portfolio process* $\pi = \{\pi(t) = (\pi_1(t), \ldots, \pi_d(t))^T, \mathscr{F}_t; 0 \le t \le T\}$ is a measurable, adapted process for which

$$(8.11) \qquad \sum_{i=1}^{d} \int_0^T \pi_i^2(t)\,dt < \infty, \quad \text{a.s.}$$

A *consumption process* $C = \{C_t, \mathscr{F}_t; 0 \le t \le T\}$ is a measurable, adapted process with values in $[0, \infty)$ and

$$(8.12) \qquad \int_0^T C_t\,dt < \infty, \quad \text{a.s.}$$

8.3 Remark. We note that any component of $\pi(t)$ may become negative, which is to be interpreted as short-selling a stock. The amount invested in the bond,

$$\pi_0(t) \triangleq X_t - \sum_{i=1}^{d} \pi_i(t),$$

may also be negative, and this amounts to borrowing at the interest rate $r(t)$.

8.4 Remark. Conditions (8.11), (8.12) ensure that the stochastic differential equation (8.10) has a unique strong solution. Indeed, formal application of Problem 6.15 leads to the formula

$$(8.13) \qquad X_t = e^{\int_0^t r(s)\,ds}\left\{x + \int_0^t e^{-\int_0^s r(u)\,du}[\pi(s)^T(b(s) - r(s)\underset{\sim}{1}) - C_s]\,ds \right.$$
$$\left. + \int_0^t e^{-\int_0^s r(u)\,du}\pi^T(s)\sigma(s)\,dW_s\right\}; \quad 0 \le t \le T,$$

where $\underset{\sim}{1}$ is the d-dimensional vector with every component equal to 1. All vectors are column vectors, and transposition is denoted by the superscript T. The verification that under (8.11), (8.12), the process X given by (8.13) solves (8.10) is straightforward.

8.5 Definition. A pair (π, C) of portfolio and consumption processes is said to be *admissible for the initial endowment* $x \ge 0$ if the wealth process (8.13) satisfies

$$(8.14) \qquad X_t \ge 0; \quad 0 \le t \le T, \quad \text{a.s.}$$

If $b(t) = r(t)\underset{\sim}{1}$ for $0 \le t \le T$, then the discount factor $e^{-\int_0^t r(s)\,ds}$ exactly offsets the rate of growth of all assets and (8.13) shows that

$$(8.15) \qquad M_t \triangleq X_t e^{-\int_0^t r(s)\,ds} - x + \int_0^t e^{-\int_0^s r(u)\,du} C_s\,ds$$

is a stochastic integral. In other words, the process consisting of current wealth plus cumulative consumption, both properly discounted, is a local martingale. When $b(t) \ne r(t)\underset{\sim}{1}$, (8.15) is no longer a local martingale under P, but becomes one under a new measure $\tilde{P}$ which removes the drift term $\pi(t)^T(b(t) - r(t)\underset{\sim}{1})$ in (8.10). More specifically, recall from Problem 8.1 that the process

(8.16) $$\theta(t) \triangleq (\sigma(t))^{-1}(b(t) - r(t)\underset{\sim}{1})$$

is bounded, and set

(8.17) $$Z_t = \exp\left\{-\sum_{j=1}^{d} \int_0^t \theta_j(s)\,dW_s^{(j)} - \frac{1}{2}\int_0^t \|\theta(s)\|^2\,ds\right\}.$$

Then $Z = \{Z_t, \mathscr{F}_t; 0 \le t \le T\}$ is a martingale (Corollary 3.5.13); the new probability measure

(8.18) $$\tilde{P}(A) \triangleq E[Z_T 1_A]; \quad A \in \mathscr{F}_T$$

is such that P and $\tilde{P}$ are mutually absolutely continuous on $\mathscr{F}_T$, and the process

(8.19) $$\tilde{W}_t \triangleq W_t + \int_0^t \theta(s)\,ds; \quad 0 \le t \le T,$$

is a d-dimensional Brownian motion under $\tilde{P}$ (Theorem 3.5.1). In terms of $\tilde{W}$, (8.13) may be rewritten as

(8.20) $$X_t e^{-\int_0^t r(s)\,ds} + \int_0^t e^{-\int_0^s r(u)\,du} C_s\,ds = x + \int_0^t e^{-\int_0^s r(u)\,du} \pi^T(s)\sigma(s)\,d\tilde{W}_s;$$
$$0 \le t \le T, \text{ a.s.}$$

For an admissible pair (π, C) the left-hand side of (8.20) is nonnegative and the right-hand side is a $\tilde{P}$-local martingale. It follows that the left-hand side, and hence also $X_t e^{-\int_0^t r(s)\,ds}$, is a nonnegative supermartingale under $\tilde{P}$ (Problem 1.5.19). Let

(8.21) $$\tau_0 = T \wedge \inf\{t \in [0, T]; X_t = 0\}.$$

According to Problem 1.3.29,

$$X_t = 0; \ \tau_0 \le t \le T \quad \text{holds a.s. on } \{\tau_0 < T\}.$$

If $\tau_0 < T$, we say that *bankruptcy* occurs at time τ_0.

From the supermartingale property in (8.20) we obtain

(8.22) $$\tilde{E}\left[X_T e^{-\int_0^T r(s)\,ds} + \int_0^T C_t e^{-\int_0^t r(s)\,ds}\,dt\right] \le x,$$

whence the following *necessary condition for admissibility*:

(8.23) $$\tilde{E}\int_0^T e^{-\int_0^t r(s)\,ds} C_t\,dt \le x.$$

This condition is also sufficient for admissibility, in the sense of the following proposition.

8.6 Proposition. *Suppose $x \geq 0$ and a consumption process C are given so that (8.23) is satisfied. Then there exists a portfolio process π such that the pair (π, C) is admissible for the initial endowment x.*

PROOF. Let $D \triangleq \int_0^T C_t e^{-\int_0^t r(s)\,ds}\,dt$, and define the nonnegative process

$$\xi_t \triangleq \tilde{E}\left[\int_t^T C_s e^{-\int_t^s r(u)\,du}\,ds \,\middle|\, \mathscr{F}_t\right] + (x - \tilde{E}D)\,e^{\int_0^t r(s)\,ds},$$

so that

$$\xi_t = e^{\int_0^t r(s)\,ds}\left\{x + m_t - \int_0^t C_s e^{-\int_0^s r(u)\,du}\,ds\right\}, \tag{8.24}$$

where

$$m_t \triangleq \tilde{E}[D|\mathscr{F}_t] - \tilde{E}D = \frac{E[DZ_T|\mathscr{F}_t]}{Z_t} - E(DZ_T)$$

from the Bayes rule of Lemma 3.5.3. Thanks to Theorem 1.3.13, we may assume that P-a.e. path of the martingale

$$N_t \triangleq E(DZ_T|\mathscr{F}_t),\ \mathscr{F}_t;\quad 0 \leq t \leq T,$$

is RCLL, and so, by Problem 3.4.16, there exists a measurable, $\{\mathscr{F}_t\}$-adapted, $\mathbb{R}^d$-valued process Y with

$$\int_0^T \|Y(t)\|^2\,dt < \infty \quad \text{and} \tag{8.25}$$

$$N_t = E(DZ_T) + \sum_{j=1}^d \int_0^t Y_j(s)\,dW_s^{(j)};\quad 0 \leq t \leq T, \tag{8.26}$$

valid a.s. P. Now $m_t = u(N_t, Z_t) - E(DZ_T)$, where $u(x, y) = (x/y)$, and from Itô's rule we obtain with $\varphi(t) \triangleq (Y(t) + N_t\theta(t))/Z_t$:

$$m_t = \sum_{j=1}^d \int_0^t \varphi_j(s)\,d\tilde{W}_s^{(j)};\quad 0 \leq t \leq T. \tag{8.27}$$

We have used the relations $dZ_t = -Z_t\theta^T(t)\,dW_t$ and (8.19). Now define

$$\pi(t) \triangleq e^{\int_0^t r(s)\,ds}(\sigma^T(t))^{-1}\varphi(t), \tag{8.28}$$

so that (8.24) becomes (8.20) when we make the identification $\xi = X$. Condition (8.11) follows from (8.4), (8.25), the boundedness of θ, and the path continuity of Z and N (the latter being a consequence of (8.26)). □

Remark. The representation (8.27) cannot be obtained from a direct application of Problem 3.4.16 to the $\tilde{P}$-martingale $\{m_t, \mathscr{F}_t; 0 \leq t \leq T\}$. The reason is that the filtration $\{\mathscr{F}_t\}$ is the augmentation (under P or $\tilde{P}$) of $\{\mathscr{F}_t^W\}$, *not* of $\{\mathscr{F}_t^{\tilde{W}}\}$.

B. Option Pricing

In the context of the previous subsection, suppose that at time $t = 0$ we sign a contract which gives us the option to buy, at a specified time T (called *maturity* or *expiration date*), one share of stock 1 at a specified price q, called the *exercise price*. At maturity, if the price $P_T^{(1)}$ of stock 1 is below the exercise price, the contract is worthless to us; on the other hand, if $P_T^{(1)} > q$, we can exercise our option (i.e., buy one share at the preassigned price q) and then sell the share immediately in the market for $P_T^{(1)}$. This contract, which is called an *option*, is thus equivalent to a payment of $(P_T^{(1)} - q)^+$ dollars at maturity. Sometimes the term *European option* is used to describe this financial instrument, in contrast to an *American option*, which can be exercised at any time between $t = 0$ and maturity.

The following definition provides a generalization of the concept of option.

8.7 Definition. A *contingent claim* is a financial instrument consisting of:

(i) a *payoff rate* $g = \{g_t, \mathscr{F}_t; 0 \le t \le T\}$, and
(ii) a *terminal payoff* f_T at maturity.

Here g is a nonnegative, measurable, and adapted process, f_T is a nonnegative, $\mathscr{F}_T$-measurable random variable, and for some $\mu > 1$ we have

$$E\left[f_T + \int_0^T g_t \, dt\right]^\mu < \infty. \tag{8.29}$$

8.8 Remark. An option is a special case of a contingent claim with $g \equiv 0$ and $f_T = (P_T^{(1)} - q)^+$.

8.9 Definition. Let $x \ge 0$ be given, and let (π, C) be a portfolio/consumption process pair which is admissible for the initial endowment x. The pair (π, C) is called a *hedging strategy against the contingent claim* (g, f_T), provided

(i) $C_t = g_t$; $0 \le t \le T$, and
(ii) $X_T = f_T$

hold a.s., where X is the wealth process associated with the pair (π, C) and with the initial condition $X_0 = x$.

The concept of hedging strategy is introduced in order to allow the solution of the *contingent claim valuation problem*: What is a fair price to pay at time $t = 0$ for a contingent claim? If there exists a hedging strategy which is admissible for an initial endowment $X_0 = x$, then an agent who buys at time $t = 0$ the contingent claim (g, f_T) for the price x could instead have invested the wealth in such a way as to duplicate the payoff of the contingent claim. Consequently, the price of the claim should not be greater than x. Could one begin with an initial wealth strictly smaller than x and again duplicate the payoff of the contingent claim? The answer to this question may be affirmative, as shown by the following exercise.

8.10 Exercise. Consider the case $r \equiv 0$, $d = 1$, $b_1 \equiv 0$ and $\sigma \equiv 1$. Let the contingent claim $g \equiv 0$ and $f_T \equiv 0$ be given, so obviously there exists a hedging strategy with $x = 0$, $C \equiv 0$, and $\pi \equiv 0$. Show that for each $x > 0$, there is a hedging strategy with $X_0 = x$.

The *fair price* for a contingent claim is the smallest number $x \geq 0$ which allows the construction of a hedging strategy with initial wealth x. We shall show that under the condition (8.3) and the assumptions preceding it, every contingent claim has a fair price; we shall also derive the explicit Black & Scholes (1973) formula for the fair price of an option.

8.11 Lemma. *Let the contingent claim (g, f_T) be given and define*

$$Q = e^{-\int_0^T r(u)\,du} f_T + \int_0^T e^{-\int_0^s r(u)\,du} g_s\,ds. \tag{8.30}$$

Then $\tilde{E}Q$ is finite and is a lower bound on the fair price of (g, f_T).

Proof. Recalling that r is uniformly bounded in t and ω, we may write $Q \leq L(f_T + \int_0^T g_s\,ds)$, where L is some nonrandom constant. From (8.17) we have for every $\nu \geq 1$:

$$\begin{aligned} Z_T^\nu = &\exp\left\{-\sum_{j=1}^d \int_0^T \nu\theta_j(s)\,dW_s^{(j)} - \frac{1}{2}\int_0^T \|\nu\theta(s)\|^2\,ds\right\} \\ &\times \exp\left\{\frac{\nu(\nu-1)}{2}\int_0^T \|\theta(s)\|^2\,ds\right\}, \end{aligned}$$

and because $\|\theta\|$ is bounded by some constant K, it follows that

$$EZ_T^\nu \leq \exp\left\{\frac{\nu(\nu-1)}{2}K^2T\right\} < \infty.$$

With μ as in (8.29) and ν given by $(1/\nu) + (1/\mu) = 1$, the Hölder inequality implies

$$\begin{aligned} \tilde{E}Q &\leq LE\left[Z_T\left(f_T + \int_0^T g_s\,ds\right)\right] \\ &\leq L(EZ_T^\nu)^{1/\nu}\left[E\left(f_T + \int_0^T g_s\,ds\right)^\mu\right]^{1/\mu} < \infty. \end{aligned}$$

Now suppose that (π, C) is a hedging strategy against the contingent claim (g, f_T), and the corresponding wealth process is X with initial condition $X_0 = x$. Recalling the Definition 8.9 and (8.30), we rewrite (8.22) as $\tilde{E}Q \leq x$.

□

8.12 Theorem. *Under condition* (8.3) *and the assumptions preceding it, the fair price of a contingent claim* (g, f_T) *is* $\tilde{E}Q$. *Moreover, there exists a hedging strategy with initial wealth* $x = \tilde{E}Q$.

PROOF. Define

$$\xi_t \triangleq e^{\int_0^t r(s)\,ds}\left[\tilde{E}Q + m_t - \int_0^t e^{-\int_0^s r(u)\,du} g_s\,ds\right], \tag{8.31}$$

where $m_t = \tilde{E}[Q|\mathscr{F}_t] - \tilde{E}Q$. Proceeding exactly as in the proof of Proposition 8.6 with D replaced by Q, we define π by (8.28) and $C \equiv g$, so that (8.31) becomes (8.20) with the identifications $X = \xi, x = \tilde{E}Q$. But then (8.31) can also be cast as

$$X_t = \tilde{E}\left[e^{-\int_t^T r(u)\,du} f_T + \int_t^T e^{-\int_t^s r(u)\,du} g_s\,ds \,\middle|\, \mathscr{F}_t\right]; \quad 0 \le t \le T, \tag{8.32}$$

whence $X_t \ge 0$; $0 \le t \le T$ and $X_T = f_T$ are seen to hold almost surely. □

8.13 Exercise. Show that the hedging strategy constructed in the proof of Theorem 8.12 is essentially (in the sense of meas $\times$ P-a.e. equivalence) the only hedging strategy corresponding to initial wealth $x = \tilde{E}Q$. In particular, the process X of (8.32) gives the unique wealth process corresponding to the fair price; it is called the *valuation process* of the contingent claim.

8.14 Example (Black & Scholes (1973) option valuation formula). In the setting of Remark 8.8 with $d = 1$ and constant coefficients $r(t) \equiv r > 0$, $\sigma_{11}(t) \equiv \sigma > 0$, the price of the bond is

$$P_0(t) = p_0 e^{rt}; \quad 0 \le t \le T,$$

and the price of the stock obeys

$$dP_1(t) = b_1(t)P_1(t)\,dt + \sigma P_1(t)\,dW_t = rP_1(t)\,dt + \sigma P_1(t)\,d\tilde{W}_t.$$

For the option to buy one share of the stock at time T at the price q, we have from (8.32) the valuation process

$$X_t = \tilde{E}[e^{-r(T-t)}(P_1(T) - q)^+ | \mathscr{F}_t]; \quad 0 \le t \le T. \tag{8.33}$$

In order to write (8.33) in a more explicit form, let us observe that the function

(8.34)

$$v(t,x) \triangleq \begin{cases} x\Phi(\rho_+(T-t,x)) - qe^{-r(T-t)}\Phi(\rho_-(T-t,x)); & 0 \le t < T, x \ge 0, \\ (x-q)^+; & t = T, x \ge 0 \end{cases}$$

with

$$\rho_{\pm}(t,x) = \frac{1}{\sigma\sqrt{t}}\left[\log\frac{x}{q} + t\left(r \pm \frac{\sigma^2}{2}\right)\right], \quad \Phi(x) = \frac{1}{\sqrt{2\pi}}\int_{-\infty}^x e^{-z^2/2}\,dz,$$

satisfies the Cauchy problem

$$(8.35)\qquad \begin{aligned} -\frac{\partial v}{\partial t} + rv &= \frac{1}{2}\sigma^2 x^2 \frac{\partial^2 v}{\partial x^2} + rx\frac{\partial v}{\partial x}; \quad \text{on } [0, T) \times (0, \infty) \\ v(T, x) &= (x - q)^+; \quad x \geq 0, \end{aligned}$$

as well as the conditions of Theorem 7.6. We conclude from that theorem and the Markov property applied to (8.33) that

$$(8.36)\qquad X_t = v(t, P_1(t)); \quad 0 \leq t \leq T, \quad \text{a.s.}$$

We thus have an explicit formula for the value of the option at time t in terms of the current stock price $P_1(t)$, the time-to-maturity $T - t$, and the exercise price q.

8.15 Exercise. In the setting of Example 8.14 but with $f_T = h(P_1(T))$, where $h: [0, \infty) \to [0, \infty)$ is a convex, piecewise C^2 function with $h(0) = h'(0) = 0$, show that the valuation process for the contingent claim $(0, f_T)$ is given by

$$(8.37)\qquad X_t = \tilde{E}[e^{-r(T-t)} h(P_1(T)) | \mathscr{F}_t] = \int_0^\infty h''(q) v_{q,T}(t, P_1(t))\, dq.$$

We denote here by $v_{q,T}(t, x)$ the function of (8.34).

C. Optimal Consumption and Investment (General Theory)

In this subsection we pose and solve a stochastic *optimal control problem* for the economics model of Subsection A. Suppose that, in addition to the data given there, we have a measurable, adapted, uniformly bounded *discount process* $\beta = \{\beta(s), \mathscr{F}_s; 0 \leq s \leq T\}$ and a strictly increasing, strictly concave, continuously differentiable *utility function* $U: [0, \infty) \to [0, \infty)$ for which $U(0) = 0$ and $U'(\infty) \triangleq \lim_{c \to \infty} U'(c) = 0$. We allow the possibility that $U'(0) \triangleq \lim_{c \downarrow 0} U'(c) = \infty$. Given an initial endowment $x \geq 0$, an investor wishes to choose an admissible pair (π, C) of portfolio and consumption processes, so as to maximize

$$V_{\pi,C}(x) \triangleq E \int_0^T e^{-\int_0^s \beta(u)\,du} U(C_s)\, ds.$$

We define the *value function* for this problem to be

$$(8.38)\qquad V(x) = \sup_{(\pi,C)} V_{\pi,C}(x),$$

where the supremum is over all pairs (π, C) admissible for x. From the admissibility condition (8.23) it is clear that $V(0) = 0$.

Recall from Proposition 8.6 that for a given consumption process C, (8.23) is satisfied if and only if there exists a portfolio π such that (π, C) is admissible for x. Let us define $\mathscr{D}(x)$ to be the class of consumption processes C for which

$$\tilde{E}\int_0^T e^{-\int_0^t r(s)\,ds} C_t\,dt = x. \tag{8.39}$$

It turns out that in the maximization indicated in (8.38) we may ignore the portfolio process π and we need only consider $C \in \mathscr{D}(x)$.

8.16 Proposition. *For every $x \geq 0$ we have*

$$V(x) = \sup_{C \in \mathscr{D}(x)} E\int_0^T e^{-\int_0^t \beta(s)\,ds} U(C_t)\,dt.$$

PROOF. Suppose (π, C) is admissible for $x > 0$, and set

$$y \triangleq \tilde{E}\int_0^T e^{-\int_0^t r(s)\,ds} C_t\,dt \leq x.$$

If $y > 0$, we may define $\hat{C}_t = (x/y)C_t$ so that $\hat{C} \in \mathscr{D}(x)$. There exists then a portfolio process $\hat{\pi}$ such that $(\hat{\pi}, \hat{C})$ is admissible for x, and

$$V_{\pi,C}(x) \leq V_{\hat{\pi},\hat{C}}(x). \tag{8.40}$$

If $y = 0$, then: $C_t = 0$; a.e. $t \in [0, T]$, almost surely, and we can find a constant $c > 0$ such that $\hat{C}_t \equiv c$ satisfies (8.39). Again, (8.40) holds for some $\hat{\pi}$ chosen so that $(\hat{\pi}, \hat{C})$ is admissible for x. □

Because U': $[0, \infty] \xrightarrow{\text{onto}} [0, U'(0)]$ is strictly decreasing, it has a strictly decreasing inverse function I: $[0, U'(0)] \xrightarrow{\text{onto}} [0, \infty]$. We extend I by setting $I(y) = 0$ for $y > U'(0)$. Note that $I(0) = \infty$ and $I(\infty) = 0$. It is easily verified that

$$U(I(y)) - yI(y) \geq U(c) - yc; \quad 0 \leq c < \infty, \quad 0 < y < \infty. \tag{8.41}$$

Define a function $\mathscr{X}$: $[0, \infty] \to [0, \infty]$ by

$$\mathscr{X}(y) = \tilde{E}\int_0^T e^{-\int_0^s r(u)\,du} I\left(yZ_s e^{\int_0^s (\beta(u) - r(u))\,du}\right) ds, \tag{8.42}$$

and assume that

$$\mathscr{X}(y) < \infty; \quad 0 < y < \infty. \tag{8.43}$$

We shall have more to say about this assumption in the next subsection, where we specialize the model to the case of constant coefficients. Let us define $\bar{y} \triangleq \sup\{y \geq 0; \mathscr{X}$ is *strictly* decreasing on $[0, y]\}$.

8.17 Problem. Under condition (8.43), $\mathscr{X}$ is continuous and strictly decreasing on $[0, \bar{y}]$ with $\mathscr{X}(0) = \infty$ and $\mathscr{X}(\bar{y}) = 0$.

Let $\mathscr{Y}$: $[0, \infty] \xrightarrow{\text{onto}} [0, \bar{y}]$ be the inverse of $\mathscr{X}$. For a given initial endowment $x \geq 0$, define the processes

$$\eta_s^* \triangleq \mathscr{Y}(x) Z_s e^{\int_0^s (\beta(u) - r(u))\,du} \tag{8.44}$$

$$C_s^* \triangleq I(\eta_s^*). \tag{8.45}$$

The definition of $\mathscr{Y}$ implies $C^* \in \mathscr{D}(x)$. We show now that C^* *is an optimal consumption process.*

8.18 Theorem. *Let* $x \geq 0$ *be given and assume that* (8.43) *holds. Then the consumption process given by* (8.45) *is optimal*:

$$V(x) = E \int_0^T e^{-\int_0^t \beta(s)\,ds} U(C_t^*)\,dt. \tag{8.46}$$

PROOF. It suffices to compare C^* to an arbitrary $C \in \mathscr{D}(x)$. For such a C, we have

$$\begin{aligned} E &\int_0^T e^{-\int_0^t \beta(s)\,ds}(U(C_t^*) - U(C_t))\,dt \\ &= E \int_0^T e^{-\int_0^t \beta(s)\,ds}[(U(I(\eta_t^*)) - \eta_t^* I(\eta_t^*)) - (U(C_t) - \eta_t^* C_t)]\,dt \\ &\quad + \mathscr{Y}(x)\tilde{E} \int_0^T e^{-\int_0^t r(s)\,ds}(C_t^* - C_t)\,dt. \end{aligned}$$

The first expectation on the right-hand side is nonnegative because of (8.41); the second vanishes because both C^* and C are in $\mathscr{D}(x)$. □

Having thus determined the value function and the optimal consumption process, we appeal to the construction in the proof of Proposition 8.6 for the determination of a corresponding optimal portfolio process π^*. This does not provide us with a very useful representation for π^*, but one can specialize the model in various ways so as to obtain V, C^* and π^* more explicitly. We do this in the next subsection.

D. Optimal Consumption and Investment (Constant Coefficients)

We consider here a case somewhat more general than that originally studied by Merton (1971) and reported succinctly by Fleming & Rishel (1975), pp. 160–161. In particular, we shall assume that U is three times continuously differentiable and that the model data are constant:

$$\beta(t) \equiv \beta, \quad r(t) \equiv r, \quad b(t) \equiv b, \quad \sigma(t) \equiv \sigma, \tag{8.47}$$

where $b \in \mathbb{R}^d$ and σ is a nonsingular, $(d \times d)$ matrix. We introduce the linear, second-order partial differential operator given by

$$L\varphi(t, y) \triangleq -\varphi_t(t, y) + \beta\varphi(t, y) - (\beta - r)y\varphi_y(t, y) - \tfrac{1}{2}\|\theta\|^2 y^2 \varphi_{yy}(t, y),$$

where $\theta = \sigma^{-1}(b - r\underset{\sim}{1})$ in accordance with (8.16). Our standing assumption throughout this subsection is that θ is different from zero and there exist $C^{1,3}$ functions G: $[0, T] \times (0, \infty) \to [0, \infty)$ and S: $[0, T] \times (0, \infty) \to [0, \infty)$ such that

$$LG(t, y) = U(I(y)); \quad 0 \le t \le T, y > 0 \tag{8.48}$$

$$G(T, y) = 0; \qquad y > 0 \tag{8.49}$$

and

$$LS(t, y) = yI(y); \quad 0 \le t \le T, y > 0 \tag{8.50}$$

$$S(T, y) = 0; \qquad y > 0. \tag{8.51}$$

Here we mean that $G_t(t, y)$, $G_{ty}(t, y)$, $G_y(t, y)$, $G_{yy}(t, y)$, and $G_{yyy}(t, y)$ exist for all $0 \le t \le T$, $y > 0$, and these functions are jointly continuous in (t, y). The same is true for S. We assume, furthermore, that G, G_y, S, and S_y all satisfy polynomial growth conditions of the form

$$\max_{0 \le t \le T} H(t, y) \le M(1 + y^{-\lambda} + y^{\lambda}); \quad y > 0 \tag{8.52}$$

for some $M > 0$ and $\lambda > 0$.

8.19 Problem. Let H: $[0, T] \times (0, \infty) \to [0, \infty)$ be of class $C^{1,2}$ on its domain and satisfy (8.52). Let g: $[0, T] \times (0, \infty) \to [0, \infty)$ be continuous, and assume that H solves the Cauchy problem

$$LH(t, y) = g(t, y); \quad 0 \le t \le T, y > 0$$

$$H(T, y) = 0; \qquad y > 0.$$

Then H admits the stochastic representation

$$H(t, y) = E \int_t^T e^{-\beta(s-t)} g(s, Y_s^{(t,y)})\, ds,$$

where, with $t \le s \le T$:

$$Y_s^{(t,y)} \triangleq y e^{(\beta - r)(s-t)} Z_s^t, \tag{8.53}$$

$$Z_s^t \triangleq \exp\{-\theta^T(W_s - W_t) - \tfrac{1}{2}\|\theta\|^2(s - t)\}. \tag{8.54}$$

(*Hint*: Consider the change of variable $\ell = \log y$.)

From Problem 8.19 we derive the stochastic representation formulas

$$G(t, y) = E \int_t^T e^{-\beta(s-t)} U(I(Y_s^{(t,y)}))\, ds, \tag{8.55}$$

$$S(t, y) = yE \int_t^T e^{-r(s-t)} Z_s^t\, I(Y_s^{(t,y)})\, ds. \tag{8.56}$$

It is useful to consider the consumption/investment problem with initial times other than zero. Thus, for $0 \le t \le T$ fixed and $x \ge 0$, we define the *value function*

(8.57) $$V(t,x) = \sup_{(\pi, C)} E \int_t^T e^{-\beta s} U(C_s)\, ds,$$

where (π, C) must be *admissible for* (t, x), which means that the wealth process determined by the equation

(8.58) $$X_s = x + \int_t^s (rX_u - C_u)\, du + \sum_{i=1}^d \int_t^s (b_i - r)\pi_i(u)\, du$$
$$+ \sum_{i=1}^d \sum_{j=1}^d \int_t^s \pi_i(u)\sigma_{ij}\, dW_u^{(j)}; \quad t \le s \le T,$$

remains nonnegative. Corresponding to a consumption process C, a portfolio process π for which (π, C) is admissible for (t, x) exists if and only if (cf. Proposition 8.6)

(8.59) $$E \int_t^T e^{-r(s-t)} Z_s^t C_s\, ds \le x.$$

For $0 \le t \le T$, define a function $\mathscr{X}(t,\cdot)\colon [0,\infty] \to [0,\infty]$ by

(8.60) $$\mathscr{X}(t,y) \triangleq E \int_t^T e^{-r(s-t)} Z_s^t I(Y_s^{(t,y)})\, ds.$$

Comparison of (8.56) and (8.60) shows that

(8.61) $$y\mathscr{X}(t,y) = S(t,y) < \infty; \quad 0 < y < \infty.$$

Now $\bar{y}(t) \triangleq \sup\{t \ge 0; \mathscr{X}(t,\cdot)$ is *strictly* decreasing on $[0,y]\} = \infty$, and we have just as in Problem 8.17 that for $0 \le t < T$, $\mathscr{X}(t,\cdot)$ is strictly decreasing on $[0,\infty]$ with $\mathscr{X}(t,0) = \infty$ and $\mathscr{X}(t,\infty) = 0$. We denote by $\mathscr{Y}(t,\cdot)\colon [0,\infty] \xrightarrow{\text{onto}} [0,\infty]$ the inverse of $\mathscr{X}(t,\cdot)$:

(8.62) $$\mathscr{Y}(t, \mathscr{X}(t,y)) = y; \quad 0 \le y \le \infty, \quad 0 \le t < T.$$

If we now define for $t \le s \le T$:

(8.63) $$\eta_s^{(t,x)} \triangleq \mathscr{Y}(t,x) Z_s^t\, e^{(\beta - r)(s-t)},$$

(8.64) $$C_s^* \triangleq I(\eta_s^{(t,x)}),$$

then

(8.65) $$V(t,x) = E \int_t^T e^{-\beta s} U(C_s^*)\, ds.$$

This claim is proved as in Theorem 8.18; the new feature here is that for $y = \mathscr{Y}(t,x)$, one has $\eta^{(t,x)} = Y^{(t,y)}$, and consequently

(8.66) $$V(t,x) = e^{-\beta t}G(t, \mathscr{Y}(t,x)); \quad 0 \le t < T, \quad x > 0.$$

Thus, if we can solve the Cauchy problems (8.48), (8.49) and (8.50), (8.51), then we can express $V(t,x)$ in closed form.

8.20 Exercise. Let $U(c) = c^{\delta}$, where $0 < \delta < 1$. Show that if

$$k \triangleq \frac{1}{1-\delta}\left(\beta - r\delta - \frac{1}{2}\|\theta\|^2 \frac{\delta}{1-\delta}\right)$$

is nonzero, then

$$G(t,y) = \frac{1}{k}(1 - e^{-k(T-t)})\left(\frac{y}{\delta}\right)^{\delta/(\delta-1)}$$

$$S(t,y) = \delta G(t,y),$$

$$V(t,x) = e^{-\beta t}\left(\frac{1 - e^{-k(T-t)}}{k}\right)^{1-\delta} x^{\delta}.$$

If $k = 0$, then $G(t,y) = (T-t)(\frac{y}{\delta})^{\delta/(\delta-1)}$, $S(t,y) = \delta G(t,y)$, and $V(t,x) = e^{-\beta t}(T-t)^{1-\delta}x^{\delta}$.

Although we have the representation (8.66) for the value function in our consumption/investment problem, we have not as yet derived representations for the optimal consumption and portfolio processes in feedback form, i.e., as functions of the optimal wealth process. In order to obtain such representations, we introduce the *Hamilton-Jacobi-Bellman* (*HJB*) equation for this model. This *nonlinear*, second-order, partial differential equation offers a characterization of the value function and is the usual technique by which stochastic control problems are attacked. Because of its nonlinear nature, this equation is typically quite difficult to solve. In the present problem, we have already seen how to circumvent the *HJB* equation by solving instead the two *linear* equations (8.48) and (8.50).

8.21 Lemma (Verification Result for the HJB Equation). *Suppose* Q: $[0,T] \times [0,\infty) \to [0,\infty)$ *is continuous, is of class* $C^{1,2}([0,T) \times (0,\infty))$, *and solves the* HJB *equation*

(8.67)
$$Q_t(t,x) + \max_{\substack{c \ge 0 \\ \pi \in \mathbb{R}^d}} \left\{ [rx - c + (b - r\underset{\sim}{1})^T \pi] Q_x(t,x) + \frac{1}{2}\|\sigma^T \pi\|^2 Q_{xx}(t,x) + e^{-\beta t}U(c) \right\} = 0; \quad 0 \le t < T, \quad 0 < x < \infty.$$

Then

(8.68) $$V(t,x) \le Q(t,x); \quad 0 \le t < T, \quad 0 \le x < \infty.$$

Proof. For any initial condition $(t,x) \in [0,T) \times (0,\infty)$ and pair (π, C) admissible at (t,x), let $\{X_s; t \le s \le T\}$ denote the wealth process determined by (8.58). With

$$\tau_n \triangleq \left(T - \frac{1}{n}\right)^+ \wedge \inf\left\{s \in [t,T];\ X_s \ge n \text{ or } X_s \le \frac{1}{n} \text{ or } \int_0^s \|\pi(u)\|^2\,du = n\right\},$$

we have $E\int_t^{\tau_n} Q_x(s,X_s)\pi^T(s)\sigma\,dW_s = 0$. Therefore, Itô's rule implies, in conjunction with (8.58) and (8.67),

$$\begin{aligned}
0 &\le EQ(\tau_n, X_{\tau_n}) \\
&= Q(t,x) + E\int_t^{\tau_n} \{Q_t(s,X_s) + [rX_s - C_s + (b - r\underline{1})^T\pi(s)]Q_x(s,X_s) \\
&\quad + \frac{1}{2}\|\sigma^T\pi(s)\|^2 Q_{xx}(s,X_s)\}\,ds \le Q(t,x) - E\int_t^{\tau_n} e^{-\beta s}U(C_s)\,ds.
\end{aligned}$$

Letting $n \to \infty$ and using the monotone convergence theorem, we obtain $E\int_t^T e^{-\beta s}U(C_s)\,ds \le Q(t,x)$. Maximization of the left-hand side over admissible pairs (π, C) gives the desired result. □

A solution to the *HJB* equation may not be unique, even if we specify the boundary conditions

(8.69) $\quad Q(t,0) = 0;\quad 0 \le t \le T \quad \text{and} \quad Q(T,x) = 0;\quad 0 \le x < \infty.$

This is because different rates of growth of $Q(t,x)$ are possible as x approaches infinity. One expects the value function to satisfy the *HJB* equation, and, in light of (8.68), to be distinguished by being the smallest nonnegative solution of this equation.

8.22 Proposition. *Under the conditions set forth at the beginning of this subsection, the value function* $V: [0,T] \times [0,\infty) \to [0,\infty)$ *is continuous, is of class* $C^{1,2}([0,T) \times (0,\infty))$, *and satisfies the HJB equation* (8.67) *as well as the boundary conditions* (8.69).

Proof. If $0 < y \le U'(0)$, then

(8.70) $$\frac{d}{dy}U(I(y)) = U'(I(y))I'(y) = yI'(y);$$

if $y > U'(0)$, then $I(y) = I'(y) = 0$ and (8.70) still holds. Because of our assumption that G and S are of class $C^{1,3}$, we may differentiate (8.48), (8.50) with respect to y and observe that $\varphi_1(t,y) \triangleq -yG_y(t,y)$ and $\varphi_2(t,y) \triangleq -y^2(\partial/\partial y)(S(t,y)/y)$ both satisfy

$$\begin{aligned}
L\varphi_i(t,y) &= -y^2 I'(y); \quad && 0 \le t \le T,\ y > 0, \\
\varphi_i(T,y) &= 0; && y > 0.
\end{aligned}$$

In particular, I' is continuous at $y = U'(0)$, i.e., a necessary condition for our assumptions is $U''(0) = \infty$. Problem 8.19 implies $\varphi_1 = \varphi_2$, because both functions have the same stochastic representation. It follows that

$$G_y(t, y) = y \frac{\partial}{\partial y}\left(\frac{1}{y} S(t, y)\right) = y\mathscr{X}_y(t, y) \tag{8.71}$$

and from (8.66), (8.62) we have

$$V_x(t, x) = e^{-\beta t}\mathscr{Y}(t, x), \tag{8.72}$$

$$\mathscr{Y}_t(t, \mathscr{X}(t, y)) = -\mathscr{Y}_x(t, \mathscr{X}(t, y))\, \mathscr{X}_t(t, y). \tag{8.73}$$

Finally, (8.50) and (8.61) imply that

(8.74)

$$-\mathscr{X}_t(t, y) + r\mathscr{X}(t, y) - (\beta - r + \|\theta\|^2)y\mathscr{X}_y(t, y) - \frac{1}{2}\|\theta\|^2 y^2 \mathscr{X}_{yy}(t, y) = I(y);$$
$$0 < y < \infty, \quad 0 \le t < T.$$

We want to check now that the function $V(t, x)$ of (8.66) satisfies the *HJB* equation (8.67). With $Q = V$, the left-hand side of this equation becomes $e^{-\beta t}$ times

$$G_t(t, \mathscr{Y}(t, x)) - \beta G(t, \mathscr{Y}(t, x)) + G_y(t, \mathscr{Y}(t, x))\, \mathscr{Y}_t(t, x) \tag{8.75}$$
$$+ \max_{\substack{c \ge 0 \\ \pi \in \mathbb{R}^d}} \left[((rx - c) + (b - r\underset{\sim}{1})^T \pi)\mathscr{Y}(t, x) + \frac{1}{2}\|\sigma^T \pi\|^2 \mathscr{Y}_x(t, x) + U(c) \right].$$

The maximization over c is accomplished by setting

$$c = I(\mathscr{Y}(t, x)). \tag{8.76}$$

Because of the negativity of $\mathscr{Y}_x$, the maximization over π is accomplished by setting

$$\pi = -(\sigma\sigma^T)^{-1}(b - r\underset{\sim}{1})\frac{\mathscr{Y}(t, x)}{\mathscr{Y}_x(t, x)}. \tag{8.77}$$

Upon substitution of (8.76), (8.77) into (8.75), the latter becomes

$$G_t(t, \mathscr{Y}(t, x)) - \beta G(t, \mathscr{Y}(t, x)) + G_y(t, \mathscr{Y}(t, x))\mathscr{Y}_t(t, x) \tag{8.78}$$
$$+ rx\mathscr{Y}(t, x) - \mathscr{Y}(t, x)I(\mathscr{Y}(t, x)) - \frac{1}{2}\|\theta\|^2 \frac{\mathscr{Y}^2(t, x)}{\mathscr{Y}_x(t, x)} + U(I(\mathscr{Y}(t, x))).$$

We may change variables in (8.78), taking $y = \mathscr{Y}(t, x)$ and using (8.71), (8.73), (8.48) to write this expression as

$$G_t(t, y) - \beta G(t, y) - y\mathscr{X}_t(t, y) + ry\mathscr{X}(t, y) - yI(y) - \tfrac{1}{2}\|\theta\|^2 y^2 \mathscr{X}_y(t, y) + U(I(y))$$
$$= y[-\mathscr{X}_t(t, y) + r\mathscr{X}(t, y) - (\beta - r + \|\theta\|^2)y\mathscr{X}_y(t, y)$$
$$- \tfrac{1}{2}\|\theta\|^2 y^2 \mathscr{X}_{yy}(t, y) - I(y)],$$

which vanishes because of (8.74). This completes the proof that V satisfies the *HJB* equation (8.67). The boundary conditions (8.69) are satisfied by V by virtue of its definition (8.57) and the admissibility condition (8.59) applied when $x = 0$. □

In conclusion, we have already shown that for fixed but arbitrary $(t, x) \in [0, T) \times (0, \infty)$, there is an optimal pair (π^*, C^*) of portfolio/consumption processes. Let $\{X_s^*; t \le s \le T\}$ denote the corresponding wealth process. If we now repeat the proof of Lemma 8.21, replacing (π, C) by (π^*, C^*) and Q by V, we can derive the inequality

(8.79)
$$0 \le V(t,x) + E\int_t^T \Big\{V_t(s, X_s^*) + [rX_s^* - C_s^* + (b - r\underline{1})^T\pi^*(s)]V_x(s, X_s^*)$$
$$+ \frac{1}{2}\|\sigma^T\pi^*(s)\|^2 V_{xx}(s, X_s^*)\Big\}\,ds \le V(t,x) - E\int_t^T e^{-\beta s}U(C_s^*)\,ds.$$

We have used the monotone convergence theorem and the inequality

(8.80)
$$V_t(s, X_s^*) + [rX_s^* - C_s^* + (b - r\underline{1})^T\pi^*(s)]V_x(s, X_s^*)$$
$$+ \tfrac{1}{2}\|\sigma^T\pi^*(s)\|^2 V_{xx}(s, X_s^*) \le -e^{-\beta s}U(C_s^*) \le 0; \quad t \le s \le T,$$

which follows from the HJB equation for V. But (8.65) holds, so equality prevails in (8.79) and hence also in the first inequality of (8.80), at least for meas $\times$ P-almost every (s, ω) in $[t, T] \times \Omega$. Equality in (8.80) occurs if and only if π_s^* and C_s^* maximize the expression

$$[rX_s^* - c + (b - r\underline{1})^T\pi]V_x(s, X_s^*) + \tfrac{1}{2}\|\sigma^T\pi\|^2 V_{xx}(s, X_s) + e^{-\beta t}U(c);$$

i.e. (cf. (8.76), (8.77)),

(8.81)
$$C_s^* = I(\mathscr{Y}(s, X_s^*)),$$

(8.82)
$$\pi_s^* = -(\sigma\sigma^T)^{-1}(b - r\underline{1})\frac{\mathscr{Y}(s, X_s^*)}{\mathscr{Y}_x(s, X_s^*)},$$

where again both identities hold for meas $\times$ P-almost every $(s, \omega) \in [t, T] \times \Omega$. The expressions (8.81), (8.82) provide the optimal consumption and portfolio processes in *feedback form.*

8.23 Exercise. Show that in the context of Exercise 8.20, the optimal consumption and portfolio processes are linear functions of the wealth process X^*. Solve for the latter and show that $X_T^* = 0$ a.s.

5.9. Solutions to Selected Problems

2.7. We have from (2.10)

$$\frac{d}{dt}\left(e^{-\beta t}\int_0^t g(s)\,ds\right) = \left(g(t) - \beta\int_0^t g(s)\,ds\right)e^{-\beta t} \le \alpha(t)e^{-\beta t},$$

whence $\int_0^t g(s)\,ds \le e^{\beta t}\int_0^t \alpha(s)e^{-\beta s}\,ds$. Substituting this estimate back into (2.10), we obtain (2.11).

2.10. We first check that each $X_t^{(k)}$ is defined for all $t \ge 0$. In particular, we must show that for $k \ge 0$,

$$\int_0^t (\|b(s, X_s^{(k)})\| + \|\sigma(s, X_s^{(k)})\|^2)\,ds < \infty; \quad 0 \le t < \infty, \text{ a.s.}$$

In light of (2.13), this will follow from

$$\sup_{0\le t\le T} E\|X_t^{(k)}\|^2 < \infty; \quad 0 \le T < \infty, \tag{9.1}$$

a fact which we prove by induction. For $k = 0$, (9.1) is obvious. Assume (9.1) for some value of k. Proceeding similarly to the proof of Theorem 2.5, we obtain the bound for $0 \le t \le T$:

$$E\|X_t^{(k+1)}\|^2 \le 9E\|\xi\|^2 + 9(T+1)K^2 \int_0^t (1 + E\|X_s^{(k)}\|^2)\,ds, \tag{9.2}$$

which gives us (9.1) for $k + 1$. From (9.2) we also have

$$E\|X_t^{(k+1)}\|^2 \le C(1 + E\|\xi\|^2) + C\int_0^t E\|X_s^{(k)}\|^2\,ds; \quad 0 \le t \le T,$$

where C depends only on K and T. Iteration of this inequality gives

$$E\|X_t^{(k+1)}\|^2 \le C(1 + E\|\xi\|^2)\left[1 + Ct + \frac{(Ct)^2}{2!} + \cdots + \frac{(Ct)^{k+1}}{(k+1)!}\right],$$

and (2.17) follows.

2.11. We will obtain (2.4) by letting $k \to \infty$ in (2.16), once we show that the two integrals on the right-hand side of (2.16) converge to the proper quantities. With $T > 0$, (2.21) gives $\max_{0\le t\le T}\|X_t(\omega) - X_t^{(k)}(\omega)\| \le 2^{-k}$, $\forall\, k \ge N(\omega)$. Consequently,

$$\left\|\int_0^t b(s, X_s^{(k)})\,ds - \int_0^t b(s, X_s)\,ds\right\|^2 \le K^2T\int_0^T \|X_s^{(k)} - X_s\|^2\,ds$$

converges to zero a.s. for $0 \le t \le T$, as $k \to \infty$. In order to deal with the stochastic integral, we observe from (2.19) that for fixed $0 \le t \le T$, the sequence of random variables $\{X_t^k\}_{k=1}^\infty$ is Cauchy in $L^2(\Omega, \mathscr{F}, P)$, and since $X_t^{(k)} \to X_t$ a.s., we must have $E\|X_t^{(k)} - X_t\|^2 \to 0$ as $k \to \infty$. Moreover, (2.17) shows that $E\|X_t^{(k)}\|^2$ is uniformly bounded for $0 \le t \le T$ and $k \ge 0$, and from Fatou's lemma we conclude that $E\|X_t\|^2 \le \underline{\lim}_{k\to\infty} E\|X_t^{(k)}\|^2$ is uniformly bounded as well. From (2.12) and the bounded convergence theorem, we have

$$E\left\|\int_0^t \sigma(s, X_s^{(k)})\,dW_s - \int_0^t \sigma(s, X_s)\,dW_s\right\|^2 = E\int_0^t \|\sigma(s, X_s^{(k)}) - \sigma(s, X_s)\|^2\,ds$$

$$\le K^2\int_0^t E\|X_s^{(k)} - X_s\|^2\,ds \to 0 \quad \text{as} \quad k \to \infty; \quad 0 \le t \le T.$$

2.12. For each positive integer k, define the stopping time for $\{\mathscr{F}_t\}$:

$$T_k = \begin{cases} 0 & \text{if } \|\xi\| > k, \\ \infty & \text{if } \|\xi\| \le k, \end{cases}$$

and set $\xi_k = \xi 1_{\{\|\xi\| \le k\}}$. We consider the unique (continuous, square-integrable) strong solution $X^{(k)}$ of the equation

$$X_t^{(k)} = \xi_k + \int_0^t b(s, X_s^{(k)})\,ds + \int_0^t \sigma(s, X_s^{(k)})\,dW_s; \quad 0 \le t < \infty.$$

For $\ell > k$, we have that $(X_t^{(k)} - X_t^{(\ell)})1_{\{\|\xi\| \le k\}}$ is equal to

$$\int_0^{t\wedge T_k} \{b(s, X_s^{(k)}) - b(s, X_s^{(\ell)})\}\,ds + \int_0^{t\wedge T_k} \{\sigma(s, X_s^{(k)}) - \sigma(s, X_s^{(\ell)})\}\,dW_s$$

$$= \int_0^t \{b(s, X_s^{(k)}) - b(s, X_s^{(\ell)})\}1_{\{s\le T_k\}}\,ds + \int_0^t \{\sigma(s, X_s^{(k)}) - \sigma(s, X_s^{(\ell)})\}1_{\{s\le T_k\}}\,dW_s.$$

By repeating the argument which led to (2.18), we obtain for $0 \le t \le T$:

$$E\left[\max_{0\le s\le t} \|X_s^{(\ell)} - X_s^{(k)}\|^2 1_{\{\|\xi\|\le k\}}\right] \le L\int_0^t E\left[\max_{0\le s\le t} \|X_s^{(\ell)} - X_s^{(k)}\|^2 1_{\{\|\xi\|\le k\}}\right] dt,$$

and Gronwall's inequality (Problem 2.7) now yields

$$\max_{0\le s\le T} \|X_s^{(\ell)} - X_s^{(k)}\| = 0, \quad \text{a.s. on } \{\|\xi\| \le k\}.$$

We may thus define a process $\{X_t; 0 \le t < \infty\}$ by setting $X_t(\omega) = X_t^{(k)}(\omega)$, where k is chosen larger that $\|\xi(\omega)\|$. Then

$$\begin{aligned}
X_t 1_{\{\|\xi\|\le k\}} &= X_t^{(k)} 1_{\{\|\xi\|\le k\}} \\
&= \xi_k + \int_0^{t\wedge T_k} b(s, X_s^{(k)})\,ds + \int_0^{t\wedge T_k} \sigma(s, X_s^{(k)})\,dW_s \\
&= \xi_k + \int_0^{t\wedge T_k} b(s, X_s)\,ds + \int_0^{t\wedge T_k} \sigma(s, X_s)\,dW_s \\
&= \left[\xi + \int_0^t b(s, X_s)\,ds + \int_0^t \sigma(s, X_s)\,dW_s\right]1_{\{\|\xi\|\le k\}}.
\end{aligned}$$

Since $P[\bigcup_{k=1}^\infty \{\|\xi\| \le k\}] = 1$, we see that X satisfies (2.4) almost surely.

3.15. We use the inequality for $p > 0$;

$$(9.3) \quad |a_1|^p + \cdots + |a_n|^p \le n(|a_1| + \cdots + |a_n|)^p \le n^{p+1}(|a_1|^p + \cdots + |a_n|^p).$$

We shall denote by $C(m,d)$ a positive constant depending on m and d, not necessarily the same throughout the solution. From (iv) and (9.3) we have, almost surely:

$$\|X_t\|^{2m} \le C(m,d)\left[\|X_0\|^{2m} + \left\|\int_0^t b(u,X)\,du\right\|^{2m} + \left\|\int_0^t \sigma(u,X)\,dW_u\right\|^{2m}\right],$$

and the Hölder inequality provides the bound

$$\begin{aligned}
\left\|\int_0^t b(u,X)\,du\right\|^{2m} &= \left[\sum_{i=1}^d \left(\int_0^t b_i(u,X)\,du\right)^2\right]^m \le t^m\left[\int_0^t \|b(u,X)\|^2\,du\right]^m \\
&\le t^{2m-1}\int_0^t \|b(u,X)\|^{2m}\,du.
\end{aligned}$$

The stopping times $S_k \triangleq \inf\{t \geq 0; \|X_t\| \geq k\}$ tend almost surely to infinity as $k \to \infty$, and truncating at them yields

$$\max_{0 \leq s \leq t \wedge S_k} \|X_s\|^{2m} \leq C(m,d)\left[\|X_0\|^{2m} + t^{2m-1}\int_0^{t \wedge S_k} \|b(u,X)\|^{2m}\,du + \max_{0 \leq s \leq t}\left\|\int_0^{s \wedge S_k} \sigma(u,X)\,dW_u\right\|^{2m}\right], \quad \text{a.s.}$$

Remark 3.3.30 and Hölder's inequality provide a bound for the last term, to wit,

$$E\left[\max_{0 \leq s \leq t}\left\|\int_0^{s \wedge S_k} \sigma(u,X)\,dW_u\right\|^{2m}\right] \leq C(m,d)\,E\left[\int_0^{t \wedge S_k} \|\sigma(u,X)\|^2\,du\right]^m$$
$$\leq C(m,d)t^{m-1}E\int_0^{t \wedge S_k} \|\sigma(u,X)\|^{2m}\,du.$$

Therefore, upon taking expectations, we obtain

$$E\left(\max_{0 \leq s \leq t \wedge S_k} \|X_s\|^{2m}\right)$$
$$\leq C(m,d)\left[E\|X_0\|^{2m} + C(T)\,E\int_0^{t \wedge S_k} (\|b(u,X)\|^{2m} + \|\sigma(u,X)\|^{2m})\,du\right]$$

for every $0 \leq t \leq T$, where $C(T)$ is a constant depending on T, m, and d. But now we employ the linear growth condition (3.16) to obtain finally,

$$\psi_k(t) \triangleq E\left(\max_{0 \leq s \leq t \wedge S_k} \|X_s\|^{2m}\right) \leq C\left[1 + E\|X_0\|^{2m} + \int_0^t \psi_k(u)\,du\right];$$
$$0 \leq t \leq T,$$

and from the Gronwall inequality (Problem 2.7):

$$\psi_k(t) \leq C(1 + E\|X_0\|^{2m})e^{Ct}; \quad 0 \leq t \leq T.$$

Now (3.17) follows from Fatou's lemma. On the other hand, if we fix $s < t$ in $[0,T]$ and start from the inequality $\|X_t - X_s\|^{2m} \leq C(m,d)(\|\int_s^t b(u,X)\,du\|^{2m} + \|\int_s^t \sigma(u,X)\,dW_u\|^{2m})$, we may proceed much as before, to obtain

$$E\|X_t - X_s\|^{2m} \leq C(t-s)^{m-1}\int_s^t \left[1 + E\left(\max_{0 \leq \theta \leq u} \|X_\theta\|^{2m}\right)\right]du$$
$$\leq C(1 + Ce^{CT})(1 + E\|X_0\|^{2m})(t-s)^m.$$

3.21. We fix $F \in \mathscr{B}_t(C[0,\infty)^d)$. The class of sets G satisfying (3.25) is a Dynkin system. The collection of sets of the form (3.26) is closed under pairwise intersection and generates the σ-field $\mathscr{B}(\mathbb{R}^d) \otimes \mathscr{B}(C[0,\infty)^r)$. By the Dynkin System Theorem, it suffices to prove (3.25) for G of the form (3.26). For such a G, we have

$$\int_G Q_j^t(x,w;F)\mu(dx)P_*(dw)$$
$$= E_*\left[\int_{G_1} Q_j^t(x,\cdot;F)\mu(dx)\cdot 1_{\varphi_t^{-1}G_2}\cdot P_*\{\sigma_t^{-1}G_3 \mid \mathscr{B}_t(C[0,\infty)^r)\}\right]$$

$$= E_*\left[\int_{G_1} Q_j^t(x,\cdot\,; F)\mu(dx)\cdot 1_{\varphi_t^{-1}G_2}\right]\cdot P_*[\sigma_t^{-1}G_3]$$

$$= P_j[G_1 \times \varphi_t^{-1}G_2 \times F]\cdot P_*[\sigma_t^{-1}G_3].$$

The crucial observation here is that, under Wiener measure P_*, $\sigma_t^{-1}G_3$ is independent of $\mathscr{B}_t(C[0,\infty)^r)$. From (3.21), we have

$$P_*[\sigma_t^{-1}G_3] = P_j[(x,w,y)\in\Theta;\ \sigma_t w\in G_3] = \nu_j[\sigma_t W^{(j)}\in G_3],$$

$$P_j[G_1 \times \varphi_t^{-1}G_2 \times F] = \nu_j[X_0^{(j)}\in G_1, \varphi_t W^{(j)}\in G_2, Y^{(j)}\in F].$$

Therefore,

$$\begin{aligned}\int_G Q_j^t(x,w;F)\mu(dx)P_*(dw) &= \nu_j[X_0^{(j)}\in G_1, \varphi_t W^{(j)}\in G_2, Y^{(j)}\in F]\cdot\nu_j[\sigma_t W^{(j)}\in G_3]\\ &= \nu_j[X_0^{(j)}\in G_1, \varphi_t W^{(j)}\in G_2, \sigma_t W^{(j)}\in G_3, Y^{(j)}\in F]\\ &= \nu_j[(X_0^{(j)}, W^{(j)})\in G, Y^{(j)}\in F] = P_j[G\times F],\end{aligned}$$

because $\{X_0^{(j)}\in G_1, \varphi_t W^{(j)}\in G_2, Y^{(j)}\in F\}$ belongs to the σ-field $\mathscr{F}_t^{(j)}$, and $\{\sigma_t W^{(j)}\in G_3\}$ is independent of it.

3.22. The essential assertion here is that $Q_j(x,w;\cdot)$ assigns full measure to a singleton, which is the same for $j=1,2$. To see this, fix $(x,w)\in\mathbb{R}^d\times C[0,\infty)^r$ and define the measure $Q(x,w;dy_1,dy_2) \triangleq Q_1(x,w;dy_1)Q_2(x,w;dy_2)$ on $(S,\mathscr{S}) \triangleq (C[0,\infty)^d\times C[0,\infty)^d, \mathscr{B}(C[0,\infty)^d)\otimes\mathscr{B}(C[0,\infty)^d))$. We have from (3.23)

$$(9.4)\quad P(G\times B) = \int_G Q(x,w;B)\mu(dx)P_*(dw);\quad B\in\mathscr{S}, G\in\mathscr{B}(\mathbb{R}^d)\otimes\mathscr{B}(C[0,\infty)^r).$$

With the choice $B = \{(y_1,y_2)\in S;\ y_1 = y_2\}$ and $G = \mathbb{R}^d\times C[0,\infty)^r$, relations (3.24) and (9.4) yield the existence of a set $N\in\mathscr{B}(\mathbb{R}^d)\otimes\mathscr{B}(C[0,\infty)^r)$ with $(\mu\times P_*)(N) = 0$, such that $Q(x,w;B) = 1$ for all $(x,w)\notin N$. But then from Fubini,

$$1 = Q(x,w;B) = \int_{C[0,\infty)^d} Q_1(x,w;\{y\})\,Q_2(x,w;dy);\quad (x,w)\notin N,$$

which can occur only if for some y, call it $k(x,w)$, we have $Q_j(x,w;\{k(x,w)\}) = 1$; $j=1,2$. This gives us (3.27). For $(x,w)\notin N$, and any $B\in\mathscr{B}(C[0,\infty)^d)$, we have $k(x,w)\in B \Leftrightarrow Q_j(x,w;B) = 1$. The $\mathscr{B}(\mathbb{R}^d)\otimes\mathscr{B}(C[0,\infty)^r)/\mathscr{B}([0,1])$-measurability of $(x,w)\mapsto Q_j(x,w;B)$ implies the $\mathscr{B}(\mathbb{R}^d)\otimes\mathscr{B}(C[0,\infty)^r)/\mathscr{B}(C[0,\infty)^d)$-measurability of k. The $\hat{\mathscr{B}}_t/\mathscr{B}_t(C[0,\infty)^d)$-measurability of k follows from a similar argument, which makes use of Problem 3.21. Equation (3.28) is a direct consequence of (3.27) and the definition (3.23) of P.

4.7. For $(\rho,\sigma)\in D$, let $A = \rho\rho^T = \sigma\sigma^T$. Since A is symmetric and positive semidefinite, there is an orthogonal matrix Q such that $QAQ^T = \left[\begin{array}{c|c}\Lambda & 0\\ \hline 0 & 0\end{array}\right]$, where Λ is a $(k\times k)$ diagonal matrix whose diagonal elements are the nonzero eigenvalues of A. Since

$$QAQ^T = (Q\rho)(Q\rho)^T = (Q\sigma)(Q\sigma)^T,$$

$Q\rho$ must have the form $Q\rho = \left[\begin{array}{c}Y_1\\ \hline 0\end{array}\right]$, where $Y_1Y_1^T = \Lambda$. Likewise, $Q\sigma = \left[\begin{array}{c}Z_1\\ \hline 0\end{array}\right]$, where $Z_1 Z_1^T = \Lambda$. We can compute an orthonormal basis for the $(d-k)$-dimensional subspace orthogonal to the span of the k rows of Y_1. Let Y_2 be the $(d-k) \times d$ matrix consisting of the orthonormal row vectors of this basis, and set $Y = \begin{bmatrix} Y_1 \\ \hline Y_2 \end{bmatrix}$. Then $(Q\rho)Y^T = \left[\begin{array}{c|c} \Lambda & 0 \\ \hline 0 & 0 \end{array}\right]$, $YY^T = \left[\begin{array}{c|c} \Lambda & 0 \\ \hline 0 & I \end{array}\right]$, where I is the $(d-k) \times (d-k)$ identity matrix. In the same way, define $Z = \begin{bmatrix} Z_1 \\ \hline Z_2 \end{bmatrix}$ so that $(Q\sigma)Z^T = (Q\rho)Y^T$, $ZZ^T = YY^T$. Then $\sigma = \rho Y^T(Z^T)^{-1}$, so we set $R = Y^T(Z^T)^{-1}$. It is easily verified that $RR^T = I$. All of the steps necessary to compute R from ρ and σ can be accomplished by Borel-measurable transformations.

4.13. Let us fix $0 \le s < t < \infty$. For any integer $n > (t-s)^{-1}$ and every $A \in \mathscr{G}_{s+(1/n)}$, we can find $B \in \mathscr{B}_{s+(1/n)}$ such that $P(A \triangle B) = 0$ (Problem 2.7.3). The martingale property for $\{M_u^f, \mathscr{B}_u; 0 \le u < \infty\}$ implies that

$$E\left[\left\{f(y(t)) - f\left(y\left(s+\frac{1}{n}\right)\right) - \int_{s+(1/n)}^t (\mathscr{A}_u^{\prime} f)(y)\,du\right\}1_A\right] \tag{9.5}$$

$$= E\left[\left\{f(y(t)) - f\left(y\left(s+\frac{1}{n}\right)\right) - \int_{s+(1/n)}^t (\mathscr{A}_u^{\prime} f)(y)\,du\right\}1_B\right] = 0.$$

If follows that the expectation in (9.5) is equal to zero for every $A \in \mathscr{F}_s = \mathscr{G}_{s+}$. We can then let $n \to \infty$ and obtain the martingale property $E[(M_t^f - M_s^f)1_A] = 0$ from the bounded convergence theorem.

4.25. Suppose $\int_{\mathbb{R}^d} \varphi(x)\mu_1(dx) = \int_{\mathbb{R}^d} \varphi(x)\mu_2(dx)$ for every $\varphi \in C_0^\infty(\mathbb{R}^d)$, and take $\psi \in C_0(\mathbb{R}^d)$. Let $\rho \in C_0^\infty(\mathbb{R}^d)$ satisfy $\rho \ge 0$, $\int_{\mathbb{R}^d} \rho(x)\,dx = 1$, and set $\varphi_n(x) \triangleq \int_{\mathbb{R}^d} \psi(x + (y/n))\rho(y)\,dy = n\int_{\mathbb{R}^d} \psi(z)\rho(nz - nx)\,dz$. Then $\varphi_n \in C_0^\infty(\mathbb{R}^d)$ and $\varphi_n(x) \to \psi(x)$ for every $x \in \mathbb{R}^d$. It follows from the bounded convergence theorem that $\int_{\mathbb{R}^d} \psi(x)\mu_1(dx) = \int_{\mathbb{R}^d} \psi(x)\mu_2(dx)$, for every $\psi \in C_0(\mathbb{R}^d)$. Now suppose $G \subset \mathbb{R}^d$ is open and bounded. Let $\psi_n(x) = 1 \wedge \inf_{y \notin G} n\|y - x\|$. Then $\psi_n \in C_0(\mathbb{R}^d)$ for all n, and $\psi_n \uparrow 1_G$. It follows from the monotone convergence theorem that $\mu_1(G) = \mu_2(G)$ for every bounded open set G. The collection of sets $\mathscr{C} = \{B \in \mathscr{B}(\mathbb{R}^d); \mu_1(B) = \mu_2(B)\}$ form a Dynkin system, and since every bounded, open set is in $\mathscr{C}$, the Dynkin System Theorem 2.1.3 shows that $\mathscr{C} = \mathscr{B}(\mathbb{R}^d)$.

5.3. For $t \ge 0$, let $E_t = \{\int_0^{t \wedge S} \sigma^2(X_s)\,ds = \infty\}$. Using the method of Solution 3.4.11, we can show that

$$\overline{\lim_{n\to\infty}} X_{t\wedge S_n} = X_0 + \overline{\lim_{n\to\infty}} \int_0^{t\wedge S_n} \sigma(X_s)\,dW_s = \infty, \quad \text{a.s. on } E_t,$$

$$\underline{\lim_{n\to\infty}} X_{t\wedge S_n} = X_0 + \underline{\lim_{n\to\infty}} \int_0^{t\wedge S_n} \sigma(X_s)\,dW_s = -\infty, \quad \text{a.s. on } E_t.$$

But X_t is continuous in the topology of the extended real numbers, so $P(E_t) = 0$. Consequently, $X_{t\wedge S} = X_0 + \int_0^{t\wedge S} \sigma(X_s)\,dW_s$ is real-valued a.s., for every $t \ge 0$, so $S = \infty$ a.s.

5.27. For $\varepsilon > 0$ and $c + \varepsilon \le x < r$, we have

$$v(x) = \int_c^x p'(y) \int_c^y \frac{2\,dz}{p'(z)\sigma^2(z)}\,dy \ge [p(x) - p(c+\varepsilon)] \int_c^{c+\varepsilon} \frac{2\,dz}{p'(z)\sigma^2(z)},$$

and (5.73) follows. A similar argument works for (5.74).

6.2. It suffices to show that if $Q(t)$ is a locally bounded, measurable, $(d \times r)$-matrix-valued function of t, then $Y_t = \int_0^t Q(s)\,dW_s$ is a Gaussian process. This is certainly true if the components of Q are simple, and the general Q can be approximated by simple matrix functions $Q^{(n)}$ so that for each fixed t, we have $\lim_{n\to\infty} E\|\int_0^t [Q^{(n)}(s) - Q(s)]\,dW_s\|^2 = 0$. Because the L^2 limit of normal random vectors must be normal, we have the desired result.

6.11. The jointly normal random variables $Z_1, Z_2, \ldots, Z_n$ are uncorrelated, as one can see from (6.25) or by observing from (6.26) that

$$Z_i = \frac{b(t_i - t_{i-1})}{(T - t_i)(T - t_{i-1})} + \int_{t_{i-1}}^{t_i} \frac{dW_s}{T - s}; \quad i = 1, \ldots, n.$$

It is also apparent that

$$EZ_i = \frac{b(t_i - t_{i-1})}{(T - t_i)(T - t_{i-1})}, \quad \mathrm{Var}(Z_i) = \frac{t_i - t_{i-1}}{(T - t_i)(T - t_{i-1})}.$$

Now write down the joint density of $(Z_1, \ldots, Z_n)$ and make the change of variables $z_i = x_i/(T - t_i) - x_{i-1}/(T - t_{i-1})$ to obtain the desired result.

6.15. Observe that

$$dZ_t = Z_t\left[A(t)\,dt + \sum_{j=1}^{r} S_j(t)\,dW_t^{(j)}\right],$$

and apply Itô's rule to the right-hand side of (6.32) to see that X defined by this formula satisfies (6.30). Uniqueness follows from Theorem 2.5.

In the case of constant coefficients for (6.33), the solution becomes $X_t = X_0 \exp(Y_t)$, where $Y_t \triangleq \mu t + \sigma B_t$, $\sigma \triangleq \sqrt{\sum_{j=1}^r S_j^2}$, $\mu = A - \sigma^2/2$ and $B_t \triangleq (1/\sigma)\sum_{j=1}^r S_j W_t^{(j)}$ is Brownian motion (by the P. Lévy characterization, Theorem 3.3.16). The strong law of large numbers (Problem 2.9.3) shows that $\lim_{t\to\infty}(Y_t/t) = \mu$, and hence also $\lim_{t\to\infty} X_t = 0$, a.s.

7.3. Proceeding as in Solution 4.2.25, we show that

$$M_t \triangleq u(X_{t\wedge\tau_D})\exp\left\{-\int_0^{t\wedge\tau_D} k(X_s)\,ds\right\} + \int_0^{t\wedge\tau_D} g(X_s)\exp\left\{-\int_0^s k(X_\theta)\,d\theta\right\}ds; \quad 0 \le t < \infty$$

is a uniformly integrable martingale under P^x. The identity $E^x M_0 = E^x M_\infty$ is (7.8).

8.1. For a $(d \times d)$ matrix Γ, define the *operator norm* $\|\Gamma\| = \sup_{\xi\neq 0}(\|\Gamma\xi\|/\|\xi\|)$. We show that $\|\Gamma\| = \|\Gamma^T\|$. According to the Cauchy-Schwarz inequality,

$$\|\eta^T \Gamma^T \xi\| = \|\xi^T \Gamma \eta\| \le \|\xi\|\,\|\Gamma\eta\| \le \|\Gamma\|\,\|\xi\|\,\|\eta\|.$$

Now set $\eta = \Gamma^T\xi$ to obtain $\|\Gamma^T\xi\| \le \|\Gamma\|\,\|\xi\|$, and thus by definition $\|\Gamma^T\| \le \|\Gamma\|$. We obtain the opposite inequality by reversing the roles of Γ and Γ^T.

Now (8.3) implies that $\sigma^T(t)$ is nonsingular, for otherwise we could find a nonzero vector ξ which makes $\xi^T a(t)\xi = \|\sigma^T(t)\xi\|^2 = 0$. Letting $\xi = (\sigma^T(t))^{-1}\eta$, we may rewrite (8.3) as

$$\|(\sigma^T(t))^{-1}\eta\| \le \frac{1}{\sqrt{\varepsilon}}\|\eta\|; \quad \forall \eta \in \mathbb{R}^d, 0 \le t \le T,$$

a.s., which is equivalent to $\|(\sigma^T(t))^{-1}\| \le (1/\sqrt{\varepsilon})$. Because $\|(\sigma(t))^{-1}\| = \|(\sigma^T(t))^{-1}\|$, we have the equivalence of (8.3) and (8.5). We have already proved (8.4), and (8.6) is established similarly.

8.17. Because I is strictly decreasing on $[0, U'(0)]$ and identically zero on $[U'(0), \infty)$, we see that $\mathscr{X}$ is nondecreasing and for $0 < y_1 < y_2 < \infty$:

$$\mathscr{X}(y_1) = \mathscr{X}(y_2) \Leftrightarrow P\left[\min_{0 \le t \le T} (Z_t e^{\int_0^t (\beta(s) - r(s))\,ds}) < \frac{U'(0)}{y_1}\right] = 0,$$

in which case $\mathscr{X}(y_1) = \mathscr{X}(y_2) = 0$. The equality $\mathscr{X}(\bar{y}) = 0$ follows. The equality $\mathscr{X}(0) = \lim_{y \downarrow 0} \mathscr{X}(y) = \infty$ is a consequence of the monotone convergence theorem. Continuity of $\mathscr{X}$ on $(0, \bar{y}]$, and hence its surjectivity, follows from (8.43) and the dominated convergence theorem.

8.19. Let $\mathscr{H}(t, \ell) = H(t, e^\ell)$, so $\mathscr{H}$ is defined and of class $C^{1,2}$ on $[0, T] \times \mathbb{R}$. A bit of computation shows that $\mathscr{H}$ solves the Cauchy problem

$$-\mathscr{H}_t + \beta\mathscr{H} - \left(\beta - r - \frac{1}{2}\|\theta\|^2\right)\mathscr{H}_\ell - \frac{1}{2}\|\theta\|^2\mathscr{H}_{\ell\ell} = g(t, e^\ell); \quad 0 \le t \le T, \quad \ell \in \mathbb{R}$$

$$\mathscr{H}(T, \ell) = 0; \qquad \ell \in \mathbb{R}.$$

Condition (8.52) on H implies that $\mathscr{H}$ satisfies (7.20) for every $\mu > 0$ and $M > 0$ depending on μ. It follows from Problem 7.7 that

$$\mathscr{H}(t, \ell) = E\left[\int_t^T e^{-\beta(s-t)} g(s, e^{L_s})\,ds \,\middle|\, L_t = \ell\right],$$

where $dL_s = (\beta - r - \|\theta\|^2/2)\,ds - \theta^T\,dW_s$. This is the stochastic differential equation satisfied by $\log Y_s^{(t,y)}$, and thus

$$H(t, y) = \mathscr{H}(t, \log y) = E\left[\int_t^T e^{-\beta(s-t)} g(s, Y_s^{(t,y)})\,ds\right].$$

5.10. Notes

Section 5.1: The term *stochastic differential equation* was actually introduced by S. Bernštein (1934, 1938) in the limiting study of a sequence of Markov chains arising in a stochastic difference scheme. He was only interested in the distribution of the limiting process and showed that the latter had a density satisfying the Kolmogorov equations. However, according to Gihman & Skorohod (1972), it would be an exaggeration to consider Bernštein the founder of this theory.

Independently of Itô's work, I. I. Gihman (1947, 1950) developed a theory of stochastic differential equations, complete with results on existence, uniqueness, smooth dependence on the initial conditions, and Kolmogorov's equations for the transition density.

Since the early work of Itô and Gihman, the interest in the methodology and the mathematical theory of stochastic differential equations has enjoyed remarkable successes. The constructive and intuitive nature of the concept, as well as its strong physical appeal, have been responsible for its popularity among applied scientists; for instance, the "dW" term on the right-hand side of (1.9) has an important interpretation as *white noise* in statistical communication theory. Apart from its original and continued relevance to physics (see the notes to Section 6, as well as Nelson (1967), Freidlin (1985)), the new methodology became gradually indispensable in fields such as *stochastic systems* (Arnold & Kliemann (1983)) and *stability theory* (Friedman (1976), Khas'minskii (1980)), *stochastic control* and *game theory* (Fleming & Rishel (1975), Friedman (1976), Krylov (1980)), *filtering* (Liptser & Shiryaev (1977), Kallianpur (1980)), *communication and dynamical systems* (Wong & Hajek (1985)), *mathematical economics* (cf. Section 5.8), *mathematical biology and population genetics* (cf. examples in Chapter XV of Karlin & Taylor (1981)). Stochastic partial differential equations arise in filtering (Pardoux (1979)) and neurophysiology (Walsh (1984, 1986)).

On the other hand, the analytical approach to diffusions and, more generally, Markov processes, has been further developed in conjunction with the Hille-Yosida theory of semigroups; see, for instance, the lecture notes of Itô (1961a) and Chapter I of Ethier & Kurtz (1986).

Section 5.2: Theorems 2.5, 2.9 are standard; they are due to Itô (1942a, 1946, 1951) and can be found in several monographs such as those by Skorohod (1965), Mc Kean (1969), Gihman & Skorohod (1972, 1979), Arnold (1973), Friedman (1975), Liptser & Shiryaev (1977), Stroock & Varadhan (1979), Kallianpur (1980), Ikeda & Watanabe (1981). These sources, as well as Friedman (1976), should be consulted for further reading on the subject matter of this chapter and some of its applications.

The study of comparison results for stochastic differential equations started with Skorohod (1965). The article by Ikeda & Watanabe (1977) contains important refinements of Proposition 2.18 and Exercise 2.19, with applications to stochastic control and to tests for explosions; see also Chapter VI in Ikeda & Watanabe (1981), Yamada & Ogura (1981), Hajek (1985).

Doss (1977) and Sussmann (1978) were the first authors who studied the possibility of *pathwise solutions* to stochastic differential equations, via an appropriate reduction to an ordinary differential equation in the spirit of Proposition 2.21. Extension of the latter to several dimensions has Lie-algebraic ramifications; see Ikeda & Watanabe (1981), pp. 107–110. The modeling issue of Subsection 5.2.D was first raised by Wong & Zakai (1965a), who obtained Proposition 2.24 under more restrictive conditions. The one-dimensional equation (2.34) was studied in detail by Chitasvili & Toronjadze (1981).

Stochastic differential equations driven by general semimartingales, instead of just Brownian motion, have been studied by Doléans-Dade (1976) and Protter (1977, 1984) among others; see also Elliott (1982), Chapter XIV. Stochastic integral equations of the Volterra type have also been considered; see, for instance, Kleptsyna & Veretennikov (1984) and Protter (1985).

Section 5.3: The methodology that leads to Proposition 3.20 and Corollary 3.23 was developed by Yamada & Watanabe (1971); we follow Ikeda & Watanabe (1981) in our exposition.

Further instances of one-dimensional equations $dX_t = \sigma(X_t)\,dW_t$ admitting no strong solution, in the spirit of Example 3.5 but with *continuous* dispersion coefficients $\sigma(\cdot)$, have been discovered by Barlow (1982).

Section 5.4: The book by Stroock & Varadhan (1979) should be consulted for further reading on martingale problems and multidimensional diffusions. For the martingale approach to more specialized questions on diffusions, such as the support theorem, boundary behavior, degeneracy, and the construction of diffusion processes on manifolds, the reader is referred to the articles by Stroock & Varadhan (1970, 1971, 1972) and the lecture notes of Priouret (1974), respectively.

The "martingale problem" approach of this section can also be employed to characterize more general Markov processes in terms of their corresponding infinitesimal operators; a good part of Chapter IV in Ethier & Kurtz (1986) is devoted to this subject.

Section 5.5: Material for the first two subsections was drawn primarily from the papers of Engelbert & Schmidt (1981, 1984, 1985). The use of local time to prove pathwise uniqueness in Theorem 5.9 is due to Perkins (1982c), although Perkins's result has been sharpened by the use of the Engelbert and Schmidt zero-one law. Proposition 5.17 is a time-homogeneous version of a more general result due to Zvonkin (1974); according to the latter, the equation (2.1) with $d = 1$ has a unique strong solution, provided that

(i) the drift $b(t, x)$ is bounded and Borel-measurable,
(ii) the dispersion $\sigma(t, x)$ is bounded (both above and away from the origin), is continuous in (t, x), and satisfies a Hölder condition of the type (2.24) with $h(u) = Ku^\alpha$; $\alpha \geq (1/2)$.

Zvonkin's results were extended to the multidimensional case by Veretennikov (1979, 1981, 1982), who showed in particular that the equation (3.2) with $d \geq 1$ has a unique strong solution for any bounded, Borel-measurable drift $b(t, x)$.

In another important development, Nakao (1972) showed that pathwise uniqueness holds for the equation (5.1), provided that the coefficients b, σ are bounded and Borel-measurable, and σ is bounded below by a positive constant and is of bounded variation on any compact interval. For further extensions of this result (to time-dependent coefficients), see Veretennikov (1979), Nakao (1983), and Le Gall (1983).

The material of Subsection C is fairly standard; we relied on sources such as McKean (1969), Kallianpur (1980), and Ikeda & Watanabe (1981), particularly the latter. A generalization of the Feller test to the multi-dimensional case is due to Khas'minskii (1960) and can be found in Chapter X of Stroock & Varadhan (1979), together with more information about explosions.

A complete characterization of strong Markov processes with continuous sample paths, including the classification of their boundary behavior, is possible in one dimension; it was carried out by Feller (1952–1957) and appears in Itô & Mc Kean (1974) and Dynkin (1965), Chapters XV–XVII. See also Meleard (1986) for an approach based on stochastic calculus. The recurrence and ergodic properties of such processes were investigated by Maruyama & Tanaka (1957); see also §18 in Gihman & Skorohod (1972), as well as Khas'minskii (1960) and Bhattacharya (1978) for the multi-dimensional case.

Section 5.6: Langevin (1908) pioneered an approach to the Brownian movement that centered around the "dynamical" equation (6.22), instead of relying on the parabolic (Fokker-Planck-Kolmogorov) equation for the transition probability density. In (6.22), X_t represents the velocity of a free particle with mass m in a field consisting of a frictional and a fluctuating force, α is the coefficient of friction, and $\sigma^2 = 2\alpha kT/m$, where T denotes (absolute) temperature and k the Boltzmann constant. Langevin's ideas culminated in the Ornstein-Uhlenbeck theory for Brownian motion; long considered a purely heuristic tool, unsuitable for rigorous work, this theory was placed on firm mathematical ground by Doob (1942). Chapters IX and X of Nelson (1967) contain a nice exposition of these matters, including the Smoluchowski equation for Brownian movement in a force field.

Section 5.7: The monograph by Freidlin (1985) offers excellent follow-up reading on the subject matter of this section, as well as on degenerate and quasi-linear partial differential equations and their probabilistic treatment.

In the setting of Theorem 7.6 with $k \equiv 0, g \equiv 0$ it is possible to verify *directly*, under appropriate conditions, that the function

$$u(t,x) = Ef(X_T^{(t,x)}) \tag{10.1}$$

on the right-hand side of (7.15) possesses the requisite smoothness and solves the Cauchy problem (7.12), (7.13). We followed such an approach in Chapter 4 for the one-dimensional heat equation. Here, the key is to establish "smoothness" of the solution $X^{(t,x)}$ to (7.1) in the initial conditions (t,x) so as to allow taking first and second partial derivatives in (10.1) under the expectation sign; see Friedman (1975), p. 124, for details.

Questions of dependence on the initial conditions have been investigated extensively. The most celebrated of such results is the *diffeomorphism theorem* (Kunita (1981), Stroock (1982)), which we now outline in the context of the stochastic integral equation (4.20). Under Lipschitz and linear growth conditions as in Theorem 2.9, this equation has, for every initial position $x \in \mathbb{R}^d$, a unique strong solution $\{X_t(x); 0 \le t < \infty\}$. Consider now the $(d+1)$-dimensional random field $\mathscr{X} = \{X_t(x,\omega); (t,x) \in [0,\infty) \times \mathbb{R}^d, \omega \in \Omega\}$. It can be shown, using the Kolmogorov-Čentsov theorem (Problem 2.2.9) in conjunction with Problems 3.3.29 and 3.15, that there exists a modification $\tilde{\mathscr{X}}$ of $\mathscr{X}$ such that:

(i) $(t, x) \mapsto \tilde{X}_t(x, \omega)$ is continuous, for a.e. $\omega \in \Omega$;
(ii) For fixed $t \geq 0$, $x \mapsto \tilde{X}_t(x, \omega)$ is a homeomorphism of $\mathbb{R}^d$ into itself for a.e. $\omega \in \Omega$.

Furthermore, if the coefficients b, σ have bounded and continuous derivatives of all orders up to $k \geq 1$, then for every $t \geq 0$,

(iii) $x \mapsto \tilde{X}_t(x, \omega)$ is a C^{k-1}-diffeomorphism for a.e. $\omega \in \Omega$.

For an application of these ideas to the modeling issue of Subsection 5.2.D, see Kunita (1986).

Malliavin (1976, 1978) pioneered a probabilistic approach to the questions of existence and smoothness for the probability densities of Brownian functionals, such as strong solutions of stochastic differential equations. The resulting "functional" stochastic calculus has become known as the *stochastic calculus of variations*, or the *Malliavin calculus*; it has found several exegeses and applications beyond its original conception. See, for instance, Watanabe (1984), Chapter V in Ikeda & Watanabe (1981), and the review articles of Ikeda & Watanabe (1983), Zakai (1985) and Nualart & Zakai (1986). For applications of the Malliavin calculus to partial differential equations, see Stroock (1981, 1983) and Kusuoka & Stroock (1983, 1985).

Section 5.8: The methodology of Subsection *A* is new, as is the resulting treatment of the option pricing and consumption/investment problems in Subsections *B* and *C*, respectively. Similar results have been obtained independently by Cox & Huang in a series of papers (e.g., (1986, 1987)). For Subsection *B*, the inspiration comes in part from Harrison & Pliska (1981) and Bensoussan (1984); this latter paper, as well as Karatzas (1988), should be consulted for the pricing of American options. Material for Subsection *C* was drawn from more general results in Karatzas, Lehoczky & Shreve (1987). The problem of Subsection *D* was introduced by Samuelson (1969) and Merton (1971); it has been discussed in Karatzas et al. (1986) on an infinite horizon and with very general utility functions. An application of these ideas to equilibrium analysis is presented in Lehoczky & Shreve (1986). See also Duffie (1986) and Huang (1987).

CHAPTER 6

P. Lévy's Theory of Brownian Local Time

6.1. Introduction

This chapter is an in-depth study of the Brownian local time first encountered in Section 3.6. Our approach to this subject is motivated by the desire to perform computations. This is manifested by the inclusion of the conditional Laplace transform formulas of D. Williams (Subsections 6.3.B, 6.4.C), the derivation of the joint density of Brownian motion, its local time at the origin and its occupation time of the positive half-line (Subsection 6.3.C), and the computation of the transition density for Brownian motion with two-valued drift (Section 6.5). This last computation arises in the problem of controlling the drift of a Brownian motion, within prescribed bounds, so as to keep the controlled process near the origin.

Underlying these computations is a beautiful theory whose origins can be traced back to Paul Lévy. Lévy's idea was to use Theorem 3.6.17 to replace the study of Brownian local time by the study of the running maximum (2.8.1) of a Brownian motion, whose inverse coincides with the process of first passage times (Proposition 2.8.5). This latter process is strictly increasing, but increases by jumps only, and these jumps have a Poisson distribution. A precise statement of this result requires the introduction of the concept of Poisson random measure, a notion which has wide application in the study of jump processes. Here we use it to provide characterizations of Brownian local time in terms of excursions and downcrossings (Theorems 2.21, 2.23).

In Section 6.3 we take up the study of the independent, reflected Brownian motions obtained by looking separately at the positive (negative) parts of a standard Brownian motion. These independent Brownian motions are tied together by their local times at the origin, a fact which does not violate their independence. Exactly this situation was encountered in the Discussion of

F. Knight's Theorem 3.4.13, where we observed that intricately connected processes could become independent if we time-change them separately and then forget the time changes. The first formula of D. Williams (Theorem 3.6) is a precise statement of what can be inferred about the time change from observing one of these reflected Brownian motions.

Section 6.4 is highly computational, first developing Feynman-Kac formulas involving Brownian local time at several points, and then using these formulas to perform computations. In particular, the distribution of local time at several spatial points, when the temporal parameter is equal to a passage time, is computed and found to agree with the finite-dimensional distribution of one-half the square of a two-dimensional Bessel process. This is the Ray-Knight description of local time; it allows us finally to prove the Dvoretzky-Erdös-Kakutani Theorem 2.9.13.

6.2. Alternate Representations of Brownian Local Time

In Section 3.6 we developed the concept of Brownian local time as the density of occupation time. This is but one of several equivalent representations of Brownian local time, and in this section we present two others. We begin with a Brownian motion $W = \{W_t, \mathscr{F}_t; 0 \le t < \infty\}$ where $P[W_0 = 0] = 1$ and $\{\mathscr{F}_t\}$ satisfies the usual conditions, and we recall from Theorem 3.6.17 (see, in particular, (3.6.34), (3.6.35)) that

(2.1) $$P[|W_t| = M_t - B_t, \quad 2L_t(0) = M_t; \forall\, 0 \le t < \infty] = 1,$$

where $B_t = -\int_0^t \operatorname{sgn}(W_s)\,dW_s$ is itself a Brownian motion,

(2.2) $$M_t = \max_{0 \le s \le t} B_s; \quad 0 \le t < \infty,$$

and $L_t(0)$ is the local time of W at the origin:

(3.6.2) $$L_t(0) = \lim_{\varepsilon \downarrow 0} \frac{1}{4\varepsilon} \operatorname{meas}\{0 \le s \le t; |W_s| \le \varepsilon\}.$$

Thus, a study of the local time of W at the origin can be reduced to a study of the more easily conceived process M. The idea of this reduction and much of what follows originated with P. Lévy (1939, 1948).

A. The Process of Passage Times

The process M of (2.2) is continuous, nondecreasing, and "flat" (constant) during excursions of the reflected Brownian motion $|W|$ away from the origin. Because the Lebesgue measure of the set $\{0 \le t < \infty; |W_t| = 0\}$ is zero, P-a.s.

(Theorem 2.9.6), the process M has time derivative almost everywhere equal to zero; it is very much like the Cantor function.

We may regard M as a new clock, which stops when $|W|$ is away from the origin and runs at an accelerated rate when $|W|$ is at the origin. In order to pass from this new clock to the original one, we introduce the right-continuous inverse of M, defined as in Problem 3.4.5:

(2.3) $$S_b \triangleq \inf\{t \geq 0; M_t > b\} = \inf\{t \geq 0; B_t > b\}; \quad 0 \leq b < \infty.$$

Each S_b is an optional (and hence also a stopping) time of the right-continuous filtration $\{\mathscr{F}_t\}$. The left-continuous inverse of the process M is simply the family of *passage times*

(2.4) $$T_b \triangleq \inf\{t \geq 0; M_t = b\} = \inf\{t \geq 0; B_t = b\}; \quad 0 \leq b < \infty.$$

Regarded as processes parametrized by the spatial variable b, $T = \{T_b; 0 \leq b < \infty\}$ and $S = \{S_b; 0 \leq b < \infty\}$ are modifications of one another (Problem 2.7.19). The existence of two inverses for M reflects the fact that M identifies the starting and ending points of the excursions of $|W|$ away from the origin. These excursions correspond to jumps of the process T, as we now elaborate.

Recall from Remark 2.8.16 that for each $t > 0$, the time in $s \in [0, t]$ for which $B_s = M_t$ is almost surely unique. Thus, for each ω in an event Ω^* of probability one, this assertion holds for every rational t. Fix $\omega \in \Omega^*$, a positive number t (not necessarily rational), and define

(2.5) $$\gamma_t(\omega) \triangleq \sup\{s \in [0, t]; W_s(\omega) = 0\} = \sup\{s \in [0, t]; B_s(\omega) = M_t(\omega)\},$$

(2.6) $$\beta_t(\omega) \triangleq \inf\{s \in [t, \infty); W_s(\omega) = 0\} = \inf\{s \in [t, \infty); B_s(\omega) = M_t(\omega)\}.$$

If $W_t(\omega) = 0$, then $\beta_t(\omega) = t = \gamma_t(\omega)$. We are interested in the case $W_t(\omega) \neq 0$, which implies

(2.7) $$\gamma_t(\omega) < t < \beta_t(\omega).$$

In this case, the maximum of $B_s(\omega)$ over $0 \leq s \leq t$ is attained uniquely at $s = \gamma_t(\omega)$, hence $T_{M_t(\omega)}(\omega) = \gamma_t(\omega)$. Similarly $T_{M_t(\omega)+}(\omega) = S_{M_t(\omega)}(\omega) = \beta_t(\omega)$, for otherwise there would be a rational $q > \beta_t(\omega)$ such that $B_{\beta_t(\omega)}(\omega) = B_{\gamma_t(\omega)}(\omega) = M_q(\omega)$, a contradiction to the choice of $\omega \in \Omega^*$. We see then that for $\omega \in \Omega^*$ and t chosen so that $W_t(\omega) \neq 0$, the size of the jump in $T_b(\omega)$ at $b = M_t(\omega)$ is the length of *the excursion interval* $(\gamma_t(\omega), \beta_t(\omega))$ *straddling* t:

(2.8) $$T_{M_t(\omega)+}(\omega) - T_{M_t(\omega)}(\omega) = \beta_t(\omega) - \gamma_t(\omega).$$

It is clear from (2.4) that T_b is strictly increasing in b, and $T_0 = 0$. It is less clear that T *grows only by jumps*. To see this, consider the zero set of $W_.(\omega)$, namely

$$\mathscr{Z}_\omega \triangleq \{0 \leq t < \infty; W_t(\omega) = 0\},$$

which is almost surely closed, unbounded, and of Lebesgue measure zero (Theorem 2.9.6). As with any open set, $\mathscr{Z}_\omega^c \cap (0, \infty)$ can be written as a countable union of disjoint open intervals

$$\mathscr{Z}_\omega^c \cap (0, \infty) = \bigcup_{\alpha \in A} I_\alpha(\omega),$$

and each of these intervals contains a number $t_\alpha(\omega)$. In the notation of (2.5), (2.6), we have

$$I_\alpha = (\gamma_{t_\alpha}, \beta_{t_\alpha}); \quad \alpha \in A.$$

Because meas$(\mathscr{Z}_\omega) = 0$ for P-a.e. $\omega \in \Omega$, we have

$$\gamma_t = \sum_{\substack{\alpha \in A \\ \beta_{t_\alpha} \le t}} (\beta_{t_\alpha} - \gamma_{t_\alpha}) = \sum_{\substack{\alpha \in A \\ \beta_{t_\alpha} \le t}} (T_{M_{t_\alpha}+} - T_{M_{t_\alpha}}) \le \sum_{x \le M_t} (T_{x+} - T_x); \quad 0 \le t < \infty,$$

almost surely. Now set $t = T_a$ to obtain

$$T_a \le \sum_{x \le a} (T_{x+} - T_x), \quad 0 \le a < \infty;$$

letting $a \uparrow b$ and using the left-continuity of T, we see finally that, except on a P-null set,

$$T_b \le \sum_{x < b} (T_{x+} - T_x), \quad 0 \le b < \infty.$$

The reverse inequality is obvious.

We summarize our observations thus far.

2.1 Theorem. *The processes $S = \{S_b; 0 \le b < \infty\}$ of (2.3) and $T = \{T_b; 0 \le b < \infty\}$ of (2.4) are modifications of one another. Being the right- (respectively, left-) continuous inverses of the process M in (2.2), they are strictly increasing. Moreover, they increase by jumps only:*

(2.9) $$S_b = \sum_{x \le b} (S_x - S_{x-}), \quad T_b = \sum_{x < b} (T_{x+} - T_x); \quad 0 \le b < \infty,$$

a.s., where $S_{0-} \triangleq 0$. These processes have stationary and independent increments, with distribution

(2.10) $$P[S_b - S_a \in dt] = P[S_{b-a} \in dt] = \frac{b-a}{\sqrt{2\pi t^3}} e^{-(b-a)^2/2t}\, dt;$$
$$0 < a < b, \quad t > 0,$$

(2.11) $$Ee^{-\alpha S_b} = e^{-b\sqrt{2\alpha}}; \quad 0 \le \alpha, b < \infty.$$

PROOF. The only new assertions are those contained in the last sentence; these follow from Proposition 2.8.5. □

It is apparent from Theorem 2.1 that T must have infinitely many jumps on any interval $[0, b]$, $b > 0$, but T can have only finitely many jumps whose size exceeds a given number $\ell > 0$. We want to know the distribution of the (random) number of such jumps. To develop a conjecture about the answer to this question, we ask another. *How large must b be in order for T to have a jump on* $[0,b]$ *whose size exceeds* ℓ? We designate by τ_1 the minimal such b, and think of it as a "waiting time" (although b is a spatial, rather than temporal, parameter for the underlying Brownian motion W). Suppose we have "waited" up to "time" c and not seen a jump exceeding size ℓ; i.e., $\tau_1 > c$. Conditioned on this, what is the distribution of τ_1? After waiting up to "time" c, we are now waiting on the reflected Brownian motion $|W_{\cdot+T_c}| = |W_{\cdot+T_c} - W_{T_c}|$ to undergo an excursion of length exceeding ℓ; thus, the conditional distribution of the remaining wait is the same as the unconditional distribution of τ_1:

$$P[\tau_1 > c + b | \tau_1 > c] = P[\tau_1 > b]. \tag{2.12}$$

This "memoryless" property identifies the distribution of τ_1 as exponential (Problem 2.2).

2.2 Problem. Let $I = [0,\infty)$ or $I = \mathbb{R}$. Show that if $G: I \to (0,\infty)$ is monotone and

$$G(b + c) = G(b)G(c); \quad b, c \in I, \tag{2.13}$$

then G is of the form $G(b) = e^{-\lambda b}$; $b \in I$, for some constant $\lambda \in \mathbb{R}$. In particular, (2.12) implies that for some $\lambda > 0$,

$$P[\tau_1 > b] = e^{-\lambda b}; \quad 0 \le b < \infty.$$

After τ_1, we may begin the wait for the next jump of T whose size exceeds ℓ, i.e., the wait for $|W_{\cdot+T_{\tau_1}}| = |W_{\cdot+T_{\tau_1}} - W_{T_{\tau_1}}|$ to have another excursion of duration exceeding ℓ. It is not difficult to show, using the strong Markov property, that the additional wait τ_2 is independent of τ_1. Indeed, the "interarrival times" $\tau_1, \tau_2, \tau_3, \ldots$ are independent random variables with the same (exponential) distribution. Recalling the construction in Problem 1.3.2 of the Poisson process, we now see that for fixed $b > 0$, the number of jumps of the process T on $[0,b]$, whose size exceeds ℓ, is a Poisson random variable. To formalize this argument and obtain the exact distributions of the random variables involved, we introduce the concept of a Poisson random measure.

B. Poisson Random Measures

A Poisson random variable takes values in $\mathbb{N}_0 \triangleq \{0,1,2,\ldots\}$ and can be thought of as the number of occurrences of a particular incident of interest. Such a concept is inadequate, however, if we are interested in recording the occurrences of several different types of incidents. It is meaningless, for example, to keep track of the number of jumps in $(0,b]$ for the process S of (2.3), because there are infinitely many of those. It *is* meaningful, though, to record the number of jumps whose size exceeds a positive threshold ℓ, but we would like to do this for all positive ℓ simultaneously, and this requires that we not only count the jumps but somehow also classify them. We can do this by letting $\nu((0,b] \times A)$ be the number of jumps of S in $(0,b]$ whose size is in $A \in \mathscr{B}((0,\infty))$, and then extending this counting measure from sets of the form $(0,b] \times A$ to the collection $\mathscr{B}((0,\infty)^2)$. The resulting measure ν on $((0,\infty)^2, \mathscr{B}((0,\infty)^2))$ will of course be *random*, because the number of jumps of $\{S_a; 0 < a \le b\}$ with sizes in A is a random variable. This random measure ν will be shown to be a special case of the following definition.

2.3 Definition. Let $(\Omega, \mathscr{F}, P)$ be a probability space, $(H, \mathscr{H})$ a measurable space, and ν a mapping from Ω to the set of nonnegative counting measures on $(H, \mathscr{H})$, i.e., $\nu_\omega(C) \in \mathbb{N}_0 \cup \{\infty\}$ for every $\omega \in \Omega$ and $C \in \mathscr{H}$. We assume that the mapping $\omega \mapsto \nu_\omega(C)$ is $\mathscr{F}/\mathscr{B}(\mathbb{N}_0 \cup \{\infty\})$-measurable; i.e., $\nu(C)$ is an $\mathbb{N}_0 \cup \{\infty\}$-valued random variable, for each fixed $C \in \mathscr{H}$. We say that ν is a *Poisson random measure* if:

(i) For every $C \in \mathscr{H}$, either $P[\nu(C) = \infty] = 1$, or else

$$\lambda(C) \triangleq E\nu(C) < \infty$$

and $\nu(C)$ is a Poisson random variable:

$$P[\nu(C) = n] = e^{-\lambda(C)} \frac{(\lambda(C))^n}{n!}; \quad n \in \mathbb{N}_0.$$

(ii) For any pairwise disjoint sets $C_1, \ldots, C_m$ in $\mathscr{H}$, the random variables $\nu(C_1), \ldots, \nu(C_m)$ are independent.

The measure $\lambda(C) = E\nu(C)$; $C \in \mathscr{H}$, is called the *intensity measure* of ν.

2.4 Theorem (Kingman (1967)). *Given a σ-finite measure λ on $(H, \mathscr{H})$, there exists a Poisson random measure ν whose intensity measure is λ.*

PROOF. The case $\lambda(H) < \infty$ deserves to be singled out for its simplicity. When it prevails, we can construct a sequence of independent random variables $\xi_1, \xi_2, \ldots$ with common distribution $P[\xi_1 \in C] = \lambda(C)/\lambda(H)$; $C \in \mathscr{H}$, as well as an independent Poisson random variable N with $P[N = n] = e^{-\lambda(H)}(\lambda(H))^n/n!$; $n \in \mathbb{N}_0$. We can then define the counting measure

$$\nu(C) \triangleq \sum_{j=1}^{N} 1_C(\xi_j); \quad C \in \mathscr{H}.$$

It remains to show that ν is a Poisson random measure with intensity λ. Given a collection $C_1, \ldots, C_m$ of pairwise disjoint sets in $\mathscr{H}$, set $C_0 = H \backslash \bigcup_{k=1}^m C_k$ so $\sum_{k=0}^m \nu(C_k) = N$. Let $n_0, n_1, \ldots, n_m$ be nonnegative integers with $n = n_0 + n_1 + \cdots + n_m$. We have

$$\begin{aligned} &P[\nu(C_0) = n_0, \nu(C_1) = n_1, \ldots, \nu(C_m) = n_m] \\ &\quad = P[N = n] \cdot P[\nu(C_0) = n_0, \nu(C_1) = n_1, \ldots, \nu(C_m) = n_m | N = n] \end{aligned}$$

$$= e^{-\lambda(H)} \frac{(\lambda(H))^n}{n!} \cdot \frac{n!}{n_0!\, n_1! \ldots n_m!} \left(\frac{\lambda(C_0)}{\lambda(H)}\right)^{n_0} \left(\frac{\lambda(C_1)}{\lambda(H)}\right)^{n_1} \cdots \left(\frac{\lambda(C_m)}{\lambda(H)}\right)^{n_m}$$

$$= \prod_{k=0}^{m} e^{-\lambda(C_k)} \frac{(\lambda(C_k))^{n_k}}{n_k!},$$

and the claim follows upon summation over $n_0 \in \mathbb{N}_0$.

2.5 Problem. Modify the preceding argument in order to handle the case of σ-finite λ.

C. Subordinators

2.6 Definition. A real-valued process $N = \{N_t; 0 \le t < \infty\}$ on a probability space $(\Omega, \mathscr{F}, P)$ is called a *subordinator* if it has stationary, independent increments, and if almost every path of N is nondecreasing, is right-continuous, and satisfies $N_0 = 0$.

A prime example of a subordinator is a Poisson process or a positive linear combination of Poisson processes. A more complex example is the process $S = \{S_b; 0 \le b < \infty\}$ described in Theorem 2.1. P. Lévy (1937) discovered that S, and indeed any subordinator, can be thought of as a superposition of Poisson processes.

2.7 Theorem (Lévy (1937), Hinčin (1937), Itô (1942b)). *The moment generating function of a subordinator $N = \{N_t; 0 \le t < \infty\}$ on some $(\Omega, \mathscr{F}, P)$ is given by*

$$Ee^{-\alpha N_t} = \exp\left[-t\left\{m\alpha + \int_{(0,\infty)} (1 - e^{-\alpha\ell})\mu(d\ell)\right\}\right]; \quad t \ge 0, \alpha \ge 0, \tag{2.14}$$

for a constant $m \ge 0$ and a σ-finite measure μ on $(0, \infty)$ for which the integral in (2.14) *is finite. Furthermore, if $\tilde{\nu}$ is a Poisson random measure on* $(0, \infty)^2$ *(on a possibly different space $(\tilde{\Omega}, \tilde{\mathscr{F}}, \tilde{P})$) with intensity measure*

$$\lambda(dt \times d\ell) = dt \cdot \mu(d\ell), \tag{2.15}$$

then the process

$$\tilde{N}_t \triangleq mt + \int_{(0,\infty)} \ell\tilde{\nu}((0, t] \times d\ell); \quad 0 \le t < \infty \tag{2.16}$$

is a subordinator with the same finite-dimensional distributions as N.

2.8 Remark. The measure μ in Theorem 2.7 is called the *Lévy measure* of the subordinator N. It tells us the kind and frequency of the jumps of N. The simplest case of a Poisson process with intensity $\lambda > 0$ corresponds to $m = 0$

and μ which assigns mass λ to the singleton $\{1\}$. If μ does not have support on a singleton but is finite with $\lambda = \mu((0, \infty))$, then N is a *compound Poisson process*; the jump times are distributed just as they would be for the usual Poisson process with intensity λ, but the jump sizes constitute a sequence of independent, identically distributed random variables (with common distribution $\mu(d\ell)/\lambda$), independent of the sequence of jump times. The importance of Theorem 2.7, however, lies in the fact that it allows μ to be *σ-finite*; this is exactly the device we need to handle the subordinator $S = \{S_b; 0 \le b < \infty\}$, which has infinitely many jumps in any finite interval $(0, b]$.

PROOF OF THEOREM 2.7. We first establish the representation (2.14). The stationarity and independence of the increments of N imply that for $\alpha > 0$, the nonincreasing function $\rho_\alpha(t) \triangleq Ee^{-\alpha N_t}$; $0 \le t < \infty$, satisfies the functional equation $\rho_\alpha(t + s) = \rho_\alpha(t)\rho_\alpha(s)$. It follows from Problem 2.2 that

$$Ee^{-\alpha N_t} = e^{-t\psi(\alpha)}; \quad t \ge 0, \alpha \ge 0, \tag{2.17}$$

holds for some continuous, nondecreasing function $\psi: [0, \infty) \to [0, \infty)$ with $\psi(0) = 0$. Because the function $g(x) \triangleq xe^{-\alpha x}$; $x \ge 0$, is bounded for every $\alpha > 0$, we may differentiate in (2.17) with respect to α under the expectation sign to obtain the existence of ψ' and the formula

$$\psi'(\alpha)e^{-t\psi(\alpha)} = \frac{1}{t}E[N_t e^{-\alpha N_t}] = \frac{1}{t}\int_{[0,\infty)} \ell e^{-\alpha\ell} P[N_t \in d\ell]; \quad \alpha > 0, t > 0.$$

Consequently, we can write

$$\psi'(\alpha) = \lim_{k\to\infty} kc_k \int_{[0,\infty)} (1 + \ell)e^{-\alpha\ell} p_k(d\ell), \tag{2.18}$$

where

$$c_k \triangleq E\left[\frac{N_{1/k}}{1 + N_{1/k}}\right], \quad p_k(d\ell) \triangleq \frac{1}{c_k}\frac{\ell}{1 + \ell}P[N_{1/k} \in d\ell].$$

If $N \equiv 0$ a.s., the theorem is trivially true, so we may assume the contrary and choose $a > 0$ so that $\varepsilon \triangleq E(N_1 1_{\{N_1 \le a\}})$ is positive. For c_k we have the bound

$$\begin{aligned} c_k &\ge E\left[\frac{N_{1/k}}{1 + N_{1/k}} \cdot 1_{\{N_{1/k} \le a\}}\right] \ge \frac{1}{1 + a}E[N_{1/k}1_{\{N_{1/k} \le a\}}] \\ &= \frac{1}{k(1 + a)}\sum_{j=1}^{k} E[(N_{j/k} - N_{(j-1)/k})1_{\{N_{j/k} - N_{(j-1)/k} \le a\}}] \\ &\ge \frac{\varepsilon}{k(1 + a)}. \end{aligned} \tag{2.19}$$

We can establish now the *tightness* of the sequence of probability measures $\{p_k\}_{k=1}^{\infty}$. Indeed,

$$P[N_1 \le \ell] \le P[N_{j/k} - N_{(j-1)/k} \le \ell;\ j = 1, \ldots, k]$$
$$= (P[N_{1/k} \le \ell])^k,$$

and thus, using (2.19), we may write

$$(2.20)\quad p_k((\ell, \infty)) \le \frac{1+a}{\varepsilon} kP[N_{1/k} > \ell] \le \frac{k(1+a)}{\varepsilon}\{1 - (P[N_1 \le \ell])^{1/k}\}.$$

Because $\lim_{k\to\infty} k(1 - \xi^{1/k}) = -\log \xi;\ 0 < \xi < 1$, we can make the right-hand side of (2.20) as small as we like (uniformly in k), by taking ℓ large. Prohorov's Theorem 2.4.7 implies that there is a subsequence $\{p_{k_j}\}_{j=1}^{\infty}$ which converges weakly to a probability measure p on $([0, \infty), \mathscr{B}([0, \infty)))$. In particular, because the function $\ell \mapsto (1 + \ell)e^{-\alpha\ell}$ is bounded for every positive α, we must have

$$\lim_{j\to\infty} \int_{[0,\infty)} (1 + \ell)e^{-\alpha\ell} p_{k_j}(d\ell) = \int_{[0,\infty)} (1 + \ell)e^{-\alpha\ell} p(d\ell).$$

Combined with (2.18), this equality shows that $k_j c_{k_j}$ converges to a constant $c \ge 0$, so that

$$(2.21)\qquad \psi'(\alpha) = c \int_{[0,\infty)} (1 + \ell)e^{-\alpha\ell} p(d\ell)$$
$$= cp(\{0\}) + c \int_{(0,\infty)} (1 + \ell)e^{-\alpha\ell} p(d\ell);\quad 0 < \alpha < \infty.$$

Note, in particular, that ψ' is continuous and decreasing on $(0, \infty)$. From the fundamental theorem of calculus, the Fubini theorem, (2.21), and $\psi(0) = 0$, we obtain now

$$(2.22)\quad \psi(\alpha) = \alpha cp(\{0\}) + c \int_{(0,\infty)} \frac{1+\ell}{\ell}(1 - e^{-\alpha\ell}) p(d\ell);\quad 0 < \alpha < \infty.$$

The representation (2.14) follows by taking $m = cp(\{0\})$ and

$$(2.23)\qquad \mu(d\ell) = \frac{c(1+\ell)}{\ell} p(d\ell);\quad \ell > 0,$$

the latter being a σ-finite measure on $(0, \infty)$. In particular, we have from (2.22):

$$(2.24)\qquad \psi(\alpha) = m\alpha + \int_{(0,\infty)} (1 - e^{-\alpha\ell})\mu(d\ell) < \infty;\quad 0 \le \alpha < \infty.$$

We can now use Theorem 2.4 with $H = (0, \infty)^2$ and $\mathscr{H} = \mathscr{B}(H)$, to construct on a probability space $(\tilde{\Omega}, \tilde{\mathscr{F}}, \tilde{P})$ a random measure $\tilde{\nu}$ with intensity given by (2.15); a nondecreasing process $\tilde{N}$ can then be defined on $(\tilde{\Omega}, \tilde{\mathscr{F}}, \tilde{P})$ via (2.16). It is clear that $\tilde{N}_0 = 0$, and because of Definition 2.3 (ii), $\tilde{N}$ has independent increments (provided that (2.25), which follows, holds, so the increments are defined). We show that $\tilde{N}$ *is a subordinator* with the same finite-dimensional distributions as N. Concerning the stationarity of increments, note that for a nonnegative simple function $\varphi(\ell) = \sum_{i=1}^n a_i 1_{A_i}(\ell)$ on $(0, \infty)$, where $A_1, \ldots, A_n$ are pairwise disjoint Borel sets, the distribution of

$$\int_{(0,\infty)} \varphi(\ell)\, \tilde{\nu}((t, t+h] \times d\ell) = \sum_{i=1}^n a_i \tilde{\nu}((t, t+h] \times A_i)$$

is a linear combination of the independent, Poisson (or else almost surely infinite) random variables $\{\tilde{\nu}((t, t+h] \times A_i)\}_{i=1}^n$ with respective expectations $\{h\mu(A_i)\}_{i=1}^n$. Thus, for any nonnegative, measurable φ, the distribution of $\int_{(0,\infty)} \varphi(\ell)\tilde{\nu}((t, t+h] \times d\ell)$ is independent of t. Taking $\varphi(\ell) = \ell$, we have the stationarity of the increment $\tilde{N}_{t+h} - \tilde{N}_t$.

In order to show that $\tilde{N}$ is a subordinator, it remains to prove that

$$\tilde{N}_t < \infty; \quad 0 \le t < \infty \tag{2.25}$$

and that $\tilde{N}$ is right-continuous, almost surely. Right-continuity will follow from (2.25) and the dominated convergence theorem applied in (2.16), and (2.25) will follow from the relation

$$\tilde{E}e^{-\alpha \tilde{N}_t} = e^{-t\psi(\alpha)}; \quad \alpha \ge 0, t \ge 0, \tag{2.26}$$

where ψ is as in (2.24). With $n \ge 1$, and $\ell_j^{(n)} \triangleq j2^{-n}$, $I_j^{(n)} = (\ell_{j-1}^{(n)}, \ell_j^{(n)}]$; $1 \le j \le 4^n$, we have from the monotone and bounded convergence theorems:

(2.27)

$$\tilde{E}\exp\left\{-\alpha \int_{(0,\infty)} \ell\, \tilde{\nu}((0,t] \times d\ell)\right\} = \lim_{n\to\infty} \tilde{E}\exp\left\{-\alpha \sum_{j=2}^{4^n} \ell_{j-1}^{(n)} \tilde{\nu}((0,t] \times I_j^{(n)})\right\}.$$

But the random variables $\tilde{\nu}((0,t] \times I_j^{(n)})$; $2 \le j \le 4^n$ are independent, Poisson, with expectations $t\mu(I_j^{(n)})$; $2 \le j \le 4^n$, and these quantities are finite because the integral in (2.14) is finite. The expectation on the right-hand side of (2.27) becomes

$$\prod_{j=2}^{4^n} \tilde{E}\exp\{-\alpha \ell_{j-1}^{(n)} \tilde{\nu}((0,t] \times I_j^{(n)})\} = \exp\left\{-t \sum_{j=2}^{4^n} (1 - e^{-\alpha \ell_{j-1}^{(n)}})\mu(I_j^{(n)})\right\},$$

which converges to $\exp\{-t\int_{(0,\infty)}(1 - e^{-\alpha\ell})\mu(d\ell)\}$ as $n \to \infty$. Relation (2.26) follows and shows that for each fixed $t \ge 0$, $\tilde{N}_t$ has the same distribution as N_t. The equality of finite-dimensional distributions is a consequence of the independence and stationarity of the increments of both processes. □

Theorem 2.7 raises two important questions:

(I) Are the constant $m \ge 0$ and the Lévy measure μ unique?
(II) Does the original subordinator N admit a representation of the form (2.16)?

One is eager to believe that the answer to both questions is affirmative; for the proofs of these assertions we have to introduce the space of RCLL functions, where the paths of N belong.

2.9 Definition. The *Skorohod space* $D[0, \infty)$ is the set of all RCLL functions from $[0, \infty)$ into $\mathbb{R}$. We denote by $\mathscr{B}(D[0, \infty))$ the smallest σ-field containing all finite-dimensional cylinder sets of the form (2.2.1).

2.10 Remark. The space $D[0, \infty)$ is metrizable by the *Skorohod metric* in such a way that $\mathscr{B}(D[0, \infty))$ is the smallest σ-field containing all open sets (Parthasarathy (1967), Chapter VII, Theorem 7.1). This fact will not be needed here.

2.11 Problem. Suppose that P and $\tilde{P}$ are probability measures on $(D[0, \infty), \mathscr{B}(D[0, \infty)))$ which agree on all finite-dimensional cylinder sets of the form

$$\{y \in D[0, \infty);\ y(t_1) \in \Gamma_1, \ldots, y(t_n) \in \Gamma_n\},$$

where $n \geq 1$, $0 \leq t_1 < t_2 < \cdots < t_n < \infty$, and $\Gamma_i \in \mathscr{B}(\mathbb{R})$; $i = 1, \ldots, n$. Then P and $\tilde{P}$ agree on $\mathscr{B}(D[0, \infty))$. (*Hint*: Use the Dynkin System Theorem 2.1.3.)

2.12 Problem. Given a set $C \subseteq (0, \infty)^2$, let $n(\cdot\,; C)\colon D[0, \infty) \to \mathbb{N}_0 \cup \{\infty\}$ be defined by

$$n(y; C) \triangleq \#\{(t, \ell) \in C;\ |y(t) - y(t-)| = \ell\}, \tag{2.28}$$

where $\#$ denotes cardinality. In particular, $n(y; (t, t+h] \times (\ell, \infty))$ is the number of jumps of y during $(t, t+h]$ whose sizes exceed ℓ. Show that $n(\cdot\,; C)$ is $\mathscr{B}(D[0, \infty))$-measurable, for every $C \in \mathscr{B}((0, \infty)^2)$. (*Hint*: First show that $n(\cdot\,; (0, t] \times (\ell, \infty))$ is finite and measurable, for every $t > 0$, $\ell > 0$.)

Returning to the context of Theorem 2.7, we observe that the subordinator N on $(\Omega, \mathscr{F}, P)$ may be regarded as a measurable mapping from $(\Omega, \mathscr{F})$ to $(D[0, \infty), \mathscr{B}(D[0, \infty)))$. The fact that $\tilde{N}$ defined on $(\tilde{\Omega}, \tilde{\mathscr{F}}, \tilde{P})$ by (2.16) has the same finite-dimensional distributions as N implies (Problem 2.11) that N and $\tilde{N}$ induce the same measure on $D[0, \infty)$:

$$P[N \in A] = \tilde{P}[\tilde{N} \in A];\ \forall\, A \in \mathscr{B}(D[0, \infty)).$$

We say that N under P and $\tilde{N}$ under $\tilde{P}$ *have the same law*. Consequently, for $C_1, C_2, \ldots, C_m$ in $\mathscr{B}((0, \infty)^2)$, the distribution under P of the random vector $(n(N; C_1), \ldots, n(N; C_m))$ coincides with the distribution under $\tilde{P}$ of $(n(\tilde{N}; C_1), \ldots, n(\tilde{N}; C_m))$. But

$$n(\tilde{N}; C) = \tilde{\nu}(C);\quad C \in \mathscr{B}((0, \infty) \times (0, \infty)) \tag{2.29}$$

is the Poisson random measure (under $\tilde{P}$) of Theorem 2.7, so

$$\nu(C) \triangleq n(N; C) = \#\{(t, \ell) \in C;\ N_t - N_{t-} = \ell\};\quad C \in \mathscr{B}((0, \infty)^2) \tag{2.30}$$

is a Poisson random measure (under P) with intensity given by (2.15).

We observe further that for $t > 0$, the mapping $\varphi_t \colon D[0,\infty) \to [0,\infty]$ defined by

$$\varphi_t(y) \triangleq \int_{(0,\infty)} \ell n(y; (0,t] \times d\ell)$$

is $\mathscr{B}(D[0,\infty))/\mathscr{B}([0,\infty])$-measurable, and

$$\varphi_t(N) = \int_{(0,\infty)} \ell \nu((0,t] \times d\ell), \quad \varphi_t(\tilde{N}) = \int_{(0,\infty)} \ell \tilde{\nu}((0,t] \times d\ell).$$

It follows that the differences $\{N_t - \varphi_t(N); 0 \le t < \infty\}$ and $\{\tilde{N}_t - \varphi_t(\tilde{N}); 0 \le t < \infty\}$ have the same law. But $\tilde{N}_t - \varphi_t(\tilde{N}) = mt$ is deterministic, and thus $N_t - \varphi_t(N) = mt$ as well. We are led to the representation

$$N_t = mt + \int_{(0,\infty)} \ell \nu((0,t] \times d\ell); \quad 0 \le t < \infty. \tag{2.31}$$

We summarize these remarks as two corollaries to Theorem 2.7.

2.13 Corollary. *Let $N = \{N_t; 0 \le t < \infty\}$ be a subordinator with moment generating function* (2.14). *Then N admits the representation* (2.31), *where ν given by* (2.30) *is a Poisson random measure on* $(0,\infty)^2$ *with intensity* (2.15).

2.14 Corollary. *Let $N = \{N_t; 0 \le t < \infty\}$ be a subordinator. Then the constant $m \ge 0$ and the σ-finite Lévy measure μ which appear in* (2.14) *are uniquely determined.*

PROOF. According to Corollary 2.13, $\mu(A) = E\nu((0,1] \times A)$; $A \in \mathscr{B}(0,\infty)$, where ν is given by (2.30); this shows that μ is uniquely determined. We may solve (2.31) for m to see that this constant is also unique. □

2.15 Definition. A subordinator $N = \{N_t; 0 \le t < \infty\}$ is called a *one-sided stable process* if it is not almost surely identically zero and, to each $\alpha \ge 0$, there corresponds a constant $\beta(\alpha) \ge 0$ such that $\{\alpha N_t; 0 \le t < \infty\}$ and $\{N_{t\beta(\alpha)}; 0 \le t < \infty\}$ have the same law.

2.16 Problem. Show that the function $\beta(\alpha)$ of the preceding definition is continuous for $0 \le \alpha < \infty$ and satisfies

$$\beta(\alpha\gamma) = \beta(\alpha)\beta(\gamma); \quad \alpha \ge 0, \gamma \ge 0 \tag{2.32}$$

as well as

$$\psi(\alpha) = r\beta(\alpha); \quad \alpha \ge 0, \tag{2.33}$$

where $r = \psi(1)$ is positive and ψ is given by (2.17), or equivalently, (2.24). The unique solution to equation (2.32) is $\beta(\alpha) = \alpha^{\varepsilon}$, and from (2.33) we see that for a one-sided stable process N,

(2.34) $$Ee^{-\alpha N_t} = \exp\{-t\psi(\alpha)\} = \exp\{-tr\alpha^\varepsilon\}; \quad 0 < \alpha < \infty.$$

The constants r, ε are called the *rate* and the *exponent*, respectively, of the process. Because ψ is increasing and concave (cf. (2.21)), we have necessarily $0 < \varepsilon \le 1$. The choice $\varepsilon = 1$ leads to $m = r$, $\mu = 0$ in (2.14). For $0 < \varepsilon < 1$, we have

(2.35) $$m = 0, \quad \mu(d\ell) = \frac{r\varepsilon}{\Gamma(1-\varepsilon)} \frac{d\ell}{\ell^{1+\varepsilon}}; \quad \ell > 0.$$

D. The Process of Passage Times Revisited

The subordinator S of Theorem 2.1 is *one-sided stable with exponent* $\varepsilon = (1/2)$. Indeed, for fixed $\alpha > 0$, $(1/\alpha)S_{b\sqrt{\alpha}}$ is the first time the Brownian motion (Lemma 2.9.4 (i)) $W^* = \{W_t^* \triangleq (1/\sqrt{\alpha})W_{\alpha t}; 0 \le t < \infty\}$ reaches level b; i.e.,

$$\frac{1}{\alpha} S_{b\sqrt{\alpha}} = S_b^* \triangleq \inf\{t \ge 0; W_t^* > b\}.$$

Consequently, $\{\alpha S_b; 0 \le b < \infty\}$ has the same law as $\{\alpha S_b^* = S_{b\sqrt{\alpha}}; 0 \le b < \infty\}$, from which we conclude that $\beta(\alpha)$ appearing in Definition 2.15 is $\sqrt{\alpha}$. Comparison of (2.11) and (2.34) shows that the rate of S is $r = \sqrt{2}$, and (2.35) gives us the Lévy measure

$$\mu(d\ell) = \frac{d\ell}{\sqrt{2\pi\ell^3}}; \quad \ell > 0.$$

Corollary 2.13 asserts then that

$$S_b = \int_{(0,\infty)} \ell\nu((0,b] \times d\ell); \quad 0 \le b < \infty,$$

where, in the notation of (2.1)–(2.4), (2.30)

(2.36) $$\begin{aligned} \nu(C) &= \#\{(b,\ell) \in C; S_b - S_{b-} = \ell\} \\ &= \#\{(b,\ell) \in C; |W| \text{ has an excursion of duration } \ell \text{ starting at time } T_b\}; \ C \in \mathscr{B}((0,\infty)^2), \end{aligned}$$

is a Poisson random measure with intensity measure $(dt\, d\ell/\sqrt{2\pi\ell^3})$. In particular, for any $I \in \mathscr{B}((0,\infty))$ and $0 < \delta < \varepsilon \le \infty$, we have

(2.37) $$E\nu(I \times [\delta,\varepsilon)) = \text{meas}(I) \int_\delta^\varepsilon \frac{d\lambda}{\sqrt{2\pi\lambda^3}} = \text{meas}(I) \sqrt{\frac{2}{\pi}} \left(\frac{1}{\sqrt{\delta}} - \frac{1}{\sqrt{\varepsilon}} \right).$$

For $0 < \delta < \varepsilon \le \infty$ and $b \ge 0$, let us consider the random variables

(2.38) $$\begin{aligned} N_b^{\delta,\varepsilon} &\triangleq \nu((0,b] \times [\delta,\varepsilon)) \\ &= \text{Number of jumps of size } \ell \in [\delta,\varepsilon) \text{ suffered by } \{S_a; 0 \le a \le b\}, \end{aligned}$$

(2.39) $$\mathcal{L}_b^{\delta,\varepsilon} \triangleq \int_{[\delta,\varepsilon)} \ell\nu((0,b] \times d\ell)$$

= Total length of jumps of size $\ell \in [\delta, \varepsilon)$ suffered by $\{S_a; 0 \le a \le b\}$.

We also agree to write

$$N_b^\delta \triangleq N_b^{\delta,\infty} = \nu((0,b] \times [\delta,\infty)), \quad \mathcal{L}_b^\varepsilon \triangleq \mathcal{L}_b^{0,\varepsilon} = \int_{(0,\varepsilon)} \ell\nu((0,b] \times d\ell).$$

The process $N^{\delta,\varepsilon} = \{N_b^{\delta,\varepsilon}; 0 \le b < \infty\}$ is Poisson with intensity $\sqrt{(2/\pi)}((1/\sqrt{\delta}) - (1/\sqrt{\varepsilon}))$; the process $\mathcal{L}^{\delta,\varepsilon} = \{\mathcal{L}_b^{\delta,\varepsilon}; 0 \le b < \infty\}$ is a subordinator which grows only by jumps (see the last paragraph of the proof of Theorem 2.7). Furthermore,

(2.40) $$Ee^{-\alpha\mathcal{L}_b^{\delta,\varepsilon}} = \exp\left[-b\int_{(\delta,\varepsilon)} (1 - e^{-\alpha\ell})\frac{d\ell}{\sqrt{2\pi\ell^3}}\right],$$

and these assertions concerning $\mathcal{L}_b^{\delta,\varepsilon}$ hold even if $\delta = 0$.

The behavior of N_b^δ as $\delta \downarrow 0$ merits some attention. Of course, $\lim_{\delta\downarrow 0} N_b^\delta = \infty$ a.s., and any meaningful statement will require some normalization.

2.17 Proposition. *For almost every $\omega \in \Omega$,*

(2.41) $$\lim_{\delta\downarrow 0} \sqrt{\frac{\pi\delta}{2}} N_b^\delta(\omega) = b; \quad \forall 0 \le b < \infty.$$

PROOF. The process

$$Q_t \triangleq \nu\left((0,b] \times \left[\frac{1}{t^2},\infty\right)\right); \quad 0 \le t < \infty,$$

has nondecreasing, right-continuous paths and independent increments. For $0 \le s < t$, the increment $Q_t - Q_s = \nu((0,b] \times [t^{-2}, s^{-2}))$ is Poisson with expectation

$$E(Q_t - Q_s) = b\int_{t^{-2}}^{s^{-2}} \frac{d\ell}{\sqrt{2\pi\ell^3}} = b\sqrt{\frac{2}{\pi}}(t - s).$$

We conclude that Q is a Poisson process, for which the strong law of large numbers of Remark 1.3.10 gives $\lim_{t\to\infty}(Q_t/t) = b\sqrt{(2/\pi)}$ a.s. By definition we have $N_b^\delta = Q_{1/\sqrt{\delta}}$, so (2.41) holds a.s. for each fixed b. Except for ω in a null set $A \subseteq \Omega$, this relation must hold for all rational, nonnegative b. The monotonicity in b of sides of (2.41) then gives us its validity for $\omega \in \Omega\backslash A$, $0 \le b < \infty$. □

As $\varepsilon \downarrow 0$, the dominated convergence theorem shows that $\mathcal{L}_b^\varepsilon \downarrow 0$ a.s. In other words, since S_b is finite, the total length of its jumps of size less than ε must approach zero with ε. We thus normalize $\mathcal{L}_b^\varepsilon$ in a manner "opposite" to the normalization of (2.41).

2.18 Proposition. *For almost every* $\omega \in \Omega$,

$$\lim_{\varepsilon \downarrow 0} \sqrt{\frac{\pi}{2\varepsilon}} \mathcal{L}_b^\varepsilon(\omega) = b; \quad \forall 0 \le b < \infty. \tag{2.42}$$

PROOF. Given $\varepsilon > 0$ and $0 < \rho < 1$, we have for all $n \ge 1$:

$$\sum_{k=1}^{n} \sqrt{\varepsilon \rho^{2k}} \nu((0, b] \times [\varepsilon\rho^{2k}, \varepsilon\rho^{2(k-1)})) \le \frac{1}{\sqrt{\varepsilon}} \mathcal{L}_b^\varepsilon \tag{2.43}$$
$$\le \sum_{k=1}^{\infty} \sqrt{\varepsilon\rho^{2(k-1)}} \nu((0, b] \times [\varepsilon\rho^{2k}, \varepsilon\rho^{2(k-1)})).$$

The left-hand side of (2.43) may be written as

$$\sum_{k=1}^{n} \{\rho^k \sqrt{\varepsilon\rho^{2k}} N_b^{\varepsilon\rho^{2k}} - \rho^{k+1} \sqrt{\varepsilon\rho^{2(k-1)}} N_b^{\varepsilon\rho^{2(k-1)}}\},$$

which converges as $\varepsilon \downarrow 0$ (see (2.41)) to

$$b\sqrt{\frac{2}{\pi}} \sum_{k=1}^{n} (\rho^k - \rho^{k+1}) = b\sqrt{\frac{2}{\pi}} (\rho - \rho^{n+1}).$$

It follows that $\underline{\lim}_{\varepsilon \downarrow 0} (1/\sqrt{\varepsilon}) \mathcal{L}_b^\varepsilon \ge b\sqrt{(2/\pi)}(\rho - \rho^{n+1})$, a.s., and letting first $n \to \infty$ and then $\rho \uparrow 1$, we obtain

$$\underline{\lim}_{\varepsilon \downarrow 0} \frac{1}{\sqrt{\varepsilon}} \mathcal{L}_b^\varepsilon \ge b\sqrt{\frac{2}{\pi}} \quad \text{a.s.}$$

Now let $\eta(\varepsilon) \triangleq \sup_{0 < \delta \le \varepsilon} |\sqrt{\delta} N_b^\delta - b\sqrt{(2/\pi)}|$, so that $\lim_{\varepsilon \downarrow 0} \eta(\varepsilon) = 0$ a.s. (Proposition 2.17). The right-hand side of (2.43) is subject to the bound

$$\sum_{k=1}^{\infty} \sqrt{\varepsilon\rho^{2(k-1)}} \nu((0, b] \times [\varepsilon\rho^{2k}, \varepsilon\rho^{2(k-1)}))$$
$$= \sum_{k=1}^{\infty} \{\rho^{k-2} \sqrt{\varepsilon\rho^{2k}} N_b^{\varepsilon\rho^{2k}} - \rho^{k-1} \sqrt{\varepsilon\rho^{2(k-1)}} N_b^{\varepsilon\rho^{2(k-1)}}\}$$
$$\le \sum_{k=1}^{\infty} \left\{\rho^{k-2}\left(b\sqrt{\frac{2}{\pi}} + \eta(\varepsilon)\right) - \rho^{k-1}\left(b\sqrt{\frac{2}{\pi}} - \eta(\varepsilon)\right)\right\}$$
$$= \frac{b}{\rho}\sqrt{\frac{2}{\pi}} + \eta(\varepsilon)\left(\frac{1}{\rho} + \frac{2}{1-\rho}\right).$$

It follows that $\overline{\lim}_{\varepsilon \downarrow 0} (1/\sqrt{\varepsilon}) \mathcal{L}_b^\varepsilon \le (b/\rho)\sqrt{(2/\pi)}$. Letting $\rho \uparrow 1$, we obtain

$$\overline{\lim}_{\varepsilon \downarrow 0} \frac{1}{\sqrt{\varepsilon}} \mathcal{L}_b \le b\sqrt{\frac{2}{\pi}}, \quad \text{a.s.}$$

This proves that (2.42) holds a.s. for each fixed b. We conclude as in Proposition 2.17. □

2.19 Exercise. Show that for fixed, positive δ, the process $X_b(\delta) \triangleq N_b^\delta - b\sqrt{2/\pi\delta}$; $0 \le b < \infty$, is a right-continuous, square-integrable martingale (with respect to the filtration $\{\mathscr{F}_b^{X(\delta)}\}$). This process has stationary, independent increments and

$$\langle X(\delta)\rangle_b = b\sqrt{\frac{2}{\pi\delta}}; \quad 0 \le b < \infty.$$

Furthermore, we have the law of large numbers

$$\lim_{b\to\infty} \frac{1}{b} N_b^\delta = \sqrt{\frac{2}{\pi\delta}}, \quad \text{a.s.}$$

2.20 Exercise. Show that for fixed, positive ε, the process $V_b(\varepsilon) \triangleq \mathcal{L}_b^\varepsilon - b\sqrt{(2\varepsilon/\pi)}$; $0 \le b < \infty$ is a right-continuous, square-integrable martingale (with respect to the filtration $\{\mathscr{F}_b^{V(\varepsilon)}\}$). This process has stationary, independent increments and

$$\langle V(\varepsilon)\rangle_b = \frac{2b}{3}\sqrt{\frac{\varepsilon^3}{2\pi}}. \tag{2.44}$$

Establish the representation

$$\mathcal{L}_b^\varepsilon = -\varepsilon N_b^\varepsilon + \int_{(0,\varepsilon)} N_b^\ell \, d\ell; \quad 0 \le b < \infty, \quad \text{a.s.,} \tag{2.45}$$

as well as the law of large numbers

$$\lim_{b\to\infty} \frac{1}{b} \mathcal{L}_b^\varepsilon = \sqrt{\frac{2\varepsilon}{\pi}}, \quad \text{a.s.} \tag{2.46}$$

(*Hint*: Obtain the moment generating function

$$Ee^{-\alpha V_b(\varepsilon)} = \exp\left[\alpha b\sqrt{\frac{2\varepsilon}{\pi}} - b\int_{(0,\varepsilon)} (1 - e^{-\alpha\ell})\frac{d\ell}{\sqrt{2\pi\ell^3}}\right].\Bigg) \tag{2.47}$$

E. The Excursion and Downcrossing Representations of Local Time

Let us now discuss the significance of Propositions 2.17, 2.18, for Brownian local time. Returning to the context of (2.1)–(2.4) and with ν the Poisson random measure given by (2.36), we may recast these propositions as

$$P\left[\lim_{\varepsilon\downarrow 0} \sqrt{\frac{\pi\varepsilon}{2}} \nu((0, M_t] \times [\varepsilon, \infty)) = M_t; \forall\, 0 \le t < \infty\right] = 1, \tag{2.48}$$

$$(2.49)\quad P\left[\lim_{\varepsilon\downarrow 0}\sqrt{\frac{\pi}{2\varepsilon}}\int_{(0,\varepsilon)}\ell\nu((0,M_t]\times d\ell)=M_t;\ \forall\, 0\le t<\infty\right]=1.$$

But, by (2.36),

$$\nu((0,M_t]\times[\varepsilon,\infty)) = \#\{\text{jumps of size} \ge \varepsilon \text{ suffered by } \{S_b;\, 0\le b\le M_t\}\}$$
$$= \#\left\{\begin{matrix}\text{excursion intervals away from the origin,}\\ \text{of duration} \ge\varepsilon, \text{ made by } \{W_u;\, 0\le u\le S_{M_t}\}\end{matrix}\right\}.$$

As observed earlier, if we have $W_t \neq 0$ so that (2.7) holds, then $S_{M_t} = \beta_t$ is the time of conclusion of the excursion of W straddling t, and

$$(2.50a)\quad \nu((0,M_t]\times[\varepsilon,\infty)) = \#\left\{\begin{matrix}\text{excursion intervals away from}\\ \text{the origin, of duration} \ge\varepsilon,\\ \text{initiated by } W \text{ before } t\end{matrix}\right\}.$$

On the other hand, if $W_t = 0$, then

$$(2.50b)\quad \nu((0,M_t]\times[\varepsilon,\infty)) = \#\left\{\begin{matrix}\text{excursion intervals away from the}\\ \text{origin, of duration} \ge\varepsilon, \text{ initiated}\\ \text{by } W \text{ at or before time } t\end{matrix}\right\}.$$

Instead of the expressions on the right-hand side of (2.50a, b), it is perhaps easier to visualize the "number of excursion intervals away from the origin, of duration $\ge \varepsilon$, *completed* by W at or before time t." This expression differs from $\nu((0,M_t]\times[\varepsilon,\infty))$ by at most *one* excursion, and such a discrepancy in counting would be of no consequence in formulas (2.48), (2.49): in the former, it would be eliminated by the factor $\sqrt{\varepsilon}$ as $\varepsilon\downarrow 0$; in the latter, the effect on the integral would be at most ε, and even after being divided by $\sqrt{\varepsilon}$, the effect would be eliminated as $\varepsilon \downarrow 0$. Recalling the identifications (2.1), we obtain the following theorem.

2.21 Theorem (P. Lévy (1948)). *The local time at the origin of the Brownian motion W satisfies*

$$L_t(0) = \lim_{\varepsilon\downarrow 0}\sqrt{\frac{\pi\varepsilon}{8}}\cdot\#\left\{\begin{matrix}\textit{excursion intervals away from the origin, of}\\ \textit{duration} \ge\varepsilon, \textit{ completed by } \{W_s;\, 0\le s\le t\}\end{matrix}\right\}$$
$$= \lim_{\varepsilon\downarrow 0}\sqrt{\frac{\pi}{8\varepsilon}}\cdot\left[\begin{matrix}\textit{Total duration of all excursion intervals away}\\ \textit{from the origin of individual duration} <\varepsilon,\\ \textit{completed by } \{W_s;\, 0\le s\le t\}\end{matrix}\right];$$
$$\forall\, 0\le t<\infty, \text{ a.s.}$$

2.22 Remark. The notion of local time $2L_t(0)$ as occupation density suggests that this quantity is determined by the behavior of the Brownian path W *near*, rather than *at*, the origin. Theorem 2.21 shows that local time is actually determined by the way in which Brownian motion spends time *at* the origin, for both representations in that theorem can be computed from knowledge of the zero set $\mathscr{Z}_\omega = \{0 \le t < \infty; W_t = 0\}$ alone.

A third representation of local time in the spirit of Theorem 2.21, the *downcrossings theorem*, was conjectured by Lévy (1959) and proved by Itô & McKean (1974). We offer a proof taken from Stroock (1982).

Recalling the notation introduced immediately before Theorem 1.3.8, let

$$D_t(\varepsilon) = D_{[0,t]}(0, \varepsilon; |W|)$$

be the number of downcrossings of the interval $[0, \varepsilon]$ by the Brownian motion $\{|W_s|; 0 \le s \le t\}$.

2.23 Theorem (P. Lévy's Downcrossings Representation of Local Time). *The local time at the origin of the Brownian motion W satisfies*

$$2L_t(0) = \lim_{\varepsilon \downarrow 0} \varepsilon D_t(\varepsilon); \quad 0 \le t < \infty, \text{ a.s.} \tag{2.51}$$

PROOF. For fixed $\varepsilon > 0$, let us define recursively the stopping times $\tau_0 \equiv 0$ and

$$\sigma_n \triangleq \inf\{t \ge \tau_{n-1}; |W_t| = \varepsilon\}, \quad \tau_n \triangleq \inf\{t \ge \sigma_n; |W_t| = 0\}$$

for $n \ge 1$. With $\eta_n = \sigma_n - \tau_{n-1}$, $\xi_n = \tau_n - \sigma_n$, we have from the strong Markov property as expressed by Theorem 2.6.16 that the pairs $(\eta_1, \xi_1), (\eta_2, \xi_2), \ldots$ are independent and identically distributed. Moreover, Problem 2.8.14 asserts that

$$E\eta_1 = \varepsilon^2, \tag{2.52}$$

and we also have

$$\{j \le D_t(\varepsilon)\} = \{\tau_j \le t\} \text{ is independent of } \{(\eta_k, \xi_k)\}_{k=j+1}^\infty. \tag{2.53}$$

Suppose that $t \in [\sigma_n, \tau_n)$ for some $n \ge 1$; then

$$\begin{aligned}\sum_{j=1}^\infty \{|W_{t\wedge\tau_j}| - |W_{t\wedge\sigma_j}|\} &= \sum_{j=1}^{n-1} \{|W_{\tau_j}| - |W_{\sigma_j}|\} + |W_t| - \varepsilon \\ &= -\varepsilon D_t(\varepsilon) + |W_t| - \varepsilon.\end{aligned}$$

If $t \notin \bigcup_{n=1}^\infty [\sigma_n, \tau_n)$, then $\sum_{j=1}^\infty \{|W_{t\wedge\tau_j}| - |W_{t\wedge\sigma_j}|\} = -\varepsilon D_t(\varepsilon)$. In either case,

$$\sum_{j=1}^\infty \{|W_{t\wedge\tau_j}| - |W_{t\wedge\sigma_j}|\} = -\varepsilon D_t(\varepsilon) + (|W_t| - \varepsilon) \sum_{j=1}^\infty 1_{[\sigma_j,\tau_j)}(t). \tag{2.54}$$

On the other hand, the local time $L_\cdot(0)$ is flat on $\bigcup_{n=1}^\infty [\sigma_n, \tau_n)$ (cf. Problem 3.6.13 (ii)), and thus from the Tanaka formula (3.6.13) we obtain, a.s.:

$$\sum_{j=1}^\infty \{|W_{t\wedge\tau_j}| - |W_{t\wedge\sigma_j}|\} = |W_t| - 2L_t(0) - \sum_{j=1}^\infty \int_{t\wedge\tau_{j-1}}^{t\wedge\sigma_j} \operatorname{sgn}(W_s)\, dW_s. \tag{2.55}$$

From (2.54), (2.55) we conclude that

$$\varepsilon D_t(\varepsilon) - 2L_t(0) = -|W_t| \sum_{j=1}^{\infty} 1_{[\tau_{j-1}, \sigma_j)}(t) - \varepsilon \sum_{j=1}^{\infty} 1_{[\sigma_j, \tau_j)}(t) + \sum_{j=1}^{\infty} \int_{t \wedge \tau_{j-1}}^{t \wedge \sigma_j} \operatorname{sgn}(W_s)\, dW_s, \quad \text{a.s.} \tag{2.56}$$

2.24 Problem. Conclude from (2.56) that, for some positive constant $C(t)$ depending only on t, we have

$$E|\varepsilon D_t(\varepsilon) - 2L_t(0)|^2 \le C(t)\varepsilon. \tag{2.57}$$

Čebyšev's inequality and (2.57) give

$$P[|n^{-2}D_t(n^{-2}) - 2L_t(0)| \ge n^{-1/4}] \le C(t)n^{-3/2},$$

and this, coupled with the Borel-Cantelli lemma, implies

$$\lim_{n\to\infty} n^{-2}D_t(n^{-2}) = 2L_t(0), \quad \text{a.s.}$$

But for every $0 < \varepsilon < 1$, one can find an integer $n = n(\varepsilon) \ge 1$ such that $(n+1)^{-2} \le \varepsilon < n^{-2}$, and obviously

$$(n+1)^{-2}D_t(n^{-2}) \le \varepsilon D_t(\varepsilon) \le n^{-2}D_t((n+1)^{-2})$$

holds. Thus, (2.51) holds for every fixed $t \in [0, \infty)$; the general statement follows from the monotoncity in t of both sides of (2.51) and the continuity in t of $L_t(0)$. □

2.25 Remark. From (2.51) and (2.1), (2.2) we obtain the identity

$$\lim_{\varepsilon \downarrow 0} \varepsilon D_{[0,t]}(0, \varepsilon; M(\omega) - B(\omega)) = M_t(\omega); \quad \forall\, 0 \le t < \infty \tag{2.58}$$

for P-a.e. $\omega \in \Omega$. The gist of (2.58) is the "miraculous fact," as Williams (1979) puts it, that the maximum-to-date process M of (2.2) can be reconstructed from the paths of the reflected Brownian motion $M - B$, in a nonanticipative way. As Williams goes on to note, "this reconstruction will not be possible for any picture you may draw, because it depends on the violent oscillation of the Brownian path." You should also observe that (2.58) offers just one way of carrying out this reconstruction; other possibilities exist as well. For instance, we have from Theorem 2.21 that

$$\lim_{\varepsilon\downarrow 0} \sqrt{\frac{\pi\varepsilon}{2}} \cdot \# \left\{ \begin{array}{l} \text{excursion intervals away from the} \\ \text{origin, of duration} \ge \varepsilon, \text{ completed} \\ \text{by } \{M_s - B_s; 0 \le s \le t\} \end{array} \right\} = M_t; \quad \forall\, 0 \le t < \infty,$$

$$\lim_{\varepsilon\downarrow 0} \sqrt{\frac{\pi}{2\varepsilon}} \cdot \left[\begin{array}{l} \text{total duration of all excursion} \\ \text{intervals, away from the origin,} \\ \text{of individual duration} < \varepsilon, \\ \text{completed by } \{M_s - B_s; 0 \le s \le t\} \end{array} \right] = M_t; \quad \forall\, 0 \le t < \infty,$$

hold almost surely.

6.3 Two Independent Reflected Brownian Motions

Our intent in this section is to show how one can create two *independent*, reflected Brownian motions by piecing together the positive and negative excursions of a standard, one-dimensional Brownian motion. This result has important consequences, and we develop some of them, most notably the *first formula of D. Williams* (Theorem 3.6). Our basic tools will be the F. Knight Theorem 3.4.13 and the theory of Brownian local time as developed in Section 3.6; we shall retain the setting, assumptions, and notation of that section.

A. The Positive and Negative Parts of a Brownian Motion

We start by examining the Tanaka formulas (3.6.11–12) a bit more closely. With $a = z = 0$, $L(t) \triangleq L_t(0)$, and

$$I_+(t) \triangleq -\int_0^t 1_{(0,\infty)}(W_s)\,dW_s, \quad I_-(t) \triangleq \int_0^t 1_{(-\infty,0]}(W_s)\,dW_s, \tag{3.1}$$

these formulas read

$$W_t^{\pm} = -I_{\pm}(t) + L(t); \quad 0 \le t < \infty \tag{3.2}$$

a.s. P^0. The processes in (3.1) are continuous, square-integrable martingales, with quadratic variations

$$\langle I_+\rangle(t) = \Gamma_+(t) \triangleq \text{meas}\{0 \le s \le t;\, W_s > 0\} \tag{3.3}$$

$$\langle I_-\rangle(t) = \Gamma_-(t) \triangleq \text{meas}\{0 \le s \le t;\, W_s \le 0\} \tag{3.4}$$

and cross-variation equal to zero:

$$\langle I_+, I_-\rangle(t) = \int_0^t 1_{(0,\infty)}(W_s)\cdot 1_{(-\infty,0]}(W_s)\,ds = 0. \tag{3.5}$$

We also have

$$\lim_{t\to\infty} \Gamma_{\pm}(t) = \infty, \quad \text{a.s. } P^0 \tag{3.6}$$

from Problem 3.6.30.

On the other hand, $W^{\pm}$ are nonnegative processes which satisfy

$$\int_0^\infty 1_{(0,\infty)}(W^{\pm}(s))\,dL(s) = 0 \tag{3.7}$$

a.s. P^0 (Problem 3.6.13 (ii)). It becomes evident then from (3.2), (3.7) that the pairs $(L, W^{\pm})$ are solutions to the Skorohod equation of Lemma 3.6.14 for the functions $-I_{\pm}$, respectively. From the explicit form (3.6.32) of the solution to this equation, we deduce

$$L(t) = \max_{0\le s\le t} I_{\pm}(s), \quad W_t^{\pm} = \max_{0\le s\le t} I_{\pm}(s) - I_{\pm}(t); \quad 0 \le t < \infty, \quad \text{a.s. } P^0. \tag{3.8}$$

Now let us introduce the right-continuous inverses of the occupation times $\Gamma_\pm$ of (3.3), (3.4), namely

$$\Gamma_\pm^{-1}(\tau) = \inf\{t \geq 0; \Gamma_\pm(t) > \tau\}; \quad 0 \leq \tau < \infty. \tag{3.9}$$

Theorem 3.4.13 asserts, in conjunction with (3.3)–(3.6), that the processes

$$B_\pm(\tau) \triangleq I_\pm(\Gamma_\pm^{-1}(\tau)); \quad 0 \leq \tau < \infty \tag{3.10}$$

are independent, standard, one-dimensional Brownian motions under P^0, and from Theorem 3.4.6 we also have the representations

$$I_\pm(t) = B_\pm(\Gamma_\pm(t)); \quad 0 \leq t < \infty. \tag{3.11}$$

Here then is the fundamental result of this section.

3.1 Theorem. *The processes*

$$W_\pm(\tau) \triangleq \pm W_{\Gamma_\pm^{-1}(\tau)}; \quad 0 \leq \tau < \infty \tag{3.12}$$

are independent, reflected Brownian motions under P^0.

PROOF. We start by introducing

$$L_\pm(\tau) \triangleq L(\Gamma_\pm^{-1}(\tau)) \tag{3.13}$$

and observing that because of (3.8), (3.11), and Problem 3.4.5 (ii), (iii) we have, a.s. P^0:

$$L_\pm(\tau) = \max_{0 \leq s \leq \Gamma_\pm^{-1}(\tau)} I_\pm(s) = \max_{0 \leq \Gamma_\pm(s) \leq \tau} B_\pm(\Gamma_\pm(s)) = \max_{0 \leq u \leq \tau} B_\pm(u) \tag{3.14}$$

for all $0 \leq \tau < \infty$; in particular, $L_\pm$ are independent, continuous nondecreasing processes.

Now for each $\tau \geq 0$, $\Gamma_\pm(\Gamma_\pm^{-1}(\tau)) = \tau < \Gamma_\pm(\Gamma_\pm^{-1}(\tau) + \delta)$ for all $\delta > 0$, and consequently $W_{\Gamma_+^{-1}(\tau)} \geq 0$, $W_{\Gamma_-^{-1}(\tau)} \leq 0$ hold a.s. P^0. It follows then that

$$W_\pm(\tau) = W^\pm_{\Gamma_\pm^{-1}(\tau)} = L_\pm(\tau) - B_\pm(\tau) = \max_{0 \leq u \leq \tau} B_\pm(u) - B_\pm(\tau) \tag{3.15}$$

also hold a.s. P^0, first for a fixed $\tau \geq 0$, and then by continuity, for all $\tau \geq 0$ simultaneously. From Theorem 3.6.17, each of the processes $W_\pm$ is a reflected Brownian motion starting at the origin; $W_\pm$ are independent because $B_\pm$ are. □

3.2 Remark. Theorem 3.6.17 also yields that the pairs $\{(W_\pm(\tau), L_\pm(\tau)); 0 \leq \tau < \infty\}$ have the same law as $\{(|W_t|, 2L_t(0)); 0 \leq t < \infty\}$, under P^0.

In particular, we obtain from (3.6.36) and (3.14), (3.15):

$$L_\pm(\tau) = \lim_{\varepsilon \downarrow 0} \frac{1}{2\varepsilon} \operatorname{meas}\{0 \leq \sigma \leq \tau; W_\pm(\sigma) \leq \varepsilon\}, \quad \text{a.s. } P^0 \tag{3.16}$$

for every $\tau \in [0, \infty)$. The processes $L_\pm$ are thus adapted to the augmentations of the filtrations $\mathscr{G}_\tau^\pm \triangleq \sigma(W_\pm(u); 0 \leq u \leq \tau)$.

3.3 Remark. As suggested by the accompanying figure, the construction of $W_+(\cdot) = W_{\Gamma_+^{-1}(\cdot)}$ amounts to discarding the negative excursions of the Brownian path and shifting the positive ones down on the time-scale in order to close up the gaps thus created. A similar procedure, with a change of sign, gives the construction of $W_-(\cdot) = W_{\Gamma_-^{-1}(\cdot)}$. Theorem 3.1 asserts then, roughly speaking, that *the motions of W on the two half-lines* $(0, \infty)$ *and* $(-\infty, 0)$ *are independent.*

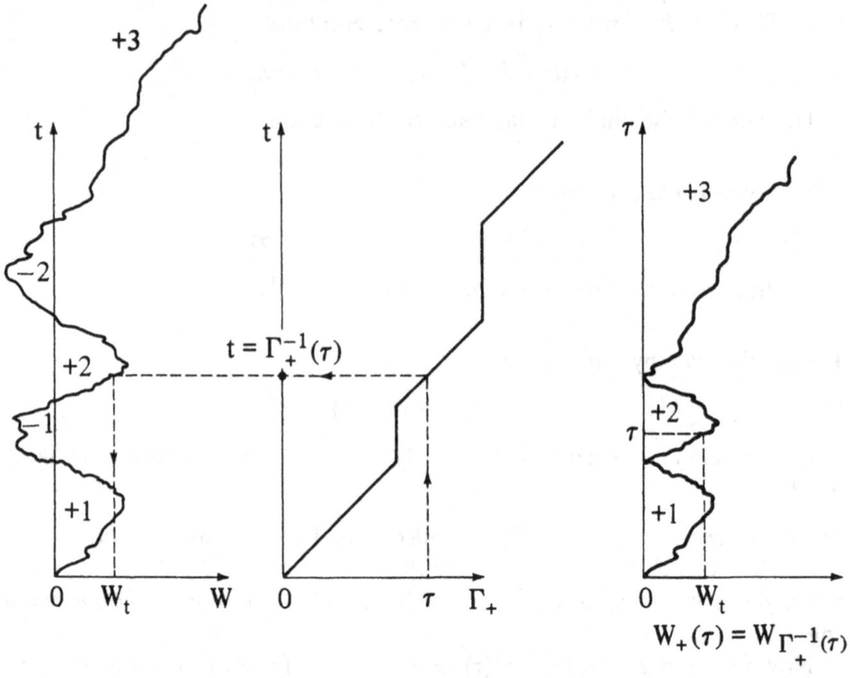

3.4 Problem. Derive the bivariate densities

$$P^0[W_t \in da;\, 2L_t(0) \in db] = \frac{b + |a|}{\sqrt{2\pi t^3}} \exp\left\{-\frac{(b + |a|)^2}{2t}\right\} da\, db; \quad a \in \mathbb{R} \tag{3.17}$$

$$P^0[|W_t| \in da;\, 2L_t(0) \in db] = \frac{2(a + b)}{\sqrt{2\pi t^3}} \exp\left\{-\frac{(a + b)^2}{2t}\right\} da\, db; \quad a > 0 \tag{3.18}$$

for $b > 0$.

3.5 Exercise. Show that the right-hand side of (5.3.13), Exercise 5.3.12, is given by $\int_\Gamma q_t(a)\, da$, where

$$q_t(a) = \frac{1}{\sqrt{2\pi t}}\left[\exp\left\{-\frac{(|a| + t)^2}{2t}\right\} + e^{-2|a|} \int_{|a|}^{\infty} \exp\left\{-\frac{(v - t)^2}{2t}\right\} dv\right]. \tag{3.19}$$

B. The First Formula of D. Williams

The processes $L_\pm$ of (3.13) are continuous and nondecreasing (cf. (3.14)), with right-continuous inverses

$$(3.20)\quad L_\pm^{-1}(b) \triangleq \inf\{\tau \ge 0; L_\pm(\tau) > b\} = \inf\left\{\tau \ge 0;\ \max_{0\le u\le\tau} B_\pm(u) > b\right\}.$$

For every fixed $b \in [0, \infty)$, $L_\pm^{-1}(b)$ is P^0-a.s. equal to the passage time

$$(3.21)\qquad T_b^\pm \triangleq T_b^{B_\pm} = \inf\{t \ge 0; B_\pm(t) \ge b\}$$

of the Brownian motion $B_\pm$ to the level b; cf. Problem 2.7.19 (i).

3.6 Theorem (D. Williams (1969)). *For every fixed $\alpha > 0$, $\tau > 0$ we have*

$$(3.22)\qquad E^0[e^{-\alpha\Gamma_+^{-1}(\tau)}|W_+(u); 0 \le u < \infty] = e^{-\alpha\tau-\sqrt{2\alpha}L_+(\tau)};\quad a.s.\ P^0.$$

PROOF. The argument hinges on the important identity

$$(3.23)\qquad \Gamma_+^{-1}(\tau) = \tau + L_-^{-1}(L_+(\tau)) = \tau + T_{L_+(\tau)}^-;\quad \text{a.s. } P^0$$

which expresses the inverse occupation time $\Gamma_+^{-1}(\tau)$ as τ, plus the passage time of the Brownian motion B_- to the level $L_+(\tau)$. But $L_+(\tau)$ is a random variable measurable with respect to the completion of $\sigma(W_+(u); 0 \le u < \infty)$, and hence *independent of the Brownian motion B_-*. It follows from Problem 2.7.19 (ii) that

$$(3.24)\qquad L_-^{-1}(L_+(\tau)) = T_{L_+(\tau)}^-,\quad \text{a.s. } P^0,$$

and this takes care of the second identity in (3.23). The first follows from the string of identities (see Problem 3.7)

$$\begin{aligned}(3.25)\qquad L_-^{-1}(L_+(\tau)) &= \inf\{t \ge 0; L_-(t) > L_+(\tau)\}\\ &= \inf\{t \ge 0; L(\Gamma_-^{-1}(t)) > L(\Gamma_+^{-1}(\tau))\}\\ &= \inf\{t \ge 0; \Gamma_-^{-1}(t) \ge \Gamma_+^{-1}(\tau)\}\\ &= \Gamma_-(\Gamma_+^{-1}(\tau)) = \Gamma_+^{-1}(\tau) - \Gamma_+(\Gamma_+^{-1}(\tau))\\ &= \Gamma_+^{-1}(\tau) - \tau,\quad \text{a.s. } P^0.\end{aligned}$$

Now the independence of $\{B_-(u); 0 \le u < \infty\}$ and $\{W_+(u); 0 \le u < \infty\}$, along with the formula (2.8.6) for the moment generating function for Brownian passage times, express the left-hand side of (3.22) as

$$e^{-\alpha\tau}E^0 e^{-\alpha T_b^-}|_{b=L_+(\tau)} = e^{-\alpha\tau-\sqrt{2\alpha}L_+(\tau)},\quad \text{a.s. } P^0. \qquad \square$$

3.7 Problem. Establish the third and fourth identities in (3.25).

Following McKean (1975), we shall refer to (3.22), or alternatively (3.23), as *the first formula of D. Williams.* This formula can be cast in the equivalent forms

$$(3.26)\quad P^0[\Gamma_+^{-1}(\tau) < t \mid W_+(u); 0 \le u < \infty] = P^0[T_b^- < t - \tau]|_{b=L_+(\tau)}$$
$$= 2\left[1 - \Phi\left(\frac{L_+(\tau)}{\sqrt{t-\tau}}\right)\right]$$
$$= \int_\tau^t L_+(\tau)\frac{e^{-L_+^2(\tau)/2(\theta-\tau)}}{\sqrt{2\pi(\theta-\tau)^3}}\,d\theta$$

a.s. P^0, which follow easily from (3.23) and (2.8.4), (2.8.5). We use the notation $\Phi(z) = (1/\sqrt{2\pi})\int_{-\infty}^z e^{-u^2/2}\,du$.

We offer the following interpretation of Williams's first formula. The reflected Brownian motion $\{W_+(u); 0 \le u < \infty\}$ has been observed, and then a time τ has been chosen. We wish to compute the distribution of $\Gamma_+^{-1}(\tau)$ based on our observations. Now W_+ consists of the positive part of the original Brownian motion W, but W_+ is run under a new clock which stops whenever W becomes negative. When τ units of time have accumulated on this clock corresponding to W_+, $\Gamma_+^{-1}(\tau)$ units of time have accumulated on the original clock. Obviously, $\Gamma_+^{-1}(\tau)$ is the sum of τ and the occupation time $\Gamma_-(\Gamma_+^{-1}(\tau))$.

Because W_- is independent of the observed process W_+, one might surmise that nothing can be inferred about $\Gamma_-(t)$ from W_+. However, the independence between W_+ and W_- holds only when they are run according to their respective clocks. When run in the original clock, these processes are intimately connected. In particular, they accumulate local time at the origin at the same rate, a fact which is perhaps most clearly seen from the appearance of the same process L in both the plus and minus versions of (3.2). After the time changes (3.12) which transform $W^\pm$ into $W_\pm$, this equal rate of local time accumulation finds expression in (3.13). (From (3.16) we see that $L_\pm$ is the local time of $W_\pm$.) In particular, when we have observed W_+ and computed its local time $L_+(\tau)$, and wish to know the amount of time W has spent on the negative half-line before it accumulated τ units of time on the positive half-line, we have a relevant piece of information: the time spent on the negative half-line was enough to accumulate $L_+(\tau)$ units of local time.

Suppose $L_+(\tau) = b$. How long should it take the reflected Brownian motion W^- to accumulate b units of local time? Recalling from Theorem 3.6.17 that the local time process for a reflected Brownian motion has the same distribution as the maximum-to-date process of a standard Brownian motion, we see that our question is equivalent to: How long should it take a standard Brownian motion starting at the origin to reach the level b? The time required is the passage time T_b^- appearing in (3.23), (3.26). Once $L_+(\tau) = b$ is known, nothing else about W_+ is relevant to the computation of the distribution of $\Gamma_-(\Gamma_+^{-1}(\tau))$.

3.8 Exercise. Provide a new derivation of P. Lévy's arc-sine law for $\Gamma_+(t)$ (Proposition 4.4.11), using Theorem 3.6.

C. The Joint Density of $(W(t), L(t), \Gamma_+(t))$

Here is a more interesting application of the first formula of D. Williams. With $\tau > 0$ fixed, we obtain from (3.26):

$$(3.27)\quad P^0[\Gamma_+^{-1}(\tau) \in dt \mid W_+(\tau) = a;\, L_+(\tau) = b] = \frac{be^{-b^2/2(t-\tau)}}{\sqrt{2\pi(t-\tau)^3}}\,dt; \quad \tau < t < \infty,$$

for almost every pair (a, b) of positive numbers. Remark 3.2 and the bivariate density (3.18) give

$$P^0[W_+(\tau) \in da;\, L_+(\tau) \in db] = \frac{2(a+b)}{\sqrt{2\pi\tau^3}} e^{-(a+b)^2/2\tau}\,da\,db; \quad a > 0, b > 0,$$

and in conjunction with (3.27):

$$(3.28)\quad P^0[W_+(\tau) \in da;\, L_+(\tau) \in db;\, \Gamma_+^{-1}(\tau) \in dt] = f(a, b; t, \tau)\,da\,db\,dt;$$
$$a > 0, \quad b > 0, \quad t > \tau$$

where

$$(3.29)\qquad f(a, b; t, \tau) \triangleq \frac{b(a+b)}{\pi\tau^{3/2}(t-\tau)^{3/2}} \exp\left\{ -\frac{b^2}{2(t-\tau)} - \frac{(a+b)^2}{2\tau} \right\}.$$

We shall employ (3.28) in order to derive, at a given time $t \in (0, \infty)$, the trivariate density for the location W_t of the Brownian motion; its local time $L(t) = L_t(0)$ at the origin; and its occupation time $\Gamma_+(t)$ of $(0, \infty)$ as in (3.3), up to t.

3.9 Proposition. *For every finite $t > 0$, we have*

$$(3.30)\quad P^0[W_t \in da;\, L(t) \in db;\, \Gamma_+(t) \in d\tau]$$
$$= \begin{cases} f(a, b; t, \tau)\,da\,db\,d\tau; & a > 0, \quad b > 0, \quad 0 < \tau < t, \\ f(-a, b; t, t-\tau)\,da\,db\,d\tau; & a < 0, \quad b > 0, \quad 0 < \tau < t, \end{cases}$$

in the notation of (3.29).

3.10 Remark. Only the expression for $a > 0$ need be established; the one for $a < 0$ follows from the former and from the observation that the triples $(W_t, L(t), \Gamma_+(t))$ and $(-W_t, L(t), t - \Gamma_+(t))$ are equivalent in law.

Now in order to establish (3.30) for $a > 0$, one could write formally

$$\frac{1}{dt} P^0[W_+(\tau) \in da;\, L_+(\tau) \in db;\, \Gamma_+^{-1}(\tau) \in dt]$$

$$= \frac{1}{d\tau} P^0[W_t \in da;\, L(t) \in db;\, \Gamma_+(t) \in d\tau]; \quad a > 0, \quad b > 0, \quad 0 < \tau < t,$$

and then appeal to (3.28). On the left-hand side of this identity, τ is fixed and we have a density in (a, b, t); on the right-hand side, t is fixed and we have a density in (a, b, τ). Because the two sides are uniquely determined only up to sets of Lebesgue measure zero in their respective domains, it is not clear how this identity should be interpreted.

We offer now a rigorous argument along these lines; we shall need to recall the random variable β_t of (2.6), as well as the following auxiliary result.

3.11 Problem. For $a \in \mathbb{R}$, $t > 0$, $\varepsilon > 0$ we have

$$P^0\left[\max_{t \le s \le t+h} W_s \ge a + \varepsilon;\ \min_{t \le s \le t+h} W_s \le a - \varepsilon\right] = o(h) \tag{3.31}$$

and for $a > 0$, $\tau > 0$:

$$P^0[W_+(\tau) > a;\ t \le \Gamma_+^{-1}(\tau) < t + h;\ \beta_t < t + h] = o(h) \tag{3.32}$$

$$P^0[W_t > a;\ \beta_t < t + h] = o(h) \tag{3.33}$$

as $h \downarrow 0$.

Proof of Proposition 3.9. For arbitrary but fixed $a > 0$, $b > 0$, $t > 0$, $\tau \in (0, t)$ we define the function

$$F(a, b; t, \tau) \triangleq \int_b^\infty \int_a^\infty f(\alpha, \beta; t, \tau)\, d\alpha\, d\beta \tag{3.34}$$

which admits, by virtue of (3.28), (3.32), the interpretation

(3.35) $F(a, b; t, \tau)$

$$= \lim_{h \downarrow 0} \frac{1}{h} P^0[W_+(\tau) > a;\ L_+(\tau) > b;\ t \le \Gamma_+^{-1}(\tau) < t + h]$$

$$= \lim_{h \downarrow 0} \frac{1}{h} P^0[W_+(\tau) > a;\ L_+(\tau) > b;\ t \le \Gamma_+^{-1}(\tau) < t + h;\ \beta_t \ge t + h].$$

For every $h > 0$ we have

$$\Gamma_+(s) = \Gamma_+(t) + s - t, \quad L(s) = L(t); \quad \forall s \in [t, t + h]$$

on the event $\{W_t > 0,\ \beta_t \ge t + h\}$. Therefore, with $0 < \varepsilon < a$ and

$$A \triangleq \{L(t) > b;\ \tau - h < \Gamma_+(t) \le \tau;\ \beta_t \ge t + h\}, \tag{3.36}$$

we obtain

(3.37)

$$P^0[W_+(\tau) > a + \varepsilon;\ L_+(\tau) > b;\ t \le \Gamma_+^{-1}(\tau) < t + h;\ \beta_t \ge t + h] - P^0[W_t > a;\ A]$$

$$\le P^0\left[\max_{t \le s \le t+h} W_s > a + \varepsilon;\ A\right] - P^0\left[\min_{t \le s \le t+h} W_s > a;\ A\right] = o(h),$$

by virtue of (3.31). Dividing by h in (3.37) and then letting $h \downarrow 0$, we obtain in conjunction with (3.35), (3.32):

$$(3.38) \quad F(a + \varepsilon, b; t, \tau) \leq \varliminf_{h \downarrow 0} \frac{1}{h} P^0[W_t > a; L(t) > b; \tau - h < \Gamma_+(t) \leq \tau].$$

Similarly,

(3.38)

$$P^0[W_t > a; A] - P^0[W_+(\tau) > a - \varepsilon; L_+(\tau) > b; t \leq \Gamma_+^{-1}(\tau) < t + h; \beta_t \geq t + h]$$

$$\leq P^0\left[\max_{t \leq s \leq t+h} W_s > a; A\right] - P^0\left[\min_{t \leq s \leq t+h} W_s > a - \varepsilon; A\right] = o(h),$$

and we obtain from (3.33):

$$(3.39) \quad F(a - \varepsilon, b; t, \tau) \geq \varlimsup_{h \downarrow 0} \frac{1}{h} P^0[W_t > a; L(t) > b; \tau - h < \Gamma_+(t) \leq \tau].$$

Letting $\varepsilon \downarrow 0$ in both (3.38), (3.39) we conclude that

$$F(a, b; t, \tau) = \lim_{h \downarrow 0} \frac{1}{h} P^0[W_t > a; L(t) > b; \tau - h < \Gamma_+(t) \leq \tau],$$

from which (3.30) for $a > 0$ follows in a straightforward manner. □

3.12 Remark. From (3.30) one can derive easily the bivariate density

$$(3.40) \quad P^0[2L_t(0) \in db; \Gamma_+(t) \in d\tau] = \frac{bte^{-tb^2/8\tau(t-\tau)}}{4\pi\tau^{3/2}(t - \tau)^{3/2}} db\, d\tau; \quad b > 0, 0 < \tau < t$$

as well as the arc-sine law of Proposition 4.4.11.

The reader should not fail to notice that for $a < 0$, the trivariate density of (3.30) is the same as that for (W_t, M_t, θ_t) in Proposition 2.8.15, for $M_t = \max_{0 \leq s \leq t} W_s$, $\theta_t = \sup\{s \leq t; W_s = M_t\}$. This "coincidence" can be explained by an appropriate *decomposition of the Brownian path* $\{W_s; 0 \leq s \leq t\}$; cf. Karatzas & Shreve (1987).

6.4. Elastic Brownian Motion

This section develops the concept of elastic Brownian motion as a tool for computing distributions involving Brownian local time at one or several points. This device allows us to study local time parametrized by the spatial variable, and it is shown that with this parametrization, local time is related to a Bessel process (Theorem 4.7). We use this fact to prove the Dvoretzky-Erdös-Kakutani Theorem 2.9.13.

We employ throughout this section the notation of Section 3.6. In particular, $W = \{W_t, \mathscr{F}_t; 0 \le t < \infty\}$, $(\Omega, \mathscr{F})$, $\{P^x\}_{x \in \mathbb{R}}$ will be a one-dimensional Brownian family with local time

$$(4.1) \quad L_t(a) = \lim_{\varepsilon \downarrow 0} \frac{1}{4\varepsilon} \operatorname{meas}\{0 \le s \le t; |W_s - a| \le \varepsilon\}; \quad 0 \le t < \infty, a \in \mathbb{R}.$$

On a separate probability space $(\Omega', \mathscr{F}', P')$, let $R_1, \ldots, R_n$ be independent, exponential random variables with (positive) parameters $\gamma_1, \ldots, \gamma_n$, respectively, i.e.,

$$P'(R_1 \in dr_1, \ldots, R_n \in dr_n) = \prod_{i=1}^n \gamma_i e^{-\gamma_i r_i} dr_i.$$

We consider n distinct points $a_1, \ldots, a_n$ on the real line and define a new process $\hat{W}$, which is the old Brownian motion W "killed" when local time at any of these points a_i exceeds the corresponding level R_i. More precisely, with

$$\tau_r(a_i) = \inf\{t \ge 0; L_t(a_i) > r\}; \quad r \ge 0, \quad \text{and}$$

$$(4.2) \quad \zeta \triangleq \inf\{t \ge 0; L_t(a_i) > R_i \text{ for some } 1 \le i \le n\} = \min_{1 \le i \le n} \tau_{R_i}(a_i),$$

we define the new process

$$\hat{W}_t \triangleq \begin{cases} W_t; & 0 \le t < \zeta, \\ \Delta; & t \ge \zeta, \end{cases}$$

and call it *elastic Brownian motion* with *lifetime* ζ. Here, Δ is a "cemetery" state isolated from $\mathbb{R}$. We may regard $\hat{W}$ as a process on $\hat{\Omega} \triangleq \Omega \times \Omega'$, $\hat{\mathscr{F}} = \mathscr{F} \otimes \mathscr{F}'$, $\hat{P}^x = P^x \times P'$.

A. The Feynman-Kac Formulas for Elastic Brownian Motion

Our intent is to study the counterpart

$$(4.3) \quad u(x) = \hat{E}^x \int_0^\infty f(\hat{W}_t) \exp\left\{-\alpha t - \int_0^t k(\hat{W}_s)\,ds\right\} dt$$
$$= \int_\Omega \int_{\Omega'} \int_0^\zeta f(W_t) \exp\left\{-\alpha t - \int_0^t k(\hat{W}_s)\,ds\right\} dt\, dP'\, dP^x.$$

of the function z in (4.4.14). The constant α is positive, and the functions f and k are piecewise continuous on $\mathbb{R}$ (Definition 4.4.8), mapping $\mathbb{R} \cup \{\Delta\}$ into $\mathbb{R}$ and $[0, \infty)$, respectively, and with

$$(4.4) \quad f(\Delta) = k(\Delta) = 0.$$

Hereafter, we will specify properties of functions restricted to $\mathbb{R}$; condition

(4.4) will always be an unstated assumption. In order to obtain an analytical characterization of the function u, we put it into the more convenient form

$$u(x) = \int_\Omega \int_0^\infty \cdots \int_0^\infty \int_0^{\min_{1\le i\le n} \tau_{r_i}(a_i)} f(W_t) e^{-\alpha t - \int_0^t k(W_s)\,ds} \prod_{i=1}^n \gamma_i e^{-\gamma_i r_i}\, dt\, dr_1 \ldots dr_n\, dP^x$$

$$= \int_\Omega \int_0^\infty \int_{L_t(a_1)}^\infty \cdots \int_{L_t(a_n)}^\infty f(W_t) e^{-\alpha t - \int_0^t k(W_s)\,ds} \prod_{i=1}^n \gamma_i e^{-\gamma_i r_i}\, dr_1 \ldots dr_n\, dt\, dP^x,$$

whence

$$u(x) = E^x \int_0^\infty f(W_t) \exp\left\{-\alpha t - \int_0^t k(W_s)\,ds - \sum_{i=1}^n \gamma_i L_t(a_i)\right\} dt. \tag{4.5}$$

4.1 Theorem. *Let $f\colon \mathbb{R} \to \mathbb{R}$ and $k\colon \mathbb{R} \to [0, \infty)$ be piecewise continuous functions, and let $D = D_f \cup D_k$ be the union of their discontinuity sets. Assume that, for some $\alpha > 0$,*

$$E^x \int_0^\infty e^{-\alpha t} |f(W_t)|\, dt < \infty; \quad \forall x \in \mathbb{R}, \tag{4.6}$$

and that there exists a function $\tilde{u}\colon \mathbb{R} \to \mathbb{R}$ which is continuous on $\mathbb{R}$, C^1 on $\mathbb{R}\backslash\{a_1, \ldots, a_n\}$, C^2 on $\mathbb{R}\backslash(D \cup \{a_1, \ldots, a_n\})$, and satisfies

$$(\alpha + k)\tilde{u} = \frac{1}{2}\tilde{u}'' + f \quad \text{on } \mathbb{R}\backslash(D \cup \{a_1, \ldots, a_n\}), \tag{4.7}$$

$$\tilde{u}'(a_i+) - \tilde{u}'(a_i-) = \gamma_i \tilde{u}(a_i); \quad 1 \le i \le n. \tag{4.8}$$

If f and $\tilde{u}$ are bounded, then $\tilde{u}$ is equal to the function u of (4.3), (4.5); *if f and $\tilde{u}$ are nonnegative, then $\tilde{u} \ge u$.*

Proof. An application of the generalized Itô rule (3.6.53) to the process

$$X_t \triangleq \tilde{u}(W_t) \exp\left\{-\alpha t - \int_0^t k(W_s)\,ds - \sum_{i=1}^n \gamma_i L_t(a_i)\right\}; \quad 0 \le t < \infty,$$

yields

$$E^x \int_0^{n \wedge S_n} f(W_t) \exp\left\{-\alpha t - \int_0^t k(W_s)\,ds - \sum_{i=1}^n \gamma_i L_t(a_i)\right\} dt = \tilde{u}(x) - E^x X_{n \wedge S_n},$$

where

$$S_n = \inf\{t \ge 0; |W_t| \ge n\}. \tag{4.9}$$

If f and $\tilde{u}$ are bounded, we may let $n \to \infty$ and use the bounded convergence theorem to obtain $u = \tilde{u}$. If f and $\tilde{u}$ are nonnegative, then $E^x X_{n \wedge S_n} \ge 0$ and we obtain $\tilde{u} \ge u$ from the monotone convergence theorem.

4.2 Remark. Theorem 4.1 is weaker than its counterpart Theorem 4.4.9 because the former assumes the existence of a solution to (4.7), (4.8) rather than asserting that u is a solution. The stronger version of Theorem 4.1, which asserts that u defined by (4.5) satisfies all the conditions attributed to $\tilde{u}$, is also true, but the proof of this result is fairly lengthy. In our applications, the function u with the required regularity will be explicitly exhibited. These comments also apply to Theorem 4.3.

A variation of Theorem 4.1 can be obtained by stopping the Brownian motion W when it exits from the interval $[b,c]$, where $-\infty < b < c \le \infty$ are fixed constants not in $\{a_1,\dots,a_n\}$. With the convention $T_\infty = \infty$, define

(4.10)

$$v(x) = E^x\left[1_{\{T_b<T_c\}}\exp\left\{-\alpha T_b - \int_0^{T_b} k(W_s)\,ds - \sum_{i=1}^n \gamma_i L_{T_b}(a_i)\right\}\right]; \quad b \le x \le c.$$

4.3 Theorem. *Let* $k: [b,c] \to [0,\infty)$ *be piecewise continuous,* $\alpha \ge 0$, *and assume that there exists a function* $\tilde{v}: [b,c] \to \mathbb{R}$ *which is continuous on* $[b,c]$, C^1 *on* $(b,c)\backslash\{a_1,\dots,a_n\}$, C^2 *on* $(b,c)\backslash(D_k \cup \{a_1,\dots,a_n\})$, *and satisfies*

$$(4.11) \qquad (\alpha + k)\tilde{v} = \frac{1}{2}\tilde{v}'' \quad \text{on } (b,c)\backslash(D_k \cup \{a_1,\dots,a_n\}),$$

$$(4.12) \qquad \tilde{v}'(a_i+) - \tilde{v}'(a_i-) = \gamma_i \tilde{v}(a_i); \quad 1 \le i \le n,$$

$$(4.13) \qquad \tilde{v}(b) = 1,$$

$$(4.14) \qquad \tilde{v}(c) = 0,$$

(*except that* (4.13) *should be omitted if* $c = \infty$). *If* $\tilde{v}$ *is bounded, then* $\tilde{v}$ *is the function* v *of* (4.10); *if* $\tilde{v}$ *is nonnegative, then* $\tilde{v} \ge v$.

PROOF. With

$$Y_t \triangleq \tilde{v}(W_t)\exp\left\{-\alpha t - \int_0^t k(W_s)\,ds - \sum_{i=1}^n \gamma_i L_t(a_i)\right\}; \quad 0 \le t < \infty,$$

the generalized Itô rule (3.6.53) yields $\tilde{v}(x) = E^x X_{T_n \wedge T_c \wedge n \wedge S_n}$; $b \le x \le c$, where S_n is given by (4.9). If $\tilde{v}$ is bounded, we may let $n \to \infty$ to conclude that $\tilde{v}(x) = E^x 1_{\{T_b<T_c\}} Y_{T_b} = v(x)$. If $\tilde{v}$ is nonnegative, Fatou's lemma gives

$$\tilde{v}(x) \ge E^x 1_{\{T_b<T_c\}} Y_{T_b} = v(x). \qquad \square$$

The following exercises illustrate the usefulness of Theorem 4.3 in computations of distributions.

4.4 Problem. For any positive numbers α, β, γ, b, justify the formula

$$(4.15)\quad E^x \exp(-\alpha T_b - \beta\Gamma_+(T_b) - \gamma L_{T_b}(0))$$

$$= \begin{cases} \dfrac{\dfrac{\gamma+\sqrt{2\alpha}}{\sqrt{2(\alpha+\beta)}}\sinh(x\sqrt{2(\alpha+\beta)}) + \cosh(x\sqrt{2(\alpha+\beta)})}{\dfrac{\gamma+\sqrt{2\alpha}}{\sqrt{2(\alpha+\beta)}}\sinh(b\sqrt{2(\alpha+\beta)}) + \cosh(b\sqrt{2(\alpha+\beta)})}; & 0 \le x \le b, \\[2ex] \dfrac{e^{x\sqrt{2\alpha}}}{\dfrac{\gamma+\sqrt{2\alpha}}{\sqrt{2(\alpha+\beta)}}\sinh(b\sqrt{2(\alpha+\beta)}) + \cosh(b\sqrt{2(\alpha+\beta)})}; & x < 0. \end{cases}$$

With $x = 0$, we obtain in the limit as $\alpha \downarrow 0$, $\beta \downarrow 0$:

$$(4.16)\qquad E^0 \exp(-\gamma L_{T_b}(0)) = \frac{1}{1+\gamma b}; \quad \gamma > 0.$$

In other words, $L_{T_b}(0)$ *under* P^0 *is an exponential random variable with expectation* $E^0 L_{T_b}(0) = b$. On the other hand, as $\alpha \downarrow 0$, $\gamma \downarrow 0$, (4.15) becomes

$$(4.17)\qquad E^x e^{-\beta\Gamma_+(T_b)} = \frac{\cosh(x\sqrt{2\beta})}{\cosh(b\sqrt{2\beta})}; \quad 0 \le x \le b,$$

and we recover (4.4.23) by setting $x = 0$. Can you also derive (4.4.23) from (2.8.29) and Theorem 3.1 without any computation at all?

4.5 Exercise. Let R_b be the first time that the Brownian path W_t falls $b > 0$ units below its maximum-to-date $M_t \triangleq \max_{0 \le s \le t} W_s$; i.e.,

$$R_b = \inf\{t \ge 0;\ M_t - W_t = b\}.$$

Show that

$$(4.18)\quad E^0 \exp(-\alpha R_b - \gamma M_{R_b}) = \frac{\sqrt{2\alpha}}{\sqrt{2\alpha}\cosh(b\sqrt{2\alpha}) + \gamma\sinh(b\sqrt{2\alpha})}$$

holds for every $\alpha > 0$, $\gamma > 0$. Deduce the formula

$$(4.19)\qquad E^0 \exp(-\gamma M_{R_b}) = \frac{1}{1+\gamma b}; \quad \gamma > 0,$$

which shows that M_{R_b} *is an exponential random variable under* P^0 *with expectation* $E^0 M_{R_b} = b$. (*Hint*: Recall Theorem 3.6.17.)

The following exercise provides a derivation based on Theorem 4.1 of the joint density of Brownian motion, its local time at the origin, and its occupation time of $[0, \infty)$. This density was already obtained from D. Williams's first formula in Proposition 3.9.

4.6 Exercise.

(i) Use Theorem 4.1 to justify the Laplace transform formula

$$(4.20)\quad E^0 \int_0^\infty 1_{[a,\infty)}(W_t) e^{-\alpha t - \beta \Gamma_+(t) - \gamma L_t(0)}\, dt$$
$$= \frac{2e^{-a\sqrt{2(\alpha+\beta)}}}{\sqrt{2(\alpha+\beta)}\,[\gamma + \sqrt{2\alpha} + \sqrt{2(\alpha+\beta)}]}$$

for positive numbers α, β, γ, and a.

(ii) Use the uniqueness of Laplace transforms, the Laplace transform formula in Remark 4.4.10, the formula

$$\int_0^\infty e^{-\lambda t} \frac{b}{\sqrt{2\pi t^3}} \exp\left\{-\frac{b^2}{2t}\right\} dt = \exp\{-b\sqrt{2\lambda}\}; \quad b > 0, \lambda > 0,$$

and (4.20) to show that for $a > 0$, $b > 0$, $0 < \tau < t$:

$$(4.21)\quad P^0[W_t \geq a;\, L_t(0) \in db;\, \Gamma_+(t) \in d\tau]$$
$$= \frac{b}{\pi \tau^{1/2}(t-\tau)^{3/2}} \exp\left\{-\frac{b^2}{2(t-\tau)} - \frac{(a+b)^2}{2\tau}\right\} db\, d\tau.$$

(iii) Use (4.21) to prove (3.30).

B. The Ray-Knight Description of Local Time

We now formulate the Ray-Knight description of local time, evaluated at time $t = T_b$, and viewed as a process in the spatial parameter.

4.7 Theorem (Ray (1963), Knight (1963)). *Let* $\{R_t, \mathscr{F}_t; 0 \leq t < \infty\}$, $(\tilde{\Omega}, \tilde{\mathscr{F}})$, $\{\tilde{P}^r\}_{r \geq 0}$ *be a Bessel family of dimension* 2, *and let* $b > 0$ *be a given number. Then* $\{\frac{1}{2}R_t^2; 0 \leq t \leq b\}$ *under* $\tilde{P}^0$ *has the same law as* $\{L_{T_b}(b-t); 0 \leq t \leq b\}$ *under* P^0.

In order to prove Theorem 4.7, it is necessary to characterize the finite-dimensional distributions of $\{R_t^2; 0 \leq t \leq b\}$. Toward that end, we define recursively

$$(4.22)\quad f_0 = 1,$$

and for $n \geq 1$,

$$(4.23)\quad f_n(t_1, \gamma_1; t_2, \gamma_2; \ldots; t_n, \gamma_n) = f_{n-1}(t_2 - t_1, \gamma_2; t_3 - t_1, \gamma_3; \ldots; t_n - t_1, \gamma_n)$$
$$+ t_1 \sum_{k=1}^{n} \gamma_k f_{n-k}(t_{k+1} - t_k, \gamma_{k+1}; t_{k+2} - t_k, \gamma_{k+2}; \ldots; t_n - t_k, \gamma_n).$$

Although we will not need this fact, the reader may wish to verify that

$$\begin{aligned} f_n(t_1, \gamma_1; t_2, \gamma_2; \ldots; t_n, \gamma_n) &= 1 + \sum_{1 \le i \le n} \gamma_i t_i + \sum\sum_{1 \le i < j \le n} \gamma_i \gamma_j t_i (t_j - t_i) \\ &+ \sum\sum\sum_{1 \le i < j < k \le n} \gamma_i \gamma_j \gamma_k t_i (t_j - t_i)(t_k - t_j) + \cdots \\ &+ \gamma_1 \gamma_2 \cdots \gamma_n t_1 (t_2 - t_1) \cdots (t_n - t_{n-1}). \end{aligned}$$

We will, however, need the relation

(4.24) $f_{n+1}(t_1, \gamma_1; \ldots; t_n, \gamma_n; t_{n+1}, \gamma_{n+1})$

$$= [1 + \gamma_{n+1}(t_{n+1} - t_n)] f_n\left(t_1, \gamma_1; \ldots; t_{n-1}, \gamma_{n-1}; t_n, \gamma_n + \frac{\gamma_{n+1}}{1 + \gamma_{n+1}(t_{n+1} - t_n)}\right),$$

valid for $n \ge 1$, which is easily proved from (4.22), (4.23) by induction.

4.8 Lemma. *For $0 \le t_1 < t_2 < \cdots < t_n < \infty$ and positive numbers $\gamma_1, \ldots, \gamma_n$, the Laplace transform of the finite-dimensional distributions of the two-dimensional Bessel process is given by*

$$\tilde{E}^0\left[\exp\left\{-\frac{1}{2}\sum_{i=1}^n \gamma_i R_{t_i}^2\right\}\right] = \frac{1}{f_n(t_1, \gamma_1; \ldots; t_n, \gamma_n)}. \tag{4.25}$$

Proof. The straightforward computation

$$E^x e^{-\gamma W_t^2/2} = \frac{1}{\sqrt{1 + \gamma t}} \exp\left\{-\frac{\gamma x^2}{2(1 + \gamma t)}\right\}; \quad x \in \mathbb{R}$$

gives

$$\tilde{E}^r e^{-\gamma R_t^2/2} = \frac{1}{1 + \gamma t} \exp\left\{-\frac{\gamma r^2}{2(1 + \gamma t)}\right\}; \quad \gamma > 0, \quad t > 0, \quad r \ge 0, \tag{4.26}$$

which proves (4.25) for $n = 1$. Assume that (4.25) holds for some value of n, and choose $0 \le t_1 < \cdots < t_n < t_{n+1} < \infty$ and positive numbers $\gamma_1, \ldots, \gamma_n, \gamma_{n+1}$. From the strong Markov property, (4.26), and (4.24), we obtain

$$\begin{aligned} &\tilde{E}^0\left[\exp\left\{-\frac{1}{2}\sum_{i=1}^{n+1} \gamma_i R_{t_i}^2\right\}\right] \\ &= \int_{[0,\infty)^n} \exp\left\{-\frac{1}{2}\sum_{i=1}^n \gamma_i r_i^2\right\} \tilde{E}^0\left[\exp\left(-\frac{1}{2}\gamma_{n+1} R_{t_{n+1}}^2\right)\middle| R_{t_n} = r_n\right] \\ &\qquad \times \tilde{P}^0[R_{t_1} \in dr_1, \ldots, R_{t_n} \in dr_n] \\ &= \int_{[0,\infty)^n} \exp\left\{-\frac{1}{2}\sum_{i=1}^n \gamma_i r_i^2\right\} \tilde{E}^{r_n}\left[\exp\left(-\frac{1}{2}\gamma_{n+1} R_{t_{n+1}-t_n}^2\right)\right] \\ &\qquad \times \tilde{P}^0[R_{t_1} \in dr_1, \ldots, R_{t_n} \in dr_n] \end{aligned}$$

$$= \int_{[0,\infty)^n} \frac{1}{1+\gamma_{n+1}(t_{n+1}-t_n)} \exp\left\{-\frac{1}{2}\sum_{i=1}^{n}\gamma_i r_i^2 - \frac{\gamma_{n+1}r_n^2/2}{1+\gamma_{n+1}(t_{n+1}-t_n)}\right\}$$
$$\times \tilde{P}^0[R_{t_1}\in dr_1,\ldots,R_{t_n}\in dr_n]$$

$$= \frac{1}{f_{n+1}(t_1,\gamma_1;\ldots;t_n,\gamma_n;t_{n+1},\gamma_{n+1})}. \qquad \square$$

4.9 Lemma. *With* $0=\xi_1<\cdots<\xi_n\le b$ *and positive numbers* $\delta_1,\ldots,\delta_n$, *we have the Laplace transform formula*

(4.27)
$$E^0\left[\exp\left\{-\sum_{i=1}^{n}\delta_i L_{T_b}(\xi_i)\right\}\right] = \frac{1}{f_n(b-\xi_n,\delta_n;b-\xi_{n-1},\delta_{n-1};\ldots;b-\xi_1,\delta_1)}.$$

PROOF. Theorem 4.3 implies that the function

$$v(x) \triangleq E^x\left[\exp\left\{-\sum_{i=1}^{n}\delta_i L_{T_b}(\xi_i)\right\}\right]; \quad -\infty<x\le b$$

can be found by seeking a bounded, continuous function $\tilde{v}$ on $(-\infty,b)$ which is linear in each of the intervals $(-\infty,0),(0,\xi_2),\ldots(\xi_{n-1},\xi_n),(\xi_n,b)$ and which satisfies

$$\tilde{v}'(\xi_i+)-\tilde{v}'(\xi_i-)=\delta_i\tilde{v}(\xi_i); \quad 1\le i\le n,$$
$$\tilde{v}(b)=1.$$

Thus, $\tilde{v}$ must be of the form

$$\tilde{v}(x)=C_i+L_ix \quad \text{on } (\xi_i,\xi_{i+1}], \quad 0\le i\le n,$$

where $\xi_0=-\infty$, $\xi_{n+1}=b$, and we must have

$$L_0=0, \tag{4.28}$$

$$C_{i-1}+\xi_iL_{i-1}=C_i+\xi_iL_i; \quad 1\le i\le n, \tag{4.29}$$

$$L_i-L_{i-1}=\delta_i(C_{i-1}+\xi_iL_{i-1}); \quad 1\le i\le n, \tag{4.30}$$

$$C_n+bL_n=1. \tag{4.31}$$

These $2n+2$ equations can be solved as follows. Let $B_{i-1}=(C_{i-1}+\xi_iL_{i-1})/C_0$; $1\le i\le n+1$, so $B_0=1$. We have

$$B_i=B_{i-1}+(\xi_{i+1}-\xi_i)\frac{L_i}{C_0}; \quad 1\le i\le n, \tag{4.32}$$

from (4.29), as well as

$$L_i=L_{i-1}+\delta_iC_0B_{i-1}; \quad 1\le i\le n, \tag{4.33}$$

from (4.30). This last identity, along with (4.28), gives

$$L_i = C_0 \sum_{j=1}^{i} \delta_j B_{j-1}; \quad 1 \le i \le n,$$

which, substituted in (4.32), yields

$$(4.34) \qquad B_i = B_{i-1} + (\xi_{i+1} - \xi_i) \sum_{j=1}^{i} \delta_j B_{j-1}; \quad 1 \le i \le n.$$

The recursion (4.34) determines $B_1, B_2, \ldots, B_n$, and (4.31) together with the definition of B_n gives $C_0 = (1/B_n)$. The constants $L_1, L_2, \ldots, L_n$ are now determined by (4.33), and $C_1, \ldots, C_n$ can be found from (4.29).

Having thus solved the equations (4.28)–(4.31), we may conclude from Theorem 4.3 that

$$C_0 = \frac{1}{B_n} = \tilde{v}(0) = v(0) = E^0\left[\exp\left\{-\sum_{i=1}^{n} \delta_i L_{T_b}(\xi_i)\right\}\right].$$

But comparison of the recursion (4.34) with (4.23) shows that

$$B_i = f_i(\xi_{i+1} - \xi_i, \delta_i; \xi_{i+1} - \xi_{i-1}, \delta_{i-1}; \ldots; \xi_{i+1} - \xi_1, \delta_1); \quad 1 \le i \le n. \quad \square$$

Proof of Theorem 4.7. We simply make the identifications $t_i = b - \xi_{n+1-i}$, $\gamma_i = \delta_{n+1-i}$ in Lemmas 4.8 and 4.9. $\square$

Armed with the description of local time in Theorem 4.7, one can provide a very simple proof for the Dvoretzky-Erdös-Kakutani Theorem 2.9.13 concerning the absence of points of increase on the Brownian path.

Proof of Theorem 2.9.13. We first show that

$$(4.35) \qquad P^0[\omega \in \Omega; W_{\cdot}(\omega) \text{ has a point of strict increase}] = 0.$$

The event in (4.35) is equal to $\bigcup_{\substack{r,p \in Q \\ 0 \le r < p}} A_{r,p}$, where

$$A_{r,p} \triangleq \{\omega \in \Omega; \exists\, \theta \in [r,p), \text{ such that } W_u(\omega) < W_\theta(\omega) < W_v(\omega),$$
$$\forall\, u \in [r,\theta), \forall\, v \in (\theta,p]\},$$

and thus it suffices to prove, for given rationals $0 \le r < p$, that $P^0(A_{r,p}) = 0$. Considering, if necessary, the Brownian motion $\{W_{r+t} - W_r, \mathscr{F}_{r+t}; 0 \le t < \infty\}$, we may assume that $r = 0$.

Because of the continuity of Brownian local time, Problem 3.6.13(ii), Theorem 4.7, and Proposition 3.3.22, we may choose an event $\Omega^* \subseteq \Omega$ with $P^0(\Omega^*) = 1$ such that for every $\omega \in \Omega^*$:

$$(4.36) \qquad \text{the mapping} \quad (t,a) \mapsto L_t(a,\omega) \quad \text{is continuous,}$$

$$(4.37) \qquad \text{for every } b \in Q, \quad \int_0^\infty 1_{\mathbb{R}\setminus\{b\}}(W_s(\omega))\, dL_s(b,\omega) = 0, \quad \text{and}$$

$$(4.38) \qquad \text{for every } b \in Q \cap (0,\infty), \quad L_{T_b(\omega)}(a,\omega) > 0; \quad \forall\, 0 \le a < b.$$

Suppose that $\omega \in \Omega^* \cap A_{0,p}$ for some $p > 0$, let θ be as in the definition of $A_{0,p}$, and choose a (positive) rational number $b \in (W_\theta(\omega), W_p(\omega))$. Let $\{b_n\}_{n=1}^\infty$ be a nondecreasing sequence of nonnegative rational numbers with $\lim_{n\to\infty} b_n = W_\theta(\omega)$. From (4.38) we have

$$(4.39) \quad 0 < L_{T_b(\omega)}(W_\theta(\omega), \omega) = [L_{T_b(\omega)}(W_\theta(\omega), \omega) - L_{T_b(\omega)}(b_n, \omega)] \\ + [L_{T_b(\omega)}(b_n, \omega) - L_\theta(b_n, \omega)] \\ + [L_\theta(b_n, \omega) - L_{T_b(\omega)}(b_n, \omega)]; \quad \forall n \geq 1,$$

thanks to (4.37). From the nature of θ, the second difference on the right-hand side of (4.39) vanishes for all $n \geq 1$, and we have $\lim_{n\to\infty} T_{b_n}(\omega) = \theta$.

We can now let $n \to \infty$ in (4.39), and use the joint continuity property (4.36), to arrive at the contradiction $0 < L_{T_b(\omega)}(W_\theta(\omega), \omega) = 0$. This shows that $P^0(A_{0,p}) = 0$, and leads ultimately to (4.35).

Consider now the process $\tilde{W}_t(\omega) \triangleq W_t(\omega) + t$; $0 \leq t < \infty$, $\omega \in \Omega$. From Corollary 3.5.2, there exists a probability measure $\tilde{P}$ on $(\Omega, \mathscr{F}_\infty^W)$ with

$$\tilde{P}(A) = E^0\left[1_A \exp\left\{-W_T - \frac{1}{2}T^2\right\}\right]; \quad \forall A \in \mathscr{F}_T^W$$

holding for every finite $T > 0$, and such that $\tilde{W}$ is standard Brownian motion under $\tilde{P}$. For every $\omega \in \Omega$, the points of increase of $W_.(\omega)$ become points of *strict* increase of $\tilde{W}_.(\omega)$, and (4.35) shows that

$$\tilde{P}[\omega \in \Omega; W_.(\omega) \text{ has a point of increase}] = 0.$$

But the two measures P^0 and $\tilde{P}$ are equivalent on $\mathscr{F}_T^W$, and consequently

$$P^0[\omega \in \Omega; W_.(\omega) \text{ has a point of increase on } [0, T]] = 0$$

for every fixed $T \in (0, \infty)$. The assertion of the theorem follows easily. □

4.10 Exercise. Verify the Cameron & Martin (1945) formula

$$E^0 \exp\left\{-\beta \int_0^b W_t^2\, dt\right\} = \frac{1}{\sqrt{\cosh(b\sqrt{2\beta})}}; \quad \beta > 0, b > 0$$

for the Laplace transform of the integral $\int_0^b W_t^2\, dt$. (*Hint*: Use (4.17).)

C. The Second Formula of D. Williams

The first formula of D. Williams (relation (3.22)) tells us how to compute the distribution of the total time elapsed $\Gamma_+^{-1}(\tau)$, given that W_+ is being observed and that τ units of time have elapsed on the clock corresponding to W_+. The relevant information to be gleaned from observing W_+ is its local time $L_+(\tau)$. The *second* formula of D. Williams (relation (4.43)) assumes the same observations, but then asks for the distribution of the local time $L_{\Gamma_+^{-1}(\tau)}(b)$ of W at a point $b \le 0$. Of course, if $b = 0$, the local time in question is just $L_+(\tau)$, which is known from the observations. If $b < 0$, then $L_{\Gamma_+^{-1}(\tau)}(b)$ is not known, but as in Williams's first formula, its distribution depends on the observation of W_+ only through the local time $L_+(\tau)$ in the manner given by (4.43).

4.11 Problem. Under P^0, the process $\{L_{T_b}(0); 0 \le b < \infty\}$ has independent increments, and

$$E^0 \exp\{-\alpha L_{T_b}(0)\} = \frac{1}{1 + \alpha|b|}; \quad \alpha > 0, b \in \mathbb{R}. \tag{4.40}$$

4.12 Lemma. *Consider the right-continuous inverse local time at the origin*

$$\rho_s \triangleq \inf\{t \ge 0; L_t(0) > s\}; \quad 0 \le s < \infty. \tag{4.41}$$

For fixed $b \ne 0$, *the process* $N_s \triangleq L_{\rho_s}(b)$; $0 \le s < \infty$ *is a subordinator under* P^0, *and*

$$E^0 e^{-\alpha N_s} = \exp\left\{-\frac{\alpha s}{1 + \alpha|b|}\right\}; \quad \alpha, s > 0. \tag{4.42}$$

Proof. Let $L_t = L_t(0)$, and recall from Problem 3.6.18 that $\lim_{t\to\infty} L_t = \infty$, P^0 a.s. The process N is obviously nondecreasing and right-continuous with $N_0 = 0$, a.s. P^0. (Recall from Problem 3.6.13 (iii) that $\rho_0 = 0$ a.s. P^0.) We have the composition property $\rho_t = \rho_s + \rho_{t-s} \circ \theta_{\rho_s}$; $0 \le s < t$, which, coupled with the additive functional property of local time, gives P^0-a.s.:

$$N_t - N_s = L_{\rho_{t-s}\circ\theta_{\rho_s}}(b) \circ \theta_{\rho_s}; \quad 0 \le s < t.$$

Note that $W_{\rho_s} = 0$ a.s. According to the strong Markov property as expressed in Theorem 2.6.16, the random variable $L_{\rho_{t-s}\circ\theta_{\rho_s}}(b) \circ \theta_{\rho_s}$ is independent of $\mathscr{F}_{\rho_s}$ and has the same distribution as $L_{\rho_{t-s}}(b) = N_{t-s}$. This completes the proof that N is a subordinator.

As for (4.42), we have from Problem 3.4.5 (iv) for $\alpha > 0$, $\beta > 0$:

$$\begin{aligned} q \triangleq E^0 \int_0^\infty \exp\{-\beta s - \alpha L_{\rho_s}(b)\}\, ds &= E^0 \int_0^\infty \exp\{-\beta L_t - \alpha L_t(b)\}\, dL_t \\ &= E^0 \int_0^{T_b} e^{-\beta L_t}\, dL_t \\ &\quad + E^0\left[e^{-\beta L_{T_b}} \int_0^\infty \exp\{-\beta L_t \circ \theta_{T_b} - \alpha L_t(b) \circ \theta_{T_b}\}\, d(L_t \circ \theta_{T_b})\right]. \end{aligned}$$

The first expression is equal to $[1 - E^0 e^{-\beta L_{T_b}}]/\beta$, and by the strong Markov property, the second is equal to $E^0 e^{-\beta L_{T_b}}$ times

$$E^b \int_0^\infty \exp\{-\beta L_t - \alpha L_t(b)\}\, dL_t$$

$$= E^b \int_{T_0}^\infty \exp\{-\beta L_t - \alpha L_t(b)\}\, dL_t$$

$$= E^b[e^{-\alpha L_{T_0}(b)} \int_0^\infty \exp\{-\beta L_t \circ \theta_{T_0} - \alpha L_t(b) \circ \theta_{T_0}\}\, d(L_t \circ \theta_{T_0})$$

$$= E^b[e^{-\alpha L_{T_0}(b)}] \cdot E^0 \int_0^\infty \exp\{-\beta L_t - \alpha L_t(b)\}\, dL_t$$

$$= q \cdot E^b[e^{-\alpha L_{T_0}(b)}] = q \cdot E^0[e^{-\alpha L_{T_b}}].$$

Therefore,

$$q = \frac{1}{\beta}[1 - E^0 e^{-\beta L_{T_b}}] + q \cdot E^0[e^{-\beta L_{T_b}}] \cdot E^0[e^{-\alpha L_{T_b}}],$$

and (4.40) allows us to solve for

$$q = \int_0^\infty e^{-\beta s} E^0(e^{-\alpha N_s})\, ds = \left(\beta + \frac{\alpha}{1 + \alpha|b|}\right)^{-1}.$$

Inversion of this transform leads to (4.42). □

4.13 Remark. Recall the two independent, reflected Brownian motions $W_\pm$ of Theorem 3.1, along with the notation of that section. If $b < 0$ in Lemma 4.12, then the subordinator N is a function of W_-, and hence is independent of W_+. To see this, recall the local time at 0 for W_-: $L_-(\tau) \triangleq L_{\Gamma_-^{-1}(\tau)}(0)$, and let

$$L_-^b(\tau) \triangleq L_{\Gamma_-^{-1}(\tau)}(b)$$

be the local time of W_- at b. Both these processes can be constructed from W_- (see Remark 3.2 for L_-), and so both are independent of W_+. The same is true for

$$\Gamma_-(\rho_s) = \inf\{\tau \ge 0 : L_-(\tau) > s\},$$

and hence also for

$$L_-^b(\Gamma_-(\rho_s)) = L_{\Gamma_-^{-1}(\Gamma_-(\rho_s))}(b).$$

But $W_{\rho_s} = 0$ a.s., and so $\Gamma_-(\rho_s + \varepsilon) > \Gamma_-(\rho_s)$ for every $\varepsilon > 0$ (Problem 2.7.18 applied to the Brownian motion $W \circ \theta_{\rho_s}$). It follows from Problem 3.4.5 (iii) that $\Gamma_-^{-1}(\Gamma_-(\rho_s)) = \rho_s$, and so $L_-^b(\Gamma_-(\rho_s)) = N_s$.

A comparison of (4.42) with (2.14) shows that for the subordinator N, we have $m = 0$ and Lévy measure $\mu(d\ell) = b^{-2} e^{-l/b}\, d\ell$.

4.14 Proposition (D. Williams (1969)). *In the notation of Section* 3, *and for fixed numbers* $\alpha > 0$, $\tau > 0$, $b \le 0$, *we have a.s.* P^0:

(4.43) $$E^0[\exp\{-\alpha L_{\Gamma_+^{-1}(\tau)}(b)\}|W_+(u); 0 \le u < \infty] = \exp\left\{-\frac{\alpha L_+(\tau)}{1+\alpha|b|}\right\}.$$

PROOF. This is obvious for $b = 0$, so we consider the case $b < 0$. We have from Problems 3.4.5 (iii) and 3.6.13 (iii) that with ρ as in (4.41),

$$\begin{aligned}\rho_{L_{\Gamma_+^{-1}(\tau)}(0)} &= \sup\{s \ge \Gamma_+^{-1}(\tau); L_s(0) = L_{\Gamma_+^{-1}(\tau)}(0)\}\\ &= \inf\{s \ge \Gamma_+^{-1}(\tau); W_s = 0\}, \quad \text{a.s. } P^0.\end{aligned}$$

Because $W_u \ge 0$ for $\Gamma_+^{-1}(\tau) \le u \le \rho_{L_{\Gamma_+^{-1}(\tau)}(0)} = \rho_{L_+(\tau)}$, the local time $L_u(b)$ must be constant on this interval. Therefore, with N the subordinator of Lemma 4.12, $N_{L_+(\tau)} = L_{\rho_{L_+(\tau)}}(b) = L_{\Gamma_+^{-1}(\tau)}(b)$, a.s. P^0. Remark 4.13, together with relation (4.42), gives

$$\begin{aligned}E^0[\exp\{-\alpha N_{L_+(\tau)}\}|W_+(u); 0 \le u < \infty] &= E^0[\exp\{-\alpha N_t\}]|_{t=L_+(\tau)}\\ &= \exp\left\{-\frac{\alpha L_+(\tau)}{1+\alpha|b|}\right\}, \quad \text{a.s. } P^0.\end{aligned}$$

4.15 Exercise.

(i) Show that for $\beta > 0$, $a > 0$, $b < 0$, we have

$$E^0[\exp\{-\beta L_{T_a}(b)\}|W_+(u); 0 \le u < \infty\}] = \exp\left\{-\frac{\beta L_{T_a}(0)}{1+\beta|b|}\right\}, \quad \text{a.s. } P^0.$$

(ii) Use (i) to prove that for $\alpha > 0$, $\beta > 0$, $a > 0$, $b < 0$,

$$E^b[\exp\{-\beta L_{T_a}(b) - \alpha L_{T_a}(0)\}] = \frac{1}{f_2(a, \alpha; a + |b|, \beta)},$$

where f_2 is given by (4.22), (4.23). (This argument can be extended to provide an alternate proof of Lemma 4.9; see McKean (1975)).

6.5. An Application: Transition Probabilities of Brownian Motion with Two-Valued Drift

Let us consider two real constants $\theta_0 < \theta_1$, and denote by $\mathscr{U}$ the collection of Borel-measurable functions $b(t, x)$: $[0, \infty) \times \mathbb{R} \to [\theta_0, \theta_1]$. For every $b \in \mathscr{U}$ and $x \in \mathbb{R}$, we know from Corollary 5.3.11 and Remark 5.3.7 that the stochastic integral equation

(5.1) $$X_t = x + \int_0^t b(s, X_s)\,ds + W_t; \quad 0 \le t < \infty$$

has a weak solution (X, W), $(\Omega, \mathscr{F}, \tilde{P}^x)$, $\{\mathscr{F}_t\}$ which is unique in the sense of probability law, with finite-dimensional distributions given by (5.3.11) for $0 \le t_1 < t_2 < \cdots < t_n = t < \infty$, $\Gamma \in \mathscr{B}(\mathbb{R}^n)$):

$$\text{(5.2)} \quad \tilde{P}^x[(X_{t_1}, \ldots, X_{t_n}) \in \Gamma]$$
$$= E^x\left[1_{\{(W_{t_1}, \ldots, W_{t_n}) \in \Gamma\}} \cdot \exp\left\{\int_0^t b(s, W_s)\,dW_s - \frac{1}{2}\int_0^t b^2(s, W_s)\,ds\right\}\right].$$

Here, $\{W_t, \mathscr{F}_t\}$, $(\Omega, \mathscr{F})$, $\{P^y\}_{y \in \mathbb{R}}$ is a one-dimensional Brownian family.

We shall take the point of view that the drift $b(t, x)$ is an element of control, available to the decision maker for influencing the path of the Brownian particle by "pushing" it. The reader may wish to bear in mind the special case $\theta_0 < 0 < \theta_1$, in which case this "pushing" can be in either the positive or the negative direction (up to the prescribed limit rates θ_1 and θ_0, respectively). The goal is to keep the particle as close to the origin as possible, and the decision maker's efficacy in doing so is measured by the expected discounted quadratic deviation from the origin

$$J(x; b) = \tilde{E}^x \int_0^\infty e^{-\alpha t} X_t^2\,dt$$

for the resulting diffusion process X. Here, α is a positive constant. The *control problem* is to choose the drift $b_* \in \mathscr{U}$ for which $J(x; b)$ is minimized over all $b \in \mathscr{U}$:

$$\text{(5.3)} \qquad J(x; b_*) = \min_{b \in \mathscr{U}} J(x; b), \quad \forall\, x \in \mathbb{R}.$$

This simple stochastic control problem was studied by Beneš, Shepp & Witsenhausen (1980), who showed that the optimal drift is given by $b_*(t, x) = u(x)$; $0 \le t < \infty$, $x \in \mathbb{R}$ and

$$\text{(5.4)} \quad u(x) \triangleq \begin{Bmatrix} \theta_1; & x \le \delta \\ \theta_0; & x > \delta \end{Bmatrix}, \quad \delta \triangleq \frac{1}{\sqrt{\theta_1^2 + 2\alpha} + \theta_1} - \frac{1}{\sqrt{\theta_0^2 + 2\alpha} - \theta_0}.$$

This is a sensible rule, which says that one should "push as hard as possible to the right, whenever the process Z, solution of the stochastic integral equation

$$\text{(5.5)} \qquad Z_t = x + \int_0^t u(Z_s)\,ds + W_t; \quad 0 \le t < \infty,$$

finds itself to the left of the critical point δ, and vice versa." Because there is no explicit cost on the controlling effort, it is reasonable to push with full force up to the allowed limits. If $\theta_1 = -\theta_0 = \theta$, the situation is symmetric and $\delta = 0$.

Our intent in the present section is to use the trivariate density (3.30) in order to compute, as explicitly as possible, the transition probabilities

$$\text{(5.6)} \qquad \tilde{p}_t(x, z)\,dz = \tilde{E}^x[Z_t \in dz]$$

of the process in (5.5), which is a *Brownian motion with two-valued, state-dependent drift*. In this computation, the switching point δ need not be related to θ_0 and θ_1. We shall only deal with the value $\delta = 0$; the transition probabilities for other values of δ can then be obtained easily by translation.

The starting point is provided by (5.2), which puts the expression (5.6) in the form

$$(5.7)\quad \tilde{p}_t(x, z)\,dz = E^x\left[1_{\{W_t \in dz\}} \exp\left\{\int_0^t u(W_s)\,dW_s - \frac{1}{2}\int_0^t u^2(W_s)\,ds\right\}\right].$$

Further progress requires the elimination of the stochastic integral in (5.7). But if we set

$$f(z) \triangleq \int_0^z u(y)\,dy = \begin{Bmatrix} \theta_1 z; & z < 0 \\ \theta_0 z; & z \geq 0 \end{Bmatrix}$$

we obtain a piecewise linear function, for which the generalized Itô rule of Theorem 3.6.22 gives

$$f(W_t) = f(W_0) + \int_0^t u(W_s)\,dW_s + (\theta_0 - \theta_1)L(t)$$

where $L(t)$ is the local time of W at the origin. On the other hand, with the notation (3.3) we have

$$\int_0^t u^2(W_s)\,ds = \theta_1^2 t + (\theta_0^2 - \theta_1^2)\Gamma_+(t),$$

and (5.7) becomes

$$(5.8)\quad \tilde{p}_t(x; z)\,dz = \exp\left[f(z) - f(x) - \frac{t}{2}\theta_1^2\right]$$
$$\cdot \int_{b=0}^{\infty}\int_{\tau=0}^{t} \exp\left\{b(\theta_1 - \theta_0) + \frac{\tau}{2}(\theta_1^2 - \theta_0^2)\right\}$$
$$\cdot P^x[W_t \in dz;\ L(t) \in db;\ \Gamma_+(t) \in d\tau].$$

It develops then that we have to compute the joint density of $(W_t, L(t), \Gamma_+(t))$ under P^x, for every $x \in \mathbb{R}$, and not only for $x = 0$ as in (3.30). This is accomplished with the help of the strong Markov property and Problem 3.5.8; in the notation of the latter, we recast (3.30) for $b > 0$, $0 < \tau < t$ as

$$P^0[W_t \in da;\ L(t) \in db;\ \Gamma_+(t) \in d\tau]$$
$$= \begin{cases} 2h(\tau; b, 0)h(t - \tau; b - a, 0)\,da\,db\,d\tau; & a < 0 \\ 2h(t - \tau; b, 0)h(\tau; b + a, 0)\,da\,db\,d\tau; & a > 0, \end{cases}$$

and then write, for $x \geq 0$ and $a < 0$:

(5.9) $$\begin{aligned} P^x[W_t \in da; L(t) \in db; \Gamma_+(t) \in d\tau] &= P^x[W_t \in da; L(t) \in db; \Gamma_+(t) \in d\tau; T_0 \le \tau] \\ &= \int_0^\tau P^x[W_t \in da; L(t) \in db; \Gamma_+(t) \in d\tau \mid T_0 = s] \cdot P^x[T_0 \in ds] \\ &= \int_0^\tau P^0[W_{t-s} \in da; L(t-s) \in db; \Gamma_+(t-s) \in d\tau - s] \cdot h(s; x, 0)\, ds \\ &= 2h(\tau; b + x, 0) h(t - \tau; b - a, 0)\, da\, db\, d\tau. \end{aligned}$$

For $x \ge 0$, $a > 0$ a similar computation gives

(5.10)

$$P^x[W_t \in da; L(t) \in db; \Gamma_+(t) \in d\tau] = 2h(t - \tau; b, 0) h(\tau; b + a + x, 0)\, da\, db\, d\tau,$$

and in this case we have also the singular part

(5.11) $$\begin{aligned} P^x[W_t \in da; L(t) = 0, \Gamma_+(t) = t] = P^x[W_t \in da; T_0 > t] &= p_-(t; x, a)\, da \\ &\equiv [p(t; x, a) - p(t; x, -a)]\, da \end{aligned}$$

(cf. (2.8.9)). The equations (5.9)–(5.11) characterize the distribution of the triple $(W_t, L(t), \Gamma_+(t))$ under P^x. Back in (5.8), they yield after some algebra:

(5.12) $\tilde{p}_t(x, z) =$

$$= \begin{cases} 2\displaystyle\int_0^\infty \int_0^t e^{2b\theta_1} h(t - \tau; b - z, -\theta_1) h(\tau; x + b, -\theta_0)\, d\tau\, db; & x \ge 0, z \le 0 \\ 2\displaystyle\int_0^\infty \int_0^t e^{2(b\theta_1 + z\theta_0)} h(t - \tau; b, -\theta_1) h(\tau; x + b + z, -\theta_0)\, d\tau\, db & \\ \quad + \dfrac{1}{\sqrt{2\pi t}} \left[\exp\left\{ -\dfrac{(x - z + \theta_0 t)^2}{2t} \right\} - \exp\left\{ -\dfrac{(x + z - \theta_0 t)^2}{2t} - 2\theta_0 x \right\} \right]; & \\ & x \ge 0, z > 0. \end{cases}$$

Now the dependence on θ_0, θ_1 has to be invoked explicitly, by writing $\tilde{p}_t(x, z; \theta_0, \theta_1)$ instead of $\tilde{p}_t(x, z)$. The symmetry of Brownian motion gives

(5.13) $$\tilde{p}_t(x, z; \theta_0, \theta_1) = \tilde{p}_t(-x, -z; -\theta_1, -\theta_0),$$

and so for $x \le 0$ the transition density is obtained from (5.12) and (5.13). We conclude with a summary of these results.

5.1 Proposition. *Let $u\colon \mathbb{R} \to [\theta_0, \theta_1]$ be given by* (5.4) *for arbitrary real δ, and let Z be the solution of the stochastic integral equation* (5.5). *In the notation of* (5.6), *$\tilde{p}_t(x + \delta, z + \delta)$ is given for every $z \in \mathbb{R}$, $0 \le t < \infty$ by*

(i) *the right-hand side of* (5.12) *if $x \ge 0$, and*

(ii) *the right-hand side of* (5.12) *with* $(x, z, \theta_0, \theta_1)$ *replaced by* $(-x, -z, -\theta_1, -\theta_0)$, *if* $x \le 0$.

5.2 Remark. In the special case $\theta_1 = -\theta_0 = \theta > 0 = \delta$, the integral term in the second part of (5.12) becomes

$$2\int_0^\infty \int_0^t e^{2\theta(b-z)} h(t-\tau; b, -\theta) h(\tau; x+b+z, \theta)\, d\tau\, db$$

$$= 2\int_0^\infty \int_0^t e^{-2\theta z} h(t-\tau; b, \theta) h(t; x+b+z, \theta)\, d\tau\, d\theta$$

$$= 2\int_0^\infty e^{-2\theta z} h(t; x+2b+z, \theta)\, db$$

$$= \frac{1}{\sqrt{2\pi t}}\left[\exp\left\{-\frac{(x+z+\theta t)^2}{2t} - 2\theta x\right\}\right.$$

$$\left. + \theta e^{-2\theta z}\int_{x+z}^\infty \exp\left\{-\frac{(v-\theta t)^2}{2t}\right\} dv\right],$$

where we have used Problem 3.5.8 again. A similar computation simplifies also the first integral in (5.12); the result is

$$(5.14)\quad \tilde{p}_t(x, z) =$$

$$= \begin{cases} \dfrac{1}{\sqrt{2\pi t}}\left[\exp\left\{-\dfrac{(x-z-\theta t)^2}{2t}\right\} + \theta e^{-2\theta z}\displaystyle\int_{x+z}^\infty \exp\left\{-\dfrac{(v-\theta t)^2}{2t}\right\} dv\right]; \\ \qquad\qquad\qquad\qquad\qquad\qquad\qquad\qquad x \ge 0,\ z > 0. \\ \dfrac{1}{\sqrt{2\pi t}}\left[\exp\left\{2\theta x - \dfrac{(x-z+\theta t)^2}{2t}\right\} + \theta e^{2\theta z}\displaystyle\int_{x-z}^\infty \exp\left\{-\dfrac{(v-\theta t)^2}{2t}\right\} dv\right]; \\ \qquad\qquad\qquad\qquad\qquad\qquad\qquad\qquad x \ge 0,\ z \le 0. \end{cases}$$

When $\theta = 1$ and $x = 0$, we recover the expression (3.19).

5.3 Exercise. When $\theta_1 = -\theta_0 = 1$, $\delta = 0$, show that the function $v(t, x) \triangleq \tilde{E}^x(Z_t^2)$; $t > 0$, $x \in \mathbb{R}$ is given by

$$(5.15)\qquad v(t, x) = \frac{1}{2} + \sqrt{\frac{t}{2\pi}}(|x| - t - 1)\exp\left\{-\frac{(|x|-t)^2}{2t}\right\}$$

$$+ \left\{(|x|-t)^2 + t - \frac{1}{2}\right\}\Phi\left(\frac{|x|-t}{\sqrt{t}}\right)$$

$$+ e^{2|x|}\left(|x| + t - \frac{1}{2}\right)\left[1 - \Phi\left(\frac{|x|+t}{\sqrt{t}}\right)\right]$$

with $\Phi(z) = (1/\sqrt{2\pi})\int_{-\infty}^{z} e^{-u^2/2}\,du$, and satisfies the equation

$$v_t = \tfrac{1}{2}v_{xx} - (\operatorname{sgn} x)v_x, \tag{5.16}$$

as well as the conditions

$$\lim_{t\downarrow 0} v(t,x) = x^2, \quad \lim_{t\to\infty} v(t,x) = \frac{1}{2}.$$

5.4 Exercise (Shreve (1981)). With $\theta_1 = -\theta_0 = 1$, $\delta = 0$, show that the function v of (5.15) satisfies the Hamilton-Jacobi-Bellman equation

$$v_t = \frac{1}{2}v_{xx} + \min_{|u|\le 1}(uv_x) \quad \text{on } (0,\infty)\times\mathbb{R}. \tag{5.17}$$

As a consequence, if X is a solution to (5.1) for an arbitrary, Borel-measurable $b: [0,\infty)\times\mathbb{R}\to[-1,1]$ and Z solves (5.5) (under $\tilde{P}^x$ in both cases), then

$$\tilde{E}^x Z_t^2 \le \tilde{E}^x X_t^2; \quad 0\le t<\infty. \tag{5.18}$$

In particular, Z is the optimally controlled process for the control problem (5.3).

6.6. Solutions to Selected Problems

2.5. If λ is σ-finite but not finite, then there exists a partition $\{H_i\}_{i=1}^\infty \subseteq \mathscr{H}$ of H with $0<\lambda(H_i)<\infty$ for every $i\ge 1$. On a probability space $(\Omega,\mathscr{F},P)$, we set up independent sequences $\{\xi_j^{(i)}; j\in\mathbb{N}_0\}_{i=1}^\infty$ of random variables, such that for every $i\ge 1$:

(i) $\xi_0^{(i)}, \xi_1^{(i)}, \xi_2^{(i)}, \ldots$ are independent,
(ii) $N_i \triangleq \xi_0^{(i)}$ is Poisson with $EN_i = \lambda(H_i)$, and
(iii) $P[\xi_j^{(i)}\in C] = \lambda(C\cap H_i)/\lambda(H_i)$; $C\in\mathscr{H}, j=1,2,\ldots$.

As before, $v_i(C) \triangleq \sum_{j=1}^{N_i} 1_C(\xi_j^{(i)})$; $C\in\mathscr{H}$, is a Poisson random measure for every $i\ge 1$, and $v_1, v_2, \ldots$ are independent. We show that $v\triangleq\sum_{i=1}^\infty v_i$ is a Poisson random measure with intensity λ. It is clear that $Ev(C) = E\sum_{i=1}^\infty v_i(C\cap H_i) = \sum_{i=1}^\infty \lambda(C\cap H_i) = \lambda(C)$ for all $C\in\mathscr{H}$, and whenever $\lambda(C)<\infty$, $v(C)$ has the proper distribution (the sum of independent Poisson random variables being Poisson). Suppose $\lambda(C)=\infty$. We set $\lambda_i = \lambda(C\cap H_i)$, so $v(C)$ is the sum of the independent, Poisson random variables $\{v_i(C)\}_{i=1}^\infty$, where $Ev_i(C)=\lambda_i$. There is a number $a>0$ such that $1-e^{-\lambda}\ge\lambda/2$; $0\le\lambda\le a$, and so with $b\triangleq 2(1-e^{-a})$:

$$\sum_{i=1}^\infty P[v_i(C)\ge 1] = \sum_{i=1}^\infty (1-e^{-\lambda_i}) \ge \frac{1}{2}\sum_{i=1}^\infty (\lambda_i\wedge b) = \infty$$

because $\sum_{i=1}^\infty \lambda_i = \lambda(C) = \infty$. By the second half of the Borel-Cantelli lemma (Chung (1974), p. 76), $P[v(C)=\infty] = P[v_i(C)\ge 1$ for infinitely many $i] = 1$. This completes the verification that v satisfies condition (i) of Definition 2.3; the verification of (ii) is straightforward.

2.12. If for some $y \in D[0,\infty)$, $t > 0$, and $\ell > 0$ we have $n(y; (0,t] \times (\ell,\infty)) = \infty$, then we can find a sequence of distinct points $\{t_k\}_{k=1}^{\infty} \subseteq (0,t]$ such that $|y(t_k) - y(t_k-)| > \ell$; $k \geq 1$. By selecting a subsequence if necessary, we may assume without loss of generality that $\{t_k\}_{k=1}^{\infty}$ is either strictly increasing or else strictly decreasing to a limit $\theta \in [0,t]$. But then $\{y(t_k)\}_{k=1}^{\infty}$ and $\{y(t_k-)\}_{k=1}^{\infty}$ converge to the same limit (which is $y(\theta-)$ in the former case, and $y(\theta)$ in the latter). In either case we obtain a contradiction.

For any interval $(t, t+h]$, the set

$$A_{t,t+h}(\ell) \triangleq \{y \in D[0,\infty); \exists s \in (t,t+h] \text{ such that } |y(s) - y(s-)| > \ell\}$$
$$= \bigcup_{k=1}^{\infty} \bigcap_{m=1}^{\infty} \bigcup_{\substack{t<r<q\leq t+h \\ q-r<(1/m), r\in Q \\ q\in Q \text{ or } q=t+h}} \left\{y \in D[0,\infty); |y(q) - y(r)| > \ell + \frac{1}{k}\right\}$$

is in $\mathscr{B}(D[0,\infty))$, as is

$$\{y \in D[0,\infty); n(y; (0,t] \times (\ell,\infty)) \geq m\} = \bigcup_{\substack{0=q_0<q_1<\cdots<q_m=t \\ \{q_1,\ldots,q_{m-1}\}\subseteq Q}} \bigcap_{i=1}^{m} A_{q_{i-1},q_i}(\ell).$$

Let us now fix $0 < t < \infty$ and $\ell > 0$ and let $\mathscr{D}_{t,\ell}$ be the Dynkin system of all sets $C \in \mathscr{B}((0,t] \times (\ell,\infty))$ for which $n(\cdot\,; C)$ is measurable. The σ-field $\mathscr{B}((0,t] \times (\ell,\infty))$ is generated by the sets of the form $(0,\tau] \times (\lambda,\infty)$; $0 < \tau \leq t$, $\ell \leq \lambda < \infty$, the collection of such sets is closed under finite intersections, and each such set belongs to $\mathscr{D}_{t,\ell}$. It follows from Theorem 2.1.3 that $n(\cdot\,; C)$ is measurable for every $C \in \mathscr{B}((0,t] \times (\ell,\infty))$. For $C \in \mathscr{B}((0,\infty)^2)$, we write C as the disjoint union

$$C = \bigcup_{k=1}^{\infty} \bigcup_{j=1}^{\infty} \left\{C \cap \left[(j-1,j] \times \left(\frac{1}{k}, \frac{1}{k-1}\right]\right]\right\}.$$

2.16. From (2.17) and Definition 2.15 we have $Ee^{-\alpha N_t} = e^{-t\beta(\alpha)\psi(1)}$, which gives (2.33) and the continuity of $\beta(\cdot)$. Because N is not identically zero, $\psi(1)$ is strictly positive. Furthermore, for every $\gamma \geq 0$,

$$e^{-t\beta(\alpha\gamma)\psi(1)} = Ee^{-N_{t\beta(\alpha\gamma)}} = Ee^{-\alpha\gamma N_t} = Ee^{-\alpha N_{t\beta(\gamma)}} = e^{-t\beta(\gamma)\psi(\alpha)}$$

and (2.32) follows from (2.33). To see that $\beta(\alpha) = \alpha^\varepsilon$, set $G(a) = \beta(e^a)$ and apply Problem 2.2. For $0 < \varepsilon \leq 1$, comparison of (2.24) and (2.33) yields

$$r\alpha^\varepsilon = m\alpha + \int_{(0,\infty)} (1 - e^{-\alpha\ell})\mu(d\ell).$$

Corollary 2.14 asserts that the constant $m \geq 0$ and σ-finite measure μ satisfying the equation are unique. If $\varepsilon = 1$, they are given by $m = r$, $\mu = 0$. If $0 < \varepsilon < 1$, we set $m = 0$, $\mu(d\ell) = r\varepsilon\, d\ell/\Gamma(1-\varepsilon)\ell^{1+\varepsilon}$ and integrate by parts to reduce the integral to a standard Laplace transform:

$$\int_{(0,\infty)} (1 - e^{-\alpha\ell})\mu(d\ell) = \frac{r\alpha}{\Gamma(1-\varepsilon)} \int_{(0,\infty)} e^{-\alpha\ell}\ell^{-\varepsilon}\, d\ell = r\alpha^\varepsilon.$$

2.24. We have from (2.56)

$$E|\varepsilon D_t(\varepsilon) - 2L_t(0)|^2 \leq 2\varepsilon^2 + 2E \sum_{j=1}^{\infty} \left(\int_{t\wedge\tau_{j-1}}^{t\wedge\sigma_j} \operatorname{sgn}(W_s)\, dW_s\right)^2.$$

This last expectation is computed as

$$E \sum_{j=1}^{\infty} \{(t \wedge \sigma_j) - (t \wedge \tau_{j-1})\} \le E \sum_{j=1}^{1+D_t(\varepsilon)} (\sigma_j - \tau_{j-1}) = E \sum_{j=0}^{\infty} 1_{\{j \le D_t(\varepsilon)\}} \eta_{j+1}$$
$$= \varepsilon^2 (ED_t(\varepsilon) + 1)$$

by virtue of (2.52), (2.53). But now the downcrossing inequality (Theorem 1.3.8 (iii)), applied to the submartingale $|W|$, gives $ED_t(\varepsilon) \le \frac{1}{\varepsilon} E|W_t| = \frac{1}{\varepsilon}\sqrt{\frac{2t}{\pi}}$.

3.7. The following are obviously valid, modulo P^0-negligible events:

$$\begin{aligned}\{t \ge 0; L_-(t) > L_+(\tau)\} &= \{t \ge 0; L(\Gamma_-^{-1}(t)) > L(\Gamma_+^{-1}(\tau))\} \\ &\subseteq \{t \ge 0; \Gamma_-^{-1}(t) \ge \Gamma_+^{-1}(\tau)\} \\ &\subseteq \{t \ge 0; L(\Gamma_-^{-1}(t)) \ge L(\Gamma_+^{-1}(\tau))\} \\ &= \{t \ge 0; L_-(t) \ge L_+(\tau)\}.\end{aligned}$$

Therefore, we have a.s. P^0:

$$\begin{aligned}T^-_{L_+(\tau)} &= \inf\{t \ge 0; L_-(t) \ge L_+(\tau)\} \le \inf\{t \ge 0; \Gamma_-^{-1}(t) \ge \Gamma_+^{-1}(\tau)\} \\ &\le \inf\{t \ge 0; L_-(t) > L_+(\tau)\} = L_-^{-1}(L_+(\tau)).\end{aligned}$$

The third identity in (3.25) follows now from (3.24). For the fourth, it suffices to observe $\Gamma_{\pm}(s) = \inf\{t \ge 0; \Gamma_{\pm}^{-1}(t) \ge s\}$; $0 \le s < \infty$, a.s. P^0, which is a consequence of Lemma 3.4.5 (v).

3.11. Using the reflection principle in the form (2.6.2) and the upper bound in (2.9.20), we obtain

$$\begin{aligned}P^0&\left[\max_{t \le s \le t+h} W_s \ge a + \varepsilon;\ \min_{t \le s \le t+h} W_s \le a - \varepsilon\right] \\ &\le \int_{-\infty}^{a} P^x\left[\max_{0 \le s \le h} W_s \ge a + \varepsilon\right] \cdot P^0[W_t \in dx] \\ &\quad + \int_{a}^{\infty} P^x\left[\min_{0 \le s \le h} W_s \le a - \varepsilon\right] \cdot P^0[W_t \in dx] \\ &\le P^a\left[\max_{0 \le s \le h} W_s \ge a + \varepsilon\right] + P^a\left[\min_{0 \le s \le h} W_s \le a - \varepsilon\right] \\ &= 2P^0\left[\max_{0 \le s \le h} W_s \ge \varepsilon\right] \\ &= 4P^0[W_h \ge \varepsilon] \le \frac{4}{\varepsilon}\sqrt{\frac{h}{2\pi}} e^{-\varepsilon^2/2h} = o(h).\end{aligned}$$

For (3.32), we notice the inequality

$$\begin{aligned}P^0&[W_+(\tau) > a; t \le \Gamma_+^{-1}(\tau) < t + h; \beta(t) < t + h] \\ &\le P^0\left[\max_{t \le s \le t+h} W_s > a;\ \min_{t \le s \le t+h} W_s \le 0\right],\end{aligned}$$

where this term is $o(h)$ as $h \downarrow 0$, thanks to (3.31); a similar argument proves (3.33).

4.11. Assume without loss of generality that $b > 0$. From the additive functional property of local time, its invariance under translation, and the composition property $T_b = T_a + T_b \circ \theta_{T_a}$; $0 < a < b$, we see that P^0-a.s.:

$$\begin{aligned} L_{T_b(\omega)}(0, \omega) - L_{T_a(\omega)}(0, \omega) &= L_{T_b(\theta_{T_a(\omega)}(\omega))}(0, \theta_{T_a(\omega)}(\omega)) \\ &= L_{T_{b-a}(\theta^*_{T_a(\omega)}(\omega))}(-a, \theta^*_{T_a(\omega)}(\omega)); \quad 0 < a < b, \end{aligned}$$

where $\theta_t^*(\omega)(s) \triangleq \omega(t + s) - \omega(t)$; $s \geq 0$. The independence of increments follows from the strong Markov property as expressed by Theorem 2.6.16. Formula (4.40) is just a restatement of (4.16).

6.7. Notes

Section 6.2: Most of the material here is due to P. Lévy (1937, 1939, 1948). The representation (2.14) is a special case of a general decomposition result for processes with stationary, independent increments (Lévy processes) into Brownian and Poisson components, obtained by P. Lévy (1937); see also Loève (1978). In our exposition of Theorem 2.7, we follow Itô & McKean (1974) and Williams (1979). Both these books, as well as McKean (1975) and Chapter VII of Knight (1981), may be consulted for further reading on Brownian local time. Chung & Durrett (1976) and Williams (1977) also deal with the subject of Theorem 2.23. Taking up a theme of Lévy, Ikeda & Watanabe (1981), pp. 123–136, show how to use local time to construct Brownian motion from a Poisson random measure on the space of excursions away from the origin. This should be read together with Chung's (1976) excellent treatise on excursions. Itô (1961b) applied Poisson random measures to the study of Markov processes. The characterizations of Theorems 2.21 and 2.23 have been generalized to Markov processes by Fristedt & Taylor (1983); see also Kingman (1973). Invariance principles are discussed by Perkins (1982a) and Csáki & Révész (1983); see also Borodin (1981). Perkins (1982b) showed that Brownian local time is a semimartingale in the spatial parameter; McGill (1982) studied its Markov properties.

Section 6.3: Theorem 3.1 appears in Itô & McKean (1974), section 2.11, and in Ikeda & Watanabe (1981), pp. 122–124. Proposition 3.9 is taken from Karatzas & Shreve (1984a); the bivariate density (3.40) was obtained by Perkins (1982b), using different methodology. The use of local time in decomposition of Brownian paths is illustrated by the work of Williams (1974) and Harrison & Shepp (1981).

Section 6.4: The strong version of Theorem 4.1 discussed in Remark 4.2, but with $n = 1$, appears in Itô & McKean (1974), pp. 45–48. For more general results along the lines of Theorems 4.1 and 4.3, the reader should consult Knight (1981), Theorem 7.4.3. Exercise 4.5 comes from Taylor (1975), where applications to finance and process control are discussed; see also Williams (1976), Lehoczky (1977), and Azéma & Yor (1979). We follow Itô & McKean (1974) in our approach to the Ray-Knight theorem 4.7 and in Lemma 4.12, and Knight (1981) for the proof of the Dvoretzky-Erdös-Kakutani Theorem 2.9.13.

Section 6.5: This material is taken from Karatzas & Shreve (1984a). The control problem of this section was introduced and solved by Beneš, Shepp, & Witsenhausen (1980) and has also been solved in the symmetric case of Exercises 5.3, 5.4 by martingale methods (Davis & Clark (1979)), stochastic comparison methods (Ikeda & Watanabe (1977)), and the stochastic maximum principle (Haussmann (1981)). See also Balakrishnan (1980). Stochastic control problems in which the optimal control process is a local time have been studied by a number of authors, including Bather & Chernoff (1967); Beneš, Shepp & Witsenhausen (1980); Chernoff (1968); Chow, Menaldi & Robin (1985); Harrison (1985); Karatzas (1983); Karatzas & Shreve (1984b, 1985); and Taksar (1985).

Bibliography

ADELMAN, O. (1985) Brownian motion never increases: a new proof to a result of Dvoretzky, Erdös and Kakutani. *Israel J. Math. 50*, 189–192.

ARNOLD, L. (1973) *Stochastic Differential Equations: Theory and Applications*. J. Wiley & Sons, New York.

ARNOLD, L. & KLIEMANN, W. (1983) Qualitative Theory of Stochastic Systems. In *Probabilistic Analysis & Related Topics* **3**, 1–79 (A. T. Bharucha-Reid, editor). Academic Press, New York.

ASH, R. B. (1972) *Real Analysis and Probability*. Academic Press, New York.

AZÉMA, J. & YOR, M. (1978) *Temps Locaux. Astérisque* **52–53**, Societé Mathématique de France.

AZÉMA, J. & YOR, M. (1979) Une solution simple au problème de Skorohod. *Lecture Notes in Mathematics* **721**, 90–115. Springer-Verlag, Berlin.

BACHELIER, L. (1900) Théorie de la spéculation. *Ann. Sci. École Norm. Sup.* **17**, 21–86. [In *The Random Character of Stock Market Prices* (Paul H. Cootner, ed.) The MIT Press, Cambridge, Mass. 1964].

BALAKRISHNAN, A. V. (1980) On stochastic bang-bang control. *Appl. Math. Optim.* **6**, 91–96.

BARLOW, M. T. (1982) One-dimensional stochastic differential equations with no strong solution. *J. London Math. Soc.* **26**, 335–347.

BATHER, J. A. & CHERNOFF, H. (1967) Sequential decisions in the control of a spaceship. *Proc. 5th Berkeley Symp. Math. Stat. & Probability* **3**, 181–207.

BENEŠ, V. E. (1971) Existence of optimal stochastic control laws. *SIAM J. Control* **9**, 446–475.

BENEŠ, V. E., SHEPP, L. A., & WITSENHAUSEN, H. S. (1980) Some solvable stochastic control problems. *Stochastics* **4**, 39–83.

BENSOUSSAN, A. (1984) On the theory of option pricing. *Acta Applicandae Mathematicae* **2**, 139–158.

BERNŠTEIN, S. (1934) Principes de la théorie des équations différentielles stochastiques. *Trudy Fiz.-Mat., Steklov Inst., Akad. Nauk.* **5**, 95–124.

BERNŠTEIN, S. (1938) Équations différentielles stochastiques. *Act. Sci. Ind.* **738**, 5–31. *Conf. Intern. Sci. Math. Univ. Genève*. Hermann, Paris.

BHATTACHARYA, R. N. (1978) Criteria for recurrence and existence of invariant measures for multidimensional diffusions. *Ann. Probability* **3**, 541–553. Correction: ibid. **8** (1980), 1194–1195.

BILLINGSLEY, P. (1968) *Convergence of Probability Measures*. J. Wiley & Sons, New York.

BILLINGSLEY, P. (1979) *Probability and Integration*. J. Wiley & Sons, New York.

BLACK, F. & SCHOLES, M. (1973) The pricing of options and corporate liabilities. *J. Polit. Economy* **81**, 637–659.

BLUMENTHAL, R. M. (1957) An extended Markov property. *Trans. Amer. Math. Soc.* **82**, 52–72.

BLUMENTHAL, R. M. & GETOOR, R. K. (1968) *Markov Processes and Potential Theory*. Academic Press, New York.

BORODIN, A. N. (1981) On the asymptotic behaviour of local times of recurrent random walks with finite variance. *Theory Probab. Appl.* **26**, 758–772.

BROSAMLER, G. A. (1970) Quadratic variation of potentials and harmonic functions. *Trans. Amer. Math. Soc.* **149**, 243–257.

BURKHOLDER, D. L. (1973) Distribution function inequalities for martingales. *Ann. Probability* **1**, 19–42.

CAMERON, R. H. & MARTIN, W. T. (1944) Transformation of Wiener integrals under translations. *Ann. Math.* **45**, 386–396.

CAMERON, R. H. & MARTIN, W. T. (1945) Evaluations of various Wiener integrals by use of certain Sturm-Liouville differential equations. *Bull. Amer. Math. Soc.* **51**, 73–90.

ČENTSOV, N. N. (1956a) Weak convergence of stochastic processes whose trajectories have no discontinuity of the second kind and the "heuristic" approach to the Kolmogorov-Smirnov tests. *Theory Probab. Appl.* **1**, 140–144.

ČENTSOV, N. N. (1956b) Wiener random fields depending on several parameters (in Russian). *Dokl. Acad. Nauk. SSSR* **106**, 607–609.

CHALEYAT-MAUREL, M., EL KAROUI, N., & MARCHAL, B. (1980) Réflexion discontinue et systèmes stochastiques. *Ann. Probability.* **8**, 1049–1067.

CHERNOFF, H. (1968) Optimal stochastic control. *Sankhyā, Ser. A*, **30**, 221–252.

CHITASHVILI, R. J. & TORONJADZE, T. A. (1981) On one-dimensional stochastic differential equations with unit diffusion coefficient; structure of solutions. *Stochastics* **4**, 281–315.

CHOW, P. L., MENALDI, J. L., & ROBIN, M. (1985) Additive control of stochastic linear systems with finite horizons. *SIAM J. Control and Optimization* **23**, 858–899.

CHOW, Y. S. & TEICHER, H. (1978) *Probability Theory: Independence, Interchangeability, Martingales*. Springer-Verlag, New York.

CHUNG, K. L. (1974) *A Course in Probability Theory*. Academic Press, New York.

CHUNG, K. L. (1976) Excursions in Brownian motion. *Ark. Mat.* **14**, 155–177.

CHUNG, K. L. (1982) *Lectures from Markov Processes to Brownian Motion*. Springer-Verlag, New York.

CHUNG, K. L. & DOOB, J. L. (1965) Fields, optionality and measurability. *Amer. J. Math.* **87**, 397–424.

CHUNG, K. L. & DURRETT, R. (1976) Downcrossings and local time. *Z. Wahrscheinlichkeitstheorie verw. Gebiete* **35**, 147–149.

CHUNG, K. L. & WILLIAMS, R. J. (1983) *Introduction to Stochastic Integration*. Birkhäuser, Boston.

CIESIELSKI, Z. (1961) Hölder condition for realizations of Gaussian processes. *Trans. Amer. Math. Soc.* **99**, 403–413.

ÇINLAR, E., JACOD, J., PROTTER, P., & SHARPE, M. J. (1980) Semimartingales and Markov processes. *Z. Wahrscheinlichkeitstheorie verw. Gebiete* **54**, 161–219.

CIREL'SON, B. S. (1975) An example of a stochastic differential equation having no strong solution. *Theory Probab. Appl.* **20**, 416–418.

COCOZZA, C. & YOR, M. (1980) Démonstration d'un théorème de F. Knight à l'aide des

martingales exponentielles. *Lecture Notes in Mathematics* **784**, 496–499. Springer-Verlag, Berlin.
COPSON, E. T. (1975) *Partial Differential Equations*. Cambridge University Press.
COURRÈGE, P. (1962/1963) Intégrales stochastiques et martingales de carré intégrable. *Seminaire Brelot-Choquet-Deny* **7**, Publ. Inst. Henri Poincaré.
COX, J. C. & HUANG, C. (1986) Optimal consumption and portfolio policies when asset prices follow a diffusion process. Submitted for publication.
COX, J. C. & HUANG, C. (1987) A variational problem arising in financial economics. *Econometrica*, to appear.
CSÁKI, E. & RÉVÉSZ, P. (1983) Strong invariance for local times. *Z. Wahrscheinlichkeitstheorie verw. Gebiete* **62**, 263–278.
DAMBIS, K. E. (1965) On the decomposition of continuous submartingales. *Theory Probab. Appl.* **10**, 401–410.
DANIELL, P. J. (1918/1919) Integrals in an infinite number of dimensions. *Annals of Mathematics* **20**, 281–288.
DAVIS, B. (1979) Brownian motion and analytic functions. *Ann. Probability*. **7**, 913–932.
DAVIS, B. (1983) On Brownian slow points. *Z. Wahrscheinlichkeitstheorie verw. Gebiete* **64**, 359–367.
DAVIS, M. H. A. & CLARK, J. M. C. (1979) On "predicted miss" stochastic control problems. *Stochastics* **2**, 197–209.
DELLACHERIE, C. (1972) *Capacités et Processus Stochastiques*. Springer-Verlag, Berlin.
DELLACHERIE, C. & MEYER, P. A. (1975/1980) *Probabilités et Potentiel*. Chaps. I–IV/V–VIII. Hermann, Paris.
DOLÉANS-DADE, C. (1976) On the existence and unicity of solutions of stochastic integral equations. *Z. Wahrscheinlichkeitstheorie verw. Gebiete* **36**, 93–101.
DOLÉANS-DADE, C. & MEYER, P. A. (1970) Intégrales stochastiques par rapport aux martingales locales. *Lecture Notes in Mathematics* **124**, 77–107. Springer-Verlag, Berlin.
DONSKER, M. D. (1951) An invariance principle for certain probability limit theorems. *Mem. Amer. Math. Soc.* **6**, 1–12.
DONSKER, M. D. & VARADHAN, S. R. S. (1975) Asymptotic evaluation of certain Markov process expectations for large time, I and II. *Comm. Pure & Appl. Math.* **28**, 1–47 and 279–301.
DONSKER, M. D. & VARADHAN, S. R. S. (1976) On the principal eigenvalue of second-order elliptic differential operators. *Comm. Pure Appl. Math.* **29**, 595–621.
DOOB, J. L. (1942) The Brownian movement and stochastic equations. *Ann. Math.* **43**, 351–369.
DOOB, J. L. (1953) *Stochastic Processes*. J. Wiley & Sons, New York.
DOOB, J. L. (1984) *Classical Potential Theory and Its Probabilistic Counterpart*. Springer-Verlag, Berlin.
DOSS, H. (1977) Liens entre équations différentielles stochastiques et ordinaires. *Ann. Inst. Henri Poincaré* **13**, 99–125.
DUBINS, L. E. & SCHWARZ, G. (1965) On continuous martingales. *Proc. Nat'l. Acad. Sci. USA* **53**, 913–916.
DUDLEY, R. M. (1977) Wiener functionals as Itô integrals. *Ann. Probability* **5**, 140–141.
DUFFIE, D. (1986) Stochastic equilibria: existence, spanning number, and the 'no expected financial gain from trade' hypothesis. *Econometrica* **54**, 1161–1183.
DUNFORD, N. & SCHWARTZ, J. T. (1963) *Linear Operators. Part I: General Theory*. J. Wiley & Sons/Interscience, New York.
DURRETT, R. (1984) *Brownian Motion and Martingales in Analysis*. Wadsworth, Belmont, California.
DVORETZKY, A. (1963) On the oscillation of the Brownian motion process. *Israel J. Math.* **1**, 212–214.
DVORETZKY, A., ERDÖS, P., & KAKUTANI, S. (1961) Nonincrease everywhere of the

Brownian motion process. *Proc. 4th Berkeley Symp. Math. Stat. & Probability* **2**, 103–116.

DYNKIN, E. B. (1965) *Markov Processes* (2 volumes). Springer-Verlag, Berlin.

DYNKIN, E. B. & YUSHKEVICH, A. A. (1969) *Markov Processes: Theorems and Problems.* Plenum, New York.

EINSTEIN, A. (1905) On the movement of small particles suspended in a stationary liquid demanded by the molecular-kinetic theory of heat. *Ann. Physik* **17** [Also in A. EINSTEIN (1959) *Investigations on the Theory of the Brownian Movement* (R. Fürth, ed.) Dover, New York].

ELLIOTT, R. J. (1982) *Stochastic Calculus and Applications.* Springer-Verlag, New York.

ELWORTHY, K. D., LI, X. M., & YOR, M. (1997) On the tails of the supremum and the quadratic variation of strictly local martingales. *Lecture Notes in Mathematics* **1655**, 113–125. Springer-Verlag, Berlin.

EMERY, M., STRICKER, C., & YAN, J. A. (1983) Valeurs prises par les martingales locales continues à un instant donné. *Ann. Probability.* **11**, 635–641.

ENGELBERT, H. J. & SCHMIDT, W. (1981) On the behaviour of certain functionals of the Wiener process and applications to stochastic differential equations. *Lecture Notes in Control and Information Sciences* **36**, 47–55. Springer-Verlag, Berlin.

ENGELBERT, H. J. & SCHMIDT, W. (1984) On one-dimensional stochastic differential equations with generalized drift. *Lecture Notes in Control and Information Sciences* **69**, 143–155. Springer-Verlag, Berlin.

ENGELBERT, H. J. & SCHMIDT, W. (1985) On solutions of stochastic differential equations without drift. *Z. Wahrscheinlichkeitstheorie verw. Gebiete* **68**, 287–317.

ETHIER, S. N. & KURTZ, T. G. (1986) *Markov Processes: Characterization and Convergence.* J. Wiley & Sons, New York.

FELLER, W. (1936) Zur theorie der stochastischen prozessen (existenz und eindeutigkeitssätze). *Math Ann.* **113**, 113–160.

FELLER, W. (1952) The parabolic differential equations and the associated semi-groups of transformations. *Ann. Math.* **55**, 468–519.

FELLER, W. (1954a) The general diffusion operator and positivity-preserving semi-groups in one dimension. *Ann. Math.* **60**, 417–436.

FELLER, W. (1954b) Diffusion processes in one dimension. *Trans. Amer. Math. Soc.* **77**, 1–31.

FELLER, W. (1955) On second-order differential operators. *Ann. Math.* **61**, 90–105.

FELLER, W. (1957) Generalized second-order differential operators and their lateral conditions. *Ill. J. Math.* **1**, 459–504.

FELLER, W. (1971) *An Introduction to Probability Theory and Its Applications, Volume II.* J. Wiley & Sons, New York.

FEYNMAN, R. P. (1948) Space-time approach to nonrelativistic quantum mechanics. *Rev. Mod. Phys.* **20**, 367–387.

FISK, D. L. (1963) Quasi-martingales and stochastic integrals. *Technical Report* **1**, Dept. of Mathematics, Michigan State University.

FISK, D. L. (1966) Sample quadratic variation of continuous, second-order martingales. *Z. Wahrscheinlichkeitstheorie verw. Gebiete* **6**, 273–278.

FLEMING, W. H. & RISHEL, R. W. (1975) *Deterministic and Stochastic Optimal Control.* Springer-Verlag, Berlin.

FOURIER, J. B. (1822) *Théorie Analytique de la Chaleur.* Paris.

FREEDMAN, D. (1971) *Brownian Motion and Diffusion.* Holden-Day, San Fransisco [Also published by Springer-Verlag, New York, 1983].

FREIDLIN, M. (1985) *Functional Integration and Partial Differential Equations.* Annals of Mathematical Studies, Vol. 109. Princeton University Press, Princeton, N.J.

FRIEDMAN, A. (1964) *Partial Differential Equations of Parabolic Type.* Prentice-Hall, Englewood Cliffs, N.J.

FRIEDMAN, A. (1975) *Stochastic Differential Equations and Applications, Volume 1.* Academic Press, New York.

FRIEDMAN, A. (1976) *Stochastic Differential Equations and Applications, Volume 2.* Academic Press, New York.

FRISTEDT, B & TAYLOR, S. J. (1983) Construction of local time for a Markov process. *Z. Wahrscheinlichkeitstheorie verw. Gebiete* **62**, 73–112.

GEMAN, D. & HOROWITZ, J. (1980) Occupation densities. *Ann. Probability* **8**, 1–67.

GIHMAN, I. I. (1947) A method of constructing random processes (in Russian). *Akad. Nauk SSSR* **58**, 961–964.

GIHMAN, I. I. (1950a) Certain differential equations with random functions (in Russian). *Uskr. Math. Ž.* **2** (3), 45–69.

GIHMAN, I. I. (1950b) On the theory of differential equations of random processes (in Russian). *Uskr. Math. Ž.* **2** (4), 37–63.

GIHMAN, I. I. & SKOROHOD, A. V. (1972) *Stochastic Differential Equations.* Springer-Verlag, Berlin.

GIHMAN, I. I. & SKOROHOD, A. V. (1979) *The Theory of Stochastic Processes, Volume III.* Springer-Verlag, Berlin.

GILBARG, D. & TRUDINGER, N. S. (1977) *Elliptic Partial Differential Equations of Second Order.* Springer-Verlag, Berlin.

GIRSANOV, I. V. (1960) On transforming a certain class of stochastic processes by absolutely continuous substitution of measures. *Theory Probab. Appl.* **5**, 285–301.

GIRSANOV, I. V. (1962) An example of non-uniqueness of the solution to the stochastic differential equation of K. Itô. *Theory Probab. Appl.* **7**, 325–331.

HAJEK, B. (1985) Mean stochastic comparison of diffusions. *Z. Wahrscheinlichkeitstheorie verw. Gebiete* **68**, 315–329.

HALE, J. (1969) *Ordinary Differential Equations.* J. Wiley & Sons/Interscience, New York.

HALL, P. & HEYDE, C. C. (1980) *Martingale Limit Theory and Its Application.* Academic Press, New York.

HARRISON, J. M. (1985) *Brownian Motion and Stochastic Flow Systems.* J. Wiley & Sons, New York.

HARRISON, J. M. & PLISKA, S. R. (1981) Martingales and stochastic integrals in the theory of continuous trading. *Stoch. Processes & Appl.* **11**, 215–260.

HARRISON, J. M. & SHEPP, L. A. (1981) On skew Brownian motion. *Ann. Probability* **9**, 309–313.

HARTMAN, P. & WINTNER, A. (1950) On the solutions of the equation of heat conduction. *Amer. J. Math.* **72**, 367–395.

HAUSSMANN, U. G. (1981) Some examples of optimal stochastic controls (or: the stochastic maximum principle at work), *SIAM Review* **23**, 292–307.

HINČIN, A. YA. (1933) Asymptotische gesetze der Wahrscheinlichkeitsrechnung. *Ergeb. Math.* **2**, No. 4 [Also published by Springer-Verlag, Berlin, 1933, and reprinted by Chelsea Publishing Co., New York, 1948].

HINČIN, A. YA. (1937) Déduction nouvelle d'une formule de M. Paul Lévy. *Bull Univ. État Moscou, Ser. Int., Sect. A, Math. et Mecan.* **1**.

HOOVER, D. N. (1984) Synonimity, generalized martingales, and subfiltrations. *Ann. Probability* **12**, 703–713.

HUANG, C. (1987) An intertemporal general equilibrium asset pricing model: the case of diffusion information. *Econometrica* **55**, 117–142.

HUNT, G. A. (1956) Some theorems concerning Brownian motion. *Trans. Amer. Math. Soc.* **81**, 294–319.

HUNT, G. A. (1957/1958) Markov processes and potentials. *Illinois J. Math.* **1**, 44–93, and 316–369; **2**, 151–213.

IKEDA, N. & WATANABE, S. (1977) A comparison theorem for solutions of stochastic differential equations and its applications. *Osaka J. Math.* **14**, 619–633.

IKEDA, N. & WATANABE, S. (1981) *Stochastic Differential Equations and Diffusion Processes.* North-Holland, Amsterdam (Kodansha Ltd., Tokyo).

IKEDA, N. & WATANABE, S. (1983) An introduction to Malliavin's calculus. *Taniguchi Symposium on Stochastic Analysis, Katata 1982* (K. Itô, editor), **1–52**. North-Holland, Amsterdam.

IMHOF, J. P. (1984) Density factorizations for Brownian motion, meander and the three-dimensional Bessel process, and applications *J. Appl. Probab.* **21**, 500–510.

ITÔ, K. (1942a) Differential equations determining Markov processes (in Japanese). *Zenkoku Shijō Sūgaku Danwakai* **1077**, 1352–1400.

ITÔ, K. (1942b) On stochastic processes II: Infinitely divisible laws of probability. *Japanese J. Math.* **18**, 261–301.

ITÔ, K. (1944) Stochastic Integral. *Proc. Imperial Acad. Tokyo* **20**, 519–524.

ITÔ, K. (1946) On a stochastic integral equation. *Proc. Imp. Acad. Tokyo* **22**, 32–35.

ITÔ, K. (1951) On stochastic differential equations. *Mem. Amer. Math. Soc.* **4**, 1–51.

ITÔ, K. (1961a) *Lectures on Stochastic Processes.* Tata Institute of Fundamental Research, Vol. 24. Bombay.

ITÔ, K. (1961b) Poisson point processes attached to Markov processes. *Proc. 6th Berkeley Symp. Math. Stat. & Probability* **3**, 225–240, University of California Press.

ITÔ, K. & MCKEAN, H. P., Jr. (1974) *Diffusion Processes and Their Sample Paths.* Springer-Verlag, Berlin.

ITÔ, K. & WATANABE, S. (1978) Introduction to stochastic differential equations. *Proc. Intern. Symp. on Stoch. Diff. Equations, Kyoto 1976* (K. Itô, ed.), i–xxx. Kinokuniya, Tokyo.

JEULIN, T. & YOR, M. (1980) Sur les distributions de certaines fonctionnelles du mouvement brownien. *Lecture Notes in Mathematics* **850**, 210–226. Springer-Verlag, Berlin.

JEULIN, T. & YOR, M. (1985) Grossissements de filtrations: exemples et applications. *Lecture Notes in Mathematics* **1118**. Springer-Verlag, Berlin.

KAC, M. (1949) On distributions of certain Wiener functionals. *Trans. Amer. Math. Soc.* **65**, 1–13.

KAC, M. (1951) On some connections between probability theory and differential and integral equations, *Proc. 2nd Berkeley Symp. on Math. Stat. & Probability*, 189–215, University of California Press.

KACZMARZ, S. & STEINHAUS, H. (1951) *Theorie der Orthogonalreihen.* Chelsea Publishing Co., New York.

KAHANE, J. P. (1976) Sur les zéros et les instants de ralentissement du mouvement brownien. *C.R. Acad. Sci. Paris* **282** (Série A), 431–433.

KAKUTANI, S. (1944a) On Brownian motion in *n*-space. *Proc. Acad. Japan* **20**, 648–652.

KAKUTANI, S. (1944b) Two-dimensional Brownian motion and harmonic functions. *Proc. Acad. Japan* **20**, 706–714.

KALLIANPUR, G. (1980) *Stochastic Filtering Theory.* Springer-Verlag, New York.

KARATZAS, I. (1980) On a stochastic representation of the principal eigenvalue of a second-order differential equation. *Stochastics* **3**, 305–321.

KARATZAS, I. (1983) A class of singular stochastic control problems. *Adv. Appl. Probability* **15**, 225–254.

KARATZAS, I. (1988) On the pricing of American options. *Appl. Math. Optimization* **17**, 37–60.

KARATZAS, I., LEHOCZKY, J. P., SETHI, S. P., & SHREVE, S. E. (1986) Explicit solution of a general consumption/investment problem. *Math. Operations Research* **11**, 261–294.

KARATZAS, I., LEHOCZKY, J. P., & SHREVE, S. E. (1987) Optimal portfolio and consumption decisions for a "small investor" on a finite horizon. *SIAM J. Control & Optimization* **25**, 1557–1586.

KARATZAS, I. & SHREVE, S. E. (1984a) Trivariate density of Brownian motion, its local and occupation times, with application to stochastic control. *Ann. Probability* **12**, 819–828.

KARATZAS, I. & SHREVE, S. E. (1984b) Connections between optimal stopping and singular stochastic control I. Monotone follower problems. *SIAM J. Control & Optimization* **22**, 856–877.

KARATZAS, I. & SHREVE, S. E. (1985) Connections between optimal stopping and singular stochastic control II. Reflected follower problems. *SIAM J. Control & Optimization* **23**, 433–451.

KARATZAS, I. & SHREVE, S. E. (1987) A decomposition of the Brownian path. *Statistics and Probability Letters* **5**, 87–93.

KARLIN, S. & TAYLOR, H. M. (1981) *A Second Course in Stochastic Processes.* Academic Press, New York.

KAZAMAKI, N. (1977) On a problem of Girsanov. *Tohōku Math. J.* **29**, 597–600.

KAZAMAKI, N. (1978) The equivalence of two conditions on weighted norm inequalities for martingales. *Proc. Intern. Symp. on Stoch. Diff. Equations, Kyoto 1976* (K. Itô, editor), 141–152 Kinokuniya, Tokyo,

KAZAMAKI, N. & SEKIGUCHI, T. (1979) On the transformation of some classes of martingales by a change of law. *Tohōku Math. J.* **31**, 261–279.

KELLOGG, O. D. (1929) *Foundations of Potential Theory.* Ungar, New York.

KHAS'MINKSII, R. Z. (1960) Ergodic properties of recurrent diffusion processes and stabilization of the Cauchy problem for parabolic equations. *Theory Probab. Appl.* **5**, 179–196.

KHAS'MINSKII, R. Z. (1980) *Stochastic Stability of Differential Equations.* Sijthoff & Noordhoff, Alphen aan den Rijn, The Netherlands.

KINGMAN, J. F. C. (1967) Completely random measures. *Pacific J. Math.* **21**, 59–78.

KINGMAN, J. F. C. (1973) An intrinsic description of local time. *J. London Math. Soc.* **6**, 725–731.

KLEPTSYNA, M. L. & VERETENNIKOV, A. YU. (1984) On strong solutions of stochastic Itô-Volterra equations. *Theory Probab. Appl.* **29**, 153–157.

KNIGHT, F. B. (1961) On the random walk and Brownian motion. *Trans. Amer. Math. Soc.* **103**, 218–228.

KNIGHT, F. B. (1963) Random walks and a sojourn density process of Brownian motion. *Trans. Amer. Math. Soc.* **107**, 56–86.

KNIGHT, F. B. (1971) A reduction of continuous, square-integrable martingales to Brownian motion. *Lecture Notes in Mathematics* **190**, 19–31. Springer-Verlag, Berlin.

KNIGHT, F. B. (1981) *Essentials of Brownian Motion and Diffusion.* Math. Surveys, Vol. 18. American Mathematical Society, Providence, R. I.

KOLMOGOROV, A. N. (1931) Über die analytischen methoden in der Wahrscheinlichkeitsrechnung. *Math. Ann.* **104**, 415–458.

KOLMOGOROV, A. N. (1933) Grundbegriffe der Wahrscheinlichkeitsrechnung. *Ergeb. Math.* **2**, No. 3 [Reprinted by Chelsea Publishing Co., New York, 1946. English translation: *Foundations of Probability Theory,* Chelsea Publishing Co., New York, 1950].

KOPP, P. E. (1984) *Martingales and Stochastic Integrals.* Cambridge University Press.

KRYLOV, N. V. (1969) On Itô's stochastic integral equations. *Theory Probab. Appl.* **14**, 330–336. Correction note: ibid. **16** (1972), 373–374.

KRYLOV, N. V. (1974) Some estimates on the probability density of a stochastic integral. *Math USSR (Izvestija)* **8**, 233–254.

KRYLOV, N. V. (1980) *Controlled Diffusion Processes.* Springer-Verlag, Berlin.

KRYLOV, N. V. (1990) Une représentation des sousmartingales positives et ses applications. *Lecture Notes in Mathematics* **1426**, 473–476. Springer-Verlag, Berlin.

KUNITA, H. (1981) On the decomposition of solutions of stochastic differential equations. *Lecture Notes in Mathematics* **851**, 213–255. Springer-Verlag, Berlin.

KUNITA, H. (1986) Convergence of stochastic flows connected with stochastic ordinary differential equations. *Stochastics* **17**, 215–251.

KUNITA, H. & WATANABE, S. (1967) On square-integrable martingales. *Nagoya Math. J.* **30**, 209–245.

KUSSMAUL, A. U. (1977) *Stochastic Integration and Generalized Martingales.* Pitman, London.

KUSUOKA, S. & STROOCK, D. W. (1983) Applications of the Malliavin calculus, I. *Taniguchi Symposium on Stochastic Analysis, Katata 1982* (K. Itô, editor), 271–306. North-Holland, Amsterdam.

KUSUOKA, S. & STROOCK, D. W. (1985) Applications of the Malliavin Calculus, II. *J. Faculty Science Univ. Tokyo* **32**, 1–76.

LAMPERTI, J. (1964) A simple construction of certain diffusion processes. *J. Math. Kyoto Univ.* **4**, 161–170.

LANGEVIN, P. (1908) Sur la théorie du mouvement brownien. *C.R. Acad. Sci. Paris* **146**, 530–533.

LAPLACE, P. S. (1809) Memoire sur divers points d'analyse. *J. École Polytech.* **8**, Cahier 15.

LEBESGUE, H. (1924) Conditions de régularité, conditions d'irrégularité, conditions d'impossibilité dans le problème de Dirichlet. *Comp. Rendus Acad. Sci.* **178**, 1, 349–354.

LE GALL, J. F. (1983) Applications des temps locaux aux équations différentielles stochastiques unidimensionnelles. *Lecture Notes in Mathematics* **986**, 15–31. Springer-Verlag, Berlin.

LEHOCZKY, J. P. (1977) Formulas for stopped diffusions with stopping times based on the maximum. *Ann. Probability* **5**, 601–607.

LEHOCZKY, J. P. & SHREVE, S. E. (1986) Explicit equilibrium so'utions for a multiagent consumption/investment problem. Technical Report 384, Stat. Dept., Carnegie Mellon University, Pittsburgh, PA.

LENGLART, E. (1977) Relation de domination entre deux processus. *Ann. Inst. Henri Poincaré* **13**, 171–179.

LÉVY, P. (1937) *Théorie de l'Addition des Variables Aléatoires.* Gauthier-Villars, Paris.

LÉVY, P. (1939) Sur certains processus stochastiques homogènes *Compositio Math.* **7**, 283–339.

LÉVY, P. (1948) *Processus Stochastiques et Mouvement Brownien.* Gauthier-Villars, Paris.

LÉVY, P. (1959) Construction du processus de W. Feller et H. P. McKean en partant du mouvement brownien. *Probability and Statistics: The Harald Cramér Volume* (U. Grenander, editor), 162–174. Almqvist and Wiksell, Stockholm.

LIPTSER, R. S. & SHIRYAEV, A. N. (1977) *Statistics of Random Processes. Vol I: General Theory.* Springer-Verlag, New York.

LOÈVE, M. (1977) *Probability Theory I* (4th edition). Springer-Verlag, New York.

LOÈVE, M. (1978) *Probability Theory II* (4th edition). Springer-Verlag, New York.

MALLIAVIN, P. (1976) Stochastic calculus of variations and hypoelliptic operators. *Proc. Intern. Symp. on Stoch. Diff. Equations, Kyoto 1976* (K. Itô, editor), 195–263. Kinokuniya, Tokyo.

MALLIAVIN, P. (1978) C^k-Hypoellipticity with degeneracy. *Stochastic Analysis* (A. Friedman and M. Pinsky, editors), 192–214. Academic Press, New York.

MANDL, P., LÁNSKÁ, V., & VRKOČ, I. (1978). Exercises in stochastic analysis. Supplement to the journal *Kybernetika* **14**, 36 pp.

MARKOV, A. A. (1906) Extension of the law of large numbers to dependent events (in Russian). *Bull. Soc. Phys. Math. Kazan* **15**, 155–156.

MARUYAMA, G. (1954) On the transition probability functions of Markov processes. *Nat. Sci. Rep. Ochanomizu Univ.* **5**, 10–20.

MARUYAMA, G. (1955) Continuous Markov processes and stochastic equations. *Rend. Circ. Matem. Palermo* **10**, 48–90.

MARUYAMA, G. & TANAKA, H. (1957) Some properties of one-dimensional diffusion processes. *Mem. Fac. Sci. Kyushu Univ., Ser. A* **13**, 117–141.

McGill, P. (1982) Markov properties of diffusion local time: a martingale approach. *Adv. Appl. Probability* **14**, 789–810.

McKean, H. P., Jr. (1969) *Stochastic Integrals*. Academic Press, New York.

McKean, H. P., Jr. (1975) Brownian local time. *Advances in Mathematics* **15**, 91–111.

Méléard, S. (1986) Application du calcul stochastique à l'étude de processus de Markov réguliers sur [0, 1]. *Stochastics* **19**, 41–82.

Merton, R. C. (1971) Optimum consumption and portfolio rules in a continuous-time model. *J. Econom. Theory* **3**, 373–413. Erratum: ibid. **6** (1973), 213–214.

Métivier, M. (1982) *Semimartingales: A Course on Stochastic Processes*. Walter de Gruyter, Berlin and New York.

Métiver, M. & Pellaumail, J. (1980) *Stochastic Integration*. Academic Press, New York.

Meyer, P. A. (1962) A decomposition theorem for supermartingales. *Illinois J. Math.* **6**, 193–205.

Meyer, P. A. (1963) Decomposition of supermartingales: the uniqueness theorem. *Illinois J. Math.* **7**, 1–17.

Meyer, P. A. (1966) *Probability and Potentials*. Blaisdell Publishing Company, Waltham, Mass.

Meyer, P. A. (1967) Intégrales stochastiques. *Lecture Notes in Mathematics* **39**, 72–162. Springer-Verlag, Berlin.

Meyer, P. A. (1971) Démonstration simplifiée d'un théorème de Knight. *Lecture Notes in Mathematics* **191**, 191–195. Springer-Verlag, Berlin.

Meyer, P. A. (1976) Un cours sur les intégrales stochastiques. *Lecture Notes in Mathematics* **511**, 245–398. Springer-Verlag, Berlin.

Millar, P. W. (1968) Martingale integrals. *Trans. Amer. Math. Soc.* **133**, 145–166.

Nakao, S. (1972) On the pathwise uniqueness of solutions of stochastic differential equations. *Osaka J. Math.* **9**, 513–518.

Nakao, S. (1983) On pathwise uniqueness and comparison of solutions of one-dimensional stochastic differential equations. *Osaka J. Math.* **20**, 197–204.

Nelson, E. (1967) *Dynamical Theories of Brownian Motion*. Princeton University Press, Princeton, N.J.

Neveu, J. (1975) *Discrete-parameter Martingales*. North-Holland, Amsterdam.

Novikov, A. A. (1971) On moment inequalities for stochastic integrals. *Theory Probab. Appl.* **16**, 538–541.

Novikov, A. A. (1972) On an identity for stochastic integrals. *Theory Probab. Appl.* **17**, 717–720.

Novikov, A. A. (1981) On estimates and the asymptotic behaviour of non-exit probabilities of a Wiener process to a moving boundary. *Math. USSR (Sbornik)* **38**, 495–505.

Nualart, D. & Zakai, M. (1986) Generalized stochastic integrals and the Malliavin calculus. *Probab. Th. Rel. Fields* **73**, 255–280.

Orey, S. (1974) *Radon-Nikodým Derivatives of Probability Measures: Martingale Methods*. Publ. Dept. Found. Math. Sciences, Tokyo University of Education.

Paley, R. E. A. C., Wiener, N., & Zygmund, A. (1933) Note on random functions. *Math. Z.* **37**, 647–668.

Papanicolaou, G. C., Stroock, D. W., & Varadhan, S. R. S. (1977) Martingale approach to some limit theorems. *Proc. 1976 Duke Conf. on Turbulence*, Duke Univ. Math. Series III.

Pardoux, E. (1979) Stochastic partial differential equations and filtering of diffusion processes. *Stochastics* **3**, 127–167.

Parthasarathy, K. R. (1967) *Probability Measures on Metric Spaces*. Academic Press, New York.

Perkins, E. (1982a) Weak invariance principles for local time. *Z Wahrscheinlichkeitstheorie verw. Gebiete* **60**, 437–451.

PERKINS, E. (1982b) Local time is a semimartingale. *Z. Wahrscheinlichkeitstheorie verw. Gebiete* **60**, 79–117.

PERKINS, E. (1982c) Local time and pathwise uniqueness for stochastic differential equations. *Lecture Notes in Mathematics* **920**, 201–208. Springer-Verlag, Berlin.

PITMAN, J. & YOR, M. (1986) Asymptotic laws of planar Brownian motion. *Ann. Probability* **14**, 733–779.

POINCARÉ, H. (1899) *Théorie du Potentiel Newtonien.* Paris, Carré & Naud.

POISSON, S. D. (1835) *Théorie Mathématique de la Chaleur.* Paris.

POLLAK, M. & SIEGMUND, D. (1985) A diffusion process and its applications to detecting a change in the drift of Brownian motion. *Biometrika* **72**, 267–280.

PORT, S. & STONE, C. (1978) *Brownian Motion and Classical Potential Theory.* Academic Press, New York.

PRIOURET, P. (1974) Processus de diffusion et équations différentielles stochastiques. *Lecture Notes in Mathematics* **390**, 38–113. Springer-Verlag, Berlin.

PROHOROV, YU, V. (1956) Convergence of random processes and limit theorems in probability theory. *Theory Probab. Appl.* **1**, 157–214.

PROTTER, P. E. (1977) On the existence, uniqueness, convergence and explosion of solutions of systems of stochastic integral equations. *Ann. Probability* **5**, 243–261.

PROTTER, P. (1984) *Semimartingales and Stochastic Differential Equations.* Lecture notes, Third Chilean Winter School of Probability & Statistics, July 1984. Tech. Report 85-25, Department of Statistics, Purdue University.

PROTTER, P. (1985) Volterra equations driven by semimartingales. *Ann. Probability* **13**, 519–530.

PROTTER, P. (1990) *Stochastic Integration and Differential Equations.* Springer-Verlag, Berlin.

PYKE, R. (1983) The Haar-function construction of Brownian motion indexed by sets. *Z. Wahrscheinlichkeitstheorie verw. Gebiete* **64**, 523–539.

RAO, K. M. (1969) On the decomposition theorems of Meyer. *Math. Scandinavica* **24**, 66–78.

RAY, D. (1956) Stationary Markov processes with continuous paths. *Trans. Amer. Math. Soc.* **82**, 452–493.

RAY, D. B. (1963) Sojourn times of diffusion processes. *Illinois J. Math.* **7**, 615–630.

REVUZ, D. & YOR, M. (1991) *Continuous Martingales and Brownian Motion.* Springer-Verlag, Berlin.

ROBBINS, H. & SIEGMUND, D. (1970) Boundary crossing probabilities for the Wiener process and sample sums. *Ann. Math. Stat.* **41**, 1410–1429.

ROBBINS, H. & SIEGMUND, D. (1973) Statistical tests of power one and the integral representation of solutions of certain parabolic differential equations. *Bull. Inst. Math. Acad. Sinica* (Taipei) **1**, 93–120.

ROGERS, L. C. G. & WILLIAMS, D. (1987) *Diffusions, Markov Processes, and Martingales. Vol. 2: Itô Calculus.* Wiley, New York.

ROSENBLATT, M. (1951) On a class of Markov processes. *Trans. Amer. Math. Soc.* **65**, 1–13.

SAMUELSON, P. A. (1969) Lifetime portfolio selection by dynamic stochastic programming. *Rev. Econom. Statist.* **51**, 239–246.

SESHADRI, V. (1988) Exponential models, Brownian motion and independence. *Can. J. Statistics* **16**, 209–221.

SHREVE, S. E. (1981) Reflected Brownian motion in the "bang-bang" control of Brownian drift. *SIAM J. Control & Optimization* **19**, 469–478.

SIMON, B. (1979) *Functional Integration and Quantum Physics.* Academic Press, New York.

SKOROHOD, A. V. (1961) Stochastic equations for diffusion processes in a bounded region. *Theory Probab. Appl.* **6**, 264–274.

SKOROHOD, A. V. (1965) *Studies in the Theory of Random Processes.* Addison-Wesley, Reading, Mass. [Reprinted by Dover Publications, New York].

SPIVAK, M. (1965) *Calculus on Manifolds.* W. A. Benjamin, New York.

SLUTSKY, E. (1937) Alcuni proposizioni sulla teoria degli funzioni aleatorie. *Giorn. Ist. Ital. Attuari* **8**, 183–199.

STROOCK, D. W. (1981) The Malliavin calculus and its applications to partial differential equations, I and II. *Math. Systems Theory* **14**, 25–65 and 141–171.

STROOCK, D. W. (1982) *Topics in Stochastic Differential Equations.* Tata Institute of Fundamental Research, Bombay and Springer-Verlag, Berlin.

STROOCK, D. W. (1983) Some applications of stochastic calculus to partial differential equations. *Lecture Notes in Mathematics* **976**, 268–382. Springer-Verlag, Berlin.

STROOCK, D. W. (1984) *An Introduction to the Theory of Large Deviations.* Springer-Verlag, Berlin.

STROOCK, D. W. & VARADHAN, S. R. S. (1969) Diffusion processes with continuous coefficients, I and II. *Comm. Pure & Appl. Math.* **22**, 345–400 and 479–530.

STROOCK, D. W. & VARADHAN, S. R. S. (1970) On the support of diffusion processes with applications to the strong maximum principle *Proc. 6th Berkeley Symp. Math. Stat. & Probability* **3**, 333–360.

STROOCK, D. W. & VARADHAN, S. R. S. (1971) Diffusion processes with boundary conditions. *Comm. Pure & Appl. Math.* **24**, 147–225.

STROOCK, D. W. & VARADHAN, S. R. S. (1972) On degenerate elliptic-parabolic operators of second order and their associated diffusions. *Comm. Pure & Appl. Math.* **25**, 651–714.

STROOCK, D. W. & VARADHAN, S. R. S. (1979) *Multidimensional Diffusion Processes.* Springer-Verlag, Berlin.

SUSSMANN, H. (1978) On the gap between deterministic and stochastic differential equations. *Ann. Probability* **6**, 19–41.

TAKSAR, M. (1985) Average optimal singular control and a related stopping problem. *Math. Operations Research* **10**, 63–81.

TANAKA, H. (1963) Note on continuous additive functionals of the 1-dimensional Brownian path, *Z. Wahrscheinlichkeitstheorie* **1**, 251–257.

TAYLOR, H. M. (1975) A stopped Brownian motion formula. *Ann. Probability* **3**, 234–246.

TROTTER, H. F. (1958) A property of Brownian motion paths. *Ill. J. Math.* **2**, 425–433.

TYCHONOFF, A. N. (1935) Uniqueness theorems for the heat equation. *Mat. Sbornik* **42**, 199–216 (in Russian).

UHLENBECK, G. E. & ORNSTEIN, L. S. (1930) On the theory of Brownian motion. *Physical Review* **36**, 823–841.

VAN SCHUPPEN, J. H. & WONG, E. (1974) Transformation of local martingales under a change of law. *Ann. Probability* **2**, 879–888.

VARADHAN, S. R. S. (1984) *Large Deviations and Applications.* SIAM Publications, Philadelphia.

VENTSEL, A. D. (1962) On continuous additive functionals of a multidimensional Wiener process. *Soviet Math. Doklady* **3**, 264–267.

VERETENNIKOV, A. YU. (1979) On the strong solutions of stochastic differential equations. *Theory Probab. Appl.* **24**, 354–366.

VERETENNIKOV, A. YU. (1981) On strong solutions and explicit formulas for solutions of stochastic integral equations. *Math USSR (Sbornik)* **39**, 387–403.

VERETENNIKOV, A. YU. (1982) On the criteria for existence of a strong solution to a stochastic equation. *Theory Probab. Appl.* **27**, 441–449.

VILLE, J. (1939) *Étude Critique de la Notion du Collectif.* Gauthier-Villars, Paris.

WALSH, J. B. (1984) Regularity properties of a stochastic partial differential equation. *Seminar on Stochastic Processes 1983*, 257–290. Birkhäuser, Boston.

WALSH, J. B. (1986) An introduction to stochastic partial differential equations. *Lecture Notes in Mathematics* **1180**, 265–439. Springer-Verlag, Berlin.

WANG, A. T. (1977) Generalized Itô's formula and additive functionals of Brownian motion. *Z. Wahrscheinlichkeitstheorie verw. Gebiete* **41**, 153–159.

WANG, M. C. & UHLENBECK, G. E. (1945) On the theory of Brownian motion II. *Rev. Modern Physics* **17**, 323–342.

WATANABE, S. (1984) *Stochastic Differential Equations and Malliavin Calculus.* Lecture Notes, Tata Institute of Fundamental Research, Bombay and Springer-Verlag, Berlin.

WENTZELL, A. D. (1981) *A Course in the Theory of Stochastic Processes.* Mc Graw-Hill, New York.

WIDDER, D. V. (1944) Positive temperatures on an infinite rod. *Trans. Amer. Math. Soc.* **55**, 85–95.

WIDDER, D. V. (1953) Positive temperatures on a semi-infinite rod. *Trans. Amer. Math. Soc.* **75**, 510–525.

WIENER, N. (1923) Differential space. *J. Math. Phys.* **2**, 131–174.

WIENER, N. (1924a) Un problème de probabilités dénombrables. *Bull. Soc. Math. France* **52**, 569–578.

WIENER, N. (1924b) The Dirichlet Problem. *J. Math. Phys.* **3**, 127–146.

WILLIAMS, D. (1969) Markov properties of Brownian local time. *Bull. Amer. Math. Soc.* **75**, 1035–1036.

WILLIAMS, D. (1974) Path decomposition and continuity of local time for one-dimensional diffusions. *Proc. London Math. Soc.* **28**, 738–768.

WILLIAMS, D. (1976) On a stopped Brownian motion formula of H. M. Taylor. *Lecture Notes in Mathematics* **511**, 235–239. Springer-Verlag, Berlin.

WILLIAMS, D. (1977) On Lévy's downcrossings theorem. *Z. Wahrscheinlichkeitstheorie verw. Gebiete* **40**, 157–158.

WILLIAMS, D. (1979) *Diffusions, Markov Processes and Martingales, Vol. 1: Foundations.* J. Wiley & Sons, Chichester, England.

WONG, E. & HAJEK, B. (1985) *Stochastic Processes in Engineering Systems.* Springer-Verlag, New York.

WONG, E. & ZAKAI, M. (1965a) On the relation between ordinary and stochastic differential equations, *Intern. J. Engr. Sci.*, **3**, 213–229.

WONG, E. & ZAKAI, M. (1965b) On the convergence of ordinary integrals to stochastic integrals. *Ann. Math. Stat.* **36**, 1560–1564.

YAMADA, T. & OGURA, Y. (1981) On the strong comparison theorems for solutions of stochastic differential equations. *Z. Wahrscheinlichkeitstheorie verw. Gebiete* **56**, 3–19.

YAMADA, T. & WATANABE, S. (1971) On the uniqueness of solutions of stochastic differential equations. *J. Math. Kyoto Univ.* **11**, 155–167.

YOR, M. (1977). Sur quelques approximations d' intégrales stochastiques. *Lecture Notes in Mathematics* **581**, 518–528. Springer-Verlag, Berlin.

YOR, M. (1978) Sur la continuité des temps locaux associés à certaines semimartingales. *Temps Locaux. Astérisque* **52–53**, 23–36.

YOR, M. (1979) Sur le balayage des semimartingales continues. *Lecture Notes in Mathematics* **721**, 453–471. Springer-Verlag, Berlin.

YOR, M. (1982) Introduction au calcul stochastique. *Séminaire Bourbaki, 34ᵉ année* (1981/1982) **590**, 1–10.

YOR, M. (1986) Sur la représentation comme intégrales stochastiques des temps d'occupation du mouvement brownien dans $\mathbb{R}^d$. *Lecture Notes in Mathematics* **1204**, 543–552. Springer-Verlag, Berlin.

ZAKAI, M. (1985) The Malliavin calculus. *Acta Applicandae Mathematicae* **3**, 175–207.

ZAREMBA, S. (1909) Sur le principe du minimum. *Bull. Acad. Sci. Cracovie.*

ZAREMBA, S. (1911) Sur le principe de Dirichlet. *Acta. Math.* **34**, 293–316.

ZVONKIN, A. K. (1974) A transformation of the state space of a diffusion process that removes the drift. *Math. USSR (Sbornik)* **22**, 129–149.

Index

D

E

F

9 780387 976556

Printed in the United States
By Bookmasters